Lecture Notes in Electrical Engineering

Volume 602

The book series *Lecture Notes in Electrical Engineering* (LNEE) publishes the latest developments in Electrical Engineering - quickly, informally and in high quality. While original research reported in proceedings and monographs has traditionally formed the core of LNEE, we also encourage authors to submit books devoted to supporting student education and professional training in the various fields and applications areas of electrical engineering. The series cover classical and emerging topics concerning:

- Communication Engineering, Information Theory and Networks
- Electronics Engineering and Microelectronics
- Signal, Image and Speech Processing
- Wireless and Mobile Communication
- Circuits and Systems
- Energy Systems, Power Electronics and Electrical Machines
- Electro-optical Engineering
- Instrumentation Engineering
- Avionics Engineering
- Control Systems
- Internet-of-Things and Cybersecurity
- Biomedical Devices, MEMS and NEMS

For general information about this book series, comments or suggestions, please contact leontina.dicecco@springer.com.

To submit a proposal or request further information, please contact the Publishing Editor in your country:

China

Jasmine Dou, Associate Editor (jasmine.dou@springer.com)

India

Aninda Bose, Senior Editor (aninda.bose@springer.com)

Japan

Takeyuki Yonezawa, Editorial Director (takeyuki.yonezawa@springer.com)

South Korea

Smith (Ahram) Chae, Editor (smith.chae@springer.com)

Southeast Asia

Ramesh Nath Premnath, Editor (ramesh.premnath@springer.com)

USA, Canada:

Michael Luby, Senior Editor (michael.luby@springer.com)

All other Countries:

Leontina Di Cecco, Senior Editor (leontina.dicecco@springer.com)

**** Indexing: The books of this series are submitted to ISI Proceedings, EI-Compendex, SCOPUS, MetaPress, Web of Science and Springerlink ****

More information about this series at http://www.springer.com/series/7818

Sumit Kundu · U. Shripathi Acharya ·
Chanchal Kr. De · Surajit Mukherjee
Editors

Proceedings of the 2nd International Conference on Communication, Devices and Computing

ICCDC 2019

Springer

Editors
Sumit Kundu
Department of Electronics and Communication Engineering
National Institute of Technology Durgapur
Durgapur, West Bengal, India

U. Shripathi Acharya
Department of Electronics and Communication Engineering
National Institute of Technology Karnataka
Mangalore, Karnataka, India

Chanchal Kr. De
Department of Electronics and Communication Engineering
Haldia Institute of Technology
Haldia, West Bengal, India

Surajit Mukherjee
Department of Electronics and Communication Engineering
Haldia Institute of Technology
Haldia, West Bengal, India

ISSN 1876-1100 ISSN 1876-1119 (electronic)
Lecture Notes in Electrical Engineering
ISBN 978-981-15-0828-8 ISBN 978-981-15-0829-5 (eBook)
https://doi.org/10.1007/978-981-15-0829-5

This Springer imprint is published by the registered company Springer Nature Singapore Pte Ltd.
The registered company address is: 152 Beach Road, #21-01/04 Gateway East, Singapore 189721, Singapore

Committee

Chief Patron

Lakshman Chandra Seth, Chairman, Haldia Institute of Technology, India

Joint Organizing Secretary

Chanchal Kr. De, Department of ECE, Haldia Institute of Technology, India
Avisankar Roy, Department of ECE, Haldia Institute of Technology, India
Tirthadip Sinha, Department of ECE, Haldia Institute of Technology, India

Joint Convener

Jagannath Samanta, Department of ECE, Haldia Institute of Technology, India
Suman Paul, Department of ECE, Haldia Institute of Technology, India

Program Committee Members

PC Members, Organization

Matteo Gaeta, University of Salerno, Fisciano (Sa), Italy
William (Bill) Buchanan, Edinburgh Napier University, UK
Victor C. M. Leung, The University of British Columbia, Canada
Lu Liu, University of Derby, UK
Paolo Bellavista, University of Bologna, Italy

Honkala Iiro, University of Turku, Finland
Biplab Sikdar, National University of Singapore, Singapore
Francisco José García Peñalvo, University of Salamanca, Spain
Gyu Myoung Lee, Liverpool John Moores University, UK
Kun Ma, University of Jinan, China
Zhili Sun, University of Surrey, UK
Chih-Shun Hsu, Shih Hsin University, Taiwan
George A. Tsihrintzis, University of Piraeus, Greece
Carlos Miguel Tavares Calafate, Universidad Politécnica de Valencia, Spain
Lei Shu, University of Lincoln, UK
Dilip Sarkar, University of Miami, Florida, USA
Sumit Kundu, NIT Durgapur, India
Bikash Kumar Dey, IIT Bombay, India
U. Shripathi Acharya, NIT Karnataka, Surathkal, India
Sunandan Bhunia, Central Institute of Technology, Kokrajhar, India
Ananda S. Chowdhury, Jadavpur University, India
Jaydeb Bhaumik, Jadavpur University, India
Animesh Maitra, University of Calcutta, India
Mainak Chatterjee, University of Central Florida, USA
Santi Prasad Maity, IIEST, Shibpur, India
Milli Pant, IIT Roorkee, India
Abhijit Biswas, University of Calcutta, India
Debashis De, MAKAUT, Kolkata, India
Prof. Roman Senkerik, Tomas Bata University in Zlin, Czech Republic
Samaresh Das, IIT Delhi, India
IvánGarcía-Magariño, University of Zaragoza, Spain
Malay Kumar Pandit, Haldia Institute of Technology, India
Kisalaya Chakrabarti, Haldia Institute of Technology, India
Sanjay Dhar Roy, NIT Durgapur, India

Technical Program Committee Members

TPC Members, Organization

Bora Döken, Istanbul Technical University, Turkey
Anjan Chakravorty, IIT Madras, Tamil Nadu, India
Susanta Kumar Parui, IIEST, Shibpur, India
Sudip Roy, IIT Roorkee, India
Somnath Ghosh, IIT Jodhpur, India
Soumya Pandit, University of Calcutta, India
Prabir Kumar Saha, NIT Meghalaya, India
Soma Barman, University of Calcutta, India
Arnab Nandi, NIT Silchar, India

Kaushik Mandal, University of Calcutta, India
Subrata Dutta, NIT Jamshedpur, India
Abhishek Sarkhel, NIT Meghalaya, India
Jishan Mehedi, Jalpaiguri Government Engineering College, India
Koushik Guha, NIT Silchar, India
Kandarpa Kumar Sarma, Gauhati University, India
Sambhu Nath Pradhan, NIT Agartala, India
Indranil Hatai, IIEST, Howrah, India
Sanjeev Kumar Metya, NIT Arunachal Pradesh, India
Pinaki Chakraborty, NIT Arunachal Pradesh, India
Bidhan Malakar, NIT Mizoram, India
Santanu Das, Jalpaiguri Government Engineering College, India
Bikash Bepari, HIT, Haldia, India
Subhankar Joardar, HIT, Haldia, India
Dilip Dey, HIT, Haldia, India
Uday Maji, HIT, Haldia, India
Shamba Chatterjee, HIT, Haldia, India
Subhranu Samanta, National University of Singapore, Singapore
Sandip Bhattacharya, Hiroshima University, Japan
Bishnu Prasad De, KIIT Bhubaneswar, India
Wriddhi Bhowmik, KIIT Bhubaneswar, India
Srijan Bhattacharya, RCC Institute of Information Technology, India
Ivy Majumdar, B. P. Poddar Institute of Management and Technology, India
Arindam Biswas, Asansol Engineering College, India
Sudipta Das, IMPS College of Engineering & Technology, Malda, India
Debashis Maji, Vellore Institute of Technology, India
Santigopal Pain, Dr. B. C. Roy Engineering College, Durgapur, India
Debasis De, Dr. B. C. Roy Engineering College, Durgapur, India
Abhijit Bhowmick, VITU, India
Sumit Kumar Jindal, VITU, India
Binod Prasad, IIT Guwahati, India
Gautam Kumar Maity, Pingla Thana Mahavidyalaya (College), India

Additional Reviewers

Reviewers, Organization

Apurba Ghosh, University Institute of Technology, Burdwan
Sudipta Sahana, JIS College of Engineering, Kalyani
Eshan Samanta, Global Institute of Science and Technology, Haldia
Aloke Saha, Dr. B. C. Roy Engineering College, Durgapur, India
Gouri Shankar Paul, Global Institute of Science and Technology, Haldia

Bidyut Das, Haldia Institute of Technology, India
Jayeeta Majumder, Haldia Institute of Technology, India
Manasija Bhattacharya, Haldia Institute of Technology, India

Advisory Committee

Sayantan Seth, Vice Chairman, HIT, Haldia
Asish Lahiri, Secretary, HIT, Haldia
M. N. Bandyopadhyay, Director, HIT, Haldia
A. K. Saha, Principal, HIT, Haldia
A. Mishra, Registrar, HIT, Haldia
A. B. Maity, Dean, SAS, HIT, Haldia
S. K. Basu, Finance Manager, HIT, Haldia
D. Das, Sr. Administrative Officer, HIT, Haldia

Organizing Committee

Surajit Mukherjee, Department of ECE, HIT, Haldia
Banibrata Bag, Department of ECE, HIT, Haldia
Santanu Maity, Department of ECE, HIT, Haldia
Tilak Mukherjee, Department of ECE, HIT, Haldia
Pallabi Pahari, Department of ECE, HIT, Haldia
Dibyendu Chowdhury, Department of ECE, HIT, Haldia
Dipak Samanta, Department of ECE, HIT, Haldia
Jayanta Kr. Bag, Department of ECE, HIT, Haldia
Kushal Roy, Department of ECE, HIT, Haldia
Asim Kumar Jana, Department of ECE, HIT, Haldia
Amit Bhattacharyya, Department of ECE, HIT, Haldia
Raj Kumar Maity, Department of ECE, HIT, Haldia
Akinchan Das, Department of ECE, HIT, Haldia
Avishek Das, Department of ECE, HIT, Haldia
Pinaki Satpathy, Department of ECE, HIT, Haldia
Razia Sultana, Department of ECE, HIT, Haldia
Moumita Jana, Department of ECE, HIT, Haldia
Sayani Ghosh, Department of ECE, HIT, Haldia
Pulak Maity, Department of ECE, HIT, Haldia
Sourav Kr. Das, Department of ECE, HIT, Haldia
Subhendu Barman, Department of ECE, HIT, Haldia
Sachindeb Jana, Department of ECE, HIT, Haldia
Atanu Pradhan, Department of ECE, HIT, Haldia
Tapan Maity, Department of ECE, HIT, Haldia

Ira Samanta, Department of ECE, HIT, Haldia
Asim Kuilya, Department of ECE, HIT, Haldia
Tanushree Bera, Department of ECE, HIT, Haldia

Preface

The 2nd International Conference on Communication, Devices and Computing (ICCDC 2019) was held at Haldia Institute of Technology, Haldia, from March 14 to 15, 2019. Haldia is a city in Purba Medinipur district of Indian state West Bengal. Haldia Institute of Technology is dedicated to the objectives of creating highly trained professional manpower in various disciplines of engineering. It has gained reputation through institutional dedication to teaching and research.

In response to call for papers of ICCDC 2019, a total of 134 papers were submitted for presentation and inclusion in the proceedings of the conference. These papers were evaluated and ranked based on their novelty, significance and technical quality by at least two reviewers per paper. After a careful and blind refereeing process, 68 papers were selected for inclusion in the proceedings. These papers cover current research in communication, signal processing, image processing, wireless network, semiconductor devices, VLSI design, antenna design and fuzzy modeling. The conference hosted three invited talks by Prof. Biplab Sikdar (National University of Singapore, Singapore), Prof. M. Sohel Rahman (Bangladesh University of Engineering and Technology, Bangladesh) and Prof. Bikash Kumar Dey (IIT Bombay, India). Also, Prof. Parthasarathi Chakrabarti, Director, IIEST, Shibpur, India, was the chief guest on this occasion.

A conference of this kind would not be possible without the full support from different committee members. The organizational aspects were looked after by the organizing committee members who spent their time and energy in making the conference a reality. We also thank all the technical program committee members and additional reviewers for thoroughly reviewing the papers submitted to the conference and sending their constructive suggestions to improve the quality of papers. Our hearty thanks to Springer for agreeing to publish the conference proceedings.

We are indebted to ISRO and Haldia Institute of Technology for sponsoring and supporting the event. Last but not least, our sincere thanks go to all speakers, participants and all authors who have submitted papers to ICCDC 2019. We sincerely hope that the readers will find the proceedings stimulating and inspiring.

Durgapur, India — Sumit Kundu
Mangalore, India — U. Shripathi Acharya
Haldia, India — Chanchal Kr. De
Haldia, India — Surajit Mukherjee

Message from the Volume Editors

It is a great pleasure for us to organize the 2nd International Conference on *Communication, Devices and Computing* (ICCDC 2019) held from March 14 to 15, 2019, at the Haldia Institute of Technology, Purba Medinipur, West Bengal, India. Our main goal is to provide an opportunity to the participants to learn about contemporary research in Communication, Devices and Computing and exchange ideas among themselves and with experts present in the conference as invited speakers. It is our sincere hope that the conference will help the participants in their research and training and open new avenues for work for those who are either starting their research or are looking for extending their area of research to a different area of current research in *Communication, Devices and Computing*.

After an initial call for papers, 134 papers were submitted for presentation at the conference. All submitted papers were sent to external referees and after refereeing, 68 papers were recommended for publication for the conference proceedings that will be published by Springer in its Lecture Notes in Electrical Engineering (LNEE) series.

We are grateful to the speakers, participants, referees, organizers, sponsors, ISRO and Haldia Institute of Technology for their support and help, without which it would have been impossible to organize the conference. We express our gratitude to the organizing committee members who work behind the scene tirelessly in taking care of the details in making this conference a success.

Contents

About the Editors

Sumit Kundu is a Professor at the Department of Electronics & Communication Engineering, National Institute of Technology, Durgapur, West Bengal, India, where he has been a member of the faculty since receiving his Ph.D. from IIT Kharagpur in 1995. He served as a loaned Assistant Professor at G. S. Sanyal School of Telecommunications, IIT Kharagpur in 2007. His research interests include cognitive radio networks, energy harvesting, wireless ad hoc and sensor networks, and physical layer security in wireless networks, and he has published more than 170 papers in various journals and conferences. He is a reviewer of several IEEE and Elsevier journals.

U. Shripathi Acharya is a Professor at the Department of Electronics and Communication Engineering, National Institute of Technology, Karnataka, India. He received his Ph.D. from the Indian Institute of Science, Bangalore in 2004. His research interests include theory and applications of error control codes, wireless communication, analog and digital communication, free space optic communication, consumer and professional electronics, and he has published more than 70 papers in various journals and conferences.

Chanchal Kr. De is an Associate Professor at the Department of Electronics and Communication Engineering, Haldia Institute of Technology, West Bengal, India. He received his Ph.D. from National Institute of Durgapur in 2015. He has 11 years of teaching experience. His research interests include radio resource management in wireless networks, OFDM technology, cooperative communication and cognitive radio networks and he has published more than 20 research articles in various journals and conferences.

Surajit Mukherjee is an Assistant Professor at the Department of Electronics and Communication Engineering, Haldia Institute of Technology, West Bengal, India. He has 9 years of teaching experience. His research interests include antenna design and RF systems design, and he has published six journal papers and five conference papers.

LDPC Code Construction from Bipartite Kneser Graphs

Tamal Chowdhury and Ankita Pramanik

Abstract Low density Parity Check (LDPC) codes show excellent characteristics in error correction. It is extensively used in the majority of the modern communication systems. Construction of LDPC code is a challenging task. In the present work, LDPC codes construction is attempted from bipartite Kneser graphs. This is the first time that Kneser graphs are used to construct LDPC codes. A special matrix consisting of very few entries of 1 is used in conjunction with the bipartite Kneser graphs to yield the final LDPC matrix. The codes obtained can be encoded by an approximate lower triangular technique. Also, the codes are four-cycle free, thus giving improved bit error rate performance. The code length and data rate of the codes obtained by the proposed method can be easily altered by choosing the parameters of the Kneser graph suitably.

Keywords Adjacency matrix · Approximate lower triangular encoding · Bipartite Kneser graph · Low density parity check codes · Special matrix

1 Introduction

Low density Parity Check (LDPC) codes are ubiquitous codes finding a vast scope of applications. LDPC codes can have entries from the binary (B) field [1] or non-binary (NB) Galois field [2]. LDPC codes defined over a Galois field (GF) are called NB-LDPC codes. Constructing, encoding, and decoding of B-LDPC codes are relatively

T. Chowdhury (✉)
Electronics and Electrical Communication Engineering Department,
Indian Institute of Technology, Kharagpur, Kharagpur, India
e-mail: tamalchowdhury27@gmail.com

A. Pramanik
Electronics and Telecommunication Engineering Department, Indian Institute of Engineering Science and Technology, Shibpur, Shibpur, India
e-mail: ankita@telecom.iiests.ac.in

S. Kundu et al. (eds.), *Proceedings of the 2nd International Conference on Communication, Devices and Computing*, Lecture Notes in Electrical Engineering 602, https://doi.org/10.1007/978-981-15-0829-5_1

easier. However, NB-codes shows better performance as compared to binary one. Performance-wise, LDPC codes are almost reaching the limit, given by Shannon [3].

The number of non zero entries is very less as compared to the number of zero entries in LDPC codes and hence the name low density. For binary LDPC codes, 1's are the only non-zero entries and for NB-LDPC codes, the entries are drawn from the GF. The codes should be so designed that it becomes four-cycle free. The presence of short cycles hampers the decoding process. It also adversely impacts the bit error rate (BER).

For representation of LDPC codes, generally, its parity check matrix (PCM) or the H matrix is used. The graphical representation of the PCM is more popular. For this, the Tanner graph of the PCM is used. The vertices of the Tanner graph are divided into two sets. In one set, there are n vertices, which are used to represent codeword bits (or variable nodes). In another set, there are m vertices, which are used to represent check nodes (or constraint nodes), and is used in the equation for parity checking. In this graph, a bit node and a check node can be connected with one edge, if that check node is used in one equation, which is used to check the parity of the bit node.

The Tanner graph may be drawn in vertical fashion also. In this case, on the left side, bite nodes are placed, and on the right side, check nodes are placed. In those cases, often the bit nodes are called as left nodes and the check nodes as right nodes. By placing message bit nodes on separate sides of the graph, one can distinguish it from the parity bits. This technique is used in a systematic code.

Multiple ways are available to construct the LDPC codes. Some of those are as follows: algebraic construction [4], random construction [5], progressive edge growth [6], quasi-cyclic [7, 8], etc. Recently graph-based approaches are becoming highly popular. LDPC codes have been constructed from voltage graphs [9], distance graphs [10], Cayley graphs [11], double cylinder graphs [12], etc. In this work, LDPC codes are constructed from bipartite Kneser graphs. To the best of authors knowledge, this is the first attempt to construct LDPC codes using Kneser graphs.

This work proposes a binary LDPC code construction. The code constructed is four-cycle free. The code is of irregular length, i.e., the no. of ones in every row and column are different.

The organization of this paper is as follows: Sect. 2 gives the literature survey on already available LDPC code construction techniques. Section 3 presents the proposed LDPC code design technique. Section 4 provides the simulation results and observations. The concluding remarks are given in Sect. 5.

2 Literature Survey

The field of LDPC codes is a matured field. Much work has been done in various aspects such as encoding [13], decoding [14] or code construction [15]. Also, much research efforts are directed towards efficient implementation of the codes [2]. Recent works also focus on the amenability of LDPC codes with various applications like space-time block codes [16], channel estimation [17], etc. The present work focuses on LDPC code construction from Kneser graphs. This section focuses on works dealing with LDPC code construction especially graph-based code construction.

In [2] a method for constructing the LDPC codes is proposed. Here the columns of the matrix are added one at a time from left to right. For obtaining the correct bit degree distribution, the column weights should be chosen accordingly. For non zero entries in each column, the rows, those are not yet full, are chosen. Degree distributions of rows for H will be imperfect, if at any point there are rows having more unfilled positions, the columns remaining are to be added there. Until the correct row degrees are obtained, the process can be started again or backtracked by a few columns.

LDPC codes belonging to the family of repeat-accumulate codes are also popular. For the last few columns of H, there are weight-2 columns in a step pattern. By this, the repeat-accumulate codes become systematic and can be easily encoded.

LDPC code construction from voltage graph was proposed in [9]. A B-matrix M is called base matrix corresponding to a tail biting LDPC block code if its parent convolution code with PCM H(P) has only monomial or zero entries and also maintain the relation as

$$M = H(P)|_{P=1} \tag{1}$$

This represent all nonzero entries in $H(P)$ being replaced by $P_0 = 1$. There may be same base matrix M for different LDPC block codes. M (base matrix) is used to represent the biadjacency matrix of the bipartite graph, which is followed by the base graph G_M. An additive group $\Delta = \{\delta\}$ is considered for voltage graph construction. The voltage graph $G_V = \{\zeta_B, \mu_B, \Delta\}$ is obtained from the base graph $G_M = \{\zeta_B, \mu_B\}$. For this a voltage value $\delta(s, \mu, \tilde{\mu})$ should be assigned to edge s which connects two vertices μ and $\tilde{\mu}$ and satisfy the property

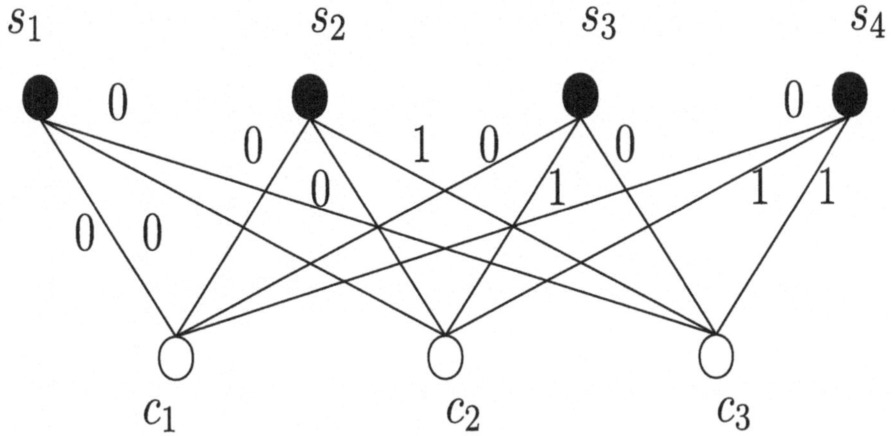

Fig. 1 Bipartite voltage graph with four symbol nodes (s_k, k = 1, 2, 3, 4) and three constraint nodes (c_l, l = 1, 2, 3)

$(s, \mu, \tilde{\mu}) = -\delta(s, \tilde{\mu}, \mu)$. The edge passes through a direction and that is responsible for the determination of the voltage of the edge. Figure 1 shows a bipartite voltage graph having four symbol nodes ($s_k, k = 1, 2, 3, 4$) and three constraint nodes ($c_l, l = 1, 2, 3$).

The work in [12] focuses on LDPC code construction from double cylinder graphs. A double cylinder type graph called the block structure graph of the PCM, denoted by $BSG(H)$ is introduced in [12]. Let, ϵ, ζ and η are positive integers and $\zeta < \eta$ and $H = (H_{i,j})_{\zeta \times \eta}$, where each $H_{i,j}$ is an $\epsilon \times \epsilon$ circulant permutation matrix or the $\epsilon \times \epsilon$ zero matrix. H is a $\zeta \times \eta$ matrix with $\epsilon \times \epsilon$ entries, which has ζ block-rows and η block-columns. In simple word, the matrix H and the quasi-cyclic LDPC code with the PCM, H are called as ϵ circulant matrix and ϵ circulant code, respectively. Let m, n and p be some non-negative integers such that $m \geq 2$, $n \geq 1$ and $p \geq 0$. A (m, n, p) double cylinder LDPC code, is the quasi cyclic LDPC code having the mother matrix $H(m, n, p)$ as given in Fig. 2.

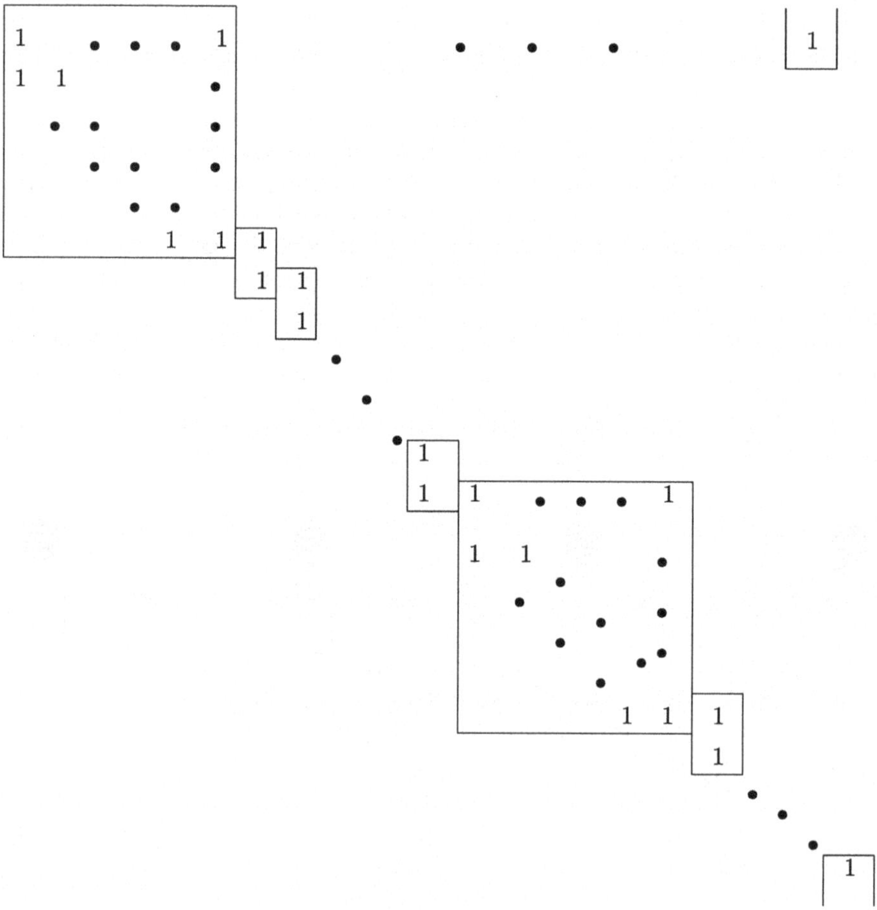

Fig. 2 LDPC code matrix from double cylinder graph

In Fig. 2, p is the number of blocks of $[1\ 1]^T$, where A^T stands for transpose of A, between each two consecutive blocks of $m \times m$ matrix $I + I^1$. The total number of blocks of the matrix $I + I^1$ is n. The final PCM is a matrix of size $(m + p - 1)n \times (m + p)n$. It must be noted that though the codes give a very high girth, they have only two 1s in every column. Thus the BER performance of the codes is very poor.

Classical LDPC codes show very good performance in different varieties of channels, and these codes can also be used by using fast and efficient algorithms. Quantum error-correcting codes can be decoded in a classical fashion. The work in [11] focuses on quantum LDPC code construction from Cayley graphs. Let, A is a subset of group R. The Cayley graph, $\Gamma(R, A)$ or $\Gamma(A)$, with vertex set R, has an edge between two vertices r and $\bar{r}$ if there exist the relation of $ra = \bar{r}$, where $a \in R$.

3 Proposed LDPC Code Construction

In this section, the proposed LDPC codes construction from bipartite Kneser graphs [18] is presented.

3.1 *Bipartite Kneser Graph*

Suppose there is a set G having n elements. Let **a**, **b** be two set of elements where set **a** contains k number of elements and set **b** contains n-k number of elements. So total number of sets containing k elements which is subset of n is $m = \binom{n}{k}$. Thus the total number of **a** sets is $\binom{n}{k} = \binom{n}{n-k}$ and total number of **b** sets is the same. If $\mathbf{a} = \{a_1, a_2, a_3, \ldots, a_m\}$ and $\mathbf{b} = \{b_1, b_2, b_3, \ldots, b_m\}$ are set of vertices of a graph then graph is called bipartite Kneser graph if the following properties are followed.

1. a_i and a_j are not connected by any edge for all i and j, similarly for **b** vertices.
2. a_i and b_j are connected by an edge if either $a_i \subset b_j$ or $b_j \subset a_i$.

The following example will help in understanding the bipartite Kneser graphs.

Example 1 Let a set $S = \{1, 2, 3, 4\}$, n = 4, k = 3, n − k = 1, so m = 4. Thus the two sets are $\mathbf{a} = \{(1, 2, 3), (1, 2, 4), (1, 3, 4), (2, 3, 4)\}$ and $\mathbf{b} = \{(1), (2), (3), (4)\}$ and the corresponding graph is given in Fig. 3.

The adjacency matrix of the constructed graph is as follows:

$$H = \begin{bmatrix} 1\,1\,1\,0 \\ 1\,1\,0\,1 \\ 1\,0\,1\,1 \\ 0\,1\,1\,1 \end{bmatrix}$$

The rows of H are elements of **a** and columns are elements of **b**.

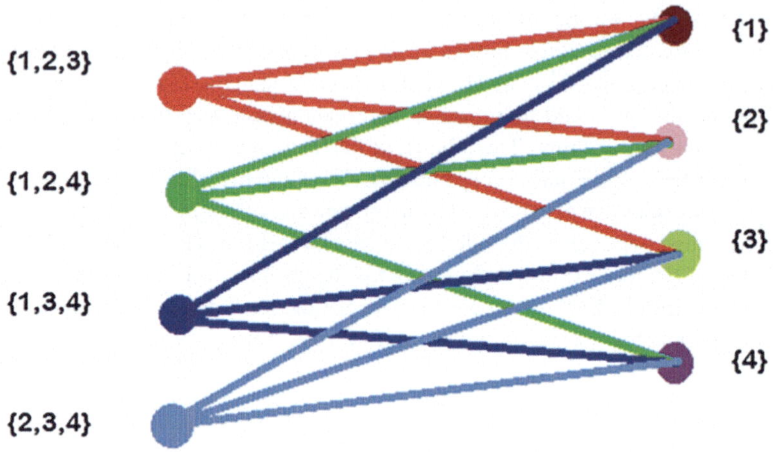

Fig. 3 Bipartite Kneser graph where n = 4, k = 3

3.2 LDPC Code Construction

The PCM of an LDPC code can be represented by a bipartite graph. So, the bipartite Kneser graph can be used as the LDPC matrix. But the graph cannot be directly used as the code matrix as four-cycle free condition needs to be ensured. Also, the bipartite Kneser graph gives rise to a square matrix. A square matrix implies a rate 0 code matrix. So, the bipartite Kneser graph cannot be directly used as the code matrix. The following proposed theorems are very useful for LDPC code construction from the bipartite Kneser graph.

Theorem 1 *A bipartite Kneser graph will be four-cycle free if n is odd and* $k = \frac{n+1}{2}$.

Proof Let n be even. Then either $k = n - k$ or $k = (n - k) + 2m$ where m is a positive integer. For the sake of simplicity, it is assumed that $k > n - k$. The results can be extended to $k < n - k$. If $k = n - k$ then the graph has no edges. For example: $S = \{1, 2, 3, 4\}$; n = 4, k = 2 = n − k; $a = \{(1, 2), (1, 3), (1, 4), (2, 3), (2, 4), (3, 4)\}$ and as $k = n - k$ so, $a = b$. As a result no vertices from a set and b set are adjacent to each other.

Now if $k = (n - k) + 2m$ then any two elements from a set are subset or superset of any elements of b. In other words if $a_i, a_j \subset a$ then there must be at least two elements $b_l, b_q \subset b$ such that either $a_i, a_j \subset b_l, b_q$ or $b_l, b_q \subset a_i, a_j$. Thus it is not four cycle-free.

If n is odd then similarly, either $k \geq (n - k) + 2$ or $k = (n - k) + 1$. If $k \geq (n - k) + 2$ then due to the above reasoning, the graph is no more four-cycle free. For example $S = \{1, 2, 3, 4, 5\}$; n = 5; k = 1; $a = \{(1, 2, 3, 4), (1, 2, 3, 5), (1, 2, 4, 5), (1, 3, 4, 5), (2, 3, 4, 5)\}$; $b = \{(1), (2), (3), (4), (5)\}$. Thus the corresponding adjacency matrix is as follows:

$$H = \begin{bmatrix} 1\,1\,1\,1\,0 \\ 1\,1\,1\,0\,1 \\ 1\,1\,0\,1\,1 \\ 1\,0\,1\,1\,1 \\ 0\,1\,1\,1\,1 \end{bmatrix}$$

Now if $k = (n - k) + 1$ then no any two elements from a set are either subset or superset of any elements of b. For example $S = \{1, 2, 3, 4, 5\}$; n = 5; k = 2;
$a = \{(1, 2, 3), (1, 2, 4), (1, 2, 5), (1, 3, 4), (1, 3, 5),$
$(1, 4, 5), (2, 3, 4), (2, 3, 5), (2, 4, 5), (3, 4, 5)\}$;
$b = \{(1, 2), (1, 3), (1, 4), (1, 5), (2, 3),$
$(2, 4), (2, 5), (3, 4), (3, 5), (4, 5)\}$.
So, the corresponding adjacency matrix is

$$H = \begin{bmatrix} 1\,1\,1\,0\,0\,0\,0\,0\,0\,0 \\ 1\,0\,0\,1\,1\,0\,0\,0\,0\,0 \\ 0\,1\,0\,1\,0\,1\,0\,0\,0\,0 \\ 0\,0\,0\,1\,0\,1\,1\,0\,0\,0 \\ 1\,0\,0\,0\,0\,0\,1\,1\,0\,0 \\ 0\,1\,0\,0\,0\,0\,1\,0\,1\,0 \\ 0\,0\,1\,0\,0\,0\,0\,1\,1\,0 \\ 0\,0\,0\,1\,0\,0\,1\,0\,0\,1 \\ 0\,0\,0\,0\,1\,0\,0\,1\,0\,1 \\ 0\,0\,0\,0\,0\,1\,0\,0\,1\,1 \end{bmatrix}$$

Thus, it can be seen that the matrix obtained is four cycle free.

After obtaining a four-cycle free bipartite Kneser graph matrix [18], a special matrix (SM) [19] is appended to it to obtain the final LDPC matrix. Also all zero, all one, and identity matrices are also appended. The matrices are so appended that the final PCM is in approximate lower triangular (ALT) format. Thus, the code can easily be encoded by the method in [13].

The SM used is of the structure as show in Fig. 4. The number of rows in SM is $n + 1$ and the number of columns is $\frac{n(n+1)}{2}$. The SM is denoted as *SM(n)*, as its size is a function of the positive integer n.

Let the LDPC PCM be denoted as H. The approximate lower triangular format of H is given as follows:

$$H = \begin{bmatrix} A & B & T \\ \tilde{C} & \tilde{D} & \mathbf{0} \end{bmatrix} \tag{2}$$

where A, B, T, $\tilde{C}$ and $\tilde{D}$ are sub-matrices of H and $\mathbf{0}$ denotes an all zero sub-matrix of suitable size.

For LDPC code construction, the bipartite Kneser graph is then split into two parts: *B.K.-A* and *B.K.-B*. *B.K.-A* is suitably appended to SM and an identity matrix to generate the A sub-matrix. *B.K.-B* is appended to an all-zero matrix to form $\tilde{C}$. The

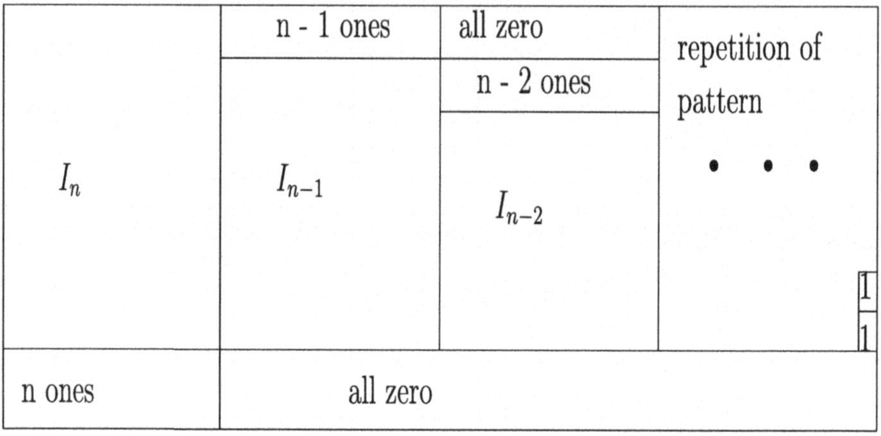

Fig. 4 Special matrix SM (n)

rest of the sub-matrices is generated by using the identity matrix, all-zero matrix, and SM.

The construction procedure can be explained by the following example.

Example 2 Let the bipartite Kneser graph having the parameters $n = 11$ and $k = 5$ be considered. In the first step of LDPC code construction, the bipartite Kneser graph is constructed. The adjacency matrix of the Kneser graph contains 462 rows and 462 columns. Now first 421 rows are included in B.K.-A and last 41 in B.K.-B. SM(60) and SM(40) are used for final LDPC code construction. The various sub-matrices of the H matrix are constructed as follows:

The structure of sub-matrix A of the proposed LDPC code is given in Fig. 5. The rest of the sub-matrices are given as follows:

$$B = \begin{bmatrix} \mathbf{0}_{481\times 41} \\ SM(40)^T \end{bmatrix} \tag{3}$$

$$\tilde{C} = \begin{bmatrix} B.K. - B \ \mathbf{0}_{41\times 1010} \end{bmatrix} \tag{4}$$

$$\tilde{D} = \begin{bmatrix} I_{41} \end{bmatrix} \tag{5}$$

$$T = \begin{bmatrix} I_{1301} \end{bmatrix} \tag{6}$$

$$\mathbf{0} = \begin{bmatrix} \mathbf{0}_{41\times 1301} \end{bmatrix} \tag{7}$$

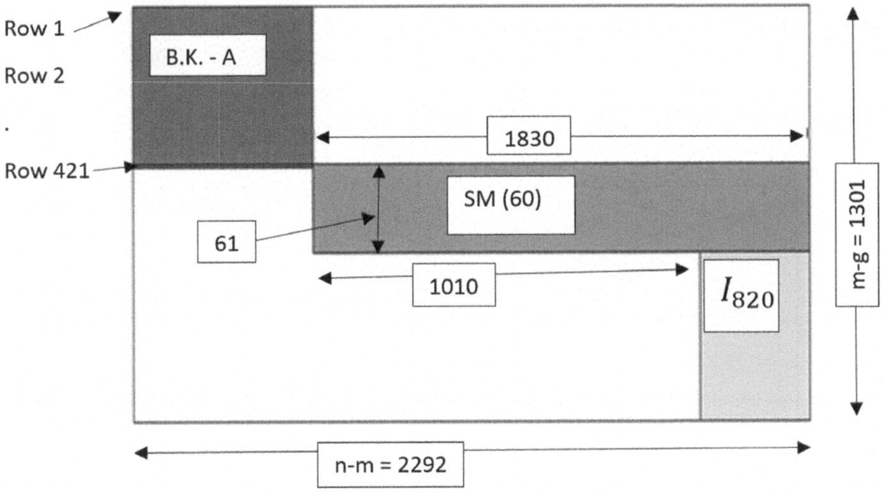

Fig. 5 Sub-matrix A of the LDPC code matrix (2292 × 1301 matrix)

It can be seen that in the obtained LDPC matrix no two columns have an entry of 1 in the same rows. Thus, the obtained code is four cycle-free.

4 Simulation Results and Discussions

The LDPC code matrix obtained from the proposed construction technique is shown in Fig. 6. After observing Fig. 6, it is clear that the designed LDPC matrix is highly

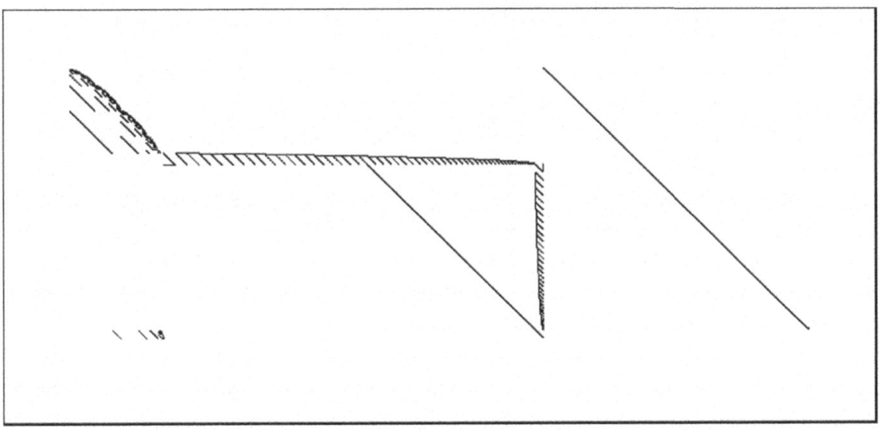

Fig. 6 Proposed LDPC code matrix from bipartite Kneser graph (dark portions are showing ones, else zeroes)

Table 1 Degree (D) distribution of the columns

D = 1	D = 2	D = 3	D = 6	D = 41
1301	1010	820	462	41

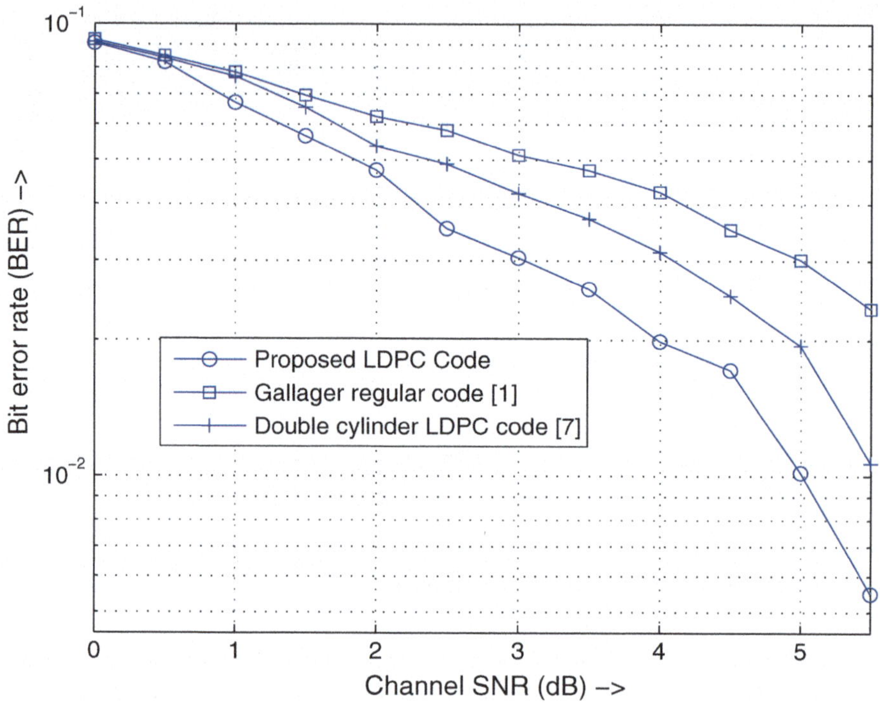

Fig. 7 BER performance of the proposed code

sparse. Such a sparse matrix can be encoded very efficiently. The degree distribution of the columns of the obtained LDPC code is shown in Table 1.

The BER performance of the obtained code in example 2 is given in Fig. 7. Thus, the constructed LDPC code is of size 1342×3634. It has a rate equal to 0.6307. The performance of the proposed code is compared with a regular Gallager LDPC [1] code and an LDPC code obtained from double cylinder graphs [12]. The Gallager code is a half-rate LDPC code of length 4000. It is taken from [20]. The code is converted into the ALT format [13] for efficient encoding. A rate $\frac{1}{9}$ code of length 2475 obtained by the method in [12] is chosen for simulation purposes.

It can be observed from Fig. 7 though the proposed LDPC code is of higher rate, it outperforms the existing codes. The proposed code structure from Kneser graphs, give better performance. Though the rate of the code as obtained from [12] is extremely low, it does not give good BER performance as it is a column weight 2 only code. For better BER performance, at least few columns of the LDPC code matrix should have column weight more than 2.

5 Conclusions

In this work, a novel LDPC code construction technique from the bipartite Kneser graph is presented. The code obtained is four-cycle free. it can be easily encoded by the ALT method. Also, the codes give better BER performance than the existing LDPC codes. Future work may be undertaken to construct LDPC codes giving further improvement in BER performance. This work is a first attempt to use Kneser graph. Investigations into amenability of the presented LDPC code matrix as a compressed sensing sampling matrix can also be taken up as a future work. The design of LDPC code matrices by customizing SM is introduced in this work. Improved custom design matrix can be designed in future to improve the performance of the designed LDPC code.

References

1. Gallager, R.G.: Low-density parity-check codes. IRE Trans. Inf. Theory **8**(1), 21–28 (1962)
2. Davey, M.C., MacKay, D.J.C.: Low density parity check codes over GF (q). In: IEEE Information Theory Workshop, pp. 70–71 (1998)
3. MacKay, D.J.C., Neal, R.M.: Near Shannon limit performance of low density parity check codes. Electron. Lett. **32**(18), 1645–1646 (1996)
4. Li, J., Liu, K., Lin, S., Abdel-Ghaffar, K.: Algebraic quasi-cyclic LDPC codes: construction, low error-floor, large girth and a reduced-complexity decoding scheme. IEEE Trans. Commun. **62**(8), 2626–2637 (2014)
5. Paolini, E., Fossorier, M.P., Chiani, M.: Generalized and doubly generalized LDPC codes with random component codes for the binary erasure channel. IEEE Trans. Inf. Theory **56**(4), 1651–1672 (2010)
6. Han, W., Huang, J.: A block-PEG construction method for LDPC codes. In: IEEE 4th International Conference on Electronics Information and Emergency Communication (ICEIEC), 2013, pp. 274–277 (2013)
7. Liu, K., El-Khamy, M., Lee, J.: Finite-length algebraic spatially-coupled quasi-cyclic LDPC codes. IEEE J. Sel. Ares Commun. **34**(2), 329–344 (2016)
8. Khodaiemehr, H., Kiani, D.: Construction and encoding of QC-LDPC codes using group rings. IEEE Trans. Inf. Theory **63**(4), 2039–2060 (2017)
9. Bocharova, I.E., Hug, F., Johannesson, R., Kudryashov, B.D., Satyukov, R.V.: Some voltage graph-based LDPC tailbiting codes with large girth. In: IEEE International Symposium on Information Theory (ISIT), 2011, pp. 732–736 (2011)
10. Malema, G., Liebelt, M.: High girth column-weight two LDPC codes based on distance graphs. EURASIP J. Wirel. Commun. Netw. **2007**(1), 6–6 (2007)
11. Couvreur, A., Delfosse, N., Zmor, G.: A construction of quantum LDPC codes from Cayley Graphs. IEEE Trans. Inf. Theory **59**(9), 6087–6098 (2013)
12. Samadieh, M., Gholami, M.: Double Cylinder Cycle codes of Arbitrary Girth (2014). arXiv:1403.0804v1 [cs.IT]

13. Richardson, T., Urbanke, R.: Modern Coding Theory. Cambridge University Press (2008)
14. Liu, X., Yuanbin, Z., Cui, R.: Variable-node-based dynamic scheduling strategy for belief-propagation decoding of LDPC codes. IEEE Commun. Lett. **19**(2), 147–150 (2015)
15. Babar, Z., Botsinis, P., Alanis, D., Ng, S.X., Hanzo, L.: Construction of quantum LDPC codes from classical row-circulant QC-LDPCs. IEEE Commun. Lett. **20**(1), 9–12 (2016)
16. Pramanik, A., Sarkar, S., Maity, S.P.: Performance Comparison of Orthogonal Complex MIMO STBC with ML Decoding and Soft Decision Decoding, accepted & presented in WISPNET (2017)
17. Pramanik, A., Farheen, Z., Maity, S.P.: On compressed sensing based channel estimation in massive MIMO. In: IEEE(EDS) Sponsored 5th International Conference on Computing. Communication and Sensor Network, Kolkata (2016)
18. Godsil, C., Royele, G.: Algebraic Graph Theory, vol. 207. Springer Science & Business Media, New York (2013)
19. Bose, R.: Information Theory, Coding and Cryptography. Tata McGraw-Hill Education (2008)
20. McKay, D.: Inference group home page. http://www.inference.org.uk/mackay/codes/data.html

Performance Analysis of CRC-Aided Polar Codes with SCL Decoding Algorithm for the Binary Deletion Channel

Tirthadip Sinha and Jaydeb Bhaumik

Abstract The performance of the polar coding scheme based on CRC-aided Successive Cancelation List (CA-SCL) decoding algorithm for practical transmission channel applications with deletion properties has been evaluated in this paper. The original information is pre-coded with Cyclic Redundancy Check (CRC) bits which helps to sense the position of deletions with high accuracy. SCL decoding considers these symbols as erasures and retrieves all information bits as well as deleted positions that agree with the CRC. A modification on polar code constructional algorithm is done to apply it in the binary deletion channels (BDC) as conventional polar codes cannot be used directly for channels with memory. The scheme is also applicable to concatenated binary discrete memoryless channels (BDMC) with the d-deletions. Results indicate that the original information can be recovered intact using the proposed scheme and the overall error probability decreases as block size N increases or polar code rate decreases for the variable list sizes. Also, the effect of CRC codelength and generator polynomials on the Block Error Rate (BLER) performance of proposed polar codes in BDC is investigated in this work.

Keywords Polar codes · SCL · CRC · BDC · BLER · CA-SCL · Deletion

1 Introduction

Arikan's (2009) polar codes [1] are recognized as the first channel coding family proven to attain the symmetric capacity of Binary-input Discrete Memoryless (BDMS) channels with low encoding–decoding complexity. These codes are also applicable practically in 5G wireless communication systems [2, 3]. Already, several efforts have made to construct polar codes for other channel models [4–6], e.g.,

T. Sinha (✉)
Haldia Institute of Technology, Haldia, West Bengal, India
e-mail: tirthadip.sinha@gmail.com

J. Bhaumik
Jadavpur University, Kolkata, West Bengal, India
e-mail: bhaumik.jaydeb@gmail.com

S. Kundu et al. (eds.), *Proceedings of the 2nd International Conference on Communication, Devices and Computing*, Lecture Notes in Electrical Engineering 602, https://doi.org/10.1007/978-981-15-0829-5_2

BSC and AWGN, providing a roadmap to polar code construction algorithms for practical transmission channels with memory [7–9] as well. The successive cancelation (SC) decoding algorithm of polar decoders is not performed well at large finite block lengths and that leads to several alternative SC decoding algorithms [6, 10] with low computational power by researchers later. The efficient implementation of Successive Cancelation List (SCL) decoding algorithm, invented by Tal and Vardy [11], outperformed LDPC and Turbo codes and is useful since then. The SCL decoding algorithm is also suitable for the information pre-coded with CRC bits. The polar SCL decoder can eradicate the codeword symbols mismatched with CRC information and attain reduced error rate even in channel imperfections by selecting correct received vectors. An optimum CRC detection algorithm for polar codes using SCL decoder is presented in [12].

In a communication system, synchronization anomalies caused by imperfect sampling outcomes dropping a few symbols of the received vector or sometimes implanting a few unwanted random symbols among the received ones. An example of a nonstationary and non-ergodic channel with memory is the Binary Deletion Channel (BDC) where information bits or symbols from the sender are either sent to the receiver (with deletion probability p) or deleted without notifying the receiver (with probability 1-p). Symbols can be deleted randomly whereas receiver is unfamiliar with deletion locations. The characteristics of a deletion channel are not similar to an erasure channel where bits may get erased or scrambled due to channel imperfections. For example, if 01010101 was transmitted to the channel, the receiver would acquire 00011 if the second, fourth, and seventh bits were deleted, and would acquire 0?0?01?1 if the bits were erased. Mitzenmacher [13] surveyed and explained deletion channels in detail. There have been several findings regarding upper and lower limits on deletion channel's capacity [13, 14] but the Shannon capacity of deletion channels is not fully explored yet [13]. The practical channels corrupted by these types of insertion or deletion errors [13] have memory and the polar codes used for memoryless channels cannot be applied straightway as the polarization of a channel with memory has not fully inspected. In a few recent works [7–9, 15–17], polarization theorems are exploited for processes with memory.

Dolecek et al. [21] showed that substitutions alongside a single deletion can be corrected using Reed-Muller (RM) codes. Due to similar algebraic constructions, polar code is a prospective candidate for correcting deletions too. Polar codes over a BDC channel can be considered as an adversarial version of the binary erasure channel (BEC) with deletion properties, and few modifications in the SCL decoding algorithm can be able to recover the original message successfully. For BEC with single deletion error, a polar decoding scheme is suggested by Thomas et al. [15] where the deleted symbols recovered by employing pre-coded CRC. First, all possible deletion patterns are enumerated and decoded for each case. The average asymptotic decoding complexity is $O(N^{d+1} \log N)$ for d-deletion symbols. Guess and check codes based upon Reed-Solomon (RS) codes are advocated by Hanna et al. [16] for deletion channel. The complexity of the decoder is $O(k^{d+2}/\log^{d} k)$, where information length is k. Tian et al. [17] proposed an alternative method which directly senses the deletion channel outputs to decode with no preprocessing and reducing the decoding complexity to $O(d^2 N \log N)$. The method can be suitable to perform CA-SCL decoding (e.g., CRC-Aided SCL) for channels with insertion and deletion errors.

In this work, a new polar decoding method exploiting CA-SCL decoding algorithm for the d-deletion channel is proposed and d could be fixed or variable. It successfully attains the symmetric information rate of BDC without any limit on the error/deletion patterns. A total of 25 different polar codes have been tested and results are plotted in respective graphs. For this test, the parameters are taken as code length $N = \{128, 256, 512, 1024, 2048\}$, Information rate $R_{info} \in \{3/7, 1/2, 5/9, 3/5, 2/3\}$ and CRC length $\in \{8, 12, 16, 23, 32\}$ with list size $|L| \geq 1$.

The rest of this paper is arranged in the given fashion. The polar code construction, SC decoding with a list, binary deletion channel, and CRC pre-coding with variable List sizes are discussed in Sect. 2. Next, the implementation of the CA-SCL decoder for deletion channel is explained based upon problem formulation. Simulation results are presented in graphs and analysis is done in Sect. 4. Complexed theoretical proofs and equations are omitted and possible references are given for in detail study. Lastly, the conclusion is drawn in Sect. 5.

2 Polar Coding Scheme

The CA-SCL Polar codes are basically binary polar codes concatenated with an external binary CRC code at the transmitter and SCL decoding together with CRC detection at the receiver. First, external CRC encoder encodes k information bits and produces $K = k + r$ bits, where r parity bits are affixed to identify precise information accurately during decoding. The internal polar encoder encoded this output K bits to create the N-length binary codeword and transmitted over BDC channel. The likelihood is determined from the received bits at receiver and delivered to the list-CRC decoder, where L survived candidate paths from the list are verified from the utmost probable paths. The suitable candidate codeword is selected for correction and corresponding k data bits are viewed as desired data bits.

2.1 Polar Code Construction

In coding theory, popular channel codes have universal structural design and architectures are not depend on the channel characteristics, but polar codes are distinct in this scenario. Polar Codes are simply linear block codes, proven to attain the symmetric capacity of BDMC. The basic idea is to create N different channels $W_N^{(i)}$ linearly from N independent copies of a BDMC channel W, $1 \leq i \leq N$. These $W_N^{(i)}$ channels are displayed as polarization as N rises. The term "polarization" means that transformed channels are either perfect/noiseless (i.e., mutual information $\longrightarrow 1$) or imperfect/noisy (i.e., mutual information $\longrightarrow 0$), asymptotically. Mutual information I(W) of a BDMC is defined as:

$$I(W) \triangleq \sum_{y \in Y} \sum_{x \in X} W(y|x) \log \frac{W(y|x)}{\frac{1}{2}W(y|0) + \frac{1}{2}W(y|1)} \tag{1}$$

where input $X = \{0, 1\}$, output = Y and transmission probabilities = $W(y|x)$.

The capacity of symmetric BDMC is identical to the mutual information of channel input–output and reliable transmission are feasible at any rate up to I(W). Optimal polar code construction algorithm is hard, therefore, many sub-optimal polar code construction has been proposed [1] at different computation complexities. The earliest polar code construction algorithm is from Arıkan using Bhattacharyya parameters [1] which are defined as:

$$Z(W) \triangleq \sum_{y \in Y} W(y|0)W(y|1) \tag{2}$$

Arıkan proved that Z(W) of BEC had a pair of upper limits as $\{z, z\} \longrightarrow \{2z - z^2, z^2\}$ on the probability of maximum likelihood (ML) decision error at each polarizing transform as Z(W) which is a measure of reliability of the channel.

In (N, k) polar code, $N = 2^n$ represents blocksize and k signifies the information length. Given an information vector u_1^N, a codeword x_1^N is generated from the equation $x_1^N = u_1^N B_N G_2^{\otimes n}$, where $G_2^{\otimes n}$ is the n-th Kronecker product of G_2 (The kernel $G_2 = \begin{bmatrix} 1 & 0 \\ 1 & 1 \end{bmatrix}$) and B_N is a bit reversal permutation matrix, as published in [1]. The vector x_1^N is sent over N independent replicas of a BDMC and the distinct channels begin polarizing as n advances. Some channels become absolutely noise-free and others become fully noisy. Noise-free channels incline to the capacity C(W). The polarization characteristics recommend the noise-free channels to send information bits and noisy channels to frozen bits known to the decoder. A message vector u_1^N contains both information bits and frozen bits (typically set to zero). The polar code constructional algorithm as proposed by Tal and Vardy in [18] can be considered as most precise and works at lower rates. The complexity of polar encoder is $\mathcal{O}(N \log N)$.

2.2 SC Decoding

SC decoding scheme for polar codes is introduced by Arikan [1] considers the channel output as y_1^N and the i-th synthesized subchannel with input u_i and output (y_1^N, u_1^{i-1}) by transition probability matrix $W_N^{(i)}$ for $i = 1, \ldots .N$. For the given y_1^N and the estimates $\hat{u}_1^{i-1}$ of u_1^{i-1}, the SC decoding algorithm evaluates u_i. The logarithmic likelihood ratios (LLR) are applied for estimation of each u_i where $i = 1, \ldots N$.

$$L_N^{(i)}\left(y_1^N, \hat{u}_1^{i-1}\right) = \log \frac{W_N^{(i)}\left(y_1^N, \hat{u}_1^{i-1} \middle| u_i = 0\right)}{W_N^{(i)}\left(y_1^N, \hat{u}_1^{i-1} \middle| u_i = 1\right)} \tag{3}$$

The sign of the LLRs controls the estimation of an unfrozen bit u_i so that $\hat{u}_i = 0$ if $L_N^{(i)}\left(y_1^N, \hat{u}_1^{i-1}\right) \geq 0$ and $\hat{u}_i = 1$ otherwise. The complexity of SC decoder is $\mathcal{O}(N \log N)$.

2.3 SCL Decoding

SCL decoder checks qualified paths related to the input bits in a tree-diagram and hold best L contenders in parallel [11] where $|L|$ signifies the size of the list. The path with the maximum likelihood is selected as the utmost reliable path and CRC detector verifies its information sequence. When the CRC detector identifies the selection as wrong, the verification of the contenders with the next probable likelihood done till any of the L contenders passes the CRC test. The corresponding flowchart of SCL decoding algorithm is presented in Fig. 1.

2.4 List-CRC Detection in CRC-Aided SCL Decoding

In Fig. 2, the block diagram of polar codes using CRC-aided SCL decoding scenario is shown. Instead of setting all $N-k$ frozen bits to zero, $N-k-r$ frozen bits can be set to zero and r bit CRC value (or the parity bits) are applied. CRC polynomial of degree r contains $r + 1$ coefficients and r coefficient bits are affixed after k information bits to produce $(K = k + r)$ bit [11, 19]. The polynomial division is performed to confirm the accuracy of the received message bits if the remainder is zero. The next message with highest LLRs selected from the list with predefined list size L and all vectors deleted if failed to pass CRC verification. The capacity of the BEC is attained by the BDC because $|L| \leq N^d$ is still sub-exponential although $d = \mathcal{O}\left(\frac{N}{\log N}\right)$.

The r bit CRC is affixed after the k bit information as an external code with code rate $k/(k + r)$ so that the actual polar code rate is increased to $R_{polar} = (k + r)/N$ keeping original information rate $R_{info} = (k/N)$ intact. Hence, polar code rate increases and error correction declines as r gets incremented to get improved error detection performance. The redundancy introduced by the outer CRC code should be planned realistically, particularly, if block size (N) is small. It can falsely identify the wrong one as right candidate if improper CRC length is applied.

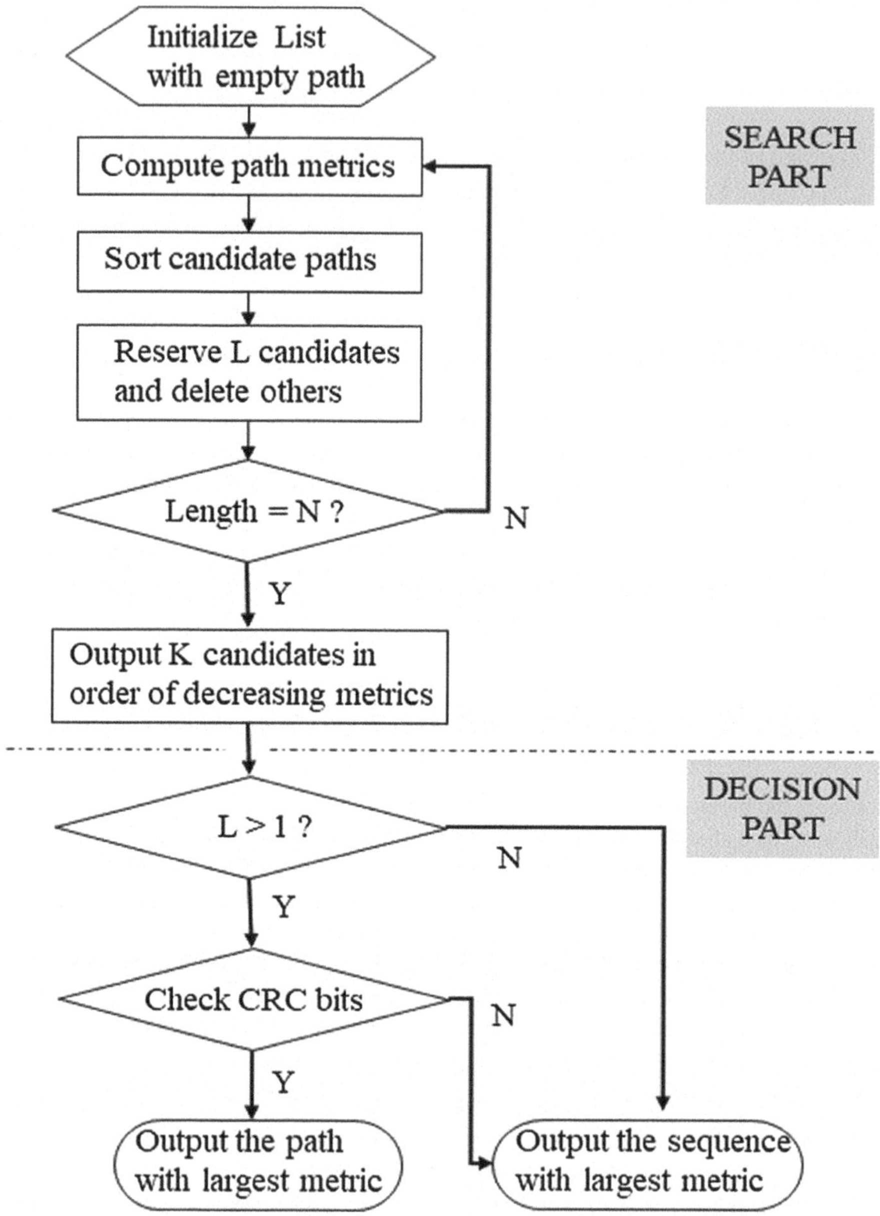

Fig. 1 SCL decoding algorithm with CRC bits

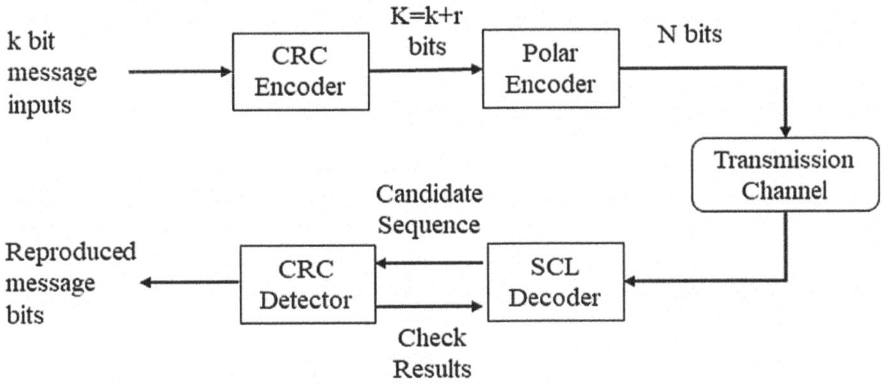

Fig. 2 Polar codes using CRC-aided SCL decoding

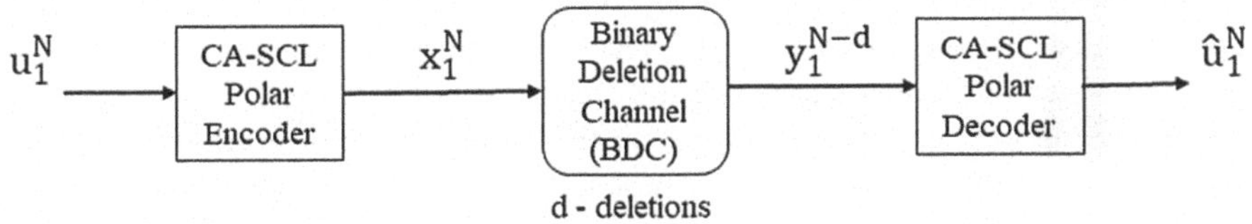

Fig. 3 Binary Deletion Channel model with d-deletions where receiver only succeed to receive N-d coded blocks instead of N

2.5 *Binary Deletion Channel (BDC)*

The block diagram of BDC with d-deletions is presented in Fig. 3. The CA-SCL decoder gets $N - d$ bits as channel output in a d-deletion channel for N transmitted bits though the receiver is unaware of deleted bit locations [13].

In some works, it is suggested that an individually coded bit can be deleted with a deletion probability $P_d \in (0, 1)$. Though the capacity of BDC is still not fully known, the upper bound capacity can be taken as $C(W) = (1 - P_d)$ if the deletion locations are considered as known to receiver and the capacity of BDC will become similar to the capacity of BEC.

3 Implementation and Problem Formulation

Considering the capacity of BDC is the same as the capacity of BEC with deletion probability $P_d \in (0, 1)$ for d-deletions, a message vector u_1^N is encoded using CRC-aided polar encoder and is transmitted through the channel. The CRC generator polynomials [2, 20] used in this work are given below as

CRC-8: $z^8+z^7+z^6+z^4+z^2+1$

CRC-12: $z^{12}+z^{11}+z^3+z^2+z+1$
CRC-16: $z^{16}+z^{15}+z^2+1$
CRC-23: $z^{23}+z^9+z^7+z^5+z^3+1$
CRC-32: $z^{32}+z^{26}+z^{23}+z^{22}+z^{16}+z^{12}+z^{11}+z^{10}+z^8+z^7+z^5+z^4+z^2+z+1$

The channel permits erasures along with d-deletions. The CA-SCL polar decoder is designed and capable to retrieve the original message from the received vector with a list |L| of linear size in N. The list contains the estimation of $\hat{u}_1^N$ for the transmitted message vector u_1^N. All the scaling parameters involved in encoding–decoding process can be saved in memory to reduced realtime decoding complexity further.

If L represents the candidate codewords formed by SCL decoder with list $L = \{1, 2, \ldots L\}$ and ℓ^* represent the list index which compares to the right codeword. Without loss of generality, it can be anticipated those smaller indices signify more probable contenders. If the list has no correct codeword, i.e., $\ell^* \notin L$, it can be assumed that $\ell^* > L$. Furthermore, let $P_{l^*}(\gamma, L)$ = the probability adapted on γ, i.e., $P_r\{\ell^\star = l|\gamma\}$ denote the distribution of ℓ^* which is the probability that ℓ^* concurs with l at a stated Signal-to-Noise Ratio over a transmission channel.

Considering the perfect situation where the right codeword in the list is recognized accurately, then the overall error probability $P_{e,id}(\gamma, L)$ of the decoder can be expressed as

$$P_{e,id}(\gamma, L) = 1 - \sum_{l=1}^{L} P_{l^*}(\gamma, L) \tag{4}$$

The error correction ability of CA-SCL decoding mainly depends on the proper utilization of CRC polynomials regardless of whether the length of CRC codes is the same.

4 Result Analysis

The effectiveness of the offered CA-SCL decoding algorithm for the BDC is shown by MATLAB simulation results in this section. Here, deletion errors with erasures are considered. However, the same algorithm might be utilized for the combination of deletion error with additive noise and with symmetric flips in case of AWGN channel and BSC, respectively. Since BDC differs from BDMC, the conventional constructional algorithm cannot be used straightway and frozen bits cannot be considered to zero always. Thus, small modifications in the polar code constructional algorithm are done and large computational complexity is required to attain high precision results. In fact, frozen bits are taken randomly and frozen bit information is fully known to the respective encoder-decoder module. Deletion bits are also taken randomly with the fixed deletion probability as high as 0.3 and deletion positions are not known to decoder. CRC polynomial is selected according to [2, 20]. More specifically, CRC-8,

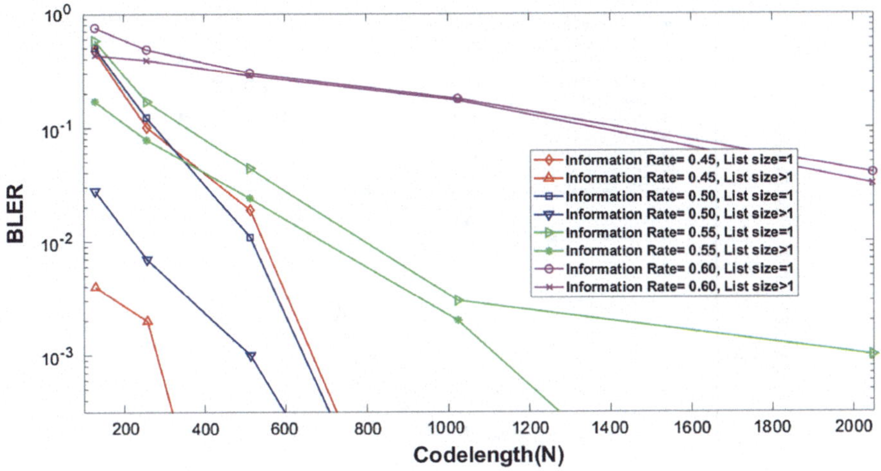

Fig. 4 Plot of Block Error Rate (BLER) against Codelength (N) with fixed deletion probability and variable CRC length

CRC-12, CRC-16, CRC-23, and CRC-32 are used with codelengths $N_1 = 128$, $N_2 = 256$, $N_3 = 512$, $N_4 = 1024$, and $N_5 = 2048$, respectively.

In Fig. 4, the relation between Codelength N and overall Block error (BLER) is demonstrated for the proposed CA-SCL polar code of different information rates $R_{info} \in \{3/7, 1/2, 5/9, 3/5, 2/3\}$ with CRC length $\in \{8, 12, 16, 23, 32\}$ and List size $|L| \geq 1$. As expected, polar codes perform better with lower codelengths N and show poor performance at a high information rate.

In Fig. 5, the relation between Polar Code Rate R_{polar} and overall error probabilities is illustrated for the proposed CA-SCL polar code of different code lengths or Block sizes $N \in \{128, 256, 512, 1024, 2048\}$ with CRC length $\in \{8, 12, 16, 23, 32\}$ and List size $|L| \geq 1$. Results show that polar codes perform better with lower code rates applying different CRC sizes and able to retrieve original messages accurately. At lower blocksize (or codelength), polar codes exhibit high error probability even in low code rates.

5 Conclusion

The error correction ability of CRC-aided polar codes is investigated in this paper for the binary deletion channel. A compromising relationship between the BLER performance and CRC codelength is observed when the polar code rate is the same. If the CRC length is lengthier than essential for a specified BLER, the dreadful performance is anticipated as a result of the comparative increment in the polar code rate. In fact, the larger size of the list and the codelength leads to longer computation

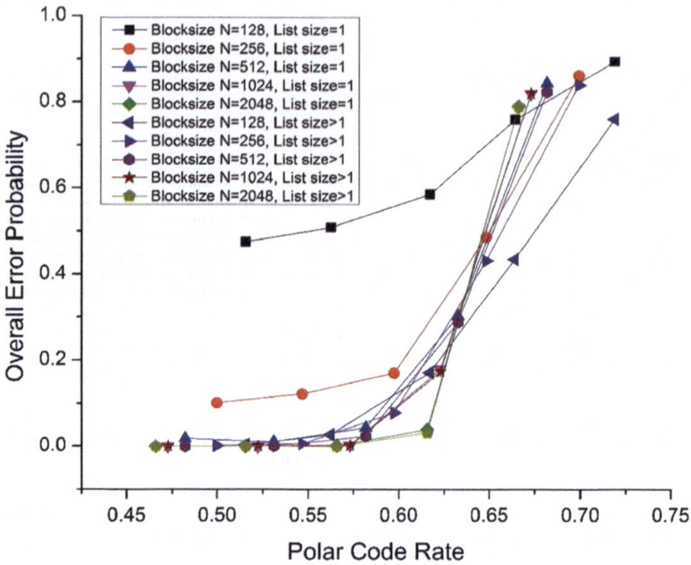

Fig. 5 Plot of Overall Error Probabilities against Polar Code Rates [$R_{polar} = (k + r)/N$] with fixed deletion probability and variable CRC length

time. The precise performance analysis of CA-SCL decoding is still challenging because of the correlation between codeword bits.

References

1. Arıkan, E.: Channel polarization: a method for constructing capacity achieving codes for symmetric binary-input memoryless channels. IEEE Trans. Inf. Theory **55**(7), 3051–3073 (2009)
2. 5G NR, Multiplexing and channel coding (3GPP TS 38.212 version 15.3.0 release 15), ETSI TS 138 212 V15.3.0 (2018-10)
3. Hui, D., Sandberg, S., Blankenship, Y., Andersson, M., Grosjean, L.: Channel coding in 5G new radio. IEEE Veh. Technol. Mag. **13**(4), 60–69 (2018)
4. Abbe, E., Barron, A.: Polar coding schemes for the AWGN channel. In: Proceedings of ISIT, pp. 194–198 (2011)
5. Sasoglu, E.: Polar coding theorems for discrete systems, Ecole Polytechnique Federale de Lausanne. Thesis no. 5219 (2011)
6. Vangala, H., Hong, Y., Viterbo, E.: Efficient algorithms for systematic polar encoding. IEEE Commun. Lett. **20**(1), 17–20 (2016)
7. Wang, R., Honda, J., Yamamoto, H., Hou, Y., Liu, R.: Construction of polar codes for channels with memory. In: Proceedings of ITW, Jeju Island, South Korea, pp. 187–191 (2015)
8. Sasoglu, E., Tal, I.: Polar coding for processes with memory. In: IEEE International Symposium on In Information Theory, pp. 225–229 (2016)
9. Shuval, B., Tal, I.: Fast polarization for processes with memory. IEEE Trans. Inf. Theory, 1–17 (2018). (Early Access)

10. Trifonov, P.: Efficient design and decoding of polar codes. IEEE Trans. Commun. **60**(11), 3221–3227 (2012)
11. Tal, I., Vardy, A.: List decoding of polar codes. IEEE Trans. Inf. Theory **61**(5), 2213–2226 (2015)
12. Zhang, Q., Liu, A., Pan, X., Pan, K.: CRC code design for list decoding of polar codes. IEEE Commun. Lett. **21**(6), 1229–1232 (2017)
13. Mitzenmacher, M.: A survey of results for deletion channels and related synchronization channels. Probab. Surv. **6**, 1–33 (2009)
14. Venkataramanan, R., Tatikonda, S., Ramchandran, K.: Achievable rates for channels with deletions and insertions. IEEE Trans. Inf. Theory **59**(11), 6990–7013 (2013)
15. Thomas, E.K., Tan, V.Y., Vardy, A., Motani, M.: Polar coding for the binary erasure channel with deletions. IEEE Commun. Lett. **21**(4), 710–713 (2017)
16. Hanna, S.K., Rouayheb, S.E.: Guess & check codes for deletions, insertions, and synchronization. IEEE Trans. Inf. Theory **65**(1), 3–15 (2019)
17. Tian, K., Fazeli, A., Vardy, A., Liu, R.: Polar codes for channels with deletions. In: Proceedings of 55th Annual Allerton Conference on Communication, Control, and Computing, pp. 572–579 (2017)
18. Tal, I., Vardy, A.: How to construct polar codes. IEEE Trans. Inf. Theory **59**(10), 6562–6582 (2013)
19. Niu, K., Chen, K.: CRC-aided decoding of polar codes. IEEE Commun. Lett. **16**(10), 1668–1671 (2012)
20. Wasserman, D.: Observations on Polar Coding with CRC-aided List Decoding. Technical Report 3041. SSC Pacific, San Diego, CA 921525001 (2016)
21. Dolecek, L., Anantharam, V.: Using reed–müller RM (1,m) codes over channels with synchronization and substitution errors. IEEE Trans. Inf. Theory **53**(4), 1430–1443 (2007)

Circular Antenna Array Synthesis Considering the Mutual Coupling Using Opposition Based DE

Avishek Das, D. Mandal and R. Kar

Abstract In this paper a circular antenna array (CAA) is designed for the improved far-field radiation pattern synthesis using a robust search algorithm called Opposition based Differential Evolution (ODE). ODE algorithm is employed here to find out the optimum current excitation weights and the inter-element separation between the array element to suppress the Side Lobe Level (SLL) and 3-dB beamwidth considering the mutual coupling between the array elements. The ODE algorithm-based results are compared with the uniform radiation pattern and with those of the reported literature with well-established algorithms like Genetic Algorithm (GA) and Particle Swarm Optimization (PSO) to show the superiority of ODE based approach.

Keywords Circular antenna array · Side Lobe Level · 3-dB beamwidth · Mutual coupling · Opposition based differential evolution

1 Introduction

The modern communication system deals with an accurate antenna array design. During the last few decades, several researches have been carried out on the antenna design for the improvement of radiation patterns. The two principal criteria of an improved radiation pattern synthesis mainly depend on the low SLL and narrow 3-dB beamwidth which is essential for avoiding the interference with other systems and for high directivity, respectively [1, 2]. But to design the antenna array maintaining these two principal criteria is very challenging because an array will never produce a low SLL and a narrow 3-dB beamwidth at the same time [1, 3].

The radiation characteristics of the antenna array degrade due to mutual coupling [4, 5]. But a very few attempts were made to suppress the mutual coupling [6] in

A. Das (✉)
Department of ECE, HIT, Haldia 721657, India
e-mail: avishek.uit0408@gmail.com

D. Mandal · R. Kar
Department of ECE, NIT Durgapur, Durgapur 713209, India

S. Kundu et al. (eds.), *Proceedings of the 2nd International Conference on Communication, Devices and Computing*, Lecture Notes in Electrical Engineering 602, https://doi.org/10.1007/978-981-15-0829-5_3

antenna arrays. The mutual coupling is compensated here by optimizing the driving admittance. The driving admittance is also known as mutual coupling [3]. The performance of detecting and transmitting the electromagnetic signal in CAA degrades heavily due to the mutual coupling [4]. This paper introduces an efficient approach for mutual coupling suppressing by adjusting the driving admittance of each array element. The driving admittance of the array is suppressed by optimizing the current excitation weights of each antenna of the array using the ODE algorithm.

The main focus of the paper is to design a CAA having a low SLL and a narrow 3-dB beamwidth considering the mutual coupling. The above-mentioned requirements of CAA are obtained by optimizing the current excitation weights and the inter-element separation keeping the phase difference zero. The simulation results for the CAA design using different algorithms GA [7–9], PSO [10–12], and ODE [13, 14] are compared in this paper. This paper is arranged as follows: Sect. 2, discuss the design of CAA. Section 3, presents the employed algorithm for CAA design. Section 4, shows the comparison of results achieved by using different algorithms. And Sect. 5, shows the conclusion of the paper.

2 Design Equation

A CAA of N elements with a radius ‘a’ is placed in the x-y plane and scanning a far-field point PP, as shown in Fig. 1. The radiation pattern of this array is represented by its array factor (AF) only as shown in (1) because all the elements are isotropic here [1].

$$AF(\phi, I, d) = \sum_{n=1}^{N} I_n \exp(j[ka \sin\theta \cos(\phi - \phi_n) + \alpha_n]) \tag{1}$$

where

$$ka = \frac{2\pi a}{\lambda} = \sum_{i=1}^{N} d_i \tag{2}$$

$$\phi_n = \left(\frac{2\pi}{ka}\right) \sum_{i=1}^{n} d_i \tag{3}$$

where the current excitation of the nth element is represented by I_n whereas the inter-element separation is represented by d_i; α_n is the phase angle; $k = \frac{2\pi}{\lambda}$, the operative wavelength is represented by λ, θ and ϕ is the elevation and the azimuth angle, respectively whereas ϕ_n is the position of the nth element of the CAA.

The Cost Function (CF) for suppressing the SLL and 3-dB beamwidth considering the mutual coupling is given in (4).

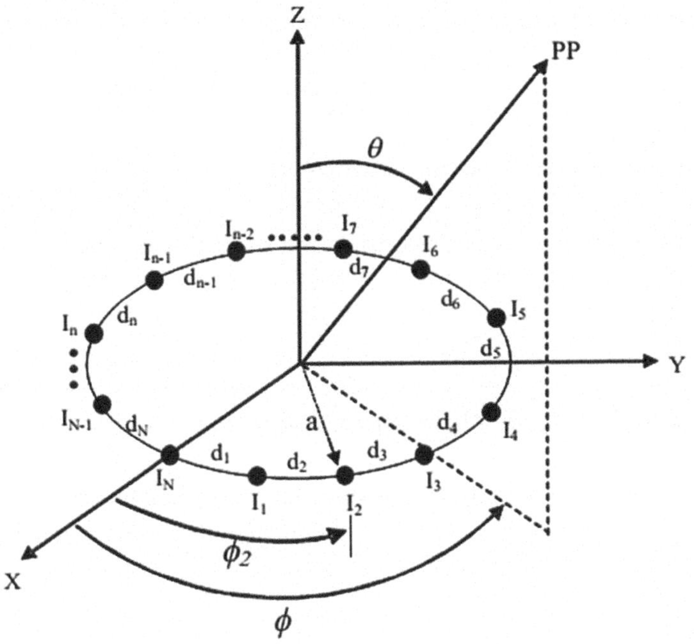

Fig. 1 Geometry of the CAA

$$CF = W_1 \times \frac{|AF(\theta_{ms1}, I_{n,i}) + AF(\theta_{ms2}, I_{n,i})|}{|AF(\theta_0, I_{n,i})|} + W_2 \times \left(\frac{1}{Z_D}\right) + W_3 \times \left[3 - dBBW_{\text{Computed}} - 3 - dBBW(I_{n,i} = 1)\right] \quad (4)$$

where

$$Z_D = \sum_{n=1}^{N} \left(\frac{\sum_{i=1}^{N} I_{n,i}}{I_{n,i}}\right); \quad (5)$$

Here W_1, W_2 and W_3 are the weighing factors and the peak angle of main beam is θ_0; θ_{ms1} and θ_{ms2} is the side lobe angle where the maximum value is attained in the lower and upper band, respectively; The current excitation weights of (*n*, *i*)th element is represented by $I_{n,i}$. 3-dB beamwidth is the angular spacing between the half power points. Hence, BW_{Computed} and 3-*dBBW* ($I_{n,i} = 1$) is the computed 3-dB beamwidth and the uniform excitation beamwidth, respectively. The 3-*dBBW* in the *CF* is considered only when 3-$dBBW_{Computed}$ > 3-*dBBW* ($I_{n,i} = 1$).

The first term in the *CF* is employed for SLL suppression; the second and third term in the *CF* is employed to reduce the mutual coupling and the 3-dB beamwidth, respectively. Here, a CAA is designed to reduce the SLL, 3-dB beamwidth, and mutual coupling with an optimal set of current excitation weights and inter-element separation.

3 Evolutionary Technique Employed

Evolutionary algorithms are characterized as robust and stochastic search methods to determine the complex optimization problem. ODE [13, 14] algorithm is applied here for CAA design.

3.1 *Opposition Based Differential Evolution*

ODE is a metaheuristic optimization method. The opposition based learning (OBL) method [15] is applied in ODE. The ODE algorithm is discussed in detail in [13, 14].

4 Result and Discussion

A comprehensive study on 8 and 10 element CAA design have been performed using ODE algorithm to find out the optimal current excitation weights and inter-element separation. The best results are achieved by using ODE algorithm after 50 trial run using MATLAB 7.5, in Intel Core (TM) 2 processor @3.00 GHz with 3 GB RAM.

4.1 *Numerical Results*

The results obtained by using ODE in the presence of mutual coupling effect for CAA design is compared with the uniformly excited radiation pattern and also with those of published reported literature having equal sets of element where mutual coupling effect is not considered.

The results of the uniform radiation pattern of 8 and 10 elements CAA as shown in Table 1 are plotted as the uniform pattern in the presented work. The SLL value for the uniform pattern of 8 and 10 elements CAA is −4.17 dB and −3.597 dB,

Table 1 Different parameters of uniform pattern for 8 and 10 elements CAA

No. of element	SLL (dB)	3-dB beamwidth (°)	Mutual coupling
8	−4.17	32.32	0.0156
10	−3.597	25.84	0.0100

Table 2 Results obtained by using ODE compared with other algorithms for N = 8 elements

Algorithms	Currents		Distance		SLL (dB)	3-dB beamwidth (°)	Mutual coupling
GA [10]	0.3289	0.2537	0.1739	0.3144	−9.811	32	0.0118
	0.7849	1.0000	0.6620	0.7425			
	0.9171	0.5183	0.6297	0.8929			
	0.6176	0.4612	0.4633	0.5267			
PSO [15]	0.7765	0.3928	0.3590	0.5756	−10.799	32	0.0080
	0.6069	0.8446	0.2494	0.7638			
	1.0000	0.7015	0.6025	0.8311			
	0.9321	0.3583	0.7809	0.3308			
ODE	0.2599	0.0004	0.3601	0.5838	−15.83	32.76	0.0146
	0.1463	0.3219	0.7784	0.9017			
	0.3342	0.1466	0.5433	0.9029			
	0.2625	0.1946	1.3859	0.3428			

respectively, whereas the 3-dB beamwidth of the above-mentioned arrays is 32.32 deg and 25.84 deg and the mutual coupling values are 0.0156 and 0.0100. Table 1 depicts clearly that 3-dB beamwidth and mutual coupling values of the uniform pattern for 8 and 10 elements CAA decreases gradually with the increment of array elements in the uniform pattern of CAA design.

The GA [9], PSO [12], and ODE based results for nonuniformly excited and nonuniformly spaced 8 elements circular antenna array design are given in Table 2. The table contains the parametric study obtained by using different algorithms for 8 elements CAA design.

The results achieved by using different algorithms for 8 elements CAA design are plotted in Fig. 2. Figure 2 reveals that ODE yields the lowest SLL value and lowest 3-dB beamwidth as compared with other well-established algorithms. The figure depicts the superiority of ODE over other algorithms.

The ODE, GA [9], and PSO [12] based results are compared in Table 3. The table shows the different parameters obtained by using different algorithms for 10 elements CAA design. The results of Table 3 show that the authors obtained much-improved result of SLL and 3-dB beamwidth by using ODE compared with other well-established algorithms.

Figure 3 depicts the radiation patterns achieved by using different algorithms for 10 elements CAA design. The plots of Fig. 3 show visibly that ODE algorithm-based results are much superior to the results obtained by using other well-established algorithms.

All the simulation results and their graphical plots of far-field radiation patterns expose the superiority of ODE over the other well-established algorithms.

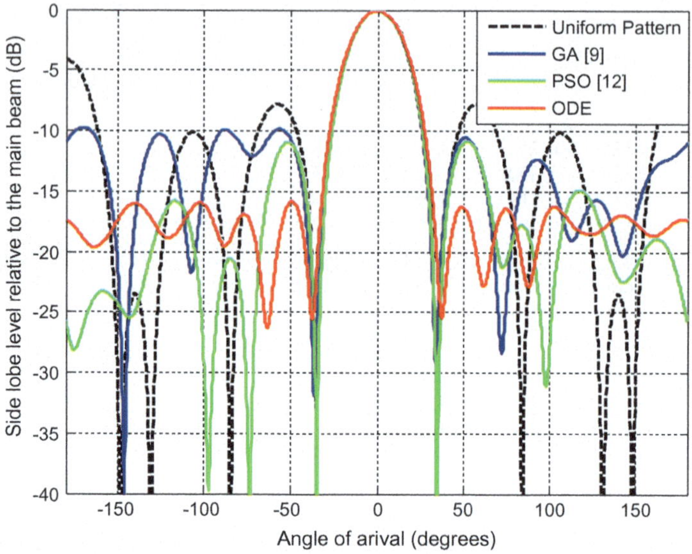

Fig. 2 Radiation patterns of different algorithms for the 8-element CAA design

Table 3 Results obtained by using ODE compared with other algorithms for N = 10 elements

Algorithms	Currents		Distance		SLL (dB)	3-dB beamwidth (°)	Mutual coupling
GA [10]	0.9545 0.3392 0.8086 0.5634 0.7045	0.4283 0.9074 0.4533 0.6015 0.5948	0.3641 0.2750 0.6902 0.4657 0.6456	0.4512 1.6373 0.9415 0.2898 0.3282	−9.811	32	0.0094
PSO [15]	1.0000 0.7519 0.5062 0.7501 1.0000	0.7529 1.0000 1.0000 0.7524 0.5067	0.3170 0.3859 0.3185 0.9657 0.9650	0.9654 0.9654 0.3164 0.3862 0.3174	−12.307	24.34	0.0063
ODE	0.3200 0.2062 0.4937 0.2671 0.3287	0.3452 0.5044 0.1448 0.0457 0.4791	0.6801 1.8986 0.5530 1.5077 0.8508	1.0279 1.5754 1.1362 0.8021 0.6074	−14.94	16.55	0.0153

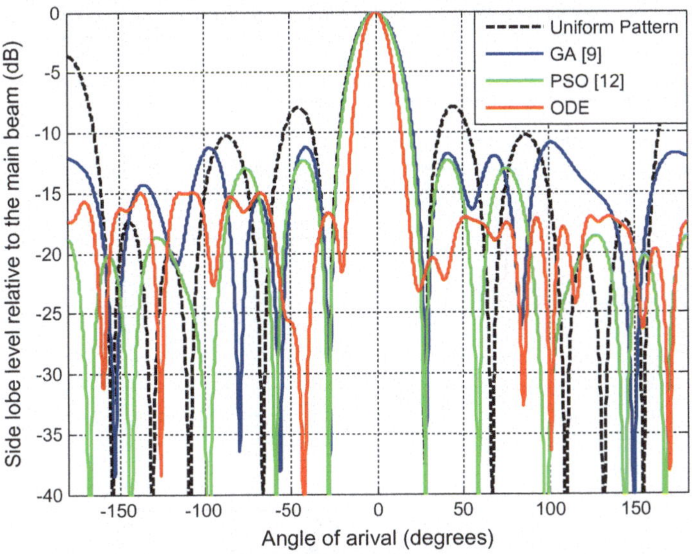

Fig. 3 Radiation pattern of different algorithms for the 10-element CAA design

4.2 Convergence Profiles of ODE

The minimum cost function value with the iteration cycles are plotted in the convergence profiles. The convergence of the ODE algorithm for 8 and 10 elements CAA design in the presence of mutual coupling are plotted in Fig. 4.

The convergence plots expose the excellent performance of the ODE algorithm in terms of precision and convergence speed.

5 Conclusion

This paper illustrates the design method of a circular antenna array for maximum SLL and 3-dB beamwidth reduction considering the mutual coupling effect by using Opposition based Differential Evolution (ODE). The ODE algorithm-based results prove the outstanding performance for the reduction of SLL and 3-dB beamwidth as compared with the published results of some well-established algorithms like GA and PSO. Thus, the ODE algorithm-based results confirm the superiority of the algorithm for the synthesis and pattern modeling of a nonuniform single ring CAA.

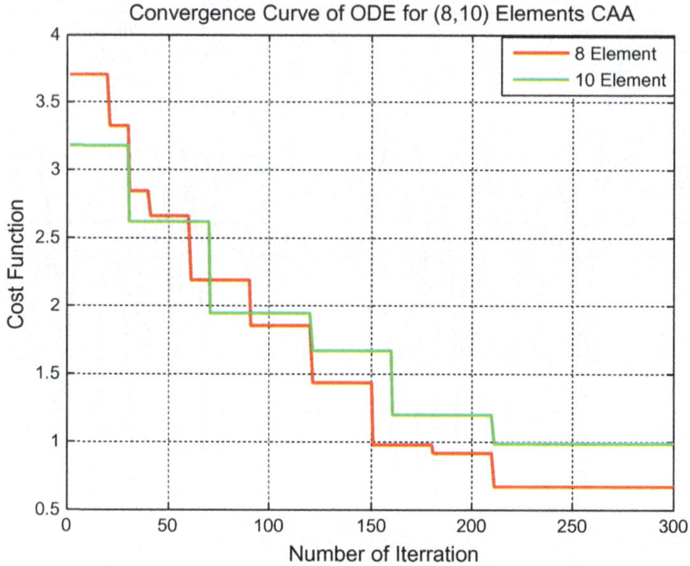

Fig. 4 Convergence profiles of 8 and 10 elements CAA in presence of mutual coupling

References

1. Ballanis, C.A.: Antenna Theory Analysis and Design, 2nd edn. Willey, New York (1997)
2. Mandal, D., Ghoshal, S.P., Bhattacharjee, A.K.: Wide null control of symmetric linear antenna array using novel particle swarm optimization. Int. J. RF Microwav. Comput. Aided Eng. **21**(4), 376–382 (2011)
3. Elliott, R.S.: Antenna Theory and Design, Revised edn. IEEE Press, New Jersey (2003)
4. Lui, H.S., Hui, H.T., Leong, M.S.: A note on the mutual-coupling problems in transmitting and receiving antenna arrays. IEEE Antennas Propag. Mag. **51**(5), 171–176 (2009)
5. Gupta, I.J., Ksienski, A.A.: Effect of mutual coupling on the performance of adaptive arrays. IEEE Trans. Antennas Propag. **ap-31**(5), 785–791 (1983)
6. Chakravorty, P., Mandal, D.: Radiation pattern correction in mutually coupled antenna arrays using parametric assimilation technique. IEEE Trans. Antennas Propag. **64**(9), 4092–4095 (2016)
7. Haupt, R.L., Werner, D.H.: Genetic Algorithms in Electromagnetics. IEEE Press and Wiley-Interscience (2007)
8. Haupt, R.L.: An Introduction to genetic algorithm for electromagnetics. IEEE Antenna Propag. Mag. **37**(2), 7–15 (1995)
9. Panduro, M., Mendez, A., Dominguez, R., Romero, G.: Design of non-uniform antenna arrays for side lobe reduction using the method of genetic algorithm. Int. J. Electron. Commun. (AEU) **60**(7), 713–717 (2006)
10. Kennedy, J., Eberhart, R.: Particle swarm optimization. In: Proceedings of the IEEE International Conference on Neural Network, Perth, pp. 1942–1948 (1995)
11. Boeringer, D.W., Werner, D.H.: Particle swarm optimization versus genetic algorithm for phased array synthesis. IEEE Trans. Antenna Propag. **52**(3), 771–779 (2004)
12. Sahib, M., Najjar, Y., Dib, N., Khodier, M.: Design of non-uniform circular antenna arrays using the particle swarm optimization. J. Electr. Eng. **59**(4), 216–220 (2008)

13. Li, X., Yin, M.: An opposition-based differential evolution algorithm for permutation flow shop scheduling based on diversity measure. Adv. Eng. Softw. **55**, 10–31 (2013)
14. Rahnamayan, S., Tizhoosh, H.R., Salama, M.M.A.: Opposition-based differential evolution. IEEE Trans. Evol. Comput. **12**(1), 64–79 (2008)
15. Tizhoosh, H.R.: Opposition-based learning: a new scheme for machine intelligence. In: Proceedings of the International Conference on Computational Control and Automation, vol. 1, pp. 695–701 (2005)

Compact Dual Band Microstrip Antenna Using Square Loop Slot for GSM 1800 and HiperLAN-2 Applications

Avisankar Roy, Sunandan Bhunia, Debasree Chanda Sarkar and Partha Pratim Sarkar

Abstract A compact dual-frequency dual-band square microstrip patch antenna using square loop slots on the patch and a rectangular slot on finite ground plane have been designed and presented in this paper. The total size of the designed antenna is very small which is only $15 \times 15\ \text{mm}^2$. The proposed fabricated antenna resonates at 1.8 and 5.5 GHz which belong to L and C band. Almost 93% compactness has been achieved with respect to the conventional microstrip antenna resonating at 1.8 GHz. The designed antenna may be suitable for GSM 1800 and HiperLAN-2 applications.

Keywords Compact · Dual-frequency · Dual-band

1 Introduction

The investigation on the design of compact dual-frequency microstrip antenna is on-demand now a days to keep wireless communication devices portable and suitable for multiple applications as the microstrip antennas are one of the most important parts of a wireless system [1–3]. Dual-band operation of a microstrip antenna is more appreciable as very close frequencies may produce interference to each other. Various techniques like cutting slots on patch, cutting slots on ground, using parasitic elements, utilizing shorting pin, etc., have been adopted to design compact dual-frequency dual-band microstrip antennas.

A. Roy (✉)
Department of ECE, Haldia Institute of Technology, East Medinipur, Haldia, WB, India
e-mail: avisankar.roy@gmail.com

S. Bhunia
Department of ECE, Central Institute of Technology, Kokrajhar, Assam, India
e-mail: snb.hit@gmail.com

D. C. Sarkar · P. P. Sarkar
DETS, University of Kalyani, Nadia, Kalyani, WB, India
e-mail: dsarkar70@gmail.com

P. P. Sarkar
e-mail: parthabe91@yahoo.co.in

S. Kundu et al. (eds.), *Proceedings of the 2nd International Conference on Communication, Devices and Computing*, Lecture Notes in Electrical Engineering 602, https://doi.org/10.1007/978-981-15-0829-5_4

Compact dual-frequency microstrip antenna design by cutting slots on the patch has been reported in [4, 5]. In [5], some unequal finger-like rectangular slots have been cut in the radiating edge of the microstrip patch to achieve dual-frequency operation with 67% compactness. Das et al. [6] have proposed a design of slotted microstrip antenna with slotted ground plane which exhibits triple-band operation and the designed antenna has been reduced by 74% by introducing the slots on ground. A modified E shaped microstrip patch antenna with shorting pin loading has been presented in [7]. The antenna operates in dual-frequency for utilizing the shorting pins in the longer arm of the patch where this technique has transformed a $\lambda/2$ resonant antenna to an $\lambda/4$ antenna. Thus, the pins have separated the two resonant frequencies. H. M. Chen has proposed a microstrip fed dual-frequency triangular monopole antenna in [8]. In that article various frequency ratios within the range of 1.37–1.75 GHz have been obtained by varying the trapezoidal slit on the triangular patch. A miniaturized microstrip patch antenna using single slotted ground plane has been presented in [9]. It has been discussed in this article that due to slotted ground plane resonant frequency has been lowered considerably and the antenna size has been reduced by 90%. In the year 2017, S. Yan et al. has proposed a planar inverted F antenna for dual-band application and presented in [10]. A dual-band circular microstrip patch antenna has been designed in [11] for energy harvesting applications. The design consists of a circular patch with direct feeding and a circular slotted ground plane which radiates by capacitive coupling between the patch and the ground plane.

In this article, a design of a compact dual-band square microstrip patch antenna using loop slots and modified ground plane has been presented. The proposed antenna has been simulated in IE3D software and measured in vector network analyzer and standard microwave test bench. The total fabricated antenna size is only $15 \times 15\,\text{mm}^2$ and it radiates at 1.8 and 5.5 GHz, i.e., L and C band. With respect to first resonant frequency 93% compactness has been achieved and the designed antenna may be suitable for GSM 1800 and HiperLAN-2 applications.

2 Antenna Design and Structure

The dimension of the square patch of reference antenna has been taken as $14 \times 14\,\text{mm}^2$ with substrate (Arlon AD300A) having dielectric constant $\varepsilon_r = 3$, thickness $h = 1.524\,\text{mm}$, and loss tangent $\tan\delta = 0.002$. The dimension of the square ground plane of the antenna has been taken as $18 \times 18\,\text{mm}^2$ to make the antenna very small. Coaxial feeding has been used at optimum location to match with 50 Ω characteristics impedance. The structure of the reference antenna which is defined as Antenna 1 is shown in Fig. 1. Two loops like slots have been introduced on the patch and a rectangular slot has been cut on the ground plane of the proposed antenna which is defined as Antenna 2. The size of the ground plane of proposed antenna has been reduced to $15 \times 15\,\text{mm}^2$ to make smaller antenna. The structure of slotted patch is shown in Fig. 2 and structure of slotted finite ground plane is shown in Fig. 3.

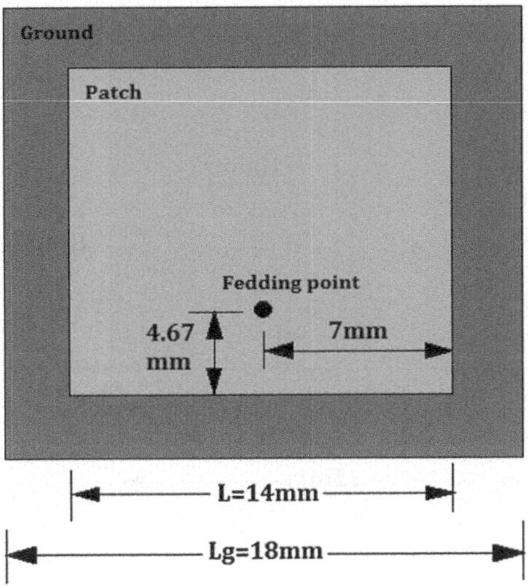

Fig. 1 Structure of reference antenna (Antenna 1)

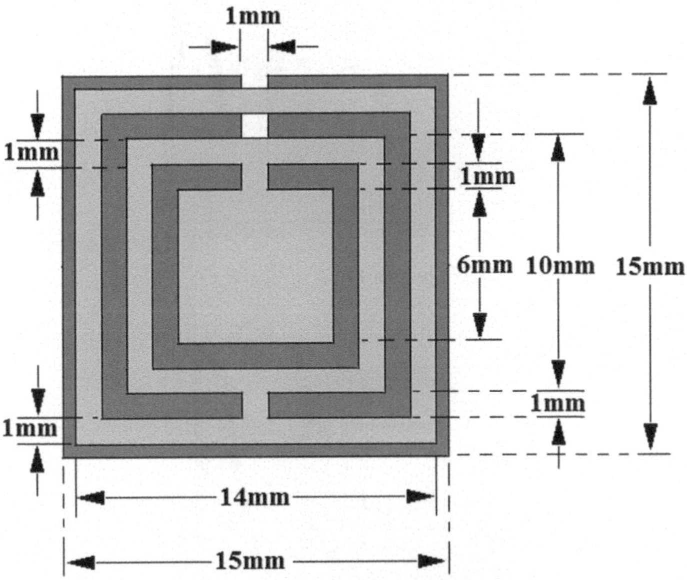

Fig. 2 Structure of the slotted patch of the proposed antenna (Antenna 2)

The fabricated Antenna 1 structure is depicted in Fig. 4. The patch of the fabricated Antenna 2 structure is illustrated in Fig. 5 and the ground plane of fabricated Antenna 2 is shown in Fig. 6.

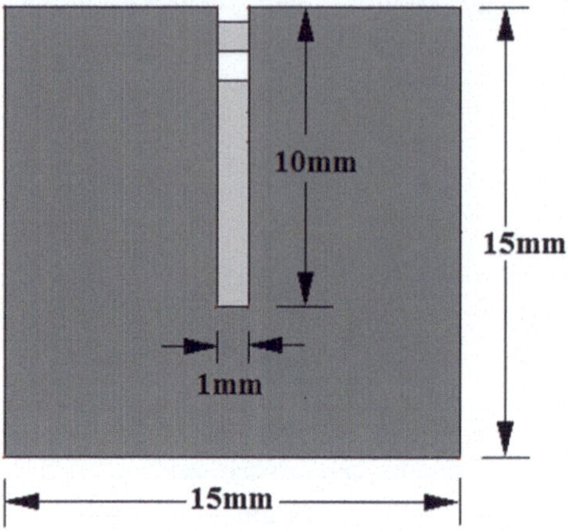

Fig. 3 Structure of the slotted ground plane of the proposed antenna (Antenna 2)

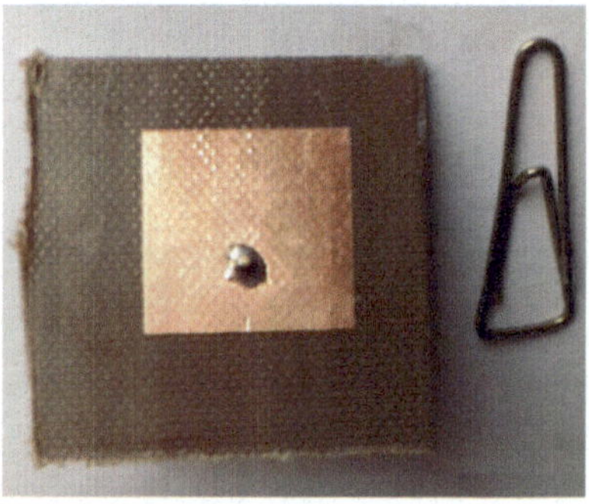

Fig. 4 Fabricated reference antenna (Antenna 1)

Fig. 5 Fabricated patch of the proposed antenna (Antenna 2)

Fig. 6 Fabricated ground plane of the proposed antenna (Antenna 2)

3 Results and Discussion

The antenna structures have been simulated using the method of moment (MOM) based software IE3D and fabricated antenna structures have been measured using vector network analyzer and standard microwave test bench. The simulated and measured results of reference Antenna 1 and proposed Antenna 2 are shown in Table 1 and Table 2, respectively.

The fabricated reference antenna (Antenna 1) resonates at 5.7 GHz with 35 dB return loss, 230 MHz (4%) bandwidth, and 5.5 dBi gain. The fabricated proposed antenna (Antenna 2) resonates at 1.8 and 5.5 GHz with 25 and 20 dB return losses. The antenna also exhibits 90 MHz (5%) and 140 MHz (2.5%) bandwidth for 1.8 GHz and 5.5 GHz frequencies and gains have been found as 2 dBi and 2.5 dBi, respectively. The frequency versus return loss graph of Antenna 1 and Antenna 2 has been illustrated

Table 1 Simulated and measured results of reference Antenna 1

	Resonant frequency (GHz)	Return loss (dB)	Bandwidth (MHz, %)	Gain (dBi)
Simulated	5.75	36	210,3.7	5.2
Measured	5.7	35	230,4	5.5

Table 2 Simulated and measured results of proposed Antenna 2

	Resonant frequency (GHz)	Return loss (dB)	Bandwidth (MHz, %)	Gain (dBi)
Simulated	2	15	90,4.5	1.5
	5.6	13	120,2.14	1.6
Measured	1.8	25	90,5	2
	5.5	20	140,2.5	2.5

in Figs. 7 and 8. The simulated current distribution of proposed Antenna 2 at 2 and 5.6 GHz has been depicted in Figs. 9 and 10.

The measured E plane radiation pattern of reference Antenna 1 at 5.7 GHz is shown in Fig. 11. The measured E plane radiation patterns of proposed Antenna 2 at 1.8 and 5.5 GHz are shown in Figs. 12 and 13. The 3 dB beamwidth for all radiation patterns have been found within 40°–60°.

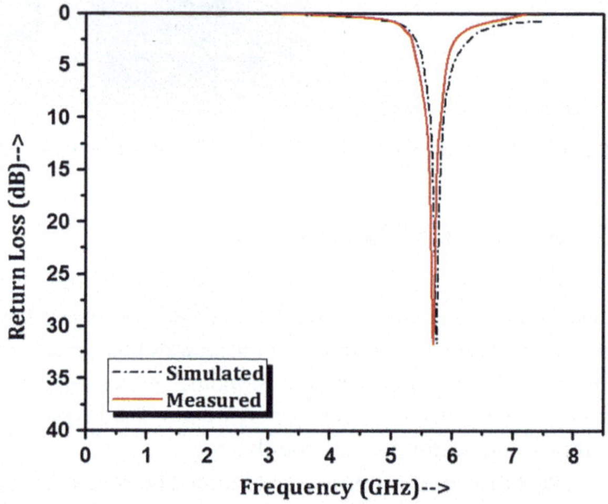

Fig. 7 Frequency versus return loss graph of reference antenna (Antenna 1)

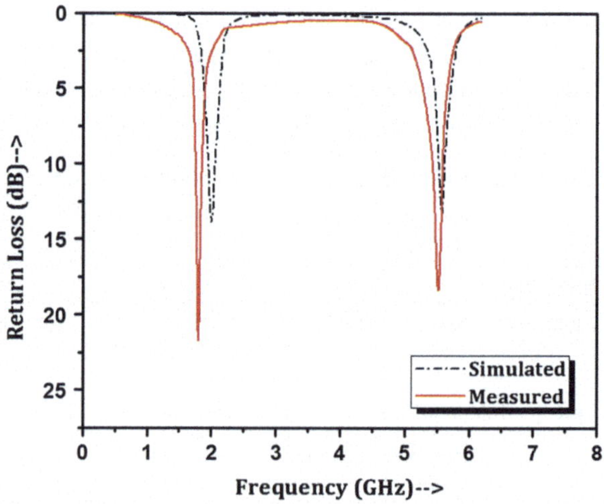

Fig. 8 Frequency versus return loss graph of the proposed antenna (Antenna 2)

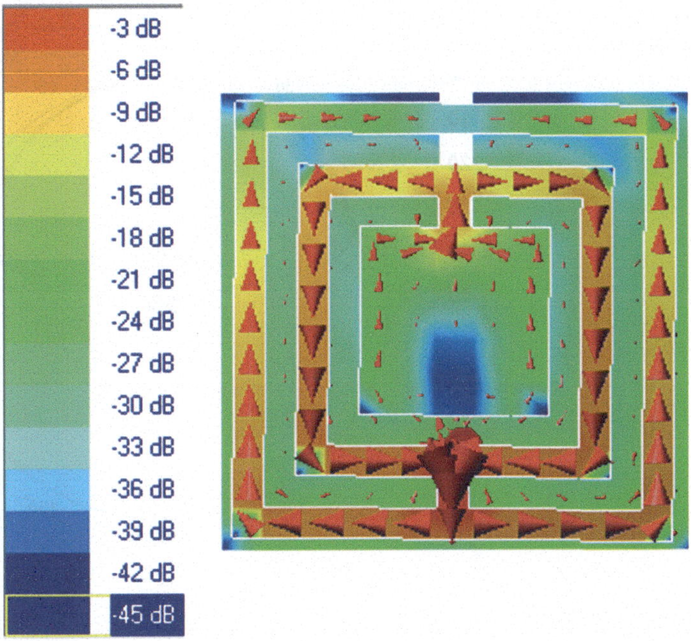

Fig. 9 Current distribution of Antenna 2 at 2 GHz

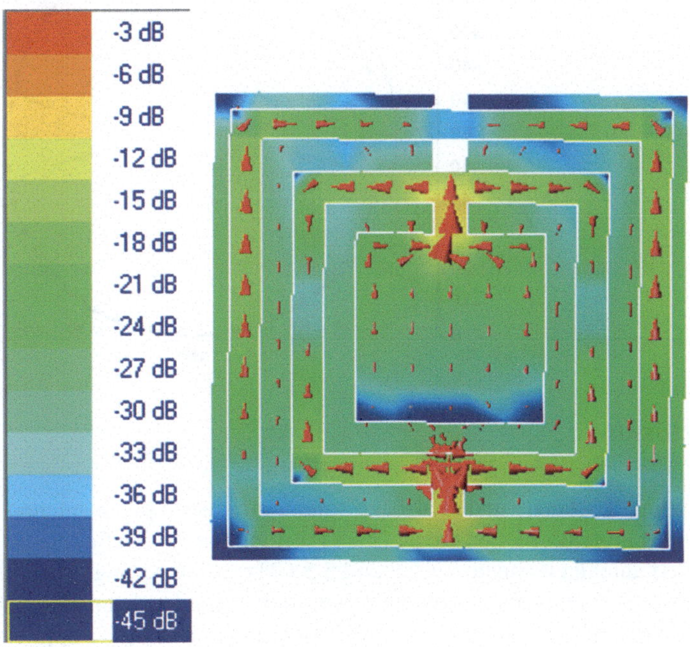

Fig. 10 Current distribution of Antenna 2 at 5.6 GHz

Fig. 11 Radiation pattern of Antenna 1 at 5.7 GHz

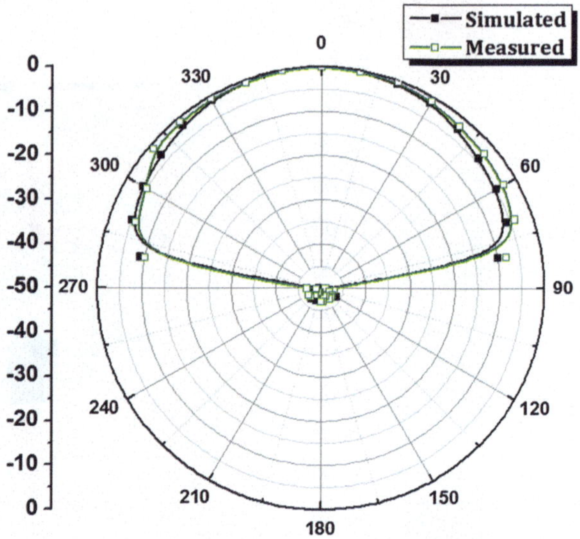

Fig. 12 Radiation pattern of Antenna 2 at 1.8 GHz

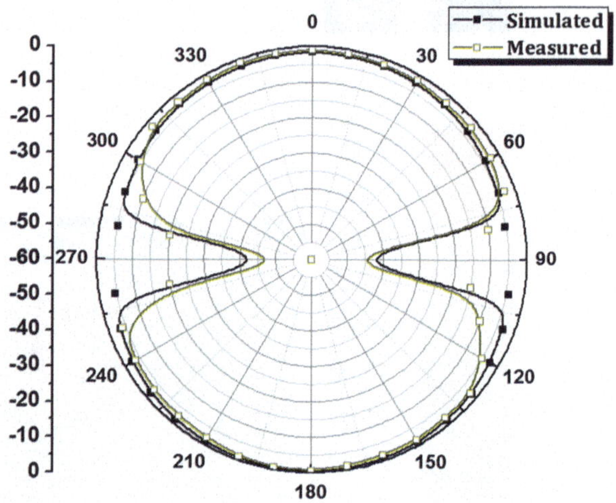

4 Conclusion

A compact dual-frequency and dual-band square microstrip patch antenna has been designed using loop like slot on the patch and rectangular slot on finite ground plane. The proposed antenna resonates at 1.8 and 5.5 GHz which belong to L and C band. With respect to first resonant frequency 93% compactness has been achieved and the designed antenna may be suitable for GSM 1800 and HiperLAN-2 applications.

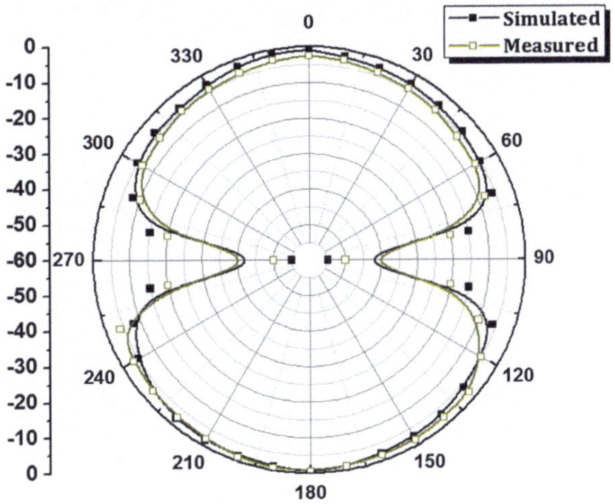

Fig. 13 Radiation pattern of Antenna 2 at 5.5 GHz

References

1. Bahl, I.J., Bhartia, P.: Microstrip Antennas. Artech House, Dedham, MA (1980)
2. James, J.R., Hall, P.S., Wood, C.: Microstrip Antennas: Theory and Design. Peter Peregrinus, London, UK (1981)
3. Wong, K.L.: Compact and Broadband Microstrip Antennas. Wiley, New York (2002)
4. Lu, J.H., Wong, K.L.: Slot-loaded, meandered rectangular microstrip antenna with compact dual frequency operation. Electron. Lett. **34**(11), 1048–1050 (1998)
5. Bhunia, S., Sarkar, P.P.: Reduced sized dual frequency microstrip antenna. Indian J. Phys. **83**(10), 1457–1461 (2009)
6. Das, S., Sarkar, P.P., Chowdhury, S.K.: Design and analysis of a compact triple band slotted microstrip antenna with modified ground plane for wireless communication applications. Prog. Electromagn. Res. B **60**, 215–225 (2014)
7. Peng, L., Ruan, C.L., Wu, X.H.: Design and operation of dual/triple-band asymmetric M-shaped microstrip patch antennas. IEEE Antennas Wirel. Propag. Lett. **9**, 1069–1072 (2010)
8. Chen, H.M.: Microstrip-fed dual-frequency printed triangular monopole. Electron. Lett. **38**(13), 619–620 (2002)
9. Sarkar, S., Majumdar, A.D., Mondal, S., Biswas, S., Sarkar, D., Sarkar, P.P.: Miniaturization of rectangular microstrip patch antenna using optimized single-slotted ground plane. Microw. Optic. Technol. Lett. **53**(1), 111–115 (2011)
10. Yan, S., Volskiy, V., Vandenbosch, G.A.E.: Compact dual-band textile PIFA for 433-MHz/2.4-GHz ISM bands. IEEE Antennas Wirel. Propag. Lett. **16**, 2436–2439 (2017)
11. Aboualalaa, M., Abdel-Rahman, A.B., Allam, A., Elsadek, H., Pokharel, R.K.: Design of a dual-band microstrip antenna with enhanced gain for energy harvesting applications. IEEE Antennas and Wireless Propag. Lett. **16**, 1622–1626 (2017)

An Efficient Concentric Circular Antenna Array Synthesis Considering the Mutual Coupling

Avishek Das, D. Mandal and R. Kar

Abstract This paper illustrate an efficient approach to enhance the directivity and minimize the side lobe level (SLL) of a non-uniform concentric circular antenna array (CCAA) considering the mutual coupling using a robust search algorithm called opposition based gravitational search algorithm (OGSA). Particle swarm optimization (PSO) and differential evolution (DE) optimization methods are also employed to compare the results obtained by using OGSA. An improved radiation characteristic with low SLL and high directivity considering the mutual coupling is achieved using an optimal current excitation weights and inter-element separation between the array elements of CCAA. The design examples of CCAA having sets of (4,6,8) and (1,4,6,8) elements are presented and their performances are studied in this paper.

Keywords Concentric circular antenna array · Side lobe level · Directivity · Mutual coupling · Opposition based gravitational search algorithm

1 Introduction

A number of radiating elements arranged in a specific configuration forms an antenna array. A CCAA is composed with a number of concentric circular rings and all the rings contain a prefixed number of radiating elements [1–4]. The total electromagnetic field emitted from an antenna array is the sum total of all the fields radiating from each array elements [5, 6]. The fields radiated from the antenna array constructively added in the desired direction to obtain an improved directive pattern of the array [7].

Antenna array design has a major role in detecting and processing the signals in wireless communication systems. An efficient antenna array design mainly deals with low SLL and narrow first null beamwidth (FNBW) which is necessary to reduce

A. Das (✉)
Department of ECE, HIT Haldia, Haldia 721657, India
e-mail: avishek.uit0408@gmail.com

D. Mandal · R. Kar
Department of ECE, NIT Durgapur, Durgapur 713209, India

S. Kundu et al. (eds.), *Proceedings of the 2nd International Conference on Communication, Devices and Computing*, Lecture Notes in Electrical Engineering 602, https://doi.org/10.1007/978-981-15-0829-5_5

the interference in the same frequency band and for the high directivity, respectively. These two prime criteria for an efficient antenna array design are in contrast to each other and hence not improved significantly at the same time [8, 9].

Mutual coupling affects the radiation characteristics of an antenna array [10]. In this paper, the driving admittance among the array elements is calculated and is optimized using OGSA algorithm for the reduction of the mutual coupling as it affects the signal processing of the CCAA [6]. Here, the driving admittance of all the array elements is minimized for the reduction of mutual coupling. Hence, the total driving admittance of the array is minimized by the set of an optimal current excitation weights obtained by using different algorithms.

The main focus of the paper is to design a CCAA having a low SLL and high directivity value considering the mutual coupling. The simulation results for CCAA design using different algorithms PSO [11–14], DE [15, 16] and OGSA [17–19] are compared in this paper.

The paper is arranged as follows: Sect. 2, discuss on the design of a CCAA. Different optimization techniques are discussed in Sect. 3. The results obtained by using different algorithms are compared in Sect. 4.1. Section 4.2, shows the convergence profiles of the algorithms applied here. Finally Sect. 5, presents the conclusion of the paper.

2 Design Equation

A CCAA of M concentric circular rings of which every rings consist with N_m number of radiating elements is shown in Fig. 1.

As all elements of the array are isotropic elements, so its Array Factor (AF) [20] is represented by (1).

$$AF(\theta, \phi, I) = \sum_{m=1}^{M} \sum_{n=1}^{N} I_{mn} \exp[j(kr_m \sin\theta \cos(\phi - \phi_{mn}) + \alpha_{mn})] \quad (1)$$

The current excitation weight of the nth element of mth ring is denoted by I_{mn}. $k = \frac{2\pi}{\lambda}$, λ represents the signal wave length. The elevation and the azimuth angle is represented as θ and ϕ, respectively, whereas, ϕ_{mn} is the inter-element separation angle given in (2). The phase difference is represented by α_{mn}.

$$\phi_{mn} = 2\pi\left(\frac{n-1}{N_m}\right);\ m = 1, 2, \ldots, M;\ n = 1, 2, \ldots, N_m. \quad (2)$$

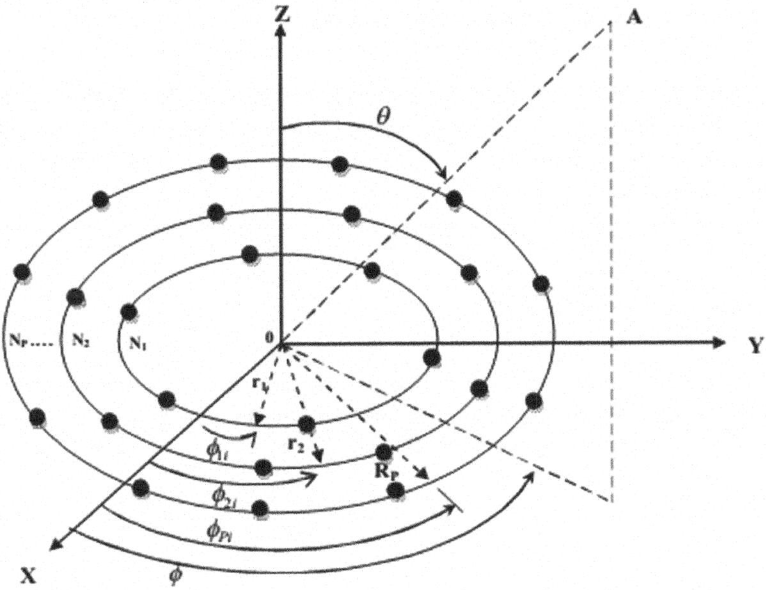

Fig. 1 Concentric circular array antenna

The Cost Function (CF) for suppressing the SLL and 3-dB beamwidth considering the mutual coupling is given in (3).

$$CF = W_1 \times \frac{|AF(\theta_{msl1}, I_{mn}) + AF(\theta_{msl2}, I_{mn})|}{|AF(\theta_0, I_{mn})|} + W_2 \times (\frac{1}{Z_D}) + W_3 \times \frac{1}{DIR} \tag{3}$$

$$Z_D = \sum_{m=1}^{M}\sum_{n=1}^{N}\left(\frac{\sum_{m=1}^{M}\sum_{n=1}^{N} I_{mn}}{I_{mn}}\right); \quad DIR = \frac{4\pi|AF_{\max}|^2}{\int_0^{\pi}\int_0^{2\pi}|AF(\theta, \phi)|^2 \sin\theta d\theta d\phi} \tag{4}$$

W_1, W_2 and W_3 are the weighing factors and θ_0 is the peak angle of the principal lobe; θ_{ms1} and θ_{ms2} is the side lobe angle where maximum value is attained in the lower band and upper band, respectively. The first and second term of CF is employed for SLL and mutual coupling reduction whereas the third term of the CF is employed for enhancing the directivity of the array. Here, OGSA algorithm is applied to minimize the CF given in (3).

3 Evolutionary Technique Employed

Evolutionary algorithms are characterized as robust and flexible search method to determine the complex optimization problem. PSO [11–14], DE [15, 16] and OGSA [17–19] algorithms are employed here for CCAA design to confirm the superiority of OGSA algorithm.

4 Result and Discussion

A comprehensive study on (1,4,6,8) and (4,6,8) element CCAA design has been performed using OGSA algorithm. The best results is achieved by using OGSA algorithm after 50 trial run using MATLAB 7.5, in Intel Core (TM) 2 processor @3.00 GHz with 3 GB RAM.

4.1 Numerical Results

The results obtained by using OGSA in the presence of mutual coupling effect for CCAA design with and without centre element are compared with the uniform pattern and with some of well-established algorithm like PSO and DE.

Table 1 shows the SLL, the FNBW and the directivity value in presence of mutual coupling for uniformly excited and uniformly spaced CCAA.

Table 2 shows a comparative analysis of the optimal results obtained by using PSO, DE and OGSA algorithm for the synthesis of (1,4,6,8) elements CCAA. The table contains the current excitation weight of each element of the array starting from the centre element toward the elements in outer most rings, inter-element spacing from ring-1 to ring-3 of CCAA and SLL, FNBW, Mutual Coupling, Directivity values. The results of Table 2 confirm the superiority of the OGSA algorithm with respect to PSO and DE.

The PSO, DE and OGSA algorithm-based results for the design of (1,4,6,8) elements CCAA considering the mutual coupling in the *CF* are plotted in Fig. 2. In this figure the plots obtained by using different algorithms are compared with the uniform pattern.

Table 1 SLL, FNBW, mutual coupling and directivity of uniformly excited and uniformly spaced CCAA with and without the central element

No. of element	SLL (dB)	FNBW (deg)	Mutual coupling	Directivity (dB)
1,4,6,8	−13.6680	95.40	0.0013	11.1271
4,6,8	−11.2332	90.36	0.0016	10.9698

Table 2 Different parameters of (1,4,6,8) elements CCAA achieved by using PSO, DE and OGSA algorithm

Algorithm	Current excitations			Inter-element spacing (λ)	SLL (dB)	FNBW (deg)	Mutual coupling	Directivity (dB)
PSO	0.2702, 0.6622, 0.6575, 0.5477, 0.5307, 0.4473, 0.1220	0.5590, 0.4810, 0.7264, 0.6894, 0.4454, 0.6161,	0.6434 0.5823 0.6996 0.4074 0.0812 0.3975,	0.5003 0.5666 0.7393	−33.048	85.14	0.0003 2658	10.8456
DE	0.3735, 0.7375, 0.9303, 0.8940, 0.8531, 0.5163, 0.2129	0.7701, 0.5155, 0.9228, 0.9219, 0.5253, 0.8032,	0.5160 0.7998 0.9211 0.5133 0.1919 0.5235	0.5000 0.5312 0.7545	−34.026	87.12	0.0004 3605	12.0608
OGSA	0.4837, 0.6622, 0.8879, 0.7994, 0.8295, 0.4774, 0.2318	0.7051, 0.6730, 0.7556, 0.8318, 0.5054, 0.7550,	0.7610, 0.8272, 0.8865, 0.5207, 0.1591, 0.5053,	0.5391 0.5517 0.7536	−36.755	84.06	0.0004 6969	12.0807

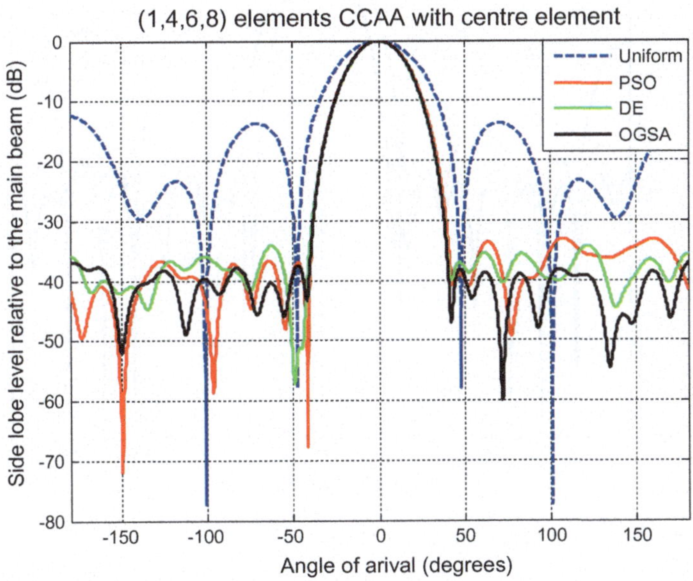

Fig. 2 Radiation pattern of (1,4,6,8) elements CCAA

Table 3 Different parameters of (4,6,8) elements CCAA achieved by using PSO, DE and OGSA algorithm

Algorithm	Current excitations			Inter-element spacing (λ)	SLL (dB)	FNBW (deg)	Mutual coupling	Directivity (dB)
PSO	0.7274	0.6426	0.7003	0.5000	−30.664	74.88	0.0006 2243	11.9997
	0.3761	0.7886	0.8876	0.5363				
	0.9279	0.8547	0.6499	0.7717				
	0.9756	0.5004	0.9544					
	0.4929	0.2048	0.5819					
	0.8993	0.5793	0.2539					
DE	0.6589	0.8050	0.5936	0.5138	−31.282	84.78	0.0007 2290	11.8518
	0.8730	0.6185	0.7652	0.5853				
	0.9712	0.4488	0.7586	0.7671				
	0.9397	0.4122	0.9373					
	0.5190	0.3356	0.5627					
	0.8161	0.5729	0.2920					
OGSA	0.6062	0.7042	0.5821	0.5000	−34.709	81.36	0.0007 2141	11.6679
	0.6631	0.5728	0.6769	0.5774				
	0.8878	0.5664	0.6190	0.7650				
	0.9325	0.4772	0.8273					
	0.4988	0.2964	0.4881					
	0.7593	0.4803	0.2675					

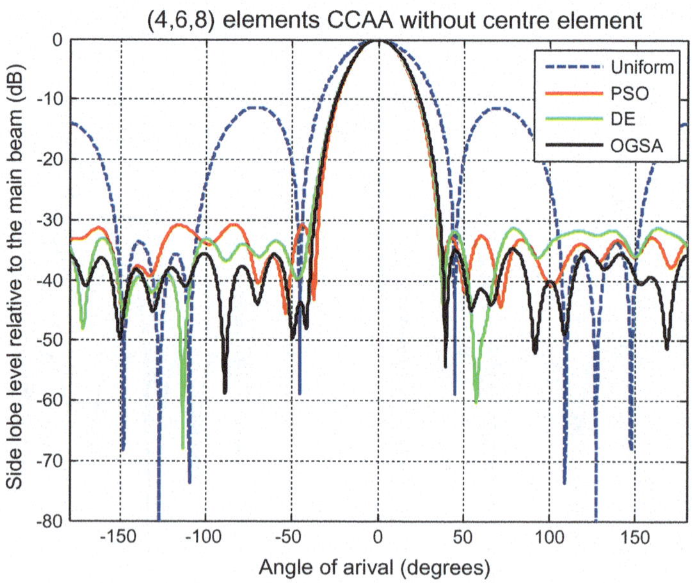

Fig. 3 Radiation pattern of (4,6,8) elements CCAA

Table 3 shows the comparison of the results achieved by using PSO, DE and OGSA algorithm for (4,6,8) elements CCAA design which shows that OGSA algorithm outperforms the PSO and DE.

Figure 3 depicts that OGSA based results have the lowest SLL and FNBW value in comparison with PSO and DE algorithm.

4.2 *Convergence Profiles Plot of OGSA*

The minimum *CF* value with the iteration cycles is plotted in the convergence profile. Figs. 4 and 5 show the convergence plot of the PSO, DE and OGSA algorithms employed for the design of CCAA with (1,4,6,8) and (4,6,8) elements in presence of mutual coupling.

The convergence plots expose the excellent performance of the OGSA algorithm in terms of precision and convergence speed.

5 Conclusion

This paper illustrates the design method of CCAA with and without centre element for the reduction of SLL as well as enhancement of directivity considering the mutual coupling effect by using OGSA. The OGSA algorithm-based results prove the outstanding performance for the reduction of SLL and enhancement of directivity as compared with the results achieved by using PSO and DE algorithm. Thus, the OGSA algorithm-based results confirm the superiority of the algorithm for the synthesis of a non-uniform CCAA.

Fig. 4 Convergence profiles of (1,4,6,8) elements CCAA with centre element in the presence of mutual coupling

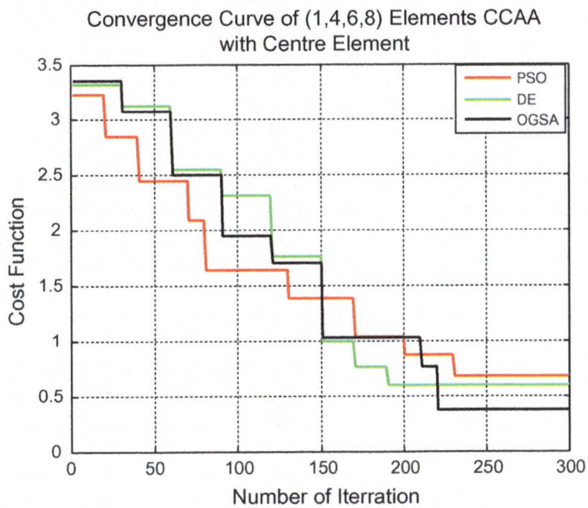

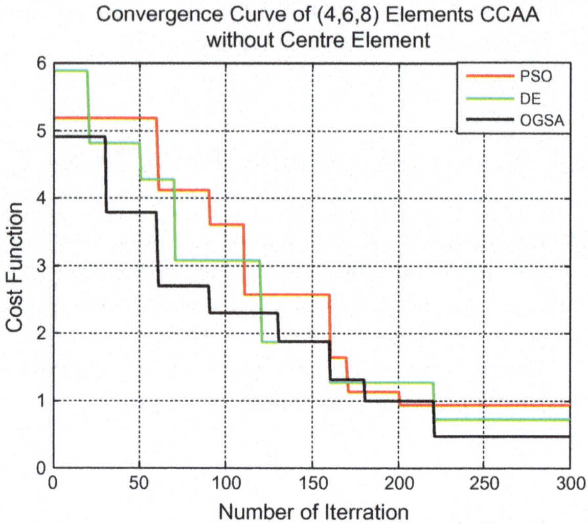

Fig. 5 Convergence profiles of (4,6,8) elements CCAA without centre element in the presence of mutual coupling

References

1. Das, R.: Concentric ring array. IEEE Trans. Antennas Propag. **14**(3), 398–400 (1966)
2. Goto, N., Cheng, D.K.: On the synthesis of concentric-ring arrays. In: Proceedings of the IEEE, vol. 58, no. 5 (1970), pp. 839–840
3. Biller, L., Friedman, G.: Optimization of radiation patterns for an array of concentric ring sources. IEEE Trans. Audio Electroacoust. **21**(1), 57–61 (1973)
4. Huebner, M.D.A.: Design and optimization of small concentric ring arrays. In: Proceedings of the IEEE AP-S Symposium (1978), pp. 455–458
5. Ballanis, C.A.: Antenna Theory Analysis and Design, 2nd edn. Wiley, New York (1997)
6. Elliott, R.S.: Antenna Theory and Design, Revised edn., New Jersey: IEEE Press (2003)
7. Collin, R.E.: Antenna and radio-wave propagation. McGraw-Hill, New York (1985)
8. Yan, K.K., Lu, Y.: Sidelobe reduction in array-pattern synthesis using genetic algorithm. IEEE Trans. Antenna Propag. **45**(7), 1117–1122 (1997)
9. Holtrup, M.G., Margulnaud, A., Citerns, J.: Synthesis of electronically steerable antenna arrays with element on concentric rings with reduced sidelobes. In: Proceedings of the IEEE AP-S Symposium (2001), pp. 800–803
10. Lui, H.S., Hui, H.T., Leong, M.S.: A note on the mutual-coupling problems in transmitting and receiving antenna arrays. IEEE Antennas Propag. Mag. **51**(5) (2009)
11. Gie, D., Rahmat-Samii, R.Y.: Particle swarm optimization for reconfigurable phased differentiated array design. Microw. Opt. Technol. Lett. **38**(3), 168–175 (2003)
12. Boeringer, D.W., Werner, D.H.: Particle swarm optimization versus genetic algorithm for phased array synthesis. IEEE Trans. Antenna Propag. **52**(3), 771–779 (2004)
13. Robinson, J., Samii, R.Y.: Particle swarm optimization in electromagnetics. IEEE Trans. Antenna Propag. **52**(2), 397–407 (2004)
14. Ciuprina, G., Ioan, D., Munteanu, I.: Use of intelligent-particle swarm optimization in electromagnetics. IEEE Trans. Mag. **38**(2), 1037–1040 (2004)
15. Storn, R., Price, K.: Differential evolution-a simple and efficient adaptive scheme for global optimization over continuous spaces, Technical Report, International Computer science Institute, Berkley (1995)
16. Karaboga, N.: Digital IIR filter design using differential evolution algorithm. EURASIP J. Appl. Signal Process. **2005**(8), 1269–1276 (2005)

17. Rashedi, E., Pour, N.H., Saryazdi, H.S.: GSA: A gravitational search algorithm. Inf. Sci. **179**(13), 2232–2248 (2009)
18. Tizhoosh, H.R.: Opposition-based learning: a new scheme for machine intelligence. In: Proceedings of the International Conference on Computational Intelligence for Modelling Control and Automation, vol. 1, pp. 695–701 (2005)
19. Ram, G., Mandal, D., Kar, R., Ghoshal, S.P.: Opposition-based gravitational search algorithm for synthesis circular and concentric circular antenna array. Scientia Iranica D **22**(6), 2457–2471 (2015)
20. Haupt, R.L.: Optimized element spacing for low sidelobe concentric ring arrays. IEEE Trans. Antenna Propag. **56**(1), 266–268 (2008)

Traffic Aggregation in Elastic Optical and Wireless Networks

Deepa Naik, Nikita, Anupama Bauri and Tanmay De

Abstract This paper proposes light trail-based algorithms for routing and resource allocation in Hybrid Networks. In hybrid networks Elastic Optical Network is back end while WiMAX complements as access network. Various network configurations coupled with different permutations of source–destination pairs are used for simulations. Algorithms proposed here found to perform better in terms of bandwidth blocking ratio in all the given networks coupled with minimum network resource utilization.

Keywords BVT (Bandwith Variable Transponder) · POLT (Programmable Optical Line Terminal) · PONU (Programmable Optical Network Unit) · Traffic grooming · Light trail · WiMAX (World Interoperability for Microwave Access Network) · WDM-PON (Wavelength Division Multiplexed Passive Optical Network)

1 Introduction

Hybrid networks have become successful in meeting high bandwidth demands [1, 2]. WiMAX network is the most preferred wireless broadband access technology [3]. The traditional wavelength routed optical networks have two limitations. They are of fixed grid and coarse granularity. Hence they are inflexible, resources therein are poorly

D. Naik (✉) · Nikita · A. Bauri · T. De
Department of Computer Science and Engineering, National Institute of Technology, Durgapur, India
e-mail: dn.12cse1103@phd.nitdgp.ac.in

Nikita
e-mail: nikitaniki027@gmail.com

A. Bauri
e-mail: anupamabauri@gmail.com

T. De
e-mail: tanmayd12@gmail.com

S. Kundu et al. (eds.), *Proceedings of the 2nd International Conference on Communication, Devices and Computing*, Lecture Notes in Electrical Engineering 602, https://doi.org/10.1007/978-981-15-0829-5_6

utilized and have high equipment cost. Further, even if the traffic request needs only a small bandwidth the entire wavelength channel is allocated to serve the request. The Elastic Optical Network (EON) [4] with its flexibility minimizes bandwidth wastage. The EON is a high-speed network, which is flexible in assigning the just enough spectrum to the traffic demands. Hence, traditional Routing and Wavelength Assignment (RWA) algorithms are not suitable for EON. In EON finding, unfilled spectrum resources for establishing a light path is called the Routing and Spectrum Allocation (RSA) [5, 6].

Traffic grooming efficiently utilizes the channel capacities by combining several low capacity demand traffic streams into high capacity wavelength channels as proposed by Dutta et al. in [7]. Traffic grooming also eliminates the necessity of guard bands. In traffic grooming there are two techniques: one is light path and another is the light trail approach. In the light path approach, traffic is transferred between source and destination nodes. This leads to inefficient utilization of network resources. In our work, we have used the light trail [8] approach where each node between the source and destination node can communicate in the data transfer without violating the channel capacities and wavelength continuity constraint. The requests are established using already active transponders. So the network resource gets utilize efficiently. In the wireless network, guard bands are used to eliminate interference from other channels. Again the traffic aggregation (grooming in optical) overcomes this hurdle by aggregating the requests together (which share the common path) without violating the capacity of wireless links. This aggregation leads to minimization of the network resources used to route the requests and also guard band required at the individual connection requests.

To the best of our knowledge, the integration of elastic optical network with WiMAX using traffic grooming approaches is not considered. The drawbacks of traditional WDM network as a backhaul motivated us to use the elastic optical network as backhaul (due to the flexibility) and WiMAX network as front end along with relay nodes to enhance the coverage area. The WDM-PON technique is used by Programmable OLT (POLT) to assign individual wavelength to each PONU integrated with the WiMAX base station. The wireless network uses the Multiple Input and Multiple Output (MIMO) OFDM technique to serve the requests. Most of the works consider the traffic requests that originate from wireless routers to central office or between any pair of wireless routers but the traffic grooming in hybrid network remained untouched. The traffic grooming approach minimizes the network resource consumption by establishing the requests on the already active transponders. So the more bandwidth is available for access networks. This leads to minimizing the bandwidth blocking ratio in the given network.

Following are the various sections of paper: Sect. 2 presents the previous work. Section 3 includes the proposed approaches and their complexity analysis. The simulation results are analyzed in Sect. 4. Finally, conclusion and future work is drawn in Sect. 5.

2 Previous Work

2.1 Integration of Optical and Wireless Network (WOBAN)

Many researchers are getting involved in different hybrid network configurations due to their flexibility in deployment, cost effectiveness, easy upgrading facility, and scalablility in achieving braodband connectivity to large number of users. In [9–12], the authors discussed the issues related to optical network, integrated with wireless network to provide broadband connectivity to mobile users. High bandwidth capable optical access networks have geographical limitations.Wireless access networks have high coverage area and lower deployment cost. They suffer from limited available spectrum [13]. The hybrid Wireless Optical Broadband Access Network (WOBAN) is a powerful combination of optical backhaul and wireless frontend. The details related to this integration are provided in our previous work [14].

The PON uses different schemes to assign the wavelengths to ONU. These schemes are Time Division Multiplexing (TDM), Wavelength Division Multiplexing (WDM), Code Division Multiplexing (CADM), and Orthogonal Frequency Division Multiplexing (OFDM). Application of Orthogonal Frequency Division Multiplexing (OFDM) in PON enhances the data rate up to 100 Gbps [15]. The TDM,WDM, AND OFDM are widely used in PON for catering access networks [16, 17].

3 Proposed Approach

Given the network topology $G(V, E)$, where V is the set of optical/wireless nodes, and E is the set of bidirectional links (elastic optical/wireless links between the links). In the network, each link capacity is divided into slots with a granularity of 6.25 GHz. The problem is to assign the frequency slots to the set of traffic demands without violating the elastic optical and wireless constraints. Here, we have considered three types of networks, namely, pure optical, pure wireless, and hybrid network (where source and destination nodes are in pure optical network or pure wireless network or it may be a hybrid request with source in the optical network and destination in the wireless network). Traffic grooming based on light trails is used for minimizing bandwidth blocking ratio of a hybrid network.

Definition of light trail: In a light path, the intermediate nodes between the source and destination nodes do not take part in communication. The light trail is a generalization of light path where the intermediate nodes are also able to access the resources (already active transponders) along the path and hence they can either add or drop data. The light trail reserves the frequency slots on all the links along path between the node pair. This is known as Routing and Spectrum Allocation (RSA) in EON and traffic aggregation in wireless network. EONs have the ability to dynamically adjust the network resources (transponder, bandwidth, and modulation type) according to

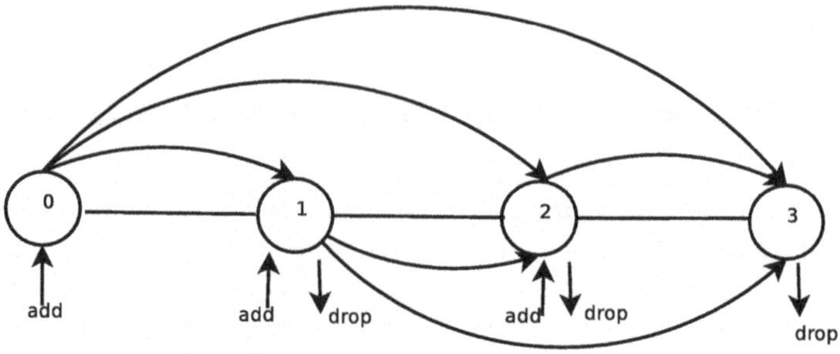

Fig. 1 Pure optical network

the demand from each connection request. In a trail, there is a Carrier Sense Multiple Access protocol (CSMA) that works for the static/dynamic demands. The control signals are sent on the path traversed by node pair. The coupler and shutter are ON and OFF indicating that the traffic should be dropped/added at intermediate nodes (Fig. 1).

In the hybrid network, we have used centralized scheme, where each relay station is under the coverage of one base station. Communication between relay stations is through the base station. Each PONU is served by dedicated wavelength. The wavelengths are divided into frequency slots. The PONU may be assigned more than one wavelength depending upon the traffic at the access nodes. The POLT is responsible for allocating the bandwidth to each PONU. To route the traffic along the selected route the Routing and spectrum Allocation (RSA) constraints in EON and transponder availability in the wireless network are applied [14]. We have proposed the different traffic grooming algorithms for minimizing bandwidth blocking ratio using minimum network resources. The wireless access network (relay station) is capable of data rate up to 240 Megabits per second (Mbps). The multiple input multiple output OFDM technique is applied to serve the requests in wireless network. We have presented three algorithms for hybrid networks. They are the Longest Hop count grooming algorithm (LHCG), Maximum Traffic Grooming algorithm (MTG) and Minimum Hop Count Grooming (MHCG) which are presented below.

For a given traffic demand matrix *T*, all possible paths are computed offline and stored in path matrix *p*. In the LHCG algorithm, all possible paths between the source and destination are sorted in descending order of their hop count. The topmost request is selected to establish the connection. The control signals are sent between the source, destination pair to know the status of the intermediate nodes willing to take part in communication along with active transponders. The source and destination of the traffic may be in the optical network or wireless network or in the hybrid network. The longest route can accommodate more traffic along the intermediate nodes and traffic gets established on the already active transponders. When a route does not have sufficient network resources for establishing the connection, next longest route is used. The intermediate requests between the selected node pair get established depending on the resources available to serve these requests. This process is repeated for all possible paths, till the resources (transponder and frequency slots) are available on the selected route to serve the requests. In case of conflict, where two routes have the same hop count, the route which serves the maximum traffic is selected to establish the connection.

Minimum Hop Count Grooming (MHCG): algorithm is just opposite of Longest Hop Count Grooming (LHCG). The route with the minimum hop distance is selected to route the requests. The traffic requests are sorted in ascending order of their hop count and stored in the path matrix. The path with the minimum distance between the source to destination is selected to route the requests. When hop count is same for two paths between given node pair, to avoid the conflict the path with the maximum traffic used to serve the requests. The intermediate requests between the selected route are served depending on the resource availability to serve these requests.

Maximum Traffic Grooming (MTG): algorithm selects the route which can pack maximum traffic along the selected route. When traffic is same for two paths between the given node pair, then the route with minimum hop count between the node pair is selected to avoid the conflict. The intermediate requests are also sorted depending up to the traffic demands. The intermediate requests with huge traffic demands are served first. If the selected route is unable to establish the connection requests then next maximum traffic route is selected to route the requests. The process of finding the route is repeated for all possible paths between the node pair.

Steps involved in **Longest Hop Count traffic Grooming (LHCG)** the hybrid network are explained below

1. Generate the traffic matrix between the node pair, find all possible paths for the non zero traffic demands. Then the requests are sorted in descending order of their hop count.
2. Choose the topmost request, If the two routes have same hop count, to resolve the conflict, select the route which accommodates maximum traffic is chosen.
3. While serving the intermediate requests between the node pair, the intermediate requests are sorted in descending order of their hop count to establish the connection requests. The requests are established depending on the availability of network resources.
4. As transponder is available at the source, destination, and intermediate nodes.
5. For serving the requests common slot index should be available along the selected links, if all conditions (2,4) satisfied then.
6. Assign slots, guard bands, and the traffic matrix is updated, update the status of the resources of hybrid network.
7. When a selected route does not have sufficient network resources,then next longest route is selected for routing (process repeats for all possible paths.)
8. The request may get blocked if network resources are not available to route, after traversing all possible paths.

The Minimum Hop Count Grooming (MHCG) and Maximum Traffic Grooming (MTG) algorithms are carried out similar to LHCG. Only the variation is in the path

Table 1 Traffic requests

Index	Source	Destination	Traffic demand/main/sub trail
s0	0	5	3-main trail
s1	0	1	1-sub trail
s2	1	5	1-sub trail
s3	1	2	2-sub trail
s4	2	5	3-sub trail
s5	0	2	3-sub trail

Table 2 Longest hop grooming

Index	Source destination	Path	Hop count	Traffic demand	Main/sub trail
s0	0–5	0–1–2–5	3	3	Main trail
s5	0–2	0–1–2	2	5	Sub trail
s2	1–5	1–2–5	2	2	Sub trail
s4	1–2	1–2	1	3	Sub trail
s3	2–5	2–5	1	3	Sub trail
s1	0–1	0–1	1	1	Sub trail

Table 3 Minimum hop count hop grooming

Index	Source destination	Path	Hop count	Traffic demand	Main/sub trail
s0	0–5	0–1–2–5	3	3	Main trail
s4	1–2	1–2	1	3	Sub trail
s3	2–5	2–5	1	3	Sub trail
s1	0–1	0–1	1	1	Sub trail
s5	0–2	0–1–2	2	5	Sub trail
s2	1–5	1–2–5	2	2	Sub trail

selection at the beginning (step 2) and at the intermediate node (step 3) depending upon the algorithms used.

Example: An example as depicted in Table 1 is to illustrate the working principle of LHCG (depicted in Table 2), MHCG (depicted in Table 3), and MTG (depicted in Table 4). Here, we have generated six requests between the source (0) and destination nodes (4) with the bandwidth requirement of three Frequency Slots (FS). Let us assume there are five intermediate requests between the source (0) and destination (5) nodes, 0–1, 0–2, 1–2, 2–5, 1–5. The bandwidth requirement of intermediate requests is 1, 5, 3, 3, and 2 FS, respectively. In LHCG, the path with the longest hop count between the source and destination node pair is selected to route the requests.

Table 4 Maximum traffic grooming

Index	Source destination	Path	Traffic demand	Main/sub trail
s0	0–5	0–1–2–5	3	Main trail
s5	0–2	0–1–2	5	Sub trail
s4	1–2	1–2	3	Sub trail
s3	2–5	2–5	3	Sub trail
s2	1–5	1–2–5	2	Sub trail
s1	0–1	0–1	1	Sub trail

In MTG, the path which accommodates maximum traffic between the node pair is selected and MHCG chooses the path with minimum distance between the node pair to route the requests. The intermediate requests between the node pair are also sorted according to the algorithm used to route the requests. When a selected route does not have sufficient network resources, then next count path is selected to establish the connection requests. Traffic requests to groom are given below in Table 1.

In the hybrid network, various scenarios were considered. They are both source and destination nodes in pure optical network or both nodes in the wireless network or hybrid network nature, i.e., either of two nodes in optical and remaining in the wireless network. The detailed optical network, wireless network constraints, and assumptions are explained in the previous work [14]. Static subscribers are assumed to be connected to relay station through wireless links. Here, only base station recognizes its relay nodes and stores path information. The available bandwidth is distributed among relay stations. The guard bands are used between the individual requests to eliminate the interference among the signals. A guard band consumes the spectrum resources. To overcome this the traffic aggregation (grooming) is used in the wireless network. So the traffic requests which share a common links are groomed together. The intermediate requests are dropped at the destination nodes and remaining is carried to next destination along the selected route. This eliminates the guard bands between each individual request. The requests are bidirectional (up link/down link diction).

The illustration of hybrid requests is depicted in Fig. 2.

Consider, a hybrid request between the nodes (Node1-RS) and to establish the connection between them the following route is used (Node0-Node1-BS-RS). The traffic requests originate at Node0, Node1, and BS are optical requests and requests between the BS and RS are wireless requests. Both the optical and wireless requests are groomed along with the connection request Node1 to RS. For the optical requests from Node1 to BS, RSA constraints are checked (contiguous slots and continuity of slots along with the selected links). From base station to relay station path slots availability and followed by transponders availability is checked before the connection is established. The intermediate requests are added/dropped at their respective destination nodes.

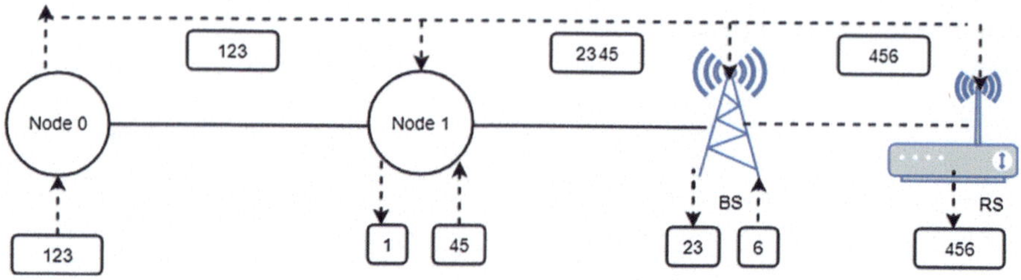

Fig. 2 Traffic grooming using light trail in elastic hybrid network

3.1 Complexity

The hybrid network consists of total n number of requests. The c presents the number of slots availability per link. The total number of paths between source to destination nodes are represented as k. In the initial step, the algorithm computer k alternate paths for node pair, which are having non zero traffic demands. So the algorithm takes $O(n^2)$. The algorithm takes the time complexity of $O(n^2)$ to generate the traffic matrix of $(n * n)$. The algorithm takes the time complexity of $O(n^2)$ to compute path at intermediate nodes. To assign the slots according to bandwidth demanded (checking the slots and network resources) the time complexity is $O(Kn^3)$. The proposed algorithm takes total time complexity of $O(n^3)$.

4 Analysis of Results

In this section, performance evaluation of the proposed algorithms is done through simulation. We have used 8 node network, NSF network (14 nodes), and German network (17 nodes) as back end network connected to WiMAX as depicted in Figs. 3, 4 and 5. Performance is compared between traffic grooming approach and without grooming approach, for Longest Hop Count Grooming (LHCG) and Minimum Hop Count Grooming (MHG), and Maximum Traffic Grooming algorithms (MTG) algorithms.

Figure 6 demonstrates the relationship between the number of slots (FS) in each link and the bandwidth blocking ratio results of LHCG, MHCG and MTG compared with the Longest Hop Without Grooming (LHWG), Minimum Hop Without Grooming (MHWG) and Maximum Traffic Without Grooming (MTWG).

In Fig. 7, the relationship between the number of transponder consumed in each node and the bandwidth blocking ratio results of LHCG, MHCG, and MTG are compared with the Longest Hop Without Grooming (LHWG), Minimum Hop Without Grooming (MHWG), and Maximum Traffic Without Grooming (MTWG). The MTG ad MHCG algorithms are consuming similar amount of transponders to serve the requests. MTWG algorithm has worst performance compared to other algorithms.

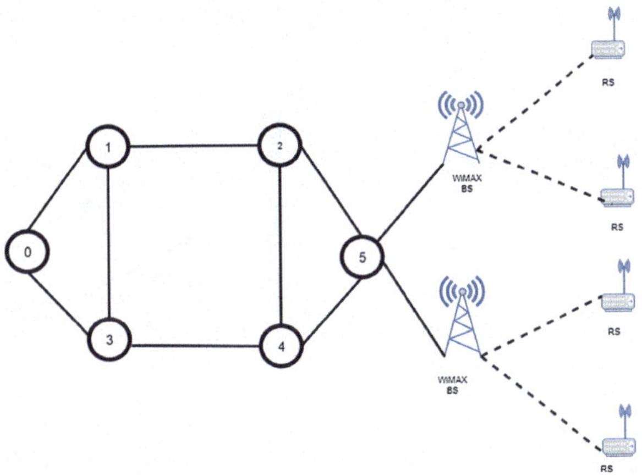

Fig. 3 8 NODE optical network connected as back end to wireless network

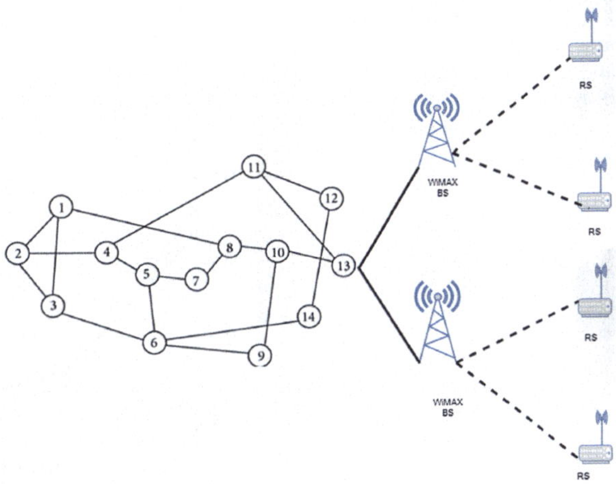

Fig. 4 NSF network connected as back end to wireless network

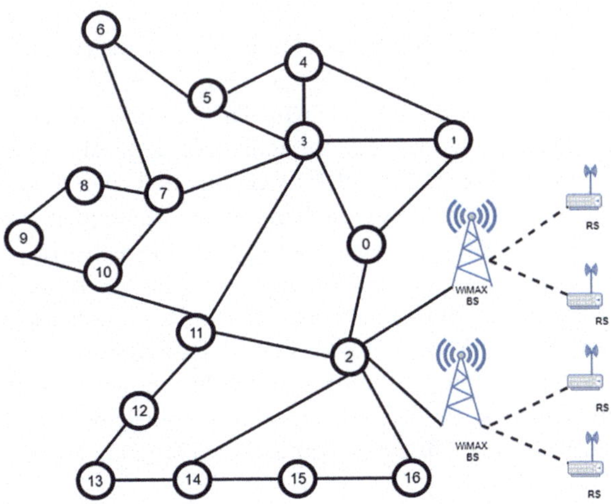

Fig. 5 GERMAN network connected as back end to wireless network

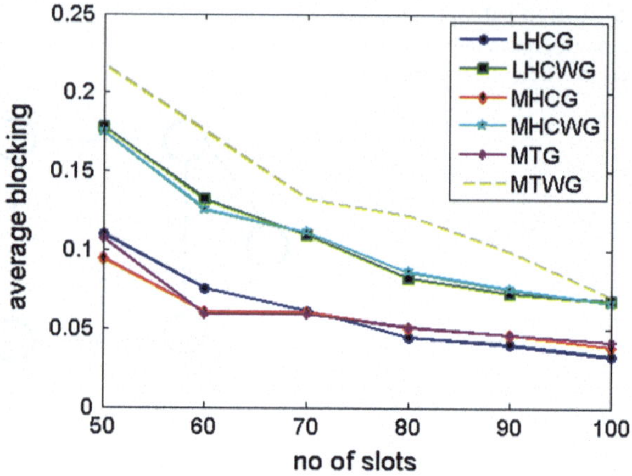

Fig. 6 Relationship between total number of slots (in each link) and bandwidth blocking ratio (8-node optical network is connected as back end to WiMAX networks)

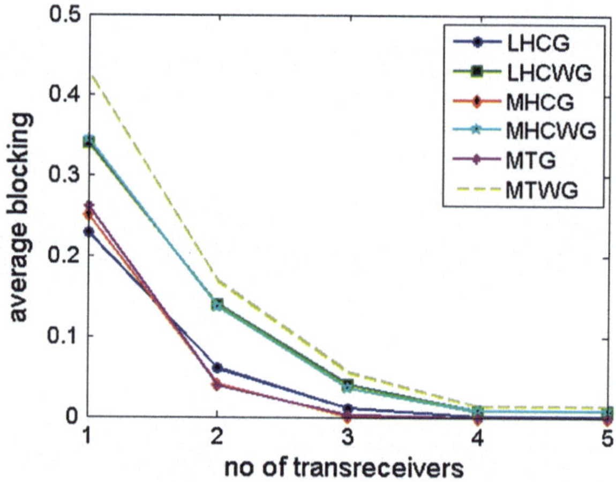

Fig. 7 Relationship between total transponders consumed and bandwidth blocking ratio (8-node optical network is connected as back end to WiMAX networks)

The relationship between the capacity of the network and slots remaining after serving the requests are shown in Fig. 8. The MTG algorithm tries to pack maximum requests between the node pair. This also minimizes number of frequency slots used as guard bands between the individual requests. The MHCG algorithm selects the minimum hop count path between the node pair to serve the requests, so it traverses the less number of links and minimizes the frequency slots used to serve the requests.

Relationship between capacity of the network, total capacity consumed, and capacity utilized as guard bands are depicted in Fig. 9. The MTG and MHCG consume less amount of network resources to serve the requests. The LHCWG algorithm consumes maximum number of frequency slots to serve the requests as it traverses through the longest route and individual requests are guarded by guard bands to serve the requests. The MTWG algorithms choose the path that accommodates maximum traffic between the node pair and the path may be longest or it may be shortest path.

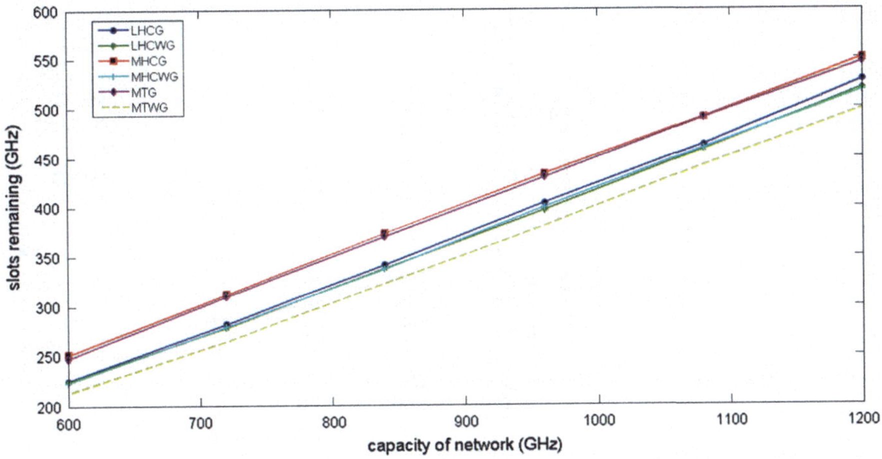

Fig. 8 Relationship between capacity of the network and slots remaining after serving the requests (8-NODE optical network is connected as back end to WiMAX networks)

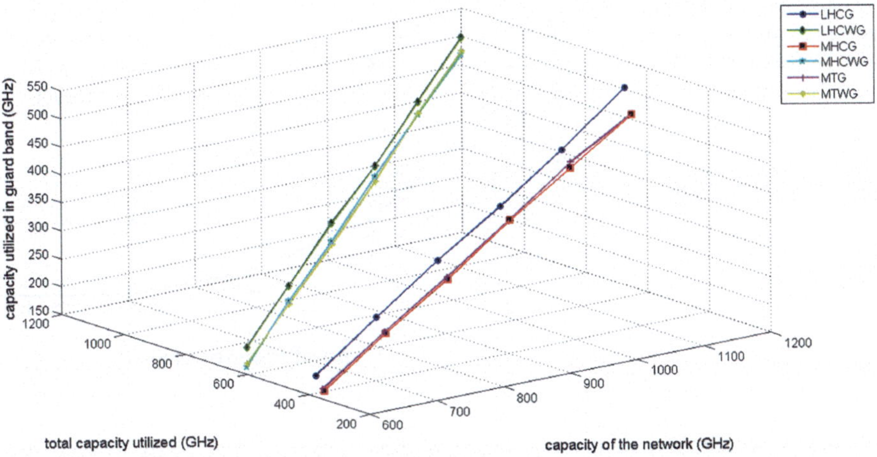

Fig. 9 Relationship between capacity of the network, total capacity consumed, capacity utilized as guard bands (8-node optical network is connected as back end to WiMAX networks

The individual requests are guarded by guard bands so it consumes maximum amount of frequency slots to serve the requests.

The simulations are also conducted on NSF network and German network to validate the performance of proposed algorithms. The Figs. 10 and 14 show relationship between total number of slots (in each link) and bandwidth blocking ratio. In both the NSF and German networks the MHCG algorithm show minimum blocking ratio as compared to other algorithms. In the 8-node network, the MHCG and MTG grooming algorithms show similar results. The main reason behind this is the 8-node

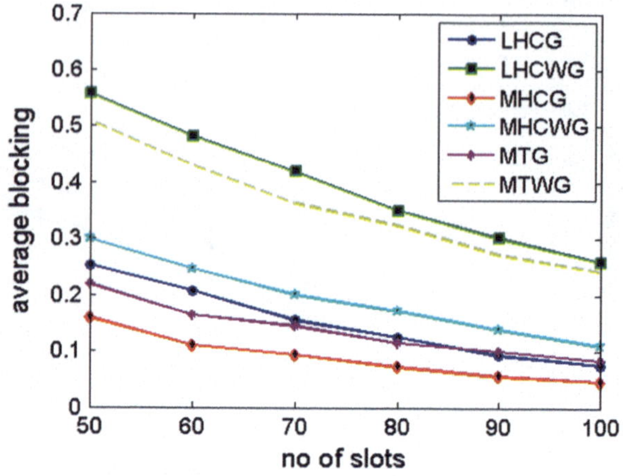

Fig. 10 Relationship between total number of slots (in each link) and bandwidth blocking ratio (14-node NSF network is connected to WiMAX networks)

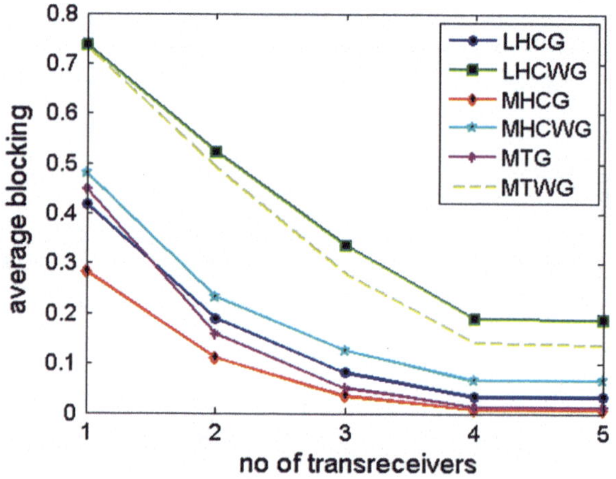

Fig. 11 Relationship between total transponders consumed and bandwidth blocking ratio (14-node NSF network is connected as back end to WiMAX networks)

network is not densely connected and same links are used repeatedly to serve the requests.

Figures 11 and 15 depict the relationship between the transponder and the bandwidth blocking ratio in the NSF and German network. The MHCG algorithm showed better performance in minimizing the bandwidth blocking ratio.

The relationship between capacity of the network and slots remaining after serving the requests are shown in Figs. 13 and 16 for NSF and German network connected as back end. The MHCG algorithm shows the maximum number of slots remaining after serving the requests.

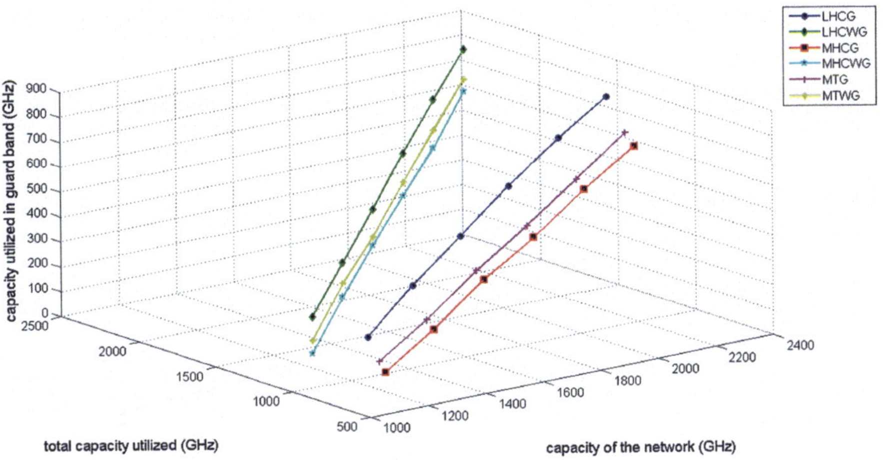

Fig. 12 Relationship between capacity of the network, total capacity consumed, capacity utilized as guard bands (14 node NSF optical network is connected as back end to WiMAX networks)

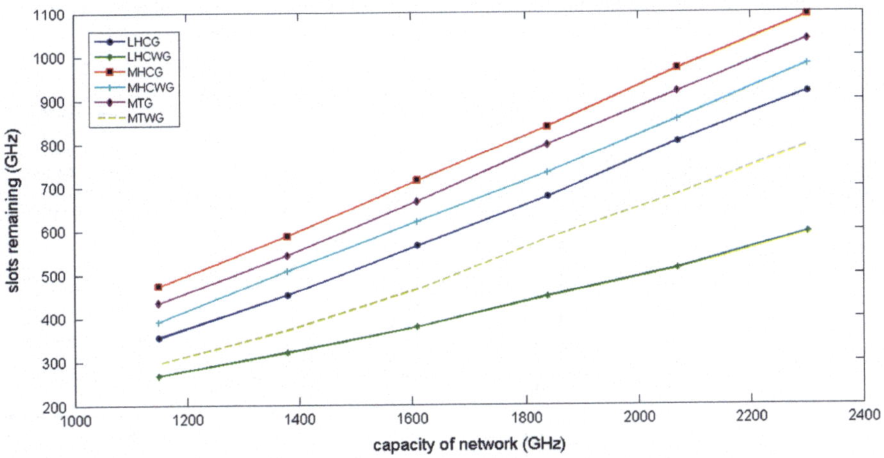

Fig. 13 Relationship between capacity of the network and slots remaining after serving the requests (14-node NSF optical network is connected as back end to WiMAX networks)

Relationship between capacity of the network, total capacity consumed, capacity utilized as guard bands are observed in Figs. 12 and 17 (for the NSF and German network). The LHCG consumes maximum number of guard bands to serve the requests, as it traverses the longest route. The MHCG performed better performance in both networks.

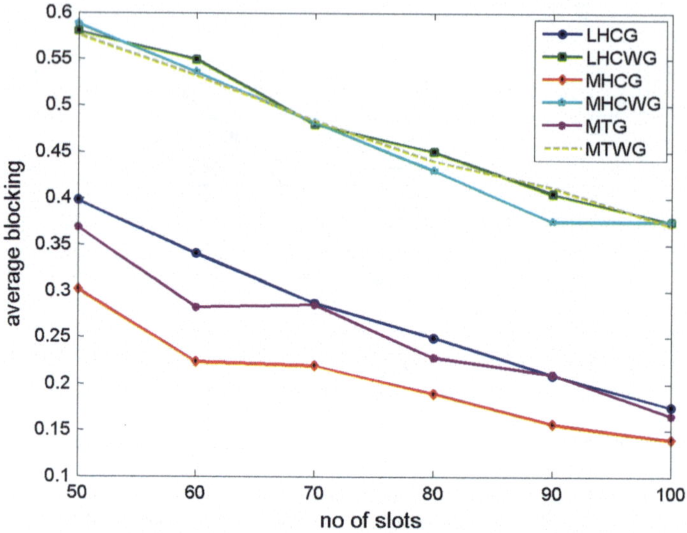

Fig. 14 Relationship between total number of slots (in each link) and bandwidth blocking ratio (17-node German optical network is connected as back end to WiMAX networks)

Fig. 15 Relationship between total transponders consumed and bandwidth blocking ratio (17-node German optical network is connected as back end to WiMAX networks)

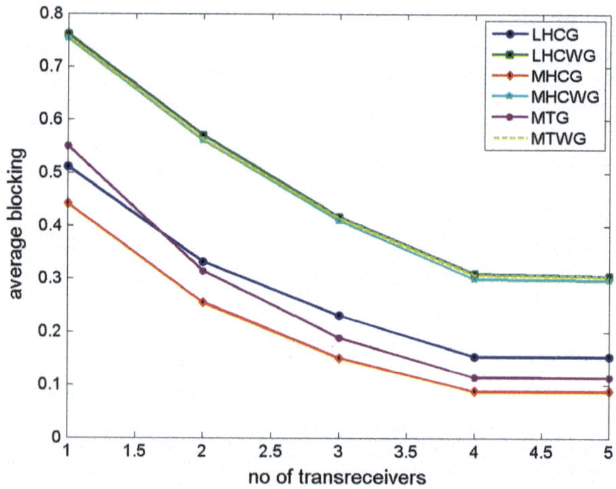

5 Conclusion

Traffic Grooming routing and spectrum assignment algorithms for integrated Elastic Optical and WiMAX network are proposed here. These algorithms perform better in terms of minimization of bandwidth blocking ratio and reduction in network resources consumption. In future, we will be addressing energy aware traffic Grooming Algorithms for such hybrid networks.

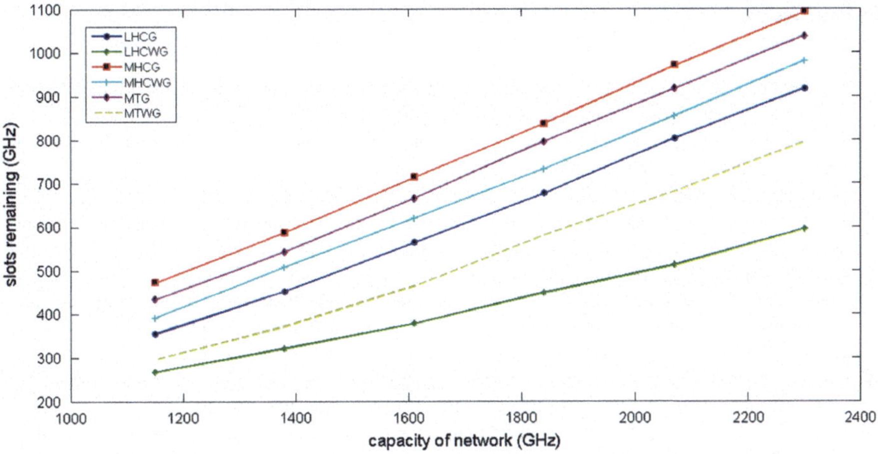

Fig. 16 Relationship between capacity of the network and slots remaining after serving the requests (17-node German optical network is connected as back end to WiMAX networks)

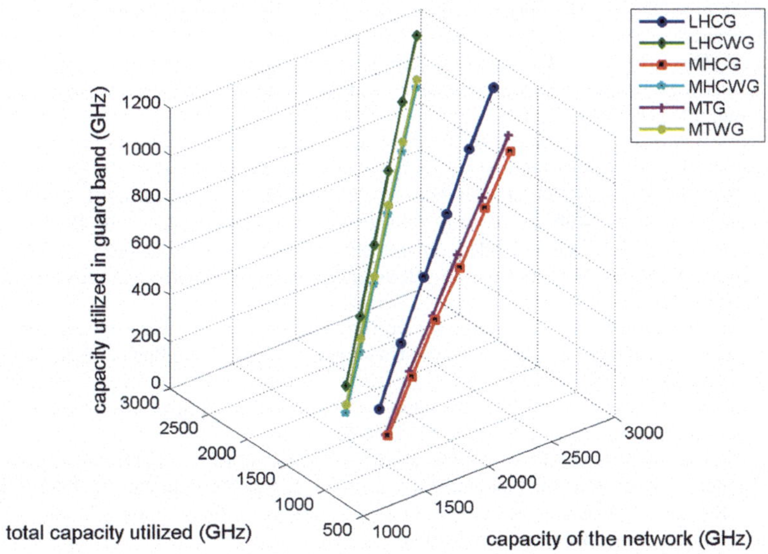

Fig. 17 Relationship between capacity of the network, total capacity consumed, capacity utilized as guard bands (17-node German optical network is connected as back end to WiMAX networks)

References

1. Ahmed, A., Shami, A.: Rpr-epon-wimax hybrid network: a solution for access and metro networks. J. Opt. Commun. Netw. **4**(3), 173–188 (2012)
2. Maier, M., Levesque, M., Ivanescu, L.: Ng-pons 1&2 and beyond: the dawn of the uber-fiwi network. IEEE Netw. **26**(2) (2012)
3. Anderson, M.: Wimax for smart grids. IEEE Spectr. **47**(7), 14–14 (2010)
4. Jinno, M., Takara, H., Kozicki, B., Tsukishima, Y., Sone, Y., Matsuoka, S.: Spectrum-efficient and scalable elastic optical path network: architecture, benefits, and enabling technologies. IEEE Commun. Mag. **47**(11) (2009)
5. Walkowiak, K., Klinkowski, M., Rabiega, B., Goścień, R.: Routing and spectrum allocation algorithms for elastic optical networks with dedicated path protection. Opt. Switch. Netw. **13**, 63–75 (2014)
6. Chatterjee, B.C., Sarma, N., Oki, E.: Routing and spectrum allocation in elastic optical networks: a tutorial. IEEE Commun. Surv. Tutor. **17**(3), 1776–1800 (2015)
7. Dutta, R., Rouskas, G.N.: Traffic grooming in wdm networks: past and future. Netw. IEEE **16**(6), 46–56 (2002)
8. Gumaste, A., Chlamtac, I.: Light-trails: a novel conceptual framework for conducting optical communications. In: Workshop on High Performance Switching and Routing, 2003, HPSR. (June 2003) 251–256
9. Shi, L., Chowdhury, P., Mukherjee, B.: Saving energy in long-reach broadband access networks: architectural approaches. Commun. Mag. IEEE **51**(2), S16–S21 (2013)
10. Yang, K., Ou, S., Guild, K., Chen, H.H.: Convergence of ethernet pon and ieee 802.16 broadband access networks and its qos-aware dynamic bandwidth allocation scheme. IEEE J. Sel. Areas Commun. **27**(2) (2009) 101–116
11. Ali, N.A., Dhrona, P., Hassanein, H.: A performance study of uplink scheduling algorithms in point-to-multipoint wimax networks. Comput. Commun. **32**(3), 511–521 (2009)
12. Reaz, A., Ramamurthi, V., Sarkar, S., Ghosal, D., Mukherjee, B.: Hybrid wireless-optical broadband access network (woban): capacity enhancement for wireless access. In: IEEE GLOBECOM 2008–2008 IEEE Global Telecommunications Conference, IEEE, pp. 1–5 (2008)
13. Ghazisaidi, N., Maier, M.: Fiber-wireless (fiwi) access networks: challenges and opportunities. IEEE Netw. **25**(1) (2011)
14. Naik, D., Nikita, De, T.: Congestion aware traffic grooming in elastic optical and wimax network. In: 2018 Technologies for Smart-City Energy Security and Power (ICSESP). IEEE, pp. 1–9 (2018)
15. Cvijetic, N., Qian, D., Hu, J.: 100 gb/s optical access based on optical orthogonal frequency-division multiplexing. IEEE Commun. Mag. **48**(7) (2010)
16. Bock, C., Prat, J., Walker, S.D.: Hybrid wdm/tdm pon using the awg fsr and featuring centralized light generation and dynamic bandwidth allocation. J. Light. Technol. **23**(12), 3981 (2005)
17. Davey, R., Kani, J., Bourgart, F., McCammon, K.: Options for future optical access networks. IEEE Commun. Mag. **44**(10), 50–56 (2006)

Reduction of the Mutual Coupling in Patch Antenna Arrays Based on Dollar Shaped Electrical Resonator

Santimoy Mandal and Chandan Kumar Ghosh

Abstract Mutual coupling is an inherent characteristic within the antenna array which degrades the antenna performances. DGS and EBG structure can help in the reduction of mutual coupling due to their capability of suppressing the surface wave propagation in a defined frequency range, but in DGS due to high back radiation it reduces the E- and H-plane radiation pattern of the antenna arrays. In the case of EBG structure due to parasitic loading effect and cavity effect will affect the bandwidth. In this paper a new method is adapted to diminish the effect of mutual coupling in an antenna array without degrading antenna characteristics. It shows that the proposed typically shaped microstrip resonator with mutual coupling reduction of more than 43 dB is achieved at the resonant frequency of 5.749GHZ.

Keywords Mutual coupling · Microstrip patch · Antenna element · Microstrip resonator

1 Introduction

Mutual coupling is an inevitable occurrence in multi-element antenna systems, which disgrace the antenna performance characteristics like antenna gain, antenna efficiency, antenna bandwidth, etc. For microstrip antenna array the electromagnetic coupling increases [1], as the distance between the patches is less than λ0/2. Therefore, to conquer such difficulty, different methods for decoupling such as defected

S. Mandal (✉)
R.V.S. College of Engineering & Technology, Edalbera, Bhilai Pahari, Jamshedpur 831012, India
e-mail: santimoy.mondal@gmail.com

C. K. Ghosh
Dr. B.C. Roy Engineering College, Jemua Road, Fuljhore, Durgapur 713206, India
e-mail: chandan.ghosh@bcrec.ac.in

S. Kundu et al. (eds.), *Proceedings of the 2nd International Conference on Communication, Devices and Computing*, Lecture Notes in Electrical Engineering 602, https://doi.org/10.1007/978-981-15-0829-5_7

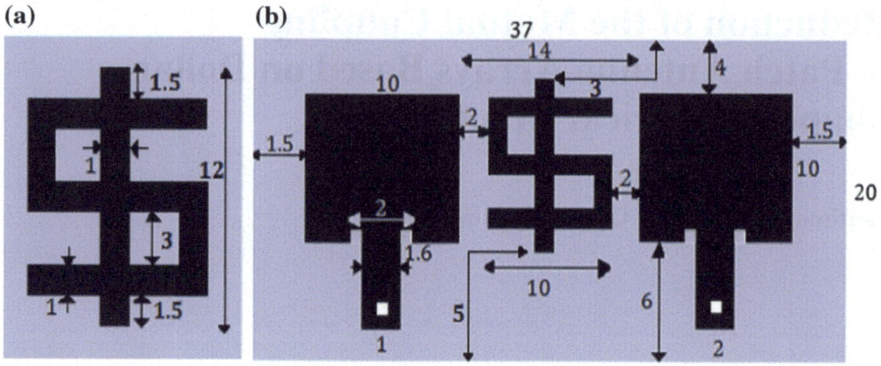

Fig. 1 **a** The dollar type resonator, **b** schematic of patch antennas with dollar shape resonator (all dimension are in mm)

ground structures (DGS) [2, 3], microstrip resonators [4, 5], and electromagnetic band-gap (EBG) [6, 7] structures have been already introduced in the literature. Episodic structures such as EBG have the ability to reduce surface wave propagation over the designed frequency band. DGS structures were preferred to improve the back radiation, ensuing in reduced front to back ratio. Multilayer dielectric substrates have some disadvantages as it increases the weight of the antenna arrays. Some other methods are also available in the literature by using metamaterial slab [8], slotted complementary split ring resonator (SCSRR) [9] and slot combined complementary split ring resonator (SCCSSR) [10] structure for suppressing the Mutual Coupling between the patch elements, but they still suffer from design complexity. The proposed technique has been simulated by the process of numerical simulation using the method of moment (MOM) based IE3D electromagnetic simulator and the electrical characteristics of the same have been studied. The schematic diagram of the proposed antenna design is shown in Figs. 1a, b.

2 Antenna Configuration and Results

A 37 × 20 mm^2 FR4 substrate with permittivity (ε_r) 4.4, thickness 1.6 mm and loss tangent 0.02 is used for designing this antenna. The schematic diagram of the proposed antenna integrated with dollar shaped resonator is shown in Fig. 1a. Simulation has been done using Method of moment (MOM) based IE3D Electromagnetic simulator. Simulated *S*-parameter of the proposed antenna shown in Fig. 2 with and without resonator which clearly reflects the mutual coupling suppression of more than 43 dB is achieved. We have optimized the dimension of dollar type resonator as shown in Fig. 1a after a good number of simulations.

After integrating a dollar shaped resonator we investigate the electrical characteristics of the antenna element (return loss S_{11} and transmission characteristic S_{21}) which is shown in Fig. 2. It is clearly evident from the given Fig. 2 that we have achieved MC (mutual coupling) reduction of more than 43 dB.

We have also studied the electrical characteristic of the proposed resonator with ports (return loss S_{11} and transmission characteristic S_{21}) in Fig. 3b. In Fig. 3b it

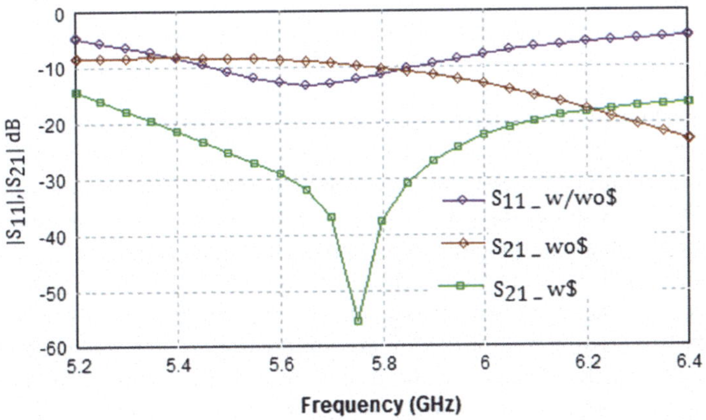

Fig. 2 Simulated *S*-parameter with and without dollar shaped resonator

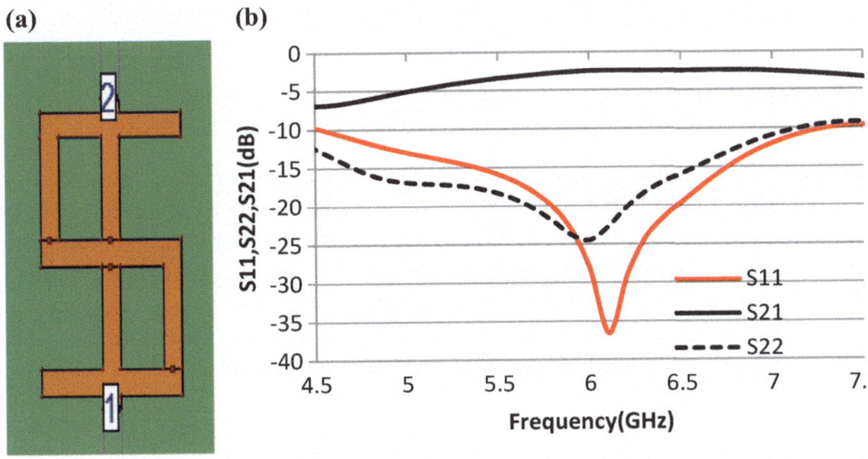

Fig. 3 **a** Dollar shaped resonator with ports, **b** simulated electrical characteristics of the resonator

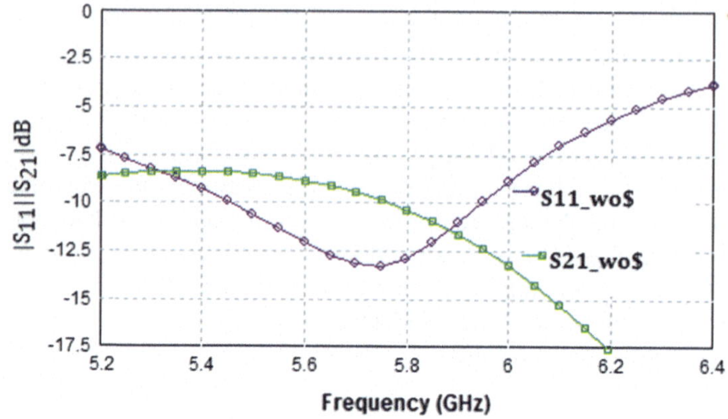

Fig. 4 Simulated electrical characteristics (w/o$) without dollar shaped resonator

is shown the filtering behavior and the stop band properties of the dollar shaped microstrip resonator.

We have also studied the electrical characteristic (return loss S_{11} and transmission characteristic S_{21}) without resonator in Fig. 4 which reflects how antenna elements are mutually coupled without embedding, properly placed resonator.

We have studied the simulated current distribution of the antenna element with dollar shaped resonator at resonant frequency in Fig. 5a. It is clearly seen from the given Fig. 5a that the suppression of mutual coupling is more at the resonant frequency 5.749 GHz. As we have removed the dollar shaped resonator the suppression of mutual coupling became less as shown in Fig. 5b.

The E-plane radiation pattern shown in Fig. 6 shows that the presence of dollar shaped resonator does not degrade the pattern in broadside direction as well as it does not produce any back radiation.

3 Conclusion

Mutual coupling reduction using dollar shaped resonator structure is presented in this paper. More than 43 dB mutual coupling suppression has been achieved between contiguous patch antennas for the designed frequency band. Microstrip patch antenna integrated with and without dollar shaped resonator has been experimentally studied and concluded that after integrating dollar shaped resonator it does not degrade the antenna performance characteristics like antenna gain, antenna bandwidth, etc. This antenna can be suitably used for different WiMAX applications.

(a)

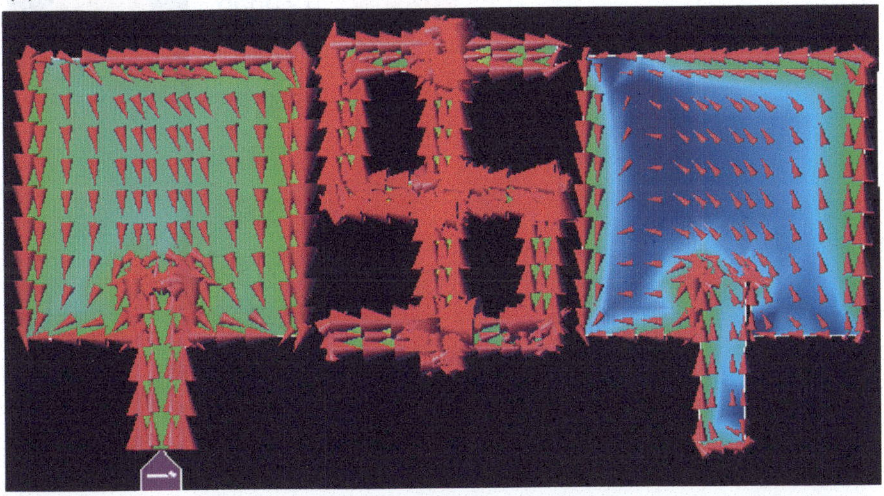

(b)

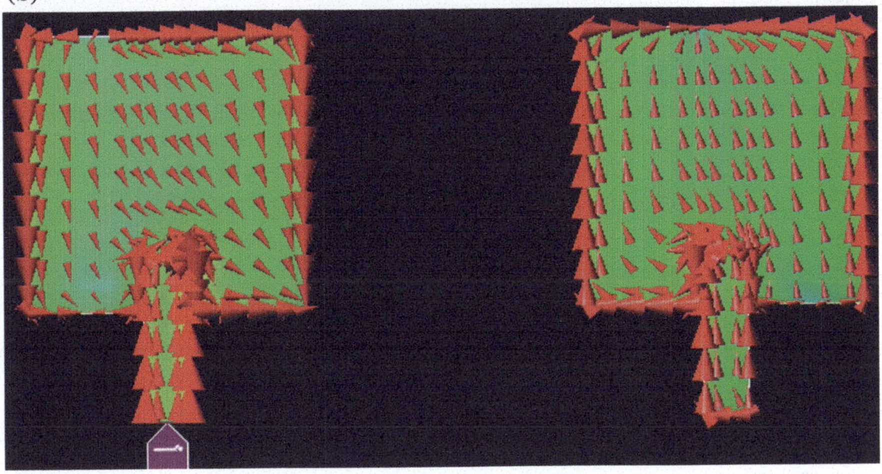

Fig. 5 **a** Simulated current distribution of the array with dollar shaped resonator, **b** simulated current distribution of the array without dollar shaped resonator

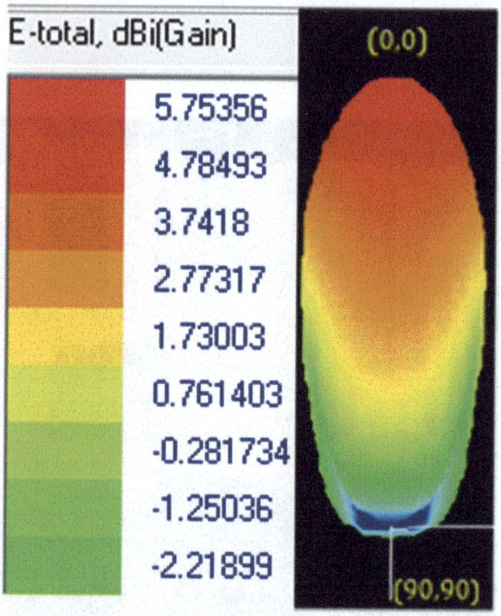

Fig. 6 Elevation pattern gain with dollar shaped resonator

References

1. Balanis, C.A.: Antenna theory: Analysis and design. Wiley Interscience, Hoboken, NJ (2005)
2. Zhu, F.G., Xu, J.-D., Xu, Q.: Reduction of mutual coupling between closelypacked antenna elements using defected ground structure. Electron Lett. **45**, 601–602 (2009)
3. Guha, D., Biswas, S., Joseph, T., Sebastian, M.T.: Defected ground structure to reduce mutual coupling between cylindrical dielectric resonator antennas. Electron Lett. **44**, 836–837 (2008)
4. Ghosh, C.K., Rana, B., Parui, S.K.: Reduction of Cross-polarization of Slotted Microstrip Antenna Array Using Spiral-Ring Resonator, Microwave and Optical Technology Letter (MOTL), vol. 55, issue 9, pp. 2083–2088. Wiley (2013)
5. Ghosh, C.K.: A compact 4-channel microstrip MIMO antenna with reduced mutual coupling. Int. J. Electron. Commun. (Elsevier) **70**, 873–879 (2016)
6. Farahani, H., Veysi, S., Kamyab, M., Tadjalli, M.A.: Mutual coupling reduction in patch antenna arrays using a UC-EBG superstrate. IEEE Antennas Wirel. Propag. Lett. **9**, 57–59 (2010)
7. Jiang, T., Jiao, T., Li, Y.: Array mutual coupling reduction using L-loading EShaped electromagnetic band gap structures. Int. J. Antennas Propag. (2016)
8. Farsi, S.D., Schreurs, Nauwelaers, B.: Mutual coupling reduction of planar antenna by using a simple microstrip u-section. IEEE Antennas Wirel. Propag. Lett. **11**, 1501–1503 (2012)
9. Ghosh, C.K., Parui, S.K.: Reduction of mutual coupling between E-shaped microstrip antennas by using a simple microstrip I-section,". Microw. Opt. Technol. Lett. **55**(11), 2544–2549 (2013)
10. Suwailam, M.M.B., Siddiqui, O.F., Ramahi, O.M.: Mutual coupling reduction between microstrip patch antennas using slotted-complementary split-ring resonators. IEEE Antennas Wirel. Propag. Lett. **9**, 876–878 (2010)

Design of Dual-Band 1×2 Antenna Array Using Wide Band Wilkinson Power Divider

Pratistha Brahma and Banani Basu

Abstract In this paper, A wideband Wilkinson Power Divider has been designed for C band application and provides a bandwidth from 970 MHz to 9.5 GHz. Two rectangular microstrip patch antennas have been designed and fed to this proposed power divider. Two frequency notches have been found at 4 GHz and 6.4 GHz and the consecutive gains for these frequencies are 5.6 dB and 10.11 dB, respectively.

Keywords Antenna array · Wilkinson power divider · Gain · Dual notch

1 Introduction

Antenna arrays are widely used nowadays in mobile communication. Usually, the array provisions of microstrip patch antennas are used to improve gain, directivity, and efficiency. As a single antenna radiates widely in pattern with large beam angle, so it is not effective for point-to-point communication. So, the beam should be more directive in nature which leads the need to use antenna array. A rectangular microstrip patch antenna array has been designed in [1] at 5.8 GHz frequency and 9.82 dB heights gain has been found. Ammar Ali et al. have designed a 4-port ultra wideband power divider and found a bandwidth of 1 GHz–11.5 GHz [2]. A 2-port Wilkinson Power Divider has been designed at 2.4 GHz frequency for Radar S-band application [3]. Yu [4] has presented a broadband Wilkinson power divider. A wideband four-element linear antenna array has been designed at multiple resonating frequencies as 2.4 GHz, 3.5 GHz, and 5.2 GHz [5]. Samudrala et al. have presented a planar array at 7.8 GHz operating frequency for C band application in [6]. Peters et al. [7] presented a four-element linear array at 60 GHz and a millimetre wave Wilkinson power divider. A low-profile array has been designed at 4.5 GHz center frequency

P. Brahma (✉) · B. Basu
Department of Electronics and Communication Engineering, National Institute of Technology, Silchar, Cachar 788010, Assam, India
e-mail: pratistha.titli07@gmail.com

B. Basu
e-mail: banani@ece.nits.ac.in

S. Kundu et al. (eds.), *Proceedings of the 2nd International Conference on Communication, Devices and Computing*, Lecture Notes in Electrical Engineering 602, https://doi.org/10.1007/978-981-15-0829-5_8

and provided 20 dBi gain [8]. Wang et al. have designed a U-slotted microstrip patch antenna array at 5.65 GHz and found a gain of 11.5 dBi [9].

This paper presents a rectangular microstrip patch antenna array for C band application. The range of the C band is 4–8 GHz. The designed antenna and antenna array are resonating at 4 GHz and 6.4 GHz. A wideband Wilkinson Power Divider (WPD) has been also designed at 4 GHz center frequency having bandwidth 960 MHz to 9.5 GHz which is covering the entire S and C band. The patch antenna and antenna array have two frequency notches at 4 GHz and 6.4 GHz frequencies. The gain has been found for single antenna as 3.9 dB and 7.8 dB at 4 GHz and 6.4 GHz frequency, respectively. The gain for antenna array has been found at 4 GHz and 6.4 GHz is 5.7 dB and 10.11 dB, respectively.

2 Proposed Wilkinson Power Divider and Antenna Array Geometry and Simulated Results

In the present investigation, a high gain Rectangular Microstrip Patch Antenna (RMPA) [10–13], a wideband 2:1 Wilkinson Power Divider and an array of two RMPA elements has been designed.

2.1 Rectangular Microstrip Patch Antenna

A RMPA has been designed to be operated at 4 GHz frequency. Rogers RO3003 having dielectric constant $\varepsilon_r = 3$ again has been chosen for this design. The detailed dimensions of the antenna structure are given in Fig. 1. The height of the substrate has been chosen as 1.6 mm.

Table 1 represents the dimensions of the rectangular patch. The lower corners of the patch are cut according to Fig. 1 to get dual notches and operating at two different frequencies. For the proposed array, as current distribution changes at those two resonating frequencies, so, we get dual notches in proposed array. The reflection coefficient and gain graph are shown in Figs. 2 and 3 and compared with conventional patch.

Since the antenna is resonation at 4 GHz and 6.4 GHz frequency, so it is compatible with the WPD shown in Fig. 2. At 4 GHz and 6.4 GHz two notches are obtained having $S11$ of -20 dB and -18 dB, respectively. Figure 3 shows the gain of the RMPA at those two consecutive resonation frequencies. At 4 GHz, the gain is 3.9 dB and at 6.4 GHz the gain is 7.8 dB, respectively.

Figure 4 shows the radiation pattern of single antenna.

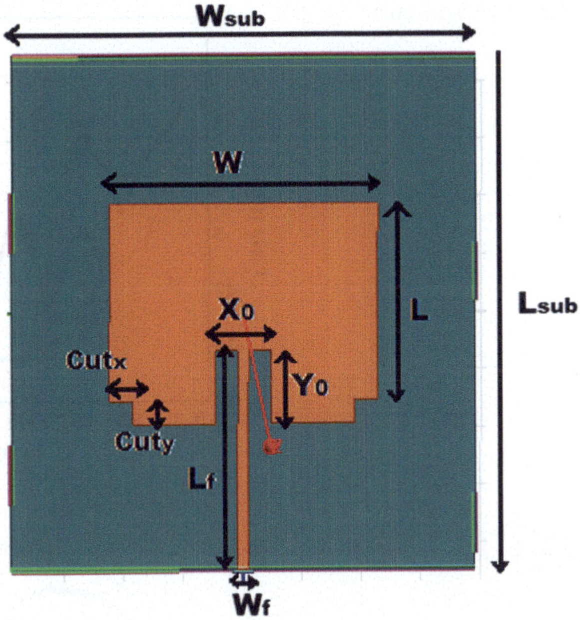

Fig. 1 Structure of rectangular patch antenna top side

Table 1 Dimension parameters and their values of rectangular MPA

Parameter name	Parameter value (mm)	Parameter name	Parameter value (mm)	Parameter name	Parameter value (mm)
Wsub	78	L	18.2	X_0	4.85
Lsub	89	W_f	1.8	Y_0	9.35
W	28	L_f	18.08	Cut_X, Cut_y	4,4

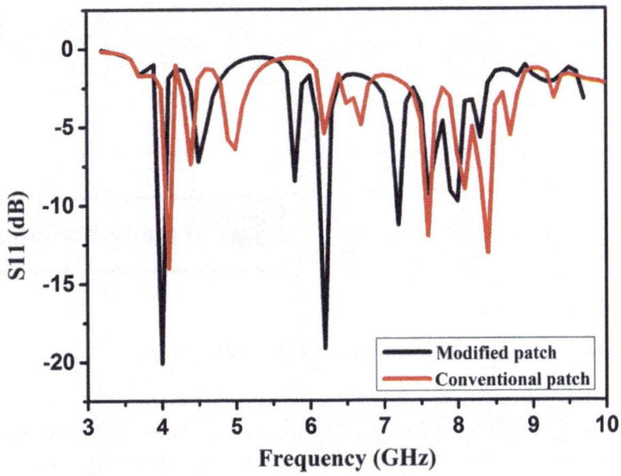

Fig. 2 $S11$ for RMPA

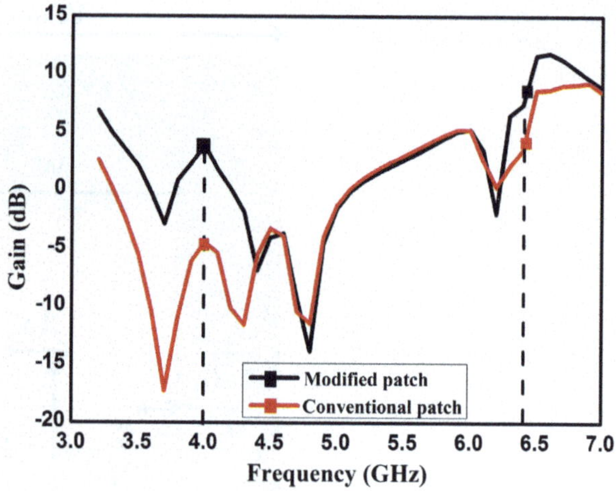

Fig. 3 Gain for RMPA

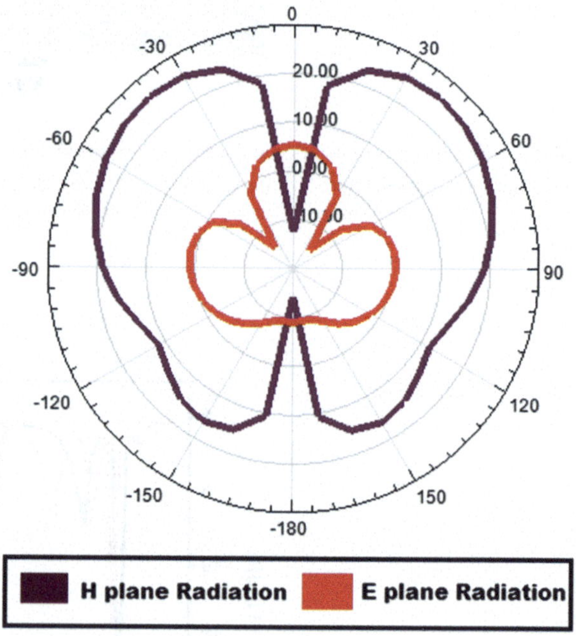

Fig. 4 Radiation pattern of single antenna

2.2 2:1 Wilkinson Power Divider

A 2:1 Wilkinson Power Divider (WPD) has been designed at 4 GHz frequency and its dimensions are shown in Fig. 5. The dielectric substrate has been chosen as Roger RO3003 ($\varepsilon_r = 3$) and height of the substrate is 1.6 mm [14–17].

Table 2 shows the dimensions of the parameters of the WPD.

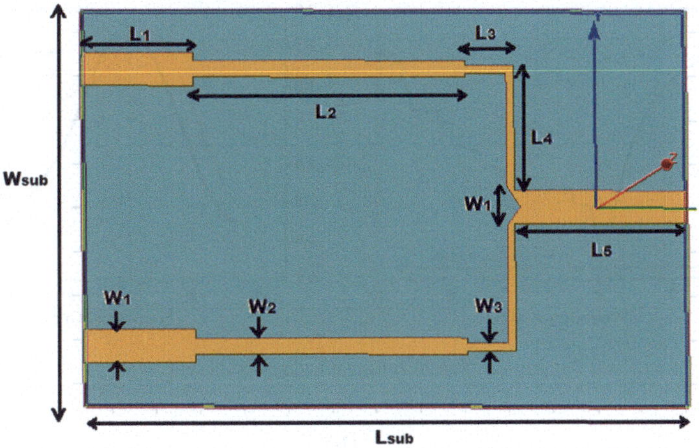

Fig. 5 Structure of WPD top side

Table 2 Dimension parameters and their values of WPD

Parameter name	Parameter value (mm)	Parameter name	Parameter value (mm)	Parameter name	Parameter value (mm)
Wsub	18	W_3 (for 100 Ω)	0.15	L_3	2
Lsub	27.33	L_1	5	L_4	6
W_1 (for 50 Ω)	1.8	L_2	12.33	L_5	8
W_2 (for 70.07 Ω)	0.43				

Figure 6 shows *S*-parameter of the WPD. Figure 6a represents a wideband reflection coefficient ($S11$) of the PD from 960 MHz to 9.5 GHz. The value of $S11$ is −31.5 dB at 4 GHz operation frequency. Figure 6b shows the transmission coefficient ($S21$, $S31$) of the PD.

Figure 7a and b shows phase of reflection coefficient and transmission coefficient of WPD. Figs. 6 and 7 ensuring that when the power is feeding at Port 1, power getting at Port 2 and Port 3 have equal magnitude and similar phase. It is seen from Fig. 7 that, at 4 GHz, the reflection and transmission phase values of $S11$, $S21$, and $S31$ are 136.480, 134.500, and 132.160, respectively.

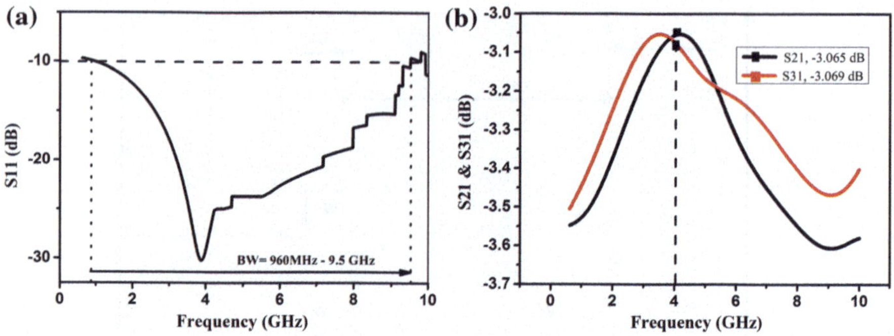

Fig. 6 *S*-parameter **a** reflection coefficient **b** transmission coefficient

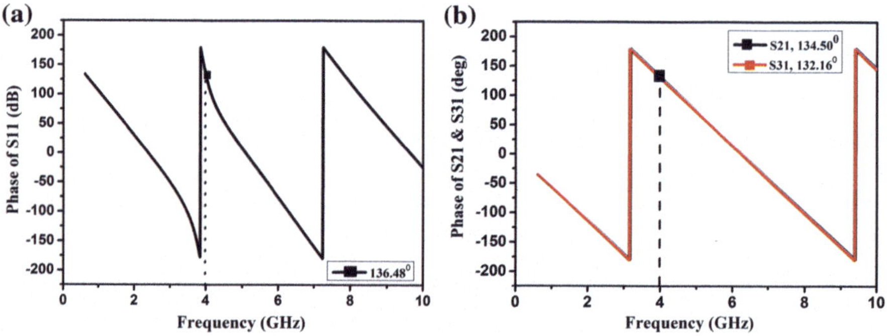

Fig. 7 Phase graph of *S*-parameters of WPD **a** *S*11 **b** *S*21 and *S*31

2.3 Rectangular Microstrip Patch Antenna Array

Two RMPAs have been feed to the wideband WPD that has been studied in previous to get higher gain. The spacing between two elements has been optimized and found as 65.5 mm as shown in Fig. 8.

Figure 8 shows the detailed dimensions of the antenna array. Table 3 represents the dimension of the array.

Figure 9 represents *S*11 for antenna array. It shows that the array is resonation within the wideband of the WPD and the minimum values of *S*11 have been found at 4 GHz (−18.01 dB), 6.4 GHz (−17.99 dB).

Figure 10 represents the gain of the antenna array. At 4 GHz resonation frequency, the gain is 5.7 dB and at 6.4 GHz frequency, the gain is 10.11 dB, respectively.

Figure 11 shows the corresponding radiation pattern of the array. Figure 12 shows the comparison of SLL for conventional and proposed array design. Co-polarization and cross-polarization of the array have not been studied yet. It will be done for

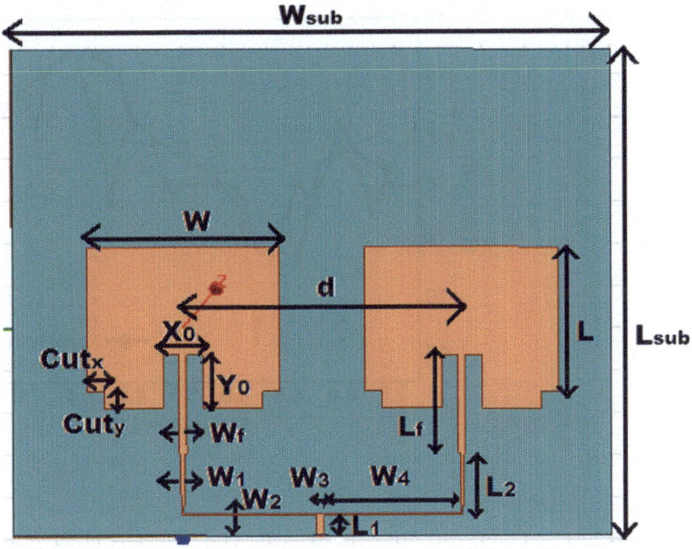

Fig. 8 Top view of antenna array

Table 3 Parameter values for modified RMPA design

Parameter names	Parameter values (mm)	Parameter names	Parameter values (mm)	Parameter names	Parameter values (mm)
Wsub	140	L_f	15.08	W_2	0.15
Lsub	115	Cut_x, Cut_y	4,4	W_3	1.8
W	28	d	65.5	W_4	31.75
L	18.2	X_0, Y_0	4.85,9.35	L_1	8
W_f	1.8	W_1	0.43	L_2	12.33

Fig. 9 *S*11 for antenna array

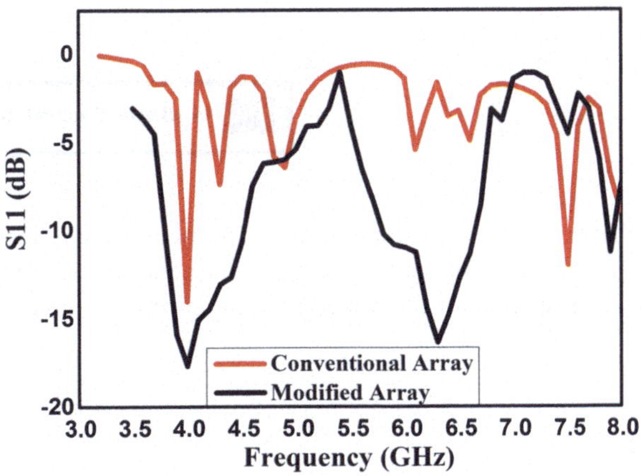

Fig. 10 Gain of antenna array

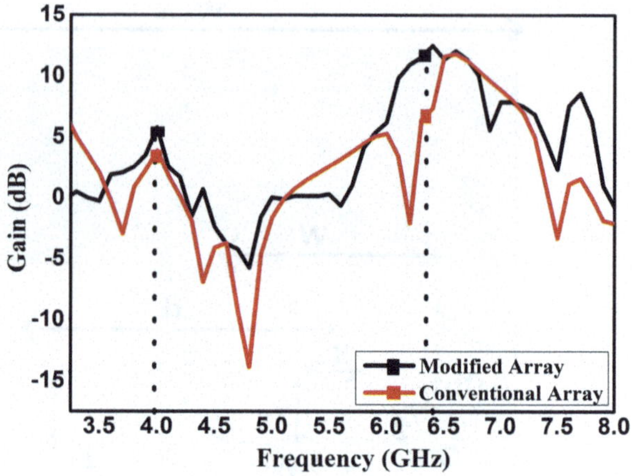

Fig. 11 Radiation pattern of antenna array

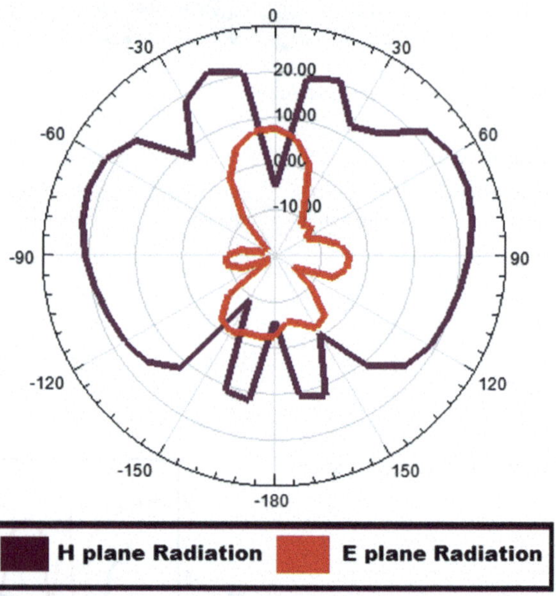

further study. The radiation losses in transmission lines and antenna can be reduced by decreasing the spacing between the conductors. All the simulations are done using High Frequency Electromagnetic Field Simulation (HFSS) Software.

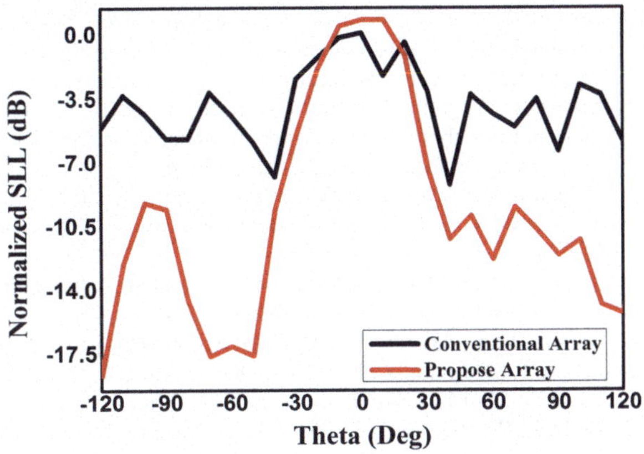

Fig. 12 SLL of conventional and proposed antenna array

3 Conclusion

In this paper, a wideband Wilkinson Power Divider, a dual notch RMPA and a dual notch two-element antenna array have been studied. Compared to conventional array, this proposed array performs better and provides better gain and dual notches. The antenna array provides high gain and minimizes the SLL which is also compared to conventional one.

References

1. Makkawi, H.M.M.: Performance evaluation of 1×2 patch antenna array based on power divider characteristics. In: 2016 Conference of Basic Sciences and Engineering Studies (SGCAC) Khartoum, pp. 6–12 (2016)
2. Ali, A.H., Abd-Alhameed, R.A., Noras, J.M. Child M.B.: An ultra wide band power divider for antenna array feeding network, pp. 1–4 (2016)
3. Taufiqqurrachman, D.P., Kurniadi, D.P.: Design and realization wilkinson power divider at frequency 2400 MHz for Radar S-band. IOSR J. Electron. Commun. Eng. (IOSR-JECE) 26–30 (2012)
4. Yu, T.: A broadband wilkinson power divider based on the segmented structure. IEEE Trans. Microw. Theory Tech. 1902–1911 (2018)
5. Sonkki, M., Pfeil, D., Hovinen, V., Dandekar, K.R.: Wideband planar four-element linear antenna array. IEEE Antennas Wirel. Propag. Lett. 1663–1666 (2014)
6. Samudrala, S., Palle, H.C.: Rectangular microstrip patch antenna array for C-band scatterometer and MBI. In: IEEE Annual India Conference (INDICON), pp. 1–4 (2016)
7. Peters, F.D.L., Hammou, D., Tatu, S.O., Denidni, T.A.: Modified millimeter-wave wilkinson power divider for antenna feeding networks. Prog. Electromagn. Res. Lett. 11–18 (2010)
8. Mak, K.M., Wong, H., Luk, K.M.: A wideband low-profile microstrip antenna and array. Microw. Opt. Technol. Lett. 729–730 (2006)
9. Wang, H., Huang, X.B., Fang, D.G.: A single layer wideband U-slot microstrip patch antenna array. IEEE Antennas Wirel. Propag. Lett. 9–12 (2008)

10. Mishra, R.D., Singhal, P.K.: Gain enhancement of rectangular microstrip patch antenna designed for exposure system using microstrip array. Int. J. Signal Process. Image Process. Pattern Recognit. 417–430 (2016)
11. Lakrit, S., Ammor, H., Matah, S., Karli, R., Saadi, A., Terhzaz, J., Tribak, A.: A new small high-gain wideband rectangular patch antenna for X and Ku bands applications. J. Taibah Univ. Sci. Taylor and Francis, pp. 202–207 (2018)
12. Rabbani, M.S., Shiraz, H.G.: Improvement of microstrip patch antenna gain and bandwidth at 60 GHz and X bands for wireless applications. IET Microw. Antennas Propag. 1167–1173 (2016)
13. Awad, N.M., Abdelazeez, M.K.: Multislot microstrip antenna for ultra-wide band applications. J. King Saud Univ.–Eng. Sci., Elsevier 38–45 (2015)
14. Balanis, C.A.: Antenna Theory Analysis and Design, 3rd edn., Wiley (2005)
15. Elliott, R.S.: Antenna Theory and Design. 4th edn., Wiley (2003)
16. Pozar, D.M.: Microwave Engineering, 3rd edn., Pearson Education Inc. (2009)
17. Ludwig, R., Bogdanov, G.: RF Circuit Design: Theory and Applications, Pearson Education Inc. (2009)

Design of Low-Profile AMC Antenna for SAR Reduction

Bidisha Hazarika and Banani Basu

Abstract A multilayered Artificial Magnetic Conductor (AMC) antenna is proposed in this article for medical body area network band. The AMC comprises a 2 × 2 array of rectangular unit-elements which provides a zero reflection phase. The AMC is placed below the circular monopole antenna to increase the gain and reduce the backward electromagnetic radiations. The operating frequency band ranges from 1.9 to 2.44 GHz. The incorporation of the AMC lowers the average 1-g SAR to 0.675 W/kg at the specific band of the flat antenna when it is 1 mm away from the antenna. The proposed antenna gives peak antenna gain of 4.2 dB, and front to back radiation ratio of 8.94 dB at 2.23 GHz.

Keywords Artificial magnetic conductor (AMC) · Specific absorption rate (SAR) · Unit-element

1 Introduction

Different Wireless Body Area Network (WBAN) applications necessitate accurate specifications for the human tissues to the Electromagnetic (EM) radiation. Two necessary instructions are introduced to restrict the EM radiation. Those are—the International Commission on Non-Ionizing Radiation Protection (ICNIRP) and the US regulatory Federal Communications Commission (FCC) [1]. Nowadays, researchers and scientists are redesigning the similar pattern of electronic technology for different medical applications which increases the safety of patients in medical domain [2]. It is noticed in the literature that by introducing AMC in the design, the Specific Absorption Rate (SAR) is reduced significantly [3]. In military applications, real-time information technology has been presented for the Future Warrior Forecast (FFW) [4]. In [5], a PIFA AMC antenna is designed to reduce SAR up to 43.3% at the designated frequency band. International guidelines for whole-body EM absorption also

B. Hazarika (✉) · B. Basu
Department of Electronics and Communication Engineering, National Institute of Technology, Silchar, Silchar, Cachar 788010, Assam, India
e-mail: bidisha322@gmail.com

S. Kundu et al. (eds.), *Proceedings of the 2nd International Conference on Communication, Devices and Computing*, Lecture Notes in Electrical Engineering 602, https://doi.org/10.1007/978-981-15-0829-5_9

utilize the Whole-Body Averaged Specific Absorption Rate (WBA-SAR) to measure the SAR value. Liu and Wang [6] presents a smaller metamaterial-based monopole antenna without calculating SAR. Therefore, by reducing the SAR absorption the human body can be guarded from harmful RF transmissions utilizing AMC.

In this article, we have proposed a double-layered AMC integrated antenna for reducing SAR in the respective frequency band. The work presents the design methodology of AMC integrated antenna and characterized by zero reflection phase. The circular antenna is situated at the top of upper substrate and separated by a 10 mm air gap. The integrated antenna proffers impedance matching at 2.23 GHz and achieves miniaturization. By placing the AMC below the circular monopole, the surface waves at the designated frequency have been reduced and increases gain significantly. The results of the proposed antenna show impedance matching of −30.5 dB, gain of 4.2 dB, and FB ratio of 8.94 dB at 2.23 GHz. By using AMC, the back lobe radiation also minimizes greatly. The integrated antenna reduces the SAR up to 0.675 W/kg, while it is 1 mm apart from the artificial tissue model. This further proves the superiority of AMC integrated antenna for body-centric applications.

2 Design of Antenna and AMC Structure

2.1 Design of AMC Reflector

The design of the proposed rectangular-shaped AMC unit-element is presented in Fig. 1a. The AMC is designed with Roger RO3003 material with $\varepsilon_r = 3$, $\tan \delta = 0.0013$ and a thickness of 1.5 mm. The AMC antenna behaves as a Perfect Magnetic Conductor (PMC) in a particular frequency region, which furnishes zero reflection phase. In this design, the zero reflection phase is obtained at 2.4 GHz band with a bandwidth of 0.25 GHz (1.85–2.45 GHz) within ±90° phase values. The interferences betwixt the incident wave and the reflected wave are occurred in that frequency region. This characterization is normally analyzed by illustrating a planar wave and introducing an infinite number of AMC unit-elements array. To create infinite unit-element array, two master/slaver boundaries and floquet ports are utilized in our proposed structure. Considering the size of feed network and radiation performance of antenna, we chose 2 × 2 array in our design. The top surface consists of rectangular-shaped patch and the bottom surface is the metal ground. To simplify the design process, the size of AMC unit-element is controlled at 55 mm × 73 mm.

In electromagnetic fields, the nature of design is characterized by its permittivity and permeability. Scattering parameter retrieval [7] is an effectual technique to extract material properties of unit-element. In addition, by giving periodic boundaries on a single unit-element, the features of an infinite number of AMC unit-elements can be definitely extricated. This is because of the reason that the permittivity and permeability equations are based on the procedure derived in [8]. However, by observing at

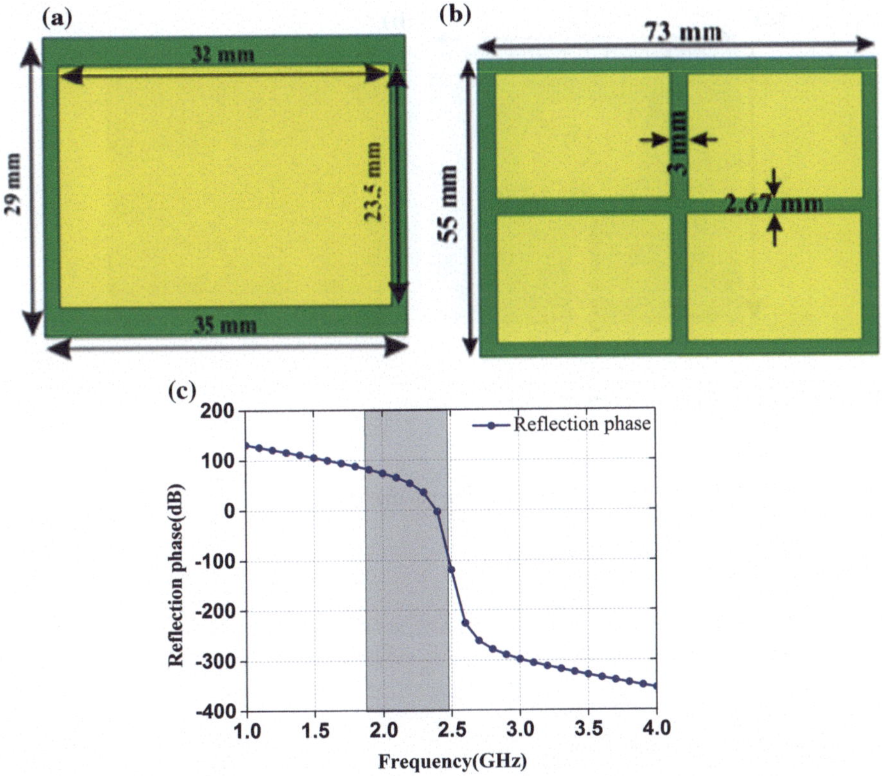

Fig. 1 Design of **a** single unit-element, **b** 2 × 2 AMC unit-element array and **c** zero phase reflection of single unit-element

the S-parameters, the features of the substrate material cannot be explained. Therefore, by applying S_{11} and S_{21} values, the permittivity (ε) and permeability (μ) are extracted which is not presented here because of massive contents.

2.2 *Antenna Design*

The top and bottom layers of circular monopole are presented in Fig. 2. The circular antenna is designed with FR4_epoxy substrate with relative dielectric permittivity of 4.4, loss tangent of 0.02, and thickness of 1.6 mm. The circular monopole is simulated using the ANSYS HFSS software. The design parameters of the antenna are carefully optimized to achieve wide bandwidth. The antenna covers 2.5 GHz frequency band. The dimensions of circular monopole are furnished in Table 1.

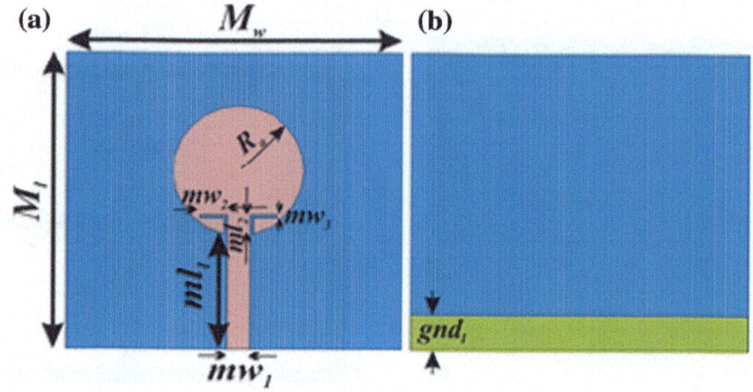

Fig. 2 **a** Top and **b** bottom layers of monopole antenna

Table 1 Antenna dimensions using HFSS simulator (Unit: mm)

Antenna design parameters	Values
M_w	45
M_l	40
mw_1	3
ml_1	16
ml_2	5
mw_2	3.2
mw_3	0.5
gnd_l	4.675
R_a	8.7

2.3 *Integration of AMC*

The AMC antenna comprises two substrates—a circular monopole, and an AMC of 2 × 2 rectangular unit-element array. The radiating element is fed by a microstrip line. Both layers are separated by a 10 mm air gap. The position of circular antenna above the AMC controls the coupling between the monopole and the AMC. The circular antenna is designed in such a way that it has a resonant response only in the parallel to x-axis. Figure 3 shows the planar configuration of AMC integrated antenna.

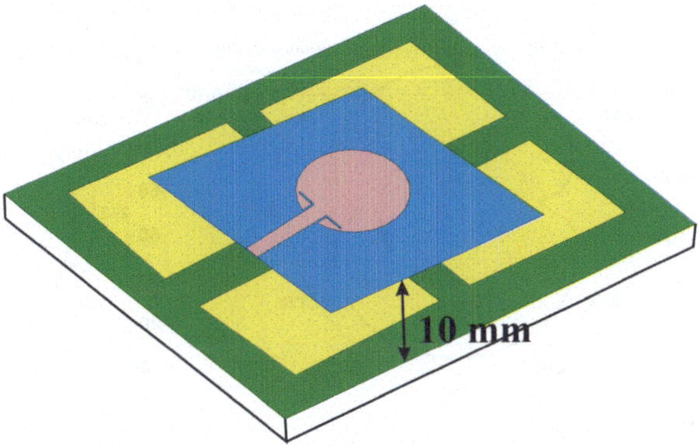

Fig. 3 Configuration of AMC integrated antenna

3 Results and Discussion

In this section, the AMC integrated antenna is analyzed in regard to S_{11}, gain, FBR, and radiation pattern. The dimension of AMC surface is optimized at 55 mm × 73 mm to resonate at specific frequency as shown in Fig. 4a. The circular monopole is situated at a height of 10 mm above the 2 × 2 array of rectangular-shaped unit-elements. Therefore, we have attained a satisfying impedance matching ($S_{11} < -10$ dB) of 1.9–2.44 GHz with a central frequency of 2.23 GHz as presented in Fig. 4a.

The aim of the portion is to examine the practical implementations of the AMC by enhancing gain, FBR, and lowering backward electromagnetic radiation in the particular frequency band. As observed in Fig. 4b, the monopole antenna offers an antenna gain of 1.53 GHz, while the AMC integrated antenna offers a broadside gain of 4.2 dB at 2.23 GHz. The radiation performance of the integrated antenna is to explain polarization features that are examined and presented in Fig. 5. The circular monopole antenna alone achieves a dipole like radiation criterion and an omnidirectional radiation criterion in E-field and H-field. Moreover, in contrast to

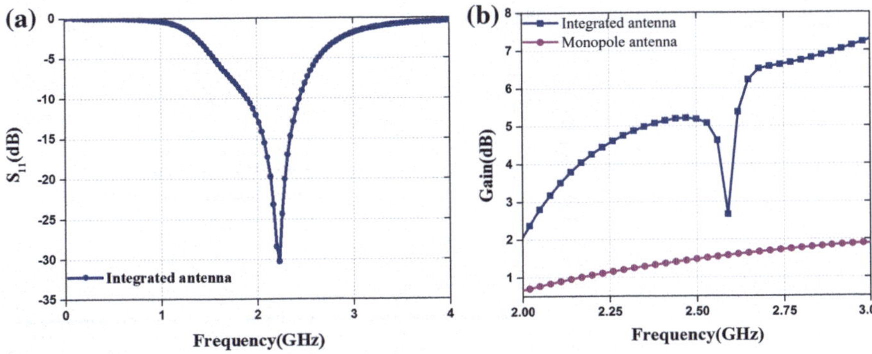

Fig. 4 Simulated **a** S_{11} of integrated antenna and **b** gain of integrated and monopole antenna

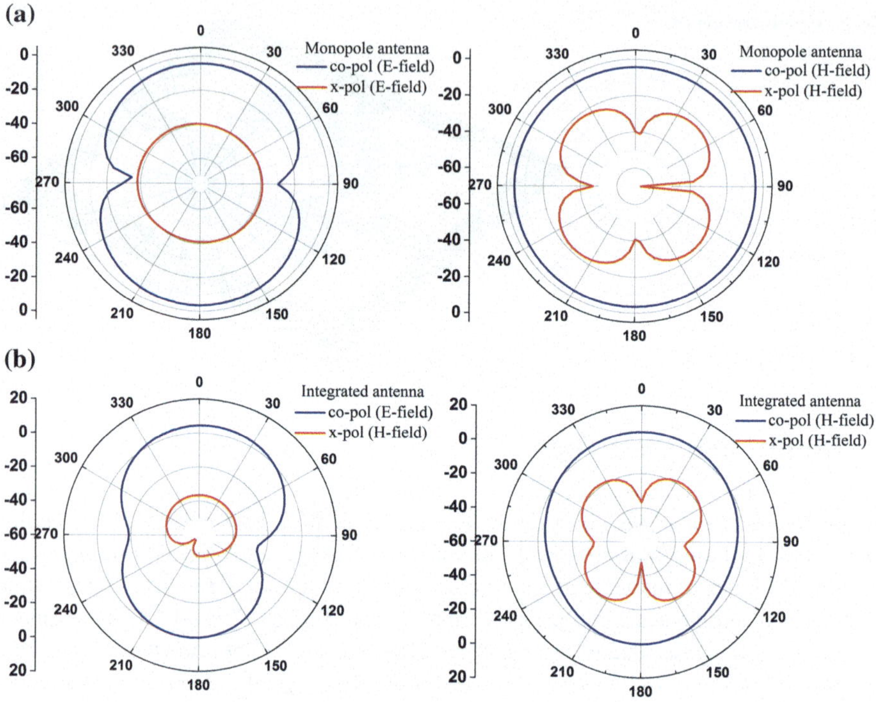

Fig. 5 Simulated E-field and H-field distributions of **a** monopole antenna and **b** integrated antenna

circular antenna, the AMC integrated antenna has lowered backward electromagnetic radiation. This indicates less backward electromagnetic energy transmission from the antenna. The AMC antenna offers a Front to Back Radiation (FBR) ratio of 8.94 dB at 2.23 GHz, while the circular antenna offers an FB ratio of 2 dB as shown in Fig. 6.

Fig. 6 Simulated FBR of monopole and AMC integrated antenna

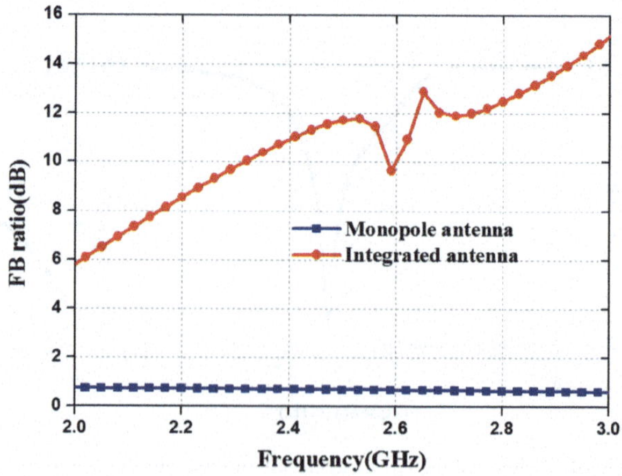

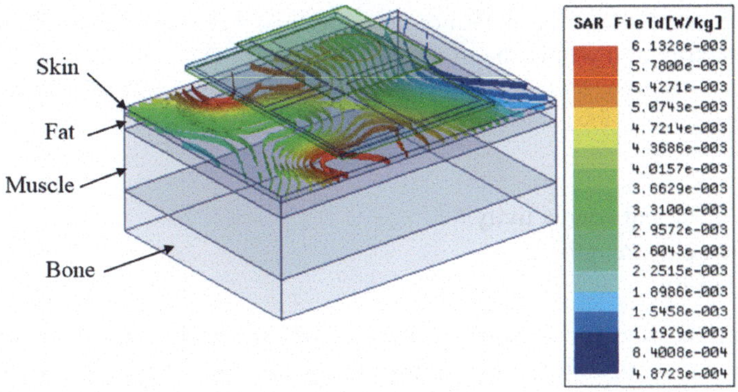

Fig. 7 Artificial body phantom integrated with AMC antenna

Table 2 Material properties of human tissue phantom

Materials	Skin	Fat	Muscle	Bone
ε_r	37.95	5.27	52.67	18.49
σ (S/m)	1.49	0.11	1.77	0.82
Density (kg/m^3)	1001	900	1006	1008
Thickness (mm)	2	5	20	13

Therefore, these analyses signify that the incorporation of the AMC with circular monopole antenna enhances the front to back radiation ratio and gain remarkably.

4 Artificial Tissue Model Analysis

An artificial human tissue phantom is designed to realize the electromagnetic behavior and effect on several parameters as shown in Fig. 7. It consists of four layers, i.e., skin, fat, muscle, and bone. Body tissues have their different material properties which are presented in Table 2. These properties of the human phantom are obtained using ANSYS HFSS software.

5 Specific Absorption Rate Simulations

The vital feature to examine the wearable antenna while it is placed near to the human bodies is SAR. SAR is a variable to know the electromagnetic absorption of human body when it is near RF devices. SAR is measured over the entire body, or 1 g or 10 g of tissue. According to the Federal Communication Commission (FCC) and IEEE

C95.1-2005, the SAR value must not be more than 1.6 W/kg. It can be measured from the electric field within the tissue as

$$\mathrm{SAR} = \frac{\sigma |E|^2}{\rho} \quad (1)$$

where σ is electrical conductivity,
E is electric field strength,
ρ is mass density.

To calculate the Specific Absorption Rate (SAR) for Body Area Network (BAN) application, the AMC antenna is positioned near skin layer. Four air gaps have been taken to investigate the biological body effect betwixt the AMC antenna and artificial human phantom. To perform the analysis, an input power of 100 mW is chosen. Table 3 describes the SAR performance of circular antenna alone. The simulated SAR performances of flat AMC integrated antenna are presented in Table 4.

Variable d_a represents the air gap among the AMC antenna and body tissues. In Table 3, the circular monopole offers a 1-g SAR of 13.6 W/kg while the antenna is placed 1 mm apart from the body tissues because of its omnidirectional behavior. Moreover, when the circular monopole is placed at a height of 5 mm, the antenna creates an SAR of 10.5 W/kg. After that, when the AMC is incorporated with circular monopole, the SAR value reduces to 0.675 W/kg at 2.23 GHz while it is only 1 mm apart from the body tissues. Furthermore, it is seen that when the AMC integrated antenna is positioned at 5 mm apart from the body tissues, the average SAR reduces to 0.00613 W/kg. The low SAR value specifies less electromagnetic power absorption toward the human body. Therefore, in contrast to circular monopole, the AMC integrated antenna has SAR reduction of 95.03% and 99.94% while it is 1 mm and 5 mm apart from human body, respectively. The performance comparison of the proposed work with previously reported papers is recorded in Table 5.

Table 3 SAR performance of flat circular monopole antenna at various d_a

d_a (mm)	SAR (W/kg)
1 mm	13.6
2 mm	12.4
3 mm	11.2
5 mm	10.5

Table 4 SAR performance of flat AMC integrated antenna at various d_a

d_a (mm)	SAR (W/kg)
1 mm	0.675
2 mm	0.634
3 mm	0.598
5 mm	0.00613

Table 5 Comparison among different antenna designs

Ref. no.	Size	Gain (dB)	FB ratio	SAR (W/kg)	SAR reduction rate (%)
[5]	$0.065\lambda \times 0.10\lambda$	n/a	n/a	0.273	43.3
[6]	$0.068\lambda \times 0.073\lambda$	3.3	n/a	n/a	n/a
Proposed work	0.064λ (circular)	4.2	8.94 dB	0.675 (1 mm apart)	95.03

6 Conclusion

The article proposed a low profile and miniaturized AMC antenna for body-centric application. The work utilizes a 2 × 2 rectangular unit-elements array for achieving zero reflection phase at 2.4 GHz. It also employs better gain, impedance bandwidth, and achieved miniaturization. The circular monopole antenna has been positioned parallel to the x-axis for efficient excitation of linearly polarized wave. The proposed AMC antenna has lowered S_{11} minima significantly and widened the impedance bandwidth. The multilayered substrate offers 4.2 dB antenna gain and front to back radiation ratio of 8.94 dB. The AMC reduces the Specific Absorption Rate (SAR) of 0.675 W/kg at designated frequency when it is 1 mm away from human body.

References

1. Trajkovikj, J., Skrivervik, A.K.: Diminishing SAR for wearable UHF antennas. IEEE Antennas Wireless Propag. Lett. **14**, 1530–1533 (2015)
2. Felício, J.M., Fernandes, C.A., Costa, J.R.: Wideband implantable antenna for body-area high data rate impulse radio communication. IEEE Trans. Antennas Propag. **64**(5), 1919–1922 (2016)
3. Hwang, J.-N., Chen, F.-C.: Reduction of the peak SAR in the human head with metamaterials. IEEE Trans. Antennas Propag. **54**(12), 3763–3770 (2006)
4. Lee, H., Tak, J., Choi, J.: Wearable antenna integrated into military berets for indoor/outdoor positioning system. IEEE Antennas Wireless Propag. Lett. **16**, 1932–1940 (2017)
5. il Kwak, S., Sim, D.-U., Kwon, J.H., Yoon, Y.J.: Design of PIFA with metamaterials for body-SAR reduction in wearable applications. IEEE Trans. Electromag. Compat. **59**, 297–300 (2017)
6. Liu, L.-Y., Wang, B.-Z.: A broadband and electrically small planar monopole employing metamaterial transmission line. IEEE Antennas Wireless Propag. Lett. **14**, 1018–1021 (2015)
7. Ahmad, B.N., Mohammad, S.S.: Extraction of material parameters for metamaterials using a full-wave simulator. IEEE Antennas Propag. Mag. **55**(5), 202–211 (2013)
8. Weir, B.W.: Automatic measurement of complex dielectric constant and permeability at microwave frequencies. Proc. IEEE **62**(1), 33–36 (1974)

Implementation of Adaptive Mobility Management Technique for Wireless Mesh Network to Handle Internet Packets

Abhishek Majumder and Sudipta Roy

Abstract Hybrid Wireless Mesh Protocol (HWMP) has been specified for routing of packets in Wireless Mesh Network (WMN) for IEEE 802.11s. But external STAs' movement is not supported by IEEE 802.11s. There are many existing techniques for managing mobility of mobile nodes. But IEEE 802.11s is not integrated with any of these techniques. The paper proposes adaptive mobility management technique to support movement of external STA in WMN. The scheme is combined with IEEE 802.11s. NS-2 has been used to simulate both the schemes. Performance comparison shows that the proposed technique performs better than HWMP in respect of average handoff cost, average end to end delay, normalized routing overhead, and throughput in different network conditions.

Keywords Wireless mesh network · Mesh gateway · External station · IEEE 802.11s · Mobility management

1 Introduction

IEEE 802.11s [1–5] was proposed for implementing IEEE 802.11 based WMN. Portals, Mesh Stations (STAs), Mesh Gates, and Access Points constitute the Mesh Basic Service Set (MBSS). A mesh node that connects the WMN to Internet is known as portal. Mesh STAs route the data traffic to destination from source. For serving the clients, mesh gate is associated with Access Point (AMPG). External STAs are served by the APMG as the proxy. In case of Internet traffic, packets are destined to a remote node which is not present in WMN. Proxy APMG of external STA routes Internet packet via portal. As external STA enters into coverage of a new APMG, handoff takes place. It will disrupt services. Mobility management scheme is desired to have

A. Majumder (✉)
Department of Computer Science & Engineering, Tripura University, Suryamaninagar, Tripura, India
e-mail: abhi2012@gmail.com

S. Roy
Department of Computer Science & Engineering, Assam University, Silchar, Cachar, India

S. Kundu et al. (eds.), *Proceedings of the 2nd International Conference on Communication, Devices and Computing*, Lecture Notes in Electrical Engineering 602, https://doi.org/10.1007/978-981-15-0829-5_10

minimum handoff cost. If handoff cost is less, the network will be less congested. It results in reduced end to end delay and increased throughput.

In IEEE 802.11s, Hybrid Wireless Mesh Protocol (HWMP) [4, 6–8] is used for routing in WMN. In HWMP, for serving external STA Basic Service Set (BSS) is formed by the APMG. Upstream Internet packets are sent to APMG of BSS by external STA, which in turn is forwarded to portal. Downstream Internet traffic is forwarded to external STA's proxy APMG by the portal. On receiving the packets, APMG forwards them to external STA. HWMP is described by IEEE 802.11s to select path for routing the packets in mesh BSS. But IEEE 802.11s [9] has not taken external STAs' movement into account. As the number of clients and their movement increase, mobility [2, 9] of external STA becomes an important issue which needs to be addressed. Adaptive mobility management scheme which utilizes the proxy tables of HWMP has been implemented in this paper to integrate mobility in IEEE 802.11s.

In this paper, existing schemes for mobility management is discussed in Sect. 2. Section 3 discusses the proposed scheme. In Sect. 4, simulation results and comparison is presented. Conclusion is presented in Sect. 5.

2 Related Work

For managing the mobility in WMN, many schemes were proposed [10–12]. In ANT [13], while Mesh Client (MC) travels from coverage of old Routing Access Point (RAP) to new one, it transmits a location update and handoff confirmation message toward location database server and old RAP, respectively. Location database server responds by sending back location update confirmation message.

In Mesh Mobility Management (M^3) [14], MC's location information is stored by Gateway (GW). MC transmits location update message periodically toward GW. As MC comes under new Mesh Router's (MR) coverage, previous MR appends forward pointer to current MR. Till GW receives location update message, the process will continue.

In iMesh [15], while the MC enters into the coverage of a MR, it transmits Host Network Association message (HNA) in whole network. Every MR of WMN updates MC's routing information. For broadcasting the HNA message OLSR is used in this scheme.

In MEMO [16], during handoff route error message is transmitted in the whole network by previous MR. The new MC proactively transmits route reply message toward GW to setup route for Internet traffic.

In SMesh [17], data group and control group are two groups to which each MC is associated with. When an MR believes that it can provide better service to the MC, it gets attached with data group. Information about the current location of MC is kept by MR in control group.

In WMM [18], each MC maintains a pair of tables: routing table and proxy table. Option field of each packet consists of four subparts: SST (timestamp of sender's host MR), ISS (IP address of sender's host MR), IRS (IP address of receiver's host MR), and RST (timestamp of receiver's host MR). When an MR forwards one packet toward some destination, it updates the proxy table and routing table corresponding to the destination.

A location server is stored at GW in Static Anchor Scheme [19]. When an MC performs handoff, previous MR appends forward pointer toward current MR. Until forward chain length reaches to k this process will continue. When chain length reaches k, MC sends location update to GW. Length of forward chain becomes 0 and current MR is set as MC's Anchor Mesh Router (AMR). In Dynamic Anchor Scheme [19] the condition for forward chain reset is same as Static Anchor Scheme. In the event of new Internet session arrival, forward chain also gets reset.

In LMMesh [20], route advertisement is periodically broadcasted by the GW. Therefore, each MR has route toward GW. When an Internet packet is transmitted by MC toward GW, option field contains location information of source MC. While packet is received, GW updates MC's routing information.

3 Proposed Scheme

Adaptive mobility management scheme [21] can handle Internet traffic only. It uses three messages in modified format: Reassociation, PREQ, and Proxy Update. External STA calculates its Session to Mobility Ratio (SMR) [22] value when it moves into coverage of new APMG. Reassociation message is used by the external STA to send its SMR value to new APMG. External STA's routing information is updated using PREQ. Proxy Update is utilized for updating the proxy information of external STA. Usage and frame format of the other frames are similar to that of HWMP.

3.1 Frame Formats

The proposed adaptive mobility management scheme uses modified forms of Reassociation, PREQ, and Proxy Update.

3.1.1 Reassociation Message

In this scheme, during handoff, the external STA sends reassociation message to new APMG. In addition to the other fields, three new fields have been introduced in this message, namely: SMR_value, STA_IP, and old_APMG_address. SMR_value is a 1-byte field which contains external STA's SMR value. STA_IP is a 4-byte field which stores the external STA's IP address. Address of external STA's old APMG is stored in Old_APMG_address field. It is a 6-byte field.

3.1.2 Path Request Frame

Inside path request frame, two new 1-bit fields have been added, namely: is_mobilestation and forward_chain. Is_mobilestation field will be 1 if the PREQ is generated for external STA. On the other hand, the field forward_chain will be 1 if PREQ is sent for extension of forward chain toward new APMG from previous APMG.

3.1.3 Proxy Update Element (PXU)

One new field, namely, PXU_station_IP has been added to the PXU element. PXU_station_IP is an 8-bit field which stores IP address of the external STA for which PXU message is generated.

3.2 Routing Table

Two fields, namely, Is_mobilestation and Part_of_forward_chain are added in the routing table. Rest of the fields in the routing table is same as HWMP. Is_mobilestation is a field of 1-bit size. It will be 1 if routing table entry is of a mobile station. The field Part_of_forward_chain will be 1 if corresponding entry is forward chain's part toward external STA.

3.3 Proxy Table

The proxy table also contains all the fields as in case of HWMP. The proposed scheme uses an additional 1-byte field, namely, Proxy_station_ip. It stores the IP address of the corresponding external STA.

3.4 Scheme Proposed

Portal transmits Route Announcement (RANN) message periodically. Therefore, every mesh STA and APMG maintain one path toward portal. Proposed scheme mainly has two parts: mobility management and routing.

3.4.1 Mobility Management

During handoff, external STA calculates its SMR value. Ratio of session departure and arrival rate's summation and rate of mobility is defined as SMR [22]. At the time of sending reassociation message to new APMG external STA sends its SMR value and IP address through the fields SMR_value and STA_IP, respectively. When reassociation message is received, APMG compares value of external STA's SMR (SMR_{APMG}) with threshold SMR value (δ_{th}).

- If $SMR_{APMG} \leq \delta_{th}$, in the routing table of new APMG external STA's entry is created. Next hop and destination address fields of the entry are set to external STA's address. Part_of_forward_chain and Is_mobiestation fields corresponding to the entry will be 1. Then PREQ is sent by new APMG toward old APMG. Is_mobilestation and Forward_chain fields of that PREQ message will be set to 1. Rest of the fields will be set using the same procedure as done in HWMP. When PREQ is received, in routing table of old APMG, external STA's entry is searched. If PREQ's originator HWMP sequence number is lesser compared to routing table entry's sequence number, it will not be updated. Else, external STA's routing table entry is updated. Part_of_forward_chain and Is_mobilestation fields corresponding to the entry will be content of Forward_chain and Is_mobilestation fields of PREQ message, respectively. Next hop address will be new APMG's address. Node address will be mesh STA's address.
- If $SMR_{APMG} > \delta_{th}$, the external STA's entry will be added into proxy table of new APMG. The proxy_station_ip field of the external STA's entry is set to the IP address of external STA. The process to fill in all the other fields is same as HWMP. The new APMG transmits PXU message toward portal. PXU_station_ip field of the PXU message is set to the IP address of the external STA. HWMP is followed to fill all the other fields of PXU message. When the intermediate APMG receives the PXU, it again transmits PXU toward next hop to reach portal. This will continue till PXU message reaches the portal. After PXU is received by the portal, it checks in its proxy table whether external STA's entry is present. If not present, it creates an entry. Proxy_station_ip of the entry is set to PXU_station_ip of PXU message. Else, it updates the proxy table.

3.5 *Routing*

Internet traffic can only be handled by the proposed scheme. Each of the APMG has a path to the portal since Route Announcement (RANN) is transmitted by portal periodically. The external STA sends uplink data packet of Internet traffic toward new APMG. The packet is forwarded by new APMG toward portal. The same procedure will also be followed by intermediate APMGs. Thus, upstream Internet packets reach the portal.

When downlink Internet packets for external STA are received by the portal, in the proxy table external STA's entry is searched through destination's IP address. MAC address of APMG of destination external STA found from that entry will be stored in the address three of MAC header. Destination external STA's MAC address corresponding to that entry will be stored in mesh address extension field of mesh control field. The portal then forwards the downstream Internet packets toward next hop for reaching destination's APMG. As mesh STA receives data packets, it checks if destination STA's entry is present in proxy table. The presence of the entry is also checked in routing table. If neither of them contains the entry, it searches entry of destination external STA's serving APMG. Then it transmits packets toward destination. Thus, data packet arrives at external STA's serving APMG. External STA's entry is searched in the proxy table when serving APMG receives data packet. If it finds an entry, the packet is forwarded to the external STA. If it does not find, the routing table will be searched for external STA's entry. Packet is forwarded to next hop which corresponds to that entry. Thus, packet reaches the external STA's current APMG. External STA then receives the packet from its current APMG.

4 Simulation and Comparison

To carry out the simulation NS-2 [23] has been used. The topology of the simulation has been shown in Fig. 1. Node 1 has been considered as portal. Random Waypoint Mobility Model is followed by external STA during movement. In the MAC layer,

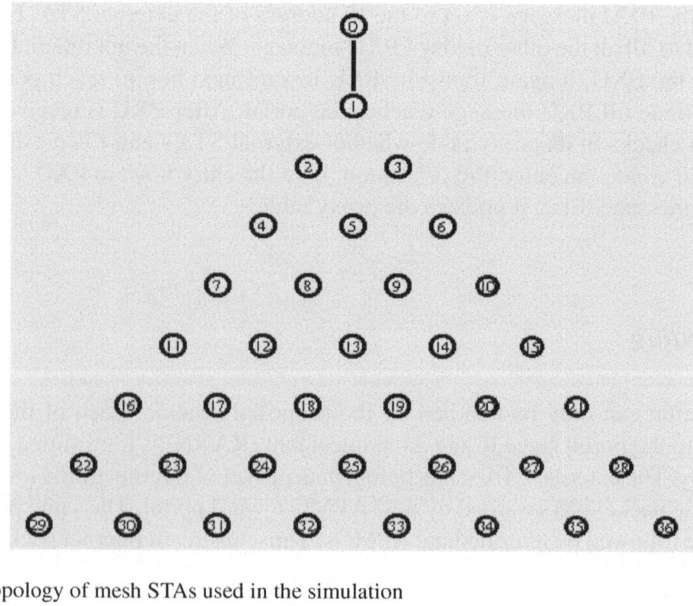

Fig. 1 Topology of mesh STAs used in the simulation

IEEE 802.11 has been used. External STA's number is varied from 10 to 100. Data rate is set to 512 kbps. 10 m/s is set as external STA's maximum speed. Simulation has been carried out 10 times for performance comparison. Parameters used in simulation are shown Table 1.

From Fig. 2 it can be seen that throughput of IEEE 802.11s is lower than that of adaptive mobility management scheme. Since PXU and PXUC are transmitted and PREQ is broadcasted in IEEE 802.11s network congestion takes place. This congestion results in reduction in throughput. In proposed scheme, PREQ's propagation is very limited and PXU message is sent only to the portal if threshold SMR is less than external STA's SMR value. Therefore, congestion is less in adaptive mobility management scheme and throughput becomes higher. In IEEE 802.11s more PREQ will be broadcasted and PXU and PXUC will be transmitted as number of external

Table 1 Value of parameters used for simulation

Parameters	Values	Parameters	Values
Maximum number of connections	10	Pause time of external STA	0 s
Simulation time	100 s	Traffic type	CBR
Range of radio	41.04 m	Number of portal	1
Two neighboring mesh STAs' distance	40 m	Size of packet	258
Number of external STA	36	Threshold session to mobility ratio in adaptive mobility management scheme (δ_{th})	0.50

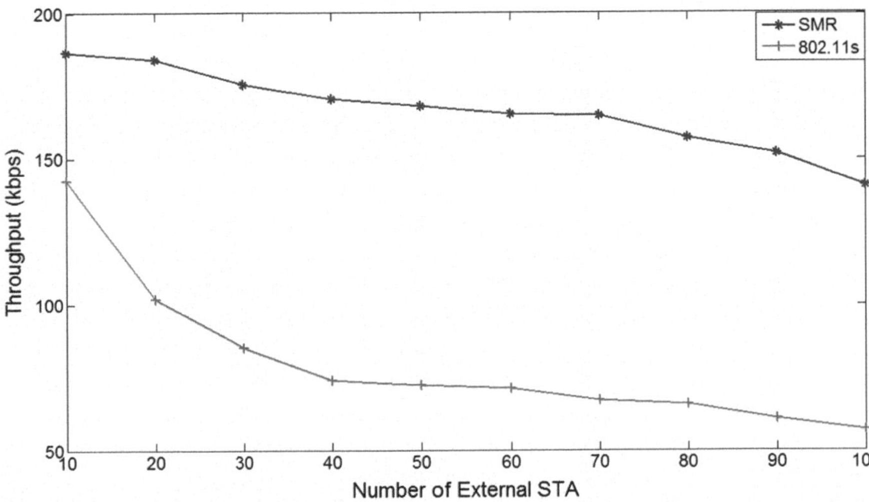

Fig. 2 Throughput versus external STA's number

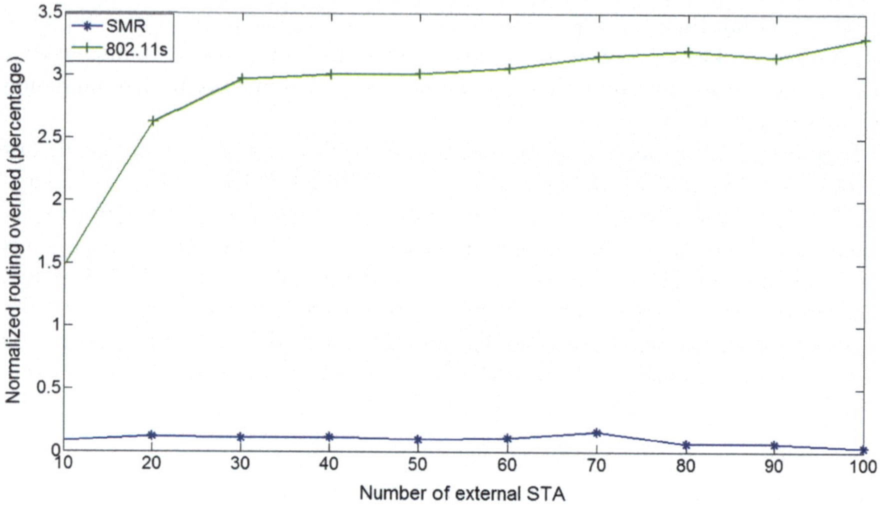

Fig. 3 Normalized routing overhead versus external STA's number

STAs increase. In the proposed scheme, more PREQ and PXU messages are transmitted as external STA's number increase. Therefore, in both schemes, with increase in external STAs' number, congestion of network becomes higher and throughput reduces.

From Fig. 3 it is observed that normalized routing overhead of adaptive mobility management scheme is lower than that of IEEE 802.11s. This is because, during every handover in IEEE 802.11s large number of PREQ, PXU and PXUC messages are transmitted. In the proposed scheme, very limited number of PREQ and PXU are transmitted. In IEEE 802.11s as external STAs' number increases, more PREQ, PXU, and PXUC are transmitted. Therefore, normalized routing overhead of IEEE 802.11s increases as external STA's number increases. In proposed technique, as external STAs' number increase very limited number of PREQ and PXU message is transmitted. Therefore, normalized routing overhead of adaptive mobility management scheme remains steady as external STA's number increase.

Figure 4 shows that proposed technique's average handoff cost is less compared to IEEE 802.11s. This is because PXU and PXUC messages are transmitted and PREQ is broadcasted in IEEE 802.11s during handoff process. As external STA's number increases more PREQ, PXU, and PXUC will be transmitted in IEEE 802.11s. This results in increase in congestion and retransmission. Therefore, in case of IEEE 802.11s, as external STA's number increase per handoff cost also increases. In the proposed scheme, propagation of PREQ and PXU is very limited. Therefore, as external STA's number increase cost per handoff remains steady.

Figure 5 shows that the proposed technique performs better compared to IEEE 802.11s with respect to average end to end delay. This is because of congestion due to transmission of PREQ, PXU, and PXUC messages in IEEE 802.11s. As external STA's number increase, number of PREQ, PXU, and PXUC transmitted also

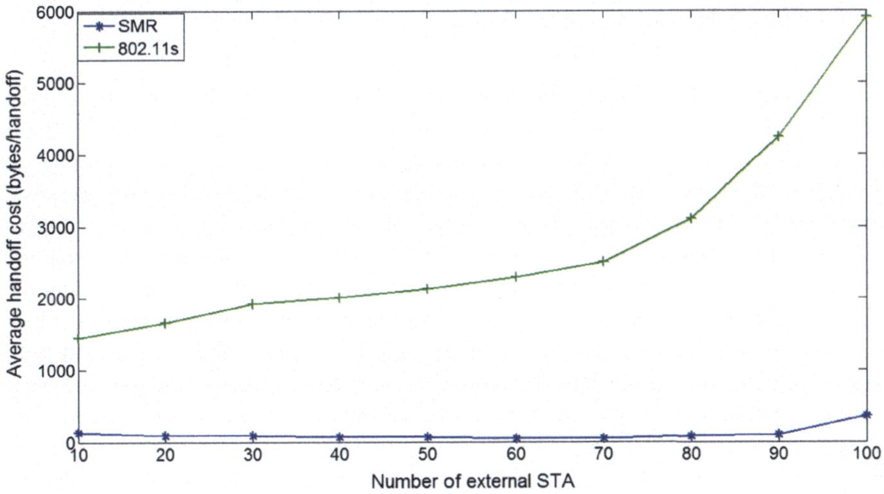

Fig. 4 Average handoff cost versus external STA's number over

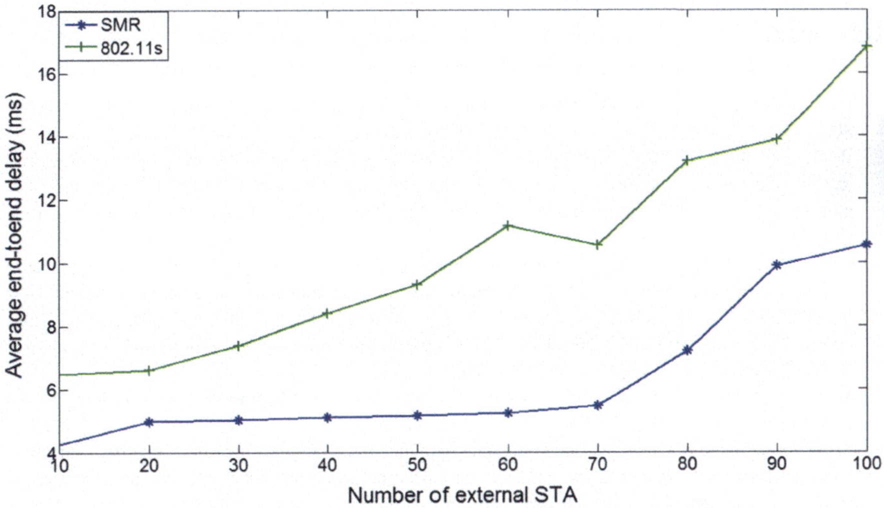

Fig. 5 Average end to end delay versus external STA's number over

increases rapidly in IEEE 802.11s. This causes congestion in the network. Therefore, in IEEE 802.11s as external STA's number increase average end to end delay also increases. In adaptive mobility management scheme also number of PREQ and PXU messages transmitted increases as external STA's number is increased. So, average end to end delay of proposed scheme becomes more as external STA's number increase.

5 Conclusion and Future Work

External STA's mobility management was the primary focus of the paper. Few of the existing techniques have been presented. But any of them is not unified with IEEE 802.11s. An IEEE 802.11s based mobility management scheme was presented in this paper. NS-2 has been used for simulation of both IEEE 802.11s and proposed technique. Performance comparison in respect of average handoff cost, average end to end delay, normalized routing overhead, and throughput show that proposed scheme outperforms HWMP. Performance analysis and comparison of the proposed scheme considering jitter, varying number of connections and TCP traffic remain as future work. Moreover, numerical analysis of proposed technique and comparison with some of existing techniques also remains as future work. The proposed technique is designed to handle only Internet traffic. It is not capable of handling Intranet traffic. Therefore, enhancing the proposed scheme to make it capable of handling intranet as well as internet traffic will be future work. The proposed scheme can also be enhanced to work in a WMN with multiple portals.

References

1. Hiertz, G.R., Denteneer, D., Max, S., Taori, R., Cardona, J., Berlemann, L., et al..: IEEE 802.11s: the WLAN mesh standard. IEEE Wirel. Commun. **17**(1), 104–111(2010)
2. IEEE Std 802.11-2012.: Information technology—telecommunications and information exchange between systems—local and metropolitan area networks—specific requirements—part 11: wireless LAN medium access control (MAC) and physical layer (PHY) specifications (2010)
3. IEEE P802.11s/D1.07-2007.: Information technology—telecommunications and information exchange between systems—local and metropolitan area networks—specific requirements—part 11: wireless medium access control (MAC) and physical layer (PHY) specifications (2007)
4. Camp, J.D., Knightly, E.W.: The IEEE 802.11s extended service set mesh networking standard. IEEE Commun. Mag. **46**(8), 120–126 (2008)
5. Wang, X., Lim, A. O.: IEEE 802.11s wireless mesh networks: Framework and challenges. Ad Hoc Netw. **6**(6), 970–984 (2008)
6. Sooriyaarachchi, S.J., Fernando, W. A.C., Gamage, C.D.: Evaluation of scalability of hybrid wireless mesh protocol in IEEE 802.11. In: Fifteenth International Conference on Advances in ICT for Emerging Regions (ICTer), pp. 152–159 (2015)
7. Bari, S.M.S., Anwar, F., Masud, M.H.. Performance study of hybrid wireless mesh protocol (HWMP) for IEEE 802.11s WLAN mesh networks. In: International Conference on Computer and Communication Engineering, pp. 712–716 (2012)
8. Bahr, M.: Proposed routing for IEEE 802.11s WLAN mesh networks. In: 2nd Annual International Workshop on Wireless Internet, p. 5 (2006)
9. Sampaio, S., Soutp, P., Vasques, F.: A review of scalability and topological stability issues in IEEE 802.11s wireless mesh networks deployments. Int. J. Commun. Syst. **29**(4), 671–693 (2015)
10. Xie, J., Wang, X.: A survey of mobility management in hybrid wireless mesh networks. IEEE Netw. **22**(6), 34–40 (2008)
11. Mase, K.: Layer 3 wireless mesh networks: mobility management issues. IEEE Commun. Mag. **49**(7), 156–163 (2011)

12. Zhang, Z., Pazzi, R.W., Boukerche, A.: A mobility management scheme for wireless mesh networks based on a hybrid routing protocol. Comput. Netw. **54**(4), 558–572 (2010)
13. Wang, H., Huang, Q., Xia, Y., Wu, Y., Yuan, Y.: A network-based local mobility management scheme for wireless mesh networks. In: IEEE Wireless Communications and Networking Conference, pp. 3792–3797 (2007)
14. Huang, R., Zhang, C., Fang, Y.: A mobility management scheme for wireless mesh networks. In: Global Telecommunication Conference, pp. 5092–5096 (2007)
15. Navda, V., Kashyap, A., Das, S.R.: Design and evaluation of iMesh: aninfrastructure-mode wireless mesh network. In: 6th IEEE International Symposia on a World Wireless Mobile Multimedia Networks, pp. 164–170 (2005)
16. Maoshen, R., Chao, L., Huizhou, Z., Tong, Z., Wei, Y.: MEMO: an applied wireless mesh network with client support and mobility management. In: IEEE Global Telecommunication Conference, pp. 5075–5079 (2007)
17. Amir, Y., Danilov, C., Hilsdale, M., Elefteri, R.M., Rivera, N.: Fast handoff for seamless wireless mesh networks. In: 4th International Conference Mobile System Application Services, pp. 85–95 (2006)
18. Huang, D.W., Lin, P., Gan, C.H.: Design and performance study for a mobility management mechanism (WMM) using location cache for wireless mesh networks. IEEE Trans. Mob. Comput. **7**(5), 546–556 (2008)
19. Li, Y., Chen, I.-R.: Design and performance analysis of mobility management schemes based on pointer forwarding for wireless mesh networks. IEEE Trans. Mob. Comput. **10**(3), 349–361 (2011)
20. Yinan, L., Ing-Ray, C.: Mobility management in wireless mesh networks utilizing location routing and pointer forwarding. IEEE Trans. Netw. Serv. Manag. **9**(3), 226–239 (2012)
21. Majumder, A., Roy, S., Dhar, K. K.: Design and analysis of an adaptive mobility management scheme for handling internet traffic in wireless mesh network. In: Annual International Conference on Emerging Research Areas, pp. 1–6 (2013)
22. Majumder, A., Roy, S.: Design and analysis of a dynamic mobility management scheme for wireless mesh network. Sci. World J. **16** (2013). Article ID 656259
23. Fall, K., Varadhan, K.: The network simulator (ns-2) (2007). http://www.isi.edu/nsnam/ns

Multi-Relay-Based Cooperative Spectrum Sensing in Cognitive Radio Network

Saradindu Mazumder, Tiyasa Mallick, Ravi Raj, Ritika Rani, Jayanta Kumar Bag and Chanchal Kr. De

Abstract In this paper, the performance of a multiple antenna-based cooperative spectrum sensing in Cognitive Radio (CR) is discussed. The system model consists of a single Primary User (PU) and several Secondary Users (SU) and one fusion center (FC) is proposed to detect the activities of PU. Here, the SUs are playing the role of a relay which operates either in Amplify and Forward (AF) or in Decode and Forward (DF) mode. Each relay is having multiple receiving antennas and only one transmitting antenna. We have reached a closed-form expression of the detection probability (P_d) and false alarm probability (P_f) in presence of Rayleigh Fading channel. The impact of number of relays and number of antennas in each relay is highlighted in our model. We have validated our analytical model by MATLAB simulation.

Keywords Cooperative communication · Amplify and forward · Decode and forward · Primary user · Secondary user · Cognitive radio · Spectrum sensing · Fusion center

1 Introduction

The revolutionary hike in wireless service users over the last decade requires more radio spectrum. The limited spectrum resource has been licensed exclusively to users who can use a particular frequency band. A research by the Federal Communications Commission (FCC) Spectrum Policy Task Force (SPTF) demonstrates that the real licensed spectrum is immensely vacant for a large amount of the time [1]. The resulting spectral scarcity has motivated the growth of CR technology. The intelligent

S. Mazumder · T. Mallick · R. Raj · R. Rani · J. K. Bag (✉) · C. Kr. De
Department of Electronics and Communication Engineering, Haldia Institute of Technology, Haldia, India
e-mail: jkbag2008@gmail.com

C. Kr. De
e-mail: chanchalkumarde@gmail.com

S. Kundu et al. (eds.), *Proceedings of the 2nd International Conference on Communication, Devices and Computing*, Lecture Notes in Electrical Engineering 602, https://doi.org/10.1007/978-981-15-0829-5_11

Cognitive Radio (CR) system is aware of its surrounding communication environment, learns from the environment, and adapts its operating parameters in real time [2].

The concept of CR is used to conquer the conflicts between spectrum scarcity and spectrum utilization. The performance of CR network is depended on the accuracy of detecting the primary user's presence by the secondary user through spectrum sensing, where the local spectrum sensing information from multiple CRs is combined to detect the PU is known as Cooperative Spectrum Sensing. To improve the communication reliability, multiple antennas in a CR are used [3]. In wireless networks, relay network is a broad class of network topology commonly use. In relay network the source and destination are interconnected by means of some nodes. They can work in Amplify and Forward (AF) or Decode and Forward (DF) mode. The received signal is decoded, re-modulated, and retransmitted by a DF relay, while AF one simply amplifies and retransmits the signal without decoding. The cooperative spectrum sensing technique using multi-antenna-based DF relay have discussed in [4]. The effect of multiple antennas in CR is quite impressive [5–8]. The authors have discussed the various spectrum sensing techniques which are used for CRs in [9–11, 13].

In this paper, we have compared the performances of relay in AF and DF mode and discussed the effect of having multiple antennas in cooperative spectrum sensing in CR. We have considered several relays to improve the performance of our system and Rayleigh Fading channels in all the links. Using MATLAB simulation, closed-form expression of detection probability and false alarm probability has been derived and implemented.

2 System Model

Our system model (shown in Fig. 1) consists of a single PU and its activities (i.e., presence or absence) are being monitored by the FC with the help of "R" number of SU continuously, which works as the relay. The relays operate either in AF or DF scheme. Each relay consists of "L" number of receiving antennas and single transmitting antenna. The receiving antennas selected the best-received signal among them by Selection Combining (SC) technique. The transmitting antenna of each relay then sends the best signal (i.e., received from PU) to the FC. All links are considered as Rayleigh Fading Channel in our system model. Here, all noise components are Additive White Gaussian Noise (AWGN) with Power Spectral Density (PSD) N_0. FC receives different copies of same signal from "R" number of relays and combines the signals using Maximum Ratio Combining (MRC) technique. The FC decides that the PU spectrum is occupied or vacant using an Energy Detector (ED). Finally, FC circulates the information to the whole network.

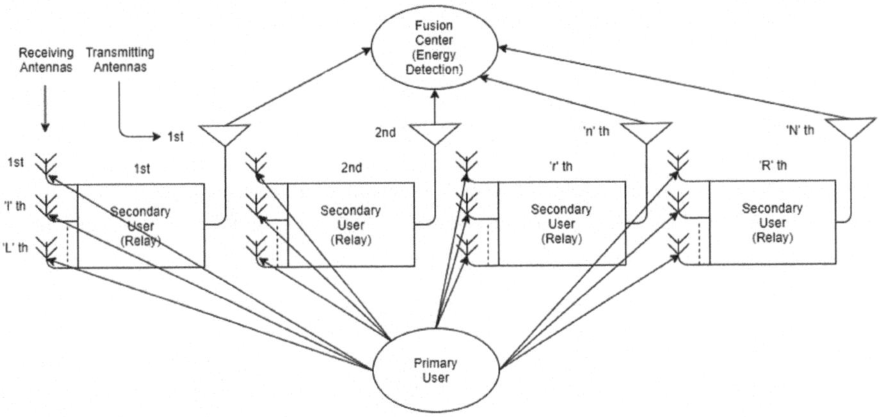

Fig. 1 System model for cooperative spectrum sensing in CR

The signal received by lth antenna of rth relay is given by

$$y_{r,l} = s.h_{p\to r,l} + n_r \tag{1}$$

where $h_{p\to r,l}$ is the channel link between PU and lth antenna of rth relay, n_r is the Additive White Gaussian Noise (AWGN) with zero mean and one-side Power Spectral Density (PSD) N_0 and s transmitted signal by the PU.

The rth relay receives L number of copies of the same signal transmitted by PU. After selection combining, the best of all the signals is selected, that can be expressed as

$$y_r = \arg \max_{l=1,2\ldots L} \{y_{r,l}\} \tag{2}$$

Substituting the value of $y_{r,l}$ into Eq. (2) we get

$$y_r = \arg \max_{l=1,2\ldots L} \{h_{p\to r,l}\}.s + n_r \tag{3}$$

The received signal at the FC from rth relay using AF mode is expressed as

$$y_{\mathrm{FC},r}^{\mathrm{AF}} = y_r.h_{r\to \mathrm{FC}}.G_r + n_{\mathrm{FC}} \tag{4}$$

where $h_{r\to \mathrm{FC}}$ is the channel links between rth relay and FC, n_{FC} is the AWGN with power spectral density N_0.

For the convenience of theoretical analysis, assume the gain of an AF relay mode $G_r = 1/h_{r\to \mathrm{FC}}$ is used to reimburse the fading of channel between the relay and FC. Substituting y_r from Eq. (1) and G_r into Eq. (2) we get

$$y_{\mathrm{FC},r}^{\mathrm{AF}} = \arg \max_{l=1,2...L} \{h_{p\to r,l}\}.s + n_r + n_{\mathrm{FC}}$$
$$= \arg \max_{l=1,2...L} \{h_{p\to r,l}\}.s + \sum_{i=1}^{2} n_i \tag{5}$$

where$n_i = n_r = n_{\mathrm{FC}}$ is the AWGN with power spectral density N_0.

FC selects the best link from R number of relays, and the received signal can be written as

$$y_{\mathrm{FC}}^{\mathrm{AF}} = \arg \max_{r=1,2,..R} \{y_{\mathrm{FC},r}^{\mathrm{AF}}\}$$
$$= h_p.s + \sum_{i=1}^{2} n_i \tag{6}$$

where $h_p = \arg \max_{r=1,2,...R} \{\arg \max_{l=1,2...L} \{h_{p\to r,l}\}\}$

Similarly, the signal received at the FC from rth relay using DF relay scheme

$$y_{\mathrm{FC},r}^{\mathrm{DF}} = h_{r\to\mathrm{FC}}.\hat{s} + n_{\mathrm{FC}}$$
$$y_{\mathrm{FC}}^{\mathrm{DF}} = \arg \max_{r=1,2,..R} \left\{y_{\mathrm{FC},r}^{\mathrm{DF}}\right\} = \arg \max_{r=1,2,..R} \left(h_{r\to\mathrm{FC}}.\hat{s}\right) + n_{\mathrm{FC}} \tag{7}$$

where $\hat{s}$ is the estimated PU signal at the DF relay.

The energy detector associated with FC decides the presence of PU on the basis of a hypothesis, which is expressed as

$$H_1 : s = E_s$$
$$H_0 : s = 0 \tag{8}$$

Let $T_{\mathrm{AF}}(H_0)$ and $T_{\mathrm{AF}}(H_1)$ be the output power of the ED of the FC under the hypothesis H_0 and H_1, respectively. From Eq. (6), we can easily obtain

$$T_{\mathrm{AF}}(H_0) = \sum_{i=1}^{2} |n_i|^2 \tag{9}$$

From Eq. (6), we can obtain

$$T_{\mathrm{AF}}(H_1) = |h_P|^2 E_s + \sum_{i=1}^{2} |n_i|^2 = |h_P|^2 E_s + 2N_0 \tag{10}$$

Corresponding to the threshold λ_{AF}, false alarm of the network can be evaluated by Eq. (9), [14]

$$P_f^{AF} = \Pr[T_{AF}(H_0) > \lambda_{AF}] = e^{-\left(\frac{\lambda_{AF}}{2N_0}\right)}\left[1 + \frac{\lambda_{AF}}{2N_0}\right] \quad (11)$$

The false alarm for corresponding to a threshold λ_{AF} using AF cooperation can be calculated by Eq. (11). We can set λ_{AF} for requirement of false alarm in a Cognitive Radio Network (CRN) and then detection probability (p_d^{AF}) of AF and corresponding false alarm P_f^{AF} are determined.

The detection probability of the PU using AF mode in the FC can be written as

$$\begin{aligned} P_d^{AF} &= \Pr[T_{AF}(H_1) > \lambda_{AF}] \\ &= \Pr\left[|h_P|^2 > \frac{\lambda_{AF} - 2N_0}{E_s}\right] \\ &= \Pr\left[\arg \max_{r=1,2,..R} \{\arg \max_{l=1,2...L} \{|h_{p\to r,l}|^2\}\} > \frac{\lambda_{AF} - 2N_0}{E_s}\right] \\ &= 1 - \left[1 - \left\{\left[1 - \exp\left(\frac{1}{\sigma^2}\max\left(\frac{\lambda_{AF} - 2N_0}{E_s}, 0\right)\right)\right]^L\right\}\right]^R \end{aligned} \quad (12)$$

Let $T_{DF}(H_0)$ and $T_{DF}(H_1)$ be the output power of the ED of the FC using the DF relay scheme under the hypothesis H_0 and H_1, respectively. From Eq. (7), we can obtain

$$T_{DF}(H_0) = |n_i|^2 \quad (13)$$

where n_i is a noise at the FC only.

$$T_{DF}(H_1) = |h_P|^2 E_s + |n_i|^2 \quad (14)$$

Corresponding to the threshold λ_{DF}, we can evaluate the false alarm based on Eq. (11) as

$$P_f^{DF} = \Pr[T_{DF}(H_0) > \lambda_{DF}] \quad (15)$$

where the threshold value of λ_{DF} can be expressed as

$$\lambda_{DF} = -2N_0 \ln(P_f^{DF}) \quad (16)$$

The detection probability of PU at the FC using DF scheme is given by

$$\begin{aligned}
P_{d,m}^{\text{DF}} &= \Pr[T_{\text{DF}}(H_1) > \lambda_{\text{DF}}] \\
&= \Pr\left[|h_P|^2 > \frac{\lambda_{\text{DF}} - N_0}{E_s}\right] \\
&= \Pr\left[\max_{n=1,2\ldots N}\{|h_P|^2\} > \frac{\lambda_{\text{DF}} - N_0}{E_s}\right] \\
&= 1 - \left[1 - \left\{\left[1 - \exp\left(\frac{1}{\sigma^2}\max\left(\frac{\lambda_{\text{DF}} - N_0}{E_s}, 0\right)\right)\right]^L\right\}\right]^R
\end{aligned} \quad (17)$$

This is the final closed-form expression obtained for our system model. We have validated our final expression and analytical using MATLAB simulation.

3 Results and Discussions

The performance of detection probability in a Cooperative Spectrum Sensing (CSS) is evaluated using MATLAB simulation test bed. This section presents the analytical and simulation results. Here, analytical results are represented by continuous curves, and simulation results are represented by discrete marks on all the figures. Fixed sensing channel SNR at 5 dB, the detection probability as a function of false alarm P_f for couple of antenna at the cognitive relay, is shown in Fig. 2. It is observed that the detection probability of the planned multi-antenna-based single CR scheme in DF mode is higher than the AF scheme over wide range of false alarm. For given channel SNR, it is observed that we try to get better the detection probability, false alarm also increases.

So, we can conclude that the performance of CR is better than the AF scheme as the detection probability is improved. For example, with 10% false alarm probability the detection probability is almost 50% in AF scheme and almost 80% in DF scheme. So, it is recommended to use the DF scheme over AF scheme in CRs but DF relaying scheme sometimes causes propagation of error in case of incorrect decoding of the information symbols [11, 12]. DF scheme makes the system more complex and costlier. In case where efficiency can be compromised AF, relaying scheme can be more preferable as it is available in low cost and simpler circuitry.

In Fig. 3 describes that with fixed probability of false alarm $(P_f) = 0.1$ the detection probability (P_d) is a function of SNR in dB with several number of antenna and couple of cognitive relay. It is observed from Fig. 2 that the detection probability of the proposed multi-antenna-based single CR scheme with couple of relay and six antennas in DF mode is higher than the AF scheme with same number of relay and couple of antennas over wide range of SNR. After increasing the number of antenna (L) from 1 to 2, with fixed false alarm probability of 0.1, detection probability is also improved to 10%. Hence, as the number of antennas is increased in CR, the detection

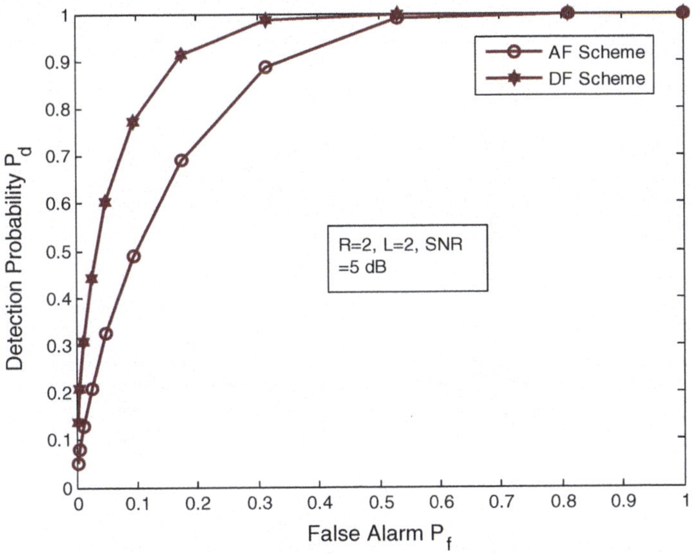

Fig. 2 Detection probability versus false alarm of AF and DF scheme of CR with same number of relay and antenna and SNR 5 db

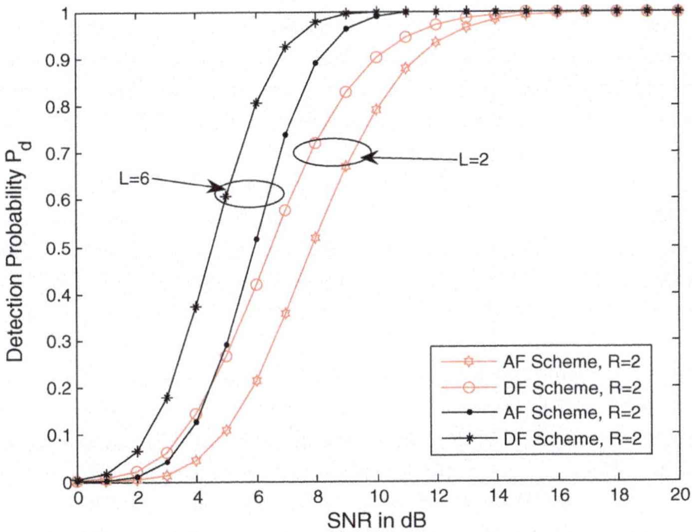

Fig. 3 Detection probability versus SNR in dB with couple of relays having different number of antennas in AF and DF scheme with false alarm $P_f = 0.01$

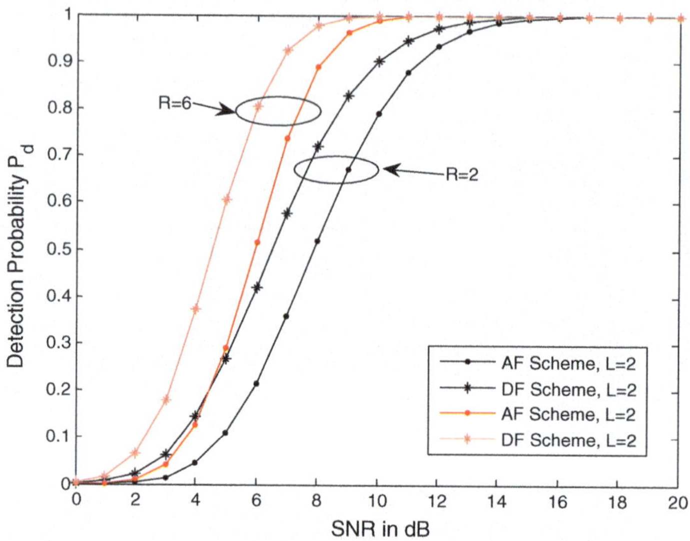

Fig. 4 Detection probability versus SNR in dB with variable number of relays each having couple of antennas in AF and DF scheme with false alarm $P_f = 0.01$

probability also increases. After increasing L from 2 to 3, Pd in CR increases around 5% increase due to diminishing return. And so as, it is essential to note that any increment more than $L = 3$ would not give any improvement. The diminishing return related to Maximum Ratio Combining technique (MRC) is the key cause behind reducing detection probability.

It is clear from Fig. 4, that the detection probability for the considered multi-antenna-based CR scheme with variable relay each having couple of antennas in DF mode is improved than the AF scheme over wide range of SNR. So we can say that the more number of relays improves the performance of a CR, as the detection probability is being increased with the increment in the number of relays in the system.

It is seen in Fig. 5 that the detection probability of our system model with increasing both the number of relays and antennas in DF scheme is higher than the AF scheme over wide range of SNR. Earlier we have seen that increase in the number of antennas in CR leaves good impact on the performance and also increase in the number of relays in the CR improves the detection probability. In this graph, we have increased both the number of relays and antennas and as per expectations we have observed a much more improved probability of detection. To increase the efficiency of CR more number of relays and antennas should be used to get the best idea of the PU's presence.

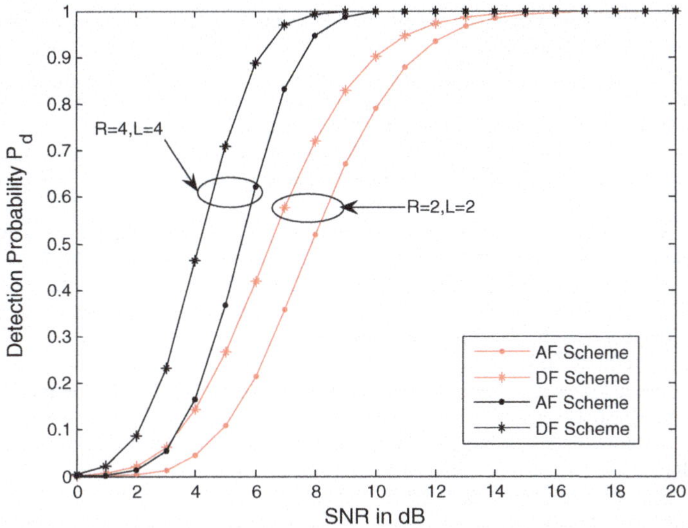

Fig. 5 Detection probability versus SNR in dB, where both the number of relays and antennas are varying in AF and DF scheme with false alarm $Pf = 0.01$

4 Conclusion and Future Work

An analytical model of a cognitive radio network with performance based on detecting the PU by the FC is presented in this paper. We have compared the characteristics of AF and DF relaying scheme in multi-antenna-based CRN for cooperative spectrum sensing. In this paper, we have reached to a closed-form expression of detection probability and false alarm probability, which we have validated using the MATLAB simulation. Based on the results, we can conclude that (1) DF scheme has higher detection probability of PU than AF scheme, (2) Increasing in the number of relays the detection probability also increases, and (3) With the increase in the number of receiving antennas CRN provides good performance. The proposed scheme with multiple receiving antennas and multiple relays helps to achieve a high detection probability of PU.

References

1. Federal Communications Commission: Spectrum Policy Task Force, Rep. ET Docket no. 02–135, (2002)
2. Haykin, S.: Cognitive radio: brain-empowered wireless communications IEEE. J. Sel. Areas Commun. **23**(2), 201–220 (2005)
3. Goldsmith, A.: Wireless Communications, 1st edn. Cambridge University Press (2005)
4. De, C.K., Sinha, R.K., Kundu, S.: Cooperative spectrum sensing with multi antenna based decode and forward relay. In: Annual IEEE India Conference (INDICON) (2014)

5. Men, J., Ge, J.: Non-orthogonal multiple access for multiple-antenna relaying networks. IEEE Commun. Lett. **19**(10) (2015)
6. Bengtsson, E.L, Karlsson, P.C, Tufvesson, F., Vieira, J., Malkowsky, S., Liu, L., Rusek, F., Ove, E.: Transmission schemes for multiple antenna terminals in real massive MIMO systems. In: IEEE Global Communications Conference (GLOBECOM) (2016)
7. Singh, A., Bhatnagar, M.R., Mallik R.K.: Cooperative spectrum sensing in multiple antenna based cognitive radio network using an improved energy detector. IEEE Commun. Lett. **16**(1) (2012)
8. Sedighi, S., Taherpour, A., Josep, S.: Spectrum sensing using correlated receiving multiple antennas in cognitive radios. IEEE Trans. Wireless Commun. **12**(11), 5754–5766 (2013)
9. Omer, A.E.: Review of spectrum sensing techniques in Cognitive Radio networks. In: 2015 International Conference on Computing, Control, Networking, Electronics and Embedded Systems Engineering (ICCNEEE), 439–446 (2015)
10. Muthumeenakshi, K., Radha, S.: Improved sensing accuracy using enhanced energy detection algorithm with secondary user cooperation in cognitive radios. Int. J. Commun. Netw. Inf. Secur. (IJCNIS) **6**(1) (2014)
11. Bagwari, A., Tomar, G.S., Verma, S.: Cooperative spectrum sensing based on two-stage detectors with multiple energy detectors and adaptive double threshold in cognitive radio networks. Can. J. Electr. Comput. Eng. **36**(4), 2278–7798 (2013)
12. Reddy, M.K.K., Aravindan, M.: A resource allocation scheme for decode and forward relay cognitive networks. Int. J. Sci. Eng. Technol. Res. (IJSETR) **3**(4) (2014)
13. Saha, A., Bhattacherjee, S. S., De, C.K., De, D.: Cooperative spectrum sharing with multi-antenna based adaptive hybrid relay in presence of multiple primary users. J. Inf. Optim. Sci. **38**(6) (2017)
14. De, C.K., Kundu, S.: Adaptive decode-and-forward protocol-based cooperative spectrum sensing in cognitive radio. Int. J. Commun. Netw. Distrib. Syst. **14**(2), 117–133 (2015)

Physical Layer Security in Cognitive Cooperative Radio Network with Energy Harvesting DF Relay Assisted with Cooperative Jamming

Shashibhushan Sharma, Sanjay Dhar Roy and Sumit Kundu

Abstract This paper presents the security in physical layer of cognitive cooperative radio network (CCRN). This network (i.e., CCRN) consists of a cognitive source (CS), a cognitive decode and forward relay (CDFR) with energy harvesting and a cognitive destination (CD). The CCRN uses the bandwidth of a primary network (PN) which contains a primary transmitter (PT) and a primary receiver (PR). The CS does not have a direct link with CD due to deep fading. An eavesdropper is present in the network which tries to decode the message of cognitive network. In this scenario, we have numerically investigated the secrecy performance by a metric named as conditional secrecy outage probability while meeting an outage constraint of PN. MATLAB based is carried out following our analytical development.

Keywords Physical layer security · Cognitive cooperative radio network · Conditional secrecy outage probability · Energy harvesting · Cooperative jamming · Decode and forward relay

1 Introduction

There has been a phenomenal increase in number of users of the communication network in recent years, which leads to scarcity of spectrum. In order to strike a balance between spectrum scarcity and spectrum underutilization, cognitive radio network [1] uses the bandwidth of the primary network (PN). In such situations, an unauthorized user (it may be called an eavesdropper) may receive the information

S. Sharma (✉) · S. D. Roy · S. Kundu
Department of Electronics and Communication Engineering, National Institute of Technology Durgapur, Durgapur, India
e-mail: ss.15ec1105@phd.nitdgp.ac.in

S. D. Roy
e-mail: sanjay.dharroy@ece.nitdgp.ac.in

S. Kundu
e-mail: sumit.kundu@ece.nitdgp.ac.in

S. Kundu et al. (eds.), *Proceedings of the 2nd International Conference on Communication, Devices and Computing*, Lecture Notes in Electrical Engineering 602, https://doi.org/10.1007/978-981-15-0829-5_12

signal of another source and decode the message, if he knows the private key to decode the message. In such scenario, security of the message is an important issue.

Traditional approach to secure the message is a cryptographic approach that may fail when the eavesdropper (EAV) knows the private key to decode the source's message from the information signal. To solve this problem, Wyner has considered the wiretap channel and proposed that the security can be maintained at the physical layer [2]. Further, in [3], the secrecy capacity is defined as the difference between the legitimate channel capacity and wiretap channel capacity. In [4], the authors have measured the secrecy in terms of secrecy outage probability (SOP). These two metrics, secrecy capacity and SOP, have been vastly used to investigate the secrecy performance of any network at physical layer [5–7]. Ahead of these metrics (secrecy capacity and SOP), the authors have investigated another metrics named as conditional secrecy outage probability (CSOP) in [8]. Further, in [8], the authors have distinguished between the SOP and the CSOP and indicated the differences between them. In [9], the authors have analyzed SOP and CSOP with jamming.

Cooperative jamming is used to prevent the EAV attacking in a network which plays an important role in physical layer security [5, 6]. The cooperative jamming reduces the signal to interference ratio at the EAV and the EAV is unable to eavesdrop the message of the source [9–11]. Energy harvesting from radio frequency (RF) sources by the communication nodes is an excellent technique to harvest the energy [9, 12] for energy-constrained nodes. There are two beneficial approaches to harvest the energy are power splitting (PS) scheme and time switching (TS) scheme [12]. The PLS has been also investigated in cognitive radio network [13, 14].

In the present paper, the work of [9] has been extended for a cognitive scenario in a single relay-based cognitive network. The proposed cognitive cooperative radio network (CCRN) contains a cognitive source (CS), an energy harvesting cognitive decode and forward relay (CDFR) and a cognitive destination (CD). The CCRN uses the bandwidth of PN which contains the primary transmitter (PT) and primary receiver (PR). We assume that the PT is present at a large distance from the CDFR and CD. Thus, the CDFR and CD do not get interference from the transmitted signal of the PT. The CS does not have a direct link with CD due to deep fading and the sharing of message is completed in two time slots. An EAV exists in the network. The EAV wants to decode the message of cognitive network. Investigating the CSOP for the above CCRN is our novel contribution of the present paper.

More precisely, our contribution can be summarized as follows:

- Evaluating CSOP in presence of a single EAV for a CCRN, where the relay harvests energy follows PS scheme.
- We indicated the impact of power splitting for harvesting energy, fraction of transmit power of the relay, spent for information transmission, and primary outage probability threshold on CSOP.

Further, the paper is organized as: system model under study is given in Sect. 2. Performance analysis is presented in Sect. 3. The numerical results discussed in Sect. 4. Finally, conclusion of the paper has been given in Sect. 5.

2 System Model

2.1 Network Model

The system model, presented in Fig. 1a, contains a cognitive source (CS), energy harvesting cognitive decode and forward relay (CDFR), and cognitive destination (CD). The CCRN uses the bandwidth of a PN which contains the primary transmitter (PT) and primary receiver (PR). We assume that the PT is present at a large distance from the CDFR and CD. Thus, the CDFR and CD do not get interference from the transmitted signal of the PT. The CS does not have any direct link with CD due to deep fading and path loss. So, communication between the CS and CD is completed in two slots. The first slot is the broadcasting slot and other one is relaying slot. The time frame is shown in Fig. 1b. An EAV, has interest to decode the message of cognitive network, is present in the network. The power assigned to the CS under outage constraint of PN while the CDFR uses transmit power under outage constraint of PN and the harvesting energy. In broadcasting slot, the broadcasted information signal of the CS is received by the relay. In the relaying slot, destination has received the forwarded information signal as well as transmitted jamming signal of the relay. The forwarded information plus jamming signal of the relay is received by the CD as well as the EAV. Thus, the EAV eavesdrops the information in relaying slot only.

2.2 Channel Model

In Fig. 1, we assume that h_{SR}, h_{RD}, h_{RE}, h_{SP}, h_{RP}, and h_{PP} are the channel coefficients of the CS to CDFR, CDFR to CD, CDFR to EAV, CS to PR, CDFR to PR, and PT to PR, respectively. All the channels are considered as Rayleigh faded and

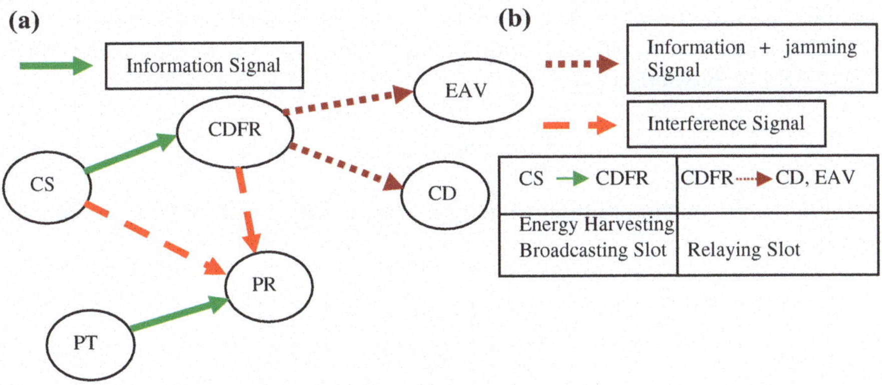

Fig. 1 **a** System model, **b** time frame

corresponding channel gains, g_i, where $i \in (SR, RD, RE, SP, RP, PP)$ are exponentially independent identically distributed (i.i.d.) random variables. The channel mean power of any link is Ω_i, where $i \in (SR, RD, RE, SP, RP, PP)$. The additive white Gaussian noise (AWGN) of the channels has zero mean and variance N_0. We assume that the CD knows the jamming signal and the channel state information (CSI) of the channel between relay and itself.

2.3 Power Allocation to the CS and CDFR, and Energy Harvesting

The power assigned to the CS and CDFR is based on outage constraint of the PN. The outage probability of the PN defined as [1]:

$$P_{OUT} = P\left\{\log_2\left(\frac{P_P g_{PP}}{P_S^C g_{SP} + N_0}\right) < R_P\right\} = 1 - \frac{\exp\left(\frac{-\gamma_{TH} N_0}{P_P \Omega_{PP}}\right)}{\gamma_{TH}\frac{P_S^C \Omega_{SP}}{P_P \Omega_{PP}} + 1} = \theta_{TH}, \quad (1)$$

where $P_{OUT} = \theta_{TH}$ is the primary outage probability threshold, R_P is the primary outage threshold rate, P_P is the primary transmit power, P_S^C transmit power of the CS under outage constraint of a PN and $\gamma_{TH} = 2^{R_P} - 1$ is the threshold signal-to-interference-plus-noise ratio. From Eq. (1), the maximum limit of transmit power $\left(P_S^C\right)$ assigned to the CS is evaluated as:

$$P_S^C = \frac{P_P \Omega_{PP}}{\gamma_{TH} \Omega_{SP}}\left\{\frac{\exp\left(\frac{-\gamma_{TH} N_0}{P_P \Omega_{PP}}\right)}{(1-\theta_{TH})} - 1\right\} \quad (2)$$

The peak device power of the CS is P_{PK}. The quality of services of PN is maintained by limiting the harmful interference by the transmitted signal of the CCRN, the CS broadcasts the signal with minimum of P_S^C and P_{PK}. The final assigned power (P_S) to CS is expressed as [1]:

$$P_S = \min\left(P_S^C, P_{PK}\right) \quad (3)$$

Similarly, the maximum limit of assigned power $\left(P_R^C\right)$ to the CDFR in relaying slot is expressed as

$$P_R^C = \frac{P_P \Omega_{PP}}{\gamma_{TH} \Omega_{RP}}\left\{\frac{\exp\left(\frac{-\gamma_{TH} N_0}{P_P \Omega_{PP}}\right)}{(1-\theta_{TH})} - 1\right\} \quad (4)$$

In the broadcasting slot, the received information signal at the relay can be expressed as

$$y_R^A = \sqrt{P_S} h_{SR} x_S + n_0 \tag{5}$$

where y_R^A is the received signal at DF relay, x_S is the unit power transmitted data of the CS and n_0 is the AWGN at the relay. In broadcasting slot, a fraction of the information signal has been used for harvesting the energy following PS scheme [9, 12]. The harvested energy is expressed as

$$E_H = \eta \beta P_S g_{SR} \tfrac{T}{2}, \tag{6}$$

where efficiency of the harvester circuit to harvest the energy is η which varies as $0 < \eta \leq 1$, β $(0 < \beta < 1)$ is the power splitting factor of information signal to harvest energy and T is the total communication time between the CS and CD in second. Since, due to low value, the AWGN power contributes negligible energy in the harvesting energy, and it is neglect [12]. In relaying slot, the possible transmit power (P_H) corresponding to the harvested energy is expressed as

$$P_H = \frac{\eta \beta P_S g_{SR} \frac{T}{2}}{\frac{T}{2}} = \eta \beta P_S g_{SR} \tag{7}$$

We assume the transmission takes place over a duration of T/2 s in the second slot. Now, the relay can use a maximum power P_H based on harvesting energy. But, due to cognitive constraint, relay transmits the signal with a power given as

$$P_R = \min\left(P_R^C, P_H\right) \tag{8}$$

The relay uses a fraction (α) of transmitted power to forward the information signal, while the rest part $(1-\alpha)$ is spent for transmitting jamming signal. The relay uses power αP_R to forward the information signal and $(1-\alpha)P_R$ to transmit the jamming signal where $0 < \alpha < 1$.

2.4 *Signal to Interference-Plus-Noise Ratio (SINR) at the Relay, EAV, and CD*

The CDFR uses β fraction of received signal for energy harvesting. The rest part of the information signal, used for forwarding the information at the CDFR, can be expressed as

$$y_R = \sqrt{(1-\beta) P_S} h_{SR} x_S + n_0. \tag{9}$$

From Eq. (9), at the DF relay, the SINR (γ_R) can be expressed as

$$\gamma_R = \frac{(1-\beta)P_S g_{SR}}{N_0}, \tag{10}$$

where N_0 is the AWGN power.

The relay forwards the information signal as well as transmit the jamming signal [10, 11] jointly. Since, α is the fraction of transmitted power of the relay, so at the CD, the received signal can be expressed as

$$y_D = \sqrt{\alpha P_R} h_{RD} x_R + \sqrt{(1-\alpha)P_R} h_{RD} x_J + n_0 \tag{11}$$

where x_R is the unit power data and x_J is the unit power jamming signal. The CD scenes and eliminates the known jamming signal as per the jamming signal and CSI knowledge [9]. The rest part of the signal, which is used to decode the message by the CD, can be expressed as

$$y_D = \sqrt{\alpha P_R} h_{RD} x_R + n_0 \tag{12}$$

But, there is no knowledge of jamming and CSI of the relay and EAV channel at the EAV (as the untrusted relay does not have this knowledge in [11]). Thus, EAV is unable to remove the jamming signal. The EAV receives the signal which can be expressed as

$$y_E = \sqrt{\alpha P_R} h_{RE} x_R + \sqrt{(1-\alpha)P_R} h_{RE} x_J + n_0 \tag{13}$$

On the basis of Eq. (12), the SINR (γ_D) at the CD can be expressed as

$$\gamma_D = \frac{\alpha P_R g_{RD}}{N_0} \tag{14}$$

From Eq. (13), the SINR (γ_E) at the EAV can be expressed as

$$\gamma_E = \frac{\alpha P_R g_{SR} g_{RE}}{(1-\alpha)P_R g_{SR} g_{RE} + N_0} = \frac{\alpha}{(1-\alpha) + \frac{N_0}{P_R g_{SR} g_{RE}}}, \tag{15}$$

For low value of N_0, Eq. (15) can be approximated as

$$\gamma_E \approx \frac{\alpha}{(1-\alpha)}. \tag{16}$$

From Eqs. (10) and (14), the end to end SINR (γ_{SD}) between the CS and CD is expressed as

$$\gamma_{SD} = \min\left(\frac{(1-\beta)P_S g_{SR}}{N_O}, \frac{\alpha P_R g_{RD}}{N_0}\right), \tag{17}$$

The end to end channel capacity (C_{SD}) is

$$C_{SD} = \frac{1}{2} \log_2(1 + \gamma_{SD}), \tag{18}$$

The relay to EAV channel capacity (C_{RE}) is

$$C_{RE} = \frac{1}{2} \log_2(1 + \gamma_E). \tag{19}$$

3 Analysis of CSOP

3.1 The Power Outage (PO)

Let us assume that the relay has a lower threshold value of transmit power $P_{R_{TH}}$ to relay the information signal. The relay must satisfy the condition which is given as

$$P_R \geq \frac{P_{R_{TH}}}{\alpha} \tag{20}$$

If the relay does not have sufficient power to transmit the signal, then the power outage occurs at the relay.

3.2 The CSOP and the PO Probability of the Cognitive Network

The secrecy capacity can be expressed as [3]:

$$C^{SEC} = [C_{SD} - C_{RE}]^+, \tag{21}$$

where $[x]^+ = \max(x, 0)$ and C^{SEC} is the secrecy capacity. The CSOP for this network is defined as [5]

$$\begin{aligned} CSOP &= P\left(C^{SEC} \leq R_{TH}^{SEC} | P_R \geq \frac{P_{R_{TH}}}{\alpha}\right) \\ &= P\left(\frac{1}{2}\log_2\left(\frac{1+\gamma_{SD}}{1+\gamma_E}\right) \leq R_{TH}^{SEC} | P_R \geq \frac{P_{R_{TH}}}{\alpha}\right) \\ &= P\left(\gamma_{SD} \leq A_1 | P_R \geq \frac{P_{R_{TH}}}{\alpha}\right) \end{aligned} \tag{22}$$

where R_{TH}^{SEC} is the threshold secrecy rate and $A_1 = \left\{2^{2R_{TH}^{SEC}}(1+\gamma_E) - 1\right\}$ which is constant as per approximation of Eq. (16). On the basis of Eq. (20), the power outage probability (POP) at the relay (i.e., relay does not have sufficient power to transmit the signal) can be expressed as

$$POP = P\left(P_R < \frac{P_{R_{TH}}}{\alpha}\right) = 1 - \exp\left(-\frac{P_{R_{TH}}}{\alpha\eta\beta P_S \Omega_{SR}}\right). \tag{23}$$

4 Numerical Results

The assigned numerical values to the physical parameter are given in Table 1.

Figures 2 and 3 show the CSOP versus power splitting to harvest energy (β) and fraction of transmitted power at the relay to relay the information (α), respectively. It is seen from the figures that we obtain optimal value $\beta = 0.8$, $\alpha = 0.4$, respectively, where the CSOP is minimum. With increase in β, in Fig. 2, harvesting energy increases before the optimal value. The increased harvesting energy provides more power to the relay for transmitting the signals. The signal strength increases at the CD and legitimate channel capacity increases but there is no effect to EAV channel capacity as per Eq. (16) for the constant value of α. Thus, secrecy capacity increases and the CSOP decreases. Beyond the optimal value, the signal strength at the relay becomes poor due to large fraction of information signal is used to harvest energy. As per scheme of DF relay, the end to end signal strength also decreases and thus secrecy capacity decreases. Corresponding to decrease in secrecy capacity, CSOP performance of cognitive network decreases.

Table 1 Assigned numerical values to different physical parameters

Physical parameters	Numerical values
Primary transmit power (P_P)	10 dBW
Primary outage threshold rate (R_P)	0.2
Primary outage probability threshold (θ_{TH})	0.2 bit/Sec/Hz
Peak device power of the CS (P_{PK})	10, 15 dBW
Power splitting factor for harvesting energy (β)	0.8
Energy conversion efficiency (η)	0.7
Fraction of transmitted power at the relay for information transmission (α)	0.4
Lower threshold value of relay transmit power ($P_{R_{TH}}$)	0.05 W
Threshold secrecy rate (R_{TH}^{Sec})	1 bit/Sec/Hz
Channel mean power all links (Ω_i)	1

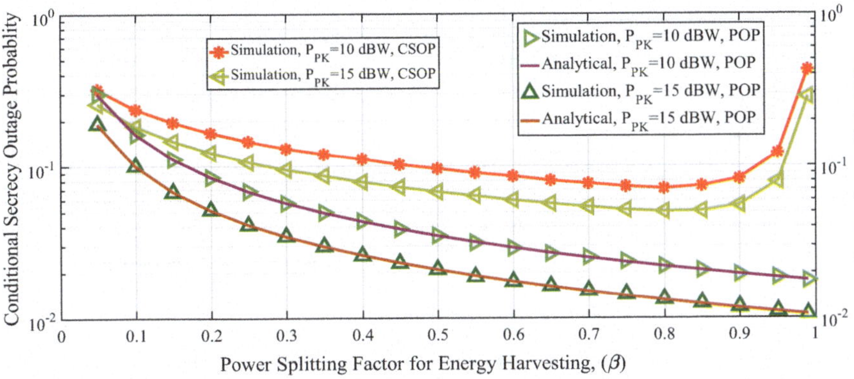

Fig. 2 The CSOP versus power splitting factor for energy harvesting (β) for different values of peak transmit power

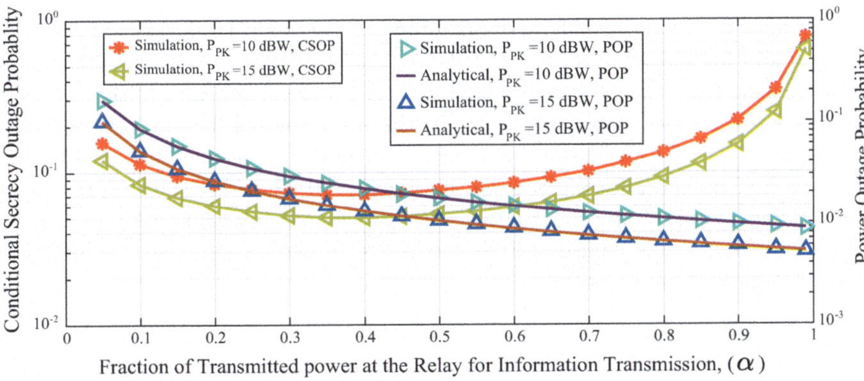

Fig. 3 The CSOP versus power splitting factor for information processing (α) for different values of peak transmit power

Further, the POP is also shown in Figs. 2, 3 and 4. The required sufficient transmit power as per Eq. (20) at relay increases with increase in $\beta\alpha$ and θ_{TH}. Correspondingly, the POP is seen to decrease.

In Fig. 3, before the optimal value of α, the signal strength at the CD increases with an increase in α. But, before the optimal value, the signal strength at the EAV does not improve as fast as at the CD due to jamming effect. Beyond the optimal value of α, with further increase in α, the jamming power at the EAV reduces, which increases the capacity of eavesdropper's channel. Resulting from this, the CSOP increases.

In Fig. 4, we have observed that the CSOP increases with an increase in primary outage probability threshold. As per Eqs. (2) and (3), the power of the CS increases with an increase in primary outage probability threshold. As transmit power of the CS increases, the signal strength at the CD also increases. But, signal strength at

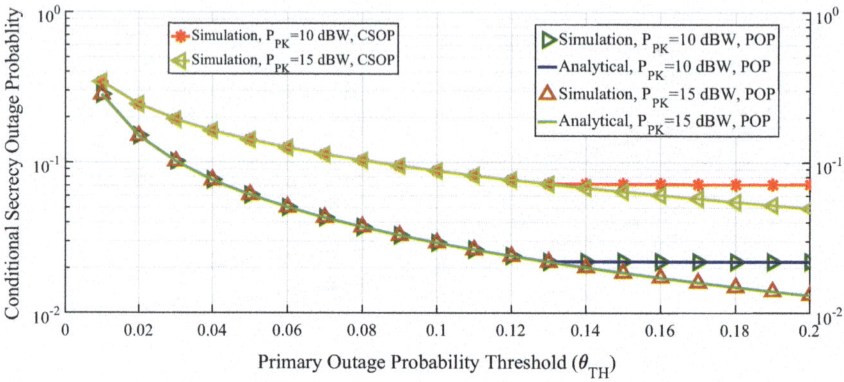

Fig. 4 The CSOP versus primary outage probability threshold (θ_{TH}) for different values of peak transmit power

the EAV does not increases due to the jamming of the relay. Thus, secrecy capacity increases and the CSOP decreases. At the $\theta_{TH} = 0.13$, the power under cognitive constraint becomes higher than the peak device power at the CS, so the transmit power of CS does not increase further. Therefore, a floor in the curve is obtained beyond the $\theta_{TH} = 0.13$. Further, it is seen that performance under two different values of P_{PK} are same for a given θ_{TH} until $\theta_{TH} = 0.13$, as the transmit power of the CS is same for both the P_{PK} till $\theta_{TH} = 0.13$, which is decided by P_S^C in Eq. (3). Moreover, in Fig. 4, the nature of POP versus θ_{TH} is also the same as that of CSOP versus θ_{TH} for the same reason.

5 Conclusion

We have investigated the CSOP of CCRN under an EAV attacking. The conditional secrecy performance has been evaluated under the situation that the relay has sufficient power to transmit the signal. The optimal value of power splitting factor to harvest energy and fraction of transmitted power at the relay for information transmission have been evaluated. Higher values of those parameters reduce the POP. It has been observed that the performance becomes better with an increase in primary outage probability threshold. Moreover, the performance also improves with an increase in peak device power of the CS. The study is useful in designing a CCRN to ensure a given level of physical layer security against an EAV attack.

Acknowledgements This research is supported by the Department of Electronics and Information Technology, Ministry of Communications and IT, Government of India under the Visvesvaraya Ph.D. Scheme administered by Media Lab Asia with Grant number PhDMLA/ 4(29)/2015-16.

References

1. Tran, H., Zepernick, H.J., Phan, H.: Cognitive proactive and reactive DF relaying schemes under joint outage and peak transmit power constraints. IEEE Commun. Lett. **17**(8), 1548–1551 (2013)
2. Wyner, A.D.: The wire-tap channel. Bell Syst. Tech. J. **54**(8), 1355–1387 (1975)
3. Leung-Yan-Cheong, S.K., Hellman, M.E.: The Gaussian wire-tap channel. IEEE Trans. Inf. Theory **24**(4), 451–456 (1978)
4. Barros, J., Rodrigues, M.R.D.: Secrecy capacity of wireless channels. In: 2006 IEEE International Symposium on Information Theory, pp. 356–360 (2006)
5. Hoang, T.M., Duong, T.Q., Vo, N.S., Kundu, C.: Physical layer security in cooperative energy harvesting networks with a friendly jammer. IEEE Wirel. Commun. Lett. **6**(2), 174–177 (2017)
6. Cao, K., Cai, Y., Wu, Y., Yang, W.: Cooperative jamming for secure communication with finite alphabet inputs. IEEE Commun. Lett. **21**(9), 2025–2028 (2017)
7. Vo, V.N., Nguyen, T.G., So-In, C., Baig, Z. A., Sanguanpong, S.: Secrecy outage performance analysis for energy harvesting sensor networks with a jammer using relay selection strategy. IEEE Access **6**, 23406–23419
8. Zhou, X., McKay, M.R., Maham, B., Hjorungnes, A.: Rethinking the secrecy outage formulation: a secure transmission design perspective. IEEE Commun. Lett. **15**(3), 302–304 (2011)
9. Sharma, S., Dhar Roy, S. Kundu, S.: Secure communication with energy harvesting multiple half-duplex DF relays assisted with jamming. Wirel. Netw. (Springer) 1–14 (2018). https://doi.org/10.1007/s11276-018-1859-0
10. Liu, Y., Li, J., Petropulu, A.P.: Destination assisted cooperative jamming for wireless physical-layer security. IEEE Trans. Inf. Forensics Secur. **8**(4), 682–694 (2013)
11. Dong, L., Yousefi, H., Jafarkhani, H.: Cooperative jamming and power allocation for wireless relay network in presence of eavesdropper. In: 2011 IEEE International Conference on Communications (ICC). Kyoto, pp. 1–5 (2011)
12. Kalamkar, S.S., Banerjee, A.: Secure communication via a wireless energy harvesting untrusted relay. IEEE Trans. Veh. Technol. **66**(3), 2199–2213 (2017)
13. Sakran, H., Shokair, M., Nasr, O., El-Rabaie, S., El-Azm, A.A.: Proposed relay selection scheme for physical layer security in cognitive radio networks. IET Commun. **6**(16), 2676–2687 (2012)
14. Tang, C., Pan, G., Li, T.: Secrecy outage analysis of underlay cognitive radio unit over Nakagami-m fading channels. IEEE Wirel. Commun. Lett. **3**(6), 609–612 (2014)

Miniaturized Multi-stopband Frequency Selective Surface for WLAN and X–Band Applications

Gouri Shankar Paul and Kaushik Mandal

Abstract In this paper, a multi-stopband frequency selective surface (FSS), covering the WLAN band (2.19–2.72 GHz), 5.89–7.05 GHz, and X-band (7.56–15.38 GHz) has been proposed. A square ring integrated with two T-shaped patches is considered as the unit cell. First square ring patch type FSS are explored, finally the design is come up with a compact square ring patch with two T-shaped single-layer FSS. It is also verified that by varying the width of the T-shaped patch resonance frequencies and bandwidth can be changed. This FSS also exhibits good angular stability performance over all the operating bands.

Keywords Frequency selective surface · Multi-stopband · Square ring · Mobile communication · WLAN · X-band

1 Introduction

Frequency selective surfaces (FSSs) are generally 2D periodic structure of metallic patches on the dielectric substrate or aperture on a metallic screen to use as suitable filtering operation as needed. Based on the filtering mechanism it can be used as band-stop or bandpass filter [1]. Use of LC filtering circuit is not suitable for microwave frequency for this reason use of FSS is very much suitable. FSSs have been investigated more than five decades. They are widely used as electromagnetic absorbers [2], radar cross section (RCS) reducing element [3], radomes [4], antenna reflectors [5], and artificial magnetic conductors [6] due to their filtering operation as band pass

G. S. Paul (✉)
Global Institute of Science & Technology, Haldia 721657, India
e-mail: gourishankar.paul@gmail.com

K. Mandal
Institute of Radiophysics & Electronics, Kolkata 700009, India
e-mail: kaushikrpe@gmail.com

S. Kundu et al. (eds.), *Proceedings of the 2nd International Conference on Communication, Devices and Computing*, Lecture Notes in Electrical Engineering 602, https://doi.org/10.1007/978-981-15-0829-5_13

or bandstop responses. Recently, the use of FSS has been tremendously increased with the quick development of wireless technology in mobile communication system, interference mitigation between adjacent wireless networks. In recent era, with the rapid development of wireless communication technology, the use of FSSs in telecommunication systems, wireless security systems, and interference mitigation between adjacent wireless networks have been investigated. FSS in recent wireless communication is used as single-band or multiband operations [7]. Single-bandstop characteristic can be achieved by incorporating modified swastika unit cell [8] with more miniaturization and angular stability but it is only for one band. Dual-band FSS consisting of a loop wire on the top metal layer and its complementary pattern on the bottom [9] is illustrated for WLAN application but unit cell dimension is 10.4 mm and its only dual-band.

Dual stopband characteristics can be achieved by using anchor-shaped loop FSS structure [10], multiple stub resonators loaded four small square slots loaded with multiple stub resonators [11] with higher design complexity, and circular loop type FSS [12]. Dual-band and miniaturized capacitive type loaded FSS is presented in [13]. Triple bandstop characteristics have been achieved using simple conducting elements on both sides [14], the design is simple with easily optimizable but the angular stability is very little only up to 30°. A complex design based on rectangular-shaped synthetic resonator [15] is also proposed for triple stopband characteristics. However, its application is limited due to the nonavailability of the synthetic resonator. Nowadays the use of Bluetooth and Wi-Fi is increasing day by day. So, it is essential to protect the users from the harmful radiations of these devices. Toward this important issue, a little bit efforts have been observed from the researchers working on FSS. Recently, an electromagnetic bandgap array [16] has been used as phone case for suppression of mobile phone radiation exposure, but EBG is much more costly than FR4. Some valid parameters like size of unit cell, number of dielectric layer, angular stability, and how much accurately covering the desired frequency bands.

There are some valid issues like size of the unit cell, number of dielectric layers, angular stability, and accurately covering the useful frequency bands are needed to be considered during design of FSSs.

Looking into the above issues, a design has been proposed which minimizes some of the issues. This paper shows that a single-side single-layer FSS with a square ring patch and two T-shaped patch as unit cell is able to show multi-bandstop characteristics covering the widely used frequency bands for the WLAN, 5.89–7.05 GHz and X-band applications. Size of the unit cell is 0.083 $\lambda \times$ 0.083 λ, where λ is the wavelength corresponding to the lowest operating frequency. This proposed design also shows stable performance over a wide range of incidence angle. Hence, the FSS is a novel multi-bandstop FSS. There are several research works on multiple stopband FSS designs but none of them has proposed such an easy and compact design. All the simulations are carried out using FEM-based Ansys HFSS EM simulator.

2 FSS Design

Both simulation and experimental investigations have been carried out extensively in a twofold manner. Two T-shaped patches are integrated with a square ring patch, as illustrated in Fig. 1. The proposed structure is designed on a single-side copper cladded FR4 substrate of dimension $(10 \times 10 \times 1.6)$ mm^3, with $\varepsilon_r = 4.4$ and loss tangent of 0.02. One square ring patch of width (W) = 0.3 mm patch is used. Two T-shaped patched with width (W1) = 2.7 mm, length (L1) = 2.85 mm, L2 = 4.25 mm, L3 = 3.95 mm and W = 0.3 mm are incorporated inside the square ring.

At first, the response of a simple square ring patch type FSS has been observed. It only resonates at a single frequency. Integration of a single T-shaped patch with the square ring provides dual stopband frequency response. Finally, after incorporation of another T-shaped patch and suitably optimizing the design parameters, triple bandstop characteristics have been achieved. These responses are summarized in Fig. 2.

This FSS can be used for a particular resonance frequency with some bandwidth. We can get different frequencies by changing the parameters of the metallic patch. The width (W1) variation of the T-shaped patches is illustrated in Fig. 3. For W1 = 2.7 mm the third operating band is showing better response.

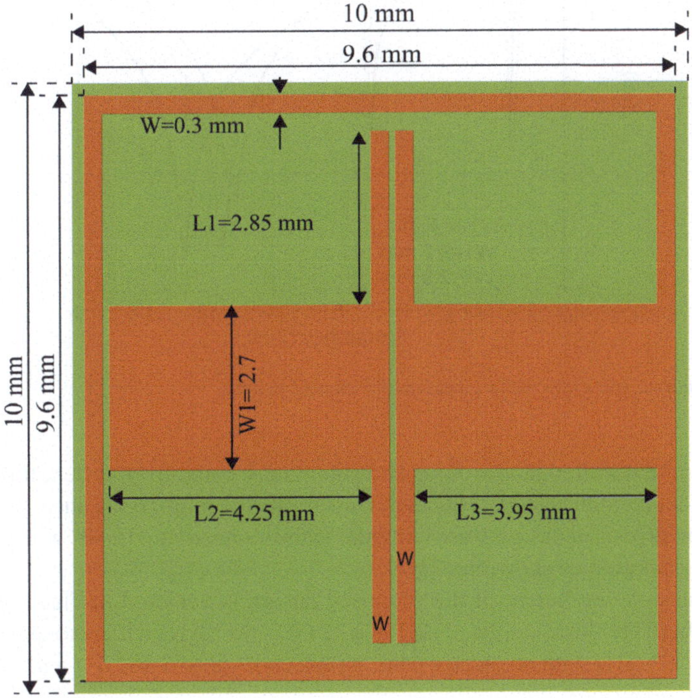

Fig. 1 Unit cell geometry of the proposed FSS

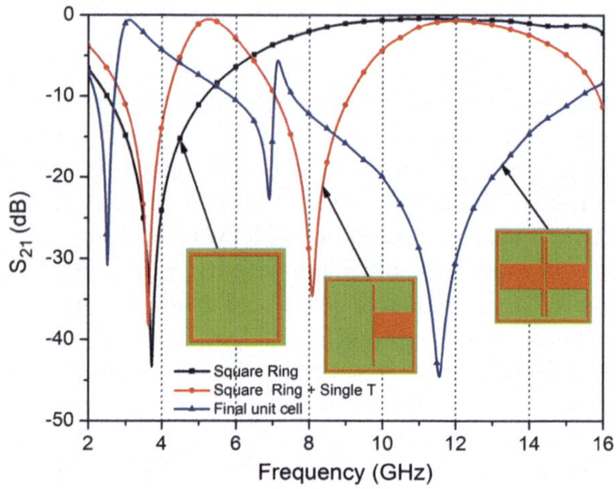

Fig. 2 Transmission coefficient (S_{21}) under different design considerations

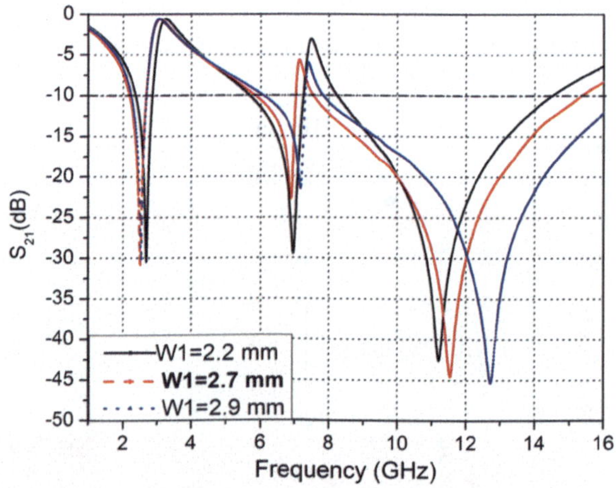

Fig. 3 S_{21} curves for width (W1) variation of T-shaped patches

FSS can be installed at any place and signal may come from different directions with different incidence angles. So, angular stability is a major issue during the design of FSS. The proposed design shows angular stability nearly up to 80° with only little frequency deviation as shown in Fig. 4.

Transmission coefficient of the proposed design is depicted in Fig. 5. It can be seen that there are three stopbands 2.19–2.72 GHz (WLAN) with resonance peak at 2.5, 5.89–7.05 GHz with resonance peak at 6.92 and 7.56–15.38 GHz with resonance

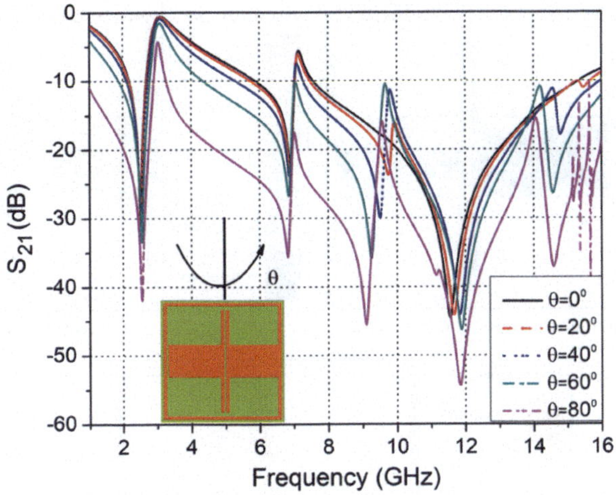

Fig. 4 Comparison of simulated S_{21} curves for different incidence angles under TE polarization

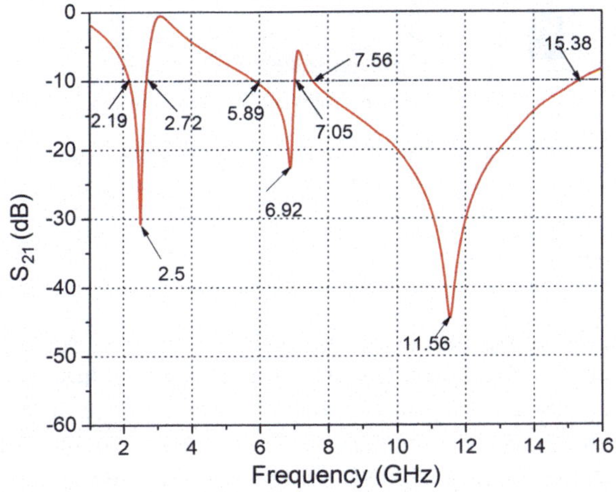

Fig. 5 S_{21} curve for the proposed FSS

peak at 11.56 GHz. The third band (7.56–15.38 GHz) is covering the entire X-band with sufficient level of attenuation.

A particular element of the proposed design is responsible for a particular resonance and these are shown in Fig. 6.

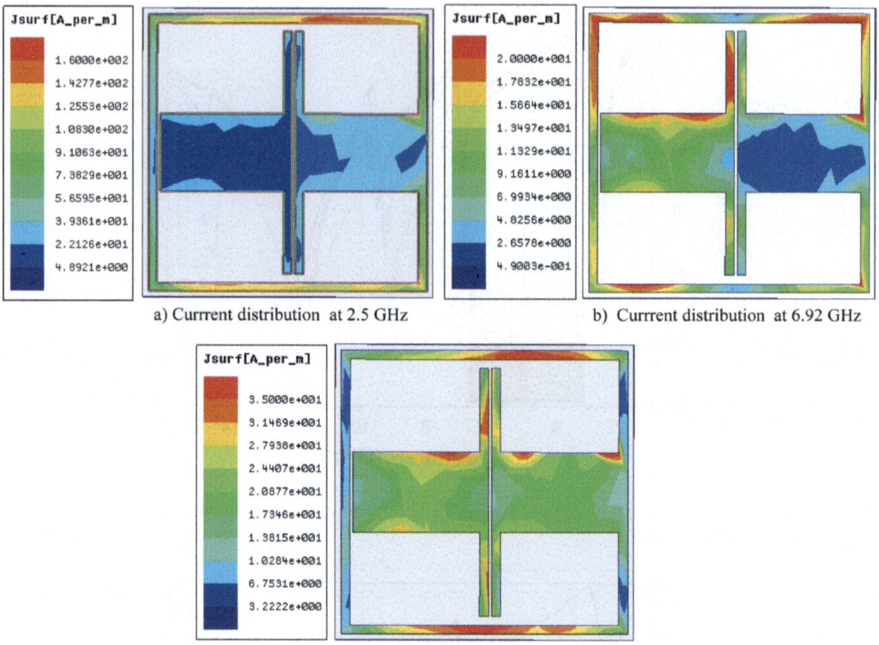

a) Currrent distribution at 2.5 GHz

b) Currrent distribution at 6.92 GHz

c) Currrent distribution at 11.56 GHz

Fig. 6 **a** Current distribution at 2.5 GHz, **b** current distribution at 6.92 GHz, **c** current distribution at 11.56 GHz

3 Conclusion

A very simple and compact multiple stopband FSS for the WLAN, 5.89–7.05 GHz, and X-band is studied rigorously. Initially, square ring FSS has been studied later two T-shaped patches are being incorporated to provide three different stopbands. The size of the unit cell is also miniaturized. Width variation of the T-shaped patches changes the operating frequency bands. Responses of this design are highly stable with different angle of incidences.

References

1. Munk, B.A.: Frequency Selective Surfaces Theory and Design. Wiley, New York (2000)
2. Luukkonen, O., Costa, F., Simovski, C.R., Monorchio, A., Tretyakov, S.A.: A thin electromagnetic absorber for wide incidence angles and both polarizations. IEEE Trans. Antennas Propag. **57**(10), 3119–3125 (2009)
3. Zahir Joozdani, M., Khalaj Amirhosseini, M., Abdolali, A.: Wideband radar cross-section reduction of patch array antenna with miniaturized hexagonal loop frequency selective surface. Electron. Lett. **52**(9), 767–768 (2016)

4. Nair, U.R., Jha, M.: Electromagnetic design and performance analysis of airborne radomes: trends and perspectives (Antenna Applications Corner). Antennas Propag. Mag. **56**(4), 276–298 (2014)
5. Mackay, A., Sanz-Izquierdo, B., Parker, E.A.: Evolution of frequency selective surfaces. Forum Electromagn. Res. Methods Appl. Technol. (FERMAT) **2**, 1–7 (2014)
6. Hiranandani, M.A., Yakovlev, A.B., Kishk, A.A.: Artificial magnetic conductors realised by frequency-selective surfaces on a grounded dielectric slab for antenna applications. IEE Proc. Microw. Antennas Propag. **153**, 487–493 (2006)
7. Bayatpur, F., Sarabandi, K.: Multipole spatial filters using metamaterial based miniaturized-element frequency-selective surfaces. IEEE Trans. Microw. Theory Tech. **56**(12), 2742–2747 (2008)
8. Natarajan, R., Kanagasabai, M., Baisakhiya, S., Sivasamy, R., Palaniswamy, S., Pakkathillam, J.K.: A compact frequency selective surface with stable response for WLAN applications. IEEE Antennas Wirel. Propag. Lett. **12**, 718–720 (2013)
9. Hu, X.-D., et al.: A miniaturized dual-band frequency selective surface (FSS) with closed loop and its complementary pattern. IEEE Antennas Wirel. Propag. Lett. **8**, 1374–1377 (2009)
10. Yan, M., Qu, S., Wang, J., Zhou, H.: A miniaturized dual-band FSS with stable resonance frequencies of 2.4 GHz/5 GHz for WLAN applications. IEEE Antennas Wirel. Propag. Lett. **13**, 895–898 (2014)
11. Rahmati, B., Hassani, H.R.: Multiband metallic frequency selective surface with wide range of band ratio. IEEE Trans. Antennas Propag. **63**, 3747–3753 (2015)
12. Döken, B., Kartal, M.: Easily optimizable dual-band frequency selective surface design. IEEE Antennas Wirel. Propag. Lett. **16**, 2979–2982 (2017)
13. Xu, R., Zhao, H., Zong, Z., Wu, W.: Dual-band capacitive loaded frequency selective surfaces with close band spacing. IEEE Microw. Wirel. Compon. Lett. **18**(12), 2008, pp. 782–784
14. Bashiri, M., Ghobadi, C., Nourinia, J., Majidzadeh, M.: WiMAX, WLAN, and X-band filtering mechanism: simple-structured triple-band frequency selective surface. IEEE Antennas Wirel. Propag. Lett. **16**, 3245–3248 (2017)
15. Kartal, M., Golezani, J.J., Doken, B.: A triple band frequency selective surface design for GSM systems by utilizing a novel synthetic resonator. IEEE Trans. Antennas Propag. **65**, 2724–2727 (2017)
16. Da, R., Yoo, H.: Application of a compact electromagnetic bandgap array in a phone case for suppression of mobile phone radiation exposure. IEEE Trans. Microw. Theory Tech. **66**, 2363–2372 (2018)

NC2H: Cooperation Among Nodes through Cluster Head in Ad Hoc Network

Abu Sufian, Anuradha Banerjee and Paramartha Dutta

Abstract In civilian data communication using mobile ad hoc networks (MANETs), all mobile nodes cannot be of homogeneous type and these nodes do not do one specific job or communication. Therefore, cooperation among these nodes is a big issue, which is very essential to successfully run a MANET for this type of data communication. Denial of service and malicious behavior of a node are the main obstacles to a secure and successful communication in this type of a network. This scheme proposed a generic idea to avoid and prevent selfish behavior of a node as well as encourage to increase cooperation among nodes by the cluster head using a single-hop clustering strategy.

Keywords Ad hoc networks · Clustering · MANETs · Networks security · Node cooperation · Selfish attack · Wireless network

1 Introduction

Mobile ad hoc network (MANET) is a kind of network consisting of a collection of mobile nodes which could quickly set up a temporary network, where nodes must be capable of transmitting and receiving radio signals. This MANET is infrastructure-less, self-organizing where nodes are mobile in nature, and could play three kinds of roles: sender, receiver, and router [1]. This type of network has many applications such as data communication in battlefield, communication for rescue operation after a natural disaster, vehicle-to-vehicle communication, communication in a large conference room, and many more [2]. There are several underlined protocols that have

A. Sufian (✉)
University of Gour Banga, Malda, West Bengal, India
e-mail: sufian.csa@gmail.com

A. Banerjee
Kalyani Government Engineering College, Kalyani, West Bengal, India

P. Dutta
Visva-Bharti University, Santiniketan, West Bengal, India

S. Kundu et al. (eds.), *Proceedings of the 2nd International Conference on Communication, Devices and Computing*, Lecture Notes in Electrical Engineering 602, https://doi.org/10.1007/978-981-15-0829-5_14

become standard to set up and exploit this type of network [3]. All these protocols can be broadly classified into three categories: one is proactive where routes are always maintained such as in DSDV [4], WRP [5], FSR [6]; the second category is reactive where a route is established when required such as AODV [7], DSR [8]; and the third category is a mix of the above two where some portion of the network is proactive and the remaining portion is reactive, such as TORA [9], EMR-PL [10], WTMR [16].

In most of the classical routing protocols of MANETs, cooperation among nodes is overlooked. It is a true fact that MANETs are not successfully applied in civilian data communication, and one of the main reasons is the lack of cooperation among participating nodes although it is very successful in specific types of communication such as data communication in battlefield and communication among team members at the time of rescue operation after a natural disaster. As we know, a mobile node has limited resources such as limited residual energy and low bandwidth, some node(s) could be selfish to save these limited resources. This node cooperation problem does not arise in military data communication or any other specific communication using MANETs, because all the nodes in such type of communication are dedicatedly designed for that communication. But in civilian data communication, mobile nodes are of different types such as cell phone, laptop, palmtop, and PDA and these are not designed for specific work. Therefore, cooperation among nodes is very much important here in order to establish a MANET using these nodes of heterogeneous type.

Here, we used a single-hop clustering strategy as a generic mechanism to increase cooperation among nodes through cluster head (CH) by giving reward and punishment in terms of trust value. We know that the clustering scheme is more stable and scalable in nature, especially single-hop clustering schemes [11, 21]. All the nodes are attached with the respective CH directly and the entire network made into different partitions called clusters in the single-hop clustering scheme. All CHs are connected to each other through some gateway nodes and a network is established. CHs will take responsibility for increasing the cooperation among nodes within networks by increasing or decreasing the trust value of the member nodes. This scheme proposes an idea to avoid and prevent the three main selfish attacks, namely **Link Breakage Attack**, **Deleting Route Cache Attack**, and **Deliberately Delay Forward Attack**.

Rest of the paper is organized as follows: in Sect. 2, we discussed the literature review, clustering strategy is briefly discussed in Sect. 3, whereas Sect. 4 explains the proposed solutions to some selfish attacks, and conclusion and future scope in Sect. 5.

2 Literature Survey

So far many node cooperation schemes have been suggested by many researchers. Some prominent schemes are as follows:

L. Buttyan and J.P. Hubaux proposed a scheme called Nuglets [12] to increase cooperation among nodes. They used virtual currency which makes cooperative nodes to increase node cooperation among the nodes in MANETs. This scheme uses two purse models: one is a Packet-Purse-Model (PPM) where a Nuglet is debited from the source node and the other is Packet-Trade-Model (PTM) where a Nuglet is debited from the destination node of the packet. These authors also explained the utilities and necessity of increasing Nuglets for a node, and also explained the required security of these Nuglets. S. Marti et.al. proposed a similar type of routing scheme based on popular routing DSR [8] to catch a misbehaving node using watchdog and add path quality labels using pathrater [13]. By this portraiture, all participating nodes are classified and then the misbehaving or malicious nodes are avoided. P. Michiardi and R. Molva suggested a scheme called CORE [14] where the reputation of a node is the main concern. This scheme invigorates selfish nodes to quit selfish behavior, for example, clipping denial of service attack. S. Zhong et.al. proposed a credit-based and cheat-proof scheme for MANETs by resolving the selfish behavior of nodes called SPRITE [15]. This is quite similar to the scheme of Nuglets and it is an incentive credit- or debit-based scheme without any tamper-proof hardware. Here, the node gets inceptive by producing acknowledgements of forwarded messages from Credit-Clearance-Service (CCS). F. Kargl et.al. proposed Advanced Detection of Selfish/Malicious Nodes in Ad hoc Networks [17]. In this scheme, activity-based overhearing, iterative probing, and unambiguous probing are used to detect malicious and selfish nodes in the MANETs. N. Kang et. al proposed a misbehaving node detection scheme [18] at the iiWAS 2010 conference. This scheme applies different Intrusion-Detection-System (IDS) equivalent to watchdog to detect malicious/selfish nodes. IDS has been used to overcome the limitation of the uses of watchdog.

E. Hernandez Orallo et.al. proposed a model called CoCoWa [19]. Here, the authors have used a collaborative contact-based watchdog to efficiently detect selfish nodes in a short period of time. The CoCoWa model uses collaborative work, based on the diffusion awareness of local selfish nodes instead of fully depending on watchdog because watchdog also could be selfish. J.M. Chang et.al. proposed another node cooperative called Cooperative Bait Detection Approach (CBDS) [20] on the concepts of DSR [8]. This scheme can be embedded in both proactive and reactive routing. A reverse tracing approach is used here to resist a collaborative attack by malicious or selfish nodes. S. Berri et.al. proposed another reputation-based node cooperation scheme [22]. They use a similar kind of approach to increase node cooperation by adding or deducting the reputation of nodes within the MANET. But here the amount of reputation, loss or increase depend on the service taker node. If a node denies giving service to a reputed node, then the reputation loss will be more compared to denying service to less-reputed nodes and vice versa.

In C3H [21], the same authors of this paper use the single-hop node clustering strategy to prevent the malicious activity of any internal node of the network through cluster heads. This scheme explains the prevention of malicious attacks by giving future scope to increase cooperation among nodes using the same strategy. The current scheme, NC2H based on the concepts of C3H and single hop clustering scheme FESC [11].

3 Clustering Strategy for Monitoring Behavior of Nodes

The two main challenges along with many challenges for routing in MANETs are lack of centralized control and high mobility of nodes. These are the reasons why classical routing protocol of infrastructure-based networks do not work for mobile ad hoc networks. Clustering routing protocols in MANETs try to get some benefit of those classical routing protocols by mimicking the topology of these classical routing protocols. Here, we have adopted FESC [11], a single-hop clustering strategy, and some modifications made to cheat proof in C3H [20] for increasing the cooperation among nodes in ad hoc networks. There are three types of nodes in this scheme, namely cluster head (CH), ordinary member node, and gateway node. This single-hop clustering scheme proposed a technique where a member node is directly attached to the elected CH and breaks the entire network into some clusters. The cluster heads (CHs) are elected temporarily among member nodes. The election is done according to high residual energy, high trust value, low mobility of nodes, and strong connectivity to downlink nodes compared to other nodes of the cluster. Electing a good candidate as the cluster head is very important as this node will monitor the activities of the other nodes; this procedure is clearly defined in our earlier scheme C3H [21].

3.1 Graphical Explanation of the Strategy

Let us consider Fig. 1 where a small MANET containing two selfish nodes *a* and *q* have been shown. These two selfish nodes could carry out three kinds of selfish attacks which are mentioned in Sect. 4.

In this scheme, selfish attacks could be prevented through CHs. The nodes *p*, *r*, and *c* mentioned in Fig. 1 become cluster heads CH1, CH2, and CH3 in Fig. 2. CH1

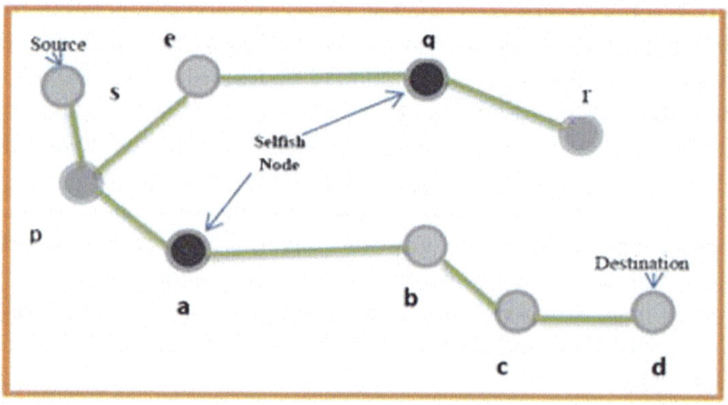

Fig. 1 A small MANET with two selfish nodes

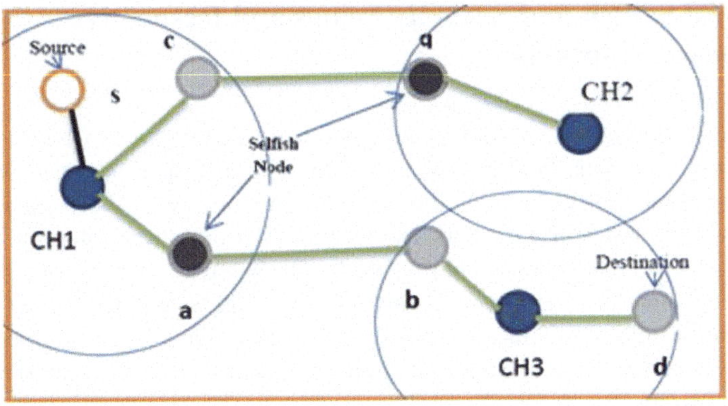

Fig. 2 Selfish nodes controlled by CHs

and CH2 could monitor the selfish nodes a and q as their respective cluster heads. The cluster heads give reward or punishment to the respective member nodes according to their participation or selfishness. The objective is clear that it is to increase node cooperation within networks and avoid selfish attacks.

3.2 Trust Value Calculation of Node

At the initial stage, that is when a node enters the network, 0.5 assigns as a default trust value, where $trust\_value$ is a variable whose values are between 0 and 1, which indicate the trust level of a node. There are two additional support variables, one is $earn\_trust$ which is a natural number starting from 2, and second one is $lose\_trust$ which is also a natural number starting from 1 up to the current $earn\_trust$.

Initial value assigned to $earn\_trust$ is 2 and $lose\_trust$ is 1 when a node joins the network for the first time. For a successful transfer of one packet, a node increases its $earn\_trust$ value by 1 unit. Similarly, if any kind of selfish behavior is shown, $lose\_trust$ value decreases by 1 unit from the responsible node. The entire activity is carried out by the cluster head. The current $trust\_value$ of a node is calculated by Eq. (1).

$$trust\_value = 1 - lose\_trust/earn\_trust \tag{1}$$

Trust value is very important for a node as this node will request for its own data packet transfer to the CH. Depending on the current $trust\_value$, a packet transfer request of a node could be granted by the respective CH.

4 Several Selfish and Security Issues and Respective Proposed Solutions

Besides many difficulties, MANETs face two serious difficulties which are selfish behavior and malicious behavior of the node. These difficulties arise within the network by some selfish or malicious node(s). According to the behavior of the mobile node, all nodes of a network can be classified into three categories, namely malicious node, selfish node, and normal node. The behavior of malicious and selfish nodes is the main concern. The malicious activity could be avoided using C3H [21], where a similar clustering tactic is used, and selfish activity can be prevented by motivating to increase the cooperation among nodes, which is the main objective of this proposed scheme.

Selfish node will try to save its resources such as energy and bandwidth as much as possible so that it could use them for its own communication in the near future. Therefore, these types of nodes try to avoid participating in communications with other nodes. Mainly, three types of selfish behavior or attack might come from selfish nodes, namely Link Breakage Attack, Deleting Route Cache Attack, and Deliberately Delay Forward Attack. These are discussed in Sects. 4.1, 4.2, and 4.3, respectively.

4.1 Link Breakage Attack

Some selfish node remains silent on receiving an RREQ packet from other nodes although it might have information about the intended destination. The objective of this kind of node is to save its own resources by avoiding to participate in the communication link of other nodes.

In this scheme, CHs identify those selfish nodes and reduce the trust values of those selfish nodes. The CHs also increase the trust value of those nodes which actually participate in communication with other nodes. Both increases and decreases in trust values are calculated by Eq. (1). As discussed earlier in this section that CHs are trustworthy nodes, link breakage attack can only arise from member nodes. Some member node selfishly tries to avoid becoming a gateway node. But in this scheme, the CH of a cluster monitors other cluster members. Whenever a route needs to be set up, the source node sends an RREQ packet to its CH, then the CH broadcasts this RREQ packet to the neighbor CHs through some member node called the gateway node (if the destination is not within the cluster). If some node remains silent after receiving the RREQ packet, the CH checks the residual energy level and the position of that node, then tries to find out the reasons why the node was unwilling to become a gateway node. If this reason is not sufficient to convince the CH, then CH reduces the trust value of that selfish node and informs the other stakeholders of this network. Similarly, CH also increases the trust value of the node for positive behavior.

4.2 Deleting Route Cache Attack

In MANETs, every node maintains a cache memory where information about the last few communications are stored. Selfish node could delete such information from its cache so that other nodes may avoid this node in future communication. In this way, the selfish node gets the chance to save its own resources.

As it was mentioned earlier, CH takes all the responsibility to store the routing information, and information not required from ordinary members. It is assumed that CHs are trusted nodes; therefore, the deleting route cache problem does not occur in this scheme. The role of the CH is very crucial, therefore, the optimum node must be the CH of a cluster at any instance. Whenever a CH hands over the charge to any other node, the routing information is also handed over.

4.3 Deliberately Delay Forward Attack

Selfish node could deliberately delay forwarding the packet of the other nodes so that those nodes could avoid this selfish node for future communication. In this way, a selfish node avoids participating in communication with other nodes.

In any instance of communication, a source node, a destination node, CH(s), and none or more gateway node(s) participate. But as mentioned earlier that CHs are the most trusted nodes and no question arises for a source and a destination node, if a delay arises, then gateway node(s) will be responsible. Any gateway node directly connects to its CH, so the gateway node either receives or delivers data packets to its CH. On the other hand, before communication starts, the route establishment phase has to complete, and at that time, the CH becomes aware of the possible time to get the data packets from the predecessor CH, and to send data packets to the successor CH. Later in communication time, if actual time does not match the estimated time approximately, then delay attack arises. The CHs then check their gateway node(s) whether the delay arises deliberately or for some other reasons. In both cases, an alternate gateway node can be selected if available, otherwise, communication will continue, but for deliberate delay, the identified gateway node gets punished by reducing the trust value of that selfish node using Eq. (1).

5 Conclusion and Future Scope

Through this NC2H single-hop clustering scheme for increasing cooperation among nodes, Link Breakage Attack, Deleting Route Cache Attack, and Deliberately Delay Forward Attack can be avoided. This NC2H could be useful to avoid other selfish attacks in ad hoc networks. The NC2H shall be implemented with standard baseline routing protocols. This scheme supposes that cluster head (CH) is the most trustworthy node, which is elected based on the single-hop clustering models, FESC, and C3H. In future work, election of CH node could be done in a more intelligent way where more parameters could be used so that the purpose of this NC2H scheme will be fulfilled more accurately.

References

1. Toh, C.K.: Ad Hoc Mobile Wireless Networks: Protocols and Systems, 1st edn. Prentice Hall PTR (2002)
2. Helen, D., Arivazhagan, D.: Applications, advantages and challenges of Ad Hoc networks. J. Acad. Ind. Res. (JAIR) **2**(8), 453–457 (2014)
3. Abolhasan, M., Wysocki, T., Dutkiewicz, E.: A review of routing protocols for mobile ad hoc networks. Ad Hoc Netw. **2**(1), 1–22 (2004)
4. Perkins C.E., Bhagwat P.: Highly Dynamic Destination-sequenced Distance-vector Routing (DSDV) for Mobile Computers, pp. 234–245. SIGCOMM ACM (1994)
5. Murthy, S., Aceves, J.J.G.L.: An efficient routing protocol for wireless networks. Mobile Netw. Appl. **1**(2), 183–197 (1996)
6. Pei, G., Gerla, M., Chen, T.W.: Fisheye state routing: a routing scheme for ad hoc wireless networks. In: IEEE International Conference on Communications. Global Convergence through Communications (2000). https://doi.org/10.1109/ICC.2000.853066.
7. Perkins C.E., Royer E.M.: Ad-Hoc On-demand Distance Vector Routing (1998). draft-ietf-manet-aodv-02.txt
8. Johnson, D.B., Hu, Y., Maltz, D.A.: The Dynamic Source Routing Protocol (DSR) for Mobile Ad Hoc Networks for IPv4. IETF Req. Comments **4728** (2007)
9. Park, V. Corson, S.: Temporary-Ordered Routing Algorithm (TORA). Internet Draft, draft-ietf-manettora-spec-04.txt, (2001)
10. Banerjee, A., Sufian, A., Duta, P.: EMR-PL: energy-efficient multipath routing based on link life prediction in ad hoc networks. J. Inform. Optim. Sci. **39**(1), 285–301 (2018). https://doi.org/10.1080/02522667.2017.1374733
11. Banerjee, A., Duta, P., Sufian, A.: Fuzzy-controlled energy-efficient single hop clustering scheme with (FESC) in Ad Hoc networks. Int. J. Inform. Technol. **10**(3), 313–327 (2018)
12. Buttyan, L., Hubaux, J. P.: Nuglets: a Virtual Currency to Stimulate Cooperation in Self-Organized Mobile Ad Hoc Networks (2001). http://infoscience.epfl.ch/record/52377
13. Marti, S., Giuli, T.J., Lai. K., Baker, M.: Mitigating Routing Misbehavior in Mobile Ad Hoc Networks, MOBICOM. ACM, USA (2000). 1-58113-197-6
14. Michiardi, P., Molva, R.: COre: a collaborative reputation mechanism to enforce node cooperation in mobile Ad Hoc Networks. Adv. Commun. Multimed. Secur., IFIP Int. Federation Inform. Process. (2002). https://doi.org/10.1007/978-0-387-35612-9_23
15. Zhong, S., Chen, J., Yang, Y.R.: SPRITE: A Simple, Cheat-Proof, Credit-Based System for Mobile Ad-Hoc Networks. IEEE INFOCOM (2003). 10-7803-7753-2/03
16. Sufian, A., Banerjee, A., Dutta, P.: Energy and velocity based tree multicast routing in mobile Ad-Hoc networks. Wirel. Personal Commun. **107**(4), 2191–2209 (2019). https://doi.org/10.1007/s11277-019-06378-y
17. Kargl, F., Klenk, A., Schlott, S., Weber, M.: Advanced detection of selfish or malicious nodes in Ad Hoc networks. Lect. Notes Comput. Sci. (2014). https://doi.org/10.1007/978-3-540-30496-8_13
18. Kang, N., Shakshuki, E. M., Sheltami, T. R.:Detecting misbehaving nodes in MANETs. In: Proceedings of the 12th International Conference on Information Integration and Web-based Applications and Services, pp. 216-222 (2010). https://doi.org/10.1145/1967486.1967522.

19. Orallo, E.H., Serrat, M.D., Olmos, Cano, J.C., Tavares De Araujo Cesariny Calafate, CM.; Manzoni, P, CoCoWa: A collaborative contact-based watchdog for detecting selfish nodes. IEEE Trans. Mobile Comput. **14**(6), 1162–1176 (2015). https://doi.org/10.1109/TMC.2014.2343627
20. Chang, J.M., Tsou, P.C., Woungang, I., Chao, H.C., Lai, C.F.: Defending against collaborative attacks by malicious nodes in MANETs: a cooperative bait detection approach. IEEE Syst. J. **9**(1) (2015)
21. Sufian, A., Banerjee, A., Dutta, P.: Cheat proof Communication through cluster head (C3H) in mobile Ad Hoc network. Pertanika J. Sci. Technol. **26**(3), 1513–1526 (2018)
22. Berri, S., Varma, V., Lasaulce, S., Radjef, M. S., Daafouz, J.: studying node cooperation in reputation based packet forwarding within mobile Ad Hoc networks. In: International Symposium on Ubiquitous Networking UNet 2017, vol. 10542. Morocco, LNCS (2017)

The Performance of Data Combining Techniques for Cooperative Leach Protocol in WSN Under Fading

Anirbaan Das, Shruthi Narayan, Pratibha Balaji and Abhijit Bhowmick

Abstract In this paper, a cluster-based wireless sensor network (WSN) is studied. The existing Low Energy Adaptive Clustering Hierarchy (LEACH) and Cooperative LEACH protocols are performed while considering channel fading and further incorporating receiver diversity techniques in the system model. The performance is investigated in terms of bit error rate (BER) and it was observed that BER reduces as the cooperation among cluster heads increases due to the introduction of gain.

Keywords WSN · LEACH · Cooperative LEACH · BER

1 Introduction

With the advent of rapid technological advancements around us, wireless sensors that operate on small and inexpensive batteries are having a substantial effect on the usage of communication-based sensors. These wireless sensor networks are required to be accurate and efficient for them to be used widely. These sensors are usually used to monitor various physical environment parameters. Due to their large networking size, constraints in terms of power supply that is available and remote deployment in harsh and difficult to access environment, the protocols based on WSN are more useful than the generally used wireless protocols. For example, the methodology used in WSN routing is that of multi-hop routing in which the principle of fairness is not followed. Instead of using point-to-point communication which is usually address

A. Das (✉) · S. Narayan · P. Balaji · A. Bhowmick
School of Electronics Engineering, Vellore Institute of Technology, Vellore, India
e-mail: anirbaan.das2015@vit.ac.in; anirbaan0307@gmail.com

S. Narayan
e-mail: shruthi.narayan2015@vit.ac.in

P. Balaji
e-mail: pratibha.balaji2015@vit.ac.in

A. Bhowmick
e-mail: abhjit.bhowmick@vit.ac.in

S. Kundu et al. (eds.), *Proceedings of the 2nd International Conference on Communication, Devices and Computing*, Lecture Notes in Electrical Engineering 602, https://doi.org/10.1007/978-981-15-0829-5_15

centric, the routing process is more data centric in WSNs. Choi et al. through his paper have established the efficiency of energy for channel aware random access for multiple parallel access channels under a collision model and have also derived the bounds on efficiency [1].

Zaman et al. have designed an energy-efficient routing protocol for clusters in the wireless sensor networks called the Position Responsive Routing Protocol which helps in reducing the amount of time a node is idle and reduces the average communication distance between the nodes [2].

The most important operation in WSN is to collect all the necessary data at the network nodes and transmit all the information to the base station where the proper analysis or processing is carried out. At present, there are various energy-efficient communication models and protocols that have been designed for specific purposes.

Tuven et al. have depicted the comparison between different gains attained by the various diversity combining techniques that are present for MIMO systems. The author has compared Maximal Ratio Combining (MRC), Selection Combining (SC), and Switch and Stay diversity techniques. The author has also stated that the performance gap present between the SC and MRC techniques tends to increase linearly as the number of transmitting antennas increases [3].

Wu et al. through their study show that the present multichannel protocols are not suitable for WSNs as not many channels are available and many real-time errors are also found. The author proposed a novel tree-based technique which allocates channels to disjoint trees and uses parallel transmissions among them [4]. Kong et al. have compared the basic LEACH protocol and the modified LEACH protocol called the Cooperative LEACH which selects N number of cluster heads in every cluster in order to increase the diversity. Using the same bandwidth, data rate, and bit error rate for both the routing protocols, Kong has observed that Cooperative LEACH protocol is more energy efficient as compared to traditional LEACH [5].

LEACH, one of the most common protocols, is hierarchical based in which most nodes transmit information to their cluster heads, which then aggregate and compress the information and send it to the base station. A stochastic algorithm is followed by the nodes, in every round, to decide whether it can become the cluster head or not in this particular round. Each node has enough radio power to be able to transmit the message directly to the base station, but doing this would lead to a waste of energy thereby reducing its efficiency. Consider P rounds, nodes that have been the cluster head once cannot become cluster heads again until P rounds. Thus, each node has a probability of 1/P to become the cluster head again. At the end of each round, each and every node that is not a cluster head selects its nearest cluster head and joins that cluster. The cluster head then decides a particular schedule for each and every node to transmit information. All the nodes that are connected and not cluster heads use TDMA to communicate with its cluster head according to a particular schedule. All of this is done by expending minimum energy. To minimize interference, the CDMA technique is also adopted in LEACH.

The LEACH protocol can be extended by using a head set instead of just a cluster head. Every time an election takes place, a head set consisting of several cluster heads is nominated. The message is to be transmitted to the distant base station by the head set. At a time, only one member of the headset is allowed to be active. Others are in a sleep mode. All the headsets uniformly transfer information to the distant base station. A cluster head is a kind of sensor node which transmits aggregated data that it receives from the other nodes in the cluster. Non-cluster head nodes just send all their collected data to the cluster head, which handles transmission from that point onward. Each cluster consists of a virtual headset. The nodes in this headset can be active only one at a time. There are two stages present in all iterations: 1. Election Phase and 2. Data Transfer Phase. There are several epochs in every data transfer phase. There are several iterations in one round.

Cooperative LEACH is obtained by just performing a simple modification in the clustering algorithm of the LEACH protocol such that virtual MIMO is put to full use based on user cooperation. Instead of selecting a single cluster head at the network layer, the Cooperative LEACH protocol enables us to obtain a diversity order equal to the number of cluster heads we choose in each cluster (say N). Due to the broadcast nature of wireless transmission cluster-heads, all the elected cluster heads, in a given cluster, will be able to simultaneously receive the data from the nodes that contain valuable information. Cooperative LEACH has been a breakthrough in cluster-based routing protocols as it saves a large amount of energy as compared to the LEACH protocol with the same data rate, bit error rate, delay, and bandwidth requirements.

Both the LEACH and the Cooperative LEACH protocols have been carried out in an ideal scenario, without considering fading in the channel. Since wireless sensor networks (WSNs) work in practical scenarios and face real-world problems, an accurate modeling of WSN transmission cannot be achieved if we consider ideal channels without allowing for real-world constraints like channel fading. Therefore, we studied a more effective and realistic way to predict and hence minimize the amount of energy required in WSN transmission by taking into account the biggest channel constraint that is the channel BER.

Motivated by the work done in [6], we consider a more realistic scenario which includes channel fading and receiver diversity techniques to combine the received information. It is considered that the channels are Rayleigh faded. In this paper, diversity techniques have been used, at the base station, to combine multiple signals from the cluster heads. The diversity technique is a powerful communication receiver technique, which is capable of providing a significant improvement in a wireless link at a nominal increase in system cost. Besides, it is helpful in saving time as it does not have the added requirement of needing to train the system. All these things make diversity the most practical solution to combat the channel fading that the signal would face, therefore our focus needs to be on being able to find the most suitable and efficient diversity techniques from Selection Combining (SC) and Maximal Ratio Combining (MRC).

We take into consideration both, the traditional LEACH algorithm as well as the Cooperative LEACH algorithm, along with diversity techniques. The algorithm used has been demonstrated in Table 1. We have investigated the bit error rate (BER) performance for several Signal-to-Noise Ratio (SNR) values and cluster size.

The performance of the diversity techniques is investigated for several cluster heads and is then compared to find the most efficient technique.

Table 1 Algorithm I

Algorithm steps
1: Form a sensor network
2: Enter the setup phase
3: Nodes decide if they can be a cluster head for the round
4: Node choose a number x
5: **if** $x <$ Threshold value ($T(n)$) **then**
6: **else**
7: Go to step 10
8: **end if**
9: Each cluster head transmits a pilot signal
10: Non-cluster head nodes receive all the pilot signals
11: Find the pilot signal with the greatest strength
12: **if** pilot signal $i >$ all other pilot signals **then**
13: Inform that cluster head of the participation in their cluster this round
14: **else**
15: Go to step 11
16: **end if**
17: Cluster head fixes TDMA slots for every node in the cluster
18: **if** it is ith non-cluster head node's TDMA slot **then**
19: ith non-cluster head can transmit the data to the cluster head
20: **else**
21: ith non-cluster heads can switch off the radio, reducing energy dissipation
22: **end if**
23: Cluster head chooses different CDMA spreading code

2 System Model

Figure 1 shows the system model of our interest where sensor nodes are distributed randomly. In every round, there are different cluster heads based on the random number selected by each node and whether or not it is lesser than the threshold value. These cluster heads then form a group of nodes around them by sending out an advertisement signal to all the other non-cluster head nodes as explained in the LEACH algorithm in the above section. The non-cluster heads receive these advertisement signal and check for their received signal strength. They then attach to the cluster head whose advertisement signal is the strongest as judged by their receiver. The attachment is completed by sending an acknowledgement signal to the cluster head. This leads to the formation of a cluster.

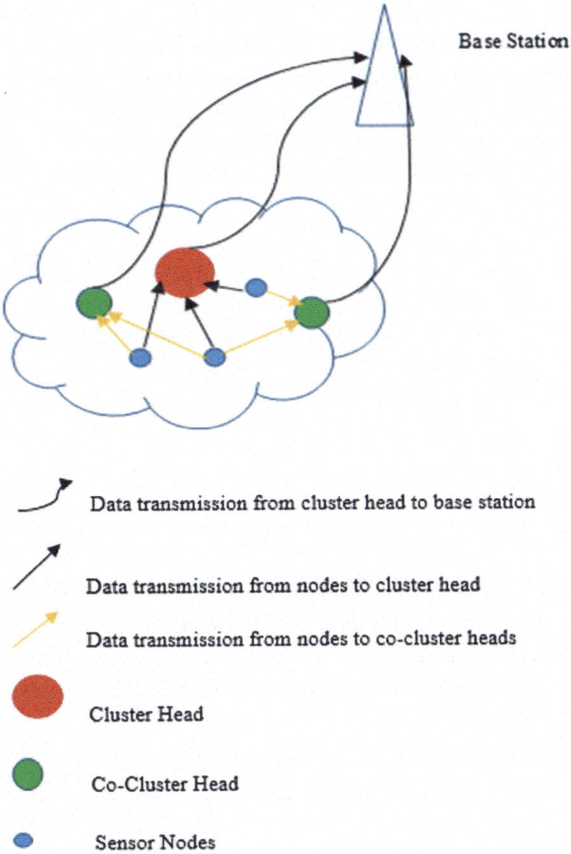

Fig. 1 System model for a hierarchical, cluster-based wireless sensor network (WSN)

Once the cluster head is aware of how many nodes are attached to it, it creates a TDMA schedule which allows it to give a dedicated time slot to receive data from each of its member nodes. The non-cluster heads can thus turn their radio off when it is not their allotted time, allowing them to reduce the energy dissipation.

In Cooperative LEACH, there can be one or more co-cluster heads which are chosen by the cluster head and allow for a diversity gain which helps improve the performance. The cooperative cluster heads are chosen in a way that they are at least half wavelength away from the position of the cluster head [7].

Therefore, in our model, we have considered both, short distance or local transmission (between the sensor nodes and their cluster heads) and long-distance transmission (between the cluster heads and the base station).

For the local transmission, the signal received by the jth cluster head in a cluster will be given by

$$y_{h,j}(t) = S \cdot h_i + n_i \tag{1}$$

where $j = 1, 2, \ldots, k$ such that $k < n$, $S = \{S_1, S_2, \ldots, S_n\}$ with S_i being the information transmitted by each individual node, h_i being the transmission channel used by the ith node and n_i being the Additive White Gaussian Noise (AWGN) in the channel, having zero mean and σ^2 variance.

For long-distance transmission, i.e., transmission between the cluster head and the base station, the signal received by the base station can be seen as a function of time as

$$y_B(t) = \beta \cdot y'_h(t) + n_j \tag{2}$$

where β is the amplification factor with which the signal is multiplied before being transmitted to the base station, n_j is the noise in the channel between the cluster head and the base station and

$$y'_h(t) = \begin{cases} \sum_{j=1}^{k} \left(y_{h,j}(t) + S_{c,j}(t)\right) w_j & ; \text{for MRC} \\ \left(y_{h,m}(t) + S_{c,m}(t)\right) w_j & ; \text{for SC} \end{cases} \tag{3}$$

where $y_{h,j}(t)$ is the combined data of all the other nodes received by the jth cluster head, $S_{c,j}(t)$ is the data of the cluster head itself, $w_j = |h_j|^2$, and $y_{h,m}(t)$ is the selected cluster head signal at the base station on the basis of the received SNR.

The bit error rate (BER) for MPSK can be expressed as [8]

$$\emptyset(\gamma) = \frac{1}{\pi} \int_0^{\frac{(M-1)\pi}{M}} \exp\left[\frac{\text{Sin}^2(\pi/M)}{\text{Sin}^2\theta}\gamma\right] d\theta \tag{4}$$

where γ is the received SNR. In our considered model, γ can be expressed as

$$\gamma = \begin{cases} \sum_{j=1}^{k} \frac{P_{c,j}(t) w_j}{\sigma^2} & ; \text{for MRC} \\ \frac{P_{c,j}(t) w_j}{\sigma^2} & ; \text{for SC} \end{cases}$$

where $P_{c,j}(t)$ is the transmit power of the jth cluster head.

3 Results and Discussions

In this section, simulation results are presented on the basis of the analytical model discussed above. The performance of the network has been investigated in terms of BER. The simulation is done in MATLAB.

In Fig. 2, BER is investigated for the SC scheme for several numbers of cluster heads. It is found that as the value of E_s/N_0 increases, BER decreases. It is also

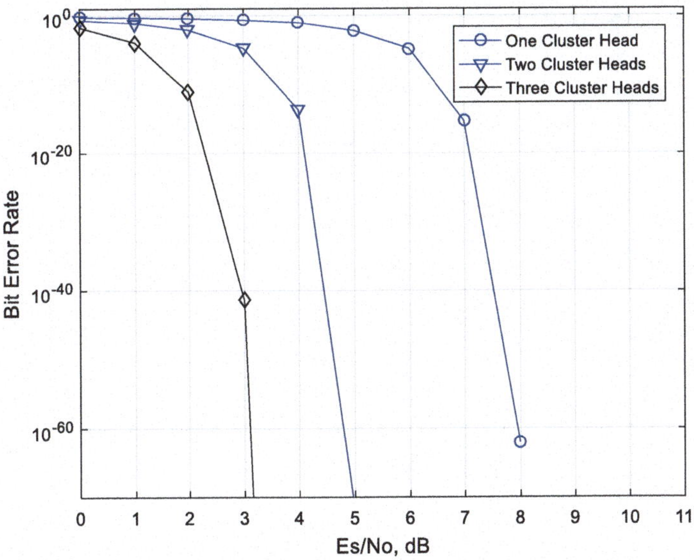

Fig. 2 BER versus E_s/N_0 for several numbers of cluster heads for SC

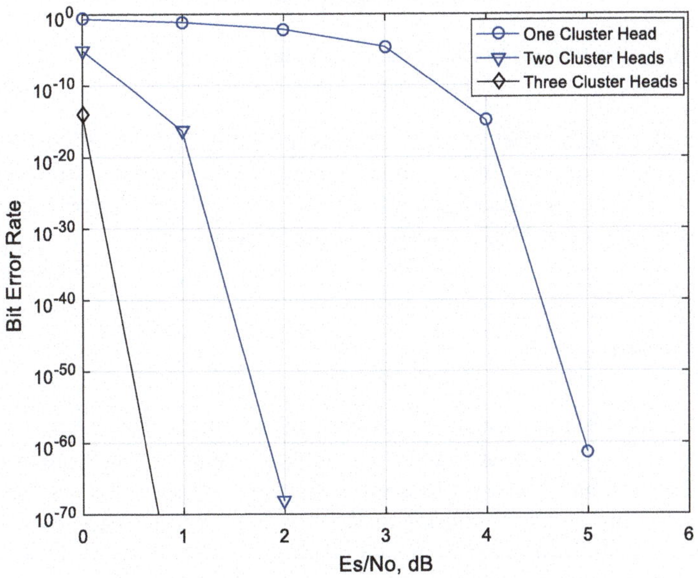

Fig. 3 BER versus E_s/N_0 for several numbers of cluster heads for MRC

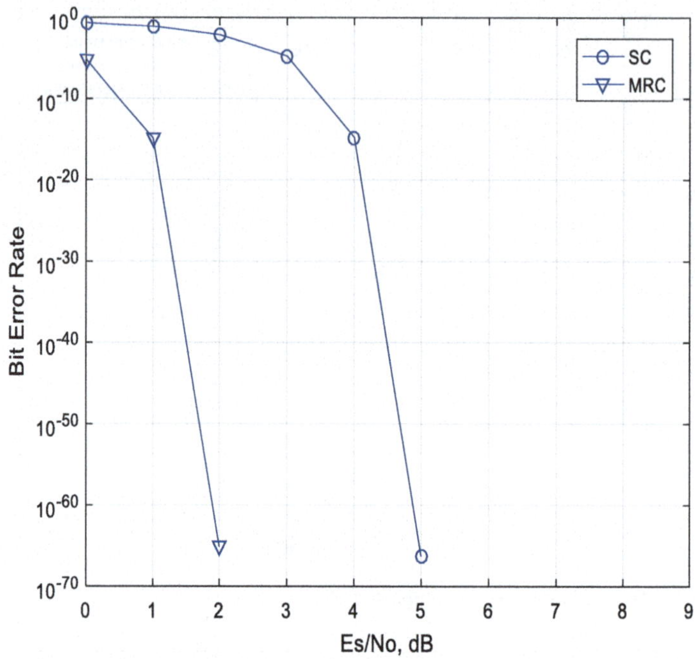

Fig. 4 A comparative BER performances for SC and MRC

observed that as the number of cluster heads increases, BER decreases for a fixed value of E_s/N_0 value. The same is investigated in Fig. 3 for the MRC scheme. A comparative BER performance for SC and MRC is investigated for two cluster heads in Fig. 4. It is found that MRC outperforms the SC scheme.

4 Conclusion

A cluster-based cooperative wireless network with multiple cluster heads is studied. In this study, the existing LEACH and Cooperative LEACH protocols are investigated under a fading channel for several receiver diversity schemes (SC and MRC). The performance is investigated in terms of BER. It is found that BER reduces as the cooperation among the number of cluster heads increases and MRC is used instead of SC for data combining. Thus, to have the best possible performance in a WSN environment, we need to have a Cooperative LEACH environment, therefore having multiple cluster heads, and in addition we need to use an optimal diversity technique like MRC at the receiver end to be able to transmit with the least amount of loss thus being the most energy-efficient method, allowing the batteries to run longer.

References

1. Choi, J.: On the energy efficiency and total bandwidth in channel-aware random access for WSNs. In: 2015 IEEE International Conference on Communications (ICC), IEEE (2015)
2. Zaman, N., Tang Jung, L., Yasin, M.M.: Enhancing energy efficiency of wireless sensor network through the design of energy efficient routing protocol. J. Sens. **2016**, Article ID 9278701, 16 (2016)
3. Tuyen, L.P., Bao, V.N.Q.: Comparison of diversity combining techniques for MIMO systems. In: 2011 17th Asia-Pacific Conference on Communications (APCC), IEEE (2011)
4. Wu, Y., Stankovic, J.A., He T., Lin, S.: Realistic and efficient multi-channel communications in wireless sensor networks. In: IEEE INFOCOM 2008-The 27th Conference on Computer Communications, Phoenix, AZ, pp. 1193–1201 (2008)
5. Kong, Hyung Yun: Energy efficient cooperative LEACH protocol for wireless sensor networks. J. Commun. Netw. **12**(4), 358–365 (2010)
6. Heinzelman, W.R., Chandrakasan, A., Balakrishnan, H.: Energy-efficient communication protocol for wireless microsensor networks. In: The 33rd Hawaii International Conference on System Sciences (2000)
7. Asaduzzaman, Y.K.: A cooperative transmission scheme for cluster based networks. In: Huang D.S., Jo K.H., Lee H.H., Kang H.J., Bevilacqua V. (eds.) Emerging Intelligent Computing Technology and Applications. With Aspects of Artificial Intelligence. ICIC 2009. Lecture Notes in Computer Science, vol. 5755. Springer, Berlin, Heidelberg (2009)
8. Zhang, L., Yang, J., Zhou, H., Jian, X.: Optimization of relay-based cooperative spectrum sensing in cognitive radio networks. In: 2011 7th International Conference on Wireless Communications, Networking and Mobile Computing, pp. 1–4. Wuhan (2011)

On Wireless Communication in Underground Mine System

Ankita Ray Chowdhury, Ankita Pramanik and Gopal Chandra Roy

Abstract Wireless communication inside the mine is much challenging as compared to the communication above the surface. The signal gets highly attenuated due to numerous attenuation and path loss components which are generally absent above the surface. The traditional rooms and pillar or longwall techniques used for mining communication, introduce multipath. Again, choosing a suitable frequency becomes paramount as the ionic air present inside the mines affects signal propagation. So, a detailed study of all the existing works is essential for developing future wireless communication systems inside the mines. Numerous existing works focus on IEEE 802.11b and IEEE 802.15.4 technologies for communication inside the mines. However, no work giving the actual range obtained with these technologies inside the mine is present. The present work compiles, classifies, and compares the different existing wireless communication technologies used in an underground mine communication system. The work also undertakes a range study in an actual coal mine. The coverage range with IEEE 802.11b and IEEE 802.15.4 is ~83% and ~62% lesser, respectively, as compared to that on the surface.

Keywords Underground mine · Operational frequencies · Underground mine environment · Radio communication · Wireless sensor network · WLAN · WPAN

A. R. Chowdhury (✉)
Department of Electrical & Electronics Engineering, IEM, Kolkata 700091, India
e-mail: ankitaraychowdhury@gmail.com

A. Pramanik
Department of Electronics and Telecommunication Engineering, IIEST, Shibpur 711103, West Bengal, India
e-mail: ankita@telecom.iiests.ac.in

G. C. Roy
Department of Mining Engineering, IIEST, Shibpur 711103, West Bengal, India
e-mail: gcr@mining.iiests.ac.in

S. Kundu et al. (eds.), *Proceedings of the 2nd International Conference on Communication, Devices and Computing*, Lecture Notes in Electrical Engineering 602, https://doi.org/10.1007/978-981-15-0829-5_16

1 Introduction

Underground coal mine has a dynamic environment. Toxic and explosive gases, namely carbon dioxide, methane, and carbon monoxide make the mine environment even more dangerous [1, 2]. The presence of these hazardous gases and/or smoke, dust, chances of increasing the temperature above tolerance limit, etc., should be monitored and communicated to the miners. In times of disaster, the establishment of communication system is the key to successful relief and rescue operations [3]. Ensuring safety of the workmen and the mine property is a big challenge for the mining company. Continuous monitoring of the underground environment through a remote system is essential to achieve safety the targeted production. Hence, proper communication infrastructure is today's need for a mining company.

In India, more than 90% of the underground coal production is from bord and pillar system of work. In this system of work, the working panel consists of many number of diverges, turns, intersects after regular intervals, and form the pillars. The working panel consists of a complex network; hence, communication through radio waves in any form becomes difficult. A typical part layout of an underground coal mine is in Fig. 1.

Designing any communication network to be used in any mine has to maintain the "permissible" level for safety purposes [4]. Permissible design typically limits the output power of any electronic devices so that the system is safe from any explosions. But limiting the output power restricts the range of communication. Moreover, electronic devices can generate heat and short circuits, making it risky.

Mine communication has to deal with soil, rock, water with electrolytes, etc., as a medium. Due to this, many problems, namely unnatural attenuation, rapid change in propagation characteristics, use of different ranges of frequencies [5], etc., will be encountered during electromagnetic (EM) wave propagation. For communication inside the mine, different waves like medium frequency (MF) [6, 7], ultrahigh

Fig. 1 A typical part layout of an underground coal mine

frequency (UHF) [8, 9], etc., have been used. Through the Earth (TTE) communication is used for reliable communication [10] since the early days. More advanced mine communication uses wireless sensor networks (WSN) [4]. Zigbee [11], Wi-Fi [12] or a combination of both [13] are used for such communication. For efficient establishment of communication networks, a study on deployment of communication networks inside mines is essential. Again, as the field of wireless communication in underground mines is nascent, a compilation of the existing works will help understand the challenges associated with it. The work in [14] has made a survey of the existing technologies in underground mining communication system. However, the work is quite old. Many newer technologies have since been developed and hence such a survey relevant for recent times is pertinent. For deployment of networks, study of signal attenuation, range obtained, etc., are important. The work in [13] gives the signal attenuation in terms of dB inside the mines. However, no work giving the actual range obtained inside the mines is available.

The main contributions of the present work are as follows:

- This work compiles the existing works so as to serve as a ready reckoner for the deployment of networks.
- Classification of the various existing works on a wireless system is done so as to make the study more effective and easier to comprehend.
- A comparative study of the performances of all the existing technologies is also provided.
- An experimental study on the range obtained with Zigbee and Wi-Fi in an actual coal mine is presented.

The rest of the paper is organized as follows: Sect. 2 explains different wireless communication types that can be used in underground mines. Section 3 presents the proposed wireless mine range testing system and the results obtained. Finally, Sect. 4 gives the conclusions and indicates future research direction in this area.

2 Different Types of Communication in Underground Mine

Underground mine communication is different from the terrestrial communication system. Due to adverse environment, the conventional equipment cannot be used for communication systems. Besides being reliable, small sized, sturdy, shock proof, and light weight are the important criteria for the equipment that can be used for mine communication [4, 9]. The early works on mine communication used wired communication systems. But as the mine size is dynamic, laying of new cables increases the cost of the infrastructure. Radio communication can be a possible solution. Most of the radio communication works with the concept of line-of-sight (LOS) but this theory fails inside the mine. As a mine is divided into different parts in terms of long wall or tunnels, the signal often gets blocked or dropped while propagating. Ionized air present inside a mine can also restrict the propagation of the signal [15]. Again, due to the electrical properties of a mine, not all the radio signals can propagate inside it. For establishing underground mine radio communication network, suitable frequency band needs to be chosen. In this section, the existing works on choosing the proper frequency is presented. Also a summary of the existing works on the different communication technologies used inside the mine is presented.

2.1 Frequencies for Underground Mine Radio Communication

The dynamic nature of the mine environment affects EM wave propagation. Different frequencies exhibit different characteristics. To establish communication networks inside the mine, the main challenge is to determine suitable operating frequency. Depending on the frequency, coverage area, propagation nature, attenuation, noise level [12], etc., will vary. In addition to these, the type of mine plays a vital role for, e.g., metallic mine helps to transmit faster than a nonmetallic mine [16].

In the earlier days, low frequency (LF) and MF signal were considered for mine communication. The main advantage of using LF is that the signal can penetrate lossy materials like pillars, rocks inside the mine with moderate power level [8]. Signals with frequency between 400 and 9000 Hz (ultra LF andvery LF) can be coupled with elongated conductors present inside mines to increase lateral range [8]. In [17], a prototype operating between 50 Hz and 6 kHz frequency to perform wireless real-time sensing for already existing vibrating wire-based geotechnical sensors to collect continuous, real-time data and to monitor the behavior of the strata remotely, has been developed. Later, MF was used for the same application. MF EM waves couple with long conductors inside a tunnel and form transmission lines through which transceivers exchange information [18]. In [6], propagation characteristics of MF waves had been studied and transmission line impedances had been measured at both ends. This gives improved coverage area as compared to LF [8, 18].

With the advancement of technologies, increased range, reduced power consumption, and bandwidth efficiency became the key criteria for network design. So the later works focused on the use of UHF and very high frequency (VHF) for mine communication. In a recent work [19], wireless underground sensor nodes communicate among themselves via the NFC/MI (near field communication/magnetic induction) technique using HF at 13.56 MHz (ISM bands). In comparison with 5.8, 2.4 GHz signals provide advantages in terms of coverage area and attenuation during propagation [12]. Paper [9] investigates the propagation characteristics and path loss incurred by EM waves while traveling inside the underground mine using frequency range in the MHz range. The major success of using UHF band for mine communication is the key for implementation of wireless local area network (WLAN) and wireless personal area network (WPAN) inside the mine. The work in [20] explained how location of each miner can be tracked inside the mine by the use of transceivers operating in UHF band. In recent years, many works have been done exploring ultrawide band (UWB) transmission for communication inside the mines. The paper in [21] proves that large bandwidth UWB communication system gives the opportunity to establish a communication infrastructure inside the mine that is capable of providing voice/video/data communication. Deterministic UWB channel models to characterize wireless channel separately in time domain and frequency domain to analyze channel parameters like delay spread, angular spread, and multipath component number was proposed in [22].

2.2 *Types of Radio Communication System*

Various types of mine communication systems are present. For ease of understanding, the present work divides the existing communication systems into three broad categories as follows:

- Wired Communication Systems
- Wireless Communication Systems
- Hybrid Systems

The classification is represented in Fig. 2. As discussed earlier, the wired communication system is cumbersome and is difficult to establish. Hybrid systems make a trade-off between the advantages and disadvantages of both systems. Among all the hybrid systems, the most popular one is leaky feeder system [4]. Due to paucity of space, the present work focuses on wireless communication system. Radio communication system inside underground mine can be further classified as follows:

1. Through the Earth (TTE) communication system,
2. In-mine radio communication system,
3. Wireless networks, and
4. Ultrawide band (UWB) systems.

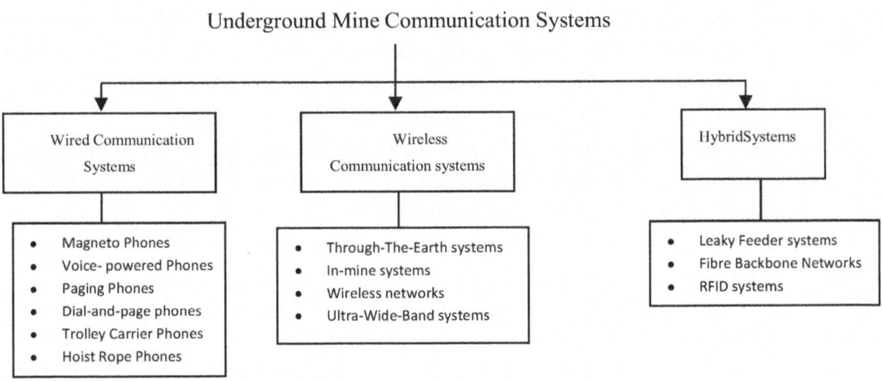

Fig. 2 Brief classification of underground mine communication systems

TTE Communication System

This communication system is based on the fact that extreme LF (ELF) signals will be coupled with elongated conductors present inside the underground mine such as railroad. Data rate offered by this system is nearly about 2.5 Kbps for real-time digitized voice. This signal then can penetrate rocks or long walls to transmit/receive information in order to establish a good communication link between the mine to the surface. TTE communication system is of two types: magnetic loop TTE, which transmits/receives signal through magnetic fields and electrode-based TTE, which transmits an electrical (current) signal through the established electric field in the mine and receiver receives a voltage signal in order to establish the link between the underground and the surface [10]. In the magnetic loop TTE radio application, the coupling effect of the signals with the elongated conductors is calculated [8]. In [16], a TTE emergency radio method for creating communication link to penetrate the earth or any other solid barrier is presented. The communication link employed a multi-carrier modulation scheme that can ascertain each carrier between the harmonic frequencies of the AC power line noise which reduces the effect of AC power line noise.

In-Mine Communication System

Underground mine communication may occur in two different regions. The first one is between the miners from the mine to the supervisors staying above the surface. This kind of communication is essential for any kind of disaster-like situation. Again establishing communication between two or more persons present inside the mine may be needed under various situations. In-mine communication system gives the infrastructure to establish a communication system by considering the radio wave propagation characteristics in underground mines. In [23], a mock mine environment was developed to conduct an experiment on channel propagation characteristics using frequency of 2.4–2.5 GHz. Linearly polarized log periodic antennas were used for transmission and reception which were placed at 1.5 m height from the tunnel floor. The result ensured that the EM wave propagation inside the mine largely is affected by the structure of the tunnel, antenna parameters, and transmission/reception frequency. Prediction of channel characteristics is very important for successful implementation of any communication system inside the mine. The work in [24] presents a channel prediction technique by a simulation method and a test was conducted where the simulated result was matched with the practical result. It divided the mine into three different regions: (a) LOS, (b) Partial-LOS (PLOS), and (c) Non-LOS (NLOS) and IEEE 802.11b system was used for communication [24]. Underground mine can be modeled as a waveguide [25]. Thus, the ray-tracing method can be used for radio wave propagation.

Wireless Networks

Presently, the deployed communication systems used in underground mines are wired systems. However, the installation of long cables all over the mine, insufficient space on laneway, etc., need to be cautiously handled. In [26], a mini intelligent methane gas sensor is designed using nRF240. nRF240 works as a sensor node to sense the presence of methane gas in the mine environment and calibrates the system autonomously. Paper [27] presents a WSN that monitors the mine to detect and identify any event and localization of miners and also roof falls. Further, [27] says that the frequency of 433 MHz is a more useful frequency in terms of path loss than 868 MHz. The work in [28] combines WSN with optical fiber sensors (OFS) to achieve more accuracy and more data rate to monitor the mine environment characteristics like temperature change, humidity level, different gas levels, etc. It investigates the possibilities of WLAN technology for providing localization capabilities inside the mine. Paper [29] adapts Owl positioning system (OwlPS), which is used for positioning technique in an indoor environment, for the inside of the mine. The obtained result is not as accurate as in the case of indoor environment but it can be concluded from the result that available general positioning systems can be implemented for low range applications inside the mines. Other than these technologies, IEEE 802.15 (WPAN) technology is also evolving. Paper [11] investigates the arrangements of deployment of sensor nodes in the WPAN network. The result shows that mesh topology gives the higher throughput and it is more reliable in comparison with cluster-tree topology. But on the other hand, cluster-tree topology is more energy efficient as a good number of nodes are in sleep mode (when they have nothing to transmit) which is useful for low energy consumption. The work in [30] monitors different system functions and applications such as temperature, humidity, illumination readings, text messaging, and controlling ventilation fans in an underground mine with the help of Zigbee (WPAN) system.

UWB Systems
The main disadvantage of having wireless network is scarcity of transmission bandwidth. The solution can be the UWB system where very low energy is used and large bandwidth is available. The UWB system is mainly deployed inside the underground mine to locate and track the miners which are very essential in any disaster situation to rescue trapped miners. In [31], a test has been performed with more advanced technique of using the UWB system with a two-way ranging (TWR) mechanism. This advanced technique gives more accuracy in underground mine communication with higher data rate transmission and larger bandwidth. A comparative study of all the radio communication systems, highlighting their advantages and disadvantages is given in Table 1.

3 Wireless Mine Range Testing System

Presently, remote monitoring of mines is being attempted by the establishing of WSN. Of all the WSNs, Zigbee-based WPAN [11] and Wi-Fi-based WLAN [12] can be chosen for communication inside the mines and also for remote monitoring due to their ease of installation and low power consumption. For measuring the range of the signal in an energy-efficient way, technology with lower data rates [27] should

Table 1 Comparative study of all radio communication systems

Name of the system	Characteristics	Benefits	Drawbacks
TTE communication systems [8, 10, 16]	• Communication between surface to underground • VLF radio waves are used	• Used in disaster control situations	• More transmitter power is required • Antenna size will be large • Dependent on frequency and depth from the surface
In-mine communication system [23–25]	• Establishes communication inside the mine, among the miners	• Helps to provide connectivity within the mine	• LOS required
Wireless networks [11, 26–30]	• Very less or no cable required • Communication established via different nodes, capable of storing and forwarding information ahead	• Less human interventions, more accuracy • Robust network • Easy installation of nodes	• More nodes needs more power • Propagation delay caused by data congestion • Poor connection beyond the network area
UWB system [20, 21, 31]	• Huge bandwidth, low power, and narrow pulses are used	• High data rate • Low power consumption • More accurate	• Range is short • Poor connection at the corners

be selected. Thus, Zigbee and Wi-Fi are the best-suited technologies to be used in such scenario. As the communication network should not be restricted just inside the mine but should reach the surface also, coverage area is the most essential measure. A detailed study to ascertain the range of the nodes is essential. The present work undertakes this study in a coal mine.

The proposed wireless mine monitoring system is presented in Fig. 3. A test signal is sent from Zigbee or the Wi-Fi transmitter. The range of the signal is measured using the corresponding receiver. The range obtained with Zigbee inside the mine is

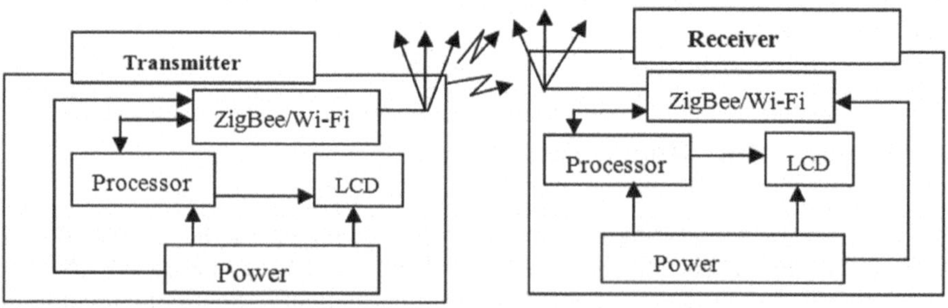

Fig. 3 Experimental setup

approximately ~13 m and that with Wi-Fi is ~17 m. To understand the amount of attenuation inside the mine, the range study is conducted with the same transceiver pair in an outdoor environment and also in an indoor environment. The ranges obtained with Zigbee in outdoor and indoor environments are ~35 m and ~18 m, respectively. The ranges obtained with Wi-Fi under the same conditions are ~95 m and ~28 m, respectively. The experimental test results are summarized in Table 2.

Thus, from Table 2, it can be seen that WPAN or WLAN, whatever the chosen technology be, the signals get highly attenuated inside the mine. Again, the WLAN signals are more affected by the mine environment (Figs. 4 and 5).

Table 2 WPAN and WLAN range test

Network	Outdoor environment (approximately)	Indoor environment (approximately)	Mine environment (approximately)	Remark
WPAN	35 m	18 m	13	• 62% less than the outdoor • 27% less than the indoor
WLAN	95 m	28 m	17	• 83% less than the outdoor • 40% less than the indoor

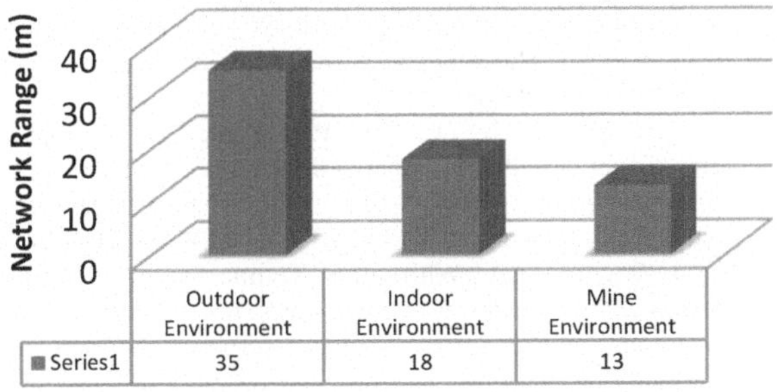

Fig. 4 Graphical representation of the result of a WPAN network range test

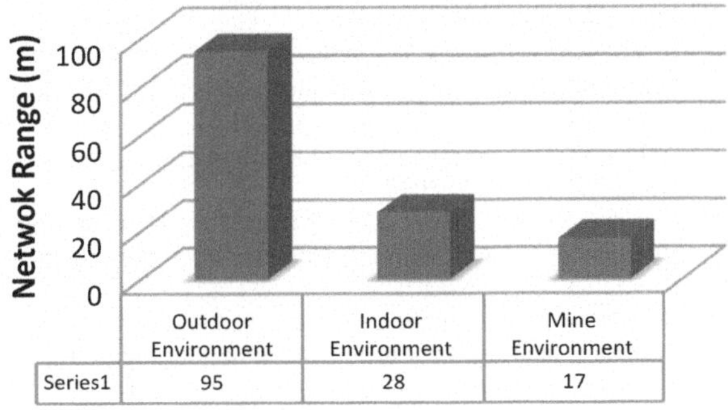

Fig. 5 Graphical representation of the result of a WLAN network range test

4 Conclusions and Scope of Future Works

From the above discussions, it can be said that no one technology can solve or address all the problems in underground mine communication as the problems faced in such a system are of varied nature. The fixed node concept used for monitoring of the mine will not help to locate the miners. Also, a centralized approach will be energy efficient but not appropriate due to the dynamic nature of the mine environment. In WPAN- and WLAN-based communication system, signal attenuation of ~27% and ~40%, respectively, compared to the indoor environment is obtained. Future works into improving the signal range in an energy-efficient manner must be developed. In recent times, software-defined radio (SDR)-based cognitive radio (CR) is used for wireless communication. CR has the capability to sense the environment intelligently, to adapt its functionality accordingly and perform likewise. If power of any transmitter/receiver decreases under any emergency situation, CR has the capability to sense it, and to regulate its operations optimally adjusting the transmission parameters so that other devices can have sufficient resources.

References

1. Misra, P., Kanhere, S., Ostry, D., Jha, S.: Safety assurance and rescue communication systems in high-stress environments: a mining case study. IEEE Commun. Mag. **48**(4) (2010)
2. Muduli, L., Mishra, D.P., Jana, P.K.: Application of wireless sensor network for environmental monitoring in underground coal mines: a systematic review. J. Netw. Comput. Appl. (2017)
3. Henriques, V., Malekian, R.: Mine safety system using wireless sensor network. IEEE Access **4**, 3511–3521 (2016)
4. Dohare, Y.S., Maity, T., Das, P.S., Paul, P.S.: Wireless communication and environment monitoring in underground coal mines–review. IETE Tech. Rev. **32**(2), 140–150 (2015)
5. Trang, H.T.H., Hwang, S.O.: Connectivity analysis of underground sensors in wireless underground sensor networks. Ad Hoc Netw. **71**, 104–116 (2018)
6. Li, J., Waynert, J.A., Whisner, B., Damiano, N.W.: Comparison of medium frequency propagation characteristics of a transmission line measured from both ends in a coal mine entry. In: Antennas and Propagation Society International Symposium (APSURSI), pp. 255–256. IEEE (2014)
7. Brocker, D.E., Sieber, P.E., Waynert, J.A., Li, J., Werner, P.L., Werner, D.H.: A hybrid approach for efficient modeling of medium-frequency propagation in coal mines. IEEE Antennas Propag. Mag. **57**(1), 164–176 (2015)
8. Ralchenko, M., Roper, M., Svilans, M., Samson, C.: Coupling of very low frequency through-the-Earth radio signals to elongated conductors. IEEE Trans. Antennas Propag. **65**(6), 3146–3153 (2017)
9. Akkaş, M.A.: Using wireless underground sensor networks for mine and miner safety. Wirel. Netw. **24**(1), 17–26 (2018)
10. Yan, L., Zhou, C., Reyes, M., Whisner, B., Damiano, N.: Mathematical modeling and measurement of electric fields of electrode-based through-the-earth (TTE) communication. Radio Sci. **52**(6), 731–742 (2017)
11. Moridi, M.A., Kawamura, Y., Sharifzadeh, M., Chanda, E.K., Wagner, M., Okawa, H.: Performance analysis of ZigBee network topologies for underground space monitoring and communication systems. Tunn. Underg. Space Technol. **71**, 201–209 (2018)

12. Kennedy, G.A., Bedford, M.D.: Underground wireless networking: a performance evaluation of communication standards for tunneling and mining. Tunn. Undergr. Space Technol. **43**, 157–170 (2014)
13. Tao, P., Xiaoyang, L.: Hybrid wireless communication system using ZigBee and WiFi technology in the coalmine tunnels. In: 2011 Third International Conference on Measuring Technology and Mechatronics Automation (ICMTMA), pp. 340–343. IEEE (2011)
14. Yarkan, S., Guzelgoz, S., Arslan, H., Murphy, R.R.: Underground mine communications: a survey. IEEE Commun. Surv. Tutor. **11**(3) (2009)
15. Jiang, Y.Z., Zhao, P., Zhai, Q., Ying, W.W., Hu, Q.L.: Signal enhancement techniques for through-the-earth communication based on multiple references and beamforming. AEU-Int. J. Electron. Commun. **86**, 86–91 (2018)
16. Roper, M.J., Kwasniok, P., Puzakov, V.: Through the earth emergency radio system. US Patent US 9,866,333 B2, issued January 9 (2018)
17. Mishra, P.K., Kumar, M., Kumar, S., Mandal, P.K.: Wireless real-time sensing platform using vibrating wire-based geotechnical sensor for underground coalmines. Sens. Actuators A Phys. (269), 212–217 (2017)
18. Li, J., Waynert, J., Whisner, B.G.: An introduction to medium frequency propagation characteristic measurement method of a transmission line in underground coal mines. Prog. Electromagn. Res. B **55**, 131–149 (2013)
19. Alshehri, A.A., Martins, C.H., Akyildiz, I.F.: Wireless FracBot (sensor) nodes: performance evaluation of inductively coupled near field communication (NFC). In: IEEE Sensors Applications Symposium (SAS), pp. 1–6 (2018)
20. Liin, J., Hunt, G.L.: Wireless underground location tracking. US Patent US2017/0006424 A1, Issued 5 Jan 2017
21. Khan, A.R., Gulhane, S.M.: A highly sustainable multi-band orthogonal wavelet code division multiplexing UWB communication system for underground mine channel. Dig. Commun. Netw. (2017)
22. Zhou, C., Jacksha, R., Yan, L., Reyes, M., Kovalchik, P.: Time domain and frequency domain deterministic channel modeling for tunnel/mining environments. Prog. Electromagn. Res. C (79), 209–223 (2017)
23. Hussain, I., Cawood, F., van Olst, R.: Effect of tunnel geometry and antenna parameters on through-the-air communication systems in underground mines: survey and open research areas. Phys. Commun. **23**, 84–94 (2017)
24. Farjow, W., Raahemifar, K., Fernando, X.: Novel wireless channels characterization model for underground mines. Appl. Math. Model. **39**(19), 5997–6007 (2015)
25. Liu, Y., Ghazal, A., Wang, C.X., Ge, X., Yang, Y., Zhang, Y.: Channel measurements and models for high-speed train wireless communication systems in tunnel scenarios: a survey. Sci. China Inf. Sci. **60**(10), 101301 (2017)
26. Xuhui, Z., Sunan, W.: Design of a wireless sensor network for methane monitoring system. In: 6th IEEE International Conference on Industrial Informatics, pp. 614–18 (2008)
27. Minhas, U.I., Naqvi, I.H., Qaisar, S., Ali, K., Shahid, S., Aslam, M.A.: A WSN for monitoring and event reporting in underground mine environments. IEEE Syst. J. **12**(1) (2018)
28. Zrelli, A., Ezzedine, T.: Design of optical and wireless sensors for underground mining monitoring system. Optik (17), 376–383 (2018)
29. Cypriani, M., Delisle, G., Hakem, N.: Wi-Fi-based positioning in underground mine tunnels. In: 2013 International Conference on Indoor Positioning and Indoor Navigation (IPIN), Oct 28, pp. 1–7. IEEE (2013)
30. Moridi, M.A., Sharifzadeh, M., Kawamura, Y., Jang, H.D.: Development of wireless sensor networks for underground communication and monitoring systems (the cases of underground mine environments). Tunn. Undergr. Space Technol. **73**, 127–138 (2018)
31. Barua, B., Kandil, N., Hakem, N.: On performance study of TWR UWB ranging in underground mine. In: Sixth International Conference on Digital Information, Networking, and Wireless Communications (DINWC), pp. 28–31. IEEE (2018)

Fuzzy-Based Session Key as Restorative Power of Symmetric Key Encryption for Secured Wireless Communication

Anirban Bhowmik, Sunil Karforma, Joydeep Dey and Arindam Sarkar

Abstract A symmetric key encryption with fuzzy-based session key has been proposed for wireless communication by satisfying the key issues like security and increase the strength of the symmetric key. In this paper, we generate "n" number of sub-keys from symmetric key using XOR operation between a random character matrix and symmetric key. Among these sub-keys, we generate a session key using fuzzy logic. Now the encryption is done by using this session key and a symmetric key. Here, we transmit the session key to the recipient end by amalgamating with the symmetric key. This amalgamated form is concatenated with ciphertext and sent to the recipient end for decryption with these keys. Here, a new authentication scheme is used with cipher text. Different types of randomness tests have been done to test the randomness of our session key. The brute-force attack analysis for this scheme and comparative study with the existing standard methods has been done with satisfactory results.

Keywords Symmetric key · Session key · Fuzzy logic · Encryption · Decryption · Brute-force

---

A. Bhowmik (✉)
Department of Computer Application, Cyber Research and Training Institute, Tinkonia, Goodshed Road, Burdwan, India
e-mail: animca2008@gmail.com

S. Karforma
Department of Computer Science, The University of Burdwan, Burdwan, India
e-mail: dr.sunilkarforma@gmail.com

J. Dey
Department of Computer Science, M.U.C. Women's College, B.C. Road, Burdwan, India
e-mail: joydeepmcabu@gmail.com

A. Sarkar
Department of Computer Science & Electronics, R.K.M. Vidyamandira, Belur Math, Belur, India
e-mail: arindam.vb@gmail.com

S. Kundu et al. (eds.), *Proceedings of the 2nd International Conference on Communication, Devices and Computing*, Lecture Notes in Electrical Engineering 602, https://doi.org/10.1007/978-981-15-0829-5_17

1 Introduction

The symmetric key encryption or single-key encryption was the only type of encryption in 1970s. A symmetric key encryption scheme has five parts—(i) plaintext (ii) encryption algorithm (iii) secret key (iv) ciphertext (v) decryption algorithm [1, 2].

For the secure use of symmetric encryption, we should focus on two requirements: (i) Strong encryption algorithm, (ii) Secret key, sender, and receiver must have the copies of this key in a secure fashion. The two basic building blocks of all encryption algorithms are substitution and transposition. There are two types of algorithms, stream cipher and block cipher and four types of algorithm modes, Electronic Code Book (ECB), Cipher Block Chaining (CBC), Cipher Feedback (CFB), Output Feedback (OFB). Many symmetric encryption algorithms are Data Encryption Standard (DES), Triple DES (3DES), International Data Encryption algorithm (IDEA), Blowfish, Advanced Encryption Standard (AES), etc. The symmetric key algorithm is also known as private key algorithm.

DES: Data Encryption Standard [3] uses a block cipher structure. It encrypts data in blocks of size 64 bits each. 64 bits of plain text is treated as the input to DES, which produces 64 bits of ciphertext. Initially, the key length is 64 bits. Before starting the DES process, every eight(8) bits of the key is discarded to produce a 56-bit key. DES is based on two main concepts of cryptography: confusion and diffusion. DES consists of 16 rounds and each round performs the step of substitution and transposition. DES results in a permutation among the 2^{64} possible arrangements of 64 bits, each of which may be either 0 or 1. Each block of 64 bits is divided into two blocks of 32 bits each, a left half block L and right half R. The DES algorithm turns 64-bit message block M into a 64-bit cipher block C. If each 64-bit block is encrypted individually, then the mode of encryption is called Electronic Code Book (ECB) mode. There are two other modes of DES encryption, namely Chain Block Coding (CBC) and Cipher Feedback (CFB), which make each cipher block dependent on all the previous message blocks through an initial XOR operation.

Double DES: Double Data Encryption Standard [2, 3] is simple to understand and it performs twice what DES does only once. Double DES uses two keys say, key1 and key2. It first performs as DES on the original plain text using key1 to get the encrypted text. It again performs as DES on the encrypted text by using the other key (that is, key2). The final output is the encryption of the encrypted text. Now for decryption, the doubly encrypted ciphertext block is first decrypted using the first key (key2) to produce the first round encrypted ciphertext. This ciphertext block is then decrypted using the key (key1) to obtain the original plain text.

RC5: RC5 [2, 3] is a symmetric key block encryption algorithm developed by Ron Rivest. The main theme of RC5 is that it is fast because it uses only primitive operations (XOR, Shift, etc.). It allows for a variable number of rounds and a variable bit-size keys to add to the flexibility. RC5 uses less memory for execution. In RC5, the word size, number of rounds, number of octets of the key, all may be of variable length but those values remain the same for a particular execution of the cryptographic algorithm. The output of RC5 is the ciphertext which is the same size as that of the input plain text.

Fuzzy Logic: Fuzzy logic [4–7] deals with fuzzy predicates and fuzzy implications made up of fuzzy predicates. It deals with how to make inferences using fuzzy predicates and fuzzy implications. A fuzzy predicate is described in terms of fuzzy sets and fuzzy implications in terms of fuzzy relations. Fuzzy relations are special kind of fuzzy sets whose domains are Cartesian products of the domain. It is also needed in the compositional form of reasoning. Fuzzy sets were introduced by Professor Lotfi A. Zadeh of University of California at Berkeley [7]. A fuzzy set on a universal domain U is defined by its membership function from U to [0, 1]. Thus, a fuzzy set on U is meant a function A: U → [0, 1]. “A” is called the membership function, A(x) is called the membership grade of x. We can write A = {x, A(x)):x€U}. It deals with reasoning with inexact or fuzzy concept. The fuzzy logic encompasses the fuzzy relations and fuzzy sets and [0, 1] is its truth value set. Most of the fuzzy logic is based on the following definitions for the logical connectives ¬¬, ∨, and ∧:

$$T(p \vee q) = \max[T(p), T(q)], T(p \wedge q) = \min[T(p), T(q)], T(p) = 1 - T(p).$$

Related Work:
Nowadays, data security has become a serious matter with the progress of communication technology. In the symmetric key encryption, DES was adopted as a national standard in 1976. Besides DES, two variations of DES have emerged which are double DES and triple DES where two keys and three keys are used to increase the robustness of encryption. IDEA [2], RC4, RC5, Blowfish, and Twofish [2] are different types of the symmetric key encryption algorithm. National Institute of Standards and Technology (NIST) announced the Advanced Encryption Standard (AES) in 2001. The AES algorithm is a symmetric block cipher with low complexity and high security level. NIST also proposed Secure Hash Algorithm (SHA) for authentication. When new encryption technique is introduced, cryptanalysts start to develop to attack. Eli Biham and Adi Samir introduced the concept of differential cryptanalysis [8]. This method looks at pairs of ciphertext whose plain texts have particular differences. Mitsuru Matsui invented the linear cryptanalysis attack [9] based on linear approximation. A timing attack is also applied to symmetric key encryption. There also exists Sensor Network Encryption Protocol (SNEP) [3] for security of sensor

network systems. Thus, many encryption algorithms are widely available and used in information security and also different types of attacks are available to break the security. In symmetric keys (or private key) encryption or secret key encryption, only one key is used to encrypt and decrypt data. DES uses one 56-bit key. Double DES uses two 56-bit keys and Triple DES (3DES) uses three 56-bit keys. While AES uses various (128,192,256) bit keys. At present, different types of concepts, logic like fuzzy logic, neural network, etc., also introduced in cryptography for increasing the robustness encryption. Our paper proposed a technique called FSKRPSKE[1] which provides a fuzzy-based session key from symmetric key using a random character key matrix. Here, session key is generated using fuzzy logic. Using these two keys, we can encrypt a file (.txt, .doc, etc.) and by the reverse way, we can decrypt the ciphertext to get plain text.

Problem Domain:
In symmetric key encryption, we can transmit huge amounts of data between sender and receiver effectively. But the whole encryption is done using a private key (symmetric key). If this private key is revealed by attackers, then the overall communication is under threat. Existing symmetric key encryption algorithm does not change their key/keys with respect to time. So the use of a single fixed key or multiple fixed keys is a problem in encryption process.

Solution Domain:
Our proposed technique provides an extra key called session key [4] from the symmetric key. This session key is generated by using a random character matrix, fuzzy logic, and the symmetric key. Since this session key may change from time to time, the use of both session key and symmetric key in encryption provides the extra robustness in our technique. Here, we deduce a novel technique by using both the symmetric key and session key for authentication proof. Thus, the use of session key with a symmetric key and an authentication cum encryption provides the added flavor in our proposed technique.

2 Methodology

Our proposed technique is composed of four parts which are (i) Session key generation, (ii) Encryption with symmetric key and session key, (iii) Authentication check and session key transpired, and (iv) Decryption. The summary of our scheme is described by a compact algorithm given below.

---

[1] FSKRPSKE means Fuzzy-Based Session Key as Restorative Power of Symmetric Key Encryption for secured wireless communication.

ALGORITHM:
Input: - plain text, symmetric key.
Output: - encrypted file with header and tailer.

Method: -

1. `Call MGA ()` // matrix is generated to create 'n' number of key population from symmetric key.
2. `Call SKG ( )` // generate session key using 'n' number of key populations and fuzzy logic.
3. `Call EP ( )` // Encryption Process i.e., cipher file is generated using two keys.
4. `Call Create_Header_Tailer ( )` // header and tailer structure is created using two keys with XOR operation.
5. `Call Concate(header, encrypted file,tailer)` // total structure is created and it is ready for transmission over network.
6. `Call AuthenticationCheck ( )` // check authentication using two keys and also generate session key using symmetric key.
7. `Call DrypPhase ( )` // plain text is generated.

All the above methods in the algorithm are described below in detail.

Session Key Generation Phase:
The session key generation process is divided into two parts. First is the predefined matrix generation and second is the session key generation from symmetric key using fuzzy logic. The predefined matrix is a square matrix and the number of columns is half the size of the symmetric key. If the key size is "n" bytes, then the number of rows and columns of the matrix is n/2.

ALGORITHM-1: Matrix Generation Algorithm (MGA)
Input: - randarr[m]: character array.
Output: - a square matrix (kmatrix[m][m]).
Method: -

```
1.Set m, i and j as integer.
2. m= half(symmetric key size).
3. kmatrix[m][m]={0}
4. for i=0 to m
5.  randarr[i]= get_randomchar();
6. end for
7. for i=0 to m
8.   for j=0 to m
9.     kmatrix [i][j]=randarr[j]
10.  end for
11. randarr[i]←rightShft(randarr[i])
12. end for
13.End.
```

ALGORITHM-2: Session Key Generation (SKG)
Input: - symmetric key and kmatrix[x][y]
Output: - session key (SK[n]).
Method: -

```
1. Set i, j, row, col,m fval as integer.
2. Set m= length (symmetric key), tmp[m/2][m/2]= {0},
   SYK[m] = symmetric key and keyarr[m/2],SK[n] as charac
   ter array.
3. row←get_row( kmatrix[m/2][m/2])& col←
  get_colmn(kmatrix[m/2][m/2]) {/*row=col=m/2*/}
 {/* step 4 to step 8 describes key population*/}
 4. for i=0 to row do
 5.    for j=0 to col do
 6.    tmp[i][j]←bitwise_XOROP( SYK[j],kmatrix[i][j])
 7.    end for
 8. end for
   {/* following part of algorithm find the fittest key
among m number of key population using fuzzy logic.*/}
 9. Set fval=0
 10. for i=0 to row do
 11.    fval+= bit_Difference (SYK [col], tmp[i][col])
 12.      if ((col/fval)<0.04) then
 13.        SK[n]←tmp[i][col]
 14.         Keyarr[col]=kmatrix[i][col]
 15.      end if
 16. end for
 17. End
```

Encryption Phase with Symmetric Key and Session Key:
Now in our proposed technique, encryption is done by using symmetric key and session key. The encryption with session key provides an extra flavor of robustness. In both cases, XOR [10] operation is executed. The encryption algorithm is given below.

ALGORITHM-3: Encryption Process (EP)
Input: - plain text, symmetric key, session key.
Output: - encrypted file.
Method: -

```
1.Set file_Plain as plain text file and file_Cipher as
  cipher text file.
2. Set file_Output as temporary file.
3. if ( !eof ) then
4. file_Output= bit_XOROP ( file_Plain , session key)
5. file_Cipher= bit_XOROP ( file_Output , symmetric
   key)
6.  end if
7.end for
8.End
```

After encryption with two keys, we create a format with Header, ciphertext, and Tailer [11] using the function *Concate()*. The result of this function is the compact form of text which is ready for transmission to the receiver end. We use Tailer part to check authentication and Header part for session key generation in the recipient end. Now the Header and Tailer structure is created using the following algorithm.

Header and Tailer Creation:

ALGORITHM-4: Create_Header_Tailer ( )
Input: - symmetric key, session key (SK[m]).
Output: - Header and Tailer.
Method: -

```
1.Set F_half, L_half and diagEl as character arrays.
2.Set key_Mat [][] as 2D character array.
3.F_half← first half of symmetric key, and L_half← last
  half of symmetric key.
4. set m= ascii valueOf(1st character of symmetric key)
5.Header← ((keyarr[] XOR L_half))<< (m mod
  length (symmetricKey)).
6.Key_Mat← Call create_matrix (F_half, session key)
7.ColmnEl← get_2ndColmn(key_Mat)
8.Tailer← bit_XOROP (ColmnEl, L_half)    // diagEl is XORed with
  L_half, bit by bit up to the last bit of L_half.
9.End
```

If the symmetric key is 16 bytes, the session key is 8 bytes and the total structure is given below which is created by calling the function Concate() which is given in the main algorithm.

Header (8 bytes)	Encrypted file	Tailer (8 bytes)

Decryption Phase:
The decryption phase occurs in the recipient end; first of all, Header section, encrypted file, and Tailer section are separated using the symmetric key. Here, we call *Create_Header_Tailer ()* function so that we can reveal the session key using the symmetric key from the Header section and we can check the authentication from the Tailer part using the function *AuthenticationCheck()*. If the authentication phase shows green signal, then plain text is generated from the encrypted file using both session key and symmetric key in reverse process of the encryption phase.

3 Results and Discussions

The encryption–decryption process is implemented in Turbo C interface in a PC of Intel Core i3 processor and 2 GB RAM. In this section, simulation results of the proposed scheme are presented. In our experiments, several sizes of files are used as plain text.

Different Types of Attacks:
There are different types of attacks that exist to recover the key in use rather than simply to recover the plain text. There are two general approaches are—(i) Cryptanalysis (ii) Brute-force attack [12, 13].

Key Space Analysis:
The size of key space is the total number of different keys in the encryption process. The brute-force attack is impractical in such crypto systems where key space is large. Now we consider a general case where secret key is k bits. There are two keys that are in our proposed scheme, first is the symmetric key with size k bits and second is the session key whose size is k/2 bits. Now for the symmetric key, the key space is 2^k and for session key, the key space is $2^{k/2,}$ and total key space is $2^{3k/2}$. Using this large key space, we discuss the following things.

Brute-Force Attack:
A good encryption technique satisfies the requirements of the resisting brute-force attack. In this attack, the attacker tries to translate the cipher text into plain text using every possible key. On average, half of all possible keys must be tried to achieve success. In most networking systems, algorithms are known to all so in this case, brute-force attack will be impossible if the algorithm uses a large number of keys. At present, the fastest super computer is Tianhe-2 having speed 33.86 petaflops, i.e., 33.86×10^{15} floating-point operations per second. Let us consider each trial requires 2000 flops to complete one check. So number of trials complete per second is 16.93×10^{12}. The number of seconds in a year is $365 * 24 * 60 * 60 = 3{,}153{,}600$ s.

Now from the above key space, the formula for breaking the keys is $2^{3k/2}/(16.93 * 10^{12} * 3{,}153{,}600) = Y$. So if k increases then Y increases. The following table and graphs shows the average time required for exhaustive key search [2] (Table 1 and Fig. 1).

Observations: From the above table, it is seen that with respect to the number of trials, our proposed technique provides good results than any standard algorithm (like DES, Triple DES, AES, etc.) with the same key size. The x-axis of the graph represents key size in bits. The above table and graph also show that our proposed technique provides good result for decryption than any standard algorithm (like DES, Triple DES, AES, etc.) with fixed decryption rate. So, it difficult for the attacker to decrypt any ciphertext using the assumed key. Thus, overall result of our technique is good with respect to any standard algorithms in brute-force attack.

Randomness Test of Session Key: In our technique, the session key is generated from a symmetric key using fuzzy logic. Now to test the randomness of the session key, we use some standard techniques such as frequency test [14], entropy [15].

Frequency Test: The frequency test is the most basic test for randomness checking. The purpose of this test is to determine whether the number of 1's and 0's in a sequence is approximately the same as would be expected for a truly random sequence. The following table shows the details (Table 2 and Fig. 2).

Table 1 Table for exhaustive key search

Symmetric key size (k bits)	No. of Trials in standard algorithms (2^k)	Time required (in years) at 16.93×10^{12} Decryption/s in standard algorithms	No. of trails in our proposed technique ($2^{(3k/2)}$)	Time required (years) at 16.93×10^{12} Decryption/s in proposed technique
56	2^{56}	0.001349	2^{84}	362,289
64	2^{64}	0.34550	2^{96}	1,483,938
128	2^{128}	6.3734×10^{18}	2^{192}	1.1775×10^{38}
168	2^{168}	7.0077×10^{30}	2^{252}	1.3555×10^{56}
192	2^{192}	1.1756×10^{38}	2^{288}	9.3148×10^{66}
256	2^{256}	2.1687×10^{57}	2^{384}	7.3799×10^{95}

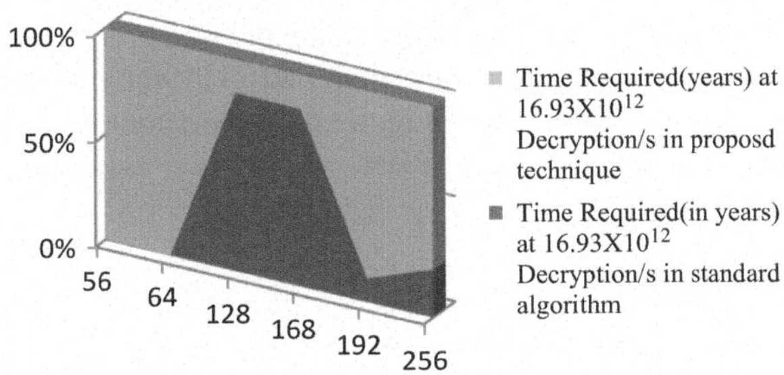

Fig. 1 3D graph of exhaustive key search

Table 2 Table for frequency test result

Symmetric key size (bits)	Frequency test result of our technique	Frequency test result of PRNG()
56	4598.255	4272.771
64	4625.148	4210.445
128	4661.584	4235.617
168	4579.129	4270.673
192	4519.966	4270.756
256	4633.421	4123.350

Fig. 2 Graph of frequency test of above table

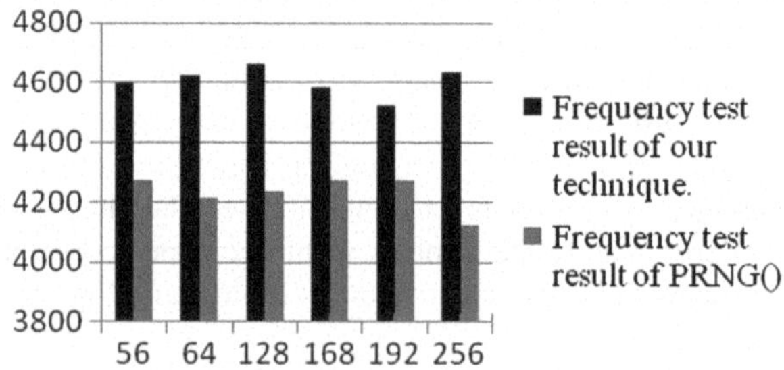

Observations: NIST SP 800-22 specifies that the randomness test must follow the three characteristics such as Uniformity, Scalability, and Consistency.

In the case of uniformity and scalability, the occurrence of a 0 or 1 is equally likely that is the probability of occurrence of 0 or 1 is half. The above table of frequency test result shows uniformity and scalability of our technique.

In case of consistency, we can say that the seed value from which we can generate the session key is a symmetric key. For cryptographic applications, the symmetric key must be secure. The session key is generated by using a random key matrix and a symmetric key. Now if the key matrix is unknown or may change from time to time and if the symmetric key is secured, then the next output bit in the sequence should be unpredictable in spite of any knowledge of the previous bits in the sequence.

It should not be feasible to determine the symmetric key from the knowledge of any generated values. There is no correlation between the symmetric key and generated values. Thus, our technique proves forward and backward unpredictability. Furthermore, from the above table and graph, it is seen that our proposed technique provides more randomness than PRNG () which is the standard technique.

Entropy Value Test: Here, we describe a comparative study between our technique and standard technique, PRNG () with the session key and symmetric key (Table 3 and Fig. 3).

Table 3 Table for entropy value

Symmetric key size (bits)	Entropy value of our technique	Entropy value of PRNG()
56	6.88	7.03
64	6.88	7.02
128	6.88	7.02
168	6.88	7.03
192	6.89	7.01
256	6.89	7.03

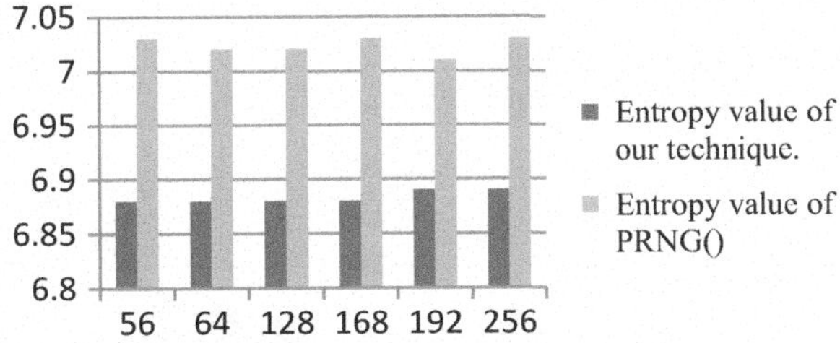

Fig. 3 2D graph of entropy value of above table

Observations: In cryptography, a cryptosystem is said to be semantically secure if it is computationally impossible for an attacker to extract any information about the plain text from ciphertext and its length. Entropy can be defined as randomness or unpredictability of information contained in a message. This randomness breaks the structure of the plain text. Entropic security in encryption is similar to semantic security when data have highly entropic distribution. The plain text entropy value is 0. Now from the comparative study of entropy value between our technique and PRNG (), it is seen that the entropy value of our technique is near to the result of PRNG (). The x-axis shows the key length. Thus, from the definition of entropic security, we say that it is very hard to predict plain text from cipher text if we use our technique to generate session key and the use of this session key and symmetric key in encryption provides robustness.

Comparative Study on Avalanche Effect:
Here, a comparative study between our technique and DES on avalanche effect [2, 16] with a fixed key is described below with a table and graph (Table 4 and Fig. 4).

Observations: A desirable property of any encryption algorithm is that a small change in either the plaintext or the key should produce a significant change in the ciphertext. In particular, one-bit change in the plaintext or one bit in the key should produce a change in many bits of the ciphertext. Thus, avalanche [2] quantifies the effect on the ciphertext when one bit changes in plaintext. An encryption algorithm that does not provide the avalanche effect can lead to an easy statistical analysis that is if the change of one bit from the input leads to the change of only one bit of the output, then it is easy to guess. The above table and graph provides comparative study between DES and our technique. In the graph, x-axis represents text size. This study

Table 4 Table for avalanche effect with fixed key

Text size (byte)	Total bit flipped in DES	Total bit flipped in our technique
17	17	17
24	25	24
32	33	32
64	63	64
92	90	90
128	126	123
144	137	141

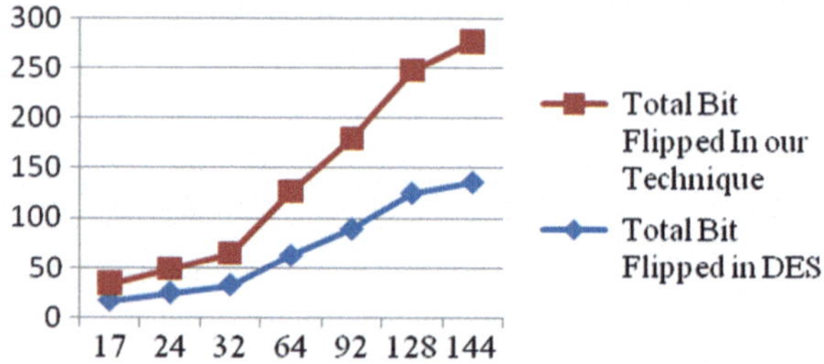

Fig. 4 Graph of avalanche effect with fixed key for the above table

tells that total number of bits flipped in our encryption technique is more than DES. Here, we use fixed size key. Thus, our technique (using fixed key) provides good results than any standard algorithm (like DES). So, our proposed scheme satisfies the desirable property for encryption algorithm.

Significance of Authentication: Authentication mechanisms [8] provide proof of identities. The authentication process ensures that the origin of the document is correctly identified, i.e., the document is coming from right person. In our scheme, we use authentication part for proof of identities. We know that symmetric key encryption provides authentication and confidentiality. But we have qualified this statement using an extra authentication scheme in our proposed technique. Here we use two structures Header and Tailer. There are complex calculations for Header and Tailer generation. Tailer structure is used for authentication checks. On the receiver side, symmetric key and particular row are used to generate session key. Now using this session key and symmetric key, we check authentication from Tailer part. Thus, our technique protects the fabrication.

4 Conclusion

We have presented encryption based on symmetric key and session key. This session key is generated from the symmetric key using some tools such as fuzzy logic, random matrix. Here, receiver decrypts the ciphertext using his or her symmetric key and session key. Our technique also provides proof of identities that enrich the robustness as well as beauty of encryption. Thus, our technique provides two contributions to symmetric key encryption. First, a new approach for extracting session key by using secured symmetric key and a random matrix. This random matrix provides randomness of our session key. Second, the encryption technique with two keys, i.e., two times encryption with session key and symmetric key provides robustness of our technique. At last, we have included an authentication mechanism in our technique. Comparative statistical tests like entropy value and frequency test between the proposed technique and standard technique proves the sturdier of our key. Finally, exhaustive key search analysis shows the acceptability of our technique. To the best of our knowledge, our proposed technique is the simplest encryption technique with symmetric key, session key with authentication mechanisms. It is practically having minimal computational overhead during encryption and decryption.

References

1. Das, A., Veni Madhavan, C.E.: Public-key Cryptography: Theory and Practice. Pearson Education (in press)
2. Stallings, W.: Cryptography and Network Security: Principles and Practice, 3rd ed. Prentice Hall (2003)
3. Boneh, D., Franklin, M.: Identity-based encryption from the Weil pairing. In: Kilian, J. (ed.) CRYPTO2001. LNCS, vol. 2139, pp. 213–229. Springer, Heidelberg (2001)
4. Fuzzy Logic: An Introduction [online]. http://www.seattlerobotics.org
5. Europe Gets into Fuzzy Logic. Electronics EngineeringTimes (1991)
6. Yager, R.R., et al.: Fuzzy Sets and Applications: Selected Papers by L.A. Zadeh. Wiley, New York (1987)
7. U.S. Loses Focus on Fuzzy Logic. Machine Design, June 21, 1990
8. Kahate, A.: Cryptography and network security. Tata McGraw-Hill Publishing Company, New Delhi (2008)
9. Stinson, D.: Cryptography: Theory and Practice, 3rd ed. Chapman & Hall/CRC (2006)
10. Oladipupo, E.T., Alade, O.A.: An approach to improve data security using modified XOR encryption algorithm. Int. J. Core Res. Commun. Eng. **1**(2)
11. Chaudhry, S.A., et al.: An improved and provably secure privacy preserving authentication protocol for SIP. Peer-to-Peer Netw. Appl. **10**(1), 1–15 (2017)
12. Agrawal, A., Gorbunov, S., Vaikuntanathan, V., Wee, H.: Functional encryption: new perspectives and lower bounds. In: Canetti, R., Garay, J.A. (eds.) CRYPTO 2013, Part II. LNCS, vol. 8043, pp. 500–518. Springer, Heidelberg (2013)
13. Buchmann, J.: Introduction to Cryptography, 2nd ed. Springer (2004)
14. Kak, A.: Lecture Notes on Computer and Network Security. Purdue University (2015) [Online]. https://engineering.purdue.edu/kak/compsec/Lectures.html
15. Zaidan, B., Zaidan, A., Al-Frajat, A., Jalab, H.: On the differences between hiding information and cryptography techniques: an overview. J. Appl. Sci. **10**, 1650–1655 (2010)
16. Delfs, H., Knebl, H.: Introduction to Cryptography: Principles and Applications. Springer (2002)

DPL-Based Novel 1-Trit Ternary Half-Subtractor

Rahul Raj, Rakesh Kumar Singh, Narendra Deo Singh, Saubhik Kumar and Aloke Saha

Abstract Subtraction is a fundamental numerical operation necessary for most arithmetic and logic computation. Ternary is a promising alternative in order to suppress demerits associated with present high-density binary (Radix-2) based digital Integrated Circuits. The present paper introduces a new strategy to design ternary (Radix-3) 1-trit half-subtractor based on Double Pass-transistor Logic (DPL) for Time-equalized/wave-pipelined digital applications. Wave-pipelining improves circuit performance without intermediate latches/flip-flops and hence can suppress difficulties of register pipelining. The proposed DPL-based ternary subtractor finds application to design low-power high-speed wave-pipelined ternary digital processing system. The complete circuit is designed and optimized with normal process E-MOS devices using BSIM3 model parameters. Ternary digit '0', '1' and '2' are represented by 0 V, 0.9 V and 1.8 V, respectively. Transient response with T-Spice simulation validates the proposed idea. The extracted speed–power performance is recorded. All the simulations are performed on TSMC 0.18 μm CMOS technology with 1.8 V supply rail and at 25 °C temperature using Tanner EDA v13.

Keywords Double Pass-Transistor Logic (DPL) · Hot-spot · Interconnect complexity · Ternary logic · Ternary subtraction · Wave-pipelining

R. Raj · R. K. Singh · N. D. Singh · S. Kumar · A. Saha (✉)
Department of Electronics & Communication Engineering, Dr. B. C. Roy Engineering College, Durgapur, India
e-mail: saha81@gmail.com

R. Raj
e-mail: rahul102136@gmail.com

R. K. Singh
e-mail: rkumarsingh35@gmail.com

N. D. Singh
e-mail: singh.narendradeo@gmail.com

S. Kumar
e-mail: saubhik.bappi@gmail.com

S. Kundu et al. (eds.), *Proceedings of the 2nd International Conference on Communication, Devices and Computing*, Lecture Notes in Electrical Engineering 602, https://doi.org/10.1007/978-981-15-0829-5_18

1 Introduction

Like addition, subtraction is also a fundamental numerical operation that finds wide applications in Digital Signal Processing (DSP), Digital Image Processing (DIP), Arithmetic Logic Unit (ALU), Central Processing Unit (CPU), Graphical Processing Unit (GPU), Digital Calculator and so on [1, 2]. In iterative computing, the subtractor plays a crucial decision-making role in order to complete the operation. Again, the speed–power performance of many critical data path elements like divider and square-root is heavily dependent on the performance of subtraction operation.

Present binary (Radix-2) based computation is gradually losing its interest due to various practical issues related to chip fabrication and/or reliability [3]. The increased number of fan-in/fan-out with complex interconnects and results in localized heat (hot-spot) makes the modern high-density SOC (System-On-Chip) unreliable. Ternary (Radix-3) [3–12] can be a feasible alternative to overcome demerits associated with binary (Radix-2)-based system design. The number of I/O pins and hence interconnect complexity can be reduced remarkably with ternary-based processing [3]. Also in communication, lesser number of communicating channels are needed to communicate ternary-based digital data as compared to conventional binary-based counterpart.

Wave-pipelining is a promising technique that eliminates intermediate latches/flip-flops of conventional register pipelining in order to improve the overall performance of digital circuit/system [13–16]. Delay equalization among all the data paths from input to output is a major task for effective wave-pipelining. Data-dependent propagation delay makes the conventional static CMOS ineffective for implementing efficient wave-pipelining. Double Pass-transistor Logic (DPL) is an emerging logic with favourable characteristics to deal with time equalization as investigated in [15, 16]. In this work, a novel ternary 1-trit half-subtractor based on DPL is proposed. The circuit has good applicability for time-equalized/wave-pipelined system design. The complete circuit is designed, optimized and validated on TSMC 0.18 μm CMOS technology with 1.8 V supply rail and at 25 °C temperature using Tanner EDA v13.

The rest of the paper is organized as follows: the operating principle of the proposed 1-trit ternary half-subtractor is explained in Sect. 2. Section 3 explores the front-end design with T-spice simulation response. The paper is concluded in Sect. 4.

2 Proposed Subtractor: Concept and Theory

The design concept of the proposed DPL-based 1-trit ternary (Radix-3) half-subtractor is explored in this section. Suppose the minuend and subtrahend 'trit' to the proposed subtractor is 'X' and 'Y', respectively. Again the Difference and Borrow output of the circuit is denoted by 'D' and 'B', respectively. Consider the circuit for Difference first. The K-Map for Difference ('D') output is shown in Fig. 1.

Fig. 1 K-Map for difference O/P ('D')

D	Y=0	Y=1	Y=2	
X=0	0	2	1	Row-1
X=1	1	0	2	Row-2
X=2	2	1	0	Row-3
	Col.-1	Col.-2	Col.-3	

Table 1 I/O relation for NTI, STI and PTI

Ternary I/P (X)	NTI O/P (X_0)	STI O/P (X_1)	PTI O/P (X_2)
0	2	2	2
1	0	1	2
2	0	0	0

The circuit construction for Difference output is explained in a step-wise manner. The inherent pass characteristic of E-NMOS and E-PMOS devices has been followed in order to achieve an optimum result. In the following explanation, the output of NTI (Negative Ternary Inverter), PTI (Positive Ternary Inverter) and STI (Simple Ternary Inverter) is denoted by X_0, X_2 and X_1, respectively, with trit input of 'X'. The I/O relation for NTI, PTI and STI is presented in Table 1. For more detail, readers are directed to [3–5].

Step-I: Consider the condition $(X, Y) = (0, 0)$ and $(X, Y) = (1, 0)$ first. To generate the corresponding microelectronic circuit, the row-3, col.-2 and col.-3 in Fig. 1 should be blocked and the output should be same as the trit value of 'X'. This can be achieved by two series connected NMOS with pass variable 'X' and with gate control variable for two NMOS as X_2 and Y_0.

Step-II: Next, to generate output for $X = 2$ and $Y = 0$, the first and second row need to be blocked. Also, the second and third column should be eliminated from consideration. This can be done by two series-connected PMOS with fixed pass voltage of 1.8 V and gate control variable for two PMOS devices as 'X_2' and $\overline{Y_0}$.

Step-III: To make the circuit for $X = 2$ and $Y = 1$ in Fig. 1, the first two rows, col.-1 and col.-3 in Fig. 1 should be eliminated. Three series-connected NMOS with pass voltage 0.9 V and gate control variables Y_2, $\overline{Y_0}$ and $\overline{X_2}$ can solve the purpose.

Step-IV: Now for the case $X = 2$ and $Y = 2$, the first two rows and first two columns in Fig. 1 should be blocked. The corresponding circuit can be achieved by two series-connected NMOS devices with a fixed pass voltage of 0 V and a control variable for two NMOS as X_2 and $\overline{Y_2}$.

Step-V: Next consider the case $X = 1$ and $Y = 2$. Here, row-1, row-2, col.-1 and col.-2 in Fig. 1 should be eliminated from consideration. Three series connected PMOS with pass voltage 1.8 V and controlling them by Y_2, X_0 and $\overline{X_2}$ can solve the problem.

Step-VI: For X = 0 and Y = 2, the row-2, row-3 and the first two column need to be blocked. This can be done by two series-connected NMOS with pass voltage 0.9 V and gate-control variable as Y_2 and X_0.
Step-VII: To generate the corresponding circuit for X = 0 and Y = 1, row-2, row-3, col.-1 and col.-3 should be eliminated. This can be achieved by three series-connected PMOS passing 1.8 V and controlling PMOS gate terminals with Y_0, $\overline{X_2}$ and $\overline{X_0}$.
Step-VIII: This is the final step to design the Difference part of the proposed 1-trit ternary half-subtractor. Consider X = 1, Y = 1. To get the corresponding circuit, row-1, row-3, col.-1 and col.-3 should be eliminated. Four NMOS devices connected in series with pass voltage 0 V and gate control variable $\overline{X_0}$, X_2, $\overline{Y_0}$ and Y_2 can be used for the purpose.

The circuit diagram to compute the difference between input trit 'X' and 'Y' based on the proposed strategy is presented in Fig. 2. The structure to generate complementary part is not included for the sake of brevity.

Next, to generate the Borrow part of the proposed subtractor, the K-map is shown in Fig. 3. The Borrow output for minuend 'X' and subtrahend 'Y' is denoted by 'B'.

The proposed strategy to generate Borrow output for the proposed 1-trit half-subtractor is explained below:

Step-I: Consider column 1 first. The output is always zero regardless of 'X' input. This can be generated by eliminating col.-2 and col.-3 of Fig. 3 and passing 0 V to

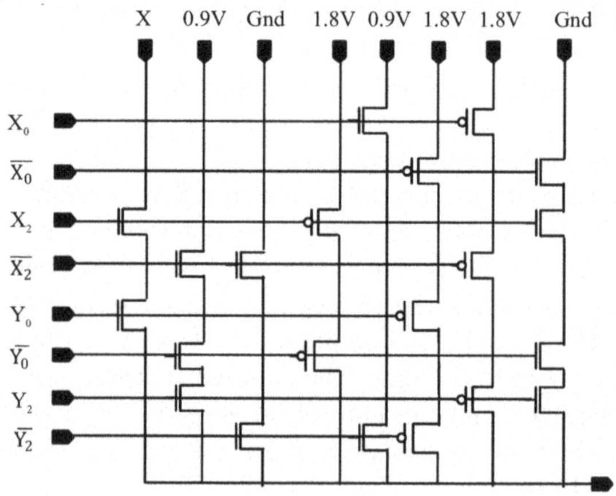

Fig. 2 Difference (D) circuit for proposed 1-trit ternary half-subtractor (minuend 'X' and subtrahend 'Y')

Fig. 3 K-Map for Borrow O/P ('B')

B	Y=0	Y=1	Y=2	
X=0	0	1	1	Row-1
X=1	0	0	1	Row-2
X=2	0	0	0	Row-3
	Col.-1	Col.-2	Col.-3	

the output. NMOS with gate control variable Y_0 and pass voltage 0 V can fulfill the purpose.

Step-II: To generate output for (X, Y) = (1,1) and (X, Y) = (2, 1), the col.-1, col.-3 and row-1 in Fig. 3 should be blocked. This can be achieved by three series-connected NMOS with pass voltage 0 V and gate control variable as X_0, $\overline{Y_0}$ and Y_2 for three NMOS devices.

Step-III: Next, consider the case X = 1, Y = 2 in Fig. 3. For this, row-1, row-3, col.-1 and col.-2 should be removed. Three series-connected NMOS with pass voltage 0.9 V and gate-control variable X_0, $\overline{Y_2}$ and X_2 for three NMOS devices can solve the problem.

Step-IV: To get output for X = 2 and Y = 2, the first two rows and first two columns of Fig. 3 should be eliminated. This can be done by two series-connected NMOS with pass voltage 0 V and control variable X_2 and $\overline{Y_2}$.

Step-V: This is the final step to generate the Borrow circuit. Consider the case (X, Y) = (0, 1) and (X, Y) = (0, 2). Here, row-2, row-3 and col.-1 should be eliminated from consideration. This can be done by two series-connected PMOS with 0.9 V pass voltage and the gate-control variable X_0 and Y_0.

The complete circuit diagram of the proposed DPL-based ternary Borrow circuit as per the aforesaid strategy is shown in Fig. 4. The same idea can be applied to generate the complementary equivalent of the Borrow output. The circuit is symmetrical in nature and hence offers balanced input capacitance that in turn reduces data-dependent delay variation at the output. The structural advantage of proposed circuit is favourable to design efficient time-equalize/wave-pipelined system.

The design and simulation result of the proposed DPL-based 1-trit ternary half-subtractor is presented in Sect. 3.

Fig. 4 Borrow circuit for proposed half-subtractor

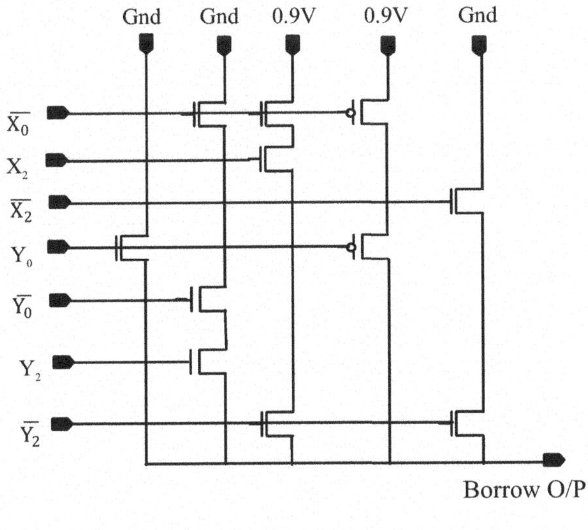

3 Design, Validation and Evaluation

The proposed 1-trit ternary half-subtractor has been designed and optimized based on BSIM3 device model on TSMC 0.18 μm CMOS technology at 25 °C temperature. The trit values 0, 1 and 2 are coded with 0 V, 0.9 V and 1.8 V, respectively. The required Negative Ternary Inverter (NTI), Positive Ternary Inverter (PTI) and Simple Ternary Inverter (STI) are designed as per the idea presented by Hang et al. in [4, 5]. The front-end schematic of the proposed subtractor is shown in Fig. 5. The schematic design for DPL-based NTI and PTI circuit are shown in Fig. 6.

In order to validate the proposed strategy, the transient response from T-Spice simulation has been observed. The ternary input to the circuit has been applied by Piece Wise Linear (PWL) input source using S-Edit of Tanner EDA v.13. The resultant wave form from the transient analysis at 25 °C temperature is presented in Fig. 7. The extracted speed–power characteristic of the proposed DPL-based 1-trit ternary half-subtractor is summarized in Table 2.

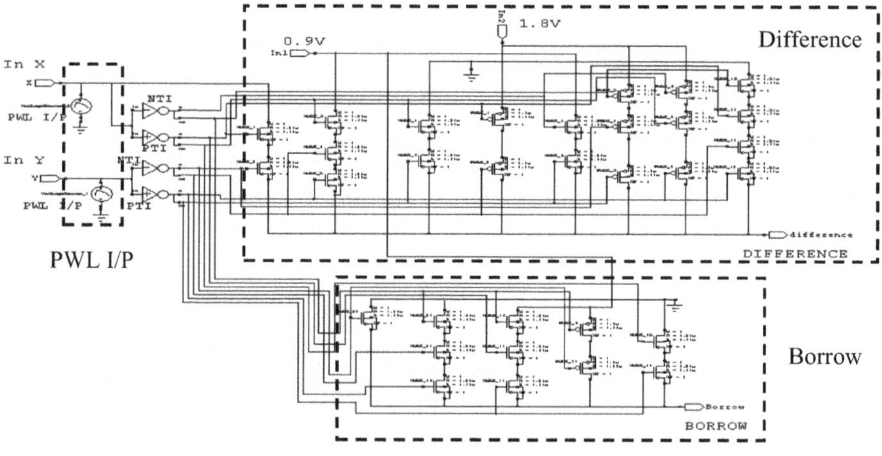

Fig. 5 Schematic design of proposed DPL-based 1-trit ternary half-subtractor

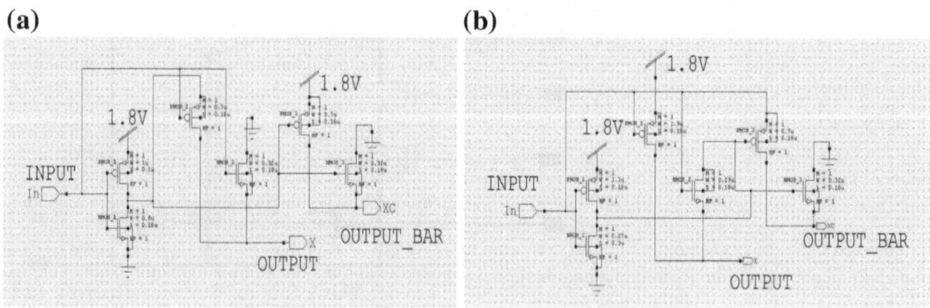

Fig. 6 Schematic design of **a** NTI **b** PTI Circuit

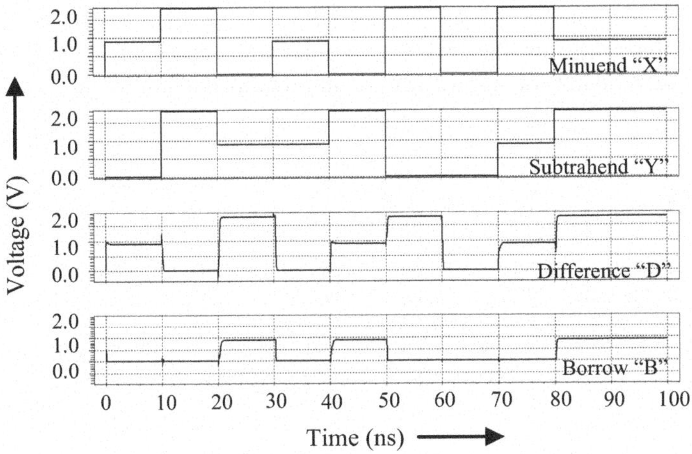

Fig. 7 Transient response of proposed 1-trit half-subtractor

Table 2 Speed–power performance of proposed subtractor

Design	Tech.	Active device count	Avg. power (μW)	Max. O/P Delay Dispersion (ps)
Proposed DPL-based 1-trit ternary half-subtractor	TSMC 0.18 μm CMOS	56	73.22	65.46

As presented in Table 2, the proposed subtractor circuit needs 56 active MOS devices to construct and dissipates 73.22 μW average power. The maximum output delay dispersion for all possible input data patterns is observed as 65.46 ps. The circuit has been designed with E-MOS devices without any threshold adjustment to reduce fabrication complexity. As per the simulation results, the proposed circuit can be favourable candidate for low-power high-speed area-efficient wave-pipelined/time-equalized system design.

4 Conclusion

The present study explores new 1-trit ternary (Radix-3) half-subtractor based on Double Pass-transistor Logic (DPL) for wave-pipelined/time-equalized applications. The balanced input capacitance property of DPL has been exploited to achieve data-independent propagation delay. The design strategy for the proposed circuit has been explained. The circuit has been designed with normal process Enhancement-type MOSFET on TSMC 0.18 μm CMOS technology with 1.8 V supply rail and at 25 °C temperature. The ternary trit values 0, 1 and 2 are coded with 0 V, 0.9 V and 1.8 V, respectively. The transient response from T-Spice simulation has been verified. As per the simulation results, the proposed circuit can be an effective choice to design low-power high-speed wave-pipelined ternary data-path element.

References

1. Thapliyal, H., Ranganathan, N.: Design of efficient reversible binary subtractors based on a new reversible gate. In: IEEE Computer Society Annual Symposium on VLSI 2009, Tampa, FL, USA, pp. 229–234, 13–15 May 2009
2. Devi, S., Suhas, N.S., Vijay, K.A.: Design of full subtractor using DPL logic and MTCMOS technique to reduce the leakage current and area. In: Second International Conference on Electrical, Computer and Communication Technologies (ICECCT) 2017, Coimbatore, India, 22–24 February 2017
3. AlokeSaha and Dipankar Pal: DPL-based novel binary-to-ternary converter on CMOS technology. AEU-Int. J. Electron. Commun. Elsevier **92**, 69–73 (2018)
4. Hang, G., Zhou, X.: Novel CMOS ternary flip-flops using double pass-transistor logic. In: IEEE International Conference on Electric Information and Control Engineering 2011, Wuhan, China, 15–17 April 2011
5. Hang, G., Zhou, X.: Novel CMOS static ternary logic using double pass-transistor logic. In: 2nd IEEE International Conference on Information Science and Engineering 2010, 4–6 December 2010
6. Vudadha, C., Surya, A., Agrawal, S., Srinivas, M.B.: Synthesis of ternary logic circuits using 2:1 multiplexers. IEEE Trans. Circ. Syst-I: Regular Papers, Article in Press, 4th June 2018
7. Vudadha, C., Srinivas, M.B.: Design of high speed and power efficient ternary prefix adders using CNFETs. IEEE Trans. Nanotechnol. **17**(4), 772–782 (2018)
8. Vudadha, C.K., Srinivas, M.B.: Design methodologies for ternary logic circuits. In: 48th IEEE International Symposium on Multiple-Valued Logic 2018, Linz, Austria, pp. 192–197, 16–18 May 2018
9. Shahrom, E., Hosseini, S.A.: A new low power multiplexer based ternary multiplier using CNTFETs. AEU-Int. J. Electron. Commun. Elsevier, **93**, 191–207 (2018)
10. Sahoo, S.K., Akhilesh, G., Sahoo, R., Muglikar, M.: High performance ternary adder using CNTFET. IEEE Trans. Nanotechnol. **16**(3), 368–374 (2017)
11. Kang, Y., Kim, J., Kim, S., Shin, S., Jang, E.S., Jeong, J.W., Kim, K.R., Kang, S.: A novel ternary multiplier based on ternary CMOS compact model. In: 47th IEEE International Symposium on Multiple-Valued Logic 2017, pp. 25–30, Novi Sad, Serbia, 22–24 May 2017
12. Srinivasu, B., Sridharan, K.: A synthesis methodology for ternary logic circuits in emerging device technologies. IEEE Trans. Circ. Syst.-I Regular Papers **64**(8), 2146–2159 (2017)
13. Burleson, W.P., Ciesielski, M., Klass, F., Liu, W.: Wave-pipelining: a tutorial and research survey. IEEE Trans Very Large Scale Integr. (VLSI) Syst. **6**(3), 464–474 (1998)
14. Saha, A., Pal, D., Chandra, M.: Low power 6-GHz wave-pipelined 8b $\times$ 8b multiplier. IET Circ. Devices Syst. **7**(3), 124–140 (2013)
15. Saha, A., Pal, D., Chandra, M.: Benchmarking of DPL based 8b $\times$ 8b novel wave-pipelined multiplier. Int. J. Electron. Lett. (IJEL), Taylor & Francis, **5**(1), 115–128 (2017)
16. Rajesh Parthasarathy, R.S., Sridhar, R.: Double pass-transistor logic for high performance wave pipeline circuits. In: Eleventh International Conference on VLSI Design: VLSI for Signal Processing, Chennai, India, pp. 495–500 (1998)

On-Chip Passive Component Optimization for RF Applications

Magnanil Goswami

Abstract Being one of the most widely used on-chip passive components, loop inductors are imperative to radio frequency (RF) applications. Optimum design of on-chip planar loop inductor for floating mode operation is presented here using a prompt and efficient semi-empirical optimization technique. Respective performance metrics comprising of physical parameters are formulated in terms of objective and constraint functions to determine globally optimal solution to this design instance. Additionally, sensitivity and trade-off analyses are also carried out toward a better insight.

Keywords Constraint · Objective · Optimization · Planar loop inductor · Semi-empirical

1 Introduction

In contrast to digital circuits which use mainly active devices, on-chip passive components are imperative adjuncts to most RF circuits [1, 2]. These components which include inductors, capacitors, resistors, etc. are well known to be cost-limiting elements in RF-integrated circuits (ICs). While these components can be realized using CMOS technology, their specific designs necessitate special consideration due to the requirement of high quality factor at relatively higher frequencies.

For low-frequency applications, passive devices can be connected externally, but as the frequency level increases, the characteristics of the passive devices would get overwhelmed by parasitic effects [3]. Consequently, on-chip passive components are preferred for RF applications. On-chip planar loop inductors in RFICs are pivotal for filtering and tuning purposes. It is due to the following reasons; planar loop inductors are the most widely used type of on-chip inductors: (i) better immunity

M. Goswami (✉)
Accendere Knowledge Management Services Pvt. Ltd., CL Educate Ltd., New Delhi 110044, India
e-mail: magnanil.goswami@accendere.co.in

S. Kundu et al. (eds.), *Proceedings of the 2nd International Conference on Communication, Devices and Computing*, Lecture Notes in Electrical Engineering 602, https://doi.org/10.1007/978-981-15-0829-5_19

to conduction losses due to minimized substrate coupling, (ii) superior shielding to substrate effects, (iii) known current return path, etc.

However, the parasitic affected planar loop inductor design space is packed with trade-offs and the existing 3-D field solver-based stochastic optimization process [4] is computationally inefficient, and hence not ideally suited for practical inductor design.

Thus, a prompt and efficient semi-empirical approach is presented herein to design on-chip planar loop inductors for use in various RFICs. This approach is based on a lucid and widely accepted inductor model [5], where expressions are in line with the proposed semi-empirical method.

After casting primary information on the semi-empirical optimization technique in Sect. 2, design of planar loop inductor-based differential resonator is formulated in Sect. 3. Interpretation of optimum performance metrics is carried out in Sect. 4. Finally, inferences based on envisaged results are arbitrated in Sect. 5.

2 Semi-empirical Optimization

The semi-empirical technique used here to carry out the optimization of on-chip planar loop inductor is orthogonal convex optimization [6]. This method, unlike classical or knowledge based or other global optimization techniques, can determine the veritable best design solution for a given set of mutually congruent design specifications. Due to the inherent nature of convex functions, this method is prompt and capable of catering vital information on sensitivity and design trade-offs, with least oversight from RFIC designers.

Orthogonal convex optimization technique is a special bracket of semi-empirical optimization, where the basic idea of modeling any practical problem starts with the formulation of design objective and constraints. Although successful formulation of each and every design aspect is not guaranteed, any duly modeled practical problem can be solved with unmatched efficiency.

Objective and constraint functions can be monomial or posynomial or positive fractional power or pointwise maximum of posynomials [6]. Standard form of such a semi-empirical optimization problem is given by

$$\begin{aligned} &\text{optimize} && f_0(x) \\ &\text{subject to} && g_i(x) = 1,\ i = (1, \ldots, p) \\ &&& f_i(x) \leq 1,\ i = (1, \ldots, m) \\ &&& x_i > 0, \quad i = (1, \ldots, n) \end{aligned} \tag{1}$$

where $g_1, \ldots, g_p$ are monomial and $f_1, \ldots, f_m$ are posynomial constraints of vector x comprising of n real positive variables. Objective function f_0 is either a posynomial to minimize or a monomial to maximize/minimize.

$g(x)$: $\mathbf{R^n} \rightarrow \mathbf{R}$ is said to be a monomial function or simply a monomial [6] if its domain is the set of vectors with positive components and its values are given by the following power law expression:

$$g(x) = cx_1^{a_1} \ldots x_n^{a_n} \tag{2}$$

where $c > 0$ is the coefficient and $a = a_1, \ldots, a_n$ is the exponent of the monomial.

$f(x)$: $\mathbf{R^n} \rightarrow \mathbf{R}$ is said to be a posynomial function or simply a posynomial [6] if its domain is the set of vectors with positive components and its values take the form of nonnegative sum of monomials

$$f(x) = \sum_{k=1}^{K} c_k g_k(x) \tag{3}$$

where $g_k(x)$ are monomials and $c_k \geq 0$ for $k = 1, \ldots, K$.

To overcome the non-convexity of monomial and posynomial functions, they are transformed into affine and convex functions, respectively, by introducing a new set of variables $y_i = \log x_i$, in lieu of x_i. Thus, Eq. (1) can be reiterated as

$$\begin{aligned} &\text{optimize} \quad f_0(y) = \log\left(\sum_{k=1}^{K_0} e^{a_{0k}^T y + b_{0k}}\right) \\ &\text{subject to } g_i(y) = a_i^T y + b_i = 0, \qquad i = (1, \ldots, p) \\ &\qquad f_i(y) = \log\left(\sum_{k=1}^{K_0} e^{a_{ik}^T y + b_{ik}}\right) \leq 0, \; i = (1, \ldots, m) \end{aligned} \tag{4}$$

From the parlance of electronic device and circuit optimization, this semi-empirical technique conforms to the generic flowchart depicted in Fig. 1.

According to Fig. 1, a set of device and circuit equations describing the electronic system are considered at the onset. Then a group of operational scenarios describing the thresholds or ranges over which these device and circuit equations remain valid are imposed on these equations as scenario-specific constraints. Based on the type of semiconductor and technology node, some scenario-independent variables related to the device physics are also introduced.

The aforesaid formulations are followed by a concurrent evaluation of the available set of equations over the entire range of user-defined constraints to determine the globally optimum solution (if any) for the design problem; otherwise, infeasibility is reported unambiguously.

3 Design Formulation

Due to the absence of ground reference, resonators with differential inputs are more immune to background electrical noise compared to single-ended input operations. The planar loop inductor shunted with a load capacitance C_L and a load resistance

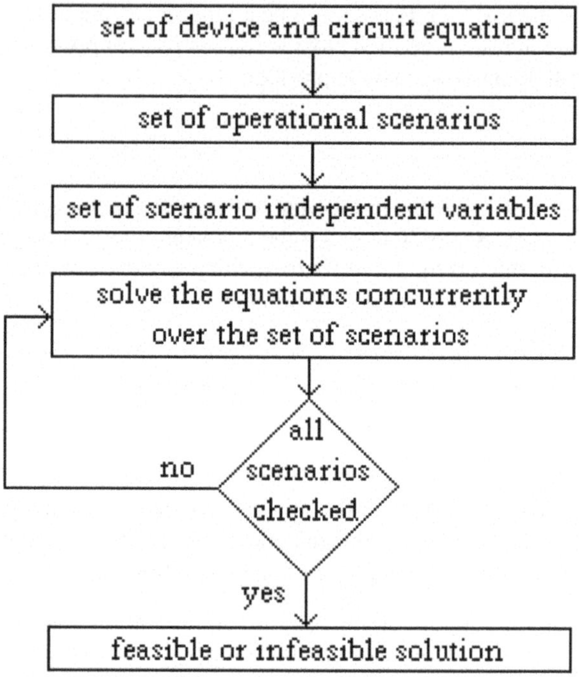

Fig. 1 Generic flowchart of the semi-empirical optimization technique

R_L forms the resonator circuit for floating (differential) mode operation, designed to resonate at operating frequency *f*. Schematic of the resonator is illustrated in Fig. 2.

Using the concepts of semi-empirical model parameter estimation [7] and simple expression for planar inductances [8, 9], compatible expressions for the lumped model of the planar loop inductor with a centerline diameter *d* and breadth *b* can be obtained with typical errors not exceeding 3%.

$$\begin{aligned} L_i &= 2.1 \times 10^{-6} d^{1.28} b^{-0.25} f^{-0.01} \\ R_s &= 0.1\tfrac{d}{b} + 3 \times 10^{-6} d b^{-0.84} f^{0.5} + 5 \times 10^{-9} d b^{-0.76} f^{0.75} + 0.02 dbf \\ C_e &= 1 \times 10^{-11} d + 5 \times 10^{-6} db \end{aligned} \quad (5)$$

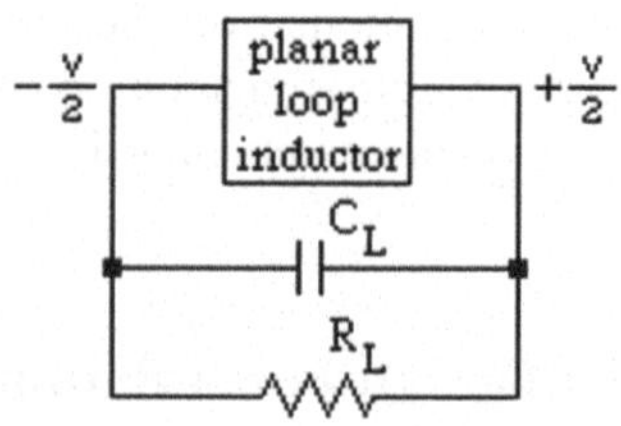

Fig. 2 Schematic of planar loop inductor-based resonator

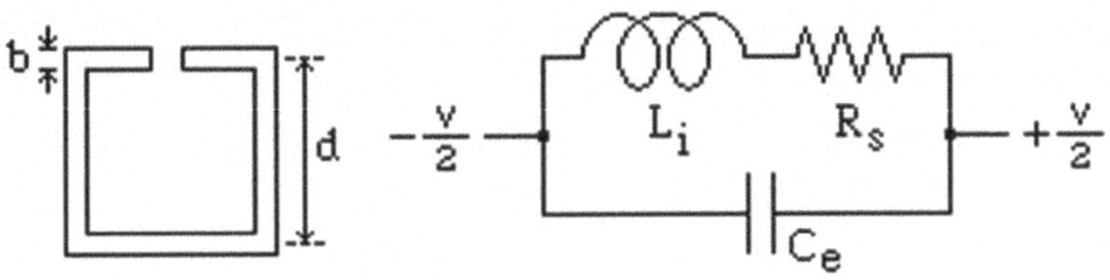

Fig. 3 Planar loop inductor and its lumped model

where L_i is inductance, R_s is series resistance, and C_e is effective capacitance of the inductor lumped model. Loss incurred from substrate capacitance is included in R_s and C_e. The planar loop inductor and its lumped model are shown in Fig. 3.

Area of the planar loop inductor A_{pli} can be formulated as

$$A_{pli} = (d + b)^2 \tag{6}$$

Total capacitance C_T of the resonator circuit is given by

$$C_T = C_e + C_L \tag{7}$$

Resonance condition for the resonator circuit is as follows:

$$4\pi^2 f^2 L_i C_T = 1 \tag{8}$$

Resulting RLC tank conductance g_t of the lumped model is

$$g_t = \frac{R_s}{4\pi^2 f^2 L_i^2} + \frac{1}{R_L} \tag{9}$$

Formally, the ratio of energy stored in the magnetic field to energy dissipated in one oscillation cycle is coined as the quality factor Q_t, which in turn is an important figure of merit for any inductor.

In case of real inductors, energy stored in the electric field due to parasitic capacitances is a loss. Hence, Q_t is proportional to the difference between the peak magnetic and electric energy. Q_t is zero at self-resonance frequency, when the peak magnetic and electric energies are equal. Also, no net magnetic energy from the inductor is available above self-resonance.

Inverse of quality factor of the RLC tank $Q_{t\text{-}inv}$ is formulated as

$$Q_{t-inv} = \frac{R_S}{2\pi f L_i} + \frac{2\pi f L_i}{R_L} \tag{10}$$

Using Eqs. (5)–(10) for the objective and constraint formulations of the semi-empirical design problem of planar loop inductor-based differential resonator circuit optimization is summarized as follows:

$$\begin{aligned}&\text{minimize } g_t\\&\text{subject to } Q_{t-inv} \leq \frac{1}{Q_{t-inv_{min}}}, \quad \left(\tfrac{d}{b}\right)_{min} \leq \tfrac{d}{b} \leq \left(\tfrac{d}{b}\right)_{max},\\&\quad d_{min} \leq d \leq d_{max}, \quad b_{min} \leq b \leq b_{max}, \quad A_{pli} \leq A_{pli-max},\\&\quad 4\pi^2 f^2 L_i C_T = 1, \quad f\text{min} \leq f \leq f\text{max}\end{aligned} \tag{11}$$

4 Result Interpretation

The optimization problem is formulated and implemented on MATLAB using ggplab toolbox. Specifications and constraints are listed in Table 1.

Optimal solution for the design instance in Eq. (11) is outlined in Table 2.

Optimum values for the lumped model of the planar loop inductor are computed from Eq. (5) by using the optimal design parameters from Table 2. Respective lumped model parameters are enlisted in Table 3.

Sensitivity of the normalized values of the area of planar loop inductor with respect to the normalized operating frequency is plotted in Fig. 4. Since the objective of this

Table 1 Constraint and specification for planar loop inductor-based resonator

Constraint	Specification
Operating frequency (GHz)	$2 \leq f \leq 6$
Load capacitance (fF)	500
Load resistance (kΩ)	1
Inductor diameter (μm)	$150 \leq d \leq 600$
Inductor breadth (μm)	$4 \leq b \leq 50$
Diameter-to-breadth ratio (–)	$10 \leq d/b \leq 100$
Inductor area (m^2)	$\leq 0.40 \times 10^{-6}$
RLC tank quality factor (–)	≥ 5
RLC tank conductance (S)	*Minimize*

Table 2 Optimal solution for planar loop inductor-based resonator

Parameter	Specification
Operating frequency (GHz)	4.9284
Inductor diameter (μm)	600
Inductor breadth (μm)	23.987
Diameter-to-breadth ratio (–)	25.0133
Inductor area (m^2)	0.38936×10^{-6}
Total capacitance of differential resonator (fF)	577.962
RLC tank quality factor (–)	6.8278
RLC tank conductance (S)	0.0026

Table 3 Optimum values for planar loop inductor lumped model

Constraint	Specification
Operating frequency (nH)	4.9284
Total capacitance of differential resonator (Ω)	577.962
RLC tank conductance (fF)	0.0026

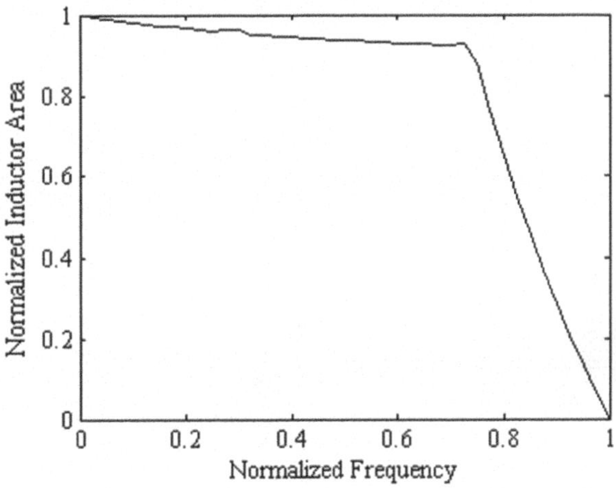

Fig. 4 Interdependence between inductor area and operating frequency

graph is to showcase the interdependence between design variables, the axes are interchangeable.

Due to compelling reasons in practical optimization instances, design variables or constraints are not really set in stone. Therefore, in order to interpret the effect of change of constraints on optimal values of design objectives, the need for trade-off analyses becomes inevitable.

In line with that, the consequences of variations in operating frequency and inductor area on normalized values of quality factor and impedance of the differential resonator are illustrated in Figs. 5, 6, 7, and 8.

These graphs are imperative measures in analyzing the effect of variation of these parameters, viz., inductor area, operating frequency on the overall performance of the resonator circuit for a trade-off packed design space bound by user-defined objective and constraints.

Figure 9 shows absolute error distribution for the lumped model of planar spiral inductor, when compared to the analytical expression of inductances computed using 3-D field solver. The graph shows that the typical errors are smaller than 3% over the entire range, which ascertains adequate level of fidelity propounded.

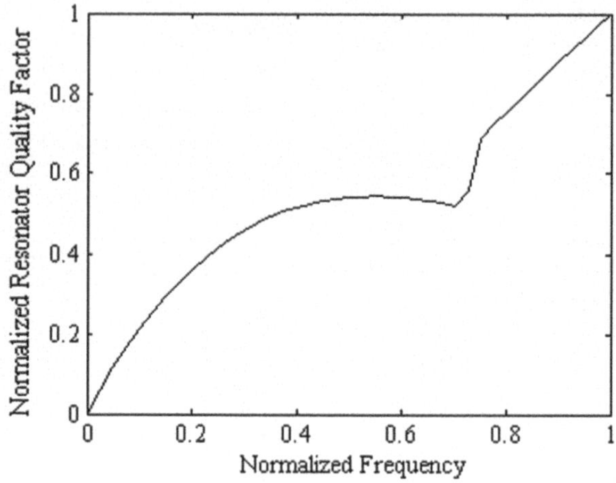

Fig. 5 Quality factor versus operating frequency trade-off

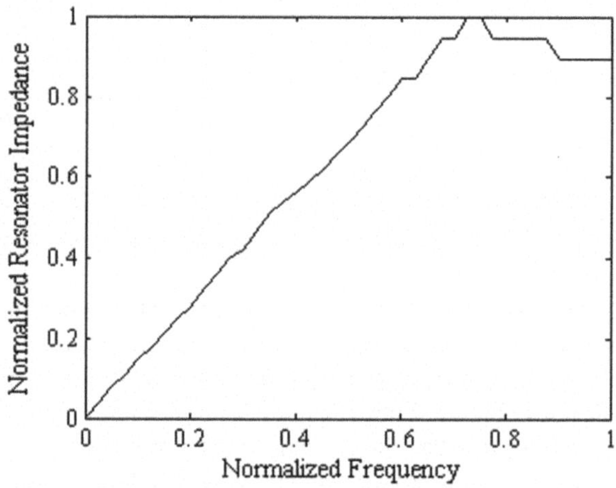

Fig. 6 Impedance versus operating frequency trade-off

5 Inference

In this discourse, a simple yet highly efficient semi-empirical optimization technique is devised to formulate and compute globally optimal design solution for an on-chip passive component (planar loop inductor)-based resonator circuit for floating mode RF applications. In connection with previous literature [7], transformation of relevant design attributes into monomial or posynomial expressions through parameter fitting is also exercised in this work.

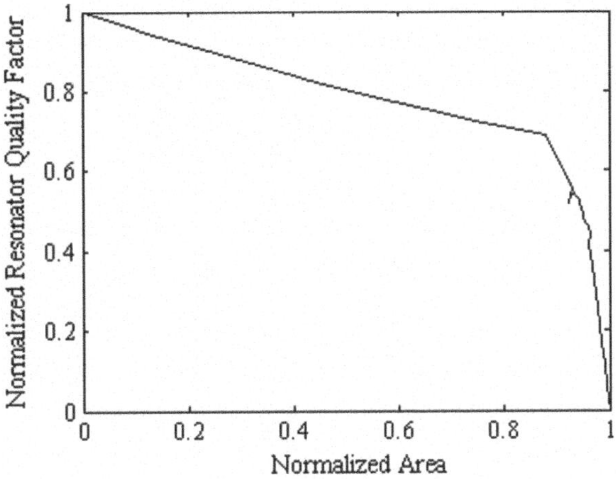

Fig. 7 Quality factor versus inductor area trade-off

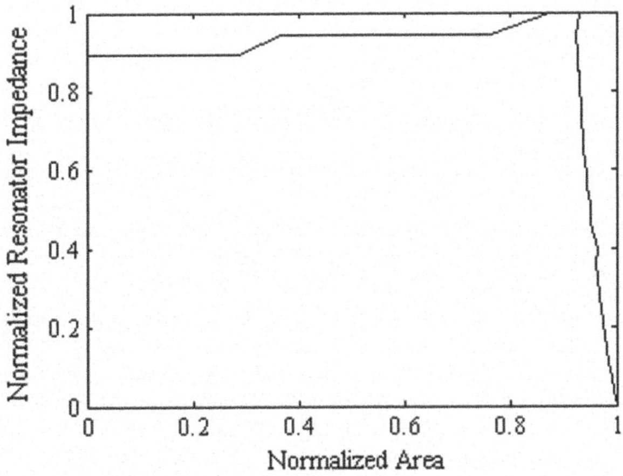

Fig. 8 Impedance versus inductor area trade-off

Apart from meeting the primary objective of determining the globally optimal design solution, prompt exploration of the design space through sensitivity and trade-off analyses is also exhibited in this discourse. Fidelity assessment of the semi-empirical lumped model of the on-chip planar spiral inductor against the inductor analytical expression reflects proximal conformity.

Based on the results obtained, it can be inferred straightaway that the proposed semi-empirical technique is capacious of optimizing on-chip passive components for RF applications with industry acceptable standard of accuracy. Exploration of more complicated RFIC design problems pertaining to the present context remains as the objective for subsequent endeavors.

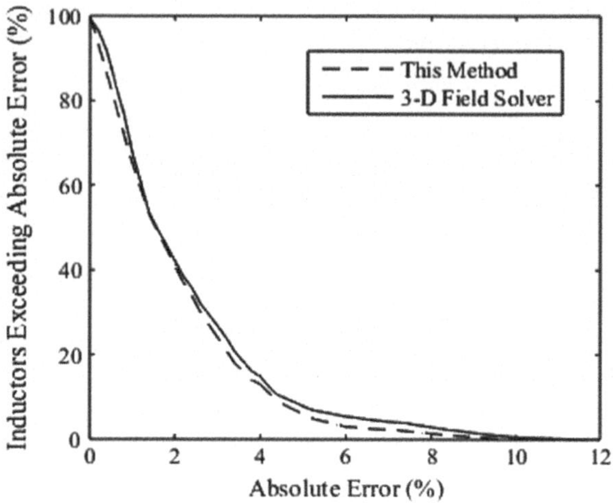

Fig. 9 Semi-empirical expression versus 3-D field solver simulation

References

1. Lee, T.H., Wong, S.S.: CMOS RF integrated circuits at 5 GHz and beyond. Proc. IEEE **88**, 1560–1571 (2000)
2. Lee, T.H.: The Design of CMOS Radio-Frequency Integrated Circuits, 2nd edn. Cambridge University Press, U.K. (2003)
3. Niknejad, A.M., Meyer, R.G.: Design, Simulation and Applications of Inductors and Transformers for Si RF ICs. Springer, USA (2000)
4. Qi, X., Wang, G., Yu, Z., Dutton, R.W., Young, T., Chang, N.: On-chip inductance modeling and RLC extraction of VLSI interconnects for circuit simulation. In: Proceedings of the IEEE Custom Integrated Circuits Conference, pp. 487–490
5. Yue, C.P., Ryu, C., Lau, J., Lee, T.H., Wong, S.S.: A physical model for planar spiral inductors on silicon. In: Proceedings of the IEEE Electron Devices Meeting. Technical Digest (1996)
6. Boyd, S., Vandenberghe, L.: Introduction to convex optimization with engineering applications, 7th edn. Stanford University Press, Stanford (2009)
7. Goswami, M., Kundu, S.: Design and analysis of semi-empirical model parameters for short-channel CMOS devices. Int J Soft Comput Eng 4(3):86–89
8. Mohan, S.S., Hershenson, M., Boyd, S.P., Lee, T.H.: Simple accurate expressions for planar spiral inductances. IEEE J Solid-State Circuits **34**, 1419–1424 (1999)
9. Yue, C.P., Ryu, C., Lau, J., Lee, T.H., Wong, S.S.: A physical model for planar spiral inductors on silicon. Tech. Digest IEEE IEDM 155–158 (1996)

Optimization of Subthreshold Parameters for Graded Channel Gate Stack Double Gate (GCGSDG) MOSFET Using Craziness-Based Particle Swarm Optimization Algorithm

D. Chowdhury, B. P. De, K. B. Maji, S. Ghosh, R. Kar, D. Mandal and S. Bhunia

Abstract In this article, the subthreshold parameters of Graded Channel Gate Stack Double Gate (GCGSDG) MOSFET are optimized using an evolutionary technique to achieve better electrical performances for the design of deep submicron CMOS digital circuits. The evolutionary algorithm employed for this study is Craziness-based Particle Swarm Optimization (CRPSO). PSO is a meta-heuristic technique. The main disadvantages of PSO are stagnation problem and premature convergence. CRPSO is the modified version of PSO to eliminate the drawbacks of PSO. The subthreshold parameters like the subthreshold swing and OFF-state current have been determined to formulate the overall objective/cost function. The overall cost function is obtained by the weighted sum approach method. CRPSO-based design shows much better results compared to the previous literature and may be considered for the efficient device design.

D. Chowdhury (✉)
Department of ECE, Haldia Institute of Technology (HIT), Haldia, India
e-mail: dibyendu.chow@gmail.com

B. P. De
School of Electronics Engineering, KIIT University, Bhubaneswar, India
e-mail: bishnu.ece@gmail.com

K. B. Maji · S. Ghosh · R. Kar · D. Mandal
Department of ECE, NIT Durgapur, Durgapur, India
e-mail: kbmaji@gmail.com

S. Ghosh
e-mail: tosg.nitd@gmail.com

R. Kar
e-mail: rajibkarece@gmail.com

D. Mandal
e-mail: durbadal.bittu@gmail.com

S. Bhunia
Department of ECE, Central Institute of Technology (CIT), Kokrajhar, Assam, India
e-mail: snb.hit@gmail.com

S. Kundu et al. (eds.), *Proceedings of the 2nd International Conference on Communication, Devices and Computing*, Lecture Notes in Electrical Engineering 602, https://doi.org/10.1007/978-981-15-0829-5_20

Keywords (GCGS) DG MOSFET · OFF-state current · Subthreshold swing · CRPSO · Optimization

1 Introduction

Double Gate (DG) MOSFETs have been the trendiest structure to its excellent potentiality in the reduction of the Short Channel Effects (SCEs). The different types of device engineering schemes lead to improvements of the device performance, which contributes to the rapid development of the semiconductor industry. Parihar et al. [1] evaluated the Short Channel Effects (SCEs) parameters for different high-k dielectric materials in Tri-gate Fin Field Effect Transistor (FinFET), Double Gate MOSFET (DGMOSFET), and Gate All Around (GAA) FinFET structures. Kumari et al. [2] have been proposed a drain current model assuming the source to drain regions is linked to the channel in the asymmetric DM-JL-DG transistor to reduce the SCEs. Faisal et al. [3] have been designed a GaN-based DG MOSFETs for the next generation logic applications. Medisetty and Dutta [4] demand that compound semiconductors are a better alternative to silicon semiconductors which reduce the SCEs in DG MOSFET. Wu et al. [5] also proposed an effective method to reduce the Band-To-Band Tunnelling (BTBT) effects by increasing the gate oxide thickness in both gate edges of the Double Gate (DG) JL-MOSFETs. Jin et al. [6] developed the physics models of electrical parameters for symmetrical GS DG strained-Si MOSFETs by solution-based two-dimensional Poisson's equation. Bentrcia et al. [7] proposed a continuous analytical model for Gate Stack (GS) Double Gate (DG) MOSFET to study the performances. The model based on both Poisson's equations and also continuity equations avoiding the charge-sheet approximation. Bendib and Djeffal [8] offered that the Multi-Objective Genetic Algorithm Optimization (MOGA) is the superior algorithm to search the optimized dimensions and the electrical parameters in DG MOSFETs for development the circuit design strategy. Djeffal et al. [9] proposed a 2D subthreshold behaviour model to improve the subthreshold performances and minimized the SCEs by introducing the high-k layer and graded channel in DG MOSFET. Aouaj et al. [10] also derived 2D analytical models with graded channel and stack gate for electrical parameters of the Dual-Material (DM) surrounding gate MOSFET. Goel et al. [11] have been exposed a method with a parabolic approximation to establish the potential distribution function by solving Poisson's equation in Graded Channel (GC) Dual-Material (DM) Double Gate (DG) MOSFETs. Swain et al. [12] have been investigated that the performances of GCGSDG MOSFET are varied with the variation of the oxide (high-k) thickness. Bendib et al. [13] show that applicability of MOGA in GCGSDG MOSFET to search the best input design parameters. Sharma et al. [14] have reported that the GCGSDG is the best device using high-speed switching function and the GSDMDG also the best device for amplification in comparison with various device design engineering.

In this work, the structural dimensions and subthreshold parameters of GCGSDG MOSFET (Fig. 1) are optimized by CRPSO [15–18] for deep submicron CMOS cir-

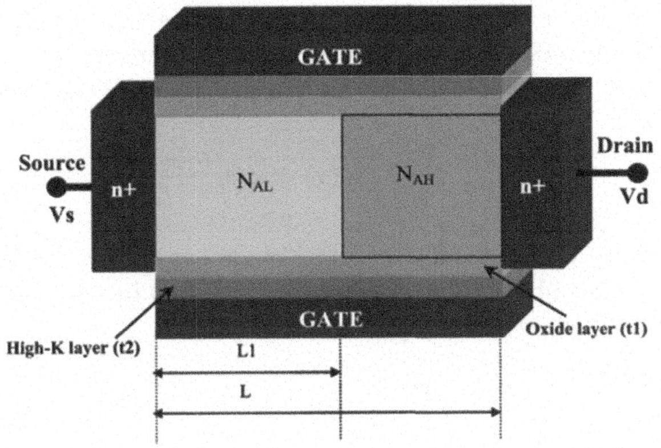

Fig. 1 A cross-sectional view of the GCGSDG MOSFET proposed an inner structure

cuits. The main contribution of this research work is the optimization of subthreshold (OFF-state current and subthreshold swing) parameters called the objective functions. All the objective functions are optimized simultaneously as a multi-objective optimization.

The article is arranged as follows: Subthreshold parameters of GCGSDG MOSFET are discussed, and the overall objective/Cost Function (CF) is formulated in Sect. 2. The CRPSO algorithm is explained in Sect. 3. Simulation results are reported in Sect. 4. Section 5 concludes the article.

2 Subthreshold Parameters for (GCGS) DG MOSFET

In this section, a compact model of subthreshold parameters for GCGSDG MOSFET is derived. The Subthreshold Swing (S) is determined from the 2D Poisson's equation [6, 9, 13] and can be represented as

$$S = S_0 \left[1 + \Delta S\right]^{-1} \tag{1}$$

where $S_0 = 60$ mV/Decade at room temperature.

ΔS is the coefficient of subthreshold swing degradation [8, 9, 13] and is given as

$$\Delta S = \frac{\left((\alpha+\beta) - \sinh\left(\frac{L_1}{\lambda}\right) - \sinh\left(\frac{L-L_1}{\lambda}\right) - \sinh\left(\frac{L}{\lambda}\right)\right)\gamma + \sinh\left(\frac{x_{\min 1}-L_1}{\lambda}\right)}{\sinh\left(L_1/\lambda\right)} \tag{2}$$

with $\alpha = \sinh\left(\frac{L_1}{\lambda}\right).\cosh\left(\frac{L_1-L}{\lambda}\right)$, $\beta = \sinh\left(\frac{L-L_1}{\lambda}\right).\cosh\left(\frac{L}{\lambda}\right)$, $\gamma = \frac{\sinh\left(\frac{x_{\min 1}}{\lambda}\right)}{\sinh\left(\frac{L}{\lambda}\right)}$, $\lambda = \sqrt{\frac{\varepsilon_{Si}.t_{oxeff}.t_{Si}}{2.\varepsilon_1}}$. Here t_1 is the SiO_2 ($\varepsilon_{ox} = \varepsilon_1$) layer thickness, t_2 is the high-k layer (ε_2) thickness, $t_{oxeff} = t_1 + \frac{\varepsilon_1}{\varepsilon_2} t_2$ is the effective oxide thickness, and $x_{\min 1}$ represent the minimum surface potential of the GCGSDG MOSFET which depends on electrical parameters and also the geometrical parameters.

In the weak inversion region, the subthreshold current density can be expressed [8, 9, 13] as

$$J_n(y) = qD_n\frac{n_{\min}(y)}{L}\left(1 - e^{-V_{ds}/V_t}\right) \tag{3}$$

where V_t denotes the thermal voltage, and D_n is the diffusion constant.

The subthreshold current can be found by integrating (3) over the entire silicon film thickness and is given as

$$I_{sub} = 2\frac{V_t}{E_s}\left(K_1\left(e^{\psi_{\min}/V_t} - e^{\psi_{s\min}/V_t}\right) + K_2\left(e^{\psi_{\min}/V_t} - e^{\psi_{s\min}/V_t}\right)\right) \tag{4}$$

where E_s denotes the constant electric field, $E_s = 2(\psi_{\min} - \psi_{s\min})/t_{si}$, $\psi_{s\min}$ is the minimum potential at (Si/SiO$_2$) interface given by $\psi_{s\min} = \psi_{\min}(x_{\min}, 0)$, $\psi_{\min}$ denotes the minimum potential, $\psi_{\min} = \psi(x_{\min}, t_{si}/2)$, and K_i is a constant defined as $K_1 = \left(q\mu_n W V_t n_i^2/L_1 N_{AL}\right)\left(1 - e^{-V_P/V_t}\right)$ and $K_2 = \left(q\mu_n W V_t n_i^2/(L - L_1) N_{AH}\right)\left(1 - e^{-(V_{ds}-V_P)/V_t}\right)$.

N_{AL} and N_{AH} denote the channel doping in a different region (Region I and Region II), respectively, with W the width of the GCGSDG MOSFET.

In the subthreshold region, the OFF-state current [8, 9, 13] is estimated by

$$I_{OFF} = I_{sub}|_{V_{gs}=0} \tag{5}$$

The multi-objective optimization problem can be formulated as follows:

Minimize I_{OFF} and ΔS

The overall cost function is obtained by 'weighted sum approach method' and is given as [13]

$$CF = w_1|\Delta S| + w_2 I_{OFF} \tag{6}$$

where w_1 and w_2 are the weight functions.

The device subthreshold swing degradation and the OFF-state current coefficient are equally significant to the device designer to get the high speed, high gain and low power dissipation. Therefore, w_1 and w_2 can be taken equal values as 1/2. The input variable vector is represented as $X = \left(t_{si}, t_1, t_2, \varepsilon_2, L, L_1, N_{AH}, N_{AL}, V_{ds}, V_g, \varphi_{MS}\right)$.

3 Craziness-Based Particle Swarm Optimization

PSO is an optimization method based on population. PSO is discussed in the different literature [15, 16]. The detail description of CRPSO algorithm is given in [17, 18].

Table 1 Best control parameters of CRPSO algorithm

Parameters	CRPSO
Population size (p)	120
The dimension of the optimization problem (q)	11
C_1	2
C_2	2
P_{cr}	0.3
$v^{craziness}$	0.0001
Iteration cycle	500

4 Simulation Results and Discussions

In this work, subthreshold parameters of GCGSDG MOSFET are optimized through the geometrical synthesis. The overall objective/cost function is optimized through CRPSO algorithm, implemented in MATLAB 7.5. The optimized dimensions (t_{si}, t_1, t_2, L, L_1) and electrical parameters $\left(\varepsilon_2, V_{ds}, V_g, \varphi_{MS}, N_{AH}, N_{AL}\right)$ are obtained from the CRPSO algorithm. The control parameters of the CRPSO algorithm are set on trial-and-error basis and are given in Table 1. For the authentication purpose, TCAD is used to design the structure of GCGSDG MOSFET for electrical simulations. To achieve the best set of design parameters, the CRPSO is run over 50 times shown in Table 2. TCAD simulation results attained from CRPSO algorithm-based optimal design of GSDG MOSFET are presented in Figs. 2, 3 and 4, respectively. CRPSO-based design results in $t_{si} = 5$ nm, $t_1 = 0.25$ nm, $t_2 = 0.30$ nm, $L = 24$ nm, $\varepsilon_2 = 25$, $\Phi_m = 4.63$ eV, $N_{AH} = 2.7 \times 10^{19}\,\text{cm}^{-3}$, $N_{AL} = 10^{18}\,\text{cm}^{-3}$, $V_{ds} = 0.5$ V, $V_{gs} = 1$ V, $I_{OFF} = 1.23 \times 10^{-11}\,\text{A}\mu m^{-1}$, $\Delta\text{S} = -6.6622 \times 10^{-4}$, $CF = 3.3311 \times 10^{-4}$. Figure 5 shows the convergence plot for CRPSO algorithm-based design of GCGSDG MOSFET. CRPSO requires 25 s to compute 60,000 parameters set with Intel Core^{TM}i5-2430M CPU@3.00 GHz. The values of on-state current (I_{ON}), the threshold voltage (V_{th}), threshold voltage roll-off ($V_{roll-off}$) and Drain-Induced Barrier Lowering (DIBL) are reported in Table 2. CRPSO shows better-optimized results as compared to MOGA [13] in terms of the device dimensions and the subthreshold parameters as shown in Table 2.

Table 2 Optimal GCGSDG MOSFET parameters

Symbol	Quantity	Random design [13]			MOGA [13]	CRPSO
		Case 1	Case 2	Case 3	Case 4	
t_{si} (nm)	Silicon thickness	8	5	10	5	5
L (nm)	Channel length	12	20	30	30	24
V_{gs} (V)	Gate voltage	0.1	0.1	0.1	0.1	1.0
V_{ds} (V)	Drain source voltage	0.5	0.5	0.5	0.5	0.5
t_1 (nm)	Thickness of the SiO_2	1.5	3	2.5	1.00	0.25
t_2 (nm)	Thickness of the high-k layer	2	1.5	2.5	1.32	0.30
Φ_m (eV)	Work function	4.33	4.33	4.33	4.33	4.63
N_{AH} (cm^{-3})	Channel doping [Region II]	10^{16}	5×10^{16}	8×10^{16}	9.28×10^{16}	2.7×10^{19}
N_{AL} (cm^{-3})	Channel doping [Region I]	10^{15}	10^{15}	2×10^{15}	5×10^{15}	10^{18}
ε_2	Permittivity of the high-k layer	40	35	20	39.58	25
L_1 (nm)	Channel length [Region I]	6	8	15	14.50	12
I_{OFF} ($A\mu m^{-1}$)	OFF-state current	3.31×10^{-3}	3.06×10^{-8}	7.01×10^{-9}	4.23×10^{-13}	1.23×10^{-11}
S (mV/dec)	Subthreshold swing	129.15	81.94	77.08	60.60	60.04
ΔS	Subthreshold swing degradation coefficient	–	–	–	–	-6.6622×10^{-4}
CF	Cost function	–	–	–	6.0495×10^{-3}	3.3311×10^{-4}
t (s)	Execution time	–	–	–	62	25
$V_{roll-off}$ (V)	Threshold voltage roll-off	−0.13	−0.036	−0.01	-1.18×10^{-5}	−0.3563
DIBL (mV/V)	Drain-induced barrier lowering	212.2	54.25	23.15	0.06	2.51
V_{th} (V)	Threshold voltage	0.45	0.46	0.53	0.54	0.58
I_{ON}/I_{OFF}	Ratio of I_{ON}/I_{OFF}	1.16	1.99×10^5	9.24×10^5	1.96×10^{10}	5.82×10^9

Fig. 2 Drain current in OFF state

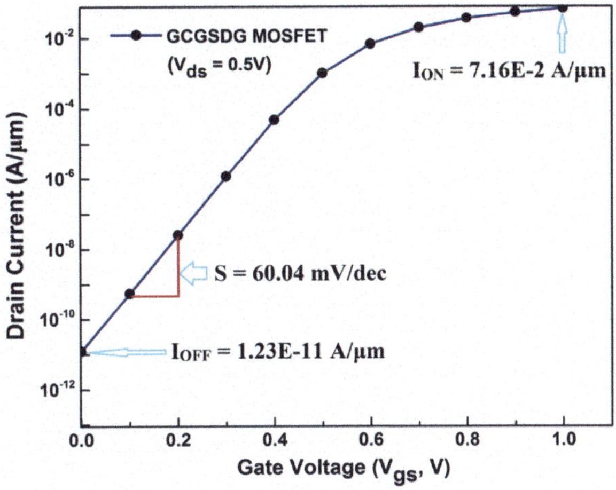

Fig. 3 The plot of drain current (I_D) versus gate voltage (V_{gs})

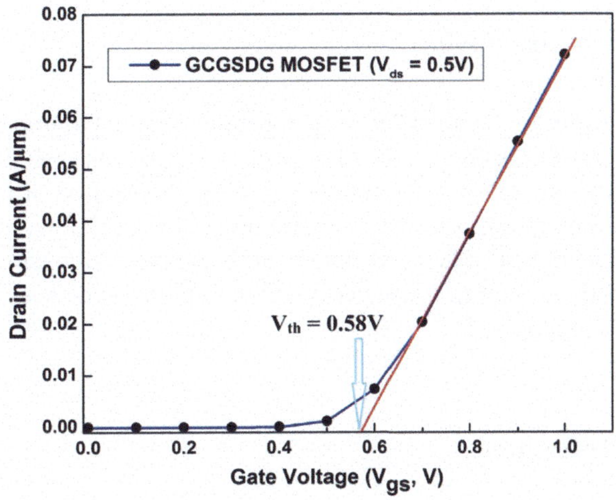

Fig. 4 The plot of I_D-ramp versus V_{gs} for GCGSGD MOSFET

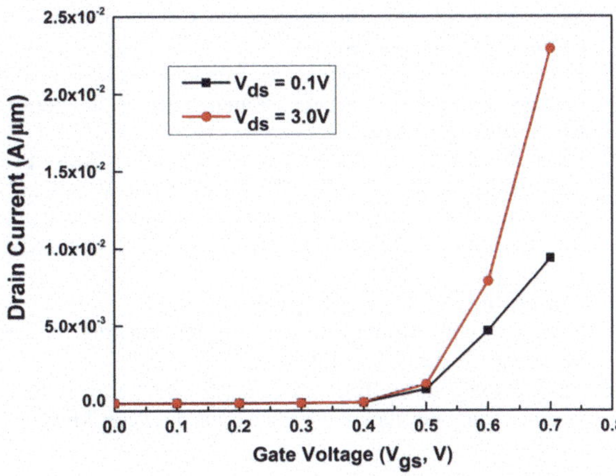

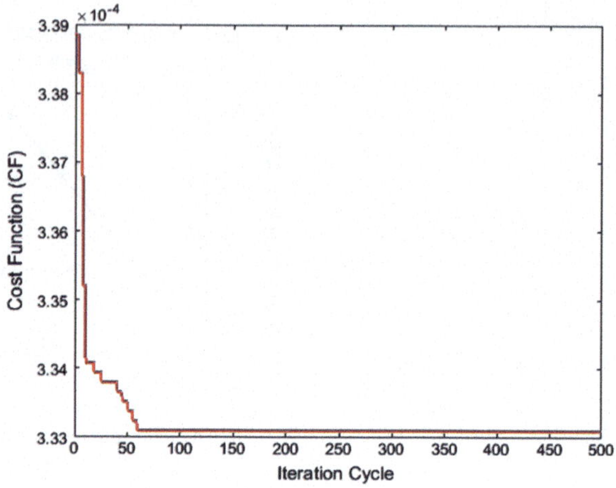

Fig. 5 The plot of convergence profile for CRPSO algorithm

5 Conclusion

In this article, the CRPSO algorithm is applied to improve the subthreshold parameters of GCGSDG MOSFET for deep submicron CMOS applications. The five objectives are taken for optimization: (1) minimization of I_{OFF}, (2) minimization of ΔS, (3) maximization of I_{ON}, (4) minimization of $V_{roll-off}$ and (5) minimization of DIBL. CRPSO has efficiently found out the optimal dimensions of the GCGSDG MOSFET and resulted in improved subthreshold performances with respect to the previous literature.

References

1. Parihar, R., Narendra, V., Mishra, R.A.: Comparative study of nanoscale FinFET structures for high-k gate dielectrics. In: International Conference on Devices, Circuits and Communications (ICDCCom), pp. 1–5 (2014)
2. Kumari, V., Modi, N., Saxena, M., Gupta, M.: Theoretical investigation of dual material junctionless double gate transistor for analog and digital performance. IEEE Trans. Electron Dev. **62**(7), 2098–2105 (2015)
3. Faisal, MdSA, Hasan, MdR, Hossain, M., Islam, M.S.: Projected performance of Sub-10 nm GaN-based double gate MOSFETs. Circ. Comput. Sci. **2**(2), 15–19 (2017)
4. Medisetty, S.J., Dutta, P.: Performance analysis of junction-less double gate MOSFET using silicon and In0.53Ga0.47As. In: International Conference on Communication and Signal Processing, pp. 1–5 (2016)
5. Wu, M., Jin, X., Kwon, H., Chuai, R., Liu, X., Lee, J.: The optimal design of junction less transistors with double-gate structure for reducing the effect of band-to-band tunnelling. J. Semicond. Technol. Sci. **13**(3), 245–251 (2013)
6. Jin, L., Hongxia, L., Bin, L., Lei, C., Bo, Y.: Study on two-dimensional analytical models for symmetrical gate stack dual gate strained silicon MOSFETs. Chin. Phys. B. **19**(10), 107302 (2010)

7. Bentrcia, T., Djeffal, F., Benhaya, A.H.: Continuous analytic IV model for GS DG MOSFETs including hot-carrier degradation effects. J. Semicond. **33**(1), 3743–3750 (2012)
8. Bendib, T., Djeffal, F.: Electrical performance optimization of nanoscale double-gate MOSFETs using multiobjective genetic algorithms. IEEE Trans. Electron Dev. **58**(11), 3743–3750 (2011)
9. Djeffal, F., Meguellati, M., Benhaya, A.: A two-dimensional analytical analysis of subthreshold behaviour to study the scaling capability of nanoscale graded channel gate stack DG MOSFETs. Phys. E **41**(11), 1872–1877 (2009)
10. Aouaj, A., Bouziane, A., Nouary, A.: Dual material gate-graded channel-gate stack (DMG-GC-Stack) surrounding gate MOSFET: analytical threshold voltage (VTH) and subthreshold swing (S) models. In: International Conference on Multimedia Computing and Systems (2011)
11. Goel, E., Kumar, S., Singh, K., Singh, B., Kumar, M., Jit, S.: 2-D analytical modeling of threshold voltage for graded-channel dual-material double-gate MOSFETs. IEEE Trans. Electron Dev. **63**(3), 966–973 (2016)
12. Swain, S.K., Adak, S., Pati, S.K., Sarkar, C.K.: Influence of channel length and high-K oxide thickness on subthreshold analog/RF performance of graded channel and gate stack DG-MOSFETs. Microelectron. Reliab. **61**, 24–29 (2016)
13. Bendib, T., Djeffal, F., Arar, D.: Subthreshold behaviour optimization of nanoscale graded channel gate stack double gate (GCGSDG) MOSFET using multi-objective genetic algorithms. J. Comput. Electron. **10**(1–2), 210–215 (2011)
14. Sharma, R.K., Gupta, M., Gupta, R.S.: TCAD assessment of device design technologies for enhanced performance of nanoscale DG MOSFET. IEEE Trans. Electron Dev. **58**(9), 2936–2943 (2011)
15. Kennedy, J., Eberhart, R.: Particle swarm optimization. In: Proceedings of IEEE International Conference On Neural Network, vol. 4, pp. 1942–1948 (1995)
16. De, B.P., Kar, R., Mandal, D., Ghoshal, S.P.: Design of symmetric switching CMOS inverter using PSOCFIWA. In: Procedings of IEEE ICCSP, pp. 1818–1824 (2014)
17. De, B.P., Kar, R., Mandal, D., Ghoshal, S.P.: Optimal analog active filter design using craziness based particle swarm optimization algorithm. Int. J. Numer. Model. Electron. Netw. Dev. Fields **28**(5), 593–609 (2015)
18. De, B.P., Kar, R., Mandal, D., Ghoshal, S.P.: Optimal high-speed CMOS inverter design using craziness based particle swarm optimization algorithm. Open Eng. **5**(1), 256–273 (2015)

Review of Single-Mode Fiber-Based Polarization-Sensitive OCT

Kisalaya Chakrabarti and Kushal Roy

Abstract Optical coherence tomography has wide applications in various fields ranging from biomedical imaging to detection of cracks in metallic object. Birefringence plays a vital role as the flow of light wave is affected by the characteristic of the material or sample under study. If the orientation of the optical crystals is aligned in the same direction in space, we cannot extract any information of the sample. In order to get complete structural information of the sample, polarization-sensitive OCT is the only choice where various intricacies of the sample body are exposed. In polarization-sensitive OCT, two different orthogonal polarized states of lights in the Stoke's vector domain emitted from a broadband laser source are impinged on the sample and the interference pattern observed in the detector arm which eradicates the problem of the orientation of light in any direction to the principal axis orientation of the sample to be examined.

Keywords PS-OCT · Time domain · Spectral domain · Stoke's vector analysis

1 Introduction

The history of OCT goes long back with the invention of interferometers when Sir Isaac Newton observed interference rings but it was introduced in medical imaging two decades ago. This is a technique of imaging an object with high resolution up to three dimensions to a depth of 2 mm [1]. This technique is noninvasive and matches with the scale of human tissue and so it is extensively used in various fields of medical investigations from retinal to cardiac imaging [2].

K. Chakrabarti (✉) · K. Roy
Department of Electronics and Communication Engineering, Haldia Institute of Technology, Haldia, W.B, India
e-mail: kisalayac@gmail.com

S. Kundu et al. (eds.), *Proceedings of the 2nd International Conference on Communication, Devices and Computing*, Lecture Notes in Electrical Engineering 602, https://doi.org/10.1007/978-981-15-0829-5_21

1.1 OCT

It is analogous to ultrasound imaging but it uses light instead of sound. The light has a frequency in the order of 10^{14}. It is based on a technique called white light interferometry, first used by Michelson where two beams of coherent light interfere. In this kind of interferometer, light emitted from a source is divided into two parts by a beam splitter, one for the sample and other for the reference arm which has a Galvo mirror and again it is mixed in the beam splitter which is detected by detector.

The depth resolution is given by the coherent length l_c which is inversely proportional to its spectral bandwidth $l_c = \frac{2c\ln 2}{\pi}\frac{1}{\Delta\nu} = \frac{2\ln 2}{\pi}\frac{\lambda_0^2}{\Delta\lambda} \approx 0.44\frac{\lambda_0^2}{\Delta\lambda}$ where $\Delta\lambda$ is the 3 dB bandwidth.

1.2 Time-Domain PS-OCT

In time-domain OCT, as is seen in Fig. 1, a rapid scanning optical delay (RSOD) line is connected in the reference arm by which the coherence length l_c necessary for the interferometers is required and is manually adjusted so that we can find interference. Then the interference pattern is detected by the two detectors. Saxer extended this concept for the imaging of the in vivo human forearm skin by polarization-sensitive OCT or PS-OCT [3]. In all PS-OCT, the main important component is a polarization modulator. It is used in the source arm after the broadband source followed by an

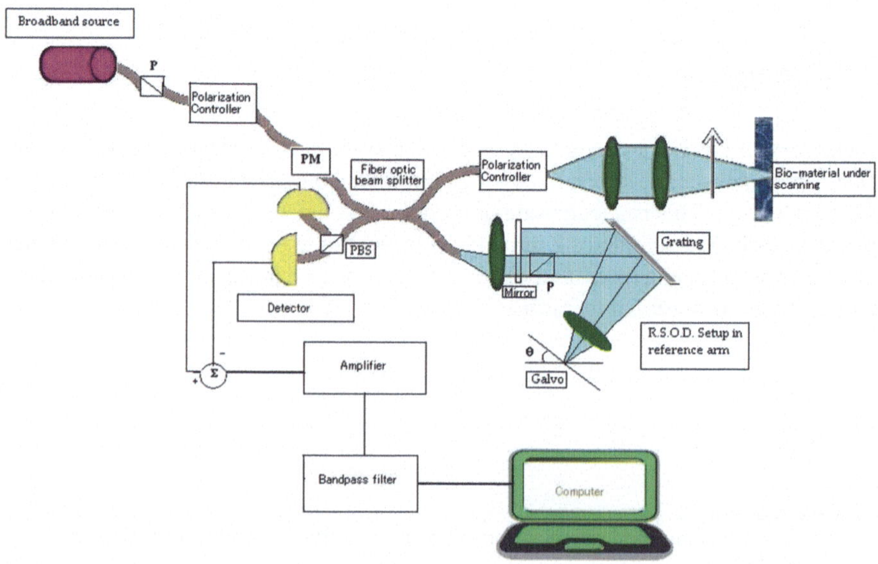

Fig. 1 Time-domain PS-OCT

isolator to counter back reflections because it will superimpose with source signal and destroy the source characteristics. The main objective of having a polarization modulator is to introduce different polarization states which are further used for the characterization of the sample and also in the reference arm. This is done to avoid the alignment of light to the axis of the orientation of the material axis. The whole idea is that if we fail to construct image by one type of polarization, the other type of polarization in the Stokes vector domain will work because it can counteract birefringence property of the sample. For this purpose, Saxer et al. used four polarization states that were orthogonal in the Poincare sphere representation. It was further extended for in vivo burn depth determination by high-speed fiber-based polarization-sensitive OCT by Hyle Park et al. [4]. He investigated the denatured collagen in skin induced by thermal injury through PS-OCT where the reduction of collagen birefringence using depth resolved changes in polarization state of the polarization state of the light which is reflected back. He used a high-speed fiber-based PS-OCT system on rat skin burned for varying period of time and for that purpose he took seven Sprague-Dawley rats and he introduced burns of 5, 20, and 30 s duration on the exposed skin on each rat using a brass rod heated to 750 °C in water bath. Park et al. as anticipated like Saxer et al. took four states of incident polarization light on Poincare's sphere which can be grouped into two pairs of polarization states comprising π radians apart or differ by a negative sign in the Stoke's vector representations. Now he averaged the intensities of the light within one pair in order to reduce noise. After averaging, he got two light intensities having two different polarization states in the Poincare's sphere. Let these be termed as I_1 and I_2. Now, after reflections from the two different depths, he received two intensities as I_1' and I_2'. The optic axis plane can be determined by I_1 and I_1' which lies between the cross product and the sum of these two polarization states. Similarly, another optic axis plane is determined by I_2 and I_2'. The intersection of these two planes gives the overall optic axis "A" of the material. Now, for degree of phase retardation calculation over the optic axis "A", Park denoted θ_1 is the degree of rotation depicted in (1), which rotates I_1 to I_1' and θ_2 in the same fashion. Due to some internal noise caused by the inhomogeneity of the sample material, these two angles are not equal, so the overall phase retardation is given by intensity weighted average of the angles as [4]

$$\theta = \frac{|I_1'|\theta_1 + |I_2'|\theta_2}{|I_1'| + |I_2'|} \tag{1}$$

These values are encoded in the gray scale in which black corresponds to a rotation of "0" and white corresponds to "π" radians. By this technique, he established a correlation between birefringence and actual burn depth.

de Boer and Milner [5] in their review paper further illustrated the same principle about PS-OCT. In their paper, the calculation of Stokes vectors of light reflected from turbid media is described and current biomedical imaging problems are also discussed. Nate J. Kemp et al. [6] obtained a novel method for high-sensitivity birefringence quantification by using enhanced polarization-sensitive optical coherence tomography (EPS-OCT) and introduces the polarimetric signal-to-noise ratio, a mathematical tool for analyzing speckle noise in polarimetry.

1.3 Spectral-Domain PS-OCT

The first in vivo spectral-domain OCT images were reported by Wojtkowski et al. [7]. In spectral-domain PS-OCT, the entire setup remained same as time-domain setup including the polarization modulator but in the detection end we have spectrometer setup around a polarization splitting medium comprised of polarization beam splitter in the form of Wollaston prism and line scan camera as shown in Fig. 2. Hyle Park et al. have published a paper in Optic Express [2] about real-time fiber-based multi-functional spectral-domain optical coherence tomography at 1–3 μm. They used two InGaAs line scan cameras capable of acquiring axial scans in 24.4 μs. In this paper, the authors discussed phase accuracy for both Doppler and spectral-domain PS-OCT. Birefringence artifact and the calibration procedure were also mentioned. In another paper by Barry Cense et al., a spectral-domain polarization-sensitive optical coherence tomography system that uses a spectrometer with a single line scan camera and a Wollaston prism in the detection arm is described. They claimed that the system is 60 times faster than a time-domain polarization-sensitive system. The sensitivity advantage of SD-OCT over TD-OCT is of twofolds, one is to shorten the acquisition times by 150 and another is improving the SNR ratio. In spectral-domain PS-OCT,

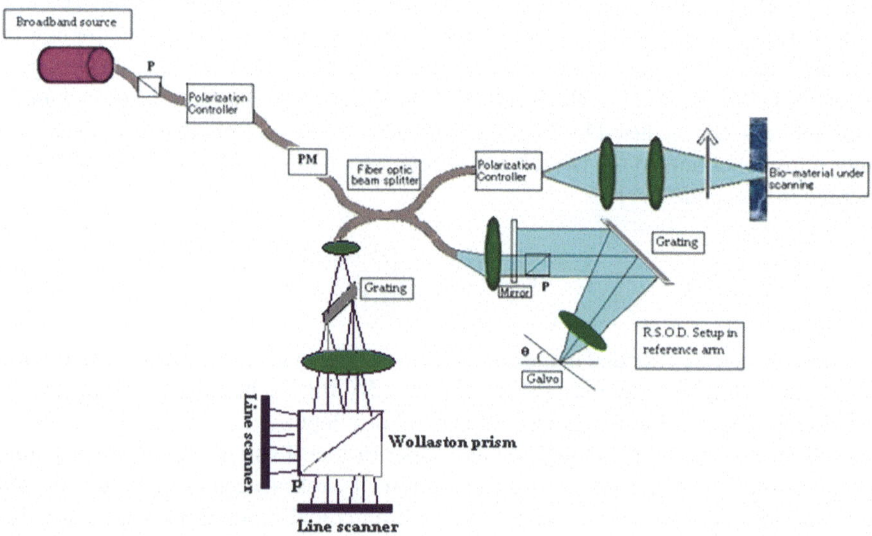

Fig. 2 Spectral-domain PS-OCT

they synchronized trigger waveforms for the line cameras and driving waveforms for the polarization modulator and fast galvanometer. In the continuation of Cense et al. work, Mujat et al. [8] published a paper on autocalibration of spectral-domain optical coherence tomography spectrometers for in vivo quantitative retinal fiber layer birefringence determination. They stated that this autocalibration method for wavelength assignment does not require separate calibration measurements and can be applied directly on the raw output data.

2 Stokes Vector Analysis

Stokes vectors play an important role in the determination of polarization states ranging from highly coherent source to unpolarized source. In the laboratory, we use quasi-monochromatic light whose eccentricity, azimuth angle, and handedness are the features by which we can characterize it. The problem encountered by measuring these parameters of unpolarized light is twofold, which is not easily measurable and the superposition principle does not hold good when two or more light waves were mixed together. In order to get rid of these problems, we use Stoke's vector.

It involves four S parameters given below:

$$S = \begin{bmatrix} S_0 \\ S_1 \\ S_2 \\ S_3 \end{bmatrix} = \begin{bmatrix} I_H + I_V \\ I_H - I_V \\ I_P + I_M \\ I_R - I_L \end{bmatrix} \tag{2}$$

where I_H and I_V are the intensities of the horizontal and vertical polarized light; I_P and I_M are the intensities of the 45° horizontal and 135° polarized light, whereas I_R and I_L are the intensities of the right circular and left circular polarized light. Therefore, the four parameters S_0, S_1, S_2, and S_3 are the total intensity, intensities of horizontal, +45° and right circular polarized light, respectively. When any of the parameters gives negative value it depicts the orthogonal polarization is predominant there. As for example, when S_1 is negative, it corresponds to polarization which is vertical rather than horizontal.

Let the electric field of a quasi-monochromatic light wave which is propagating in the z-direction and δ is the phase delay which has two components:

$$E_x(t) = E_{0x}(t)\sin\{(kz - wt) + \delta\} \tag{3}$$

$$\text{and} \quad E_y(t) = E_{0y}(t)\sin\{(kz - wt) + \delta\} \tag{4}$$

We have also the operational definitions of the Stoke's parameter given by

$$S = \begin{bmatrix} S_0 \\ S_1 \\ S_2 \\ S_3 \end{bmatrix} = \begin{bmatrix} 2I_0 \\ 2I_1 - 2I_0 \\ 2I_2 - 2I_0 \\ 2I_3 - 2I_0 \end{bmatrix} \quad (5)$$

Here I_0, I_1, I_2, and I_3 are the irradiances coming out of optical filters which are cascaded horizontally in a series and are isotropic, transparent to electric field vectors horizontally to the axis, transparent to the polarization of light at +45° with the horizontal axis of light propagation, and transparent to the right circular polarized light, respectively.

The total irradiance is given by

$$I = \varepsilon_0 c[|E_{0x}(t)|^2 + |E_{0y}(t)|^2] \quad (6)$$

As the values $E_{0x}(t)$ and $E_{0y}(t)$ slowly change with time, we can write

$$I = \frac{1}{2}\varepsilon_0 c[|E_{0x}(t)|^2 + |E_{0y}(t)|^2] \quad (7)$$

In the long time interval, one can get an average of (7) and become

$$I = \frac{1}{2}\varepsilon_0 c[\langle E_{0x}\rangle^2 + \langle E_{0y}\rangle^2] \quad (8)$$

This is nothing but $2I_0$ because both depict total irradiance. Similarly, for the second Stoke's parameter, the value of I_1 is necessary and is obtained from the irradiance detector placed in front of the horizontal electric field filter

$$I_1 = \frac{1}{2}\varepsilon_0 c[\langle E_{0x}\rangle^2 - \langle E_{0y}\rangle^2] \quad (9)$$

Then, calculating for S_1, we get

$$S_1 = \frac{1}{2}\varepsilon_0 c[\langle E_{0x}\rangle^2 - \langle E_{0y}\rangle^2] \quad (10)$$

In the same fashion, one can obtain the value of I_2 and I_3 from the irradiance detector placed on optical filters which is transparent to the electric field vector inclined +45° to the optical axis of ray propagation and right circular polarizer, respectively.

$$I_2 = \frac{1}{4}\varepsilon_0 c[\langle E_{0x}\rangle^2 + \langle E_{0y}\rangle^2 + \langle 2E_{0x}E_{0y}\cos\delta\rangle] \quad (11)$$

$$I_3 = \frac{1}{4}\varepsilon_0 c[\langle E_{0x}\rangle^2 + \langle E_{0y}\rangle^2 + \langle 2E_{0x}E_{0y}\sin\delta\rangle] \tag{12}$$

In this way, one can find that

$$S_2 = \frac{1}{2}\varepsilon_0 c\langle 2E_{0x}E_{0y}\cos\delta\rangle \tag{13}$$

and

$$S_3 = \frac{1}{2}\varepsilon_0 c\langle 2E_{0x}E_{0y}\sin\delta\rangle \tag{14}$$

Now, (5) becomes

$$S = \begin{bmatrix} S_0 \\ S_1 \\ S_2 \\ S_3 \end{bmatrix} = \frac{1}{2}\varepsilon_0 c \begin{bmatrix} \langle E_{0x}\rangle^2 + \langle E_{0y}\rangle^2 \\ \langle E_{0x}\rangle^2 - \langle E_{0y}\rangle^2 \\ \langle 2E_{0x}E_{0y}\cos\delta\rangle \\ \langle 2E_{0x}E_{0y}\sin\delta\rangle \end{bmatrix} \tag{15}$$

Now, the degree of polarization (DOP) is given by

$$P = \frac{\sqrt{S_1^2 + S_2^2 + S_3^2}}{S_0} \tag{16}$$

For a full polarized light, the value of P is 1, for the unpolarized light it is 0, and for the intermediate polarization it changes from $0 < P < 1$.

3 Spectral-Domain PS-OCT Processing

Processing of data is very important for spectral-domain data processing and for that calibration should be done in an efficient way so that data will not be lost. Mujat et al. [8] clearly explained the principle of the method for calibration of what they have undergone through it. They performed a perfect sinusoidal modulation as a function of wave vector (k) by passing the light through a microscope cover slip in the interferometer's source arm. In this cover slide, the light is reflected twice internally and created a modulated source light. Generally, for the detector arm of spectral modulator, there is a line camera where each pixel of the CCD line camera does not correspond to evenly distribute "k". The sinusoidal intensity modulation changes the wavelength assignments until it perfectly matches as a perfect sinusoid as a "k" function. Now the sinusoidal function which is generated by the cover slide of the broadband source convolved with the interference signal and subsequently bandpass filter gives rise to a perfectly sinusoid as a function of "k" in the Fourier space. So wavelength assignment is necessary (whenever there is a phase nonlinearity) which is a function of "k" space. In this way, the authors have generated a wavelength array "W" which is used to interpolate the spectral interference fringes to equally spaced "k" values. This interpolation process is improved by zero padding. Phase of this zero padded interpolated polynomial is in the order of three. This phase polynomial has a nonlinear term $\sigma(K)$ with the orders of 2 and 3 which is dependent of "k", where $K = \frac{2\pi}{W}$. Assuming the nonlinearity is caused by wrongly assigned wavelength in "W", the post-correction is needed.

A new "k" array, namely, "K'" is introduced starting from the previous value $K = \frac{2\pi}{W}$ array and $\sigma(K)$ given by $K' = K + \frac{\sigma(k)}{z(\text{peak})}$ is used for the correction of phase nonlinearity. Here, $z(\text{peak}) = 2\pi\,\text{Peak} - \text{index}/(K_{\max} - K_{\min})$. Peak index is the coherent peak to the slide index space, $K_{\max}$ and $K_{\min}$ are the two extremities in the K array. After a quite number of iterations, we find $W' = 2\pi/K'$ that agrees a linear phase as a function of K'.

In PS-OCT, the proper wavelength is much more important. In PS-OCT, the spectral interference is analyzed in two different polarization channels side by side, H (horizontal) and V (vertical). In order to avoid any kind of artifacts, the two channels are analyzed for the same depth range say "l". In this case, the authors used two microscope cover slides having two different thicknesses which will produce two coherent peaks in two different positions in "k" space. The H spectrum and K_H are kept as it is and the "k" array associated to the V spectrum, K_V is modified. Residual nonzero phase difference in between the two polarization spectrums indicates that the spectra are shifted with reference to each other. A ghost birefringence which acts as an offset to the tissue birefringence is observed due to inappropriate wavelength assignments. So Mujat et al. used two cover slides where the fringe patterns of two different polarizations are separated and are convolved together to observe best overlapping situation. This technique is used to analyze PS-OCT system.

In another paper based on PS-OCT by Cense [9], the SD-OCT system has made polarization modulator in source arm as usual but in the detection arm a Wollaston prism is introduced in between grating and lenses of the spectrometer. A high-power superluminescent diode (SLD-371-HP, Superlum, $\lambda_0 = 840$ nm, $\Delta\lambda_{FWHM} = 50$ nm) was isolated by means of broadband isolator(OFR) and at the exit of the isolator the light was linearly polarized.

With the help of the PC and dual Intel Xeon processor, a square waveform is generated and amplified which was sent to the polarization modulator which is assembled just after the isolator to prevent the back reflection. This waveform drove the polarization modulator in the frequency of 29,300 Hz which in turn fabricated two different orthogonal polarization states in a Poincare's sphere. The polarization modulator was synchronized to the line scan camera with the spectrometer in detector arm so that alternate "A" lines were acquired with two different input polarization states.

He used an 80/20 fiber coupler accomplished in the system. A "RSOD" (rapid scanning optical delay) was equipped with a polarizing beam splitter which ensures transmission of equal amount of light for both polarization states through the reference arm. Interference is observed in the detection arm and it is recorded with polarization-sensitive spectrometer. In the detection arm, light is first collimated and was diffracted with the help of transmission grating. Then, this diffracted light was channelized through a Wollaston prism made of calcite, with the separation angle of 6° which will direct both polarization states on the same line of CCD line camera with 1024 pixels each. After that, it was focused with the help of a Nikon lens (f = 105 mm) in a line scan CCD camera with 2048 pixels mounted on a 10 × 10 μm span.

Unlike M. Mujat et al., B. Cense et al. used two Fabry–Perot cavities in the source arm instead of two microscope cover slides which generated an intensity modulation of the source spectrum that is perfectly periodic in the wave vector domain. Here also as in the case of TD-PSOCT by comparing Stoke's vector at the surface of the nerve fiber layer with vectors in depth, the double pass phase retardation was calculated. It is found that the double pass phase retardation per unit depth is proportional to the birefringence (Fig. 3).

The outcomes of the heat damage are noticeably evident in the histology (purple region) and the parametric birefringence image (red region).

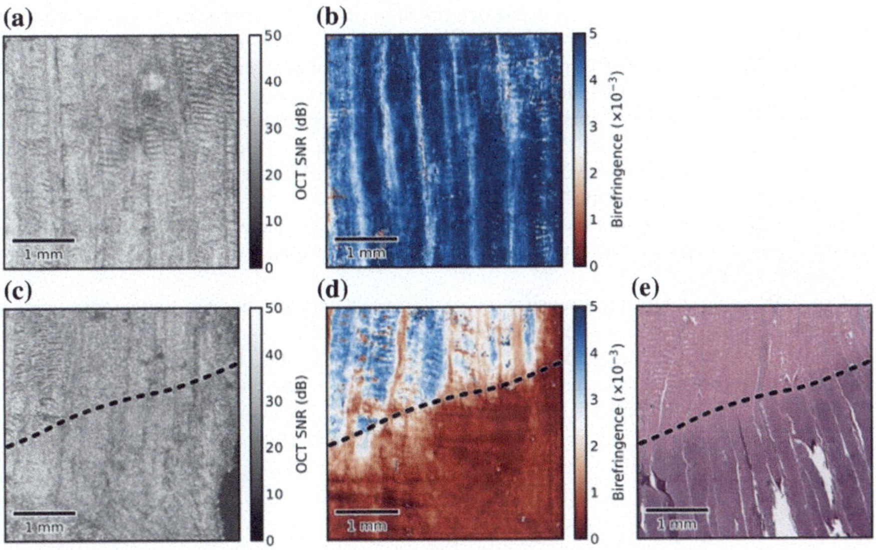

Fig. 3 Example of en face parametric birefringence imaging using PS-OCT applied to a section of heat-damaged porcine tendon (Reproduced from [10]). **a** En face OCT, **b** parametric birefringence images of the sample before heating, **c** En face OCT, **d** parametric birefringence images of the sample after the lower half of the sample was immersed into heated (69 °C) physiological Krebs solution for 5 min, **e** haematoxylin and eosin (H&E) stained histological section corresponding to the region shown in (**c**) and (**d**). The dashed lines indicate the location of the liquid–air boundary during heating

4 Conclusion

Polarization-sensitive OCT (PS-OCT) extends the perception of OCT and utilizes the information that is carried by polarized light to obtain additional information on the tissue. A number of structures in the human tissue modify the polarization state of the light and demonstrate explicitly a tissue contrast in PS-OCT images.

References

1. Huang, D., et al.: Optical coherence tomography. Science **254**, 1178–1181 (1991)
2. Park, H.B., et al.: Real-time fiber-based multi-functional spectral-domain optical coherence tomography at 1.3 μm, Opt. Express **13**(11), 3931–3944 (2005)
3. Saxer, C., et al.: High-speed fiber-based polarization-sensitive optical coherence tomography of in vivo human skin. Opt. Lett. **25**, 1355–1369 (2000)
4. Park, H.B., et al.: In vivo burn depth determination by high-speed fiber-based polarization sensitive optical coherence tomography. J. Bio-Med. Opt. **6**, 474–479 (2001)
5. de Boer, J.F., et al.: Review of polarization sensitive optical coherence tomography and stokes vector determination. J. Bio-Med. Opt. **7**, 359–371 (2002)
6. Kemp, N.J., et al.: High-sensitivity determination of birefringence in turbid media with enhanced polarization-sensitive optical coherence tomography. J. Opt. Soc. Am. A **22**(3) (2005)
7. Wojtkowski, M., et al.: In vivo human retinal imaging by Fourier domain optical coherence tomography. J. Bio-Med. Opt. **7**, 457–463 (2002)
8. Mujat, M., et al.: Autocalibration of spectral-domain optical coherence tomography spectrometers for in vivo quantitative retinal nerve fiber layer birefringence determination. J. Bio-Med. Opt. **12**, 041205 (2007)
9. Cense, B., et al.: Polarization-sensitive spectral-domain optical coherence tomography using a single line scan camera. Opt. Express **15**, 2421–2431 (2007)
10. Chin, L., Yang, X., McLaughlin, R.A., Noble, P.B., Sampson, D.D.: En face parametric imaging of tissue birefringence using polarization-sensitive optical coherence tomography. J. Biomed. Opt. **18**(6), 066005 (2013)

Performance Evaluation of Beetroot Sensitized Solar Cell Device

Debasis De, Dola Sinha and Abdul Ayaz

Abstract This paper is intended on the exploration of photo-sensitizing properties of beetroot as natural dye used in dye-sensitized solar cell (DSSC) device, where nanostructured ZnO as photoanode. The nanorod-shaped ZnO is grown on conductive oxide glass substrate by chemical technique. The beetroot (*Beta vulgaris*) contains betalain pigment of red to violet colour and can absorb visible light in the range of 500–600 nm. The beetroot dye contains anchoring groups like hydroxyl, methyl and amine which are able to attach on ZnO surface and that increases electron mobility. The fabricated DSSC device provides maximum short circuit current of 0.79 mA and open-circuit voltage of 0.475 V with cell efficiency of 0.228%.

Keywords ZnO photoanode · Beetroot sensitizer · Conductive glass substrates · Solar cell device

1 Introduction

The main motivation behind research on the dye-sensitized solar cell device is a promising and prospective renewable energy source. The cell can function well in a low-intensity light which depends on the absorption characteristics of dye sensitizer. Hence, solar cell device generates power in diffused sunlight. Natural dyes are chosen as they are very cheap and abundant in supply. So, DSSC has become cost-effective over other types of solar cells. Important component of a DSSC is dye sensitizer which is able to attach with mesoporous semiconductor metal oxide (SMO). SMO is grown on a transparent conductive oxide (TCO) based glass substrate and acts as another important part of DSSC-photoanode. The other two parts of DSSC are platinized

D. De (✉)
Energy Institute, Bangalore (a unit of R.G.I.P.T), Bangalore, Karnataka, India
e-mail: debasis.de@bcrec.ac.in; debasisd@rgipt.ac.in

D. Sinha
Department of Electrical Engineering, Dr. B. C. Roy Engineering College, Durgapur, India

A. Ayaz
Department of Electrical Engineering, Sree Ramkrishna Shilpa Vidyapith, Suri, Birbhum, India

S. Kundu et al. (eds.), *Proceedings of the 2nd International Conference on Communication, Devices and Computing*, Lecture Notes in Electrical Engineering 602, https://doi.org/10.1007/978-981-15-0829-5_22

conductive glass substrate as a counter electrode and an electrolyte. In sunlight, electrons of the dye are excited and jump into lowest unoccupied molecular orbital. These electrons finally insert into the conduction band of the SMO. The restoration of the dye occurs by electrons from electrolyte which intercepts conduction band electrons, flowing to the surface of the anode [1, 2]. All components of the DSSC lead to the efficiency of the cell. In this paper, we have grown nanostructured ZnO on florin-doped tin oxide (FTO) substrate through chemical bath deposition (CBD) [3] process. Nanosized ZnO used in the DSSC increases surface area for maximum uptake of dye molecules on the surface, which make direct flow path of electrons and increase mobility [4]. Dyes play an important role to harvest electrical energy from sunlight. Some dyes can harvest energy from UV-rays also (like *Butea monosperma*). Natural dyes can be drag out from different parts of plants, [5, 6], that may contain various kinds of pigments like anthocyanin, betacyanin, chlorophyll, etc., and can absorb light energy at different wavelengths. Different pigments consist of different chemical groups that confirm the anchoring capability of the dye with the SMO and substantiate the stability of the cell [7]. The beetroot (*Beta vulgaris*) dye composed of many anchoring groups like hydroxyl, methyl and amine [6] which can adhere firmly with SMO anode and increase the electron mobility. The paper is structured as follows: details fabrication procedure of solar cell device is explored in Sect. 2. Performance of beetroot dye as sensitizer and nanostructured ZnO as an anode in the device is evaluated. The performance of the fabricated solar cell is mentioned in Sect. 3. Conclusions are given in the end.

2 Experimental Procedure

Extraction of betanin dye sensitizer: Fresh beetroots taken from the market are washed properly in distilled water. Cleaned and peeled beetroots are kept for boiling in the water of 98 °C for 10 min to inactivate the enzymes and increase the fixation. After drying at room temperature, the beetroots are mashed into a paste using mortar. After 15 min ultrasonication at 37 Hz, the betalain dye is extracted from the paste. The dark colour of the sample was set apart using a centrifugal at 2500 rpm for 30 min. The solids residues are eliminated by mesh sieve filtration process at 13,000 rpm for 20 min in ethanolic solution and pH of prepared betalain dye solution is ~5.35.

Synthesize of nanostructured ZnO anode: FTO substrate was cleaned properly and used for photoanode of the solar cell. Nanostructured ZnO was allowed to grow slowly on FTO substrate at ~80 °C for 8 h using a chemical bath deposition (CBD) technique. The detailed CBD technique was discussed in our earlier paper [7]. The ZnO grown conducting substrate was heated at 450 °C for 2 h in air for better anchoring between ZnO and FTO. Surface morphology of ZnO is characterized by field emission scanning electron microscopy (FESEM). Light absorption and reflectance properties of natural dye anchored ZnO are analyzed by UV-Vis spectroscopy.

Fabrication of DSSC: The DSSC device was prepared with fabricated photoanode and natural betalain dye. The annealed ZnO coated FTO substrate was kept in betalain dye solution for 24 h to take up the utmost natural dye on the anode. I^-/I_3^- was taken as Redox couple which can act as an electrolyte and restore the eliminated electron from the dye. The carbon-coated FTO is used as a counter electrode (cathode) for the cell device which provides the return path of the electron. For fabricating external electrical contacts, silver paste is used. The performance of natural beetroot sensitized ZnO based solar cell device is evaluated in a solar simulator (light source of 100 mW/cm^2).

Open-circuit voltage (V_{oc}), short circuit current (I_{sc}), the voltage at maximum power point (V_{mp}) and current at maximum power point (I_{mp}) are measured from the *I–V* and *P–V* characteristics. Fill factor (*FF*) and efficiency (η) of the cell are calculated by the following Eqs. (1) and (2), respectively.

$$\text{Fill factor} = \frac{V_{mp} \times I_{mp}}{V_{oc} \times I_{sc}} \tag{1}$$

$$\text{Efficiency}(\eta) = \frac{V_{oc} \times I_{sc} \times Fill\ Factor}{P_{in}} \tag{2}$$

3 Results and Discussion

The photo-sensitizing property of the natural dye extracted from beetroot is analyzed and anchoring of dye with nanostructured ZnO anode is explored in this section. Figure 1 shows the FESEM image of ZnO deposited on FTO substrate. The figure confirms the presence of nanorod-shaped ZnO oriented randomly. The average length and diameter of the nanorods are ~4 μm and ~300 nm, respectively. All the peaks in the XRD pattern of the nanostructured ZnO coated FTO substrate are matched with standard hexagonal wurtzite ZnO structure [8].

The optical properties of beetroot were measured by UV-visible absorption spectroscopy. Figure 2 shows the absorption spectrum of the betalain dye. Broad absorption band starting from 500 to 600 nm with an absorbance maximum at $\lambda = 550$ nm is exhibited in the spectrum. This nature of spectrum is observed for red–violet betalain group. The UV–vis reflectance spectra of nanostructured ZnO and dye-loaded ZnO are investigated for the compatibility of the nanostructured ZnO with betalain dye are observed in the Fig. 2. Nanostructured ZnO has revealed full reflectance characteristics after 400 nm, but the betalain dye absorbed ZnO film shows lesser diffused reflectance, i.e., 48% reflectance due to the absorbance of light by the dye. The maximum absorption of betalain dye at 550 nm is shifted to 560 nm in dye absorbed nanostructured ZnO. The red shifting of absorption maximum is occurred due to anchoring of betalain dye with nanostructured ZnO particles. Thus, it can be concluded that ZnO and beetroot dye are chemically compatible with each other.

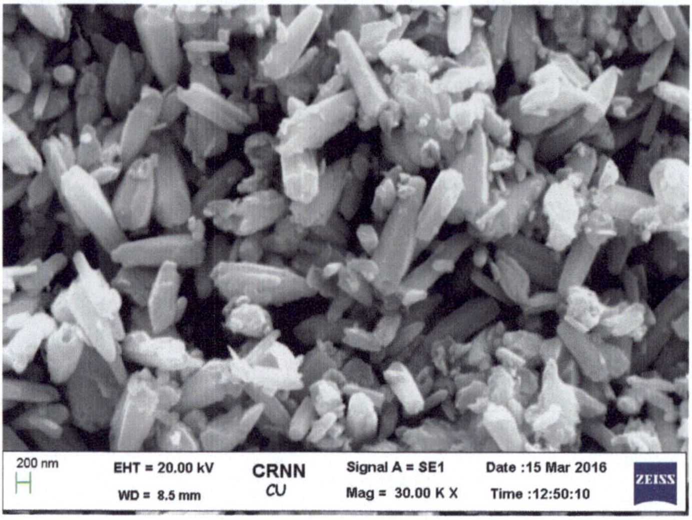

Fig. 1 FESEM image of nanostructured ZnO as photoanode

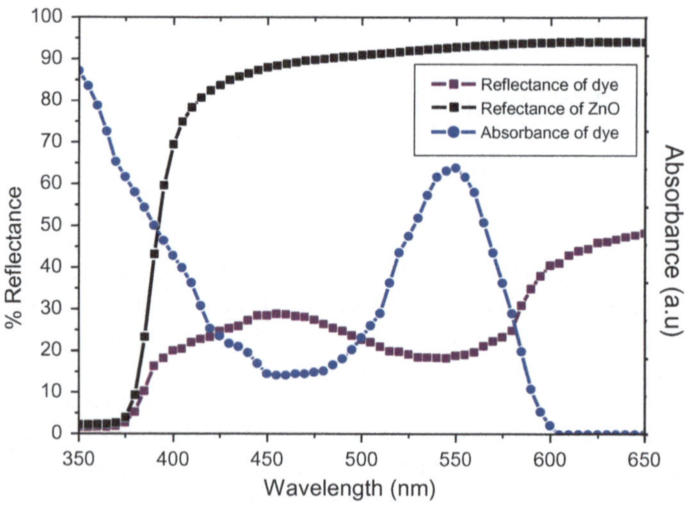

Fig. 2 UV–Vis light absorption and reflection characteristics of dye with and without ZnO

The performance of fabricated DSSC by beetroot dye is evaluated by *I–V* and *P–V* characteristics. It is shown from Fig. 3 that maximum current generated from the cell is 0.79 mA and maximum produce voltage is of 0.475 V. The calculated voltage and current at maximum power point is 0.38 V and 0.60 mA, respectively. The calculated fill factor of the device is 0.61 and light to electrical energy conversion efficiency of the device is 0.228%.

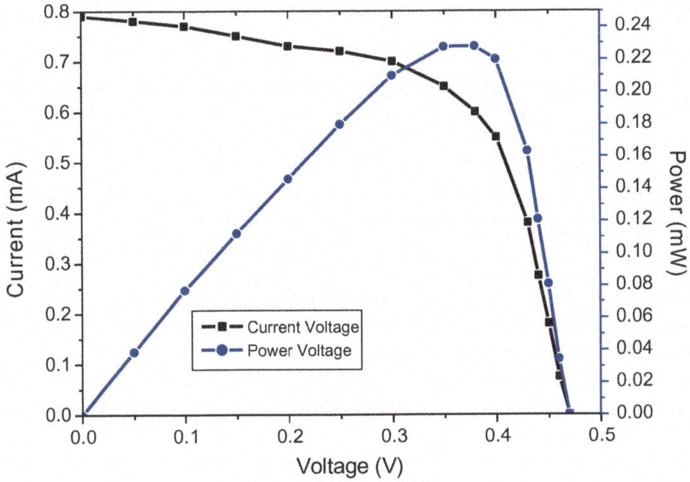

Fig. 3 *I–V* and *P–V* characteristics of the fabricated solar cell device

It is observed that efficiency is not a breakthrough. Conducting glass substrate has an effect on current generation as well as on the efficiency of the cell. It has been reported that ITO-based substrate provides less efficiency than FTO-based substrate for fabrication of DSSC with the same dye [6, 7]. FTO-based TCO is less influenced by temperature and has less sheet resistance so electron flow is high for which current generation is going high. On the other hand, the resistance of ITO will increase with higher annealing temperature. The voltage might remain same for both the cases. It is shown in literature [7] that using FTO efficiency of DSSC with turmeric enhanced 0.064% than using ITO. In this paper, efficiency enhancement technique has not been used. Efficiency can be increased by using acid treatment of dye with TiO_2 as a SMO, or metal doped ZnO/TiO_2, or using a composite SMO used as photoanode which also enhances the cost and complexity.

4 Conclusion

Dye-sensitized solar cell device is made up of ZnO nanostructure as photoanode and natural beetroot as photo-sensitizer. Nanostructured ZnO has been fabricated using CBD technique. Average length and diameter of the ZnO nanorod are 4 μm and 300 nm, respectively. Betalain is extracted from beetroot through a specific process. Natural dye has absorbed 500–600 nm broadband in the visible range. Hydroxyl, methyl and amine group present in the dye can attach firmly with ZnO anode and increase the electron mobility. The fabricated solar cell device has been achieved maximum short circuit current of 0.79 mA and open-circuit voltage of 0.475 V with a fill factor of 61% and cell efficiency of 0.228%.

References

1. Gong, J., Liang, J., Sumathy, K.: Review on dye sensitized solar cells (DSSCs): fundamental concepts and novel materials. Renew. Sustain. Energy Rev. **16**, 5848–5860 (2012)
2. Hoffmann, M.R., Martin, S.T., Choi, W., Bahnemann, D.W.: Environmental applications of semiconductor photocatalysis. Chem. Rev. **95**, 69–96 (1995)
3. De, D., Sarkar, D.K.: Superhydrophobic ZnAl double hydroxide nanostructures and ZnO films on Al and glass substrates. Mater. Chem. Phys. **185**, 195–201 (2017)
4. Afifiand, K., Tabatabaei, A.M.: Efficiency investigation of dye-sensitized solar cells based on the zinc oxide nanowires. Orient J. Chem. **30**, 155–160 (2014)
5. Goswami, D., Sinha, D., De, D.: Nanostructured ZnO and natural dye based DSSC for efficiency enhancement. In: Proceeding of IEEE 3rd International Conference on Science, Technology, Engineering and Management, pp. 556–560 (2017)
6. Sinha, D., De, D., Goswami, D., Ayaz, A.: Fabrication of DSSC with nanostructured ZnO photoanode and natural dye sensitizer. In: Proceedings of Materials Today, vol. 5, pp. 2056–2063. Elsevier (2018)
7. Sinha, D., De, D., Ayaz, A.: Performance and stability analysis of curcumin dye as a photo sensitizer used in nanostructured ZnO based DSSC. Spectrochim. Acta **193**, 467–474 (2018)
8. JCPDS# 75-1526

The Simplest Takagi–Sugeno Fuzzy Controller-Based Control Strategy for Collocated Systems

Ritu Raj

Abstract Control of multi-mode resonant systems is a challenging task. They exhibit vibrations and are lightly damped making them highly susceptible to mechanical resonance. Many inherent nonlinearities such as hysteresis and creep are exhibited by such systems. The conventional control method employed in such kind of systems is a damping controller and a tracking controller to deal with nonlinearity and resonance. The damping controller used in this paper is the Integral Resonant Controller (IRC). Integral or Proportional–Integral (PI) controllers are used as conventional tracking controller. In this paper, the simplest Takagi–Sugeno (TS) fuzzy controller is introduced for improved tracking performance with lesser vibrations in presence of hysteresis nonlinearity. The TS fuzzy controller is the simplest controller as only two input fuzzy sets are considered. Fuzzy controllers can easily handle nonlinearities due to their structure.

Keywords TS fuzzy controller · Vibration · IRC

1 Introduction

Collocated systems show a very interesting property of pole-zero interlacing. The location of poles and zeros on the imaginary axis are interchanged by simply adding a feed-through term. This guarantees the phase response to lie between 0 and $-180\,^{\circ}$. Controller design based on the collocated resonant system has provided with better performance and robustness. Vibration control of collocated structure is a significant and daunting task. Several methods such as Positive Position Feedback (PPF) [1] and integral resonant control [2] were introduced to take care of the resonant modes of collocated systems. A model of hysteresis nonlinearity present in piezoelectric actu-

R. Raj (✉)
Department of Electrical Engineering, IIT Kharagpur, Kharagpur 731202, West Bengal, India
e-mail: riturajsam@gmail.com

O P Jindal University, Raigarh 496109, Chhattisgarh, India

S. Kundu et al. (eds.), *Proceedings of the 2nd International Conference on Communication, Devices and Computing*, Lecture Notes in Electrical Engineering 602, https://doi.org/10.1007/978-981-15-0829-5_23

ators was proposed [3] using Bouc–Wen equations. Vibration control of a flexible arm having piezoelectric actuator was reported [4] using PPF and negative derivative feedback control. An output feedback position tracking of linear piezoelectric ceramic motor was studied [5] in the presence of system uncertainties and external disturbances.

Many linear control strategies have been considered in the past to damp the resonant modes of collocated systems. Recently some nonlinear methods such as incorporating fuzzy logic to curb the problem of vibration have been taken into consideration. A fuzzy controller for nanopositioning system was designed [6] to manipulate the piezoelectric actuators. The hysteresis present in piezoelectric actuators was modelled [7] by a TS fuzzy system and a feedforward controller with Proportional–Integral Derivative (PID) controller was designed for stabilization and disturbance compensation. A robust μ-synthesis and hybrid fuzzy control method was proposed [8] to compensate the effect of hysteresis present in piezoelectric actuator which is essential for precise nanopositioning. A fuzzy adaptive internal model controller was designed [9] based on TS fuzzy hysteresis model for nanoscale applications. A dual-loop control strategy was given [10] for precise nanopositioning using fuzzy logic controller.

In this paper, the simplest TS fuzzy PI controller is introduced as a tracking controller for position control of piezoelectric actuator. A dual-loop control strategy is applied where inner loop damps the resonant mode and the outer loop helps to track the reference command. An IRC damping controller is considered to damp the resonance and TS fuzzy PI controller is employed as a tracking controller. The TS fuzzy controller reduces the vibration and provides better tracking performance when compared to linear integral tracking controller.

Rest of the paper is organized as follows: Sect. 2 introduces the simplest TS fuzzy PI controller. The control strategy is discussed in Sect. 3. The problem statement is defined and simulation results are presented in Sect. 4. Section 5 concludes the paper.

2 The Simplest TS Fuzzy PI Controller

In this paper, the simplest TS Fuzzy PI controller [11] is considered as a tracking controller. The fuzzy PI controller is termed simplest as it utilizes only two input fuzzy sets, as shown in Fig. 1. The inputs of TS fuzzy PI controller are error, $e(k)$ and change of error, $\Delta e(k)$. A TS rule base is considered and the rules are given as follows:

R_1: IF ($e(k)$ is N_e AND $\Delta e(k)$ is $N_{\Delta e}$) OR ($e(k)$ is P_e AND $\Delta e(k)$ is $P_{\Delta e}$) THEN $\Delta u_s^1(k) = a^1 e(k) + b^1 \Delta e(k)$
R_2: IF ($e(k)$ is N_e AND $\Delta e(k)$ is $P_{\Delta e}$) OR ($e(k)$ is P_e AND $\Delta e(k)$ is $N_{\Delta e}$) THEN $\Delta u_s^2(k) = a^2 e(k) + b^2 \Delta e(k)$

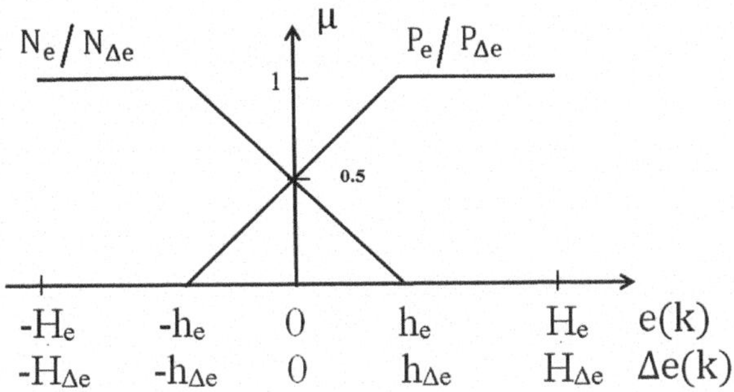

Fig. 1 Input fuzzy sets

Algebraic product AND operator and bounded sum OR operator are considered to evaluate the IF part of the rule base. Centre of Gravity (CoG) defuzzification method is considered to obtain the crisp output.

The structure [11] of the simplest TS fuzzy PI controller is given as follows:

$$\Delta u_s(k) = \gamma_1(e(k), \Delta e(k)).e(k) + \gamma_2(e(k), \Delta e(k)).\Delta e(k) \tag{1}$$

where $\gamma_1(.)$ and $\gamma_2(.)$ are defined as the gains of the fuzzy controller. The expression of gains $\gamma_1(.)$ and $\gamma_2(.)$ is given as
$\gamma_1(e(k), \Delta e(k)) = \frac{(a^1+a^2)h_e h_{\Delta e}+(a^1-a^2)e(k)\Delta e(k)}{2h_e h_{\Delta e}}$; and
$\gamma_2(e(k), \Delta e(k)) = \frac{(b^1+b^2)h_e h_{\Delta e}+(b^1-b^2)e(k)\Delta e(k)}{2h_e h_{\Delta e}}$

Hence,

$$\Delta u_s(k) = \frac{\begin{array}{c}[(a^1+a^2)h_e h_{\Delta e} + (a^1-a^2)e(k)\Delta e(k)]e(k) + \\ [(b^1+b^2)h_e h_{\Delta e} + (b^1-b^2)e(k)\Delta e(k)]\Delta e(k)\end{array}}{2h_e h_{\Delta e}} \tag{2}$$

Equation (2) is used as the fuzzy tracking controller to improve the tracking performance of multi-mode resonant systems.

3 Control Strategy

Collocated systems have multiple modes which create the problems of resonance and vibrations. The presence of nonlinearities, such as hysteresis and creep, further introduces complications. The conventional method to solve these problems is to incorporate damping controller to curb the effect of resonance, and a tracking controller is employed to reduce the position error. This control scheme is demonstrated in Fig. 2.

Generally, IRC damping controller is preferred over other damping controllers such as PPF and Positive Velocity and Position Feedback (PVPF). PPF or PVPF controller design results in high-order controller for multi-mode resonant systems. It

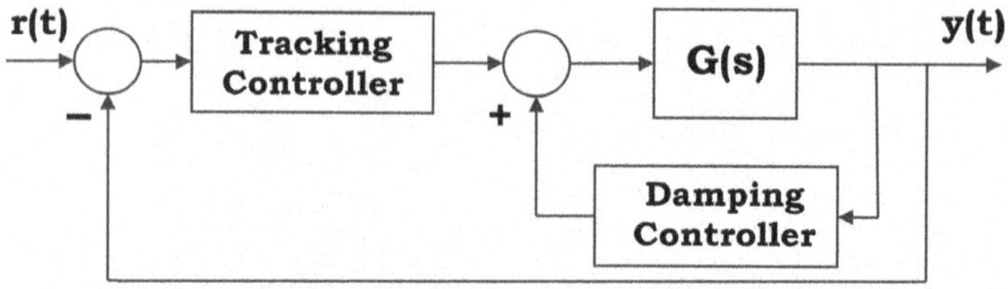

Fig. 2 A general block diagram for damping and tracking control of multi-mode resonant system

can be only applied for systems where the first mode is dominant over other modes. IRC, on the other hand, results in a low-order controller capable of damping multiple modes while retaining high stability margins [2]. A block diagram of IRC damping controller and integral tracking controller is shown in Fig. 3 where d_f is feed-through term, k_d is the damping gain and k_t is the integral tracking gain.

The conventional tracking scheme uses a linear integral controller in the forward path to reduce position error. We suggest to incorporate a TS fuzzy controller in place of linear integral controller in the forward path to improve the tracking performance. This control scheme is shown in Fig. 4. The fuzzy controller incorporated in the control scheme of Fig. 4 is the simplest fuzzy PI controller of TS type which is given by Eq. (2). It will be shown in the next section that by incorporating TS fuzzy PI controller, the tracking performance of multi-mode resonant systems is enhanced.

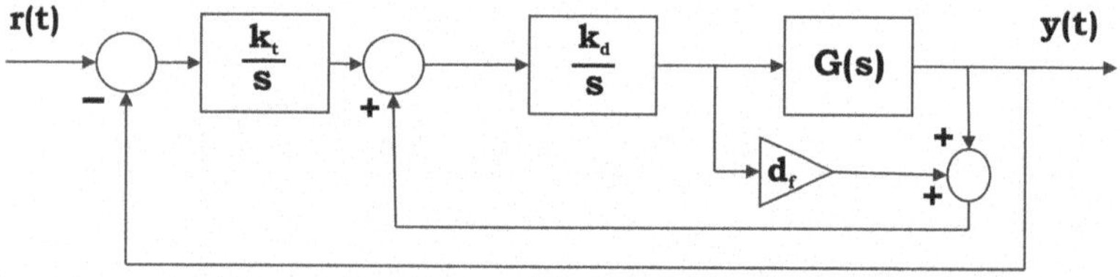

Fig. 3 Block diagram for IRC damping controller and integral tracking controller [12] with feed-through term d_f for multi-mode resonant systems

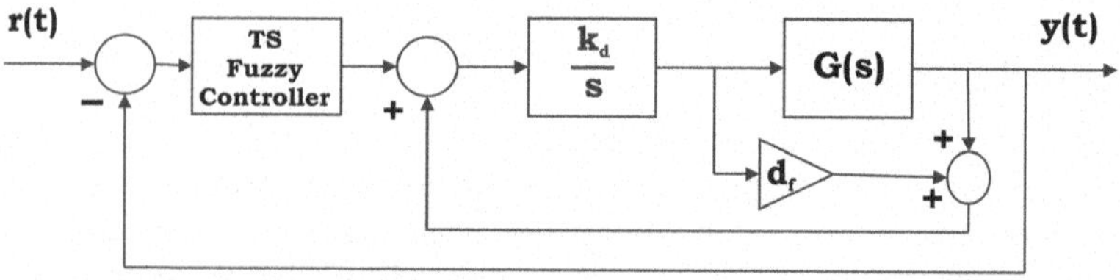

Fig. 4 Block diagram for TS fuzzy tracking control for multi-mode resonant systems

4 Problem Statement

A cantilever beam structure [2] is considered, as shown in Fig. 5, which is an example of collocated system having multiple modes. The collocated transfer function of the cantilever beam is given by Eq. (3). It is a multi-modal resonant system having the first resonant mode at 12.36 Hz.

$$\frac{Y(s)}{U(s)} = \frac{2250}{s^2 + 0.3854s + 6035} + \frac{8971}{s^2 + 1.49s + 217100} + \frac{90960}{s^2 + 3.573s + 1.697 \times 10^6} \quad (3)$$

This model of cantilever beam is having three pairs of collocated piezoelectric actuators attached to it. The piezoelectric actuator introduces hysteresis nonlinearity into the system. We consider the Bouc–Wen model [3] for the hysteresis nonlinearity which is given as follows:

$$\begin{aligned} m\ddot{x} + b\dot{x} + kx &= k(du - h) \\ \dot{h} &= \alpha d\dot{u} - \beta|\dot{u}|h - \gamma\dot{u}|h| \end{aligned} \quad (4)$$

where h represents the nonlinear relation between force and displacement. $\alpha = 0.26$, $\beta = 0.005$, $\gamma = 0.00068$, and $d = 2\,\mu\text{m}$ per volt to define the hysteresis loop's magnitude and shape. u is the applied voltage, x is the displacement of the piezoelectric actuator, m, b, k and d are the mass, damping coefficient, stiffness and piezoelectric coefficients, respectively. The hysteresis nonlinearity introduced by the Bouc–Wen model is shown in Fig. 6.

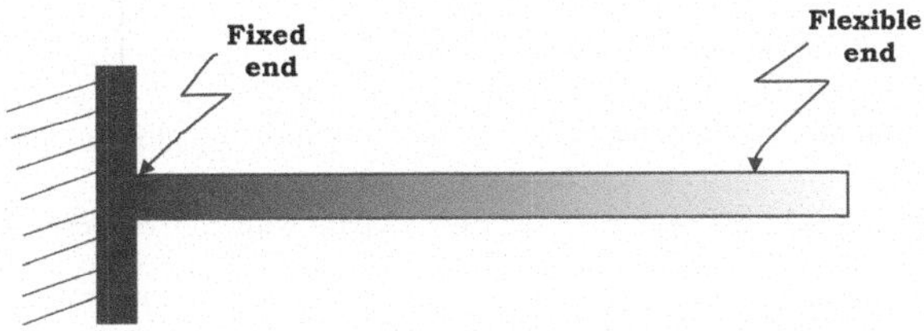

Fig. 5 A cantilever beam

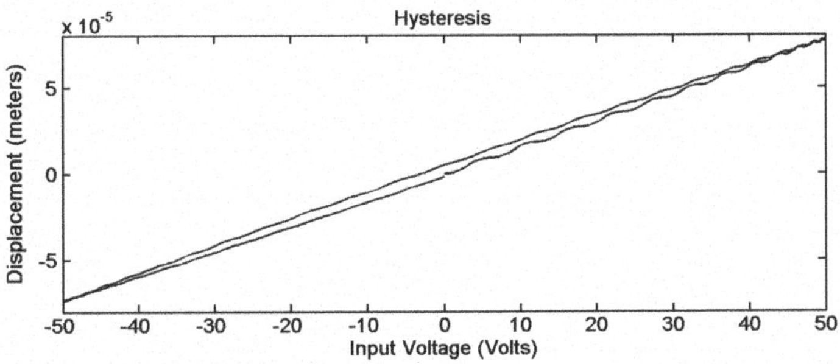

Fig. 6 Nature of hysteresis nonlinearity

The objective is to design a tracking controller to track square and sawtooth wave of amplitude 10 μm and frequency 0.1 rad/sec in absence and presence of hysteresis. The IRC damping controller is first designed [12] to damp out the resonant modes and a linear integral controller is obtained to track square and sawtooth wave. $d_f = -0.7456$ and $k_t k_d < 4046$ are obtained for IRC damping controller and integral tracking controller in absence of hysteresis. Similarly, the simplest TS fuzzy controller is designed to achieve the objectives and the fuzzy controller parameters are given as follows: $a^1 = 1.98 * 10^{-3}$, $a^2 = 0.73 * 10^{-3}$, $b^1 = 4 * 10^{-3}$, $b^2 = 1 * 10^{-3}$, $h_e = 80$, $h_{\Delta e} = 2$, and $S_{\Delta u}^{-1} = 3$. The corresponding tracking responses are shown in Figs. 7 and 8. It can be clearly seen from Figs. 7b and 8b that the vibrations have been reduced by incorporating the simplest TS fuzzy PI controller.

The controllers are redesigned for obtaining tracking responses in the presence of hysteresis. $a^1 = 7.12$, $a^2 = 1.88$, $b^1 = 0.4 * 10^{-3}$, $b^2 = 0.1 * 10^{-3}$, $h_e = 0.022$, $h_{\Delta e} = 20$ and $S_{\Delta u}^{-1} = 180$ are the fuzzy PI controller parameters. The gain product $k_t k_d$ is increased to a high value to obtain acceptable tracking performance for the linear integral tracking controller. The corresponding tracking responses are shown in Figs. 9 and 10. It is clearly evident that by incorporating the simplest TS fuzzy PI controller, the tracking response of multi-modal systems in the presence of hysteresis is enhanced.

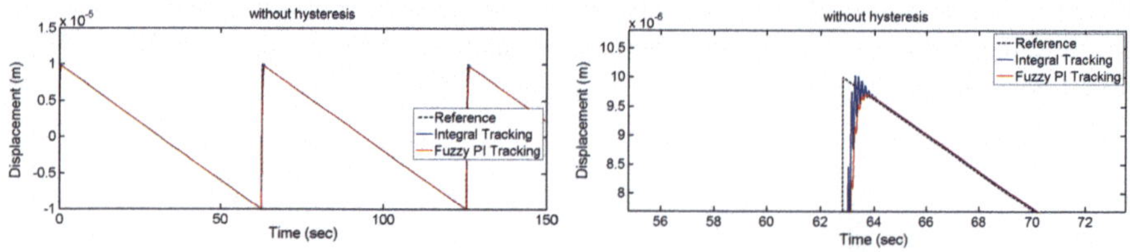

(a) Tracking sawtooth wave (b) Enlarged view of the response

Fig. 7 Sawtooth wave tracking response of cantilever beam without hysteresis nonlinearity

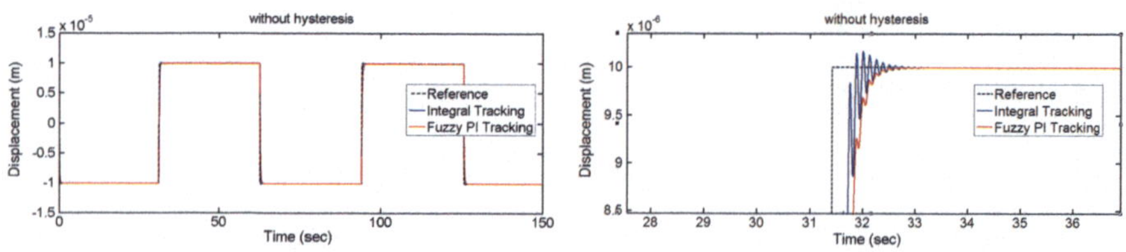

(a) Tracking square wave (b) Enlarged view of the response

Fig. 8 Square wave tracking response of cantilever beam without hysteresis nonlinearity

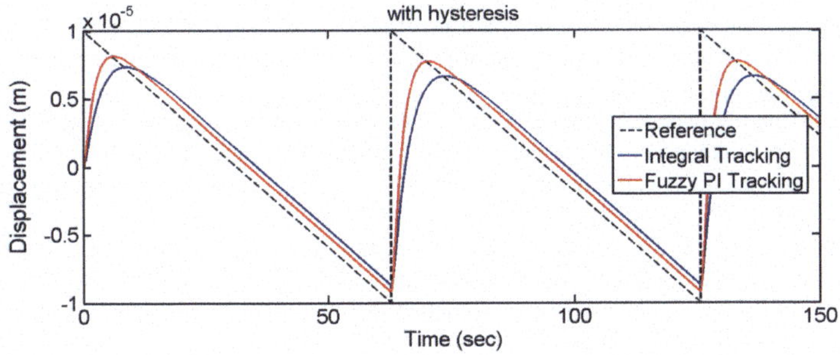

Fig. 9 Sawtooth wave tracking response of cantilever beam with hysteresis nonlinearity

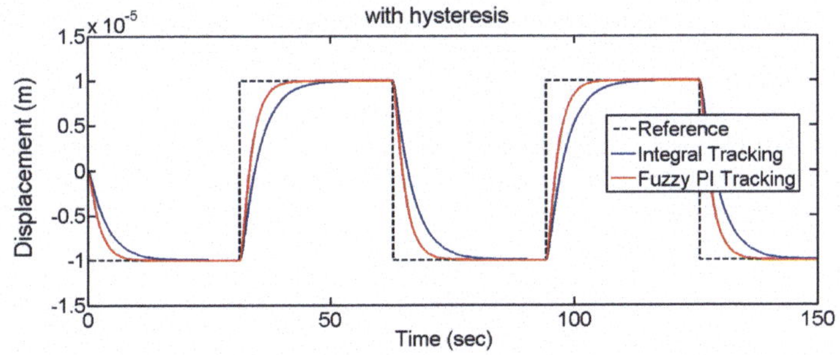

Fig. 10 Square wave tracking response of cantilever beam with hysteresis nonlinearity

5 Conclusion

In this paper, the simplest TS fuzzy PI controller is introduced to enhance the performance of multi-mode resonant systems in the presence of nonlinearities. The TS fuzzy controller is termed as the simplest controller since the input variables are fuzzified by only two input fuzzy sets. The incorporation of the TS fuzzy controller enhanced the tracking performance and reduced the vibrations in comparison to the conventional tracking method. The behaviour of fuzzy controller is nonlinear, and due to this, it can easily handle nonlinearities and resonance.

References

1. Friswell, M.I., Inman, D.J.: The relationship between positive position feedback and output feedback controllers. Smart Mater. Struct. **8**, 285–291 (1998)
2. Aphale, S.S., Fleming, A.J., Moheimani, S.O.R.: Integral resonant control of collocated smart structures. Smart Mater. Struct. **16**, 439–446 (2007)
3. Rakotondrabe, M.: Bouc-Wen modeling and inverse multiplicative structure to compensate hysteresis nonlinearity in piezoelectric actuators. IEEE Trans. Autom. Sci. Eng. **8**(2), 428–431 (2011)

4. Syed, H.H.: Comparative study between positive position feedback and negative derivative feedback for vibration control of a flexible arm featuring piezoelectric actuator. Int. J. Adv. Robot. Syst. **14**(4), 1–9 (2017)
5. Safa, A., Abdolmalaki, R.Y., Nejad, H.C.: Precise position tracking control with an improved transient performance for a linear piezoelectric ceramic motor. IEEE Trans. Ind. Electr. (2018)
6. Badr, B.M., Ali, W.G.: Nano positioning fuzzy control for piezoelectric actuators. Int. J. Eng. Technol. **10**(1), 70–74 (2010)
7. Li, P., Yan, F., Ge, C., Zhang, M.: Ultra-precise tracking control of piezoelectric actuators via a fuzzy hysteresis model. Rev. Sci. Inst. **83**(8), 85–114 (2012)
8. Abdelaziz, M., Abdelhameed, M.M., Tolbah, F.A.: A hybrid fuzzy robust control for piezo-electric actuators. J. Am. Sci. **8**(12), 483–489 (2012)
9. Li, P., Li, P., Sui, Y.: Adaptive fuzzy hysteresis internal model tracking control of piezoelectric actuators with nanoscale application. IEEE Trans. Fuzzy Syst. **24**(5), 1246–1254 (2016)
10. Altaher, M., Aphale, S.S.: Fuzzy-enhanced dual-loop control strategy for precise nanopositioning. In: Proceedings of IEEE International Symposium on Computer Science and Intelligent Controls (ISCSIC), pp. 73–81 (2017)
11. Raj, R., Mohan, B.M.: Analytical structures and stability analysis of the simplest Takagi-Sugeno fuzzy two-term controllers. Int. J. Process Syst. Eng. **5**(1), 67–92 (2019)
12. Namavar, M., Fleming, A.J., Aleyaasin, M., Nakkeeran, K., Aphale, S.S.: An analytical approach to integral resonant control of second-order systems. IEEE/ASME Trans. Mech. **19**(2), 651–659 (2014)

FPGA and ASIC Implementation of SEC-DED-DAEC Codes for SRAM Applications

Sayan Tripathi, Jhilam Jana, Jagannath Samanta, Amandeep Anand, Chandrabhan Kumar and Gautam Raj

Abstract Multiple bit upsets (MBUs) admit progressively in memory density with continual expand. The valuable information of memories are demolished due to soft errors. To avoid these errors in MBUs of memory, error correction codes are introduced. Single error correction-double error detection-double adjacent error correction (SEC-DED-DAEC) codes are used to protect SRAM devices from radiation-induced MBUs. These codes are able to correct single and double adjacent errors, and also detect double errors. In this paper, seven different SEC-DED-DAEC codes have been designed and implemented for static random access memories (SRAM) with word width of 16-bit, 32-bit, and 64-bit. All the codec architectures are simulated and synthesized both in FPGA and ASIC platforms. Performances of different SEC-DED-DAEC codecs are observed in terms of area and delay.

Keywords Static random access memory (SRAM) · Soft errors · Multiple bit upsets (MBUs) · SEC-DED-DAEC codes

---

S. Tripathi (✉) · J. Jana
Dept. of ETCE, Jadavpur University, Jadavpur, Kolkata 700032, India
e-mail: tripathysayan@gmail.com

J. Jana
e-mail: jhilamjana2014@gmail.com

J. Samanta · A. Anand · C. Kumar · G. Raj
Haldia Institute of Technology, Haldia, India
e-mail: jagannath19060@gmail.com

A. Anand
e-mail: Anandamsn99@gmail.com

C. Kumar
e-mail: chandrabhan04b@gmail.com

G. Raj
e-mail: Gautmraj.012@gmail.com

S. Kundu et al. (eds.), *Proceedings of the 2nd International Conference on Communication, Devices and Computing*, Lecture Notes in Electrical Engineering 602, https://doi.org/10.1007/978-981-15-0829-5_24

1 Introduction

Soft errors are generally caused by alpha particles radiation from package decay, from cosmic rays creating energetic neutrons and protons, thermal neutrons, and also from signal integrity problems [1]. It can change an instruction in a program or data value [2–5]. Soft errors are occurred at chip-level or system-level. Chip-level soft errors are occurred due to particles hit at the chip, whereas system-level soft errors are occurred when data being processed by hitting with a noise phenomenon. Error-correcting codes (ECCs) are commonly used in situations where there is a possibility that data may be corrupted and the consequences of such an error are unacceptable [6].

SRAM has been widely used since decades due to its high random access speeds and compatibility with CMOS logic process. It is typically used as the main building block for system registers, on-chip cache, etc. SRAM bit cells can also be optimized at the transistor level for high performance. ECCs can be used in data situations to safeguard the reading and writing of storage devices, as in ECC random access memory (RAM), or can be used in transmission situations, where data is sent over a potentially noisy channel [7]. Spaceflight is one of the former situations where spacecraft, unshielded by Earth's atmosphere, can be struck by cosmic rays which can change the data stored in memory cells. ECC methodology which is cost-efficient is used to correct the double adjacent and nonadjacent bit-errors which are occurred on MBUs. Single error correction-double error detection-double adjacent error correction with completely eliminated mis-correction up to d bits (SEC-DED-DAEC-CEM-d) codes can correct all single errors and double adjacent errors and detect all double errors and also completely eliminate mis-correction for nonadjacent double errors [8]. These codes are able to correct single error, correct double adjacent errors, detect double errors, detect y bits adjacent errors. These SEC-DED-DAEC codes can improve the data transfer efficiency. VLSI implementation of these codecs is essential to the researchers.

In this paper, we have been designed and implemented the Dutta-based (i) SEC-DED-DAEC codes [2, 9], (ii) SEC-DED-DAEC-CEM-d codes [8], and (iii) SEC-DED-DAEC codes with no mis-correction codes [10] with word length of 16-bit, 32-bit, and 64-bit.

This paper is organized for remainder portion as follows. Section 2 provides the literature review. Section 3 presents overview of SEC-DED-DAEC codes and SEC-DED-DAEC-CEM-d codes for different H-matrices. Section 4 provides the design of Dutta (22, 16) SEC-DED-DAEC code. Section 5 presents the area and delay overhead. Section 6 contains the synthesis results of SEC-DED-DAEC codes. Section 7 presents the conclusion.

2 Literature Review

The several approaches for the basic single error correction-double error detection (SEC-DED) Hamming code have been previously introduced in [11]. The codes are described by Dutta as Single error correction-double error detection-double adjacent error correction (SEC-DED-DAEC) code and Single error correction-double-error detection-double adjacent error correction with complete elimination of mis-correction with a dispersion window *d* (SEC-DED-DAEC-CEM-*d*) codes. Neale et al. described a method to generate the Single error correction-double error detection-triple adjacent error-detecting codes (SEC-DED-TAEC) which can detect up to 11-bit adjacent errors [4, 5]. They provided a tool to assist the developers to implement the SEC-DED code and also internal triple mode redundancy (TMR) using SRAM-based FPGAs platform [4, 5].

These SEC-DED-DAEC codes suffer from high degree of mis-correction probability for nonadjacent double errors which could present a significant reliability threat for memories. Several researchers tried to reduce the mis-correction probability of the nonadjacent double errors by either choosing different permutation of the columns of the H-matrix via search or by using additional columns by making use of the spare columns kept for memory repair [2, 9]. Further improvements were shown by using unused spare columns [8, 12]. Even further reduction in mis-correction probability of the random nonadjacent double errors is reported in [8, 12]. Since the overall mis-correction probability is still nonzero, the most likely nonadjacent double errors still reduce the reliability of memories employing such codes [8].

3 Overview of Dutta-Based SEC-DED-DAEC Codes and SEC-DED-DAEC-CEM-*d* Codes

In this section, basics of Dutta-based different SEC-DED-DAEC codes are described.

Firstly, Dutta-based SEC-DED-DAEC codes deal with the low area overhead of SEC-DED codes. An additional constraint on the H-matrix is added respect to Hsiao and Hamming SEC-DED codes. This constraint provides the DAEC property. Different construction rules of H-matrix are described as follows [2, 9]:

1. All columns must have null vector.
2. Entire columns should be distinct.
3. No linear dependencies exist including three columns.
4. The linear composition of any two adjacent columns does not same any of the individual columns or the linear composition of any other two adjacent columns.

In the other portion, ECC technology is used to correct the MBUs in memory applications. The SEC-DED-DAEC-CEM-*d* codes can correct all single errors, double adjacent errors and detect all double errors and eliminate mis-correction for nonadjacent double errors completely [8].

$$H = \begin{bmatrix} 1100110001110100100000 \\ 0001010011011001010000 \\ 1010100111100011001000 \\ 1001001110010110000100 \\ 0110101100001101000010 \\ 0111011000101010000001 \end{bmatrix}$$

Fig. 1 H-matrix of (22, 16) SEC-DED-DAEC code

$$H = \begin{bmatrix} 101110010101100001001010011100001000000 \\ 010101001011000100011100011000010100000 \\ 101010100110001010110000110000110010000 \\ 010100011100010101100001100001110001000 \\ 001001100000101101001011000011100000100 \\ 100001001001011010010110100111000000010 \\ 010010110010110010100101001110000000001 \end{bmatrix}$$

Fig. 2 H-matrix of (39, 32) SEC-DED-DAEC code

$$H = \begin{bmatrix} 100101011101011001010010010000101100010011001000110100001110000010000000 \\ 101110011000001010100100100001011000100110010001101000011100000101000000 \\ 011010101011101011001001000010110001001100100011010000111000001100100000 \\ 111101100101100110010010100101100010011001000110100001100000011100010000 \\ 100101100011110110100101001011000100110010001100000011010000111000001000 \\ 111011001111110111001010010110001001100000011001000110100001110000000100 \\ 011010110110111100010100101100000011000100110010001101000011100000000010 \\ 010111101110011100101001011000010110001001100100011010000111000000000001 \end{bmatrix}$$

Fig. 3 H-matrix of (72, 64) SEC-DED-DAEC code

The H-matrices of SEC-DED-DAEC codes, SEC-DED-DAEC-CEM-d codes, and SEC-DED-DAEC codes with no mis-correction have been considered and implemented in Figs. 1, 2, 3, 4, 5, 6 and 7. We have observed the simulated waveform and synthesis results in terms of area and delay in FPGA and ASIC platforms.

In Fig. 1, H-matrix of (22, 16) SEC-DED-DAEC code is mentioned. The (n) number of columns in parity check (H) matrix has (k) number of data and $(n-k)$ number of parity columns. This H-matrix consists of 16 data (d_1–d_{16}) columns and 6 parity (c_1–c_6) columns.

The H-matrix of (39, 32) SEC-DED-DAEC code is provided in Fig. 2. In this matrix, 32 numbers of data and 7 numbers of parity are presented.

The H-matrix of (72, 64) SEC-DED-DAEC code is described in Fig. 3. This matrix contains d_1–d_{64} data columns and c_1–c_8 parity columns. The $(n-k)$ rows and (n) columns are required to consist of SEC-DED-DAEC (n, k) code.

In Fig. 4, H-matrix of (22, 16) SEC-DED-DAEC-CEM-2 code is mentioned. In this H-matrix, $(n-k)$ rows and (n) columns are required. Here, 16 data columns and 6 parity columns are presented.

The H-matrix of (40, 32) SEC-DED-DAEC-CEM-5 code is presented in Fig. 5. The parity check (H) matrix contains of d_1–d_{32} data columns and c_1–c_8 parity columns.

$$H = \begin{bmatrix} 0011101010110111000001 \\ 0101111011101101000010 \\ 1100111100011011000100 \\ 1011011011011000001000 \\ 0100111110110010010000 \\ 1010110111010100100000 \end{bmatrix}$$

Fig. 4 H-matrix of (22, 16) SEC-DED-DAEC-CEM-2 code

$$H = \begin{bmatrix} 0110001000101100101101111111111100000001 \\ 0010111110010001010011000001010100000010 \\ 1100101001010111010100000001010110000100 \\ 0111011010000001001001001010010000001000 \\ 1100010100100100100000010010001000010000 \\ 0001000000101001000101010101000000100000 \\ 0101101101001000100100001001000001000000 \\ 1000000010101001001001010000001010000000 \end{bmatrix}$$

Fig. 5 H-matrix of (40, 32) SEC-DED-DAEC-CEM-5 code

$$H = \begin{bmatrix} 1010100101000100000000000001010010001010011111110111111111111111000000001 \\ 1011110101001101010001010101001011101011000000010100101001010101000000010 \\ 0010101101010101000010101000010010100001010010110000000000101011000000100 \\ 1010100000100100100001101100010100001010000000010100110100010000000001000 \\ 1000010111010110101100000101101000100100010100000000100001000010000010000 \\ 1110001010010010011010100010001001001100001000000010100100100000000100000 \\ 0001010001000010100100010000010101010001000100101000000100010000001000000 \\ 0100101010101000010101010011000010010000100001000101000000000010010000000 \\ 0101000100101001001010001000100000000001010101000100001000100000100000000 \end{bmatrix}$$

Fig. 6 H-matrix of (73, 64) SEC-DED-DAEC-CEM-5 code

Fig. 7 H-matrix of (42, 32) SEC-DED-DAEC code with no mis-correction

$$H = \begin{bmatrix} 011111111111111100000000000000001000000000 \\ 010101010000000111110000100000000100000000 \\ 110010101000000100101111010000000010000000 \\ 101010000100000010000000001100100001000000 \\ 101000000010100000010101000000000000100000 \\ 000100000010010001000100010101000000010000 \\ 000000100001010010000010000111110000001000 \\ 000001000000101101000000110101010000000100 \\ 000000001001000000011000111010010000000010 \\ 000000010100000100010101000001010000000001 \end{bmatrix}$$

The H-matrix of (73, 64) SEC-DED-DAEC-CEM-5 code is provided in Fig. 6. The H-matrix of (73, 64) SEC-DED-DAEC-CEM-5 contains of d_1–d_{64} data columns and c_1–c_9 parity columns.

In Fig. 7, H-matrix of (42, 32) SEC-DED-DAEC code with no mis-correction is described. This matrix are satisfied the construction rule of SEC-DED-DAEC codes and no mis-correction property. The d_1–d_{32} data columns and c_1–c_{10} parity columns are presented in this H-matrix.

4 Design of SEC-DED-DAEC Codec

In this section, the (22, 16) SEC-DED-DAEC code is represented and employed. The encoding and decoding techniques are described as follows.

4.1 Encoding Process

The arrangement of codeword of the (22, 16) SEC-DED-DAEC code is as follows:

$$v = (d_1 d_1 d_2 d_3 d_4 d_5 d_6 d_7 d_8 d_9 d_{10} d_{11} d_{12} d_{13} d_{14} d_{15} d_{16} c_1 c_2 c_3 c_4 c_5 c_6). \quad (1)$$

The matrix of (22, 16) SEC-DED-DAEC provides the following check-bit calculations. These calculations are shown in Eq. (2).

$$\begin{aligned}
c_1 &= d_1 \oplus d_2 \oplus d_5 \oplus d_6 \oplus d_{10} \oplus d_{11} \oplus d_{12} \oplus d_{14} \\
c_2 &= d_4 \oplus d_6 \oplus d_9 \oplus d_{10} \oplus d_{12} \oplus d_{13} \oplus d_{16} \\
c_3 &= d_1 \oplus d_3 \oplus d_5 \oplus d_8 \oplus d_9 \oplus d_{10} \oplus d_{11} \oplus d_{15} \oplus d_{16} \\
c_4 &= d_1 \oplus d_4 \oplus d_7 \oplus d_8 \oplus d_9 \oplus d_{12} \oplus d_{14} \oplus d_{15} \\
c_5 &= d_2 \oplus d_3 \oplus d_5 \oplus d_7 \oplus d_8 \oplus d_{13} \oplus d_{14} \oplus d_{16} \\
c_6 &= d_2 \oplus d_3 \oplus d_4 \oplus d_6 \oplus d_7 \oplus d_{11} \oplus d_{13} \oplus d_{15} \qquad (2)
\end{aligned}$$

The schematic diagram of (22, 16) SEC-DED-DAEC code is presented in Fig. 8.

4.2 Decoding Process

In this decoding technique, the error detection technique is computed by the syndrome value. For received codeword, there is no error, if the syndrome value is zero ($SY = 0$). On the other hand, some bit-errors occur if the syndrome value is nonzero ($SY \neq 0$). The expressions of syndrome bits are provided in equation (3) for the matrix shown in Fig. 1.

$$\begin{aligned}
s_1 &= r_1 \oplus r_2 \oplus r_5 \oplus r_6 \oplus r_{10} \oplus r_{11} \oplus r_{12} \oplus r_{14} \oplus r_{17} \\
s_2 &= r_4 \oplus r_6 \oplus r_9 \oplus r_{10} \oplus r_{12} \oplus r_{13} \oplus r_{16} \oplus r_{18} \\
s_3 &= r_1 \oplus r_3 \oplus r_5 \oplus r_8 \oplus r_9 \oplus r_{10} \oplus r_{11} \oplus r_{15} \oplus r_{16} \oplus r_{19} \\
s_4 &= r_1 \oplus r_4 \oplus r_7 \oplus r_8 \oplus r_9 \oplus r_{12} \oplus r_{14} \oplus r_{15} \oplus r_{20} \\
s_5 &= r_2 \oplus r_3 \oplus r_5 \oplus r_7 \oplus r_8 \oplus r_{13} \oplus r_{14} \oplus r_{16} \oplus r_{21} \\
s_6 &= r_2 \oplus r_3 \oplus r_4 \oplus r_6 \oplus r_7 \oplus r_{11} \oplus r_{13} \oplus r_{15} \oplus r_{22} \qquad (3)
\end{aligned}$$

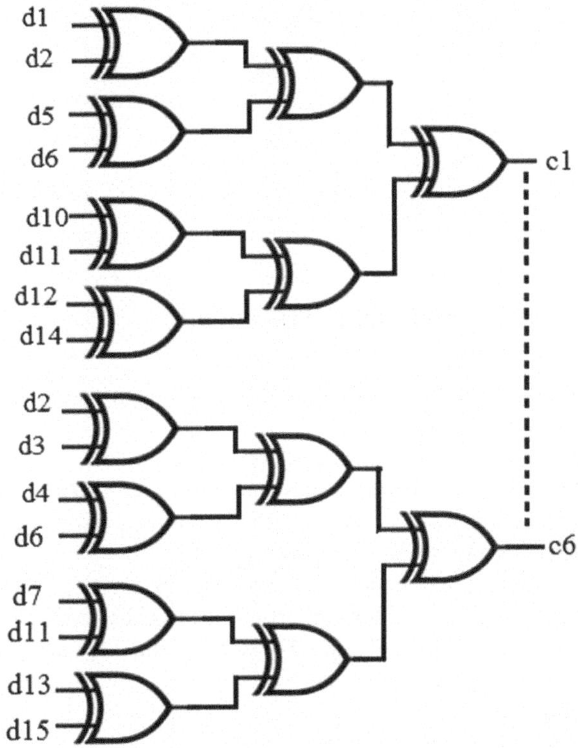

Fig. 8 Parity generation logic of (22, 16) SEC-DED-DAEC code

The matrix in Fig. 1 provides the decoder output calculations. These calculations are presented in Eq. (4).

$$
\begin{aligned}
dc_1 &= r_1 \oplus ((s_1 s_2^{'} s_3 s_4 s_5^{'} s_6^{'}) + (s_1^{'} s_2^{'} s_3 s_4 s_5 s_6))\\
dc_2 &= r_2 \oplus ((s_1 s_2^{'} s_3^{'} s_4^{'} s_5 s_6) + (s_1^{'} s_2^{'} s_3 s_4 s_5 s_6) + (s_1 s_2^{'} s_3^{'} s_4^{'} s_5^{'} s_6^{'}))\\
dc_3 &= r_3 \oplus ((s_1^{'} s_2^{'} s_3 s_4^{'} s_5 s_6) + (s_1 s_2^{'} s_3^{'} s_4^{'} s_5^{'} s_6^{'}) + (s_1^{'} s_2 s_3 s_4 s_5 s_6^{'}))\\
dc_4 &= r_4 \oplus ((s_1^{'} s_2 s_3^{'} s_4 s_5^{'} s_6) + (s_1^{'} s_2 s_3 s_4 s_5 s_6^{'}) + (s_1 s_2 s_3 s_4 s_5 s_6))\\
dc_5 &= r_5 \oplus ((s_1 s_2^{'} s_3 s_4^{'} s_5 s_6^{'}) + (s_1 s_2 s_3 s_4 s_5 s_6) + (s_1^{'} s_2 s_3 s_4^{'} s_5 s_6))\\
dc_6 &= r_6 \oplus ((s_1 s_2 s_3^{'} s_4^{'} s_5^{'} s_6) + (s_1^{'} s_2 s_3 s_4^{'} s_5 s_6) + (s_1 s_2 s_3^{'} s_4 s_5 s_6^{'}))\\
dc_7 &= r_7 \oplus ((s_1^{'} s_2^{'} s_3^{'} s_4 s_5 s_6) + (s_1 s_2 s_3^{'} s_4 s_5 s_6^{'}) + (s_1^{'} s_2^{'} s_3 s_4^{'} s_5^{'} s_6))\\
dc_8 &= r_8 \oplus ((s_1^{'} s_2^{'} s_3 s_4 s_5 s_6^{'}) + (s_1^{'} s_2^{'} s_3 s_4^{'} s_5^{'} s_6) + (s_1^{'} s_2 s_3^{'} s_4^{'} s_5 s_6^{'}))\\
dc_9 &= r_9 \oplus ((s_1^{'} s_2 s_3 s_4 s_5^{'} s_6^{'}) + (s_1^{'} s_2 s_3^{'} s_4^{'} s_5 s_6^{'}) + (s_1 s_2^{'} s_3^{'} s_4 s_5^{'} s_6^{'}))\\
dc_{10} &= r_{10} \oplus ((s_1 s_2 s_3 s_4^{'} s_5^{'} s_6^{'}) + (s_1 s_2^{'} s_3^{'} s_4 s_5^{'} s_6^{'}) + (s_1^{'} s_2 s_3^{'} s_4^{'} s_5^{'} s_6))\\
dc_{11} &= r_{11} \oplus ((s_1 s_2^{'} s_3 s_4^{'} s_5^{'} s_6) + (s_1^{'} s_2 s_3^{'} s_4^{'} s_5^{'} s_6) + (s_1^{'} s_2 s_3 s_4 s_5^{'} s_6))\\
dc_{12} &= r_{12} \oplus ((s_1 s_2 s_3^{'} s_4 s_5^{'} s_6^{'}) + (s_1^{'} s_2 s_3 s_4 s_5^{'} s_6) + (s_1 s_2^{'} s_3^{'} s_4 s_5 s_6))
\end{aligned}
$$

$$
\begin{aligned}
dc_{13} &= r_{13} \oplus ((s_1' s_2 s_3' s_4' s_5 s_6) + (s_1 s_2' s_3' s_4 s_5 s_6) + (s_1 s_2 s_3' s_4 s_5' s_6)) \\
dc_{14} &= r_{14} \oplus ((s_1 s_2' s_3' s_4 s_5 s_6') + (s_1 s_2 s_3' s_4 s_5' s_6) + (s_1 s_2' s_3 s_4' s_5 s_6)) \\
dc_{15} &= r_{15} \oplus ((s_1' s_2' s_3 s_4 s_5' s_6) + (s_1 s_2' s_3 s_4' s_5 s_6) + (s_1' s_2 s_3' s_4 s_5 s_6)) \\
dc_{16} &= r_{16} \oplus ((s_1' s_2 s_3 s_4' s_5 s_6') + (s_1' s_2 s_3' s_4 s_5 s_6) + (s_1 s_2 s_3 s_4' s_5 s_6'))
\end{aligned}
\quad (4)
$$

5 Area and Delay Overhead

This section presents the area and delay overhead which contains of area overhead and critical path delay.

In Table 1, there are area overhead comparison of SEC-DED-DAEC codes, SEC-DED-DAEC-CEM-*d* codes and SEC-DED-DAEC codes with no mis-correction codes. The number of gates that we have used for the area calculation of both the encoder and decoder is shown in Table 1.

The critical path delay comparison for SEC-DED-DAEC codes, SEC-DED-DAEC-CEM-*d* codes, and SEC-DED-DAEC codes with no mis-correction are presented in Table 2. The number of gates that we have used for this critical path calculation is shown here.

Table 1 Area overhead comparison of different SEC-DED-DAEC codes

Code	Block	XOR2	AND2	OR2	NOT	Equivalent NAND2
(22, 16) SEC-DED-DAEC	Encoder	42	-	-	-	168
	Decoder	64	235	31	128	947
(39, 32) SEC-DED-DAEC	Encoder	88	-	-	-	352
	Decoder	127	570	63	321	2158
(72, 64) SEC-DED-DAEC	Encoder	116	-	-	-	464
	Decoder	440	1337	127	782	5597
(22, 16) SEC-DED-DAEC-CEM-2	Encoder	52	-	-	-	208
	Decoder	74	235	31	116	975
(40, 32) SEC-DED-DAEC-CEM-5	Encoder	92	-	-	-	368
	Decoder	132	672	64	373	2437
(73, 64) SEC-DED-DAEC-CEM-5	Encoder	195	-	-	-	780
	Decoder	268	1528	127	831	5340
(42, 32) SEC-DED-DAEC code with no mis-correction	Encoder	86	-	-	-	344
	Decoder	128	855	63	596	3007

Table 2 Critical path delay comparison of different SEC-DED-DAEC codes

Code	XOR2	AND2	OR2	NOT	Equivalent NAND2
(22, 16) SEC-DED-DAEC	18	5	2	1	89
(39, 32) SEC-DED-DAEC	28	6	2	1	131
(72, 64) SEC-DED-DAEC	62	7	2	1	269
(22, 16) SEC-DED-DAEC-CEM-2	22	5	2	1	105
(40, 32) SEC-DED-DAEC-CEM-5	40	7	2	1	181
(73, 64) SEC-DED-DAEC-CEM-5	68	8	2	1	295
(42, 32) SEC-DED-DAEC code with no mis-correction	28	9	2	1	137

6 Synthesis Results

In this section, SEC-DED-DAEC codes (16-bit, 32-bit, and 64-bit), SEC-DED-DAEC-CEM-*d*codes (16-bit, 32-bit, and 64-bit), and SEC-DED-DAEC code with no mis-correction are simulated and synthesized using FPGA and ASIC platforms.

6.1 *FPGA-Based*

SEC-DED-DAEC codes (16-bit, 32-bit, and 64-bit), SEC-DED-DAEC-CEM-*d*codes (16-bit, 32-bit, and 64-bit), and SEC-DED-DAEC code with no mis-correction have been represented in Verilog hardware description language. All functional blocks are simulated and synthesized using Virtex-6 (XC6VCX75T) FPGA device family. The FPGA synthesis results have been presented in Table 3 in terms of area look-up tables (LUTs), input output buffers (IOBs), and delay.

6.2 *ASIC-Based*

In Table 4, ASIC-based synthesis results of SEC-DED-DAEC codes, SEC-DED-DAEC-CEM-*d* codes, and SEC-DED-DAEC codes with no mis-correction for 350 nm technology are presented. The synthesis results are observed in Leonardo Spectrum tool (Mentor Graphics). Three parameters such as number of gates, number of accumulated instances, and slack time are observed. Slack time is the difference between the required time and arrival time.

Table 3 FPGA synthesis results of different SEC-DED-DAEC codes

Code	Block	Area (LUTs)	IOBs	Delay(ns)
(22, 16) SEC-DED-DAEC	Encoder	12	22	1.179
	Decoder	44	44	2.351
(39, 32) SEC-DED-DAEC	Encoder	21	39	1.704
	Decoder	100	78	4.022
(72, 64) SEC-DED-DAEC	Encoder	45	72	1.880
	Decoder	215	144	4.555
(22, 16) SEC-DED-DAEC-CEM-2	Encoder	11	22	1.742
	Decoder	51	44	3.367
(40, 32) SEC-DED-DAEC-CEM-5	Encoder	21	40	1.590
	Decoder	118	80	3.609
(73, 64) SEC-DED-DAEC-CEM-5	Encoder	41	73	1.590
	Decoder	237	146	4.336
(42, 32) SEC-DED-DAEC code with no mis-correction	Encoder	22	42	1.477
	Decoder	135	84	4.421

Table 4 ASIC synthesis results of different SEC-DED-DAEC codes

Code	Block	Number of gates	Number of accumulated instances	Slack time
(22, 16) SEC-DED-DAEC	Encoder	64	33	8.80
	Decoder	195	131	6.90
(39, 32) SEC-DED-DAEC	Encoder	129	67	8.19
	Decoder	376	254	6.31
(72, 64) SEC-DED-DAEC	Encoder	272	142	7.97
	Decoder	731	491	4.78
(22, 16) SEC-DED-DAEC-CEM-2	Encoder	58	30	8.47
	Decoder	191	129	6.63
(40, 32) SEC-DED-DAEC-CEM-5	Encoder	131	68	8.40
	Decoder	412	284	5.90
(73, 64) SEC-DED-DAEC-CEM-5	Encoder	260	136	8.11
	Decoder	870	528	5.26
(42, 32) SEC-DED-DAEC code with no mis-correction	Encoder	132	68	8.57
	Decoder	460	322	6.25

7 Conclusion

In this paper, the different SEC-DED-DAEC codes have been designed and evaluated. We have analyzed the critical path and area complexity of seven different SEC-DED-DAEC codecs. The results are shown in terms of LUTs and delay. These SEC-DED-DAEC codes are synthesized in both FPGA and ASIC platforms. This work is valuable to the VLSI domain designers. We may implement (22, 16) SEC-DED-DAEC code on a transistor level in future. Designing a new matrix for SEC-DED-DAEC code and implementing higher order matrices are our future goal.

References

1. Dixit, A., Wood, A.: The impact of new technology on soft error rates. In: 2011 IEEE International Reliability Physics Symposium (IRPS), pp. 5B–4. IEEE (2011)
2. Dutta, A., Touba, N.A.: Multiple bit upset tolerant memory using a selective cycle avoidance based SEC-DED-DAEC code. In: 25th IEEE VLSI Test Symposium, 2007, pp. 349–354. IEEE (2007)
3. Ibe, E., Taniguchi, H., Yahagi, Y., Shimbo, K.I., Toba, T.: Impact of scaling on neutron-induced soft error in SRAMs from a 250 nm to a 22 nm design rule. IEEE Trans. Electr. Dev. **57**(7), 1527–1538 (2010)
4. Neale, A., Jonkman, M., Sachdev, M.: Adjacent-MBU-tolerant SEC-DED-TAEC-*y*AED codes for embedded SRAMs. IEEE Trans. Circuits Syst. II Express Briefs **62**(4), 387–391 (2015)
5. Neale, A., Sachdev, M.: A new SEC-DED error correction code subclass for adjacent MBU tolerance in embedded memory. IEEE Trans. Device Mater. Reliab. **13**(1), 223–230 (2013)
6. Samanta, J., Tripathi, S.: Comments on a novel approach of error detection and correction for efficient energy in wireless networks. Multimed. Tools Appl. 1–6 (2018)
7. Cha, S., Yoon, H.: Efficient implementation of single error correction and double error detection code with check bit pre-computation for memories. JSTS J. Semicond. Technol. Sci. **12**(4), 418–425 (2012)
8. Dutta, A.: Low cost adjacent double error correcting code with complete elimination of miscorrection within a dispersion window for multiple bit upset tolerant memory. In: 2012 IEEE/IFIP 20th International Conference on VLSI and System-on-Chip (VLSI-SoC), pp. 287–290. IEEE (2012)
9. Dutta, A., Cypress Semiconductor Corp (US).: Low cost adjacent double error correcting code. U.S. Patent 8,875,002 (2014)
10. Jun, H.Y., Lee, Y.S.: Single error correction, double error detection and double adjacent error correction with no miscorrection code. IEICE Electr. Express **10**(20), 20130743–20130743 (2013)
11. Hamming, R.W.: Error detecting and error correcting codes. Bell Syst. Tech. J. **29**(2), 147–160 (1950)
12. Datta, R., Touba, N.A.: Exploiting unused spare columns to improve memory ECC. In: 27th IEEE VLSI Test Symposium, 2009. VTS'09, pp. 47–52. IEEE (2009)

Effect of Pocket Intrinsic Doping on Double and Triple Gate Tunnel Field Effect Transistors

Ritam Dutta and Nitai Paitya

Abstract Band-to-band tunneling at the source–channel junction of multigate tunnel field effect transistors (TFETs) plays a major role to boost ON current for eliminating short channel effects (SCE). In this paper, the conventional double gate TFET structure (DG-TFET) is designed and modified by applying gate engineering to design triple gate TFET (TG-TFET). The work mainly focuses on drain current (I_{ON}) assessment, fast switching ratio (I_{ON}/I_{OFF}), and subthreshold swing (*SS*) by comparing between conventional DG-TFET and proposed TG-TFET with pocket intrinsic layers. The variation of drain current for different pocket intrinsic doping conditions on TFET performance has been reviewed and demonstrated with SILVACO TCAD simulator. The pocket intrinsic triple gate TFET shows a higher I_{ON} and better I_{ON}/I_{OFF} ratio of 5.51 for source and pocket doping of 1×10^{20} cm^{-3} and 1×10^{15} cm^{-3}. This provides 68.51% fast switching ratio as compared to other nonconventional tunnel FETs. But the subthreshold swing (*SS*) is limited by variation of pocket doping from 10^0 to 10^{18} cm^{-3} for TG-PI-TFET, since DG-PI-TFET provides better subthreshold swing of 23.93 mV/dec ($\ll$60 mV/dec) considering the leakage current for low-power applications.

Keywords DG-PI-TFET · I_{ON}/I_{OFF} ratio · TG-PI-TFET · Pocket intrinsic doping

---

R. Dutta (✉)
Department of Electronics and Communication Engineering, Surendra Institute of Engineering and Management, Maulana Abul Kalam Azad University of Technology, West Bengal, India
e-mail: ritam_siliguri@yahoo.com

N. Paitya
Department of Electronics and Communication Engineering, Sikkim Manipal Institute of Technology, Sikkim Manipal University, Sikkim, India
e-mail: nitai.p@smit.smu.edu.in

S. Kundu et al. (eds.), *Proceedings of the 2nd International Conference on Communication, Devices and Computing*, Lecture Notes in Electrical Engineering 602, https://doi.org/10.1007/978-981-15-0829-5_25

1 Introduction

Miniaturization of metal oxide field effect transistor (MOSFET) channel lengths has been a real challenge nowadays. Hence, the tunnel field effect transistor (TFET) is found to be motivating to raise the low leakage current and steeper subthreshold slope [1, 2].

Since the minimum subthreshold swing (*SS*) attainable in an MOSFET has been limited to 60 mV/decade, therefore an MOSFET does not behave like an ideal switch. The tunnel FET has the prospect of reducing the subthreshold swing less than the conventional limit [3–5]. However, tunnel FETs are ambipolar in nature and also produce very low drain current (I_{ON}); hence by increasing the doping concentration, the ambipolarity can be removed and the I_{ON} can be boosted by introducing gate and channel engineering [6, 7].

This paper mainly focuses on to boost I_{ON} targeting the best possible subthreshold swing below 60 mV/decade. Various multigate tunnel FET structures have been reviewed and analyzed to assess the performance improvement of tunnel FET [8–12]. The two-way tunneling junctions of multigate structure of tunnel FETs increase the tunneling current compared to single gate TFET [13, 14].

The incorporation of thin pocket intrinsic doping in the tunnel FET produces another electric field component to the intrinsic channel region of the device. As a result of which, the tunneling current of the device improves [15–17]. Due to band-to-band tunneling (BTBT) mechanism, the thin pocket intrinsic layers will raise the I_{ON} for both double gate and triple gate pocket intrinsic TFET structures. The maximum I_{ON} is recorded for TG-PI-TFET as 1.286×10^{-4} A/μm whereas for DG-PI-TFET is 1.018×10^{-4} A/μm.

In this work, using the dopant engineering thin pocket layers are created near the drain channel junction to address the poor drive current (I_{ON}). The asymmetry of the device helps the ambipolarity reduction, and the dopant pocket near the source junction helps in achieving higher performance compared to single gate TFET structures. In addition to this, the pocket intrinsic doping has been varied from 1×10^{0} cm^{-3} to 1×10^{15} cm^{-3} at supply voltage of 0.5 V in order to identify the limitation of triple gate TFET structure. In this paper, the effect of pocket doping has been thoroughly investigated by making sure that the pocket is always fully depleted and annealing schemes on best device performance for low-power applications.

2 Device Structures and Physical Models

2.1 Device Specifications

For tunnel FET the boosting of I_{ON} has been a real challenge. The conventional DG-TFET has equal source and drain regions of 20 nm length with channel length of 50 nm. In order to introduce effective area of BTBT tunneling, the conventional

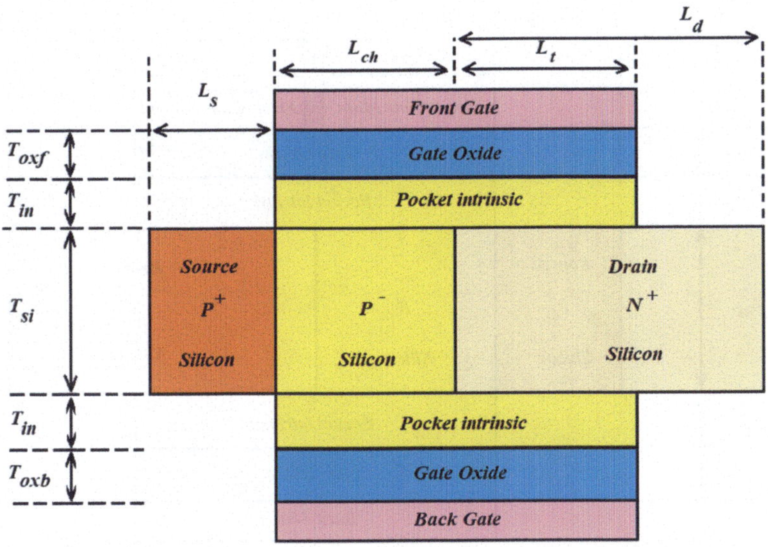

Fig. 1 Cross-sectional view of double gate pocket intrinsic tunnel FET structure (DG-PI-TFET)

double gate TFET structure has been modified by incorporating thin pocket intrinsic layers both in the front and back of silicon substrate as shown in Fig. 1. This structure is termed as double gate pocket intrinsic TFET (DG-PI-TFET). In DG-PI-TFET structure, the drain region has been extended in order to access overlapping region from 20 nm to 45 nm, which shrinks the 50 nm channel length to 25 nm, whereas the remaining 25 nm of the channel length is used as tunneling length.

The conventional DG-TFET structure with pocket intrinsic layer (DG-PI-TFET) shows a considerable increase in the ON current. But the OFF current has been an issue as for fast switching.

Therefore, by introducing gate engineering, the front gate has been redesigned into two lateral poly gates having length of 24 nm each. A narrow vertical gap of 2 nm is created in between of the two newly alienated gate named as: front gate main and front gate control. The vertical gap is filled with same gate oxide material, i.e., silicon. By introducing triple gate (i.e., front gate main, front gate control, and back gate), it enhances better control on the transposition of carriers through quantum band-to-band tunneling. This exhibits a high switching (I_{ON}/I_{OFF}) ratio for a channel length of 50 nm compared to double gate TFET structure.

Although multigate structure of tunnel FET provides better drive current (I_{ON}), but the leakage current has been an issue for obtaining better switching (I_{ON}/I_{OFF}) ratio. Keeping this limitation in view, the thin pocket intrinsic layers are placed similar to that of double gate pocket intrinsic tunnel FET structure for analyzing I_{ON}, I_{OFF}, and I_{ON}/I_{OFF} ratio.

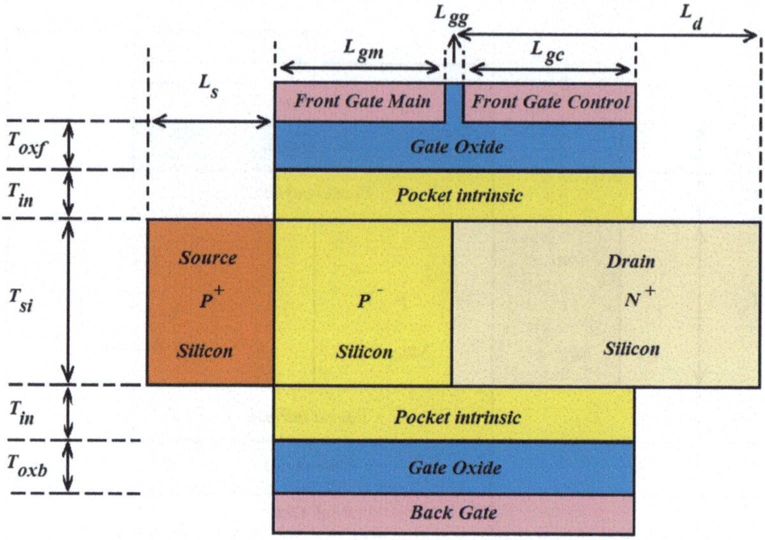

Fig. 2 Cross-sectional view of proposed triple gate pocket intrinsic tunnel FET structure (TG-PI-TFET)

Figure 2 shows the cross-sectional view of proposed triple gate structure with pocket intrinsic layer keeping the silicon film thickness and the gate length unchanged. In addition to this, the triple gate will help to control the barrier width better. This overlapping region at channel drain junction has been observed as an effective gate length or tunneling length. By using band-to-band tunneling (BTBT) phenomenon, the probability of transposition of carriers from source to drain is increased.

The proposed triple gate tunnel FET structure with pocket intrinsic layer (TG-PI-TFET) shows a considerable increase in the ON current without degrading the leakage current for obtaining better I_{ON}/I_{OFF} ratio for a pocket doping concentration of 1×10^0 cm^{-3}.

2.2 Device and Electrical Parameter Analysis

The DG-PI-TFET and TG-PI-TFET are designed by varying the device and electrical parameters to obtain better I_{ON}. The Source/Drain Junction Parameters and Doping concentrations are also varied for multigate tunnel FET Structures to analyze the leakage current (I_{OFF}). The typical parameters for the abovementioned tunnel FET structures are summarized in Tables 1 and 2.

Table 1 Typical parameters of various tunnel FET structures

Parameters	DG-TFET Ref. [18]	DG-PI-TFET	TG-PI-TFET
Gate length/ Intrinsic channel length (L_{ch})	50 nm	25 nm	24 nm
Tunneling length (L_t)	–	25 nm	24 nm
Length of source (L_S)	20 nm	20 nm	20 nm
Length of drain (L_d)	20 nm	45 nm	45 nm
Length of gate gap (L_{gg})	–	–	2 nm
Length of every separated gate (L_{gm} *and* L_{gc}) [$L_{ch} = L_{gm} + L_{gg} + L_{gc}$]	–	–	24 nm
Gate oxide thickness (T_{oxf})	2 nm	2 nm	2 nm
Gate oxide thickness (T_{oxb})	2 nm	2 nm	2 nm
Silicon film thickness (T_{si})	10 nm	10 nm	10 nm
Pocket intrinsic thickness (T_{in})	–	5 nm	5 nm
Gate voltage (V_G)	3 V	3 V	3 V($V_{GM} = V_{GC}$)
Supply voltage (V_D)	0.5 V	0.5 V	0.5 V
Gate work-function (Φ)	4.8 eV	4.8 eV	4.8 eV

Table 2 Source/drain junction parameters and doping used of various tunnel FET structures

Parameters	DG-TFET [Ref 18]	DG-PI-TFET	TG-PI-TFET
P^+ source doping concentration (N_a)	1×10^{20} cm^{-3}	1×10^{20} cm^{-3}	1×10^{20} cm^{-3}
N^+ drain doping concentration (N_d)	1×10^{18} cm^{-3}	1×10^{18} cm^{-3}	1×10^{18} cm^{-3}
Intrinsic doping concentration (N_i)	1×10^{15} cm^{-3}	1×10^{15} cm^{-3}	1×10^{15} cm^{-3}
Source p-n junction abruptness	–	2–3 nm/dec	2–3 nm/dec
Peak pocket doping	–	1×10^{0} to 1×10^{15} cm^{-3}	1×10^{0} to 1×10^{15} cm^{-3}

2.3 Simulation Framework

The double gate and triple gate tunnel FET device with their modified pocket intrinsic structures, i.e., DG-PI-TFET and TG-PI-TFET are designed in SILVACO TCAD device simulator. The results have been thoroughly compared and analyzed. In this paper, the Shockley–Read–Hall (SRH) model is used for the recombination effects at junctions. Velocity saturation in high electric field is modeled using the field-dependent mobility model (FLDMOB). The bandgap narrowing (BGN) model has been used for high doping concentration and the non-local band-to-band tunneling (BTBT) model is introduced for the tunneling in lateral direction.

3 Results and Discussions

3.1 Transfer Characteristics

The I_{ON}/I_{OFF} characteristics for all different TFET structures are performed with gate voltage (V_G) variation from –3 V to 3 V. The drain current obtained in log scale where the triple gate PI-TFET attains maximum value of I_{ON} as 1.286×10^{-4} A/μm, whereas the conventional DG-PI-TFET provides $(I_D)_{max}$ as 1.018×10^{-4} A/μm as shown in Fig. 3.

Now, to analyze the I_{OFF} characteristics for different multigate TFET structures, the gate voltage (V_G) is varied from –3 V to 0 V. The drain current obtained in log scale where the TG-PI-TFET attains minimum value of I_{OFF} as 2.335×10^{-5} A/μm, whereas the conventional DG-PI-TFET provides $(I_D)_{min}$ as 3.113×10^{-5} A/μm.

From Fig. 3, it can be seen that to boost I_{ON}, the leakage current (I_{OFF}) simultaneously plays very important role for designing fast switching device. Due to reduction in I_{OFF} for the TG-PI-TFET structure, the subthreshold slope can be increased. Table 3 provides the detail analysis and comparison of I_{ON}, I_{OFF}, and I_{ON}/I_{OFF} ratio.

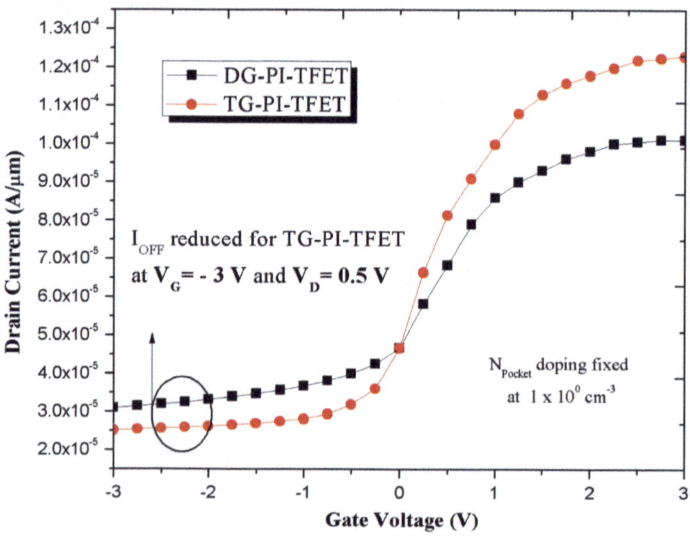

Fig. 3 I_{ON}/I_{OFF} characteristics of conventional DG-PI-TFET and TG-PI-TFET. The I_D versus V_G curves where I_D in log scale and V_G in linear scale when $V_D = 0.5$ V

Table 3 I_{ON}/I_{OFF} ratio of conventional DG-PI-TFET and TG-PI-TFET for gate voltage variation with supply voltage $V_D = 0.5$ V

Devices	I_{ON} at $V_G = 3$ V	I_{OFF} at $V_G = -3$ V	I_{ON}/I_{OFF} ratio
DG-PI-TFET	1.018×10^{-4} A/μm	3.113×10^{-5} A/μm	3.27
TG-PI-TFET	1.286×10^{-4} A/μm	2.335×10^{-5} A/μm	5.51

3.2 *Pocket Intrinsic Doping Variation*

The source doping and pocket intrinsic doping concentration variations are the area of research for designing tunnel FET as fast switching device. Since the ON state tunneling occurs at the source–channel junction, so the source doping concentration plays an important role for the ON current assessment of TFETs. At low doping concentrations, the ON current increases exponentially with increasing the source doping concentration. At high doping levels, the ON current starts to saturate [15].

Similarly, the bandgap narrowing influence tunnel barrier height and the intrinsic channel Fermi energy determining tunneling electron number are defined exclusively by the pocket intrinsic doping concentration.

In this paper, the double gate and triple gate pocket intrinsic TFET device performance is examined by varying the pocket doping concentration (N_{Pocket}) from 1×10^0 cm^{-3} to 1×10^{18} cm^{-3}, keeping source doping (N_a) at 1×10^{20} cm^{-3}. From Fig. 4, it is evident that the boosting of I_{ON} is further limited at pocket doping of 1×10^{18} cm^{-3}, without sacrificing the I_{OFF} leakage current.

Whereas for triple gate pocket intrinsic TFET although the off current was manageable for pocket doping concentration of 1×10^0 cm^{-3} at supply voltage $V_D = 0.5$ V. But with further raise in pocket doping the I_{ON} remains unaltered and the I_{OFF} is increased here (Fig. 5).

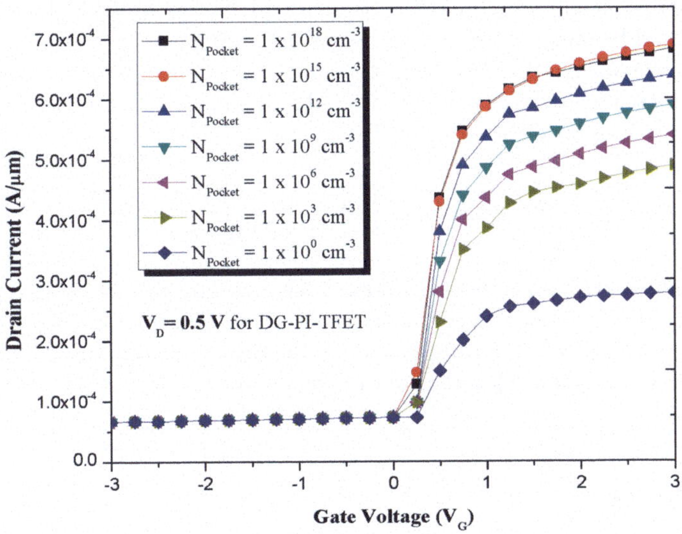

Fig. 4 The simulated I_D versus V_G curves for different pocket doping concentrations of conventional DG-PI-TFET where I_D in log scale and V_G in linear scale when $V_D = 0.5$ V

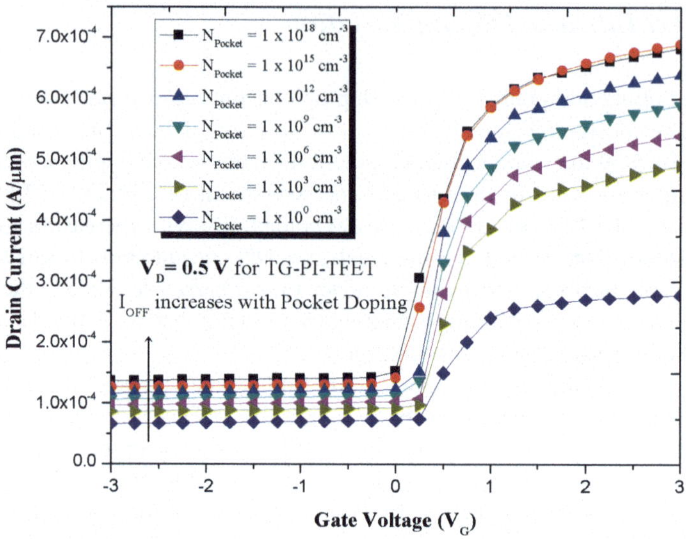

Fig. 5 The simulated I_D versus V_G curves for different pocket doping concentrations of conventional TG-PI-TFET where I_D in log scale and V_G in linear scale when $V_D = 0.5$ V

3.3 *Subthreshold Swing (SS)*

Keeping the overall gate length same as 50 nm for all multigate tunnel FET structures, the subthreshold swing (*SS*) is also measured and compared at $V_D = 0.5$ V, $T_{si} = 10$ nm, $T_{ox} = 2$ nm. Since in all the transfer characteristics curves the drain current (I_D) is taken in log scale, therefore using the formula for finding *SS* swing the comparison report is thoroughly investigated.

$$Sub - threshold\ Swing\ (SS) = \frac{dV_g}{d(\log_{10} I_d)} \text{mV/dec} \tag{1}$$

The comparison report is shown in Table 4. The subthreshold swing (*SS*) provided by the DG-PI-TFET has been recorded best as 23.93 mV/dec in comparison with TG-PI-TFET as 30.22 mV/dec for different pocket doping concentrations. 23.93 mV/dec subthreshold swing is much less than that of other non-conventional TFET structures.

4 Conclusion

In this work, we have reviewed various multigate TFET structures by using 2D TCAD simulations on the performance enhancement of I_{ON} for the low-power applications. I_{ON}/I_{OFF} and *SS* as figures of merit for DC performance are also investigated.

Table 4 Subthreshold swing (SS) of DG-PI-TFET and TG-PI-TFET for gate voltage variation V_G from –3 V to 3 V with V_D = 0.5 V

Devices	Pocket doping concentration (cm^{-3})	Subthreshold swing (*SS*) (mV/dec)
DG-PI-TFET	1×10^0	40.25
	1×10^3	30.56
	1×10^6	28.31
	1×10^9	24.83
	1×10^{12}	24.25
	1×10^{15}	23.93
	1×10^{18}	23.93
TG-PI-TFET	1×10^0	41.47
	1×10^3	30.67
	1×10^6	30.42
	1×10^9	29.66
	1×10^{12}	29.73
	1×10^{15}	30.06
	1×10^{18}	30.22

I_{ON}/I_{OFF} is significantly larger in TG-PI-TFET making the device suitable for low-power sensor applications without altering the doping concentration. But variation in pocket doping limits the subthreshold slope for TG-PI-TFET. So this shows need of optimization in source and pocket doping throughout the TFET device could pave the way to operate the TFET considering the leakage current.

References

1. Thomas, N., Philip Wong, H.S.: The end of Moore's law: a new beginning for information technology. IEEE J. Comput. Sci. Eng. **19**(2), 41–50 (2017)
2. Nikonov, D.E., Young, I.A.: Overview of beyond-CMOS devices and a uniform methodology for their benchmarking. Proc. IEEE **101**(12), 2498–2533 (2013)
3. Seabaugh,. A., Alessandri, C., Min Li, H., Paletti, P.: Steep slope transistors: tunnel FETs and beyond,.In: Proceedings of 46th European Solid State Device Research Conference, pp. 349–351. IEEE (2016)
4. Choi, W.Y., Park, B.G., Lee, J.D., King Liu, T.J.: Tunneling field-effect transistors (TFETs) with sub-threshold swing (SS) less than 60 mV/dec. IEEE Electron Device Lett. **28**(8), 743–745 (2007)
5. Zhang, Q., Zhao, W., Seabaugh, A.: Low-subthreshold-swing tunnel transistors. IEEE Electron Device Lett. **27**(4), 297–300 (2006)
6. Hraziia, Vladimirescu, A., Amara, A., Anghel, C.: An Analysis on the Ambipolar Current in Si Double-Gate Tunnel FETs, Solid-State Electronics, vol. 70, pp. 67–72. Elsevier (2012)
7. Bagga, N., Sarkhel, S., Sarkar, S.K.: Recent research trends in gate engineered tunnel FET for improved current behavior by subduing the ambipolar effects: a review. In: Proceedings of IEEE International Conference on Computing, Communication and Automation. India (2015)

8. Pal, A., Sachid, A.B., Gossner, H., Rao, V.R.: Insights Into the design and optimization of tunnel-FET devices and circuits. IEEE Trans. Electron Devices **58**(4), 1045–1053 (2011)
9. Datta, S., Liu, H., Narayanan, V.: Tunnel FET Technology: A Reliability Perspective, Microelectronics Reliability, vol. 54, pp. 861–874. Elsevier (2014)
10. Leonelli, D., Vandooren, A., Rooyackers, R., Gendt, S., Heyns, M.M., Groeseneken, G.: Drive Current Enhancement in p-Tunnel FETs by Optimization of the Process Conditions, Solid-State Electronics, vol. 65, pp. 28–32. Elsevier (2011)
11. Boucart, K., Ionescu, A.M.: Length Scaling of the Double Gate Tunnel FET with a High-K Gate Dielectric, Solid-State Electronics, vol. 51, pp. 1500–1507. Elsevier (2007)
12. Verhulst, A.S., Sorée, B., Leonelli, D., Vandenberghe, W.G., Groeseneken, G.: Modeling the single-gate, double-gate, and gate-all-around tunnel field-effect transistor. J. Appl. Phys. **107**(2), 024518 (2010)
13. Seunggyu, J.,Hyungtak, K., Hwan, C.: Characteristics of recess structure tunneling field effect transistor for high on current drivability. J. Semicond. Technol. Sci. **18**(3) (2018)
14. Wei, S., Zhang, G., Geng, L., Shao, Z., Yang, C.F.: Comparison of the Performance Improvement for the Twonovel SOI-Tunnel FETs with the Lateral Dual-Gate and Triple-Gate, Microsystem Technologies. Springer, Berlin (2018)
15. Jhaveri, R., Nagavarapu, V., Woo, J.: Effect of pocket doping and annealing schemes on the source-pocket tunnel field-effect transistor. IEEE Trans. Electron Devices **58**(1), 80–86 (2011)
16. Kumar, M.J., Ramaswamy, S.: Double gate symmetric tunnel FET: investigation and analysis. IET Circuits Devices Syst. **11**(4), 365–370 (2017)
17. Graef, M., Hosenfeld, F., Horst, F., Farokhnejad, A., Hain, F., Iñíguez, B., Kloes, A.: Advanced Analytical Modeling of Double-Gate Tunnel-FETs—A Performance Evaluation, Solid-State Electronics, vol. 141, pp. 31–39. Elsevier (2018)
18. Gracia, D., Nirmal, D.: Performance Analysis of Dual Metal Double Gate Tunnel-FETs for Ultralow Power Applications, Nanoelectronic Materials and Devices–LNEE, vol. 466, pp. 11–18. Springer, Berlin (2018)

Design and Evaluation of Neale-Based Multi-bit Adjacent Error-Correcting Codec for SRAM

Sayan Tripathi, Jhilam Jana, Jagannath Samanta, Aman Raj, Deep Ranjan and Mayank Pratap Singh

Abstract Due to scaling in CMOS technology, multiple bit upsets (MBUs) have been widely occurred in memories. As a result, multiple adjacent bits of memories are corrupted and valuable information are lost forever. To mitigate these problems, multi-bit adjacent error-correcting codes are generally employed in SRAM memories. Single error correction-double error detection-double adjacent error correction (SEC-DED-DAEC) codes are used to mitigate radiation or noise source induced MBUs to protect static random access memory (SRAM) devices. These codes are able to correct single and double adjacent errors, and also detect double errors. In this paper, three different SEC-DED-DAEC codes have been designed and implemented for SRAM memories. All functional blocks of these codecs are simulated and synthesized both in FPGA and ASIC platforms. Performances of the different SEC-DED-DAEC codes are observed in terms of area and delay.

Keywords Neale code · Error Correction Codes (ECCs) · Memories · Soft errors · SEC-DED-DAEC · Multiple Bit Upsets (MBUs)

S. Tripathi (✉) · J. Jana
Dept. of ETCE, Jadavpur University, Jadavpur, Kolkata 700032, India
e-mail: tripathysayan@gmail.com

J. Jana
e-mail: jhilamjana2014@gmail.com

J. Samanta · A. Raj · D. Ranjan · M. P. Singh
Haldia Institute of Technology, Haldia, India
e-mail: jagannath19060@gmail.com

A. Raj
e-mail: amanstm1410@gmail.com

D. Ranjan
e-mail: deepranjan.02@gmail.com

M. P. Singh
e-mail: Mayank.jsr2196@gmail.com

S. Kundu et al. (eds.), *Proceedings of the 2nd International Conference on Communication, Devices and Computing*, Lecture Notes in Electrical Engineering 602, https://doi.org/10.1007/978-981-15-0829-5_26

1 Introduction

The information are hampered due to soft errors. These errors can also affect an instruction in a system program [1–4]. Soft error can be occurred from noise, signal integrity, and cosmic rays [5]. Error-correcting codes (ECCs), including Hamming codes, are commonly used in situations where there is a possibility that data may be corrupted and the consequences of such an error are unacceptable [6]. SRAM has been widely used since decades due to its high random access speeds and compatibility with CMOS logic process. It is typically used as the main building block for system registers, on-chip cache, etc. SRAM bit cells can also be optimized at the transistor level for high performance. ECCs can be used in data situations to safeguard the reading and writing of storage devices, as in ECC random access memory (RAM), or can be used in transmission situations, where data is sent over a potentially noisy channel. Spaceflight is one of the former situations where spacecraft, unshielded by Earth's atmosphere, can be struck by cosmic rays which can change the data stored in memory cells [1, 2]. The SEC-DED codes are capable of correcting single random error and detecting all possible of random double errors. These codes have been presented to secure SRAM devices.

In this paper, we have been able to design and implement the Neale-based SEC-DED-DAEC codes [1, 2].

This paper is organized for remainder portion as follows. Section 2 provides the related work. Section 3 presents the basics of SEC-DED-DAEC code, H-matrix construction rules, and three different H-matrices. Section 4 provides the encoding and decoding techniques of optimized code. Section 5 presents the area and delay analysis. Section 6 contains the FPGA and ASIC synthesis results. Section 7 presents the conclusion.

2 Related Work

Mitigation of single bit-error is commonly accomplished with Hamming [6] and Hsiao [7] codes. These SEC-DED codes are considered to have an acceptable latency and area overhead and are found in many applications such as ECC RAM, where the error rate is low [7]. These codes can be extended to offer double error correction with an increase in the number of parity bits. The Bose–Chaudhuri–Hocquenghem (BCH) codes required additional area and latency to prevent widespread adoption [1, 8]. These codes are implemented most commonly as multi-cycle solutions, which are unacceptable in many applications such as embedded memories [1, 3, 9]. The addition of a single parity bit is sufficient to provide double error correction, but the tradeoff is that parity bit cannot be calculated until the rest of the parity bits have been calculated, resulting in a longer critical path and hence higher latency. A method of this calculation has been investigated, but this method increases the size and therefore cost of the error correction circuit. Another way that multiple bit-errors have traditionally been addressed is to use interleaving [2, 10–12]. The result of this is that a single event that upsets multiple bits will produce multiple single bit-errors instead of a single multiple bit-errors. The fundamental principle is the distribution of the bits within a single data word. This technique requires complicated wiring and in certain cases, such as register files, is not feasible. Moreover, in situations such as spaceflight, where cosmic rays with high linear energy transfer (LET) characteristics may strike the device, single events with up to thirteen bit-flips. In the next section, the basics of SEC-DED-DAEC codes have been discussed.

3 Basics of SEC-DED-DAEC Code

In this section, parity check matrices have been constructed for the proposed codes with up to 11-bits of adjacent error detection for memories with 16, 32 and 64 data bits per codeword. The matrix is divided into two separate portions (top and bottom) each responsible for providing different behavior. There are series of horizontally concatenated L identity matrices. Code bits are divided physically. H-matrix is used to calculate each of the syndrome bits. The top portion contains n/L sub-matrices, H_i. For each H_i sub-matrix, the odd column vectors are identical to one another, and likewise for the even column vectors. For example, for a four-column H_i sub-matrix, the first and third columns are identical to one another, and the same is true for the second and fourth columns. The four-column H_i sub-matrix is described by

$$H_i = (hoi,\ hei,\ hoi,\ hei) \tag{1}$$

where hoi represents the odd column vectors for the ith top portion sub-matrix and hei represents the even column vectors.

3.1 *H*-Matrix Construction Rules for SEC-DED-DAEC Code

The H-matrix is constructed by a set of constraints. The constraints for SEC-DED-DAEC codes are as follows:

1. Every column should not be zero.
2. Every column should have a distinct value.
3. XOR results of any two columns should be different with any of the individual columns.
4. XOR results of any two adjacent columns should be distinct, and separate from any column in H.
5. XOR result of any three adjacent columns should be unique and dissimilar from the single bit and double adjacent bit-error-correcting syndromes.
6. Number of syndromes between adjacent and non-adjacent errors must be decreased.

Fig. 1 H-matrix of (41, 32) IIMS SEC-DED-DAEC code

$$H = \begin{bmatrix} 000000010101101010010101000000010101011111 \\ 101010111111101010000000001010101010111111 \\ 101010000000111111101010000000001010101010 \\ 10000010000010000010000010000010000010000 \\ 01000001000001000001000001000001000001000 \\ 00100000100000100000100000100000100000100 \\ 00010000010000010000010000010000010000010 \\ 00001000001000001000001000001000001000001 \\ 00000100000100000100000100000100000100000 \end{bmatrix}$$

Fig. 2 H-matrix of (24, 16) ICB SEC-DED-DAEC code

$$H = \begin{bmatrix} 010101010000101000000000 \\ 000001010101000010100000 \\ 000010101010010100000000 \\ 000010100000000000001111 \\ 100010001000100010001000 \\ 010001000100010001000100 \\ 001000100010001000100010 \\ 000100010001000100010001 \end{bmatrix}$$

Fig. 3 H-matrix of (24, 16) optimized code

$$H = \begin{bmatrix} 111010101000000000001010 \\ 010100000000010101010100 \\ 000000000010111010101010 \\ 100001000010000100001000 \\ 010000100001000010000100 \\ 001000010000100001000010 \\ 000100001000010000100001 \\ 000010000100001000010000 \end{bmatrix}$$

The H-matrices of (41, 32) IIMS SEC-DED-DAEC code, (24, 16) ICB SEC-DED-DAEC code, and (24, 16) optimized codes have been considered and implemented in Figs. 1, 2 and 3. We are also observing the simulated waveform and corresponding output in terms of area and delay in FPGA and ASIC platform.

In Fig. 1, H-matrix of (41, 32) IIMS SEC-DED-DAEC code is mentioned. The (n) number of columns in parity check (H) matrix has (k) number of data and $(n-k)$ number of parity columns. This H-matrix consists of 32 data (d_1–d_{32}) columns and 9 parity (c_1–c_9) columns.

The H-matrix of (24, 16) ICB SEC-DED-DAEC code is provided in Fig. 2. In this parity check matrix, 16 data and 8 parity are presented.

The H-matrix of (24, 16) optimized code is described in Fig. 3. The H-matrix of (24, 16) optimized code contains of d_1–d_{16} data columns and c_1–c_8 parity columns. $(n-k)$ rows and (n) columns are required for consisting (n, k) (24, 16) optimized code.

4 Encoding and Decoding Techniques

In this section, the (24, 16) optimized code Fig. 3 is designed and implemented. The encoding and decoding techniques are described. In encoding technique, the arrangement of codeword of the (24, 16) optimized code with maximum row check-bits is as follows:

$$v = (c_1 d_1 d_2 c_2 d_3 c_4 d_4 c_6 d_5 c_8 d_6 c_5 c_3 d_7 d_8 d_9 d_{10} d_{11} d_{12} d_{13} d_{14} d_{15} d_{16} c_7) \quad (2)$$

The G_{Top} matrix in Fig. 3 provides the least significant (c_{LSBs}) check bit calculations. These calculations are shown in Eq. (3).

$$\begin{aligned} c_1 &= d_1 \oplus d_2 \oplus d_3 \oplus d_4 \oplus d_5 \oplus d_{14} \oplus d_{16} \\ c_2 &= d_1 \oplus d_7 \oplus d_9 \oplus d_{11} \oplus d_{13} \oplus d_{15} \\ c_3 &= d_6 \oplus d_7 \oplus d_8 \oplus d_{10} \oplus d_{12} \oplus d_{14} \oplus d_{16} \end{aligned} \quad (3)$$

While the G_{Bottom} matrix provides the following most significant (c_{MSBs}) check bit calculations in Eq. (4).

$$\begin{aligned} c_4 &= d_1 \oplus d_2 \oplus d_3 \oplus d_4 \oplus d_5 \oplus d_6 \oplus d_9 \oplus d_{16} \\ c_5 &= d_1 \oplus d_4 \oplus d_{10} \oplus d_{15} \\ c_6 &= d_2 \oplus d_6 \oplus d_7 \oplus d_8 \oplus d_{10} \oplus d_{11} \oplus d_{12} \oplus d_{14} \\ c_7 &= d_1 \oplus d_5 \oplus d_9 \oplus d_{11} \oplus d_{12} \oplus d_{13} \oplus d_{15} \\ c_8 &= d_3 \oplus d_8 \oplus d_{13} \end{aligned} \quad (4)$$

The schematic diagram of parity bit generator logic of (24, 16) optimized code have been implemented as follows in Fig. 4.

In this decoding technique, the detection of error is calculated by the syndrome value. For received codeword, if the syndrome value is zero ($SY = 0$), it means the received codeword has no error. On the other hand, if the syndrome value is no-zero ($SY \neq 0$), it means there are some bit-errors. The matrix in Fig. 3 provides the following syndrome calculations in Eq. (5).

$$\begin{aligned} s_1 &= r_1 \oplus r_2 \oplus r_3 \oplus r_5 \oplus r_7 \oplus r_9 \oplus r_{21} \\ s_2 &= r_2 \oplus r_4 \oplus r_{14} \oplus r_{16} \oplus r_{18} \oplus r_{20} \oplus r_{22} \\ s_3 &= r_{11} \oplus r_{13} \oplus r_{14} \oplus r_{15} \oplus r_{17} \oplus r_{19} \oplus r_{21} \oplus r_{23} \\ s_4 &= r_1 \oplus r_6 \oplus r_{11} \oplus r_{16} \oplus r_{21} \\ s_5 &= r_2 \oplus r_7 \oplus r_{12} \oplus r_{17} \oplus r_{22} \\ s_6 &= r_3 \oplus r_8 \oplus r_{13} \oplus r_{18} \oplus r_{23} \end{aligned}$$

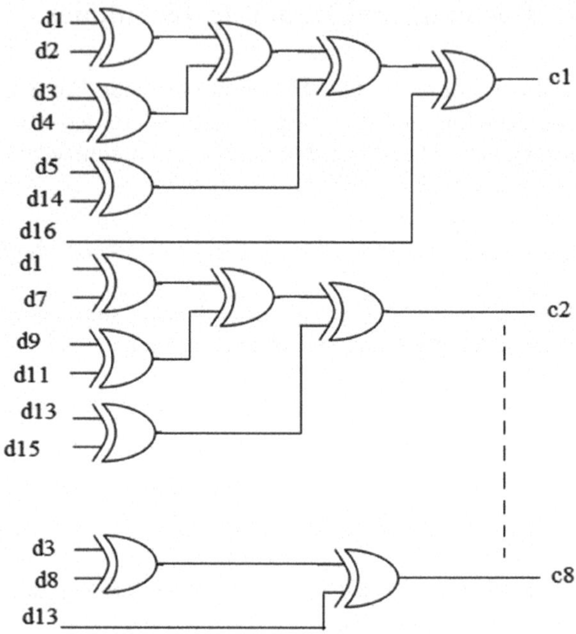

Fig. 4 Parity bit generation logic of (24, 16) optimized code

$$
\begin{aligned}
s_7 &= r_4 \oplus r_9 \oplus r_{14} \oplus r_{19} \oplus r_{24}\\
s_8 &= r_5 \oplus r_{10} \oplus r_{15} \oplus r_{20}
\end{aligned} \tag{5}
$$

The matrix in Fig. 3 provides the decoder output calculations. These calculations are presented in Eq. (6).

$$
\begin{aligned}
dc_1 &= r_2 \oplus ((s_1 s_2 s_3' s_4' s_5 s_6' s_7' s_8') + (s_1' s_2 s_3' s_4 s_5 s_6' s_7' s_8') + (s_1' s_2 s_3' s_4' s_5 s_6 s_7' s_8'))\\
dc_2 &= r_3 \oplus ((s_1 s_2' s_3' s_4' s_5' s_6 s_7' s_8') + (s_1' s_2 s_3' s_4' s_5 s_6 s_7' s_8') + (s_1 s_2 s_3 s_4' s_5' s_6 s_7 s_8'))\\
dc_3 &= r_5 \oplus ((s_1 s_2' s_3' s_4' s_5' s_6' s_7' s_8) + (s_1 s_2 s_3' s_4' s_5' s_6' s_7 s_8) + (s_1 s_2' s_3' s_4 s_5' s_6' s_7' s_8))\\
dc_4 &= r_7 \oplus ((s_1 s_2 s_3' s_4' s_5 s_6' s_7' s_8') + (s_1 s_2' s_3' s_4 s_5 s_6' s_7' s_8') + (s_1 s_2' s_3' s_4' s_5 s_6 s_7' s_8'))\\
dc_5 &= r_9 \oplus ((s_1 s_2' s_3' s_4' s_5' s_6' s_7 s_8') + (s_1 s_2' s_3' s_4' s_5' s_6 s_7 s_8') + (s_1 s_2' s_3' s_4' s_5' s_6' s_7 s_8))\\
dc_6 &= r_{11} \oplus ((s_1' s_2' s_3 s_4 s_5' s_6' s_7' s_8') + (s_1' s_2' s_3 s_4 s_5' s_6' s_7' s_8) + (s_1' s_2' s_3 s_4 s_5 s_6' s_7' s_8'))\\
dc_7 &= r_{14} \oplus ((s_1' s_2 s_3 s_4' s_5' s_6' s_7 s_8') + (s_1' s_2 s_3' s_4' s_5' s_6 s_7 s_8') + (s_1' s_2 s_3' s_4' s_5' s_6' s_7 s_8))\\
dc_8 &= r_{15} \oplus ((s_1' s_2' s_3 s_4' s_5' s_6' s_7' s_8) + (s_1' s_2 s_3' s_4' s_5' s_6 s_7 s_8') + (s_1' s_2 s_3' s_4' s_5' s_6' s_7 s_8))\\
dc_9 &= r_{16} \oplus ((s_1' s_2 s_3' s_4 s_5' s_6' s_7' s_8') + (s_1' s_2 s_3 s_4 s_5' s_6' s_7' s_8) + (s_1' s_2 s_3 s_4 s_5 s_6' s_7' s_8'))\\
dc_{10} &= r_{17} \oplus ((s_1' s_2' s_3 s_4' s_5 s_6' s_7' s_8') + (s_1' s_2 s_3 s_4 s_5 s_6' s_7' s_8') + (s_1' s_2 s_3 s_4' s_5 s_6 s_7' s_8'))\\
dc_{11} &= r_{18} \oplus ((s_1' s_2 s_3' s_4' s_5' s_6 s_7' s_8') + (s_1' s_2 s_3 s_4' s_5 s_6 s_7' s_8') + (s_1' s_2 s_3 s_4' s_5' s_6 s_7 s_8'))\\
dc_{12} &= r_{19} \oplus ((s_1' s_2' s_3 s_4' s_5' s_6' s_7 s_8') + (s_1' s_2 s_3 s_4' s_5' s_6 s_7 s_8') + (s_1' s_2 s_3 s_4' s_5' s_6' s_7 s_8))
\end{aligned}
$$

$$
\begin{aligned}
dc_{13} &= r_{20} \oplus ((s_1' s_2 s_3' s_4' s_5' s_6' s_7' s_8) + (s_1' s_2 s_3 s_4' s_5' s_6' s_7 s_8) + (s_1 s_2 s_3 s_4 s_5' s_6' s_7' s_8)) \\
dc_{14} &= r_{21} \oplus ((s_1 s_2' s_3 s_4 s_5' s_6' s_7' s_8') + (s_1 s_2 s_3 s_4 s_5' s_6' s_7' s_8) + (s_1 s_2 s_3 s_4 s_5 s_6' s_7' s_8')) \\
dc_{15} &= r_{22} \oplus ((s_1' s_2 s_3' s_4' s_5 s_6' s_7' s_8') + (s_1 s_2 s_3 s_4 s_5 s_6' s_7' s_8') + (s_1 s_2 s_3 s_4' s_5 s_6 s_7' s_8')) \\
dc_{16} &= r_{23} \oplus ((s_1 s_2' s_3 s_4' s_5' s_6 s_7' s_8') + (s_1 s_2 s_3 s_4' s_5 s_6 s_7' s_8') + (s_1 s_2' s_3 s_4' s_5' s_6 s_7 s_8'))
\end{aligned}
\tag{6}
$$

5 Area and Delay Analysis

This section presents the theoretical analysis which comprises area overhead and critical path delay.

5.1 *Area Overhead*

Theoretical area overhead comparison of (41, 32) IIMS SEC-DED-DAEC code, (24, 16) ICB SEC-DED-DAEC code and (24, 16) optimized codes are shown in Table 1. We have presented the area complexity in terms of equivalent two input NAND (NAND2) gates.

5.2 *Critical Path Delay*

In Table 2, there are critical path delay comparison of (41, 32) IIMS SEC-DED-DAEC code, (24, 16) ICB SEC-DED-DAEC code and (24, 16) optimized codes. The number of gates that we have used for this critical path delay calculation is shown in Table 2.

Table 1 Area overhead comparison of different SEC-DED-DAEC codes

Code	Block	XOR2	AND2	OR2	NOT	Equivalent NAND2
(41, 32) IIMS	Encoder	117	–	–	–	468
	Decoder	158	760	63	532	2873
(24, 16) ICB	Encoder	43	–	–	–	172
	Decoder	67	384	32	231	1363
(24, 16) optimized	Encoder	88	–	–	–	352
	Decoder	127	570	63	321	2158

Table 2 Critical path delay comparison of different SEC-DED-DAEC codes

Code	XOR2	AND2	OR2	NOT	Equivalent NAND2
(41, 32) IIMS	42	8	2	1	191
(24, 16) ICB	18	7	2	1	93
(24, 16) optimized	16	7	2	1	85

6 FPGA and ASIC Synthesis Results

6.1 FPGA-Based

The SEC-DED-DAEC (41, 32) IIMS, SEC-DED-DAEC (24, 16) ICB, and (24, 16) optimized codes have been represented in Verilog hardware description language. All functional blocks are simulated and synthesized using Virtex-6 (XC6VCX75T) FPGA device family. The FPGA synthesis results have been presented in Table 3 in terms area (look-up tables (LUTs)), input–output buffers (IOBs), and delay.

Table 3 represents the FPGA-based synthesis results for three different codes.

6.2 ASIC-Based

In this section, ASIC-based synthesis result of (41, 32) IIMS SEC-DED-DAEC code, (24, 16) ICB SEC-DED-DAEC code, and (24, 16) optimized codes for 350 nm technology in Leonardo Spectrum tool (Mentor Graphics) are presented. The number of gates and the number of accumulated instances are provided in Table 4. The slack time is the difference between time of data required and time of data arrival. This parameter is also obtained in Table 4.

Table 3 FPGA-based synthesis result

Code	Block	Area (LUTs)	IOBs	Delay (ns)
(41, 32) IIMS	Encoder	21	41	1.72
	Decoder	112	82	3.77
(24, 16) ICB	Encoder	12	24	1.17
	Decoder	48	48	2.64
(24, 16) Optimized	Encoder	13	24	1.13
	Decoder	57	48	2.77

Table 4 ASIC-based synthesis results

Code	Block	Number of gates	Number of accumulated instances	Slack time
(41, 32) IIMS	Encoder	118	61	7.63
	Decoder	387	259	5.18
(24, 16) ICB	Encoder	67	35	8.90
	Decoder	194	129	7.00
(24, 16) Optimized	Encoder	56	29	8.46
	Decoder	211	150	6.63

7 Conclusion

In this paper, the different SEC-DED-DAEC codes have been evaluated. We have analyzed the critical path and area complexity of three SEC-DED-DAEC codes architecture. The performance of three different codes is investigated in FPGA and ASIC platforms. The synthesis results are observed in terms of LUTs and delay. These works are very useful for VLSI designers. The transistor level implementation of (24, 16) SEC-DED-DAEC optimized code is our future motto.

References

1. Neale, A., Jonkman, M., Sachdev, M.: Adjacent-MBU-tolerant SEC-DED-TAEC-*y*AED codes for embedded SRAMs. IEEE Trans. Circ. Syst. II: Express Br. **62**(4), 387–391 (2015)
2. Neale, A., Sachdev, M.: A new SEC-DED error correction code subclass for adjacent MBU tolerance in embedded memory. IEEE Trans. Device Mater. Reliab. **13**(1), 223–230 (2013)
3. Ibe, E., Taniguchi, H., Yahagi, Y., Shimbo, K.I., Toba, T.: Impact of scaling on neutron-induced soft error in SRAMs from a 250 nm to a 22 nm design rule. IEEE Trans. Electron Device **57**(7), 1527–1538 (2010)
4. Dutta, A., Touba, N.A.: Multiple bit upset tolerant memory using a selective cycle avoidance based SEC-DED-DAEC code. In: 25th IEEE VLSI Test Symposium, pp. 349–354. IEEE (2007)
5. Dixit, A., Wood, A.: The impact of new technology on soft error rates. In: 2011 IEEE International Reliability Physics Symposium (IRPS), pp. 5B–4. IEEE (2011)
6. Hamming, R.W.: Error detecting and error correcting codes. Bell Syst. Tech. J. **29**(2), 147–160 (1950)
7. Hsiao, M.Y.: A class of optimal minimum odd-weight-column SEC-DED codes. IBM J. Res. Dev. **14**(4), 395–401 (1970)
8. Naseer, R., Draper, J.: Parallel double error correcting code design to mitigate multi-bit upsets in SRAMs. In: 34th European Solid-State Circuits Conference. ESSCIRC 2008, pp. 222–225 (2008)
9. Alameldeen, A.R., Wagner, I., Chishti, Z., Wu, W., Wilkerson, C., Lu, S.L.: Energy-efficient cache design using variable-strength error-correcting codes. ACM SIGARCH Comput. Archit. News **39**(3), 461–472 (2011)
10. Samanta, J., Tripathi, S.: Comments on a novel approach of error detection and correction for efficient energy in wireless networks. Multimed. Tools Appl. 1–6 (2018)

11. Neale, A., Sachdev, M.: A 0.4 V 75 kbit SRAM macro in 28 nm CMOS featuring a 3-adjacent MBU correcting ECC. In: 2014 IEEE Proceedings of the Custom Integrated Circuits Conference (CICC), pp. 1–4 (2014)
12. Ming, Z., Yi, X.L., Wei, L.H.: New SEC-DED-DAEC codes for multiple bit upsets mitigation in memory. In: 2011 IEEE/IFIP 19th International Conference on VLSI and System-on-Chip (VLSI-SoC), pp. 254–259. IEEE. (2011)

Performance of All-Optical Logic Soliton-Based AND Gate Using Reflective Semiconductor Optical Amplifier (RSOA)

K. Maji, K. Mukherjee and Asif Raja

Abstract In this chapter, we have analyzed and described all-optical soliton-based logic AND gate using refractive semiconductor optical amplifier (RSOA). We have used control signal and data signal as a soliton pulse. Input and output operations are also shown. We have found output extinction ratio, contrast ratio, relative eye-opening, and quality factor and how they vary with control pulse energy for the optimization of the performance.

Keywords AND gate · Refractive semiconductor optical amplifier (RSOA) · Soliton · Quality factor

1 Introduction

Reflective semiconductor optical amplifiers (RSOAs) are used to amplify signals in the downstream while demodulating them with end-user information in the upstream communication link [1]. Recently, RSOAs find increasing application for passive optical network (PON) applications [1–3]. RSOAs can also be used to design all-optical logic processors [4–6] and have higher gain at lower injection current compared to ordinary semiconductor optical amplifiers (SOAs). A signal double passes the RSOA length and gets amplified twice giving better performance compared to single-pass SOA. RSOAs have low noise figure and high optical gain for low drive current, and can be easily saturated at low optical power. In [5, 6], conventional SOAs along with RSOAs are used for logic gates and are complex in hardware. This limits the processing speed. But, in [4], only RSOAs have been used to design X-OR gate and processing speed of 120 Gbps has been reported. In this communication, AND gate based on only two RSOAs is designed and its performance is simulated. In [4], RSOA is modeled for cross-phase modulation and the X-OR gate is designed using

K. Maji · K. Mukherjee (✉) · A. Raja
Department of Physics, B.B. College, Asansol, India
e-mail: klmukherjee003@gmail.com

K. Maji
e-mail: kajalmaji200@gmail.com

S. Kundu et al. (eds.), *Proceedings of the 2nd International Conference on Communication, Devices and Computing*, Lecture Notes in Electrical Engineering 602, https://doi.org/10.1007/978-981-15-0829-5_27

interferometric structure. Unlike [4], in this communication, cross-gain modulation (XGM) is used to design the AND gate and RSOA is modeled for XGM without using any interferometric structure for the first time using soliton pulses as it has due advantage in long distance communication system.

2 Theoretical Modeling

The basic mechanism of working of this RSOA-based logic gate is to gain modulation of RSOA by application of control pulse. This effect of gain modulation controls the output of another signal called probe signal and the phenomenon is called cross-gain modulation. When both control signal and data signal pass through the RSOA, the gain of the RSOA decreases. So the output power of the RSOA is

$$\mathrm{OP_1} = \mathrm{RG(t)P_{in}(t)} \tag{1}$$

When the only data signal present in the RSOA in the absence of control signal passes through the RSOA, the output power of the RSOA is

$$\mathrm{OP_2} = \mathrm{R\,G_U G_U P_{in}(t)} \tag{2}$$

where R is the reflectivity, G_U is the unsaturated gain [7] of the RSOA, and $P_{in}(t)$ is the input power. When the control is injected into the RSOA, gain of the RSOA is [7]

$$\mathrm{G(t)} = \exp[2\mathrm{h(t)}] \tag{3}$$

where h(t) is calculated by following formula [7]:

$$h(t) = \ln\left[1 - \left(1 - \frac{1}{G_U}\right)\exp\left(-\frac{E_c(t)}{E_s}\right)\right] \tag{4}$$

where Es is the saturation energy [7] and $E_c(t)$ is the control pulse energy.

Now,

$$E_{cp}(t) = \int_{-\infty}^{t} P_{cp}(t')dt' \tag{5}$$

We consider the soliton pulse [4, 8, 9]

$$P_{cp}(t) = \sum_{n=1}^{n=N} a_{nA,B} P_{soli} \sec h^2\left(1.763\frac{(t - n\chi)}{\tau_{fwhm}}\right) \tag{6}$$

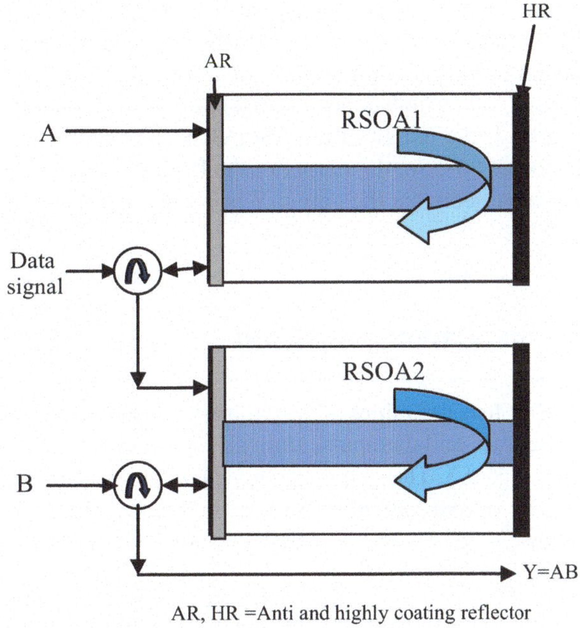

Fig. 1 AND gate using RSOA

where $P_{soli} = \left(\frac{1.763}{2\pi}\right)^2 \frac{A_{eff}\lambda^3 D}{n_2 c \tau_{fwhm}^2}$ is the soliton peak power, D is the dispersion constant, n_2 is the nonlinear coefficient, λ and c are the wavelength and velocity of light, A_{eff} is the fiber effective area, τ_{fwhm} is the full width half maximum, and E_{cp} $(t \to \infty) = P_{soli} \times \tau_{fwhm} = E_c$ is the total control pulse energy.

2.1 RSOA-Based AND Gate

Soliton-based AND gate using reflective semiconductor optical amplifier is shown in Fig. 1. It is made by two RSOA, RSOA1, and RSOA2.

2.2 Operating Principle of RSOA-Based AND Gate

Case 1. When both control signal pulse A and signal B are absent, i.e., 0. We see that in Fig. 1, there is no data signal present in RSOA2. So, the output of the Y becomes 0.
Case 2. When the control signal A becomes high, i.e., "1" and signal B is low, i.e., "0" at this time, the output of the RSOA1 becomes low but their is no data signal present in the RSOA2. Therefore, the output of the RSOA2 becomes zero, i.e., Y becomes "0".

Case 3. When the control signal A becomes low, i.e., "0" and signal B is high, i.e., "1" at this time the output of the RSOA1 becomes high. So, the output of the RSOA2 becomes low, i.e., Y becomes "0".
Case 4. When both control signal pulse A and signal B are present, i.e., 1. We see that in Fig. 1, the output of the RSOA1 becomes low, i.e., control signal of the RSOA2 becomes "0". So, the output of the RSOA2 becomes high, i.e., Y becomes "1".
We have seen that the output Y is 0 0 0 1 and thus it gives AND operation.

3 Simulation and Result

Table 1 lists the values of the parameters used in the simulation. We have used control pulse train and data pulse train as soliton pulses. The signals A and B are shown in Figs. 2 and 3. The output of the AND operation is shown in Fig. 4.

The output extinction ratio is defined as the ratio of minimum values of peak power of "1" to the maximum values of peak power of "0" [9, 10].

$$\text{E.R.} = \frac{\text{Minimum values of peak power of 1}}{\text{Maximum values of peak power of 0}} \tag{7}$$

Table 1 RSOA parameters used in simulation

Symbol	Parameters	Value
C	Velocity of light	3×10^8 m/s
R	Reflectivity	1
Γ	Confinement factor of RSOA	0.48
α_N	Differential gain	3.3×10^{-20} m^2
w	Width of the RSOA	1.5 μm
d	Depth of the RSOA	250 nm
λ	Wavelength of light	1550 nm
E_c	Control pulse energy	10 fJ
τ_{fwhm}	Full width half maximum of control pulse	1.5 ps
n_2	Nonlinear coefficient	2.6×10^{-20} m^2/w
D	Dispersion constant	1 ps/(nm–km)
A_{eff}	Fiber effective area	5×10^{-13} m^2
I	Injection current	170 mA
b_0	Optical bandwidth	3 nm
Nsp	Spontaneous noise parameter	2

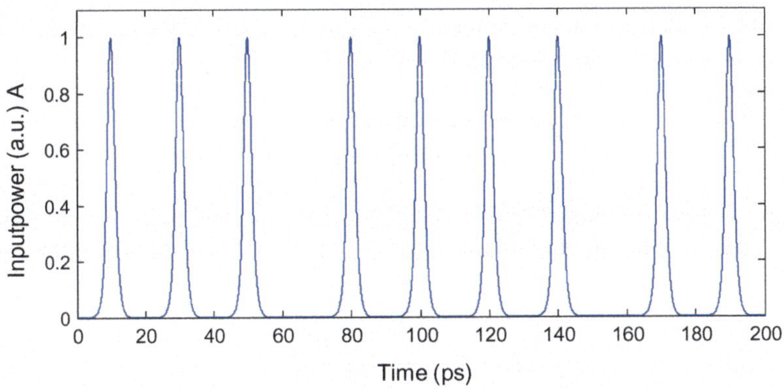

Fig. 2 Input power A

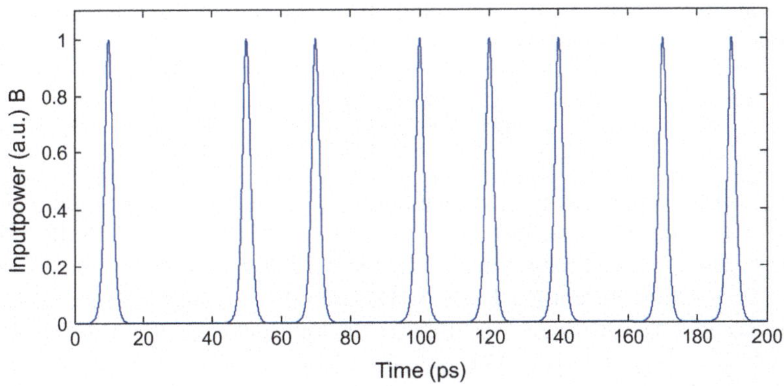

Fig. 3 Input power B

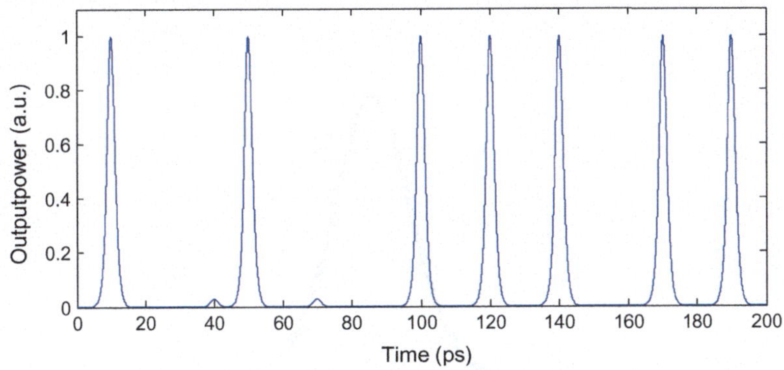

Fig. 4 Output of AND operation

The output contrast ratio is defined as the ratio of average values of peak power of “1” to the average values of peak power of “0” [7, 10].

$$\text{C.R.} = \frac{\text{Average values of peak power of 1}}{\text{Average values of peak power of 0}} \quad (8)$$

The relative eye-opening (REOP) is defined as the ratio of difference between the minimum and maximum output power at “1” and “0” to the minimum output power at “1” [7].

$$\text{REOP} = \frac{\text{P}^1_{\text{min}} - \text{P}^0_{\text{max}}}{\text{P}^1_{\text{min}}} \quad (9)$$

The output quality factor or Q value is defined as the ratio of the difference between average output power of “1” and “0” to the sum of the standard deviation of the output power “1” and “0” [7].

$$\text{Q} = \frac{P^1_{mean} - P^0_{mean}}{sd^1 + sd^0} \quad (10)$$

Figure 5 shows the “pseudo-eye-diagram” [11] of the AND operation and after optimization from Fig. 5, we have got E.R. = 15.35 dB, C.R. = 17.88 dB, relative eye-opening is 97%, and output quality factor is 77.069 at control pulse energy Ec = 10fJ and injection current I = 170 mA. The amplified spontaneous noise (ASE) is numerically added to investigate its effect on the logic gate performance and is given by [12], $P_{ASE} = N_{SP} \times h\upsilon(G_u-1)b_0$, h being Planck’s constant, υ the frequency of the input signal, G_u is the maximum gain of the RSOA, b_0 is the optical bandwidth, and N_{SP} is the ASE noise parameter. It is found that for larger maximum gain, the ASE has little effect on the quality factor or on the performance of the AND gate.

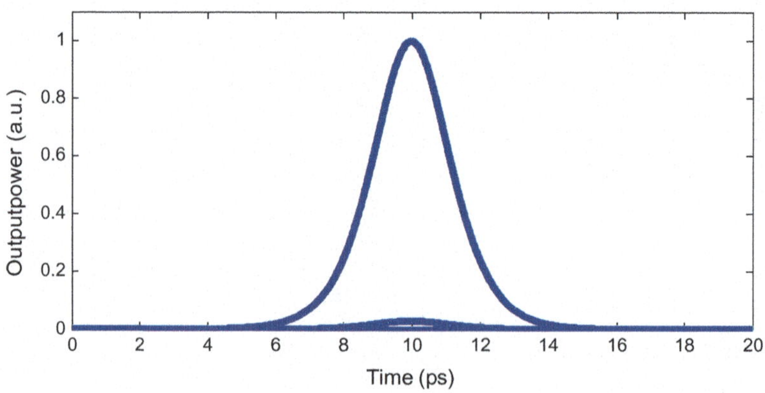

Fig. 5 “Pseudo-eye-diagram” of the AND operation

The dependence of E.R., C.R., quality factor and relative eye-opening on control pulse energy is shown below.

Figures 6 and 7 show variation of extinction ratio and contrast ratio with control pulse energy for constant injection current. It is found that E.R and C.R slightly decrease with increase in the control pulse energy because control signal's increase causes gain of the RSOA to decrease.

Figures 8 and 9 show the dependence of relative eye-opening and Q factor on control pulse energy. From Fig. 8, it is found that relative eye-opening is 97% and from Fig. 9 it is found that Q value rapidly decreases with increase in the control pulse energy. This is because greater power difference means larger Q value and when the control signal increases output power difference decreases.

Fig. 6 Variation of E.R. with control pulse energy

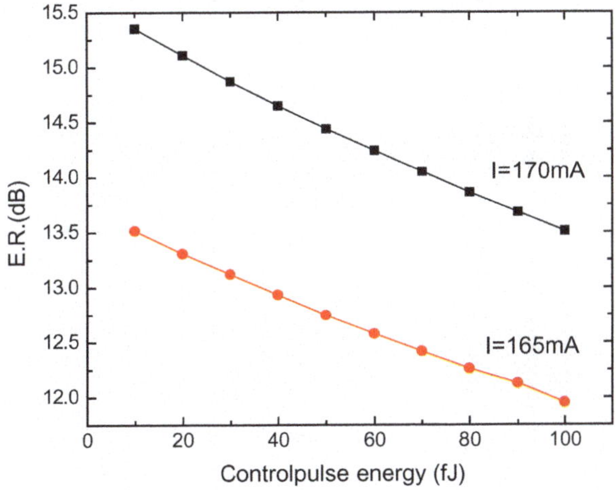

Fig. 7 Variation of C.R. with control pulse energy

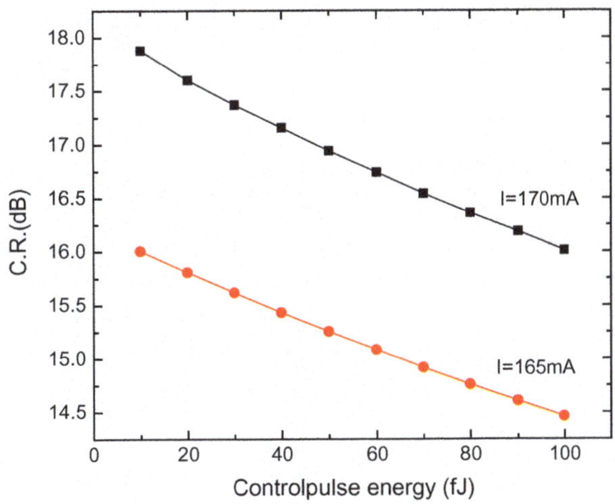

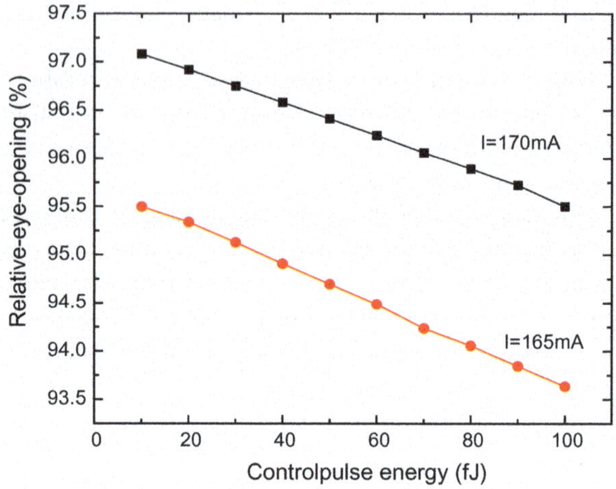

Fig. 8 Variation of REOP with control pulse energy

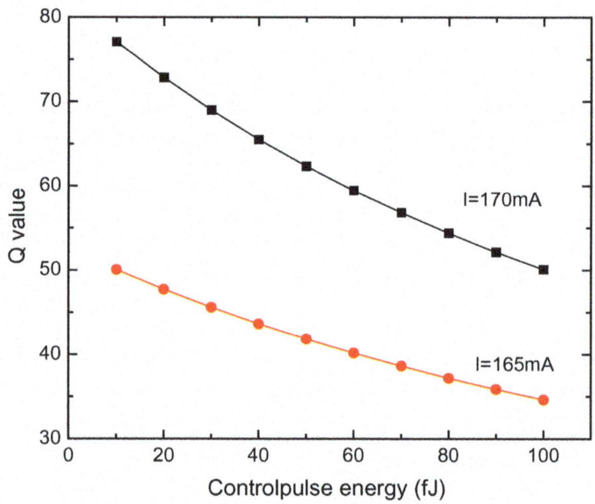

Fig. 9 Dependence Q value with control pulse energy

4 Conclusions

RSOA-based all-optical logic is proposed and analyzed using soliton pulses. The design shows better performance than the single-pass SOA with Q value nearly 80 and relative eye-opening of 97%. The values of ER, CR, and Q factor show improvement also. The simulation results confirm the practical feasibility of the gate.

References

1. Rizou, Z., Zoiros, K.: Theoretical analysis of directly modulated reflective semiconductor optical amplifier performance enhancement by microring resonator-based notch filtering. Appl. Sci. **8**(2), 223 (2018). https://doi.org/10.3390/app8020223
2. Zhan, W., Zhou, P., Zeng, Y., Mukaikubo, M., Tanemura, T., Nakano, Y.: Optimization of modulation-canceling reflective semiconductor optical amplifier for colorless wdm transmitter applications. J. Lightwave Technol. **35**, 274–279
3. Chandra Mandal, G., Mukherjee, R., Das, B., Patra, A.S.: Next-generation bidirectional triple-play services using RSOA based WDM radio on free-space optics PON. Opt. Commun. **411**, 138–142 (2018)
4. Kotb, A., Zoiros, K.E., Guo, C.: Performance investigation of 120 Gb/s all-optical logic XOR gate using dual-reflective semiconductor optical amplifier-based scheme. J. Comput. Electron (2018). https://doi.org/10.1007/s10825-018-1243-4
5. Mukherjee, K.: A novel frequency encoded all optical logic gates exploiting polarization insesensitive four wave mixing in semiconductor optical amplifier, filtering property of ADD/DROP multiplexer and non-linearity of reflective semiconductor amplifier. Optik **122**(10), 891–895 (2011)
6. Ghosh, B., Hazra, S., Haldar, N., et al.: A novel approach to realize of all optical frequency encoded dibit based XOR and XNOR logic gates using optical switches with simulated verification Opt. Spectrosc. **124**, 337 (2018). https://doi.org/10.1134/S0030400X1803013X
7. Chattoipadhyay, T.: All-optical clocked delay flip-flop using a single terahertz optical asymmetric demultiplexer–based switch: a theoretical study. Appl. Opt. **49**(28) (2010)
8. Maji, K., Mukherjee, K., Raja, A.: Frequency encoded all-optical universal logic gates using terahertz optical asymmetric demultiplexer. IJPOT **4**(3), 1–7 (2018)
9. Maji, K., Mukherjee, K., Raja, A.: Analysis of tera hertz optical asymmetric demultiplexer (TOAD) based optical switch using soliton pulse. In: 1st International Conference 2018 IEEE Electron Device Kolkata (2018 IEEEEDKCON) Nov 24–25. ISBN: 978-1-5386-6415-5
10. Roy, J.N., Gayen, D.K.: Integrated all-optical logic and arithmetic operations with the help of TOAD based interferometer device—alternative approach. Appl. Opt. **46**(22), 5304–5310 (2007)
11. Schneider, T.: Nonlinear Optics in Telecommunications. Springer, Berlin (2004). Chap. 6, Sect. 6
12. Komatsu, K., Hosaya, G., Yashima, H.: All-optical logic NOR gate using a single quantum-dot SOA-assisted an optical filter. Opt. Quant. Electron. **50**, 131 (2018)

CMOS Design and Analysis of Four-Quadrant Analog Multiplier Circuit for LF Applications

Abhishek Kumar Gond and Soumya Pandit

Abstract Low-frequency analog integrated circuit design is a challenging task considering the threat imposed by flicker noise. This paper presents design and analysis of analog multiplier circuit using p-channel MOS transistor, in contrast to the traditional approach of designing using n-channel MOS transistor. Exhaustive simulation using SCL 180 nm process technology is performed. This includes dc, transient, and harmonics analysis. We derive compact analytical model of various performance parameters and compare it with simulation results. Good matching between the two has been obtained. The application of multiplier circuit for demodulation purpose is shown.

Keywords Analog multiplier · LF applications · Four-quadrant multiplier · CMOS design

1 Introduction

Low-frequency analog signal processing has recently gained a lot of interest for biomedical and seismic applications. The primary challenge involved in analog circuits working in the low-frequency range is to mitigate the flicker noise. Mathematical analysis of flicker noise shows that the power spectral density of flicker noise is inversely proportional to [1] (i) frequency of applied input signal, i.e., f, (ii) area of the channel region ($W.L$) and directly proportional to Γ, which is a process-dependent constant. Empirically, p-channel MOS transistors have about two to five times less $1/f$ noise than n-channel MOS transistors. Therefore, for low-frequency applica-

A. K. Gond (✉) · S. Pandit
IC Design Laboratory, Institute of Radio Physics and Electronics, University of Calcutta, Kolkata, India
e-mail: abhishek.kr.gond@gmail.com

S. Pandit
e-mail: sprpe@caluniv.ac.in

S. Kundu et al. (eds.), *Proceedings of the 2nd International Conference on Communication, Devices and Computing*, Lecture Notes in Electrical Engineering 602, https://doi.org/10.1007/978-981-15-0829-5_28

tions, p-channel MOS transistors are used for input transistors and the channel area $W.L$ needs to be large.

In analog signal processing, it is often required for a circuit that takes two analog inputs x, y and produces an output Z proportional to the product, i.e., $Z = Kxy$, where K is some constant value in suitable dimension. Such circuit is termed as analog multiplier [2]. Its application with continuous-valued signals can be found in communications, instrumentation, and control system designs.

Earlier, analog multiplier circuits based on transconductance technique are shown by Gilbert [3]. Wu et al. and Cracan et al., respectively, introduce a multiplier circuit based on MOS translinear principle where transistors are biased in the weak inversion region [4, 5]. This approach has the advantage of low power consumption but its input range is very small and operation speed is low. Carvajal et al. introduce a low voltage analog cell called flipped voltage follower to realize the multiplier circuit [6]. CMOS analog multiplier using feedback techniques is introduced by Liu et al. [7]. Lopez-Martin et al. reported dual supply voltage which is not applicable for integrated circuit design [8]. Miguel et al. and Pawarangkoon et al. have shown four-quadrant multiplier architectures for various applications [9, 10]. In this work, four-quadrant multiplier architecture operating in linear and saturation region is proposed, which gives high input range and low distortion in comparison with [11].

The paper is organized as follows: Sect. 2 presents the design principle and description of the proposed circuit for analog multiplier. Section 3 describes the various simulation results and associated discussions. Finally, we conclude the paper in Sect. 4.

2 Design Principle and Circuit Analysis

2.1 Mathematical Formulation

The architecture of our circuit is formulated based on the mathematical equality shown in (1) [12].

$$\left[(X+x)(Y+y)+(X-x)(Y-y)\right]-\left[(X-x)(Y+y)+(X+x)(Y-y)\right]=4xy \quad (1)$$

where X, Y represent the DC values of the two input signals (multiplier and multiplicand) and x, y represent the corresponding small signal AC values. It may be noted that DC component of the two input signals, i.e., X and Y are eliminated in (1). A block diagram representation of architecture of a four-quadrant analog multiplier is shown in Fig. 1a.

Now we implement each of the quadrant shown in Fig. 1a, i.e., (x, y), $(-x, y)$, $(x, -y)$, and $(-x, -y)$ using p-channel MOS transistor. The circuit shown in Fig. 1b implements $(x.y)$. The transistor M_1 operates in linear mode, while M_2 operates in saturation mode. Therefore, we write

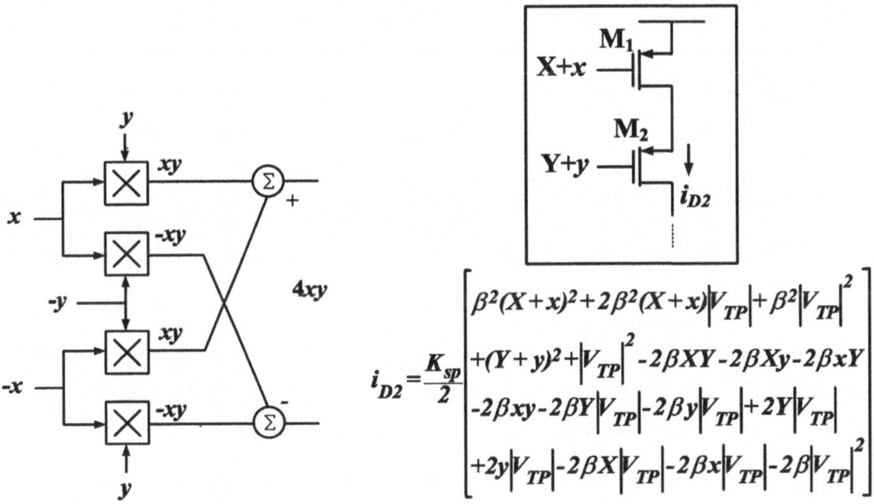

(a) Basic architecture of 4 quadrant multiplier

(b) PMOS implementation of one quadrant

Fig. 1 Block diagram representation of four-quadrant multiplier and implementation of one portion

$$v_{D1} > X + x + |V_{TP}| = \beta\,(X + x + |V_{TP}|) \tag{2}$$

where β is a positive integer, whose value is to be empirically determined and $|V_{TP}|$ is modulus of threshold voltage of a p-channel transistor. The drain current i_{D2} of M_2 is written as

$$i_{D2} = \frac{K_{sp}}{2}\left[P - 2\beta x.y\right] \tag{3}$$

where $K_{sp} = \mu_0 C_{ox}\frac{W}{L}$ is the transconductance parameter of a p-channel MOS transistor operating in saturation mode [13]. We find that (3) gives the term xy, remaining components being included in the factor P. Employing p-channel transistors similar to that shown in Fig. 1b for each quadrant, i.e., $(-x, y)$, $(-x, -y)$, and $(x, -y)$, the corresponding drain currents will have the term $\pm 2\beta xy$ with other terms to be included in factors similar to P.

Adding individual circuit implementation of each quadrant, with little intuition, the proposed circuit diagram of the four-quadrant multiplier is shown in Fig. 2.

2.2 Circuit Analysis

The circuit in Fig. 2 is fully differential arrangement of the architecture shown in Fig. 1a. Suitable DC values of the input signals for biasing the transistors, i.e., X, Y are generated using voltage divider circuit with resistive components $r_1 = 2\ \mathrm{K}\Omega$

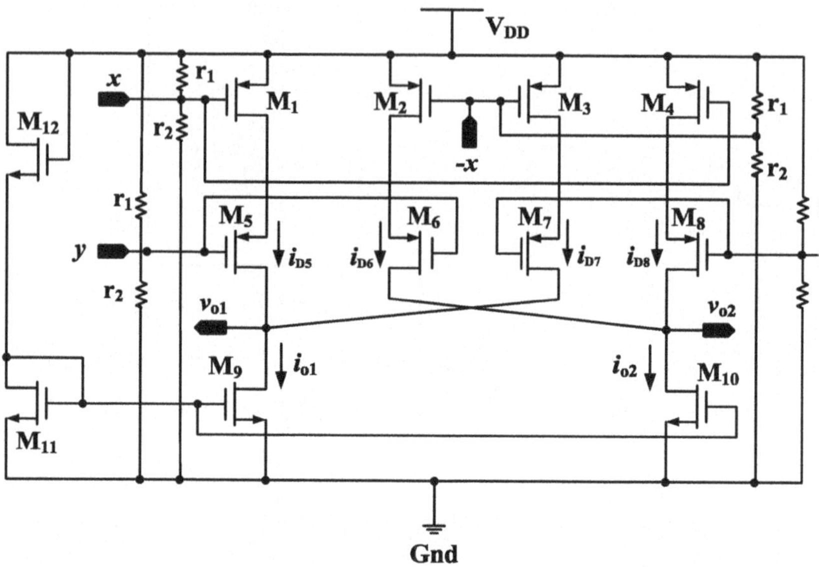

Fig. 2 Proposed circuit of four-quadrant analog multiplier circuit using p-channel MOS transistor

and $r_2 = 1.1\ \text{K}\Omega$. These resistors are implemented using MOS transistors. The bias conditions for transistors in Fig. 1b are written as

$$\begin{aligned} X &< V_{DD} - |V_{TP}| \\ V_{D1} &> X + |V_{TP}| \Rightarrow V_{D1} = \beta\,(X + |V_{TP}|) \\ Y &< V_{D1} - |V_{TP}| \\ V_{D2} &\leq Y + |V_{TP}| \end{aligned} \tag{4}$$

where V_{D1} and V_{D5} are the drain voltage of transistors M_1 and M_5, respectively. We assume that (a) threshold voltage V_T is same for transistors M_1–M_8, (b) the transistors M_1–M_4 operate in linear mode and have same transconductance parameter K_{lp}, and (c) the transistors M_5–M_8 operate in saturation mode and have same transconductance parameter K_{sp}. The difference of the output current yields

$$i_o = i_{o1} - i_{o2} = (i_{D5} + i_{D7}) - (i_{D6} + i_{D8}) \tag{5}$$

where

$$i_{D5} = \frac{K_{sp}}{2}\Big[\beta(X + x + |V_{TP}|) - (Y + y) - |V_{TP}|\Big]^2 \tag{6}$$

$$i_{D6} = \frac{K_{sp}}{2}\Big[\beta(X - x + |V_{TP}|) - (Y + y) - |V_{TP}|\Big]^2 \tag{7}$$

$$i_{D7} = \frac{K_{sp}}{2}\Big[\beta(X - x + |V_{TP}|) - (Y - y) - |V_{TP}|\Big]^2 \tag{8}$$

$$i_{D8} = \frac{K_{sp}}{2}\Big[\beta(X + x + |V_{TP}|) - (Y - y) - |V_{TP}|\Big]^2 \tag{9}$$

Therefore, the final output current i_o is written as

$$i_o = -4K_{sp}\beta xy \tag{10}$$

Consequently, the output voltage is written as

$$v_o = v_{o1} - v_{o2} = i_o R_D = -4K_{sp}\beta xy R_D \tag{11}$$

where R_D is the output load resistance.

3 Results and Discussion

3.1 Simulation Setup

The W/L ratios of M_1–M_8 transistors are 6/5, and that for M_9–M_{11} and M_{12} transistors are 5/5 and 2/5, respectively. The circuit is simulated using Semiconductor Laboratory (SCL) 180 nm CMOS technology. The supply voltage is taken to be 1.8 V. The effective mobility parameters for the n-channel and p-channel MOS transistors are extracted to be ≈218 cm^2/V.s and ≈81 cm^2/V.s, respectively, in the saturation mode. The typical oxide thickness parameters for the n-channel and p-channel MOS transistors are 4 nm. We use Spectre SPICE simulator of Cadence and BSIM4 compact model for simulation purpose. All simulations are performed assuming room temperature.

3.2 DC Analysis

Initially, DC operating conditions of the circuit are set up, so that biasing requirements in 4 of the transistors are satisfied. For this, we select $X = Y = 640$ mV and $I_{o1} = I_{o2} = 16\,\mu$A. For DC sweep analysis, we vary the small signals x and y above the bias voltages X and Y, respectively. The DC transfer characteristics of V_{o1}–V_{o2} versus x for constant values of y are shown in Fig. 3a. The simulation result for DC analysis is compared with that predicted by analytical model in (11) derived earlier. We find that reasonably good match is obtained between analytical and simulation results. From the simulation results, the empirical value of β is found out to be 9. On the other hand, DC transfer characteristics of V_{o1}–V_{o2} versus y for constant values of x

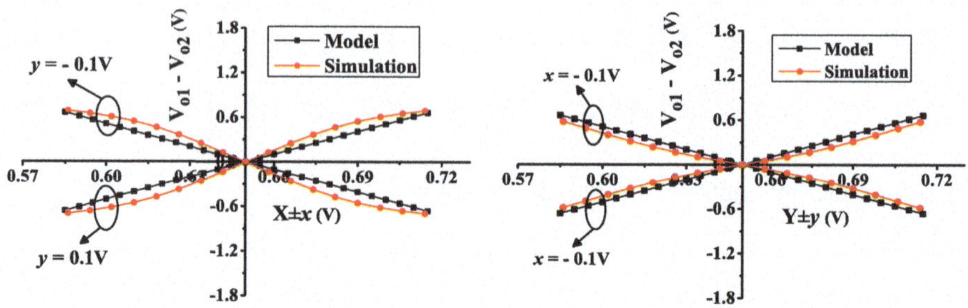

(a) V_{o1}-V_{o2} versus x (b) V_{o1}-V_{o2} versus y

Fig. 3 DC transfer characteristics of four-quadrant analog multiplier

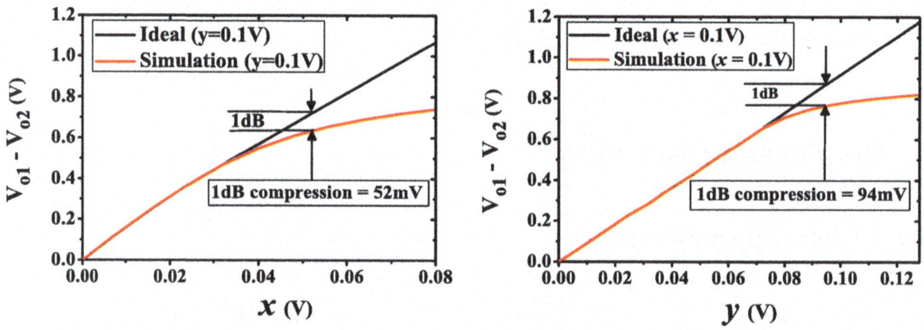

(a) V_{o1}-V_{o2} versus x (b) V_{o1}-V_{o2} versus y

Fig. 4 Linearity analysis of four-quadrant analog multiplier circuit

are shown in Fig. 3b. We observe that in this case also, the simulation results match well with analytical results.

From Fig. 3a, b, we observe that the DC transfer characteristics show some non-linearity for higher values of x and y. This necessitates analysis of the linearity of the circuit. We use 1-dB compression point technique to carry the linearity analysis. The input voltage levels above the 1-dB compression point cause distortion in the output signal. The necessary graphs are shown in Fig. 4a, b. Figure 4a indicates the the linear range for x up to 52 mV at a fixed value of $y = 0.1$ V. On the other hand, Fig. 4b indicates that the linear range for y up to 94 mV at a fixed value of $x = 0.1$ V. Thus, our subsequent simulations are due within this input range.

3.3 *Transient Analysis*

In order to perform the transient analysis, we consider two time-varying signals: a 1 KHz frequency signal (y) and another one, a 3 Hz frequency sinusoidal signal (x).

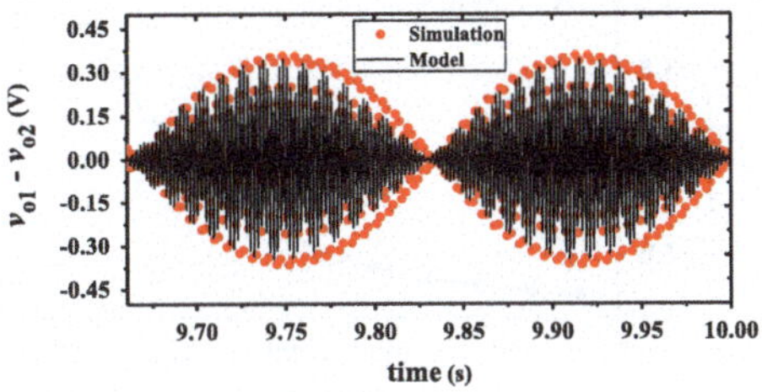

(a) Transient analysis of 4 quadrant analog multiplier circuit, showing product of two sinusoidal signals.

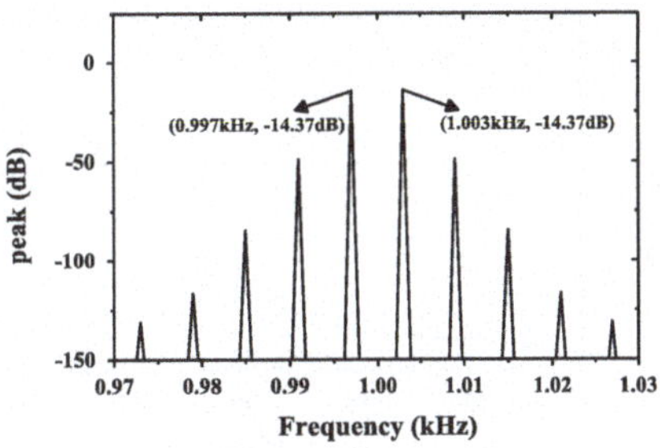

(b) Output fundamental and its harmonics, at (1KHz±3Hz)

Fig. 5 Transient analysis of four-quadrant analog multiplier circuit

The peak values of signals x and y are 30 mV and 80 mV, respectively. The output signal, shown in Fig. 5a, ensures that the circuit carries out multiplication of two sinusoidal signals, resembling a double sideband suppressed carrier (DSB-SC) waveform. We further compare the simulation results with mathematical model and find good accuracy. This illustrates that our circuit shown in Fig. 2 will work as a modulator [14].

The multiplication of two sinusoidal signals offers frequency translation. Simultaneously, it also generates harmonics of the input signal. We, therefore, carry out harmonics analysis of the product signal. This is shown in Fig. 5b. It shows spectrum response at the output of the multiplier and indicates the harmonics and the high spikes centered at the high-frequency signal of 1 KHz ± 3 Hz. The spectrum response does not have component of the high-frequency signal (1 KHz sinusoid), and the reason behind the multiplication of two sinusoidal signals is said to be suppressed carrier.

3.4 Total Harmonic Distortion (THD)

Because of nonlinear response of the circuit, for larger values of x and y, we prefer to study total harmonic distortion (THD) analysis of the circuit.

Figure 6a shows the % THD when a DC voltage is swept from 0–0.1 V for x while (1 KHz, 50 mV) sinusoidal signal is applied to y. We observe that if the value of x remains within the input range (<52 mV), %THD remains up to 1.513 for a fixed amplitude of y, while the maximum %THD shoots up to 5.114. Similarly, Fig. 6b shows the %THD when a DC voltage is swept from 0 to 0.15 V for y while a (3 Hz, 30 mV) sinusoidal signal is applied to x. We observe that if the value of y remains within the input range (<100 mV), %THD remains 2.314 for a fixed amplitude of x, while the maximum %THD shoots up to 4.24.

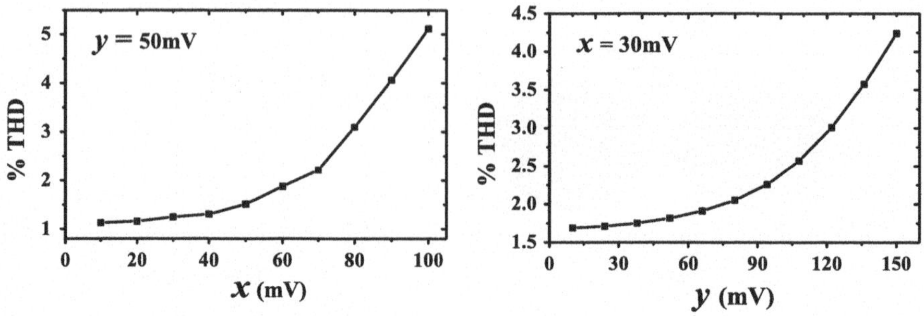

(a) THD computation for x signal (b) THD computation for y signal

Fig. 6 Total Harmonic Distortion analysis of the four-quadrant analog multiplier

Table 1 Performance comparison

Parameters	This work	Panigrahi and Paul [11]	Liu and Liu [16]
V_{DD}	1.8 V	0.5 V	1.5 V
Input	52 mV (x)	60 mV (V_{in1})	±120 mV
Range	94 mV(y)	80 mV (V_{in2})	
THD%	1.53 (x)	5.3 (V_{in1})	4.2
	2.32 (y)	7.79 (V_{in2})	
Power	4.1 mW	0.7 μW	6.7 μW
I_{bias}	16 μA	300 nA	100 nA
Process	180 nm	180 nm	350 nm
	SCL technology	IP6M CMOS	2P4M CMOS

3.5 Comparison with Existing Works

Table 1 shows a detailed comparison of the performances with the results available in earlier reported works. The existing multiplier topologies shown in [11, 16] work with PMOS transistors. We observe that the input range for our circuit shown in Fig. 2 is improved in comparison with the work shown in [11]. THD of the proposed circuit shown in this work is much better than the circuits of [11, 16]. The maximum power consumption for the proposed multiplier circuit is 4.1 mW, which is more than in comparison with that reported in [11, 16]. This is because the circuits in [11, 16] operate in weak inversion region with low current bias. Also, low supply voltage results with low power consumption.

3.6 Application as Demodulator

In order to use the analog multiplier circuit as demodulator, we generate a double sideband suppressed carrier (DSB-SC) modulated test signal using Verilog-A [15], which is shown in Fig. 7(a). This test signal is multiplied with a high-frequency signal (1 KHz) which results in the waveform of demodulated signal at the output of the multiplier. This is shown in Fig. 7(b). The harmonics response of the demodulated signal is shown in Fig. 7(c). We observe that it eliminates the interference added to the DSB-SC signal, and is showing the peaks at n times of the input frequency(3 Hz), where n is the odd integer value. In order to get back the low-frequency signal, we apply the demodulated signal to a passive RC filter, implemented with $R_f = 4$ MΩ and C_f=13 nF. Figure 7(d) shows the output signal at the output of the filter. It gives a 3 Hz sinusoidal signal, with an amplitude of 76 μV.

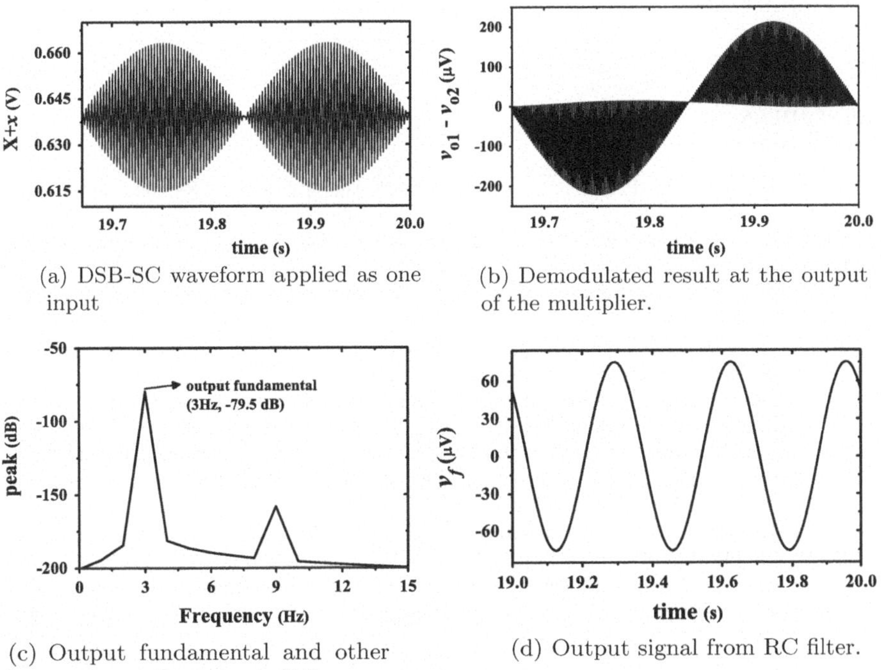

(a) DSB-SC waveform applied as one input

(b) Demodulated result at the output of the multiplier.

(c) Output fundamental and other harmonics at the output of the demodulator.

(d) Output signal from RC filter.

Fig. 7 Use of the multiplier circuit as demodulator

4 Conclusion

In this paper, we present a circuit topology for analog multiplier circuit using p-channel MOS transistors as input. The circuit is meant for low-frequency applications. We present exhaustive simulation results for dc sweep analysis, transient analysis, and THD analysis. The simulation results are compared with our derived analytical results and we can find good accuracy. We find that the input range and total harmonic distortion show better result than the same for some other topologies reported in literature. The future scope of this work involves detailed noise analysis, compensating power consumption as well as process variability analysis, and final layout design.

Acknowledgements The authors thank SMDP-C2SD project, University of Calcutta, sponsored by Ministry of Electronics and Information Technology, Govt. of India to provide research facilities. The authors further thank UGC-UPE-II project, Group B: Signal Processing, University of Calcutta for necessary support.

References

1. Pandit, S., Mandal, C., Patra, A.: Nano-Scale CMOS Analog Circuits. CRC Press, Boca Raton (2014)
2. Chen, C., Li, Z.: A low-power CMOS analog multiplier. IEEE Trans. Circ. Syst. **53**(2), 100–104 (2006)
3. Gilbert, B.: A precision four-quadrant multiplier with sub nanosecond response. IEEE J. Solid-State Circ. **3**, 353–365 (1968)
4. Wu, R., Xing, J.: MOS translinear principle based analog four-quadrant multiplier. Int. Conf. IASID **2012**, 1–4 (2010)
5. Cracan, A., Bonteanu, G., Bozomitu, R.: A weak-inversion CMOS analog multiplier/divider circuit. In: 24th IEEE SIITME (2018)
6. Carvajal, R.G., Ramirez-Angulo, J., Torralba, A., Lopez-Martin, A., Galan, J., Carlosena, A., Munoz, F.: The flipped voltage followers: a useful cell for low-voltage low-power circuit design. IEEE Trans. Circ. Syst. **52**, 1276–1291 (2005)
7. Liu, S., Hwang, Y.: CMOS four-quadrant multiplier using feedback techniques. IEEE J. Solid-State Circ. **29**(6) (1994)
8. Lopez-Martin, A.J., Blas Carlos A., Ramirez-Angulo. J., Carvajal, R.G.: Current-mode CMOS multiplier/divider circuit operating in linear/saturation regions. Analog Integr. Circ. Signal Process 299–302 (2011)
9. Algueta-Miguel, J.M., Blas Carlos, A., Lopez-Martin, A.J.: Fully differential current-mode CMOS triode translinear multiplier. IEEE Trans. Circ. Syst. II **58**(1) (2011)
10. Pawarangkoon, P., Sawigun, C.: A compact bulk-driven four quadrant analog multiplier in weak inversion. In: IEEE Asia Pacific Conference on Circuits and System (2018)
11. Panigrahi, A., Paul, P.K.: A novel bulk-input low voltage and low power four quadrant analog multiplier in weak inversion. Analog Integr. Circ. Signal Process 237–243 (2013)
12. Han, G., Sanchez-Sinencio, E.: CMOS transconductance multipliers: A tutorial. IEEE Trans. Circ. Syst. II: Analog Digit. Signal Process **45**(12) (1998)
13. Razavi, B.: Design of Analog CMOS Integrated Circuits. McGraw-Hill Inc., New York (2001)

14. Palls-Areny, R., Webster, J.G.: Analog Signal Processing. Wiley, New York (1999)
15. Kundert, K.S.: The Designers Guide Book Series. Kluwer Academic Publishers, New York (2004)
16. Liu, W., Liu, S.: Design of a CMOS low power and low voltage four quadrant analog multiplier. Analog Integr. Circ. Signal Process 307–312 (2010)

Computational Devices for Intelligent Vehicular Systems and Smart Networking

P. Raj Kumar, Ravisankar Malladi, Nishu Gupta and Sara Paiva

Abstract A novel microstrip bandpass filter with split ring resonator (SRR) defected ground structure (DGS) patch antenna for computational applications is proposed in this article. This filter and antenna design can collaboratively work as *filtenna* for improved performance by minimizing the total power consumption in the system. The investigation is carried out on width of the slit opening of SRR slot (W) and rotation of the inner square-shaped ring structure in clockwise direction. The proposed filtenna is designed to provide a tunable frequency operation. Simulation is performed on high-frequency structural simulator (HFSS) which demonstrates good output response making filtenna design suitable for intelligent vehicular communication and wireless local area network (WLAN) applications. The proposed filtenna is compact in size, has low insertion loss, exhibits high selectivity, and demonstrates excellent out-of-band performance.

Keywords DGS · DSRC · ITS · Computational devices · VANET

P. Raj Kumar
Computer Science and Engineering Department, Vaagdevi College of Engineering, Warangal, India
e-mail: rajkumar_p@vaagdevi.edu.in

R. Malladi
Computer Science and Engineering Department, Godavari Institute of Engineering & Technology, Rajahmundry, India
e-mail: mrsankar@giet.ac.in

N. Gupta (✉)
Electronics and Communication Engineering Department, Vaagdevi College of Engineering, Warangal, India
e-mail: gupta_n@vaagdevi.edu.in

S. Paiva
ARC4Digit – Instituto Politécnico de Viana do Castelo, Viana do Castelo, Portugal
e-mail: sara.paiva@estg.ipvc.pt

S. Kundu et al. (eds.), *Proceedings of the 2nd International Conference on Communication, Devices and Computing*, Lecture Notes in Electrical Engineering 602, https://doi.org/10.1007/978-981-15-0829-5_29

1 Introduction

Split ring resonator (SRR) has strong magnetic polarizability. It is electrically small in size and hence is broadly analyzed although proposed initially [1, 2]. Ong et al. [3] presented a detailed study of SRR parameters on the performance of coaxial-fed monopole antenna. Parametric study is carried out on SRR parameters and it is found that there is a change in the notch and frequency band, and hence can work as a frequency tunable notch function. Various studies are carried out on SRR and negative permittivity and permeability materials [4]. In [5], methodologies, viz., rotating the inner ring of SRR in a hybrid structure and by changing the measurements of SRR, are presented for achieving an increment in the bandwidth of SRR-based metamaterial structures. In the last decade, a lot of research works have been carried out on new structures and materials, and hence DGS structure is introduced into the microwave circuitries, viz., antenna, filter, etc. which eventually enhances suppression of high-frequency response [6, 7].

Proposed novel microstrip filtenna has center frequency at 5.9 GHz, making it suitable for vehicular application in the dedicated short-range communication (DSRC) band. Vehicular ad hoc network (VANET) adopts DSRC technology which is based on short-range wireless communication [8]. VANET evolves exponentially because of the rising demand of the user in terms of safety traffic, real-time management of the traffic, entertainment, and many other wide range of applications. VANET has emerged as a new technology that can support such emerging vehicular applications. VANET has an important role in intelligent transportation system (ITS) which makes the users aware of the surrounding path they wish to follow [9]. Hence, ITS is greatly supported by VANET and is enhanced by vehicular communication systems [10, 11].

Parametric analysis is done by studying the effects of the variations in the dimensions of the slit cut width of the SRR slot and also effects due to the rotation of the inner square-shaped ring structure. The slit cut width has been increased from 1 to 5 mm and the inner square-shaped ring structure has been rotated clockwise. Its effects are studied in terms of frequency band, return loss and insertion loss characteristics, and investigation of the SRR parameters on the performance of the bandpass filtenna has been carried out.

2 Microstrip Bandpass Filtenna with SRR DGS

The structure of the microstrip bandpass filtenna with SRR DGS is shown in Fig. 1a, b with dimensions mentioned in Table 1. The structure comprises a hairpin slot cut in the filtering circuitry and SRR DGS in the ground plane. The proposed filtenna uses 50 Ω, $\lambda_g/4$ microstrip line and is designed on the substrate of size $20 \times 19.5 \times 1.524\ mm^3$. The filtenna design is realized on NH9350 microwave substrate having dielectric constant of 3.5, loss tangent of 0.003, and thickness of 1.524 mm.

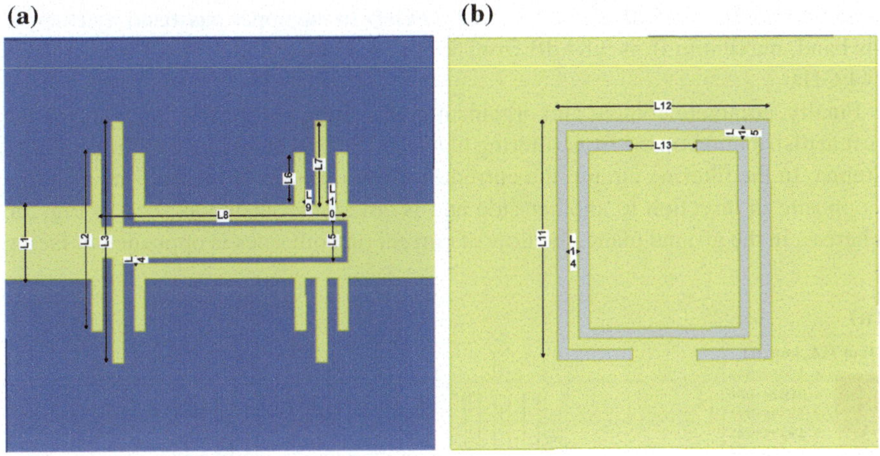

Fig. 1 Proposed bandpass filtenna showing **a** top view, **b** SRR DGS ground plane

Table 1 Dimensions of the proposed bandpass filtenna (mm)

L1	L2	L3	L4	L5	L6	L7	L8
3.4	8.4	11.4	0.3	2	2.5	4	11.5
L9	**L10**	**L11**	**L12**	**L13**	**L14**	**L15**	
0.5	0.5	11.4	10	3	0.5	0.5	

3 Simulation Results and Analysis

Parametric analysis of the SRR parameters in terms of slit cut width of the SRR slot and rotation of the inner square-shaped ring structure has been carried out. Thereafter, final design is proposed and analysis of the return loss (RL) and insertion loss (IL) of the bandpass filtenna with SRR DGS is shown in Fig. 2. The RL comes out to be −13 dB at 5.9 GHz resonant frequency with a bandwidth of 700 MHz (11.86%). The maximum IL is 4.84 dB from 1.13 to 1.36 GHz in the lower passband, whereas

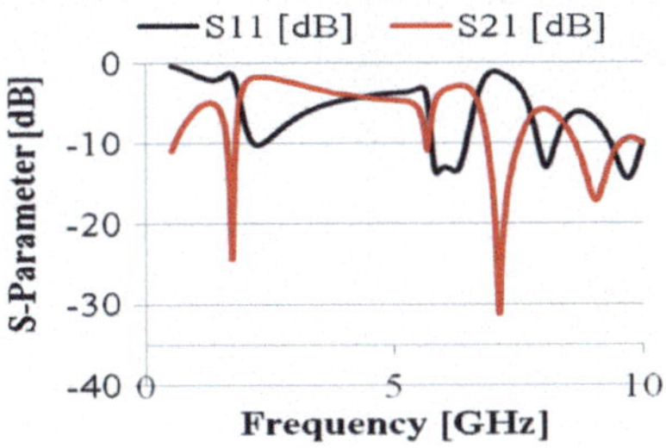

Fig. 2 Simulated S-parameter characteristics of bandpass filtenna

the maximum IL is 1.5 dB from 2.15 to 2.77 GHz in the upper passband. Further, in this band, maximum IL is 3.84 dB from 3.92 to 4.37 GHz and 3.15 dB from 5.98 to 6.44 GHz.

Finally, complete content and organizational editing before Fig. 3a, b shows the current distribution on the front filtering plane and back ground plane of the proposed filtenna. In the filtering circuit, the current flow on one side of the transmission line is opposite in direction to another side and is emerging outside as seen in Fig. 3a. Whereas, in the ground plane, the flow of current on both sides is opposite but facing

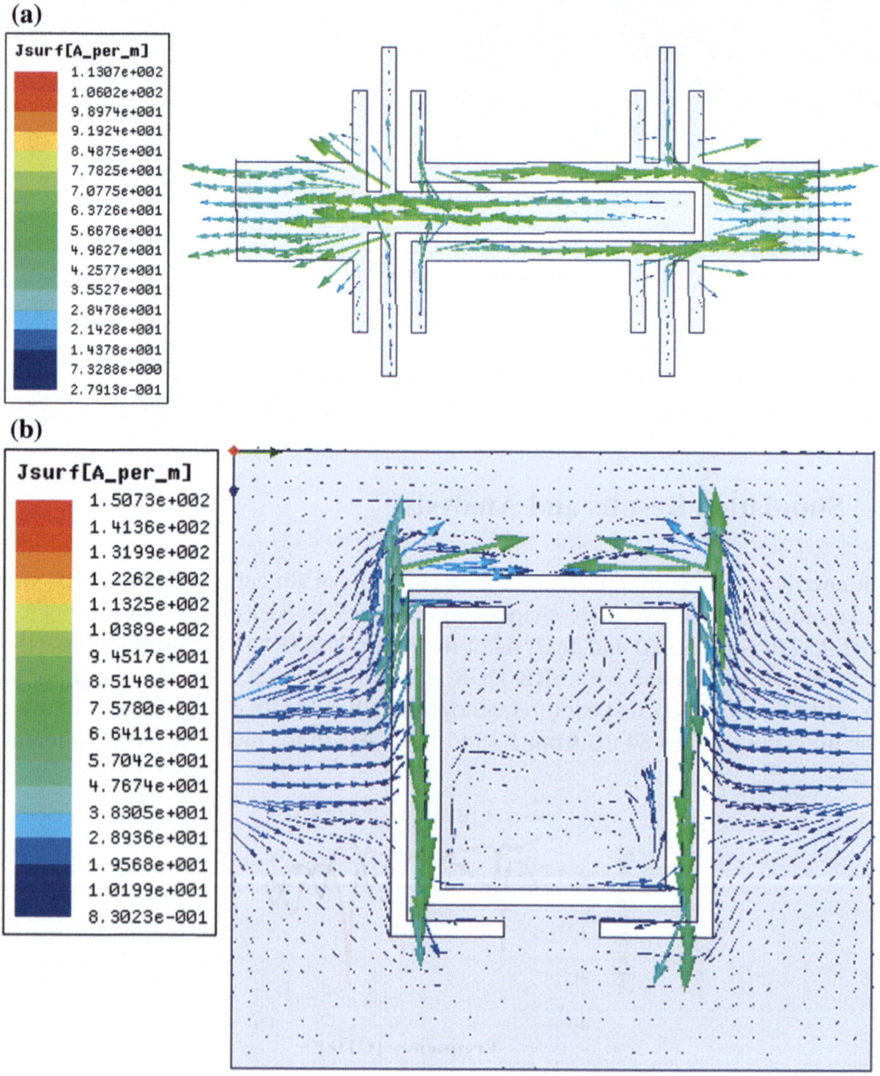

Fig. 3 Surface current density on **a** top filtering circuit, **b** ground plane

each other and is moving inside to the center as shown in Fig. 3b. Hence, 5.9 GHz resonant frequency is produced due to the energies coupled between them.

3.1 Parametric Analysis

Parametric study is performed on the SRR parameters by (i) making variations in the dimensions of the slit opening width of the SRR slot and (ii) rotation of the inner square-shaped ring of the SRR. Its effect on the scattering parameters and bandwidth is analyzed as discussed in the following subsections:

(i) **Slit opening width of the SRR slot**

The slit cut width (W) of the SRR slot is increased from 1 mm to 5 mm as shown in Fig. 4 and its effect on the RL and IL is analyzed in Figs. 5 and 6, respectively. Comparative performance in terms of the center frequency, RL, bandwidth, when

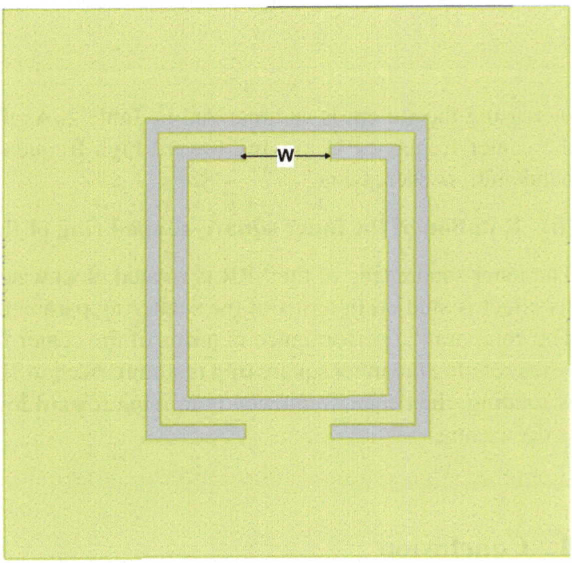

Fig. 4 SRR showing slit opening width (W)

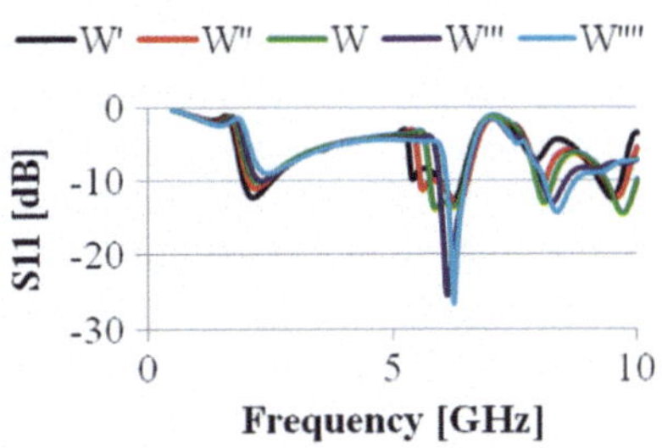

Fig. 5 Return loss of the bandpass filtenna after parametric analysis is performed on "W"

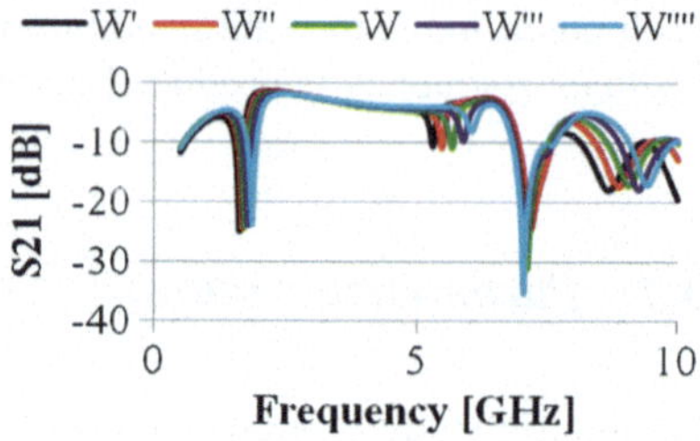

Fig. 6 Insertion loss of the bandpass filtenna after parametric analysis is performed on "W"

Table 2 Comparative performance after parametric analysis in terms of "W"

Slit cut width (W) (mm)	Center frequency (GHz)	RL (dB)	Bandwidth (MHz)
1	2.09, 6.21	−12.26, −12.52	570, 470
2	2.17, 6.28	−11, −13.24	420, 610
3	5.9	−13	700
4	6.09	−25.38	530
5	6.21	−26.45	340

increasing the slit cut is summarized in Table 2. As the slit cut width is increasing, the center frequency is moving toward high frequency, RL is improving, and the bandwidth is decreasing.

(ii) **Rotation of the inner square-shaped ring of the SRR**

The inner square ring of the SRR is rotated clockwise (CW) as shown in Fig. 7a–d. Its effect is studied in terms of the scattering parameters as shown in Figs. 8 and 9. The comparative performance in terms of the center frequency, RL, and bandwidth when rotating the inner square ring is summarized in Table 3. As the inner square ring is rotating, the center frequency is moving toward lower frequency and bandwidth is decreasing.

4 Conclusion

A novel microstrip bandpass filtenna with SRR DGS for intelligent vehicular communication and WLAN applications is reported in this paper. DGS is included in the filtenna to suppress high-frequency response and to improve the overall performance of the filtenna. Further, parametric analysis is done on the SRR parameter in terms of slit opening width and rotation of the inner square-shaped ring of the SRR. This eases performing frequency tuning operation and enhances the bandwidth and scattering parameters. The proposed filtenna is miniaturized, exhibits high selectivity, achieves

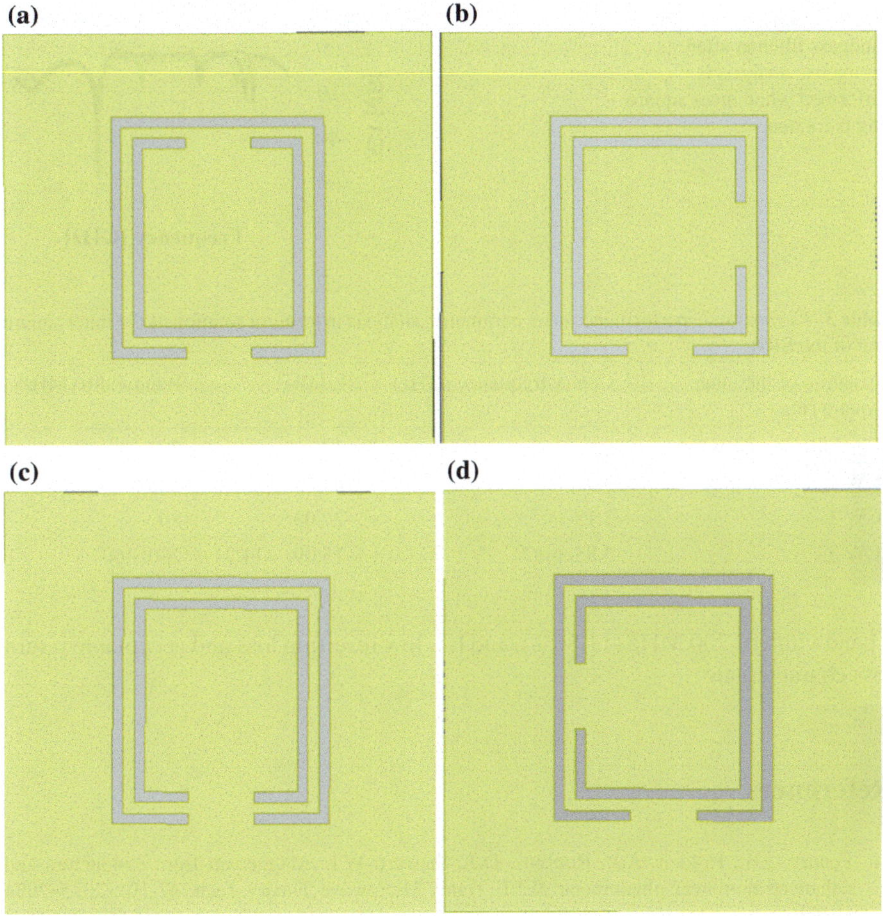

Fig. 7 Rotation of the inner square ring of the SRR in CW with **a** final filtenna design, **b** CW-1, **c** CW-2, and **d** CW-3

Fig. 8 Return loss of the bandpass filtenna after parametric analysis is performed when inner square ring is rotated

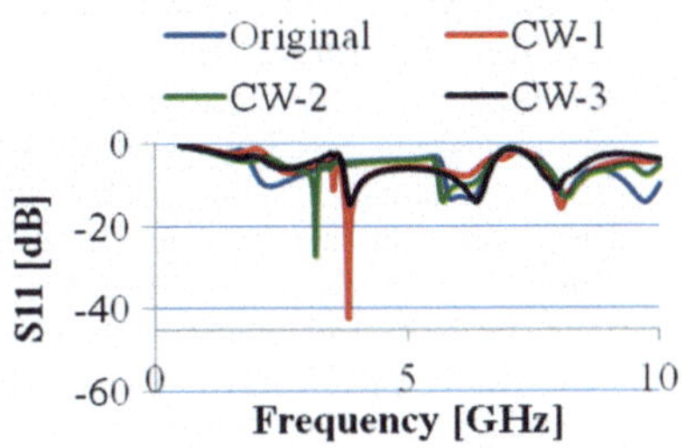

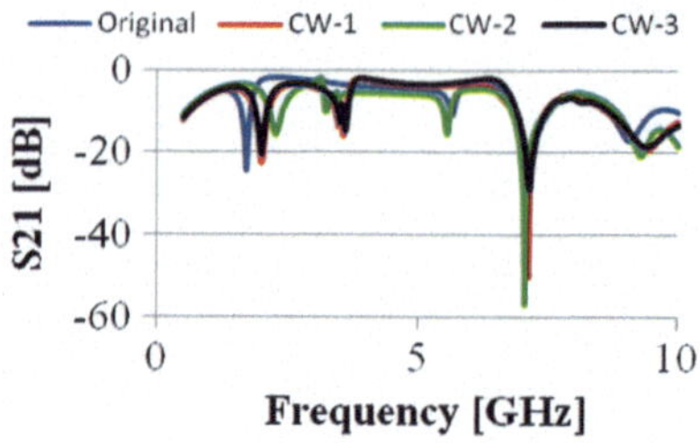

Fig. 9 Insertion loss of the bandpass filtenna after parametric analysis is performed when inner square ring is rotated

Table 3 Comparative performance after parametric analysis in terms of rotation of the inner square ring of the SRR

Rotation of the inner square ring	Center frequency (GHz)	RL (dB)	Bandwidth (MHz)
Final filtenna design	5.9	−13	700
CW-1	3.77	−41.53	320
CW-2	3.15	−27.03	80
CW-3	3.83, 6.32	−15.09, −14.24	280, 460

a bandwidth of 700 MHz (11.86%), and has low insertion loss and acceptable return loss characteristic.

References

1. Pendry, J.B., Holden, A.J., Robbins, D.J., Stewart, W.J.: Magnetism from conductors and enhanced nonlinear phenomena. IEEE Trans. Microwave Theory Tech. **47**(10), 2075–2084 (1999)
2. Pulido-Mancera, L.M., Baena, J.D.: Equivalent circuit model for thick split ring resonators and thick spiral resonators. Phys. Opt. (2014)
3. Ong, T.K., Ahmad, B.H., Aziz, M.Z.A.Abd., Othman, M.A., Suaidi, M.K., Sulaiman, H.A., Malek, F.Abd.: Investigation of SRR parameters on monopole antenna performance. ARPN J. Eng. Appl. Sci. **10**(2) (2015)
4. Veselago, V.G.: The electrodynamics of substances with simultaneously negative values of epsilon and mu. Sov. Phys. Uspekhi **10**, 509–514 (1968)
5. Rigi-Tamandani, A., Ahmadi-Shokouh, J., Tavakoli, S.: Wideband planar split ring resonator based metamaterials. Prog. Electromagn. Res. M **28**, 115–128 (2013)
6. Fang, J., Li, Q., Zhang, Y., Gong, L.: Compact printed ultra-wide band filter employing SRR-DGS. In: 2016 IEEE International Conference on Ubiquitous Wireless Broadband (ICUWB)
7. Saini, Y., Kumar, M.: Performance analysis of Chebyshev UWB bandpass filter using defected ground structure. In: International Conference on Advances in Computing, Communications and Informatics, ICACCI 2014, pp. 1364–1369 (2014)
8. Gupta, N., Prakash, A., Tripathi, R.: Adaptive beaconing in mobility aware clustering based MAC protocol for safety message dissemination in VANET, Hindawi Publishing Corporation. Wirel. Commun. Mob. Comput. **2017**, Article ID 1246172, 15 pp.

9. Gupta, N., Prakash, A., Tripathi, R.: Clustering based cognitive MAC protocol for channel allocation to prioritize safety message dissemination in vehicular ad-hoc network. Veh. Commun. **5**, 44–54 (2016)
10. Park, J., Lee, W.-H., Lee, S.-S.: An intelligent interface selection scheme for vehicular communication system using LTE and IEEE 802.11p. In: 2016 International Conference on Information and Communication Technology Convergence (ICTC)
11. Khodaei, M., (Panos) Papadimitratos, P.: The key to intelligent transportation: identity and credential management in vehicular communication systems. IEEE Veh. Technol. Mag. **10**(4), 63–69 (2015)

Load Balancing Applying BPN in a Dual-Core Processor in IP Router Running with QUEST Scheduler

Suman Paul and Malay Kumar Pandit

Abstract In this paper, we investigate on a load balancing scheme applying neural network—backpropagation network (BPN) algorithm in a dual-core processor in IP router for processing of multimedia IP traffic. Load balancing is an important challenging issue among the cores of a processor in network routers. A balanced equally distributed and a steady-state load of the cores enhances the system quality of service (QoS) and performance. In this system, each core runs with a QoS-enhanced stochastic smart packet scheduler—QUEST. For analysis, we take both the cases of moderately and heavily unbalanced initial load of the cores of a dual-core processor in router. Each core is provisioned with the QUEST scheduler and runs to attain a ratio of process utilization which is of steady-state featured for scheduling and processing of multimedia IP traffic, namely, deadline-sensitive Voice over Internet Protocol, Internet Protocol Television, and best-effort Hypertext Transfer Protocol. We observe the computed balanced load applying the BPN for different conditions. Applying BPN, the load imbalance among the cores is minimized and the results show that load of each core has attained a target steady-state balanced load.

Keywords Backpropagation network (BPN) · Dual-core processor in IP router · Load balancing · Network QoS · Scheduler

1 Introduction

In modern communication networks and internetworking devices, QoS plays a noteworthy role and it is treated as an important performance metric. The important QoS attributes are fair bandwidth allocation [1], packet loss rate (PLR), latency,

S. Paul (✉) · M. K. Pandit
Department of ECE, Haldia Institute of Technology, Haldia, West Bengal, India
e-mail: paulsuman999@gmail.com

M. K. Pandit
e-mail: mkpandit.seci@gmail.com

S. Paul
School of Engineering and Technology, West Bengal University of Technology (Maulana Abul Kalam Azad University of Technology West Bengal), Kolkata, India

S. Kundu et al. (eds.), *Proceedings of the 2nd International Conference on Communication, Devices and Computing*, Lecture Notes in Electrical Engineering 602, https://doi.org/10.1007/978-981-15-0829-5_30

jitter [2], throughput, etc. Processing with multicore processors in router has significant advantages over the single-core processors in running of time-critical applications of modern real-time multimedia IP traffic, where processing demands its completion within the specific deadline. Load balancing among the cores is a challenging issue in multicore processor in router. Load balancing is the process of reallocation of the total load of the system among its cores equally at run time and to maintain a load distribution of steady-state type over time. Efficient load balancing results in improvement of performance of the system. A steady-state balanced uniform and equal distribution of the system load among the cores enforces maximized system QoS in networking devices, i.e., core and edge routers. In routers, the cores of the processor process the incoming packets in parallel. An efficient load balancing method is capable of supervising the load of the cores and it fixes on at run time to balance from the overloaded to underutilized cores. The equal distribution of the load among the cores will increase the computing and processing speed of deadline-specific traffic. The work presented in this paper is an extension and further analysis of our earlier proposed scheduler [3]. In this paper, we have demonstrated the working principle and the analysis of run time load balancing in a dual-core processor in IP router applying backpropagation neural network (BPN) algorithm. Each core is running with a QoS-enhanced smart packet scheduler which is stochastic in nature [3]. The scheduler running in each core has the attribute of process utilization, which is of steady-state type. The ratio of process utilization, which is steady state in nature, is in the order of [0.80:0.16:0.04] for VoIP, IPTV, and web browsing HTTP traffic. The scheduler complies with processing of traffic of deadline-specific classes and enforces the utilization close to 100% of each core. Further, it avoids the starvation of low priority processes. The load imbalance in the system is minimized in an intelligent way by applying BPN algorithm with an aim to retain an equal load and a steady-state core load distribution ratio. This results in the benefit of highest system utilization, higher throughput, lower value of mean waiting time, and lower value of PLR.

2 Literature Review

Load balancing among the different cores in network router is a challenging one. In order to achieve target system QoS, it is essential to dynamically allocate the tasks and to ensure whether the distribution of the load among the cores is equal. Jadon and Yadav [4] have proposed a dynamic load balancing algorithm in real-time multicore system based on heuristic. The imbalance among the cores is minimized. The algorithm shows a lower value of migration overhead. The authors in [5] have demonstrated a deadline-aware scheduling algorithm for real-time multicore systems with an objective to save power and to balance the estimations of the loads. Wu and Wolf [6] have proposed an algorithm for run time task allocation in a multicore system which processes incoming packets. The profiling information during execution is acquired. The tasks with heavy processing requirements are duplicated. A balanced workload is obtained by applying the duplication algorithm. Tan et al. [7] have proposed an adaptive load balancer. In this scheme, the number of tasks in relation to the number of processing cores has been considered.

3 Working Principle of Proposed Load Balancing Scheme Applying QUEST Scheduler for Each Core

In this section, we first revisit our earlier proposed utilization-enforced and dynamic priority-featured scheduler [3] which is governed by the HMM in a concise for single-core processor. Figure 1 states the working principle. The two matrices define and as well control the working model of the scheduler. An initial TPM is stated in (1). Equation (2) denotes the error probability matrix. The elements of the matrix, p_{ij}, denote the probability of execution of the specific class-defined traffic (processes). The Metropolis–Hastings (MCMC algorithm) algorithm is applied to estimate the elements of "P".

$$P = \begin{pmatrix} 0.90\ 0.08\ 0.02 \\ 0.39\ 0.56\ 0.05 \\ 0.42\ 0.18\ 0.40 \end{pmatrix} \tag{1}$$

$$E = \begin{pmatrix} 0.98\ 0.9\ 0.8 \\ 0.02\ 0.1\ 0.2 \end{pmatrix} \tag{2}$$

The scheduler has been designed with an aim to achieve a ratio of process utilization (steady-state distribution of Pareto type) and with specifically focusing on

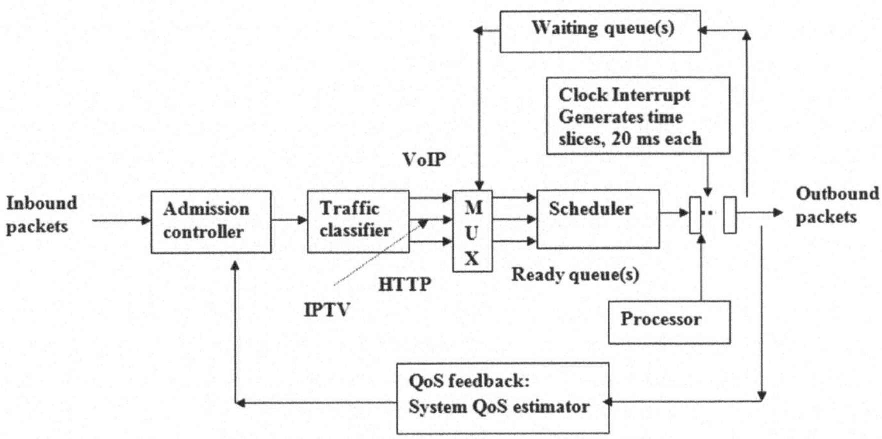

Fig. 1 Internal working flow of M/BP/1/./QUEST scheduling framework running in each core

Table 1 Parameters of the service model: arrival nature, deadline

Traffic class	Arrival characteristics	Deadline (ms)
VoIP	MMPP	20
IPTV	MMPP	100
HTTP-web browsing	MMPP	400

class-specific deadlines. Traffic of Voice over Internet Protocol (VoIP), Internet Protocol Television (IPTV), and Hypertext Transfer Protocol (HTTP) traffic run in this scheduler. The provisioned steady-state ratio (process utilization) is [0.8 0.16 0.04] for highly deadline-sensitive VoIP, moderately deadline-sensitive IPTV, and best-effort traffic HTTP traffic, respectively. The characteristics of inbound traffic are by nature Markovian type. These traffics are modulated by the Poisson processes. A traffic classifier enforces to differentiate the inbound traffic and the packets are sent to individual queue for each flow. The characteristics of the service time of the scheduler are Bounded Pareto. The processor works on incoming processes for time slots (time quantum) of 20 ms. The process (class-specific) which is executed is based on the result of a random number generated at run time. The three multimedia traffics are executed with priorities where the priorities are inversely proportional to their respective deadlines. Considering the deadlines (Table 1), the priorities of execution of the traffic are treated considering the order as HTTP < IPTV < VoIP. A uniform burst time (16 ms) and a timing jitter of 4 ms will set the required value of time quantum (slot) of 20 ms.

3.1 Load Balancing Applying Backpropagation Neural Network

Figure 2 depicts the working principle of load balancing among different cores (a dual-core system) applying backpropagation neural networks (BPN) [8, 9]. The working flow of the load balancing training algorithm has following four steps. First, the run time load of each core in the system is set as initial value for further processing through the BPN. Second, the process of feedforward is performed. After completion of the feedforward, the step which performs backpropagation of errors is executed. Finally, the load and biases are updated for each core. To analyze the performance of the load balancing scheme, different values of initial unbalanced load for the cores have been taken. The initial unbalanced target and system-estimated loads of the core have been stated in Tables 2 and 3 in Sect. 4. In this supervised learning, we set the target-balanced load for each core. First, each of the hidden units receives the value of unbalanced load from each core. This is known as feedforward. The activation function is calculated by the hidden units and the hidden units drive signals to output units. For the specified input pattern, the response of the net is formed by the activation function. The activation function is calculated by the output unit. The activation

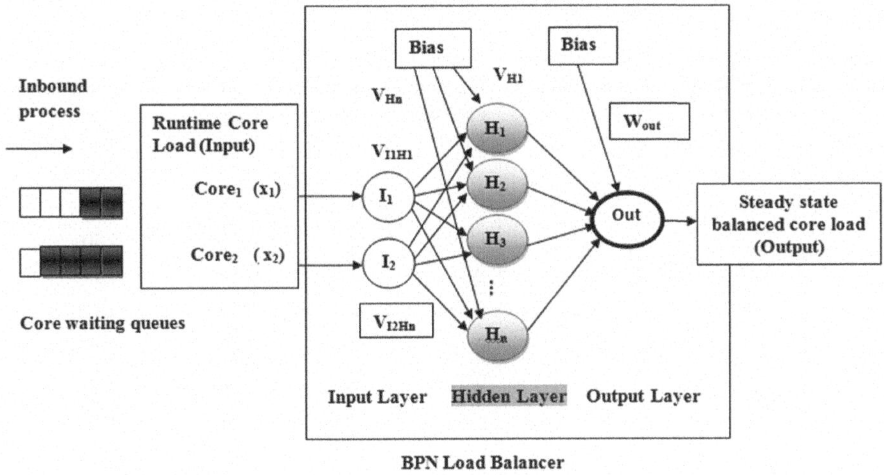

Fig. 2 Working flow of BPN-based load balancing for dual-core processor

Table 2 Observation for no. of cores 2, no. of hidden neurons 3

Learning rate	No. of hidden neurons	Converged at iteration	Initial run time core load $Core_{(1)}$	Initial run time core load $Core_{(2)}$	Target core load $Core_{(n),n=1,2}$	System calculated run time core load $Core_{(n),n=1,2}$
0.3	3	575	0.48	0.52	0.499	0.499
			0.47	0.53	0.501	0.501
			0.54	0.46	0.499	0.499
			0.53	0.47	0.499	0.499
0.5	3	412	0.48	0.52	0.499	0.499
			0.47	0.53	0.501	0.501
			0.54	0.46	0.499	0.499
			0.53	0.47	0.499	0.499
0.7	3	288	0.48	0.52	0.499	0.4989
			0.47	0.53	0.501	0.5009
			0.54	0.46	0.499	0.4991
			0.53	0.47	0.499	0.4992

function we have used in this BPN for load balancing is tanh(x). During the process of backpropagation, errors compare their computed activation with the target value. The target value is the balanced load of each core in the system. The objective is to calculate the associated errors for specific pattern with that output unit.

Table 3 Observation for no. of cores 2, no. of hidden neurons 3

Learning rate	No. of hidden neurons	Converged at iteration	Initial run time core load $Core_{(1)}$	Initial run time core load $Core_{(2)}$	Target core load $Core_{(n),n=1,2}$	System calculated run time core load $Core_{(n),n=1,2}$
0.3	3	2834	0.40	0.60	0.499	0.499
			0.39	0.61	0.501	0.501
			0.58	0.42	0.499	0.499
			0.55	0.45	0.499	0.499
0.5	3	2413	0.40	0.60	0.499	0.499
			0.39	0.61	0.501	0.501
			0.58	0.42	0.499	0.499
			0.55	0.45	0.499	0.499
0.9	3	2933	0.40	0.60	0.499	0.499
			0.39	0.61	0.501	0.501
			0.58	0.42	0.499	0.499
			0.55	0.45	0.499	0.499

Based on the calculated error, the factor is computed. The factor is used to distribute the error at the output unit back to all units which are in the previous layer. Similarly, the error factor is computed for each hidden unit. Finally, the weights and biases are updated using error factor and the activation.

4 Results and Discussion

The framework for load balancing has been implemented using MATLAB. The unbalanced loads of the cores are trained as input of the BPN-driven load balancer. We observe the system-estimated balanced load for the cores for different parameters and conditions for both the cases of moderately and heavily unbalanced system environment. The system model iteratively computes the load of the cores. It is observed that with the increased number of iterations, the load imbalance is minimized and the system reaches a target steady-state balanced load. The results for different cases are stated in Tables 2 and 3. The obtained trends of minimization of load imbalance are depicted in Figs. 3, 4, 5, 6, and 7.

Case 1: Number of cores 2, moderately unbalanced initial load.

Case 2: Number of cores 2, heavily unbalanced initial load.

From the abovementioned figures, we observe that for moderately unbalanced environment, the system converges faster to a target steady-state balanced load of each core with lower number of iterations. In this case, a faster convergence is achieved with increasing learning rate. For heavily unbalanced environment, the

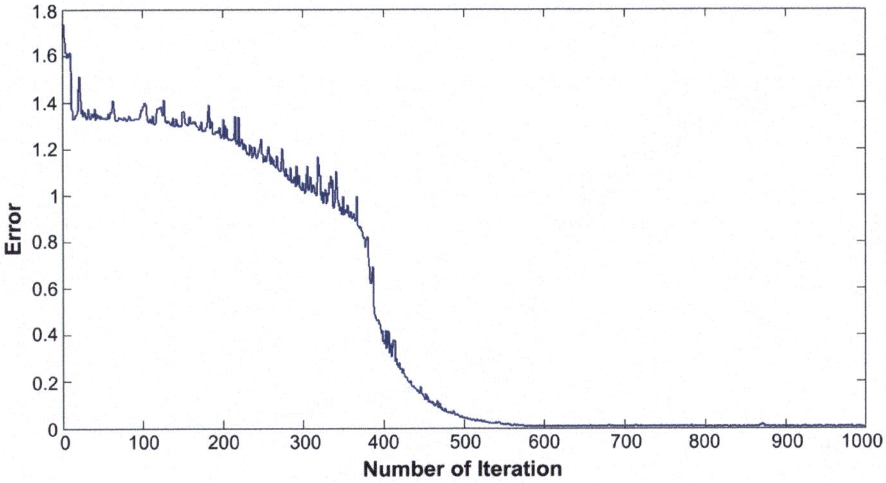

Fig. 3 Plot of error with increasing number of iteration, number of hidden neurons 3, learning rate 0.3

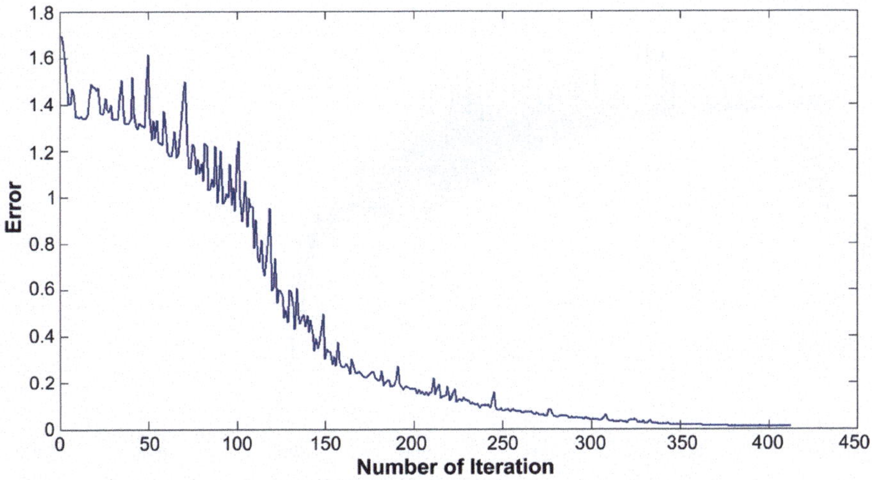

Fig. 4 Plot of error with increasing number of iteration, number of hidden neurons 3, learning rate 0.5

trend of convergence toward the target load of balanced condition takes comparatively more time and more number of iterations compared to moderately unbalanced environment.

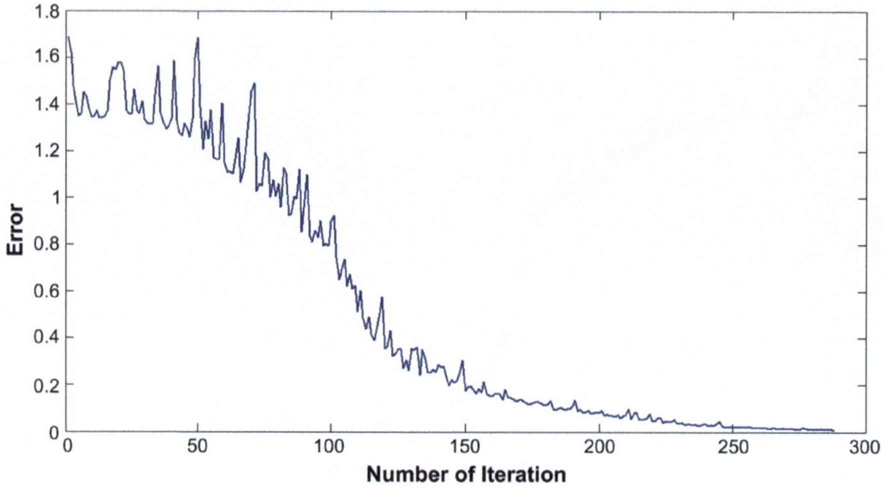

Fig. 5 Plot of error with increasing number of iteration, number of hidden neurons 3, learning rate 0.7

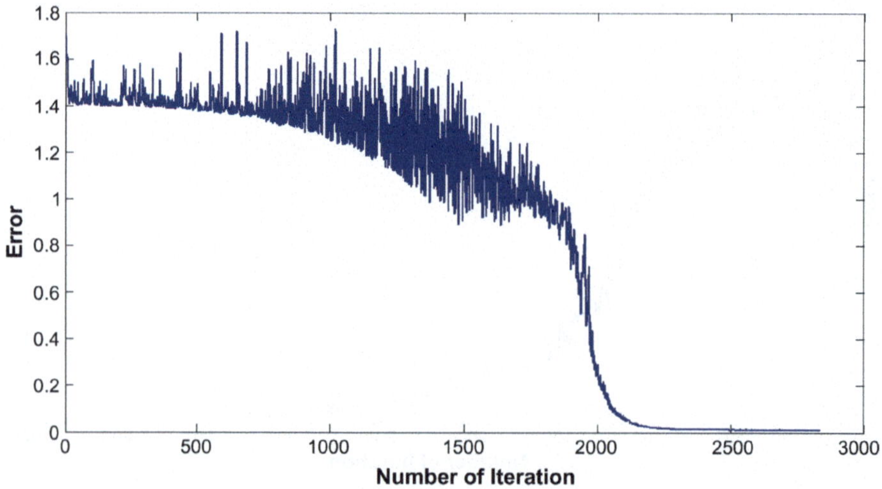

Fig. 6 Plot of error with increasing number of iteration, number of hidden neurons 3, learning rate 0.3

5 Future Work and Conclusion

In this paper, we have demonstrated a BPN-based load balancing scheme for dual-core processor in IP router, where each core is running by the QUEST scheduler. The performance of the load balancing scheme can be analyzed with increased number of cores for both cases of moderately and heavily unbalanced load among the

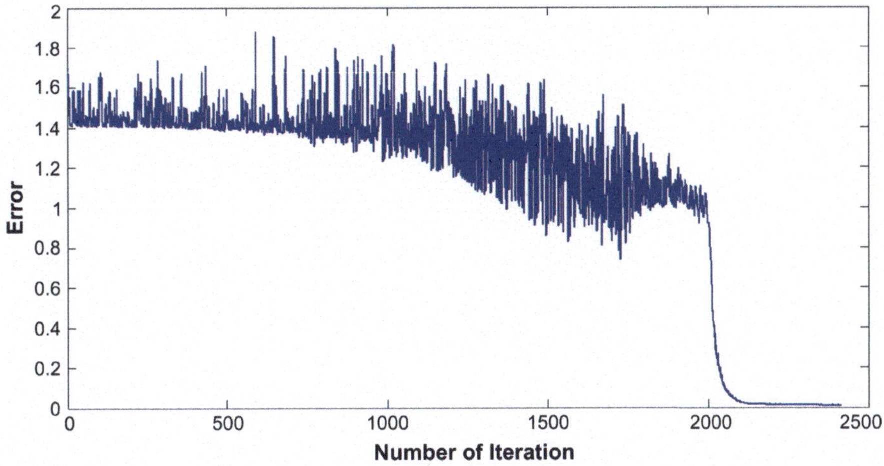

Fig. 7 Plot of error with increasing number of iteration, number of hidden neurons 3, learning rate 0.5

cores. Moreover, the proposed BPN-based load balancing mechanism can be compared with other state-of-the-art neural network algorithms—radial basis function network (RBFN), counterpropagation network, self-organizing feature maps (SOM), and recurrent neural network (RNN) for load balancing and analysis of loads among the cores.

References

1. Wang, Y., Xu, F., Chen, Z., Sun, Y., Zhang, H.: An application-level QoS control method based on local bandwidth scheduling and Haiyan Zhang. J. Electr. Comput. Eng. **2018**, 1–9 (2018). Article ID 4576245
2. Zhu, Z., Wang, F., Zhao, H.: QoS-aware resource allocation for network virtualization in an integrated train ground communication system. Wirel. Commun. Mobile Comput. **2018**, 1–12 (2018). Article ID 2653405
3. Paul, S., Pandit, M.K.: A QoS-enhanced intelligent stochastic real-time packet scheduler for multimedia IP traffic. Multimedia Tools Appl. 1–24 (2017)
4. Jadon, S., Yadav, R.S.: Load balancing in multicore systems using heuristics based approach. I. J. Intell. Syst. Appl. **12**, 56–68 (2018)
5. Cho, K.-M., Tsai, C.-W., Chiu, Y.-S., Yang, C.-S.: A high performance load balance strategy for real-time multicore systems. Sci. World J. 1–14 (2014). Article ID 101529
6. Wu, Q., Wolf, T.: Runtime task allocation in multicore packet processing systems. IEEE Trans. Parallel Distrib. Syst. **23**(10), 1934–1943 (2012)
7. Tan, I.K.T., Chai, I., Hoong, P.K.: An adaptive task-core ratio load balancing strategy for multicore processors. Int. J. Comput. Electr. Eng. **3**(5), 748–753 (2011)
8. Song, W., Zou, S., Tian, Y., et al.: Hum. Cent. Comput. Inf. Sci. **8**, 29 (2018). https://doi.org/10.1186/s13673-018-0152-7
9. Zhang, N., Zhou, K., Li, D.: Earth Sci. Inform. **11**, 553 (2018). https://doi.org/10.1007/s12145-018-0346-6

AF and DF Relay-Based Cooperative Spectrum Sensing Scheme with Dynamic Clustering

Aniket Kumar Sinha, Amishu Kumar Barnwal, Kuldeep Kumar, Ankur Kumar, Santanu Maity and Chanchal Kr. De

Abstract In this paper, we deal with the analysis of an intra-cluster-based Cooperative Spectrum Sensing (CSS) technique to improve the sensing performance of the primary user (PU) in cognitive radio network. The network model consists of one Fusion Center (FC), where FC continuously senses the spectrum of Primary User (PU) with the help of M number of Access Points (APs) and R number of clusters having N number of Cognitive Relays (CRs). The clusters are made on the basis of the quality of channel between CRs and APs. The scheme is based on two-level decisions. First decisions are made at APs using energy detector and second-level decision is made at FC using OR rule and AND rule. A closed-form expression is derived in presence of Rayleigh fading channel for the detection probability and the false alarm with respect to several sensing parameters like sensing channel SNR, number of relays in a cluster, and number of APs.

Keywords Cooperative spectrum sensing · Cognitive radio network · Energy detection · Intra-cluster · Primary user

1 Introduction

Due to improvement in wireless communication engineering, spectrum sensing plays a vital role in this emerging field. Research shows that spectrum utilization of all licensed spectrums can be unused in numerous regions in different time slots. One of the approaches to eradicate this limitation is by spectrum sensing and cognitive radio network has been much discussed due to its high spectrum efficiency [1].

A. K. Sinha (✉) · A. K. Barnwal · K. Kumar · A. Kumar · S. Maity · C. Kr. De
ECE Department, Haldia Institute of Technology, Haldia, West Bengal, India
e-mail: aniket123hit@gmail.com

S. Maity
e-mail: santanu2010@gmail.com

C. Kr. De
e-mail: chanchalkumarde@gmail.com

S. Kundu et al. (eds.), *Proceedings of the 2nd International Conference on Communication, Devices and Computing*, Lecture Notes in Electrical Engineering 602, https://doi.org/10.1007/978-981-15-0829-5_31

The cognitive radio network depicts the path for the secondary users to share the newly arrived licensed spectrum with ease, with minimal interference at some stipulated time slot, and specified geographical topology. Nevertheless, whenever the primary user returns in the operation, the secondary user should readily give access to the spectrum to prevent unwanted interference with the primary user. Thus, the unremitting spectrum sensing is necessary to perceive the presence of the primary user [2]. Furthermore, while the channel suffers from fading and shadowing, the overall result of the sensing of the cognitive user would be falling down [3]. To overcome this difficulty, cooperative spectrum sensing plays a great role by usage of multiuser diversity in sensing technique. A cluster-based CSS is studied to reduce the huge amount of transmission overhead [4]. The cluster-based CSS scheme is used to reduce the required energy and optimize the required bandwidth. A dynamic clustering-based CSS scheme is proposed in using improved energy detectors. Most of the previous work concentrates on single relay with multiple antennas or multiple relays in a network without clustering technique.

In our work, an AF and DF relay-based CSS scheme using dynamic clustering method is studied for improving the sensing performance of cognitive radio network. Closed-form expressions for false alarm and detection probabilities for a cluster-based CSS network are derived using AF and DF relay protocols over Rayleigh fading channel. The logistic results are compared with simulation results for dynamic cluster-based cognitive radio network.

2 Mathematical Modeling

In cognitive radio network, secondary users are allowed to use the licensed spectrums which are not occupied by primary users in some specific time and location. So, the important equipment of cooperative spectrum sensing scheme in cognitive radio network is to detect the presence of the Primary User (PU). In our proposed system model is shown in Fig. 1, where Fusion Center (FC) continuously senses the spectrum of PU with the help of M number of Access Points (APs) and R number of Clusters having N number of Cognitive Relays (CRs). The PU is linked to all the clusters, which is formed by adjacent relay nodes. All the clusters are also interlinked to all the APs in cognitive networks which are further linked to the FC. It is considered that the cognitive relays in a cluster receive the signals from PU and forward it to the APs without any hard decision. The relay that is used in a model consists of cooperative relaying protocols having both forwarding scheme as amplify-forward (AF) and decode-forward (DF). The AF relay works on the principle of transmitting the original signal upon amplification and forwarding it to all APs, whereas the DF relay attains to restore the original signal and forward it to all APs. The best CR is selected based on better channel coefficient among all links between the PU and CRs in a cluster and the sensing information from each cluster is sent to the APs. At each AP, the hard decision is taken, i.e., whether the band of PU is vacant (i.e., inactive) or occupied (i.e., active) with the help of Energy Detection (ED) technique of spectrum

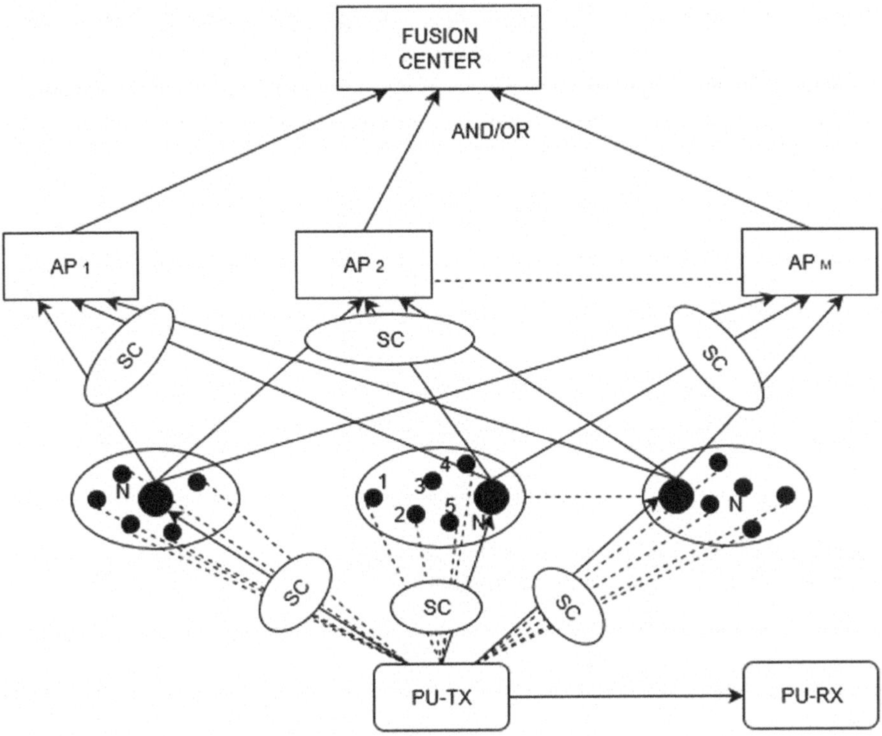

Fig. 1 System model of dynamic cluster-based CSS scheme in CR

sensing. The information received from all the APs reaches the FC which combines all decisions of the APs to make a final decision and share the final sensing decision to the whole network. The fusion rule in the FC is a hard decision fusion rules such as an OR rule or AND rule.

The signal received by nth relay of rth cluster is given by [5]

$$r_{C_r} = \max_{n=1,2\ldots N}\{h_{P,C_{r,n}}\}s + n_{C_r} \tag{1}$$

where $h_{P,C_{r,n}}$ is the channel link between PU and nth relay of rth cluster, n_{C_r} is the Additive White Gaussian Noise (AWGN) with zero-mean and one-sided Power Spectral Density (PSD) N_0 and s indicates PU signal, namely, $s = E_s$ implies the presence of PU and $s = 0$ implies its absence. Here, all channel links are Raleigh fading channel and having equal variance (i.e., $\sigma^2_{P,C_{r,n}} = \sigma^2$ for $r = 1, 2, \ldots$ R and $n = 1, 2, \ldots$ N) and the links between APs and FC are ideal and noiseless.

The signal received in mth AP from the selected relays is given by in AF mode

$$y^{AF}_{AP_m} = r_{C_r} h_{C_r,AP_n} + n_{AP_n} \tag{2}$$

where h_{C_r,AP_m} is the channel links between rth cluster and mth AP, n_{AP_n} is the AWGN noise with zero-mean and one-sided power spectral density N_0.

Let the gain of an AF relay mode $G_m = 1/h_{C_r,AP_n}$ is used to compensate the fading of channel between the relay and AP for the convenience of theoretical analysis. Substituting r_{C_r} from Eq. (1) and G_m into Eq. (2) yields

$$\begin{aligned} y_{AP_m}^{AF} &= \max_{n=1,2\ldots N}\{h_{P,C_{r,n}}\}s + n_{C_r} + n_{AP_n} \\ &= h_{P,C_r}s + \sum_{i=1}^{2} n_i \end{aligned} \tag{3}$$

where $h_{P,C_r} = \max_{n=1,2\ldots N}\{h_{P,C_{r,n}}\}$ and $n_{C_r} = n_{AP_n} = n_i$.

Similarly, the received signal at mth AP using DF relay scheme is

$$y_{AP_m}^{DF} = h_{P,C_r}\hat{s} + n_{i_n} \tag{4}$$

where $\hat{s}$ is the estimated PU signal at the DF relay.

The detector associated with AP decides on

$$\begin{aligned} &H_1 : s = E_s \\ &H_0 : s = 0 \end{aligned} \tag{5}$$

Let $T_{AF}(H_0)$ and $T_{AF}(H_1)$ be the output power of the ED of the Access Point (AP) under the hypotheses H_0 and H_1, respectively. From Eq. (3), we can easily obtain

$$T_{AF}(H_0) = \sum_{i=1}^{2} |n_i|^2 \tag{6}$$

where all the noise components are independent AWGN with equal power spectral density N_0.

From Eq. (4), we can obtain

$$T_{AF}(H_0) = \left|h_{P,C_r}\right|^2 E_s + \sum_{i=1}^{2} |n_i|^2 = \left|h_{P,C_r}\right|^2 E_s + 2N_0 \tag{7}$$

False alarm of the network can be evaluated by Eq. (6), corresponding to the threshold λ_{AF}

$$P_f^{AF} = \Pr[T_{AF}(H_0) > \lambda_{AF}] = e^{-\left(\frac{\lambda_{AF}}{2N_0}\right)}\left[1 + \frac{\lambda_{AF}}{2N_0}\right] \tag{8}$$

It can be calculated as false alarm to corresponding threshold λ_{AF} using AF cooperation in Eq. (8). We can set λ_{AF} for requirement of false alarm in a CRN and then find out detection probability (p_d^{AF}) of AF cooperation case corresponding false alarm P_f^{AF}.

The probability of detection of the PU using AF mode in the mth AP can be written as

$$\begin{aligned} P_{d,m}^{AF} &= \Pr[T_{AF}(H_1) > \lambda_{AF}] = \Pr\left[\left|h_{P,C_r}\right|^2 > \frac{\lambda_{AF} - 2N_0}{E_s}\right] \\ &= \Pr\left[\max_{n=1,2\ldots N}\left\{\left|h_{P,C_{r,n}}\right|^2\right\} > \frac{\lambda_{AF} - 2N_0}{E_s}\right] \\ &= 1 - \left[1 - \exp\left(\frac{1}{\sigma^2}\max\left(\frac{\lambda_{AF} - 2N_0}{E_s}, 0\right)\right)\right]^N \end{aligned} \tag{9}$$

Let $T_{DF}(H_0)$ and $T_{DF}(H_1)$ be the output power of the ED of the Access Point (AP) using the DF relay scheme under the hypotheses H_0 and H_1, respectively. From Eq. (4), we can easily obtain

$$T_{DF}(H_0) = |n_i|^2 \tag{10}$$

where n_i is a noise at the access point only.

$$T_{DF}(H_1) = \left|h_{P,C_r}\right|^2 E_s + |n_i|^2 \tag{11}$$

Based on Eq. (11), false alarm can be calculated from the corresponding threshold λ_{DF}

$$P_f^{DF} = \Pr[T_{DF}(H_0) > \lambda_{DF}] \tag{12}$$

where the threshold value of λ_{DF} can be written as

$$\lambda_{DF} = -2N_0 \ln(P_f^{DF}) \tag{13}$$

The detection probability of PU using DF scheme at the mth AP is given by

$$\begin{aligned} P_{d,m}^{DF} &= \Pr[T_{AF}(H_1) > \lambda_{DF}] = \Pr\left[\left|h_{P,C_r}\right|^2 > \frac{\lambda_{AF} - N_0}{E_s}\right] \\ &= \Pr\left[\max_{n=1,2\ldots N}\left\{\left|h_{P,C_{r,n}}\right|^2\right\} > \frac{\lambda_{AF} - N_0}{E_s}\right] \\ &= 1 - \left[1 - \exp\left(\frac{1}{\sigma^2}\max\left(\frac{\lambda_{AF} - N_0}{E_s}, 0\right)\right)\right]^N \end{aligned} \tag{14}$$

The information received from all the APs reaches the FC which combines all decisions of the APs to make a final decision and circulate the final sensing decision to the whole network. The fusion rule in the FC is a hard decision fusion rules such as an OR rule and AND rule [6].

The detection probability of PU in Fusion Center in using AF mode is

$$P_d^{AF} = 1 - [1 - P_{d,m}^{AF}]^M \tag{15}$$

The detection probability of PU in Fusion Center in using DF mode is

$$P_{d,FC}^{DF} = \prod_{m=1}^{M} P_{d,m}^{DF} \tag{16}$$

In OR rule, if the decision taken by any of the AP is inactive (i.e., available of PU band) then secondary user can initiate the access of the spectrum of PU and so that spectrum utility will be high for fixed false alarm. Whereas, in AND rule, if the decision taken by any of the AP is active (i.e., no available of PU band) then secondary user will not initiate the access of the spectrum and so the spectrum utility will be less.

3 Results and Discussions

The performance of detection probability in a cluster-based CSS is evaluated using MATLAB simulation test bed. Here, analytical results are shown by continuous curves, and simulation results are shown by discrete marks on all figures.

Figure 2 shows the detection probability (P_d) versus SNR of different relay protocols in dynamic cluster-based CSS scheme. It is seen that if SNR value is increased then corresponding detection probability (P_d) will also be increased for both CR schemes over a wide range of SNR values. DF scheme is always better then AF scheme. For example, consider at 10 db SNR for AF scheme, the corresponding (P_d) is 0.33, whereas for DF scheme, the corresponding (P_d) is 0.58, so (P_d) is higher for DF scheme OR rule which gives better performance in comparison to AND rule in terms of (P_d) but the problem is that the false alarm will also be increased. Although AND rule gives lesser performance but it will have lesser false alarm (P_f) than that of OR rule. And it is better to have less false alarm value.

Figure 3 depicts the detection probability of dynamic cluster-based CSS scheme versus false alarm sensing channel for different fusion rules in FC. It is seen that the increase in the detection probability leads to increase in the false alarm P_f for a fixed value of false alarm P_f. This graph also follows the same nature of curve as the previous one. Here also OR rule is better than AND rule in terms of P_d over wide range of false alarm. It is also seen that DF scheme is always better than AF scheme. For example, at a fixed false alarm value of 0.1 for OR rule P_d in DF scheme is 0.23 while in AF scheme P_d it is 0.03. Similarly, for AND rule P_d in DF scheme is higher than P_d in AF scheme at a particular value of false alarm P_f.

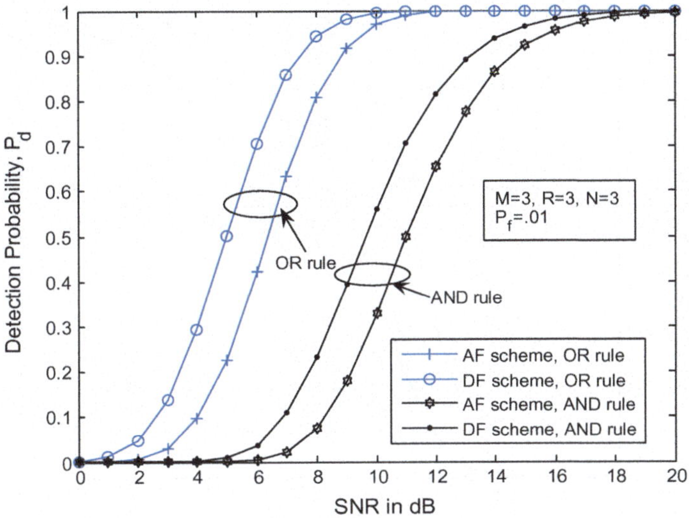

Fig. 2 Variation of P_d with SNR for different cognitive relay schemes

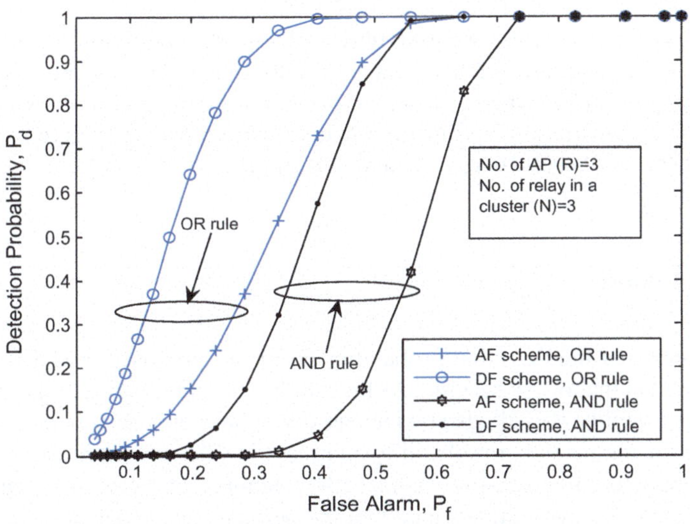

Fig. 3 Detection probability versus false alarm for different fusion rules

Figure 4 depicts the detection probability of dynamic cluster-based CSS scheme versus SNR of sensing channel for different number of CRs and APs. It is seen that the detection probability of our system model is increased with the increase of both the number of CRs and APs in both the schemes and rules, i.e., in DF and AF and OR and AND rule. In this graph, we have increased both the number of relays and

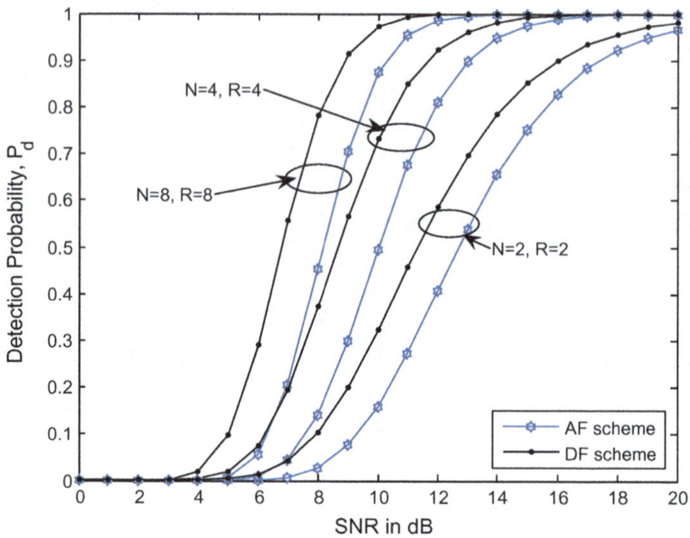

Fig. 4 Impact of number of N and R on detection probability

APs and as per expectations, we have observed a much more improved probability of detection, for example, at a fixed value of 8 db SNR in DF scheme when $N = 2$ and $R = 2$, P_d is 0.1 but when $N = 4$ and $R = 4$, P_d is 0.38, and when $N = 8$ and $R = 8$, P_d is 0.79, whereas in AF scheme when $N = 2$ and $R = 2$, P_d is 0.02 but when $N = 4$ and $R = 4$, P_d is 0.16 and when $N = 8$ and $R = 8$, P_d is 0.48.

4 Conclusion

We have seen that the detection probability P_d of our proposed model is better in DF scheme than that of AF scheme. As the number of relays and APs increases, the detection probability P_d will also be increased. We have also derived a closed-form expression for the detection probability and the false alarm with respect to several sensing parameters like sensing channel SNR, number of relays in a cluster, and number of APs. So, we can conclude that the performance of DF scheme is better than the AF scheme as the detection probability is higher. For example, with 10% false alarm probability, the detection is almost 50% in AF scheme and almost 80% in DF scheme. So, it is recommended to use the DF scheme over AF scheme in CRs, but DF relaying scheme sometimes causes a propagation of error in case of incorrect decoding of the information symbols. DF scheme makes the system more complex and costlier. In case where efficiency (P_d) can be compromised, AF relaying scheme would be more preferable as it is available in low cost and simpler circuitry. For OR and AND rule, it should be used according to the company requirements.

References

1. Haykin, S., Qu, Z., Xu, Y., Yin, S.: A novel clustering-based spectrum sensing in cognitive radio wireless sensor networks. In: 2014 IEEE 3rd International Conference on Cloud Computing and Intelligent Systems (CCIS), pp. 695–699 (Nov 2014)
2. Deng, F., Fanzi, Z., Li, R.: Clustering-based compressive wide-band spectrum sensing in cognitive radio network. In: 5th International Conference on Mobile Ad-hoc and Sensor Networks, 2009. MSN'09, pp. 218–222 (Dec 2009)
3. Roy, A., Pattepu, S., Datta, A.: Energy detection analysis of hybrid cooperative relaying mode in cognitive radio network. In: 1st IEEE International Conference on I-SMAC (Feb 2017)
4. Reisi, N., Jamali, V., Ahmadian, M., Salari, S.: Cluster-based cooperative spectrum sensing in cognitive radio networks under log-normal shadow-fading. In: 2011 19th Iranian Conference on Electrical Engineering, pp. 1–5. Tehran (2011)
5. Yadav, K., Bhowmick, A., Dhar Roy, S., Kundu, S.: Cooperative spectrum sensing based on dynamic clustering with improved energy detector. In: 3rd International Conference on Recent Advances in Information Technology I RAIT-2016
6. Sunil, P.: Performance analysis of moving multi-antenna relay cooperation with hybrid relaying scheme in cooperative wireless networks. In: 2016 International Conference on Computational Intelligence and Networks

New Compact SEC-DED-DAEC Code for Memory Applications

Raj Kumar Maity, Jagannath Samanta and Jaydeb Bhaumik

Abstract The demand of higher capacity, smaller size, and reliable memory is increasing with the continuous scaling of semiconductor technology with time. But reliability of memory is greatly influenced by soft errors caused due to radiation effects. These soft errors lead to corruption of data stored in one or multiple cells of memory. Error Correction Codes (ECCs) are frequently employed for mitigating the effects of soft errors in memories. Single Error Correction-Double Error Detection-Double Adjacent Errors Correction (SEC-DED-DAEC) code is one of the popularly known ECC schemes which is employed when Multiple Bit Upsets (MBUs) occur in memory. In this paper, a new SEC-DED-DAEC code has been proposed for memory applications. The proposed codecs have been designed and synthesized in FPGA platform for some common word lengths frequently used in memory applications. The performance of proposed codecs have been compared with other related works. The proposed codecs require lesser area compared to existing codecs.

Keywords Memory · Soft errors · ECCs · SEC-DED-DAEC · FPGA

1 Introduction

Memory is an integral part of most of the modern electronic devices and systems. Proper functionality of these electronic systems mostly depends on the reliable operation of memory. But reliability of memory is hugely affected by soft errors which alter value stored in one or multiple memory cells [1]. In [2], Chen et al. have shown

R. K. Maity (✉) · J. Samanta
Haldia Institute of Technology, Haldia, India
e-mail: hitece.raj@gmail.com

J. Samanta
e-mail: jagannath19060@gmail.com

J. Bhaumik
Department of ETCE, Jadavpur University, Kolkata, India
e-mail: bhaumik.jaydeb@gmail.com

S. Kundu et al. (eds.), *Proceedings of the 2nd International Conference on Communication, Devices and Computing*, Lecture Notes in Electrical Engineering 602, https://doi.org/10.1007/978-981-15-0829-5_32

that effect of soft errors in memory can be reduced by employing Error Correction Codes (ECCs). Single Error Correction (SEC) and Single Error Correction-Double Error Detection (SEC-DED) codes [3] are normally employed when only single bit of stored codeword in memory is affected by soft error. Due to continuous scaling of technology, occurrence of errors in multiple bits of stored codeword known as Multiple Bit Upsets (MBUs) has increased nowadays [4]. These MBUs bring forward a major challenge to existing SEC or SEC-DED codes. Interleaved version of SEC-DED code [5] can handle MBUs up to some extent in the expense of complicated memory design [6]. As an alternative, codes with multiple bit error correcting capability such as Bose–Chaudhuri–Hocquenghem (BCH) code [7] and Reed Solomon (RS) code [8] can protect MBUs in memory but these codes have more complex design of their decoders.

Due to non-negligible chances of double errors in a sequence, Single Error Correction-Double Error Detection-Double Adjacent Error Correction (SEC-DED-DAEC) codes [6, 9–13] are drawing the interest of the researchers for protecting memory against MBUs. The SEC-DED-DAEC codes presented by Dutta et al. [10] are planed for memory and designed with reduced redundancy. But main limitation of Dutta codes is higher miscorrection probability, i.e., higher chances of treating nonadjacent double errors as double adjacent error. An attempt to reduce miscorrection probability of SEC-DED-DAEC codes has been made by Ming et al. in [11] where they have proposed a new structure of SEC-DED-DAEC code design without changing the redundancy. A codec designed to correct double adjacent errors has completely nullified the chances of miscorrection against two-random errors [12]. The SEC-DED-DAEC codes have been designed by Neale et al. [6, 9] to provide scalable adjacent error detection feature with reduced miscorrection probability and moderate hardware complexity. Two schemes for hardware-efficient decoding of SEC-DED-DAEC code have been described in [13] where the first scheme of decoding makes the decoder area efficient but the second scheme provides fast decoding. So the major constraints of designing SEC-DED-DAEC code are miscorrection probability which must be lower and hardware complexity for its codec design which must be reduced for realization of compact and fast codecs.

In this paper, an area-efficient SEC-DED-DAEC code has been proposed. Subsequent are the major contributions of this work: (i) the H-matrix of a new SEC-DED-DAEC code has been proposed for some common word lengths frequently used in memory, (ii) proposed codecs have been designed and synthesized in FPGA platform, (iii) FPGA synthesis results of proposed codecs have been compared with other existing designs, and (iv) the proposed codecs show better performance in area (LUTs) compared to existing other codecs.

The remaining of this brief is ordered as follows. Section 2 provides the H-matrix construction of proposed SEC-DED-DAEC codes. Design of proposed SEC-DED-DAEC codecs is presented in Sect. 3. The FPGA-based synthesis results for the proposed codecs are presented in Sect. 4, and finally Sect. 5 concludes the paper.

2 H-matrix Construction of Proposed SEC-DED-DAEC Codes

The parity check (H) matrix of an (n, k) ECC is a matrix with $(n - k)$ number of rows and n number of columns [14]. In systematic form, the H-matrix can be expressed as follows:

$$H = [P^T \quad I_{n-k}] \tag{1}$$

where P^T and I represent transpose of parity matrix (P) and identity matrix, respectively. Encoder and decoder circuits of any ECC are designed by employing the H-matrix. The proposed SEC-DED-DAEC H-matrices are created by employing the following conditions:

1. All the columns in H-matrix are nonzero and distinct.
2. All the data columns in H-matrix have constant Hamming weight equal to three (3).
3. The XOR sum of any two adjacent columns in H-matrix is unique and has a constant Hamming weight equal to two (2).

Condition 1 provides the Single Error Correction (SEC) feature. Double Error Detection (DED) feature of the proposed codes is confirmed by condition 2 as single error produces an odd weight syndrome, whereas weight of the syndrome is even for double errors. The Double Adjacent Error Correction (DAEC) feature of the proposed codes is maintained by condition 3.

The H-matrices of proposed SEC-DED-DAEC codes have been constructed for some common word lengths used in memory, i.e., for $k = 16$, 32, and 64 bits. For the proposed (n, k) SEC-DED-DAEC codes, the length of codewords (n) and number of parity bits $(n - k)$ are determined by employing Eq. (2) and these values are listed in Table 1.

$$\binom{n-k}{3} \geq k \quad \text{and} \quad \binom{n-k}{2} \geq n - 1 \tag{2}$$

The H-matrices of proposed SEC-DED-DAEC are shown in Figs. 1, 2, and 3 for 16, 32, and 64-bit word lengths, respectively.

Table 1 Number of parity bits and codeword lengths of proposed codes

Number of data bits (k)	Number of parity bits ($n - k$)	(n, k)
16	8	(24, 16)
32	10	(42, 32)
64	13	(77, 64)

$$H = \begin{bmatrix} 000011100000011110000000 \\ 011110000001110001000000 \\ 110000001011100000100000 \\ 000000011110000100010000 \\ 000001000111000000001000 \\ 000111110000001100000100 \\ 101100000000111000000010 \\ 111000111100000000000001 \end{bmatrix}$$

Fig. 1 H-matrix of proposed (24, 16) SEC-DED-DAEC code

$$H = \begin{bmatrix} 011100000000110000000000001110111000000000 \\ 111111000000000111010001000000000100000000 \\ 110000001000000000001111010000000010000000 \\ 000111100000000010000011111110000001000000 \\ 100000110000111111000000000000010000100000 \\ 000000000011111110000100000000000000010000 \\ 000010000111000000001111000000000000001000 \\ 001000000000000100001111110000000000000100 \\ 000000011110000000111000000001110000000010 \\ 000001111100001000000000110011100000000001 \end{bmatrix}$$

Fig. 2 H-matrix of proposed (42, 32) SEC-DED-DAEC code

$$H = \begin{bmatrix} 00001111100000011111100000000010000000000000000000000000101011100000000000 \\ 00000000000000001000000000011111000100000000000000000000010000000001010000 \\ 0000000 \\ \end{bmatrix}$$

parity bits which are generated by encoder provide reliability to stored data in memory. The decoder detects or corrects error in stored codewords caused by soft errors to retrieve the original data.

The expressions for parity bits of proposed (24, 16) SEC-DED-DAEC code based on its H-matrix in Fig. 1 are presented as follows :

$$\begin{aligned}
c_1 &= d_5 \oplus d_6 \oplus d_7 \oplus d_{14} \oplus d_{15} \oplus d_{16} \\
c_2 &= d_2 \oplus d_3 \oplus d_4 \oplus d_5 \oplus d_{12} \oplus d_{13} \oplus d_{14} \\
c_3 &= d_1 \oplus d_2 \oplus d_9 \oplus d_{11} \oplus d_{12} \oplus d_{13} \\
c_4 &= d_8 \oplus d_9 \oplus d_{10} \oplus d_{11} \oplus d_{16} \\
c_5 &= d_6 \oplus d_{10} \oplus d_{11} \oplus d_{12} \\
c_6 &= d_4 \oplus d_5 \oplus d_6 \oplus d_7 \oplus d_8 \oplus d_{15} \oplus d_{16} \\
c_7 &= d_1 \oplus d_3 \oplus d_4 \oplus d_{13} \oplus d_{14} \oplus d_{15} \\
c_8 &= d_1 \oplus d_2 \oplus d_3 \oplus d_7 \oplus d_8 \oplus d_9 \oplus d_{10}
\end{aligned} \tag{3}$$

The encoder circuit of proposed (24, 16) SEC-DED-DAEC code is revealed in Fig. 4 which computes the parity bits employing Eq. (3). In Fig. 4, the logic design of first (c_1) and last parity (c_8) bits has been shown. The encoder circuit consists of only two-input XOR (XOR2) gates.

Decoder circuit in proposed SEC-DED-DAEC codec retrieves back the original data stored in memory by correcting the errors caused by soft errors. Decoder circuit for proposed SEC-DED-DAEC code performs three major tasks, namely, (i) computation of syndrome, (ii) detection of error location, and (iii) correction of error.

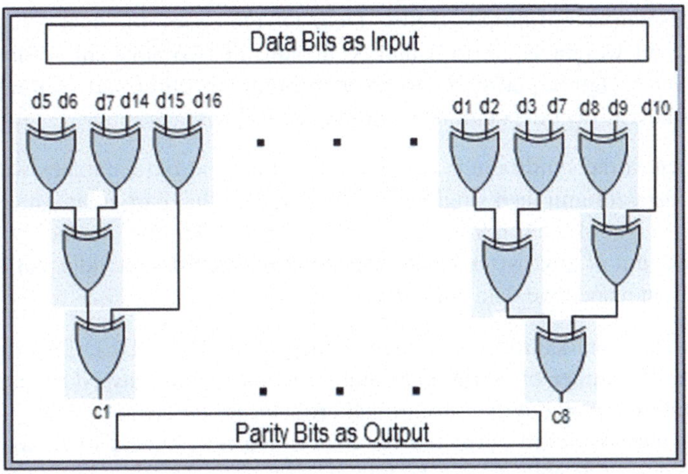

Fig. 4 Encoder circuit of proposed (24, 16) SEC-DED-DAEC code

Syndrome (S) for any ECC is computed based on the H-matrix of the code and r (stored codeword in memory) by employing equation (4).

$$S = r \cdot H^T \tag{4}$$

The syndrome expressions for proposed (24, 16) SEC-DED-DAEC code with H-matrix as indicated in Fig. 1 are as follows:

$$\begin{aligned}
S_1 &= r_5 \oplus r_6 \oplus r_7 \oplus r_{14} \oplus r_{15} \oplus r_{16} \oplus r_{17} \\
S_2 &= r_2 \oplus r_3 \oplus r_4 \oplus r_5 \oplus r_{12} \oplus r_{13} \oplus r_{14} \oplus r_{18} \\
S_3 &= r_1 \oplus r_2 \oplus r_9 \oplus r_{11} \oplus r_{12} \oplus r_{13} \oplus r_{19} \\
S_4 &= r_8 \oplus r_9 \oplus r_{10} \oplus r_{11} \oplus r_{16} \oplus r_{20} \\
S_5 &= r_6 \oplus r_{10} \oplus r_{11} \oplus r_{12} \oplus r_{21} \\
S_6 &= r_4 \oplus r_5 \oplus r_6 \oplus r_7 \oplus r_8 \oplus r_{15} \oplus r_{16} \oplus r_{22} \\
S_7 &= r_1 \oplus r_3 \oplus r_4 \oplus r_{13} \oplus r_{14} \oplus r_{15} \oplus r_{23} \\
S_8 &= r_1 \oplus r_2 \oplus r_3 \oplus r_7 \oplus r_8 \oplus r_9 \oplus r_{10} \oplus r_{24}
\end{aligned} \tag{5}$$

If all the syndrome expressions are zero then there are no errors in r to be detected. Otherwise, if syndrome corresponds to a nonzero value then error is detected. So whether error has occurred or not in r can be determined by checking the syndrome values. Locations of detected errors are also determined based on syndrome of r, i.e., if syndrome values match any of data column or XOR sum of two adjacent data columns in H-matrix then single or double adjacent errors are located in respective position of data bit in r. This is the conventional way of error location detection method for SEC-DED-DAEC codes with unequal column weight in H-matrices.

In the H-matrices of proposed SEC-DED-DAEC codes, the data columns have a constant odd weight equal to 3 and XOR sum of two adjacent columns has a constant even weight equal to 2. So error location detection part of the proposed SEC-DED-DAEC decoders can be simplified as follows:

(i) If weight of the syndrome is odd and 1's in the syndrome matches with the 1's in any data column then single error is located on that particular data column.
(ii) Else if weight of the syndrome is even and 1's in syndrome matches with the 1's in XOR sum of any two adjacent data columns then double adjacent errors are located on respective data columns.

This simple error location detection logic for proposed SEC-DED-DAEC decoders requires lesser number of AND gates and does not require any NOT gate in their implementation compared to conventional error location detection logic.

The located errors are corrected by error correction logic part of the decoder which employs XOR gates to flip the bits in errors. The decoder circuit of proposed (24, 16) SEC-DED-DAEC code has been indicated in Fig. 5. As shown in Fig. 5, the syndrome computation part of the decoder generates syndrome bits and this part is designed using only XOR2 gates. The error location detection part of the decoder

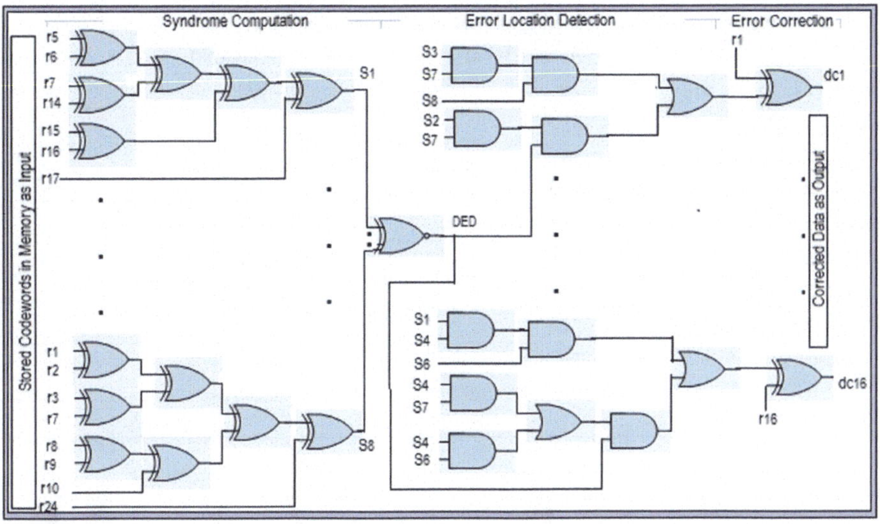

Fig. 5 Decoder circuit of proposed (24, 16) SEC-DED-DAEC code

identifies the locations of errors and this part is implemented using two-input AND (AND2) and two-input OR (OR2) gates. Finally, the located errors are corrected by error correction part which employs XOR2 gates to flip the bits in errors.

4 Synthesis Results

The proposed SEC-DED-DAEC codecs have been represented in Verilog Hardware Description Language (HDL). These codecs have been simulated and synthesized on Field Programmable Gate Array (FPGA)-based Artix7 (XC7A100T-2csg324) device family. The FPGA-based synthesis results of proposed codecs have been compared with other related codecs in [6, 9, 13]. These comparisons are summarized in Table 2. As shown in Table 2, proposed codecs require lesser number of LUTs (Lookup Tables) compared to both the existing codecs in [6, 9] and in [13] with first scheme of optimized decoding. Also, the proposed codecs are more area (LUTs) efficient than the codecs in [6, 9] compared to the codecs in [13]. The delay performance of proposed codecs is better than the codecs presented in [6, 9]. But delay of proposed codecs is worse compared to the codecs in [13]. Graphical comparison of improvements in LUTs and delay has been indicated in Fig. 6. Proposed codecs exhibit better performance in terms of LUTs compared to the existing codecs. Also, improvement in terms of LUTs for proposed codecs compared to the codecs in [6, 9] increases with increase in word length. But improvement diminishes with increase in word length against codecs in [13]. Also, delay performance of proposed codecs is better with respect to codecs in [6, 9] only and these improvements decrease with

Table 2 Comparison of FPGA-based synthesis results for proposed codecs

Codecs	Area (LUTs)	Delay (ns)
Neale (24, 16) [6, 9]	70	4.931
Reviriego (23, 16) [13]	65	3.634
Proposed (24, 16)	58	4.084
Neale (41, 32) [6, 9]	129	6.278
Reviriego (39, 32) [13]	114	4.970
Proposed (42, 32)	101	5.338
Neale (75, 64) [6, 9]	260	6.727
Reviriego (73, 64) [13]	214	5.904
Proposed (77, 64)	197	6.651

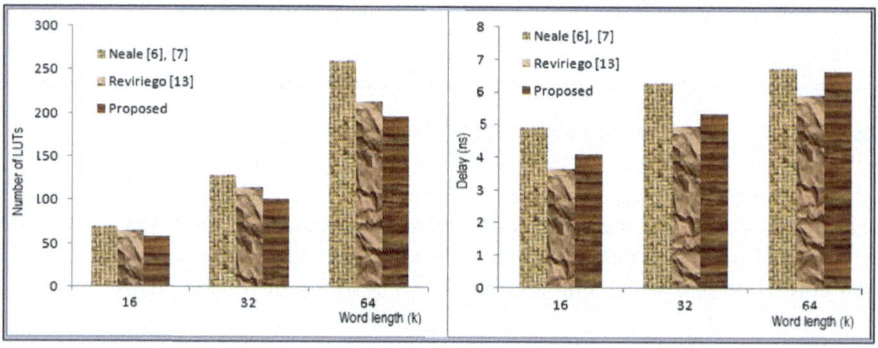

Fig. 6 Graphical comparison of improvements in LUTs and delay of proposed codecs

increase in word length. The proposed (77, 64) codec exhibits a maximum improvement of 24.23% in LUTs against Neale et al. (77, 64) codec, whereas a maximum improvement of 17.18% has been attained by proposed (24, 16) codec against Neale et al. (24, 16) codec.

5 Conclusion

In this paper, a new SEC-DED-DAEC code has been proposed for some common word length frequently applied in memory. FPGA synthesis results of proposed codecs have been compared with other two related works. The proposed codecs exhibit better performance in area and delay against one existing work and better performance in area against another existing work. So the proposed codecs can be employed in memory for protecting corruption of stored data due to soft errors.

References

1. Baumann, R.C.: Soft errors in advanced computer systems. IEEE Des. Test Comput. **22**(3), 258–266 (2005)
2. Chen, C.L., Hsiao, M.Y.: Error-correcting codes for semiconductor memory applications: a state-of-the-art review. IBM J. Res. Develop. **28**(2), 124–134 (1984)
3. Hsiao, M.Y.: A class of optimal minimum odd-weight-column SEC-DED codes. IBM J. Res. Dev. **14**(4), 301–395 (1970)
4. Ibe, E., Taniguchi, H., Yahagi, Y., Shimbo, K., Toba, T.: Impact of scaling on neutron-induced soft error in SRAMs from a 250 nm to a 22 nm design rule. IEEE Trans. Electron Devices **57**(7), 1527–1538 (2010)
5. Reviriego, P., Maestro, J.A., Baeg, S., Wen, S., Wong, R.: Protection of memories suffering MCUs through the selection of the optimal interleaving distance. IEEE Trans. Nucl. Sci. **57**(4), 2124–2128 (2010)
6. Neale, A., Sachdev, M.: A new SEC-DED error correction code subclass for adjacent MBU tolerance in embedded memory. IEEE Trans. Device Mater. Rel. **13**(1), 223–230 (2013)
7. Naseer R., Draper, J.: Parallel double error correcting code design to mitigate multi-bit upsets in SRAMs. In: Proceedings of the 34th European Solid State Circuits Conference, pp. 222–225 (2008)
8. Pontarelli, S., Reviriego, P., Ottavi, M., Maestro, J.A.: Low delay single symbol error correction codes based on reed solomon codes. IEEE Trans. Comput. **64**(5), 1497–1501 (2015)
9. Neale, A.: Design and analysis of an adjacent multi-bit error correcting code for nanoscale SRAMs, Ph.D. Thesis (2014)
10. Dutta, A., Touba, N.A.: Multiple bit upset tolerant memory using a selective cycle avoidance based SEC-DED-DAEC code. In: Proceedings 25th IEEE VLSI Test Symposium, pp. 349–354 (2007)
11. Ming, Z., Yi, X.L., Wei, L.H.: New SEC-DED-DAEC codes for multiple bit upsets mitigation in memory. In: Proceedings IEEE/IFIP 20th International Conference VLSI Systems Chip, pp. 254–259 (2011)
12. Dutta, A.: Low cost adjacent double error correcting code with complete elimination of mis-correction within a dispersion window for multiple bit upset tolerant memory. In: Proceedings IEEE/IFIP 20th International Conference VLSI Systems Chip, pp. 287–290 (2012)
13. Reviriego, P., Martínez, J., Pontarelli, S., Maestro, J.A.: A method to design SEC-DED-DAEC codes with optimized decoding. IEEE Trans. Device Mater. Reliab. **14**(3), 884–889 (2014)
14. Lin, S., Costello, D.J.: Error Control Coding: Fundamentals and Applications. Prentice-Hall (1983)

Performance Evaluation of Low-Cost RGB-Depth Camera and Ultrasonic Sensors

Akash Adhikary, Rishav Vatsa, Aman Burnwal and Jagannath Samanta

Abstract Depth data from ranging sensors are generally employed in site data acquisition, navigation, and photogrammetry. Different types of sensors have their own edges and limitations depending upon the situation, hence a fusion of these sensors has the potential to overcome the limitations of single sensor devices. To model a robust device, the thorough performance characteristic of each sensor is required. Designing a robust system is challenging for companies and researchers as the available test data is very confined. In this paper, the performance of ultrasonic ranging module and Kinect sensor in the context of spatial modeling and ranging has been evaluated. The experiments are conducted to evaluate the accuracy, range and operating conditions of these sensors. A very robust performance of ultrasonic and Kinect sensor is observed within a range of 3 m and 4 m respectively. The ultrasonic sensor is suitable for rigid objects and has an accuracy of $\pm 2.61\%$ whereas Kinect performs accurately with a wider range of objects and has an average accuracy of $\pm 0.89\%$. This result establishes a benchmark for performance expectation in similar modeling applications and exhibits a scope for designing a hybrid multisensor spatial modeling device.

Keywords Depth sensing · Kinect · Ultrasonic sensor · 3-D modeling

A. Adhikary (✉) · R. Vatsa · A. Burnwal · J. Samanta
Haldia Institute of Technology, Haldia, West Bengal, India
e-mail: akash.adhikary@hotmail.com

R. Vatsa
e-mail: rishavrishuvatsa23@gmail.com

A. Burnwal
e-mail: amankolindia@gmail.com

J. Samanta
e-mail: jagannath19060@gmail.com

S. Kundu et al. (eds.), *Proceedings of the 2nd International Conference on Communication, Devices and Computing*, Lecture Notes in Electrical Engineering 602, https://doi.org/10.1007/978-981-15-0829-5_33

1 Introduction

With the advancement of new technologies in the domain of navigation and digital signal processing (DSP), the application of spatial modeling and object tracking [1] have surged up as a very prominent tool to explore the previously unknown. Simultaneous Localization and Mapping (SLAM) [2–4], Light Detection and Ranging (LiDAR) [5], and ultrasonic and Infrared (IR) [6] are sensors of choice due to their robustness. Kinect sensor [7, 8] is a comparatively new yet popular device and is also used for this purpose. It is a low-cost RGB-depth sensor primarily developed as a motion input device for the Xbox 360 gaming console. But the sensor became more popular among researchers due to its capability of providing accurate depth data in relatively low cost. This device with the support of intelligent algorithms is contributing to various applications, such as object tracking, object recognition [9, 10], activity analysis [11], etc. Another popular low-cost sensor used for ranging is ultrasonic sensor HC-SR04 [12] which is a conventionally used ranging sensor.

The efficiency of these sensors varies analogous to reflecting surfaces, angles, and distances. As these sensors are low-cost devices, their reliability has a constraint of confined ranges. The performance evaluation of Kinect has been presented in [13–17] and performance analysis of HC SR-04 in [18–20]. In literature, very few works have been reported based on comparative performance evaluation of these two sensors.

In this paper, we have performed some experiments and set a few performance metrics upon which we have compared Microsoft Xbox Kinect and HC SR-04 ultrasonic sensor. We also analyze their accuracy under different test conditions. We have varied certain external factors like reflecting surfaces, medium, and angle of incidence which affected the performance of our sensors significantly.

The further sections of this paper are organized as follows: In Sect. 2, an overview of the device under test is presented. Section 3 gives a detailed description of the experiment and performance metrics. Section 4 shows the results and observations of carried out experiments and further analyzes the performance of the sensors. We conclude the paper in Sect. 5.

2 Overview of Ultrasonic and Kinect Sensor

The working principle of ultrasonic and Kinect sensor, employed in this work, is described in the following.

2.1 Ultrasonic Ranging Module HC-SR04

In robotics, ultrasonic sensors are widely used for contact-less, mid-range distance measurements. The ultrasonic sensor uses Time of Flight (TOF) method for distance measurement, which refers to the time taken for a pulse to travel from the transmitter to an observed object and back to the receiver. The distance of a reflecting surface from this sensor is computed as follows:

$$d = \frac{1}{2} \times v_s \times T_f \tag{1}$$

where d is the distance between the reflecting surface and the ultrasonic sensor, T_f is time taken by the ultrasonic pulse from the transmitter to the receiver, and v_s is the velocity of sound in the medium at an ambient temperature. In Fig. 1, the mechanism of distance measurement can be visualized. Here, v_s is assumed to be 343 m/s. The speed of sound does not vary significantly with respect to the variations in the temperature [22]; however, the reflecting surface and the medium of transmission of the sound waves affect its performance notably.

2.2 *Kinect Sensor*

Kinect sensor is a highly sophisticated hardware device that contains a depth vision and an RGB camera, an array of four microphones and a 2−DOF (Degree of Freedom) tilt motor ranging from −27° to +27°. The whole setup of the sensor in Fig. 2a consists of a pair of IR projector and a Complementary Metal Oxide Semiconductor (CMOS) IR camera. The firmware on the onboard prime sense [23] controller works on machine learning algorithms which are used for object and skeletal tracking [24].

The Kinect projects a pattern of Infrared (IR) dots on the surface as depicted in Fig. 3a and determines the disparities between the emitted light beam and the observed position of the light dots. The disparity is then used to calculate the distance to the reflecting object by triangulation [25] as shown in Fig. 2b. If x is the coordinate of a particular point from the projector end, then it should correspond with a coordinate x' of the same point as seen from the IR receiver end. The equation

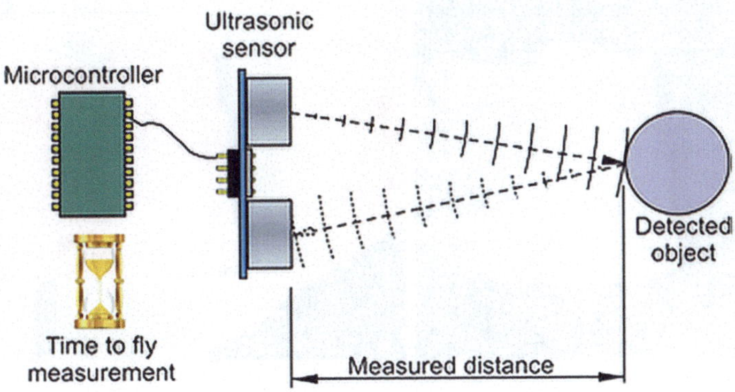

Fig. 1 Distance measurement using ultrasonic sensor [21]

governing this mechanism is written in Eq. (3).

$$\text{Disparity} = x - x' = \frac{B \times f}{z} \tag{2}$$

$$d_{\text{kinnect}} = 109150 - \frac{348 \times 10^4}{z} \tag{3}$$

In Fig. 2b, P is the position of IR projector and I is the position of IR camera which is constant throughout. In Eq. (2), z is the depth in pixels and f is a variable distance, the distance between point P and I is known as Baseline B as depicted in Fig. 2b. From this depth, distance in meters, i.e., d_{kinnect} can be calculated using Eq. (3). Once triangulation is achieved through the depth sensor, the RGB module provides the color to the pixels. This result can be clearly inferred from the output of raw depth data achieved during the course of observation in Fig. 3a.

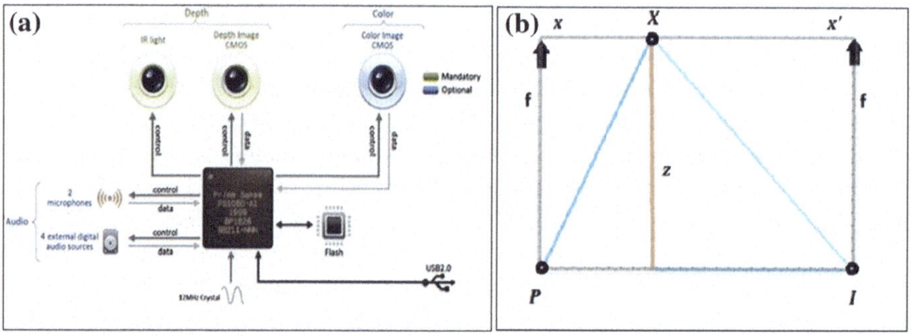

Fig. 2 **a** System architecture of Kinect sensor. **b** Triangulation in Kinect sensor

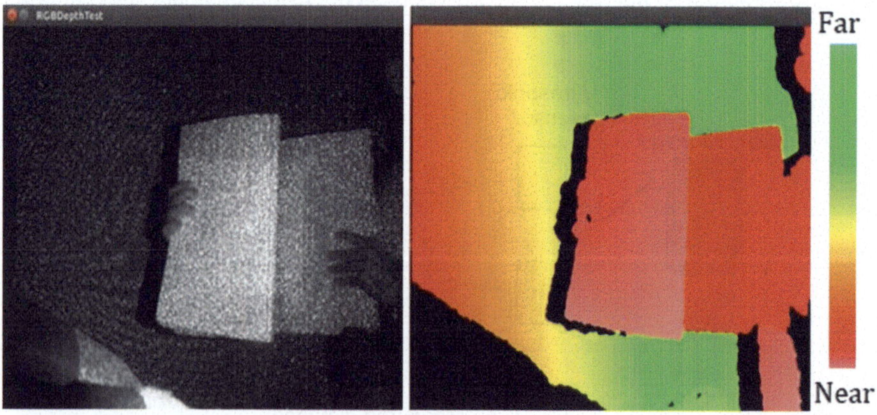

Fig. 3 **a** IR dots projected by the IR camera in Kinect. **b** Colored depth image with vertical color gradient to distinguish distances

Table 1 Combination of reflecting surfaces and medium of propagation

Sl. No.	Reflecting surfaces	Medium
1	Wall	Air
2	Sponge	
3	Metal sheet	

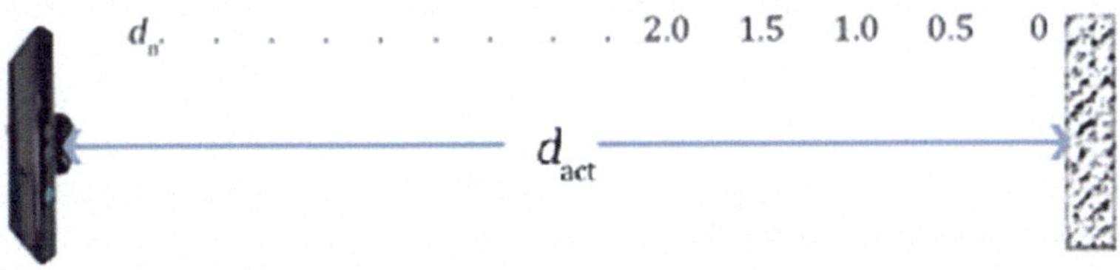

Fig. 4 Top view of the setup of experiment to measure distance error percentage

In the next section, detailed description of experiment setup and the different performance metrics for sensor evaluation are presented.

3 Experiment and Performance Metrics

The experiments are designed in such a way that all the factors affecting the performance of the sensors could be tested. In this set of experiments, the operating point and performance of the devices are to be evaluated whose threshold varies with different mediums and different reflecting surfaces. A combination of such factors has been listed in Table 1.

3.1 Experiment Setup

To evaluate performance in terms of distance error percentage, sensors are placed at a distance of d_{act} meters from the reflecting surface, $0 \leq d_{act} \leq \gamma_{max}$ where d_{act} is the actual distance between sensor and surface, γ_{max} is the threshold at which the performance of the sensors starts to degrade considerably. The sensor is moved from $d_{act} = 0$ to $d_{act} = \gamma_{max}$ in steps of 0.25m. Measured distance d_{mea} along with actual distance d_{act} is recorded at every step. Percentage error is calculated and is plotted with respect to distance in order to find the range ($\mathbb{R}$), threshold, and operating points. d_{act} is measured carefully with a measuring tape, and Fig. 4 shows a clear picture of the setup. We repeated this experiment and recorded the mean values listed in Table 2.

Table 2 Performance analysis of the sensors on wall and metal sheet

Parameter	Wall		Metal Sheet	
Metric	Kinect	Ultrasonic	Kinect	Ultrasonic
γ_{min}(m)	0.510	0.020	0.480	0.030
γ_{max}(m)	4.860	2.750	4.930	2.700
$\mathbb{R}$(m)	4.350	2.720	4.450	2.670
MPE (%)	0.900	2.700	0.870	2.530
SD (m)	0.284	0.852	0.338	0.673

Table 3 Combinations of sensor and mesh placement in front of the screen and respective pixel density of detection (P_{PDM} and P_{PDS})

$(d - \lambda)$(m) →	0.5		0.8		1.0	
λ(m) ↓	P_{PDM}(%)	P_{PDS}(%)	P_{PDM}(%)	P_{PDS}(%)	P_{PDM}(%)	P_{PDS}(%)
0.00	100.0	00.0	100.0	00.0	100.0	00.0
0.10	55.5	44.5	59.4	40.6	62.4	37.6
0.30	57.0	43.0	79.9	20.1	85.3	14.7
0.50	75.7	24.3	89.6	10.4	93.2	06.8
0.70	81.1	18.9	94.8	05.2	96.6	03.4
0.80	87.2	12.8	96.9	03.1	97.4	02.6
1.40	87.9	12.1	97.6	02.4	98.2	01.8

In the case of the mesh, we placed a screen at a distance d from the sensor. In between the sensor and screen, we put a 20 screen size nylon mesh at γ meters from the screen and $(d - \gamma)$ meters from the sensors. Here both the mesh and the screen act as reflecting surfaces. We wanted to test if these sensors could detect a highly porous mesh or if it penetrates through them to detect the object behind the mesh, i.e., screen. Both d and γ is altered in sets, as listed in Table 3. The setup is as shown in Fig. 5.

Prior to the experiment, both the sensors are given a warm-up time of 15 min at an ambient temperature of 22C with an average variation of ±5C. The actual distance is measured with the help of measuring tape which was always adjusted perpendicular to the camera plane so that the true distance can be measured. The angles are measured using a protractor. Processing 3.4 and Arduino IDE for Linux were used for acquiring raw data from Kinect and HC SR-04 module, respectively.

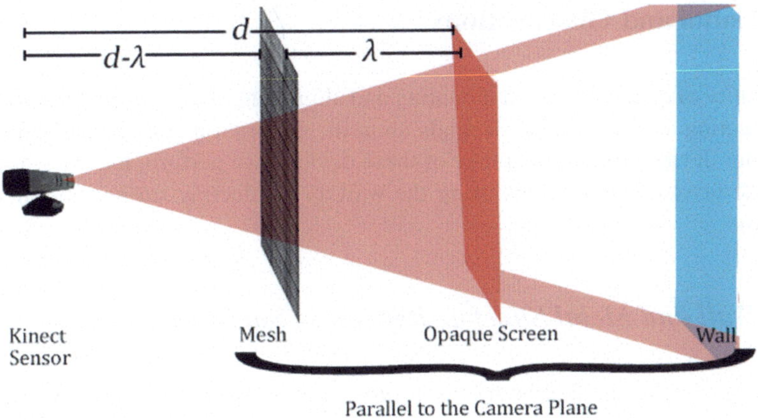

Fig. 5 Experimental setup for mesh as reflecting

3.2 *Performance Metrics*

The performance metrics used as a benchmark for the devices are as follows:

(a) **Range of Sensor** ($\mathbb{R}$) = $\{\gamma_{\min}, \gamma_{\max}\}$
The range of the sensor is defined as the maximum and minimum value of the applied parameter that can be measured. In our case, it is the maximum distance $\gamma_{\max}$ and minimum distance $\gamma_{\min}$ that can be measured by our sensors with distance error $d_{\text{err}} \leq 3\%$.

(b) **Effectual Angle** (θ_{E})
Effectual angle is the maximum incidence angle with respect to the reflecting surface at which distance error d_{err} is within the acceptable limits.

(c) **Mean Percentage Error (MPE)**
Let there be N measurements and let the ith measured distance be d_{mea_i} and actual distance be d_{act_i} then the mean percentage error will be computed as

$$\textbf{MPE} = \frac{1}{N}\sum_{i=1}^{N}\left(\frac{d_{\text{act}_i} - d_{\text{mea}_i}}{d_{\text{act}_i}}\right) \times 100\% \tag{4}$$

The mean percentage error is the computed average of percentage errors by which the output of the sensor varies from actual values.

4 Results and Observations

The quality of data depends on mounting and alignment of the respective sensor w.r.t the reflecting surface. It also depends upon the calibration and characteristics of a particular device. To find the range of these devices, we perform the experiments in the most favorable conditions using the wall as a reflecting surface and air as the medium.

4.1 Wall and Metal Sheet as Reflecting Surfaces

The range and threshold of both sensors are evaluated from the distance versus error curve in Fig. 6. Maximum allowable error is fixed to be 3%. Both the devices perform well within their operating points. The analyzed data is presented in Table 2. While testing Kinect with metal sheet and wall, its performance is exceptional, (**MPE** = 0.89%) as the opaque surface acted like a projector screen to the IR sensor of Kinect. The ultrasonic sensor also performs well on these two surfaces because the surface is rigid, and has a good echoing boundary layer. (**MPE** = 2.61%)

4.2 Sponge as Reflecting Surface

The sponge is opaque but its micro-perforated structure has an absorbing boundary layer so sound waves get absorbed by it. Kinect performed efficiently on this surface

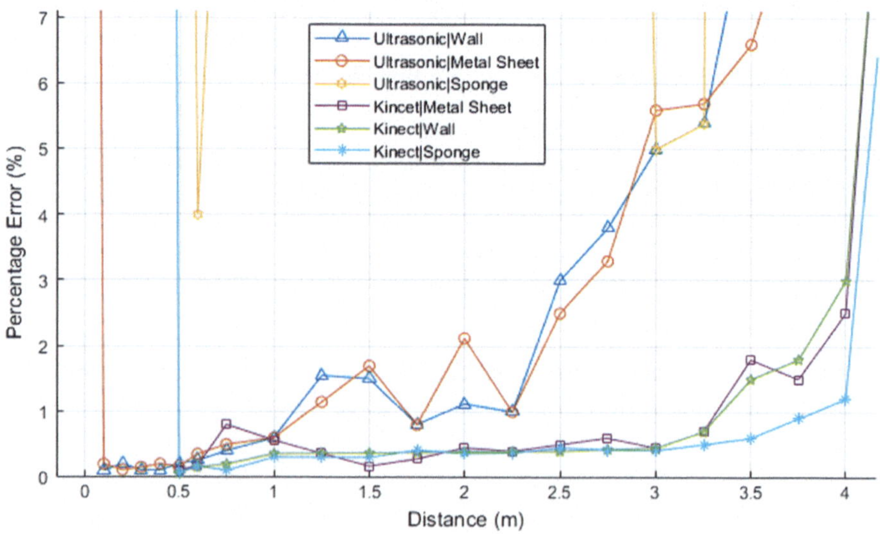

Fig. 6 Variation of distance error percentage with respect to actual distance

but ultrasonic sensors showed huge errors and random values. Figure 6 shows performance comparison of both these devices on a piece of 5-inch thick sponge. The setup of the experiment is similar to Fig. 4.

4.3 Nylon Mesh as Reflecting Surface

The behavior of both the sensors is very different while using nylon mesh as the surface. As it is neither opaque nor a reflective boundary, both the IR projections of Kinect and the pulse tray from the ultrasonic sensor pass through it. For the ultrasonic sensor, net acts as a transmissible boundary layer throughout and has no effect on the performance of sonic sensor at all even after adding five layers of mesh in between. In contrary, Kinect has a depth resolution of 640×480 at 30 frames per second (fps) and each pixel stores a depth value. Since the mesh is perforated, with a change in the relative position of the sensor and mesh, the proportion of pixels detecting the mesh changes. When sensor is close to the mesh, it acts as a transparent layer while at a distance more than 0.5 m the transparency gradually decreases. The proportion of pixels also depends upon the proximity of the object behind the mesh, i.e., the screen. We compare the ratio of pixels detecting mesh and screen for different distances. Percentage of pixels detecting the screen and mesh is called P_{PDS} and P_{PDM}, respectively. The setup is depicted in Fig. 5 and the result of the experiment is presented in Table 3 and Fig. 7. As γ increases, P_{PDS} decreases for all cases of $(d - \gamma)$ listed in Table 3. If a set of $(d - \gamma)$ for which we conducted the experiment is called $\mathbb{E}$ then we can establish the following relations:

$$\forall (d - \gamma) \in \mathbb{E}, P_{PDS} \propto \frac{1}{\gamma} \tag{5}$$

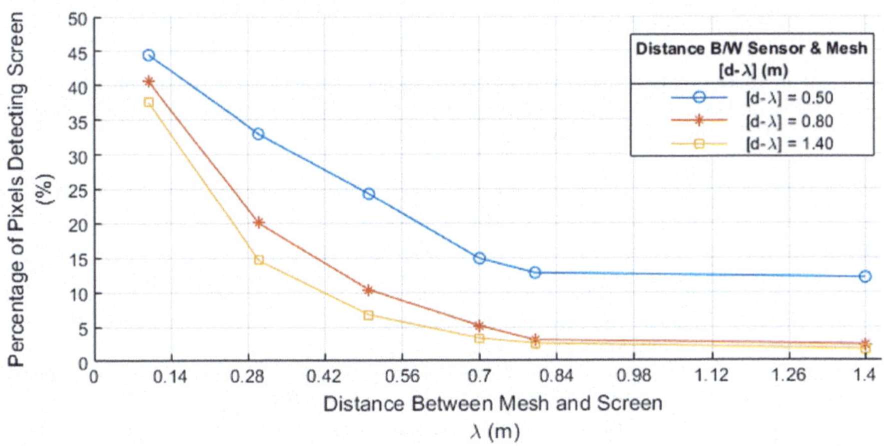

Fig. 7 Plot of relation between distance of mesh from the screen λ and percentage of pixels detecting the screen P_{PDS} at various $(d - \gamma)$

As, $P_{PDS} + P_{PDM} = 100$
Hence, using (5), we can infer

$$\forall(d-\gamma) \in \mathbb{E}, P_{PDM} \propto \gamma \tag{6}$$

Also,

$$\forall(\gamma) \in \mathbb{E}, P_{PDM} \propto d-\gamma \tag{7}$$

Equations (6) and (7) are depicted through the graph in Fig. 7. It is observed that as the distance between the mesh and the camera increases, the visibility of the objects behind the mesh decreases even if it is in the range of the Kinect sensor. Similarly, when the mesh and sensor are kept stationary, the screen behind the mesh is made to move away from the sensor, and the value of P_{PDS} decreases.

5 Conclusion

From the series of experiments conducted under controlled environment, we come to the conclusion that both the sensors have their own merits and demerits. Both the sensors have centimeter-level accuracy. Ultrasonic sensor can perform very well with rigid surfaces like metal, cemented wall, etc. The range of ultrasonic sensor is less than that of kinect but the former can measure an object as close as 0.02m, while the latter cannot detect objects closer than 0.5m. Apart from measuring distance, Kinect can also be used to record the color of the pixel which is an edge over all conventional sensors like LiDAR, etc. In future, we can use these sensors to design a hybrid multisensor device. We can expect these sensors to perform identically in similar environment under analogous circumstances. This research helps one to fuse data of ultrasonic and Kinect sensors, thus increasing the range and accuracy of the system in all operating conditions.

Acknowledgements This research was supported by the DRDO Robotics and Unmanned Systems Exposition (DRUSE). The additional tool for conducting experiments was offered by Physics laboratory at Haldia Institute of Technology.

References

1. Al Najjar, M., Ghantous, M., Bayoumi, M.: Object tracking. In: Video Surveillance for Sensor Platforms, pp. 119–146. Springer, Berlin (2014)
2. Fuentes-Pacheco, J., Ruiz-Ascencio, J., Rendón-Mancha, J.M.: Visual simultaneous localization and mapping: a survey. Artif. Intell. Rev. **43**(1), 55–81 (2015)
3. Thrun, S., Montemerlo, M.: The graph slam algorithm with applications to large-scale mapping of urban structures. Int. J. Robot. Res. **25**(5–6), 403–429 (2006)

4. Mirowski, P., Ho, T.K., Yi, S., MacDonald, M.: Signalslam: Simultaneous localization and mapping with mixed wifi, bluetooth, LTE and magnetic signals. In: International Conference on Indoor Positioning and Indoor Navigation, pp. 1–10. IEEE (2013)
5. Carter, J., Schmid, K., Waters, K., Betzhold, L., Hadley, B., Mataosky, R., Halleran, J.: Lidar 101: an introduction to lidar technology, data, and applications. National Oceanic and Atmospheric Administration (NOAA) Coastal Services Center, vol. 30, p. 2015 (2012)
6. Mohammad, T.: Using ultrasonic and infrared sensors for distance measurement. World Acad. Sci., Eng. Technol. **51**, 293–299 (2009)
7. Zhang, Z.: Microsoft kinect sensor and its effect. IEEE Multimed. **19**(2), 4–10 (2012)
8. Schmeing, M., Krauskopf, E., Jiang, X.: Real-time depth fusion using a low-cost depth sensor array. In: 2014 3DTV-Conference: The True Vision-Capture, Transmission and Display of 3D Video (3DTV-CON), pp. 1–4. IEEE (2014)
9. Spehr, J.: On hierarchical models for visual recognition and learning of objects, scenes, and activities (2015)
10. Gong, S., Liu, C., Ji, Y., Zhong, B., Li, Y., Dong, H.: Visual object tracking. In: Advanced Image and Video Processing Using MATLAB, pp. 391–428. Springer, Berlin (2019)
11. Aggarwal, J.K., Ryoo, M.S.: Human activity analysis: a review. ACM Comput. Surv. (CSUR) **43**(3), 16 (2011)
12. Indoware: Ultrasonic Ranging Module HC - SR04 Datasheet (2013)
13. Sabale, A.S., Vaidya, Y.M.: Accuracy measurement of depth using kinect sensor. In: 2016 Conference on Advances in Signal Processing (CASP), pp. 155–159. IEEE (2016)
14. Galna, B., Barry, G., Jackson, D., Mhiripiri, D., Olivier, P., Rochester, L.: Accuracy of the microsoft kinect sensor for measuring movement in people with parkinson's disease. Gait Posture **39**(4), 1062–1068 (2014)
15. Khoshelham, K.: Accuracy analysis of kinect depth data. In: ISPRS Workshop Laser Scanning. vol. 38 (2011)
16. Chow, J., Ang, K., Lichti, D., Teskey, W.: Performance analysis of a low-cost triangulation-based 3D camera: Microsoft kinect system. In: International Conference for Photogrammetry and Remote Sensing Congress (ISPRS), vol. 39, p. B5 (2012)
17. Rafibakhsh, N., Gong, J., Siddiqui, M.K., Gordon, C., Lee, H.F.: Analysis of xbox kinect sensor data for use on construction sites: depth accuracy and sensor interference assessment. In: Construction Research Congress 2012: Construction Challenges in a Flat World, pp. 848–857 (2012)
18. Planas, S., Rosell, J.R., Pomar, J., Camp, F., Solanelles, F., Gracia, F., Llorens, J., Gil, E., et al.: Performance of an ultrasonic ranging sensor in apple tree canopies. Sensors **11**(3), 2459–2477 (2011)
19. Adarsh, S., Kaleemuddin, S.M., Bose, D., Ramachandran, K.: Performance comparison of infrared and ultrasonic sensors for obstacles of different materials in vehicle/robot navigation applications. In: IOP Conference Series: Materials Science and Engineering, vol. 149, p. 012141. IOP Publishing (2016)
20. Massa, D.P.: Choosing an ultrasonic sensor for proximity or distance measurement part 1: acoustic considerations. Sensors **16**(2), 3 (1999)
21. Kelemen, M., Virgala, I., Kelemenová, T., Miková, L., Frankovský, P., Lipták, T., Lörinc, M.: Distance measurement via using of ultrasonic sensor. J. Autom. Control. **3**(3), 71–74 (2015)
22. Bohn, D.A.: Environmental effects on the speed of sound. In: Audio Engineering Society Convention 83. Audio Engineering Society (1987)
23. DiFilippo, N.M., Jouaneh, M.K.: Characterization of different microsoft kinect sensor models. IEEE Sens. J. **15**(8), 4554–4564 (2015)
24. Abhishek, K., et al.: Skeletal tracking using microsoft kinect. Methodology **1**(1), 11 (2010)
25. Koyuncu, H., Yang, S.H.: A survey of indoor positioning and object locating systems. IJCSNS Int. J. Comput. Sci. Netw. Secur. **10**(5), 121–128 (2010)

Cost-Efficient Bluetooth-Controlled Robot Car for Material Handling

Sayan Tripathi, Jhilam Jana, Sayan Mandal, Debraj Pal, Koushik Das, Asim Kumar Jana and Malay Kumar Pandit

Abstract Nowadays, material handling with the help of robotics knowledge is the emerging part in our day-to-day life. Material handling can be used to diminish the physical efforts. The cost-efficient Bluetooth-controlled robot car has made it possible for increasing number of handling tasks that is the main benefit of this paper. In this paper, we have designed and implemented the Bluetooth-controlled robot car for material handling using Arduino, pyroelectric infrared (PIR) sensor, and Bluetooth module which prove to be lower in cost than the preexisting designs. This device can be beneficial to farmers, storage godowns, educational organization, and general public for reducing their physical labors.

Keywords Pyroelectric infrared (PIR) sensor · Arduino · Bluetooth · Robotics · PCB design

S. Tripathi · J. Jana (✉)
Dept. of ETCE, Jadavpur University, Jadavpur, Kolkata 700032, India
e-mail: jhilamjana2014@gmail.com

S. Tripathi
e-mail: tripathysayan@gmail.com

S. Mandal · D. Pal · K. Das · A. K. Jana · M. K. Pandit
Haldia Institute of Technology, Haldia, India
e-mail: sayan.mandal72@gmail.com

D. Pal
e-mail: debrajpal147@gmail.com

K. Das
e-mail: koushik.das1996.kd78@gmail.com

A. K. Jana
e-mail: asimkjana@gmail.com

M. K. Pandit
e-mail: mkpandit.seci@gmail.com

S. Kundu et al. (eds.), *Proceedings of the 2nd International Conference on Communication, Devices and Computing*, Lecture Notes in Electrical Engineering 602, https://doi.org/10.1007/978-981-15-0829-5_34

1 Introduction

Today, the most common method is material handling process. This process is improving the customer service, reducing the delivery time and physical hard work. To minimize the material handling issues through robotics, the Bluetooth-controlled robot car is an emerging technique which is cheap, user-friendly, and effective. This paper represents the cost-efficient Bluetooth-controlled robot car for material handling. Here, the main motto of this work is to reduce the physical efforts during the material handling. The proposed system is very essential for domestic as well as industrial purposes. So, this proposed scheme is described to get rid of the hazardous problems. In our proposed design, any materials within 6 kg load capacity are handled and controlled. Arduino, Bluetooth module, PIR sensor, and motor driver are the main components of this design. Arduino can be interfaced with Bluetooth module to establish the connectivity and is also attached with motor driver to control the motor speeds. Another important module of this design is PIR sensor which can sense infrared (IR) radiation changes within its viewing range. Fresnel lens focuses irradiation on PIR sensor, measures the changes in the IR rate, and creates an electric potential difference corresponding to the variation in the IR radiation. This PIR module senses the living objects and generates the sound and slowdowns the robot car's motion to avoid collision. The user can control the robot car using the Android programmed app. This robot car will be moved according to the instruction through the Android application by the user.

The proposed model of robot car can be operated on any kind of surface that is another advantage. This is also cost efficient and simple than other existing systems. The proposed system is more suitable for modern society.

This paper is organized for remainder portion as follows. Section 2 provides the literature survey. Section 3 presents the proposed design. Section 4 contains the required component details. Section 5 provides mechanism of cost-efficient Bluetooth-controlled robot car. Section 6 presents the PCB design. Section 7 contains empirical result analysis. Section 8 presents the applications. Section 9 provides the conclusion.

2 Literature Survey

The main aim is to design the cost-efficient Bluetooth-controlled robot car for material handling which is eco-friendly. Many researchers had developed robotic systems to reduce human efforts and described their technologies. These robotic designs were controlled by software programs. Winter et al. developed Android-controlled robot. This system was used for transferring the information wirelessly [1]. On the other hand, the robotic system was developed using 8051 microcontroller, Bluetooth module which has attached a camera for surveillance [2]. Guardi et al. invented a communication-based robotic platform using Android and Bluetooth [3]. The robots are developed for not only reducing the human labors but are also operated in security and rescue operations [4–6].

Braun et al. described navigation system for autonomous off-road robots which depend on navigation to avoid obstacles in the map [7]. Tezel and Hangun designed and implemented a Bluetooth-controlled robot using Arduino. In this collision-free design, sensor data was got by linear interpolation [8]. The pick and drop robot was also invented. It can be very useful in defence purpose such as diffusing a bomb, land mines, etc [9]. The Android-based robotic car via Bluetooth module was also invented [10]. Ritika Pahuja and Narender Kumar designed an Android-controlled Arduino-based robot car. When the Android app is turned on, the system is connected via Bluetooth. The Android provides a user-friendly experience [11]. Other Android-based robotic car was designed that was eco-friendly automation system [12]. Avoiding human errors, the manual and autonomous driving switching techniques were described [13]. Shivaprasad B. developed a robot which can be utilized for fertilization cleaning purpose. The developed system collected the valuable information through sensor on the robot [14]. At another design, the robotic field was involved. This design was very suitable effect on military purpose [15]. Nelson et al. described the Bluetooth-controlled robot model that was controlled by sensor and Bluetooth module. This design does not affect any physical barriers like walls, doors, etc [16].

Bluetooth-controlled robotic systems for material handling have yet not obtained wide popularity and publicity. Public are still dependent on physical labors of human. So, these devices need much publicity and can be used as the best alternative to human physical labors.

3 Proposed Prototype of Cost-Efficient Bluetooth-Controlled Robot Car

Cost-efficient Bluetooth-controlled robot car for material handling is designed by using PIR sensor and Arduino microcontroller. This proposed system is applicable for material handling. As per our design, one job can be controlled at a time and any materials within 6 kg load capacity are handled. The PIR sensor senses the changes in IR rate and it creates an electric potential difference corresponding to the variation in IR radiation and detects the presence of living organism to avoid accidents. PIR sensor and Arduino UNO microcontroller module are the heart of this proposed system. The block diagram of this proposed design is described in Fig. 1. Bluetooth module, buzzer, and light-emitting diode (LED) are also required for this purpose. This proposed design is low cost, compact, and simple.

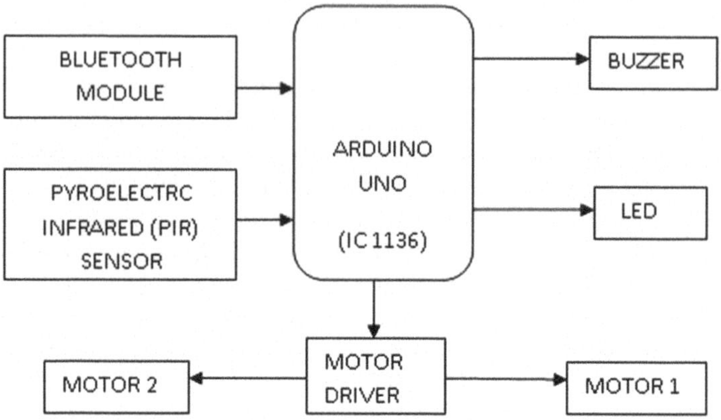

Fig. 1 General block diagram of cost-efficient Bluetooth-controlled robot car for material handling

3.1 Characteristics of Cost-Efficient Bluetooth-Controlled Robot Car

Cost-efficient Bluetooth-controlled robot car has many features. Some of the notable points are as follows:
1. Power efficient: The device consumes less power and there is no need to modify the circuit for extra separate power supply.
2. Cost-effective: The cost of the electronic device is less so that every individual and family can buy and use it.
3. Compactness: The device uses few electronic components compare to other existing systems.
4. Reducing the physical labors: This proposed design reduces the human labors to transfer the material from one end to other end.
5. Avoidance of collision: This robot car avoids the collision with other cars and living obstacles.
6. Simple circuit design: The circuit can be simply made so that mass production can be easily done.
7. The proposed robot car can be operated on any kind of surface.

4 Necessitate Components of Proposed System

1. **Arduino Uno (IC 1136)**
The main device in proposed system is Arduino Uno (IC 1136) which has a typical low power, high performance. It is manufactured by Atmega. Operating range is 6.0 to 20 V. This microcontroller provides a versatile and profitable solution to many embedded applications.

2. **Pyroelectric (PIR) Sensor**
Passive infrared sensor (PIR) is a sensor which detects the presence of living object such as human being, animals, etc. There are two important parts in this sensor, one is pyroelectric crystal which detects the heat from living organism and another is Fresnel lens which focuses on IR radiation on PIR sensor.

3. **Bluetooth Module**
The Bluetooth module is wireless communication module for this proposed system which can be operated by the property of master/slave. The functionality of this module is that it is configured only by user's commands. Bluetooth module is also programmable input/output control module and the input/output voltage range is 3.3 to 5V.

4. **Motor Driver**
Motor driver raises the motor driving IC that consists of two motors. These motors work at the same time. This module's operating range is 5.9 to 20 V. This motor driver steers the motor speed.

5. **Buzzer**
It is an audio signaling device. This device has various functionalities. PIR sensor confirms the presence of living object, and thereafter a message will be sent to the Arduino (IC 1136) microcontroller and then the buzzer will be buzzed.

5 Mechanism of Cost-Efficient Bluetooth-Controlled Robot Car for Material Handling System

Figure 2 represents the mechanism of the proposed design. Handling the material properly has always been important. So, the proposed robot car can reduce the human labor and deliver the material in time. Arduino, Bluetooth module, and PIR sensor are the heart of this design. In this paper, Arduino and Bluetooth are interfaced. The Bluetooth chip establishes the connection between Android devices and Arduino. The motor driver is also attached with Arduino to control the speed of two motors. PIR sensor module is another important module which is interfaced with Arduino. The PIR sensors are used to detect living objects movement around the robot car to avoid the accidents. Whenever a movement is detected, the car will beep a buzzer and the indicators blink.

The power supply is given to UNO board and motor driver using 14 volts rails from any power source. The power is given to PIR sensor and Bluetooth module using +5 volts rail of the UNO board. The pins RX and TX of Bluetooth module are connected to the pins TX and RX of the UNO, respectively. The VCC and ground pin are connected, respectively. The output pin of PIR sensor is connected to A1 pin of UNO board and this pin will be the input pin of UNO. The VCC and ground pins are connected accordingly. The 14 volt power supply module consists of the 14 volt 1 amp batteries. Enables (ENA) and (ENB), Inputs (IN1), (IN2), (IN3), and (IN4)

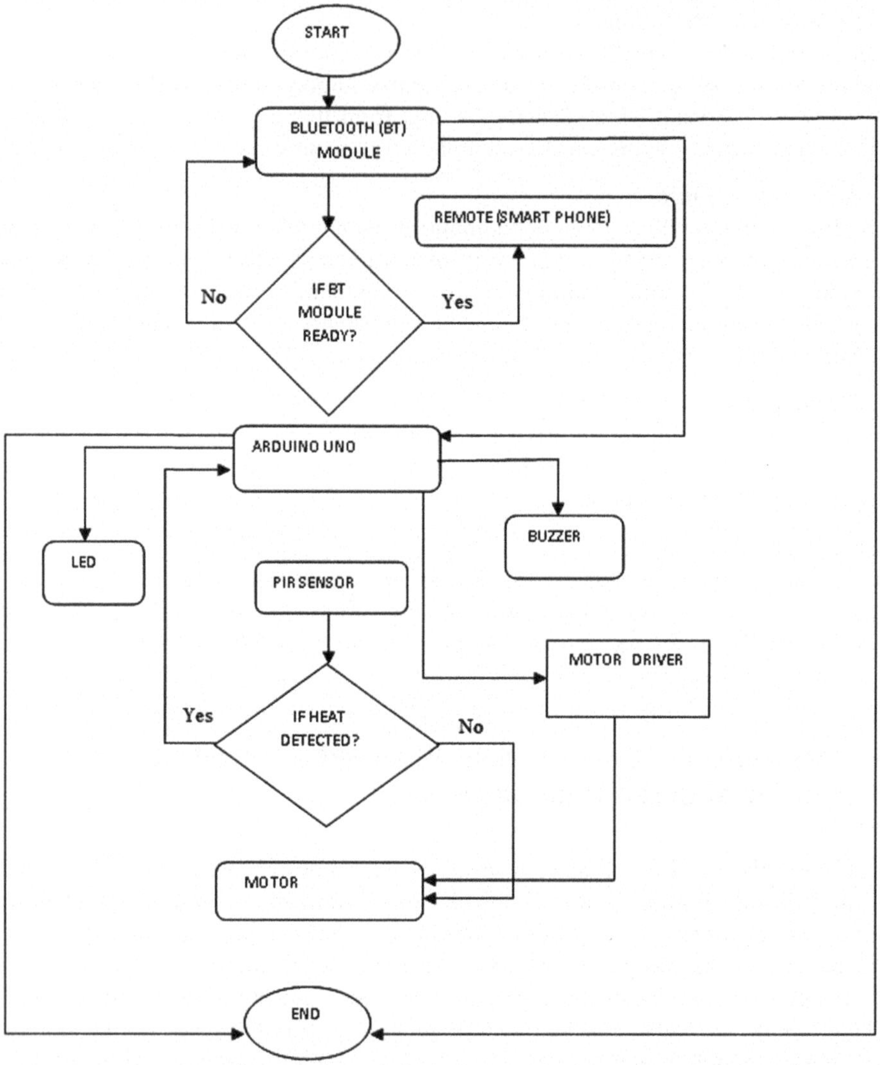

Fig. 2 Mechanism of proposed system

pins of motor driver are connected to the UNO board pins 3, 11, 4, 5, 9, and 10, respectively. The VCC and ground pins are also connected. The written assembly program burns the microcontroller IC of UNO.

On the other hand, the PCB layout is developed by PCB wizard and this PCB layout is designed and implemented using PCB fabrication steps.

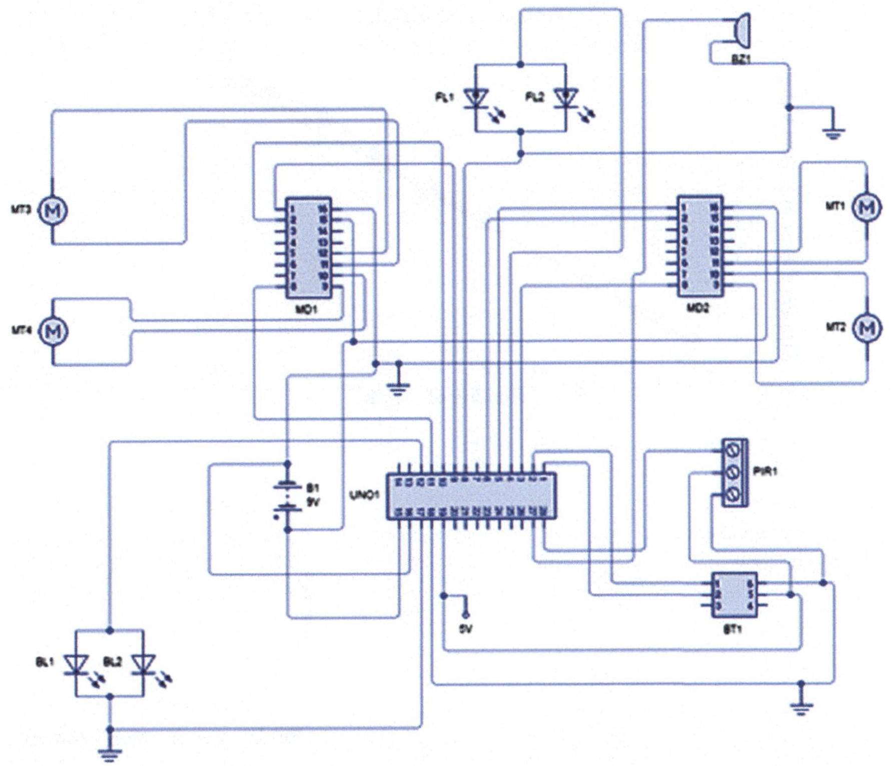

Fig. 3 Cost-efficient Bluetooth-controlled robot car for material handling circuit using PCB wizard

6 PCB Design of Proposed Scheme

There are many types of software for the PCB design like the PCB wiz, express PCB, Eagle PCB, Proteus, and Circuit wizard. PCB Wizard is used to design the proposed cost-efficient Bluetooth-controlled robot car for material handling circuit's PCB. It is very easy to use and understand this proposed design. The proposed circuit is designed in the PCB Wizard. Therefore, by clicking tools and converting, finally the PCB solder side artwork of proposed circuit is designed. Figure 3 shows the layout of PCB design of proposed circuit. This PCB design is developed using PCB fabrication steps.

7 Empirical Results Analysis

Cost-efficient Bluetooth-controlled robot car has been tested manually in different paths such as straight path, circular path, elliptical path, etc. Some errors occur to

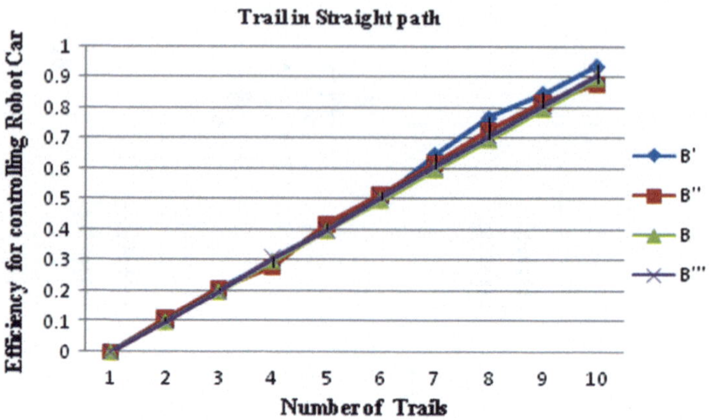

Fig. 4 Controlling efficiency and number of trails straight path

Fig. 5 Controlling efficiency and number of trails circular path

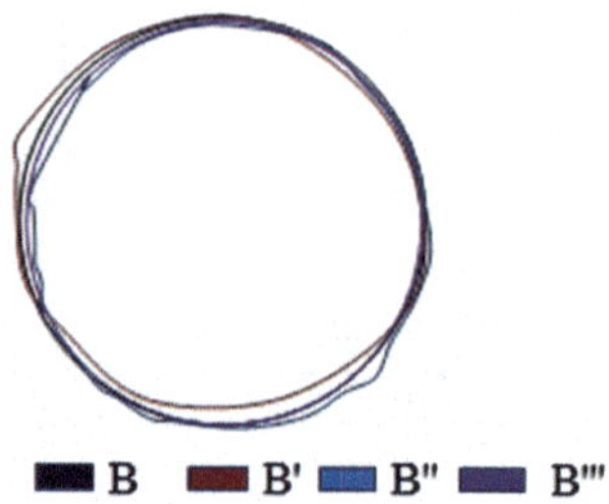

Fig. 6 Controlling efficiency and number of trails elliptical path

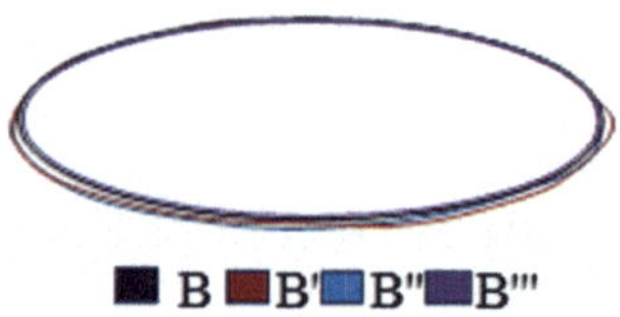

move the proposed robot car from source to destination. Figures. 4, 5, and 6 show the graphical representation of controlling efficiency of robot car and number of trails in straight path, circular path, and elliptical path, respectively. To make perfection, we have trailed number of times to reduce errors. Errors are located in these graphs like point B', point B'', and point B'''.

Figure 7 shows the graphical representation of the range of PIR sensor in terms of voltage with respect to distance. The voltage from PIR decreases according to the increase in the distance.

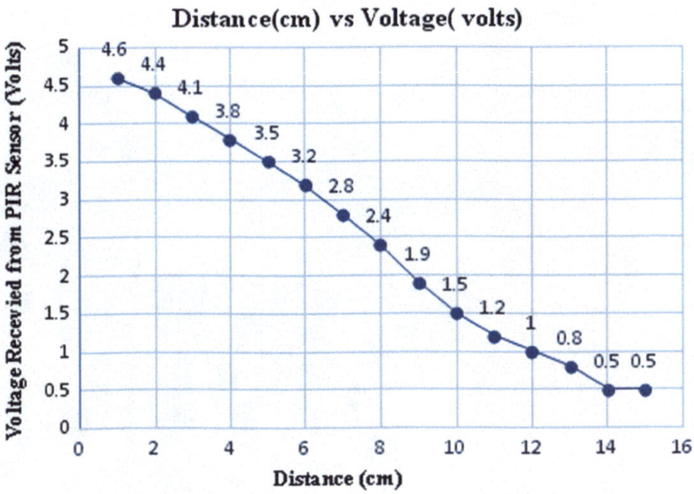

Fig. 7 The voltage received from PIR sensor versus distance graph

8 Applications

The proposed cost-efficient Bluetooth-controlled robot car is applicable in following cases:

1. Household Applications: Trying to shifting materials can be hazardous in a family. The device is used in household applications like kitchen, storeroom, etc.

2. Industrial Applications: This device is more suitable for industrial purpose. It can be used to deliver the valuable documents, flies, materials, etc.

3. Educational Applications: To transfer the important documents, papers, files, etc of school, college from source to destination, this device is very useful. This proposed device is also suitable for the laboratory and library.

4. Others Applications: This proposed device is also applicable for hospitals, banks, offices, etc.

9 Conclusion

The cost-efficient Bluetooth-controlled robot car is designed to lessen the physical labors of human being during the material handling. The performance of this proposed design could be improved with modification. This work is to increase the productivity of industry. The publicity of this design is very much needed. This device can be used in purpose of the military also. By reducing the physical labors for transferring the valuable materials of the industry, home, farms, etc, we can min-

Fig. 8 Implemented cost-efficient Bluetooth-controlled robot car

imize the adverse effects and approach the problems electronically which will be beneficial for the future generations.

The future goal of this design is to improve the efficiency using Internet of things (IoT) and wireless sensor network (WSN) and add the web camera with security features, Fig. 8.

Acknowledgements We would like to thank Dr. Malay Kumar Pandit, Professor and Asim Kumar Jana, Associate Professor, Department of Electronics and Communication Engineering, Haldia Institute of Technology, Haldia, West Bengal for their kind support, guidance, and suggestions for successful completion of this research work.

References

1. Kazacos Winter, J.: Android controlled mobile robot (2013)
2. Selvam, M.: Smart phone based robotic control for surveillance applica-tions. Int. J. Res. Eng. Technol. **3**(3), 229–232 (2014)

3. Guardi, V.M.: Design of a Bluetooth Enabled Android Application for a Microcontroller Driven Robot. Diss. Rensselaer Polytechnic Institute (2014)
4. Birk, A., Schwertfeger, S., Pathak, K.: A networking framework for teleoperation in safety, security, and rescue robotics. IEEE Wirel. Commun. **16**(1), 6–13 (2009)
5. Casper, J.L., Micire, M., Murphy, R.R.: Issues in intelligent robots for search and rescue. In: Unmanned Ground Vehicle Technology II. International Society for Optics and Photonics, vol. 4024, pp. 292–303 (2000)
6. Liu, Y., Nejat, G.: Robotic urban search and rescue: a survey from the control perspective. J. Intell. Robot. Syst. **72**(2), 147–165 (2013)
7. Braun, T., Schaefer, H., Berns, K., Topological large-scale off-road navigation and exploration RAVON at the European Land Robot Trial 2008. In IEEE/RSJ International Conference on Intelligent Robots and Systems, IROS 2009, pp. 4387–4392. IEEE (2009)
8. Tezel, C., Hangn, B.: Design and implementation of bluetooth controlled collision avoidance 4 wheel robot using arduino with linear interpolation method for determination. Int. J. Eng. Sci. Appl. **1**(4), pp. 151–156
9. Goud, R.K., Kumar, B.S.: Android based robot implementation for pick and retain of objects. Int. J. Eng. Trends Technol. (IJETT) **16**(3) (2014)
10. Sharma, A., Verma, R., Gupta, S., Bhatia, S.K.: Android phone controlled robot using bluetooth. Int. J. Electron. Electr. Eng. ISSN, pp. 0974–2174 (2014)
11. Pahuja, R., Kumar, N.: Android mobile phone controlled bluetooth robot using 8051 microcontroller. Int. J. Sci. Eng. Res. **2**(7), 14–17 (2014)
12. Maity, A., Paul, A., Goswami, P., Bhattacharya, A.: Android application based bluetooth controlled robotic car. Int. J. Intell. Inf. Syst. **6**(5), 62 (2017)
13. Sheng, W., Ou, Y., Tran, D., Tadesse, E., Liu, M., Yan, G.: An integrated manual and autonomous driving framework based on driver drowsiness detection. In: 2013 IEEE/RSJ International Conference on Intelligent Robots and Systems (IROS), pp. 4376–4381. IEEE (2013)
14. Shivaprasad, B.S., Ravishankara, M.N., Shoba, B.N.: Design and implementation of seeding and fertilizing agriculture robot. Int. J. Appl. Innov. Eng. Manag. (IJAIEM) **3**(6), 251–255 (2014)
15. Kuhnert, K.D.: Software architecture of the autonomous mobile outdoor robot AMOR. In: Intelligent Vehicles Symposium, pp. 889–894. IEEE (2008)
16. Rai, N., Rasaily, D., Wangchuk, T.R., Gurung, M., Khawas, R.K.: Bluetooth Remote Controlled Car using Arduino

Design of a Non-iterative First-Order Compensator for Type 1 Higher Order Systems

C. Ganesh, R. Shanmugasundaram and A. Singaravelan

Abstract Most of the real-time systems used in control applications are of higher order. Implementation of the controller requires dynamic and intelligent tuning to achieve the desired time response in control applications. An attempt is made in this article to extend the use of a simple first-order non-iterative compensator used in a Brushless Direct Current (BLDC) drive-based position control system to type 1 higher order systems. The compensator is found to yield the desired frequency response specifications with lesser overshoot and settling time than those of the conventional lag and lead compensators for various type 1 higher order systems.

Keywords Compensator · Non-iterative · Type 1 · Higher order

1 Introduction

Lag and lead compensators are applied in various servo applications. The lag compensator is used to improve the steady-state response, while lead compensator is applied to improve the transient response [1, 2]. Design of compensator is made for servo systems using lag and lead compensators [3–5]. Design technique [4, 6] of such compensators is iterative, laborious, and time-consuming [7–9]. Graphical tools, viz., Bode, Nyquist, and Root locus plots, are employed [1] to achieve the desired gain and phase margin for the systems with parameter uncertainty. A non-trial-and-error design method was proposed [10] for lag–lead compensator design to satisfy the desired specifications such as steady-state error, gain margin, and phase margin of a third-order system. An auto-tuning procedure was presented [6] for lag and lead compensators to achieve the desired specifications for the systems in which models cannot be perfectly anticipated. Design procedures for lag and lead compensators

C. Ganesh (✉) · A. Singaravelan
New Horizon College of Engineering, Near Marathahalli, Bengaluru 560103, India
e-mail: c.ganesh.mtech72@gmail.com

R. Shanmugasundaram
Sri Ramakrishna Engineering College, Vattamalaipalayam, Coimbatore 641022, India

S. Kundu et al. (eds.), *Proceedings of the 2nd International Conference on Communication, Devices and Computing*, Lecture Notes in Electrical Engineering 602, https://doi.org/10.1007/978-981-15-0829-5_35

were developed [5, 7, 9] in the stable operating regions of systems when comprehensive information on gain margin and phase margin was available. An iterative technique for the design of lag and lead compensators was established [8] to accomplish desired specifications in the position control system. Lead compensator was used [11] for dead-band relay sub-controller to attain very good relative stability in a Direct Current (DC) motor-based position control system. A lag–lead compensator was developed [12] for robots. A digital non-iterative controller proposed [13] for a BLDC servosystem was found to yield better transient and steady-state response.

It is observed from the literature that compensators are quite useful in servo applications. However, their design techniques are iterative, tedious, and time-consuming. A simple non-iterative first-order compensator design technique was proposed [14]. Reference [15] suggested this design technique as an alternative to performance index-based proportional plus integral plus derivative (PID) controller tuning for a BLDC drive fed position control system to achieve optimum results. Reference [16] extended this procedure for a type 1 third-order system.

The time delay in feedback control [17] has become significant nowadays with the extensive use of networked control systems. Time delays cause adverse effects on the control system dynamics and may lead to closed-loop instabilities. BLDC motor control systems [18] are stable when time delays are not considered. However, inevitable time delays may destabilize the system when a BLDC motor is controlled through a network. For this reason, time delays must be considered during a controller design. Therefore, care must be taken to compensate for instability due to time delays while extending the use of first-order compensator [15] to type 1 higher order systems with time delays.

This research article deals with the design of proposed first-order compensator for a BLDC drive system (Type 1 third order) by considering time delay. Extending the use of the compensator for other type 1 higher order systems is validated with five illustrations. The proposed method can be implemented to the problem stated in [19].

2 Design of the First-Order Compensator

A non-iterative first-order compensator [15] is found to yield comparable results to PID controller, and it fulfills the desired specifications for a BLDC drive-based position control system. The procedure is summarized below.

Step 1: The required gain (K_c) of the system is chosen to meet the steady-state requirement with better response and negligible oscillations.

Step 2: Gain crossover frequency (ω_g) of the compensated system is so chosen that it corresponds to the desired phase margin.

Step 3: Transfer function of the compensator [2, 4] to be designed is expressed by

$$G_{comp}(s) = \frac{(1 + T_a s)}{(1 + T_b s)} \tag{1}$$

where T_a and T_b are time constants of compensating zero and pole, respectively.

Step 4: Gain of the uncompensated system is determined at the desired gain crossover frequency, ω_g. Let this gain be $\left|G_p(j\omega)\right|_{\omega g}$.

Step 5: Gain of the compensated system transfer function at the desired gain crossover frequency, ω_g, is given by

$$|Gpc(j\omega)| = K_c\left|\frac{(1 + j\omega T_a)}{(1 + j\omega T_b)}\right||Gp(j\omega)| = 1 \tag{2}$$

Step 6: ωT_a is so chosen that it is greater than or equal to a constant, (say 100) to nullify the phase angle contribution by the compensator at the desired gain crossover frequency. T_a and T_b are evaluated from (2).

Step 7: The compensated system transfer function is determined by substituting T_a, T_b, and K_c in (2).

The open-loop transfer function of BLDC drive fed position control system [20–23] considered for the illustration [15] is used below:

$$\frac{\theta(s)}{E_a(s)} = \frac{0.036}{s\left[\left(s4.68 \times 10^{-6} + 4.68 \times 10^{-5}\right)\left(1.08 + s1.86 \times 10^{-3}\right) + \left(0.036^2\right)\right]} \tag{3}$$

When the BLDC drive system is controlled through a network, the time delay is introduced through a transportation lag e^{-sT} connected in cascade with the above transfer function which makes the closed-loop system unstable when $T > 0.055$ s. The proposed compensator is designed for the three cases, viz., (i) Stable BLDC drive fed position control system with 0.045 s time delay, (ii) an unstable BLDC drive fed position control system with 0.065 s time delay, and (iii) stable BLDC drive fed speed control system without time delay.

2.1 *Stable BLDC Drive System with 0.045 s Time Delay*

Choosing the desired gain and phase margins as 6 dB and 45°, respectively, gain of the uncompensated system is evaluated as 4.44 dB at 16 rad/sec. ωT_a is chosen as 100. T_a and T_b are obtained as 6.25 and 10.4206, respectively, from (2). The resultant compensator becomes $G_{comp}(s) = (1 + 6.25\ s)/(1 + 10.4206\ s)$. The unit-step responses and Bode plots of the closed-loop system are obtained using the proposed compensator, lead compensator, and lag compensator. Safety factor (ε), gain margin, phase margin, rise time, peak overshoot, and settling time are obtained from the plots [2, 4] and tabulated in Table 1.

Table 1 Specifications of the BLDC drive system with 0.045 s time delay

Method	ε (deg)	t_r (sec)	t_r (sec)	M_p (%)	GM (dB)	PM (deg)
Uncompensated	–	0.03	1.49	80.3	1.62	15.3
Proposed design	0.1	0.05	0.45	29	6.05	45
Lag compensator	9	0.06	0.62	28.2	7.39	45.4
Lead compensator	20	Unstable			−0.87	−62

Table 2 Specifications of the BLDC drive system with 0.065 s time delay

Method	ε (deg)	t_r (sec)	t_r (sec)	M_p (%)	GM (dB)	PM (deg)
Uncompensated	–	Unstable			−1.37	−15.3
Proposed design	0.2	0.07	0.64	29.2	6.04	45
Lag compensator	10	0.09	0.99	30.1	7.65	44.6
Lead compensator	18	Unstable			−0.55	−178

2.2 *Unstable BLDC Drive System with 0.065 s Time Delay*

The uncompensated system is unstable with a gain margin of −1.37 dB and a phase margin of −15.3°. The desired gain and phase margins are specified as 6 dB and 45°, respectively. The compensator is designed as $G_{comp}(s) = (1 + 8.7719\ s)/(1 + 20.6117\ s)$. The unit-step responses and bode plots of the closed-loop system are obtained using the proposed compensator, lead compensator, and lag compensator. Time- and frequency-domain specifications are obtained from the plots and tabulated in Table 2. It is seen from the Tables 1 and 2 that the proposed non-iterative compensator yields satisfactory results.

2.3 *Stable BLDC Drive System Without Time Delay*

The open-loop transfer function of BLDC drive fed position control system is given in Eq. (3). The uncompensated system is stable with infinite gain margin and phase margin of 17.1°. The desired phase margin is specified as 45°. The resultant compensator is $G_{comp}(s) = (1 + 0.1242\ s)/(1 + 0.766\ s)$. The compensated system yields a phase margin of 45°.

3 Analysis of Compensator Design for Higher Order Systems

Typical stable and unstable systems with and without time delay were considered [5, 6], for establishing their compensating techniques. In order to demonstrate the effectiveness of the proposed non-iterative first-order compensator design, few type 1 higher order systems are also considered. Five examples are discussed in this section. The first two examples consider typical stable third-order systems with and without delay. The third example is chosen to show that the approach can also be used for unstable systems. The fourth example is a fifth-order system with complex poles, and the last one is to show that the same compensating design procedure is useful for a lower order system. Desired specifications are given in frequency-domain specifications. The compensator is designed for each system using the proposed design procedure and compared with the results of lag and lead compensators.

3.1 A Stable Third-Order System with Time Delay

A servo plant given in Eq. (4) is considered [5]. Choosing the specifications of gain and phase margins as 10 dB and 60°, respectively, the resultant compensator is given in Eq. (5).

$$G(s) = \frac{5e^{-0.23s}}{s(s + 0.7)(s + 7)} \tag{4}$$

$$G_{comp}(s) = \frac{(1 + 333.13s)}{(1 + 1028s)} \tag{5}$$

A comparison is made with the lag compensator design method [2, 4] which gives $G_{lag}(s) = \frac{(1+21.978s)}{(1+119.3957s)}$. This is also compared with the lead compensator design method [2], [4] which yields $G_{lead}(s) = \frac{(1+1.7854s)}{(1+0.3763s)}$ The unit-step responses of the closed-loop system are obtained using the proposed compensator, lag compensator, and lead compensator and shown in Fig. 1. The frequency response of the plant is shown in Fig. 2. Time- and frequency-domain specifications are obtained from the plots and tabulated in Table 3.

3.2 A Stable Third-Order System Without Time Delay

A position control system with $G(s) = \frac{58390}{s(s+36)(s+100)}$ is considered [8]. The desired phase margin is chosen as 69.2°. The resultant compensator is $G_{comp}(s) = \frac{(1+10.3734s)}{(1+16.6892s)}$. This is compared with lag compensator design method which gave the

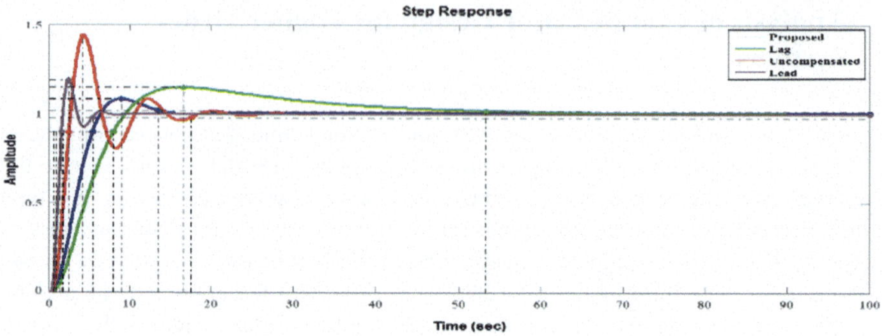

Fig. 1 Step responses for the third-order system with time delay

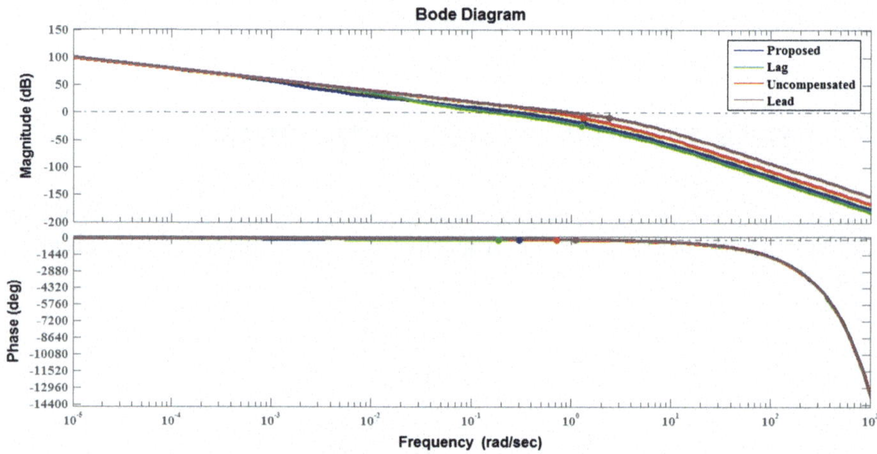

Fig. 2 Bode plots for the third-order system with time delay

Table 3 Specifications of the third-order system with time delay

Method	ε (deg)	t_r (sec)	t_r (sec)	M_p (%)	GM (dB)	PM (deg)
Uncompensated	–	1.56	17.5	43.6	8.93	29.3
Proposed design	0.56	4.05	13.5	8.35	18.7	60
Lag compensator	12	6.21	53.3	15.1	23.1	60
Lead compensator	10	0.96	5.45	19.1	8.77	49.5

compensator transfer function as $G_{lag}(s) = \frac{(1+0.7843s)}{(1+2.6884s)}$. This is also compared with lead compensator design method which yielded $G_{lead(s)} = \frac{(1+0.065s)}{(1+0.0356s)}$. The unit-step responses of the closed-loop system are obtained using the proposed compensator, lag compensator, and lead compensator and shown in Fig. 3. Bode plot of the plant

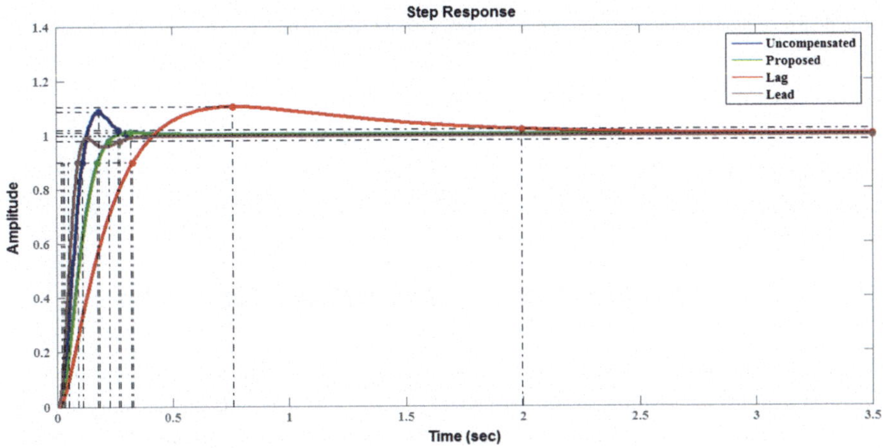

Fig. 3 Step responses for the third-order system without time delay

is shown in Fig. 4. Time- and frequency-domain specifications are obtained from the plots and tabulated in Table 4.

It is obvious from Table 4 that the proposed compensator design method has achieved the desired results. ε is varied on a trial-and-error basis for the design of lag and lead compensators, whereas it is negligible in the proposed design.

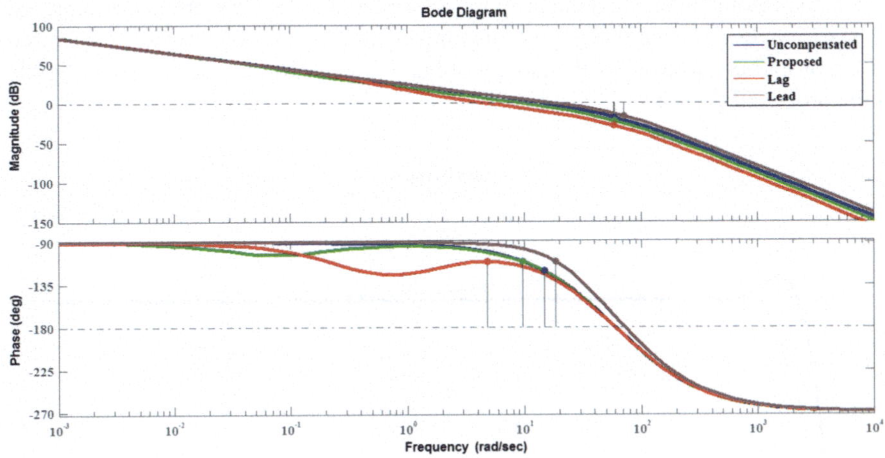

Fig. 4 Bode plots for the third-order system without time delay

Table 4 Specifications of the third-order system without time delay

Method	ε (deg)	t_r (sec)	t_r (sec)	M_p (%)	GM (dB)	PM (deg)
Uncompensated	–	0.08	0.27	8.67	18.5	59.2
Proposed design	0.4	0.14	0.23	1.21	22.6	69.2
Lag compensator	10	0.28	2	10.6	28.9	69.2
Lead compensator	7	0.07	0.27	0	17	69.2

3.3 An Unstable Third-Order System Without Time Delay

A position control system with $G(s) = \frac{5}{s(s+1)(0.5s+1)}$ is considered [6]. The uncompensated system is unstable with gain margin of −4.44 dB and phase margin of −13°. The desired gain and phase margins are specified as 10 dB and 60°, respectively. Hence, the transfer function of the proposed compensator is expressed as $G_{comp}(s) = \frac{(1+278.5515s)}{(1+3588.6s)}$. This is compared with lag compensator design method which yields the compensator transfer function as $G_{lag}(s) = \frac{(1+20.4082s)}{(1+568.5962s)}$. The unit-step responses of the closed-loop system obtained using the proposed compensator and lag compensator are shown in Fig. 5. The unit-step response of the closed-loop uncompensated system is shown in Fig. 6. Bode plot of the system is shown in Fig. 7. Time- and frequency-domain specifications are obtained from the plots and tabulated in Table 5. It is found that lead compensator design is not possible for meeting the desired specifications.

It is seen from Table 5 that the proposed compensator design method has achieved satisfactory results even though the uncompensated system is unstable. ε is varied on a trial-and-error basis for the design of lag compensator, whereas it is negligible in the proposed design.

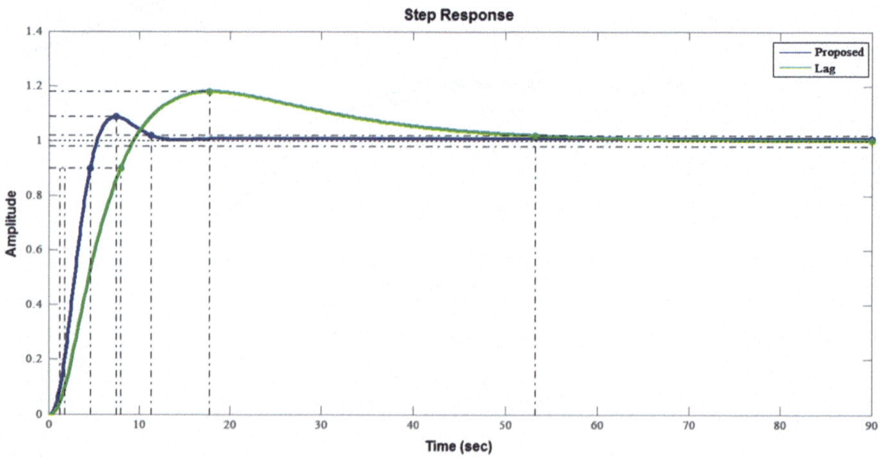

Fig. 5 Step responses for the third-order system

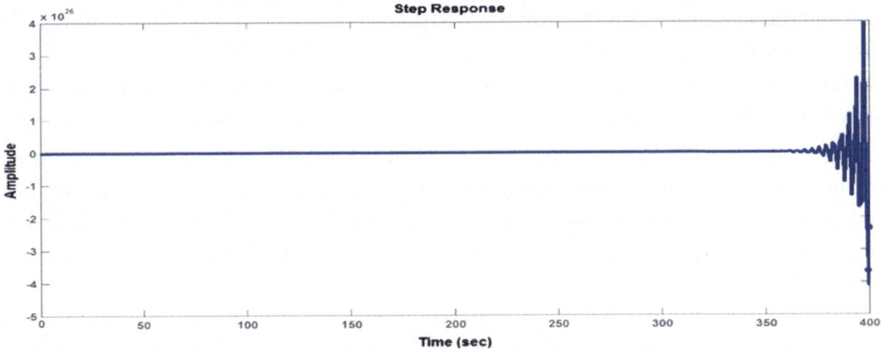

Fig. 6 Unit-step response for the uncompensated system

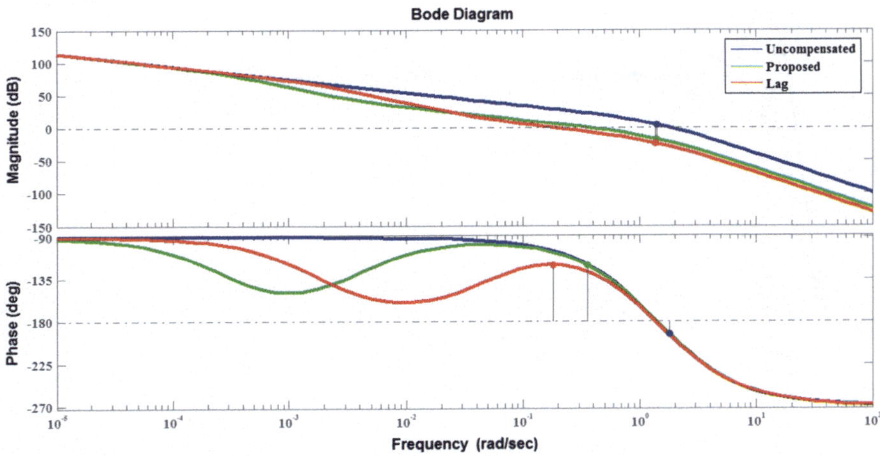

Fig. 7 Bode plots for the third-order system

Table 5 Specifications of the third-order system

Method	ε (deg)	t_r (sec)	t_r (sec)	M_p (%)	GM (dB)	PM (deg)
Uncompensated	–	Unstable			−4.44	−13
Proposed design	0.77	3.38	11.3	8.7	17.7	60
Lag compensator	14	6.24	53.2	17.9	23.8	60

3.4 *A Stable Fifth-Order System Without Time Delay*

A process with $G(s) = \frac{1}{s(s+1)^2(s^2+2s+3)}$ is considered. For the uncompensated system, the gain margin and phase margin are 8.06 dB and 43.9°, respectively. The desired gain and phase margins are specified as 10 dB and 60°, respectively. Hence, the transfer function of the proposed compensator is expressed as $G_{comp}(s) = \frac{(1+507.614s)}{(1+829.9413s)}$. This is compared with lag compensator design method which yielded the compensator transfer function as $G_{lag}(s) = \frac{(1+27.027s)}{(1+60.0203s)}$. This is also compared with lead compensator design method which yields $G_{lead}(s) = \frac{(1+3.6356s)}{(1+1.1598s)}$. The unit-step responses of the closed-loop system obtained using the proposed compensator, lag compensator, and lead compensator are shown in Fig. 8. Bode plot of the system is shown in Fig. 9. Time- and frequency-domain specifications are obtained from the plots and tabulated in Table 6.

It is seen from Table 6 that the proposed compensator design method has achieved satisfactory results, whereas the lead compensated system does not satisfy the desired

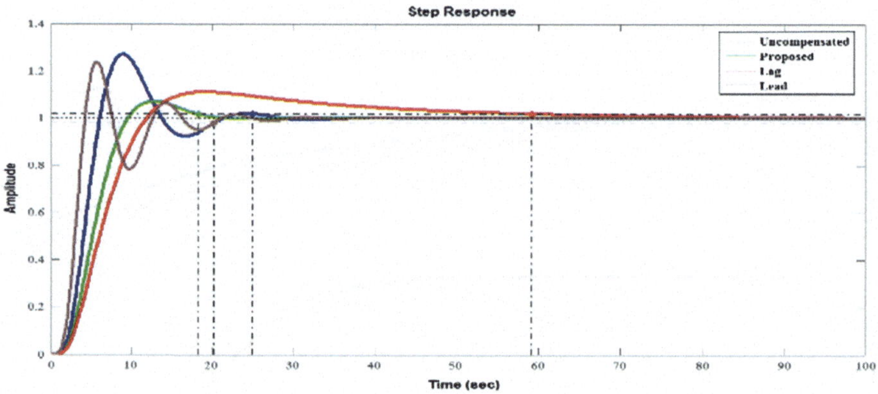

Fig. 8 Unit-step responses for the fifth-order system

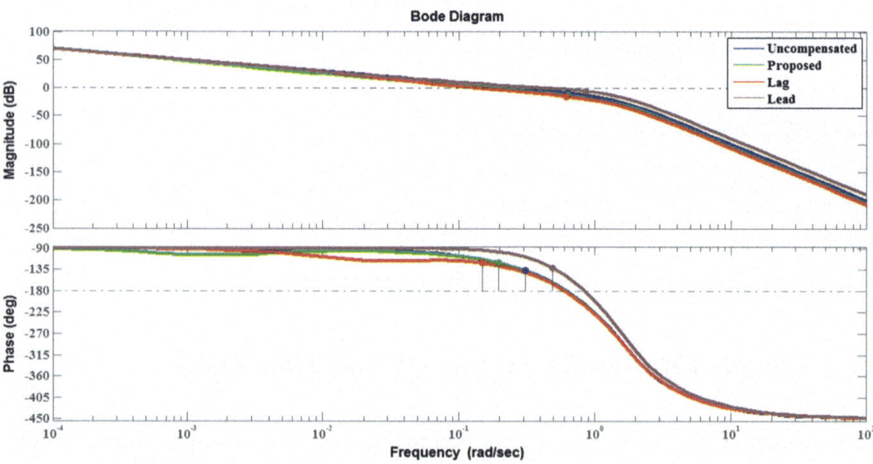

Fig. 9 Frequency responses for the fifth-order system

Table 6 Specifications of the fifth-order system

Method	ε (deg)	t_r (sec)	t_r (sec)	M_p (%)	GM (dB)	PM (deg)
Uncompensated	–	3.35	24.9	27.1	8.06	43.9
Proposed design	0.32	5.52	18.3	7.12	12.3	60
Lag compensator	8	7.13	59.2	11.1	14.7	60
Lead compensator	15	2.19	20.2	23.8	5.1	49.5

specifications. ε is varied on a trial-and-error basis for the design of lag and lead compensators, whereas ε is negligible in the proposed design.

3.5 A Stable Second-Order System Without Time Delay

A unity feedback system with $G(s) = \frac{4}{s(2s+1)}$ is considered. It is desired to design a lag compensator and obtain a phase margin of 40°. The transfer function of the compensator is obtained using the proposed compensator design as $G_{comp}(s) = \frac{(1+170.068s)}{(1+751.005s)}$. This is compared with the lag compensator design method which gave the compensator transfer function as $G_{lag}(s) = \frac{(1+10.8992s)}{(1+78.5778s)}$. A comparison is also done with lead compensator resulting in $G_{lead}(s) = \frac{(1+0.8951s)}{(1+0.3777s)}$. The unit-step responses of the closed-loop system obtained using the proposed compensator, lag compensator, and lead compensator are shown in Fig. 10. Bode plot of the system is shown in Fig. 11. ε, gain margin, phase margin, rise time, peak overshoot, and settling time obtained from the various compensated systems along with the uncompensated system are tabulated in Table 7.

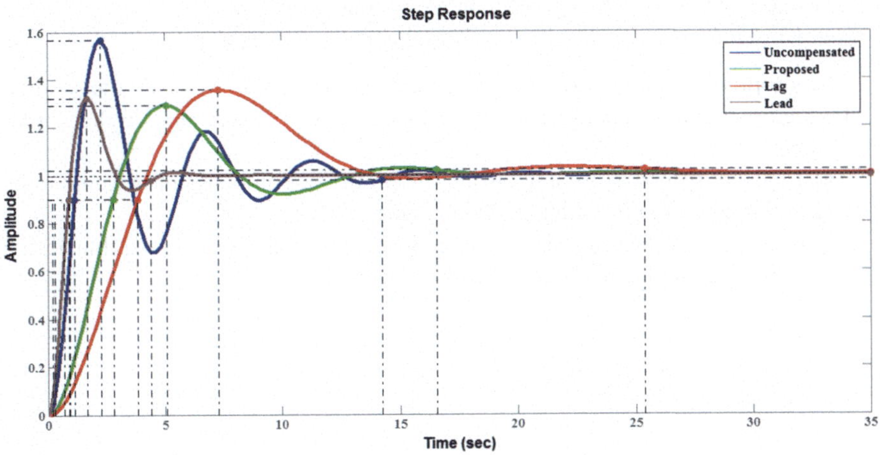

Fig. 10 Unit-step responses for the second-order system

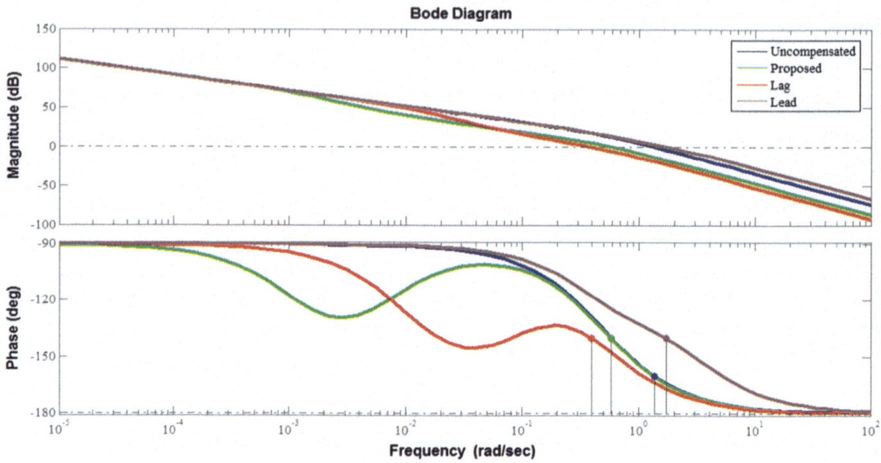

Fig. 11 Frequency responses for the second-order system

Table 7 Specifications of the second-order system

Method	ε (deg)	t_r (sec)	t_r (sec)	M_p (%)	GM (dB)	PM (deg)
Uncompensated	–	0.85	14.2	57	Infinity	20
Proposed design	0.65	2.11	16.6	29.3	Infinity	40
Lag compensator	11	2.88	25.4	35.6	Infinity	40
Lead compensator	4	0.68	4.4	32.1	Infinity	40

It is seen from Table 7 that the proposed compensator design method has achieved satisfactory results. ε is varied on a trial-and-error basis for the design of lag and lead compensators, whereas it is negligible in the proposed design.

Following inferences can be drawn from the above five examples.

1. The proposed compensator design procedure is useful to meet the desired frequency-domain specifications in various type 1 lower and higher order systems.
2. The transient response obtained using the proposed compensator is found to be better compared to that of lag and lead compensators.
3. The proposed compensator design procedure is non-iterative, whereas the design methods of lag and lead compensators are iterative.

4 Conclusions

Proposed first-order compensator is found to yield very good results in terms of time- and frequency-domain specifications for type 1 second-, third-, and higher order systems with and without time delay. Since the compensator designed is capable of giving results better than a conventional lag and lead compensator, it can also be extended as the generalized non-iterative procedure for the design of compensators to various systems.

The compensator is found to yield comparable results to PID controller for a BLDC drive fed position control system under dynamic conditions, and hence it can be employed in type 1 second-, third-, and higher order systems as an alternative to PID controller. The desired specifications can be fixed based on the results of the PID controller in such systems.

To extend the use of the proposed compensator further, a higher order system can be reduced into an equivalent type 1 third-order system and accordingly the desired specifications can be fixed. Moreover, the design procedure can be investigated to find its suitability to type 0 and type 2 systems also.

References

1. Keel, H.L., Bhattacharya, S.P.: Robust parametric classical control design. IEEE Trans. Autom. Control **39**(7), 1524–1530 (1994)
2. Norman, S.N.: Control Systems Engineering, 7th edn. Wiley, New Jersey (2015)
3. Lee, H.: A new phase-lead design method using the root locus diagrams. IEEE Trans. Autom. Control **50**(11), 1887–1891 (2005)
4. Ogata, K.: Modern Control Engineering, 5th edn. Pearson Education, London (2011)
5. Wang, Q.G., Ye, Z., Chieh Hang, C.: Tuning of phase-lead compensators for exact gain and phase margins. Automatica **42**(2), 349–352 (2006)
6. Loh, A.P., Cai, X., Tan, W.W.: Auto-tuning of phase lead/lag compensators. Automatica **40**(3), 423–429 (2004)
7. Teixeira, M.C.M., Assunção, E.: On lag controllers: design and implementation. IEEE Trans. Educ. **45**(3), 285–288 (2002)
8. Wang, F.Y.: The exact and unique solution for phase-lead and phase-lag compensation. IEEE Trans. Educ. **46**(2), 258–262 (2003)
9. Wang, D.J.: Synthesis of phase-lead/lag compensators with complete information on gain and phase margins. Automatica **45**(4), 1026–1031 (2009)
10. Yeung, K.S., Wong, K.W., Chen, K.: A non-trial-and-error method for lag-lead compensator design. IEEE Trans. Educ. **41**(1), 76–80 (1998)
11. Li, G., Tsang, K.M.: Concurrent relay-PID control for motor position servo systems. Int. J. Control Autom. Syst. **5**(3), 234–242 (2007)
12. Chen, Y.C.Y.: Implementation of a lag-lead compensator for robots. In: 27th IEEE Conference on Decision and Control, pp. 174–179. Texas, USA (1988)
13. Shanmugasundram, R., Zakariah, M.K., Yadaiah, N.: Design and digital implementation of non-iterative controller for improved performance of typical BLDC servo systems. Int. J. Power Electron. **3**(5), 510–526 (2011)
14. Ganesh, C., Jeba, S., Saranya, R., Geethu, S., Patnaik, S.K.: A non-iterative controller design for a BLDC drive system. In: 2009 International Conference on Advances in Recent Technologies in Communication and Computing, pp. 141–145. ARTCom 2009, Kottayam, India (2009)

15. Ganesh, C., Patnaik, S.K.: A simple first order compensator for brushless direct current drive based position control system. J. Vib. Control **21**(4), 647–661 (2015)
16. Haneef, H., Ganesh, C.: Investigations on the design aspects of first order controller for type 1 third order system. Int. J. Appl. Eng. Res. **10**(10), 9438–9445 (2015)
17. Ayasun, S.: Stability analysis of time-delayed DC motor speed control system. Turk. J. Electr. Eng. Comput. Sci. **21**(2), 381–393 (2013)
18. Pan, C.T., Fang, E.: A phase-locked-loop-assisted internal model adjustable-speed controller for BLDC motors. IEEE Trans. Industr. Electron. **55**(9), 3415–3425 (2008)
19. Singaravelan, A., Kowsalya, M.: Control of converter fed microgrid using fuzzy controller. In: 2013 International Conference on Energy Efficient Technologies for Sustainability, pp. 1179–1184. ICEETS 2013, Nagercoil, India (2013)
20. Ganesh, C., Patnaik, S.K.: Artificial neural network based proportional plus integral plus derivative controller for a brushless DC position control system. J. Vib. Control **18**(14), 2164–2175 (2012)
21. Shanmugasundram, R., Yadaiah, N., Ganesh, C., Poornaselvan, K.J.: Compensator design by pole-zero adjustment for a typical position control system. In: 2008 International Conference on Information and Automation for Sustainability ICIAFS 2008, pp. 19–24. Colombo, Srilanka (2008)
22. Vinodhini, R., Ganesh, C., Patnaik, S.K.: Genetic algorithm optimized on-line neuro-tuned robust position control of BLDC motor. In: 2012 IEEE Students' Conference on Electrical, Electronics and Computer Science SCEECS 2012. Bhopal, India (2012)
23. Ganesh, C., Prabhu, M., Rajalakshmi, M., Sumathi, G., Bhola, V., Patnaik, S.K.: ANN based PID controlled brushless DC drive system. Int. J. Electr. Power Eng. **03**(01), 45–48 (2012)

A Novel Optimized Nanoceramic Thermal Sensor with La Modifications for Body Temperature and Fluctuation Measurements

Kushal Roy and Ritesh Khanna

Abstract Perovskite-type materials particularly PZT have remained at the center of research due to their applicability in fabrication of sensors and actuators. The work presented in this paper focuses on fabrication of PZT (70/30) by sol–gel technique followed by La modifications in structural and dielectric properties. The structural characterization of both parent and modified materials is carried out by SEM and XRD techniques followed by detailed study on dielectric values and variations with frequency and temperature. It is hereby proposed that La-modified PZT(70/30) material with varying concentration of La can prove to be excellent material for the fabrication of thermal sensors which can measure human body temperature and fluctuations in a noninvasive and noncontactual way contrary to the existing conventional body temperature measurement systems. The fabricated sensors on the lab scale are found to be highly power efficient and sensitive. It is established in this paper that the sensitivity and range of body temperature measurement vary with the amount of La doping in parent PZT(70/30) material used to fabricate the sensor.

Keywords PZT nanoceramics · Temperature sensors · XRD · EDAX

1 Introduction

Lead Zirconium Titanate (PZT) Pb[ZrxTi(1−x)]O_3 electro-piezo ceramic system, discovered by Jaffe et al., finds vast and wide range of applications in the field of Electrical and Electronics Engineering as smart structures [1], memory devices [1, 2], piezoelectric electromechanical actuators [3], high sensitivity compact sensors

K. Roy (✉)
Department of Electronics and Communication Engineering, Haldia Institute of Technology, Haldia 721657, West Bengal, India
e-mail: kushalroy1979@gmail.com

R. Khanna
Department of Electronics and Instrumentation Engineering, CSJM University Kalyanpur, Kanpur 208025, India
e-mail: rkhanna23@gmail.com

S. Kundu et al. (eds.), *Proceedings of the 2nd International Conference on Communication, Devices and Computing*, Lecture Notes in Electrical Engineering 602, https://doi.org/10.1007/978-981-15-0829-5_36

[4], and ceramic capacitors [4, 5] of various configurations of shapes and sizes, due to its well-known piezoelectric properties for different values of x, where $x < 1$. Since its discovery about four decades back, it has always remained a point of attraction for various researchers and scientists because of its highly modifiable piezoelectric and dielectric properties, obtained by replacing either whole or a part of the host atom in ABO_3-type perovskite structure of PZT, by foreign ions [6]. Analogous to semiconductors, these foreign ions are also called dopants, classified as isovalent, acceptors, and donors. The PZT system is still under extensive research for further improvement in its piezoelectric properties. It has been reported by many authors that significant changes can be achieved by the addition of various dopants in PZT with a view to their dielectric and piezoelectric properties. La is one of the widest accepted hard dopants for PZT nanoceramic systems generally used to enhance the material's temperature sensing capacity. The synthesis of $Pb[Zr0 \cdot 7Ti0 \cdot 3]O_3$ nanoceramics is carried out by sol–gel process with the raw materials lead acetate trihydrate $Pb(CH_3COO)_2 \cdot 3H_2O$ (Merck), zirconium propoxide $Zr(C_3H_7O)4$ (Aldrich), and titanium isopropoxide $Ti[(CH_3)_2CHO]_4$ (Aldrich). The virgin PZT powder obtained from the sol–gel process is mixed with 1.5 and 2.5 weight percent of La (from La_2O_3, Merck USA) followed by calcination at 900 °C to prepare the appropriate La-modified PZT powder which are transformed into cylindrical pallets of 10 mm diameter and 1.5 mm thickness using a hydraulic press at 100 psi. The ceramic disks obtained are coated with silver on both sides by VPD process to produce a capacitive structure.

2 Methodologies, Results, and Discussions

2.1 Sensor Material Characterization

The developed virgin PZT (70/30) and the La-modified nanoceramic material are subjected to structural analysis using SEM and XRD to determine the suitability of the nanostructures for the fabrication of human body temperature sensors. The XRD characterization graph clearly shows the inclusion of La ions in correct and predetermined position of the crystal lattice as shown in Fig. 1. After a comparative analysis of the SEM micrographs of the virgin PZT and La-modified PZT, it is clearly observed that La modification induces more compactness of the domains as compared to the virgin PZT sample, thereby making it more suitable for the development of thermal sensors as shown in Figs. 2, 3. Though compactness is extremely desired for the development of appropriate sensing element, it is observed that higher concentration of La doping causes deformation of the crystal structure as noted from the peak splitting effect in the XRD graph shown in Fig. 4, which will cause the induction of nonlinearity in measurement of any physical parameter.

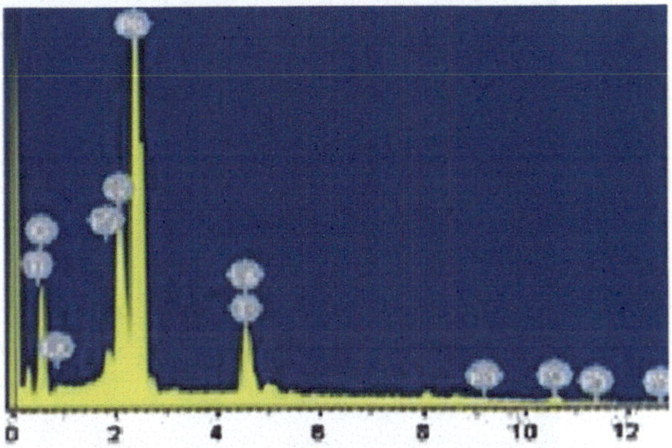

Fig. 1 EDAX graph of La-modified PZT (70/30)

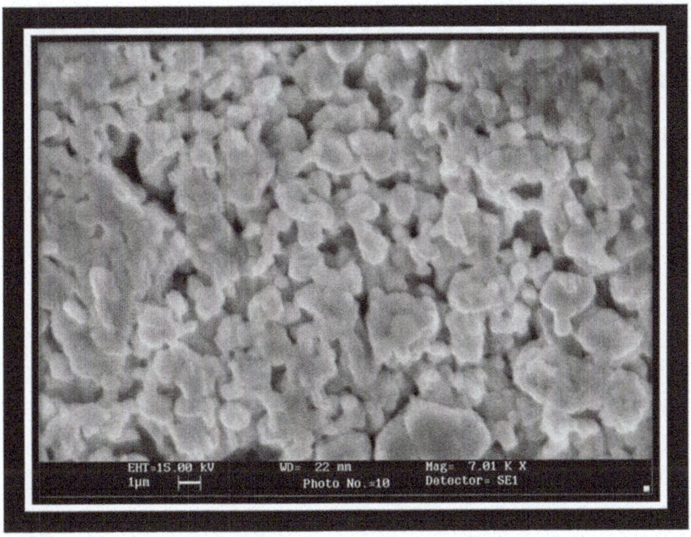

Fig. 2 SEM micrograph of PZT (70/30)

2.2 *Sensor Calibration and Temperature Detection*

Figure 5 shows the La-modified PZT capacitive configuration sensors developed on lab scale for remote detection of human body temperature and its fluctuations. The sensors are subjected to measurement of the variation in K (dielectric constant) with temperature using an isotropic thermal radiator having a controlled radiating temperature ranging from 90 to 110 °F (32.2–43.3 °C). The developed capacitive sensors are placed at a radial distance of 100 mm (10 cm) from the radiator within a controlled atmosphere lab having a temperature of 25 °C and RH <60%. The variation of "K" (dielectric constant) with radiating temperature is plotted as shown in Fig. 6. As discussed in the previous section, the variation of K shows considerable linearity

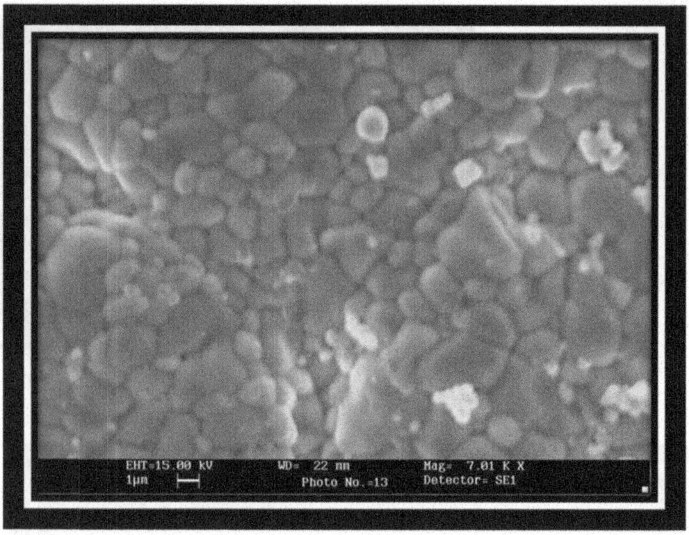

Fig. 3 SEM micrograph of 1.5% La-modified PZT (70/30)

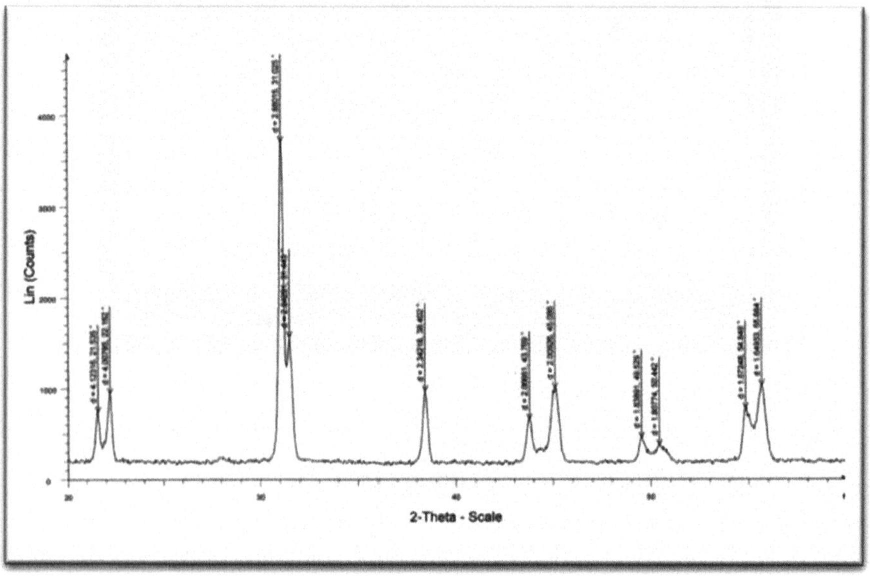

Fig. 4 XRD graph of 2.5% La-modified PZT showing peak splitting effect

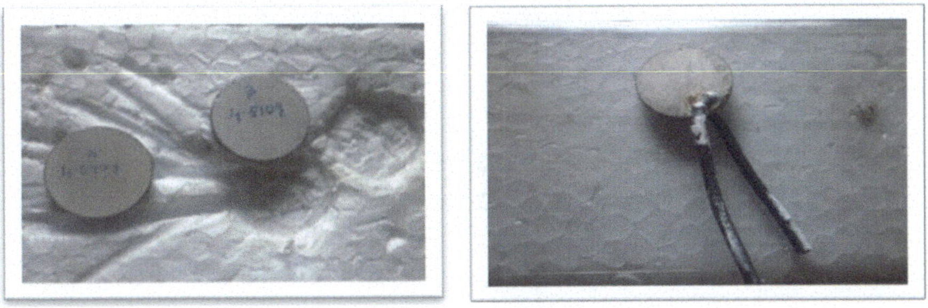

Fig. 5 Fabricated body temperature sensors in capacitive configurations

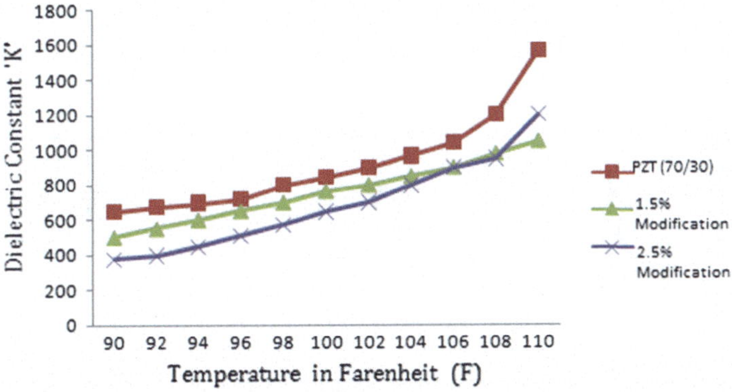

Fig. 6 Variation of dielectric constant "K" with simulated human body temperature

within the temperature range of functional human body for 1.5% La-modified PZT sensor as compared to 2.5% modified sensor, thereby making it more suitable and sensitive to body temperature fluctuations.

2.3 *Sensor Designing*

In order to measure the human body temperature, the capacitive sensors fabricated can be easily incorporated in an electronic circuit where variation in K (causing change in capacitance) resulting in change of current can be directly calibrated for change in temperature. The overall power consumption can be restricted to <2 W if the distance of 10 cm is maintained between the human body and the sensing element. Figure 7 shows the circuit diagram for the entire measurement setup. The equivalent component level circuit diagram shown in Fig. 8 represents the process of emf e_0 generated due to change in current i_{cr} (due to change in dielectric constant "K") from the sensor.

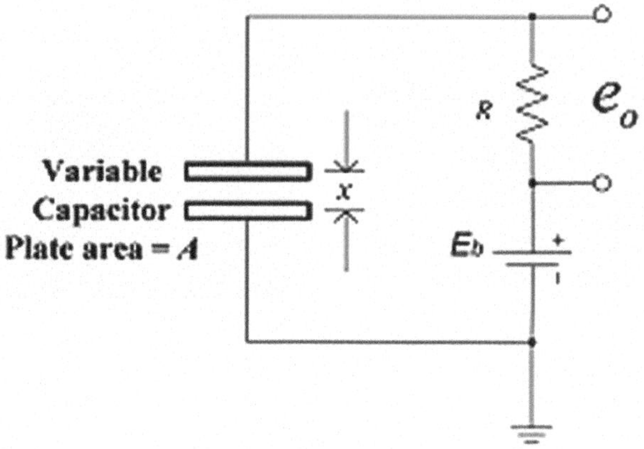

Fig. 7 The circuit diagram of incorporating the nanoceramic variable capacitive sensor

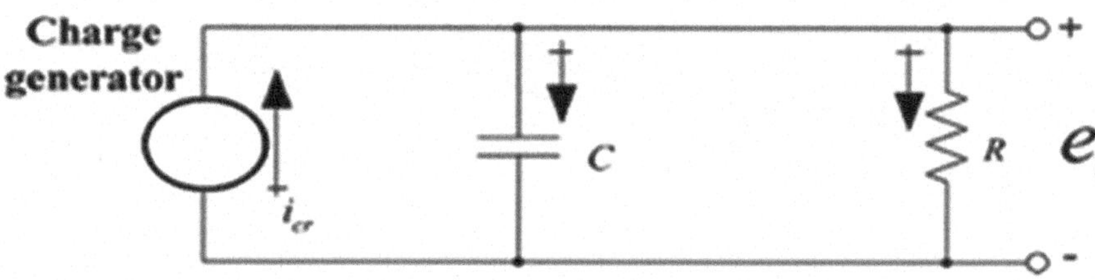

Fig. 8 The equivalent circuit showing the emf e0 generated due to the change in dielectric constant "K"

3 Conclusion

After thorough investigation, it is hereby concluded that 1.5% La-modified PZT nanoceramic material developed is one of the optimal doping levels, best suited for noncontactual remote measurement of body temperature. The lower doping level maintains the compactness of the sensing element proposed along with preserving its linearity and sensitivity to temperature changes. The proposed method eradicated the drawbacks of conventional temperature measurement systems currently used in medical purpose, and at the same time consumes extremely low electrical power and has considerable ease of design.

Acknowledgements The author would like to thank SERB (formerly DST SERC) for providing the research funding, Smart Materials Research Lab (SMR) and Institute Instrumentation Center, IIT Roorkee for providing the characterization results, and Haldia Institute of Technology for their constant support.

References

1. Tahar, R.B.H., Tahar, N.B.H., Salah, A.B., et al.: Preparation and characterization of PZT solid solutions via sol-gel process. J. Cryst. Growth. **307**, 420–432 (2007)
2. Smitha, P., Pandey, P.K., Gajbhiye, N.S., et al.: Polyol based auto-combustion synthesis of nanostructured PZT and its characterization. Mater. Chem. Phys. **109**, 500–505 (2008)
3. Chaudhari, V.A., Bichile, G.K.: Structural and impedance spectroscopic studies on $PbZrxTi1-xO_3$ ceramics. Phys. B **405**, 534–539 (2017)
4. Fasquelle, D., Carru, J.C.: Electrical characterizations of PZT ceramics in large frequency and temperature ranges. J. Eur. Ceram. Soc. **28**, 2071–2074 (2018)
5. Noimanee, S., Tunkasiri T, Siriwitayakorn, K., et al.: Design considerations for aural vital signs using PZT Piezoelectric ceramics sensor based on the computerization method. Sensors **120**, 3192–3208 (2007)
6. Kang, M-G., Jung, W-S., Kang, C-Y., et al.: Recent progress on PZT based piezoelectric energy harvesting technologies. Actuators **5**(1), 5 (2016)

On Relative Performances and Decoding of CRC Concatenated Polar Codes with Different Lists for Solid-State Drives

Tirthadip Sinha, Sudip Nayek and Jaydeb Bhaumik

Abstract The storage capacity and cost-per-bit of solid-state drives (SSDs) have improved gradually in recent years, which have driven to the extensive acceptance in data storage of modern computing systems. SSDs convey greater read and write performance, enhancement in random-access input/output (I/O) operations, and resistance to physical shock, while form factor and static power requirement are not as much of magnetic hard disk drives (HDDs). Though HDDs are progressively replaced by SSDs as a primary data storage, error correction is still severe to SSDs as NAND flash memories have worsening reliability which can diminish the life span of the flash storage devices. Presently, popular error control codes show inadequate error correction ability or poor error floor, and the growing storage density of NAND flash memory cells exhibits more data errors. All of these stimulate probing for finest channel coding to recover NAND flash reliability. Polar codes are favorable for SSDs as they are theoretically verified optimal codes with good error floor behavior. With the low computational complexity and the high error correction competency, cyclic redundancy check (CRC) concatenated polar code with list decoding can beat existing error correction codes in data communication applications. Hence, it also steered to implement in flash storage devices. In this paper, CRC concatenated polar codes with different list sizes for SSDs are designed and error correction performances are evaluated in C/C++ environment. Evaluated outcomes illustrate the decoding performance achieving preferred error floor behavior for error correction in flash storage devices. While the outcomes display that concatenated polar codes with list decoding are favorable for SSDs, some open problems are remained there for further investigations.

T. Sinha (✉) · S. Nayek
Haldia Institute of Technology, Haldia, West Bengal, India
e-mail: tirthadip.sinha@gmail.com

S. Nayek
e-mail: sudip.nayek@yahoo.com

J. Bhaumik
Jadavpur University, Jadavpur, Kolkata, India
e-mail: bhaumik.jaydeb@gmail.com

S. Kundu et al. (eds.), *Proceedings of the 2nd International Conference on Communication, Devices and Computing*, Lecture Notes in Electrical Engineering 602, https://doi.org/10.1007/978-981-15-0829-5_37

Keywords SSD · Polar code · Flash memory · CRC · List decoding · Error correction code · BLER

1 Introduction

With the necessity of massive data storage applications, SSDs have improved and accepted gradually as an efficient data storage device to support the huge storage capacity and simultaneously save the cost using NAND flash memory in modern computing systems. In comparison with magnetic HDDs, SSDs convey greater read and write performance, enhancement in random-access input/output (I/O) operations, resistance to physical shock, low form factor, and less static power consumption. With no mechanical parts, SSDs are less probable to undergo unexpected drive failure than HDDs. Though HDDs are progressively replaced by SSDs as a primary data storage, error correction is still severe to SSDs since NAND flash memories have worsening reliability which can diminish the life span of the flash storage devices. The graphical voltage distribution model of 2-bit NAND flash memory cell is presented in Fig. 1, and four separated states show all probable combinations of two bits in a NAND flash memory cell [1]. The state S_0 is termed as *erased state* and other three states (S_1–S_3) are known as *programmed state*. The voltage distribution in a flash memory cell has a Gaussian-like distribution due to read/program disturbances and different interferences [2, 3]. So, Gaussian distribution is taken in this work as a competent model to portray the flash memory cell. At the start of SSD lifespan, the noise variance is insignificant and the gaps between each distribution are able to offer adequate information to differentiate different states. But standard deviation of distributions produced by noise is considerably inflated with the rise of program and erase (P/E) cycles [1, 4], and thus stronger protections for stored data are required using error correction codes (ECCs) to retrieved data properly.

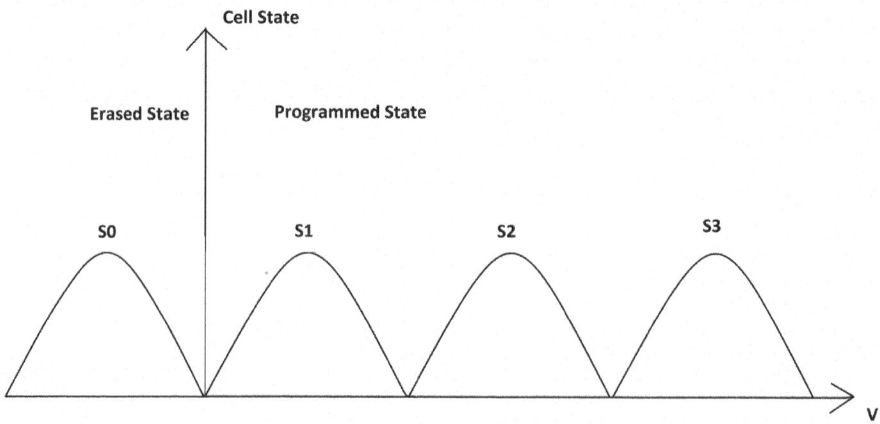

Fig. 1 The voltage distribution model of 2-bit NAND flash memory cell

Generally, memory subsystems have used Hamming code, whereas Reed Solomon (RS) code is used in HDDs and CD-ROMs. Later, Bose–Chaudhuri–Hocquenghem (BCH) code, Low-Density Parity-Check (LDPC) code, or combination of both [5] is used as ECCs. Though they exhibit inadequate error correction capability or extended decoding latency [6], polar codes, proposed by Arıkan [7], are renowned to attain channel capacity under binary-input discrete memory less channels (BDMCs) with low computational complexity of O(n log n) for hardware implementation of both encoder and decoder. Also, the performance of polar codes is better than LDPC with small code length and flexible with code rate [8]. Due to all these advantages, polar code is considered as promising ECC scheme for SSD.

Polar codes are realized by channel polarization which is produced by the method of channel splitting and combining, respectively. Then, the information bits are sent over only the good channels and bad channels are discarded. Several kinds of decoding algorithms have been suggested for polar codes and applied very effectively in many applications. Successive cancellation list (SCL) decoding, offered by Tal and Vardy [2], encourages the use of concatenated polar codes using cyclic redundancy check (CRC) codes as precoding, so that the right codeword from the list can be retrieved during decoding at the receiver [9]. Still, detecting the best CRC code length and list size combination for the concatenated polar codes has remained rather unknown [5].

The error correction competency of the polar code extremely depends upon the perfection of polar code structure. A code structure is basically design layout for input bits and built solely based upon the status of channel quality. But the status of channel quality is increased by the amount of program/erase (P/E) cycles of data blocks. Flash memory generally implements the array assembly to construct the chip for reducing the area size and design of the circuit, although the array assembly likewise brings severe interference among the attached memory cells when the read, write, or erase operation is done. Usually, a bigger unit is adopted to execute these processes [1, 10] to escape this situation. Previously, various works have been done to implement the polar codes in the field of flash drive [2, 3, 11].

In this work, therefore, attempts have been made to investigate the effect of the encoded block length and *Signal-to-Noise Ratio (SNR) per bit* (EbNo) on the subsequent block error rate (BLER) performance of the concatenated polar codes using CRC precoding and list decoding with different list sizes so that it can be applied effortlessly in an SSD. Also, relative error correction performances are evaluated in C/C++ environment when compared to existing work [10] and outcomes illustrate that the decoding performance approaches the preferred error floor behavior for error correction in flash storage devices. To identify efficient polar code construction with low BLER, totally 30 polar code combinations are investigated with code length $N = \{512, 1024, 2048\}$, code rate = ½, CRC length $|m| = \{8, 16\}$, and list size $|L| = \{1, 2, 4, 8, 16\}$.

The rest of this paper is prepared in the given fashion. The basics of polar encoding and decoding methods and channel modeling of flash cells in SSD are discussed in Sects. 2 and 3, respectively. In Sect. 4, simulation outcomes are displayed to compare the error correction performances of different polar code combinations for various

N, |c|, and |L| with fixed code rate = ½. Also, the relative performance comparison and result analysis are done with the existing work [10]. Lastly, the conclusion is drawn in Sect. 5.

2 Polar Coding Scheme

A polar code is basically a linear block error-correcting code. The code construction takes place on the basis of multiple recursive concatenation of a tiny kernel code because of which the physical channel gets converted into virtual superficial channels. When the number of recursions is extended, then the reliability of virtual channels became either high or low, and the data bits are allotted to the best reliable channels and the less reliable channels are treated as frozen bits that give a value of 0. It is the first proven channel code with an unambiguous structure to theoretically achieve the channel capacity for symmetric BDMC. Also, the reasonable encoding and decoding complexities of polar codes make them more attractive for many modern applications [10, 12].

2.1 *Polar Code Construction*

According to Shannon capacity calculation, it is well known that the noiseless channels have lower error probabilities and higher capacities in comparison with noisy channels, and thus theoretically it is always preferred to transmit the bits of information through a noiseless channel. For encoding in polar code initially, simple selection of K out of N indices is to be done where $N = 2^n$, the code length is N, and the length of information is K [13]. For channel polarization exploitation, many algorithms are proposed previously but the simple one to use is the recursion one given as $z \rightarrow \{2z - z^2, z^2\}$ as shown in Fig. 2. Then, a particular channel is chosen (e.g., AWGN, BEC, BSC). Though any channel can be taken into consideration only knowing the Bhattacharya parameter of the channel as for AWGN channel, it is $z = e^{-Ec/No}$. Then two values are created using two different values and so a chaining method is introduced until the tree has N leaves indexed from top 0,1, …., N − 1.

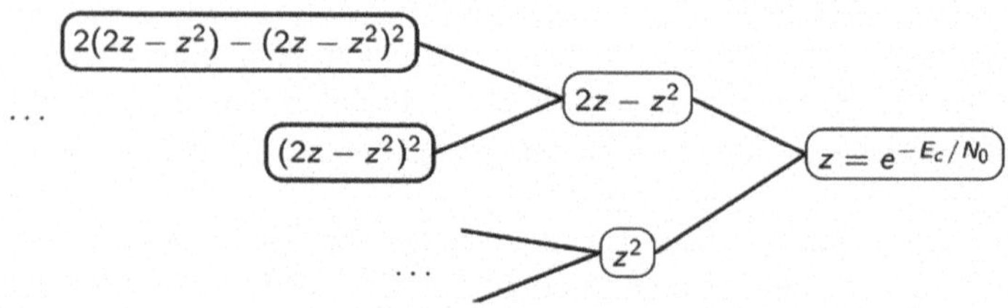

Fig. 2 Implementation of recursion algorithm for polarization of the channel

Then, the k list values are picked from the leaves and their indices are stored in a set J and are shown as the output.

An (N, K, I) polar code is selected where the code length is N after adding redundancy 2^n, K is the message length, and I is the set of the bit reversed of the set J where the indices are stored and is known as the information bit indices. The complementary set of I, Ic, is called frozen bit indices. Another process used is the use of kernel (binary) [13] given by

$$F^{\otimes n} \triangleq F \otimes F \otimes \ldots \text{ (n times)} \tag{1}$$

It does chronical product operations, that is, the dimensions grow as a power of small n.

$$F = \begin{pmatrix} 1 & 1 \\ 0 & 1 \end{pmatrix} \tag{2}$$

$$F^{\otimes 2} = \begin{pmatrix} F & F \\ 0 & F \end{pmatrix} \tag{3}$$

$$F^{\otimes 3} = \begin{pmatrix} F^{\otimes 2} & F^{\otimes 2} \\ 0 & F^{\otimes 2} \end{pmatrix} \tag{4}$$

Encoding is simply multiplying the kernel with a vector to get the encoded vector:

$$\mathrm{X} = F^{\otimes n}\mathrm{d} \tag{5}$$

where $d_{I_c} = 0$ and $\mathrm{d_I} = \mathrm{u}$ (the message). It is similar to the product of vector with generator matrix.

As this is a matrix multiplication, it has quadratic complexity of $O(N^2)$ which is not implementation friendly. Thus, a more efficient implementation is used which has complexity of O(NlogN), which is known as the butterfly structure. It is basically several XOR operations being performed on several d converting to X. Now each XOR operation represents one kernel. As a result, it will show F times of whatever input that is going to come in. Thus, cascading several F intelligently, we are going to implement the n-fold chronicle product in an efficiently computational fashion. It is an efficient implementation by which the complexity can be reduced easily. A general representation [14] of the butterfly structure is

$$x = F^{\otimes n}d,\ in\,just\left(\frac{N}{2}\log_2 N\right)XOR \tag{6}$$

The code structure is constructed by controlling the more reliable *K* bit-channels for the information and selecting the more unreliable (N − K) bit-channels for the frozen bits. If the code structure comprises some unreliable or weak bit-channels for information bits, then the error correction proficiency would be extremely worsened [15].

2.2 Successive Cancellation List (SCL) Decoding

Two-way recursive algorithm is used on N received likelihood. It takes N likelihood of the elements in X and converts to N likelihoods of the bits in d sequentially [14]. If two likelihoods are given, then they get converted into two other likelihoods as two variable functions. The second operation depends on the immediately bit decision from upper branch. The two likelihood operations are in use:

$$\begin{pmatrix} L_1 \\ L_2 \end{pmatrix} \rightarrow \begin{pmatrix} f(L_1, L_2) \\ g(L_1, L_2) \end{pmatrix} = \begin{pmatrix} \frac{L_1 L_2 + 1}{L_1 + L_2} \\ L_2 . L_1 \; or \; L_2 / L_1 \end{pmatrix} \tag{7}$$

The second function is special because it has two variants which are dependent on bit of upper branch in a way directly or indirectly the upper likelihood decides what bit is going to come in this block and the lower likelihood decides which function is to be used; if upper likelihood is zero, then it will be product and if it is 1 then it is division.

Issues usually, as polar codes, are used at higher block lengths, so numerical decision issues arises very quickly; so to solve the numerical underflows, log domain representations are used.

$$\begin{pmatrix} l_1 \\ l_2 \end{pmatrix} \rightarrow \begin{pmatrix} lnf\left(e^{l_1}, e^{l_2}\right) \\ lng\left(e^{l_1}, e^{l_2}\right) \end{pmatrix} = \begin{pmatrix} \ln\left(\frac{1+\exp(l_1+l_2)}{\exp(l_1)+\exp(l_2)}\right) \\ l_2 + l_1 \; or \; l_2 - l_1 \end{pmatrix} \tag{8}$$

Assuming there is computational tree, they are slowly converted to single function gradually by decreasing by n/2 step operations in each level. After certain level, certain levels are made as inactive since its calculated values are already available in the next stage. The penultimate stage is always full of g functions. These N trees are naturally embedded in an $N * (n + 1)$ array with order N log N. The correlation of the butterfly structure between the source bits is considered as interference in source bit domain. Successive cancellation (SC) of the *interference* produced by the previous bits increases the reliability in the recovery of source bits. Because of the regular structure of polar codes, SC decoding algorithm too can be styled as a code tree or a trellis structure.

In SCL decoding which is basically an advanced version of SC decoding [15], extra paths are involved by including the different list sizes to increase error correction performance further. Usually, SCL is functioned as a breadth-first search algorithm and permits a maximum of L contestant paths compared to SC where only one path is retained after processing at each level. SCL doubles the quantity of contestants by assigning a bit (0 or 1) to each contestant path. It also chooses a maximum of L ones with the major metrics saving in a list. In a *lazy copy* memory sharing arrangement, the SCL decoder can be applied with computational complexity of O(LNlogN), whereas direct implementation will need O(LN^2). The effective equation is given by

$$L_{i,j} = \begin{cases} 2tanh^{-1}\left[\tanh\left(\frac{L_{i+1,j}}{2}\right).\tanh\left(\frac{L_{i+1,j+2^{i-1}}}{2}\right)\right], & \left\lfloor\frac{j-1}{2^{i-1}}\right\rfloor mod2 = 0 \\ \left(1-2s_{i,j-2^{i-1}}\right)\left(L_{i+1,j-2^{i-1}}\right)+L_{i+1,j}, & otherwise \end{cases} \tag{9}$$

2.3 Concatenated Polar Code with CRC Precoding and SCL Decoding

CRC precoding and SCL decoding are introduced in conventional polar codes to further improve the error correction ability. Here, the SCL decoder outputs the contestant paths into a CRC detector, and the checking outcomes are used to identify the accurate codeword. Adaptive CRC concatenated SCL or CRC-aided SCL (CA-SCL) decoding is suggested by gradually increasing the list sizes to lower the complexity of SCL decoding. The error correction ability is substantially enhanced and outpaces than normal decoding process depending upon CRC length and generator polynomial under CRC-aided decoding structures [16, 17]. But the redundancy linked with CRC codes turns into more prominent and thus decreases the overall competency of the concatenated polar code if the code length is small. Recently, a search algorithm for polar codes is developed to ensure an optimal CRC search using SCL decoding [6, 16].

3 Channel Modeling of Flash Cells in SSD

A pictorial example of the polar code combination implementation in flash cells for polar code (8, 6) is presented in Fig. 3. Here, U signifies the bit-channels which are resulting from the code structure consistent with the input data, whereas X and Y indicate the encoded bits and the noisy bits, respectively. Y-bits could be altered/corrupted after X-bits are stored into flash cells in the presence of noise and polar code fails on decoding/correcting some noisy bits. Due to the polarization, each bit-channel may be connected to the several physical/flash cells. Based upon the code construction aimed to achieve high error correction ability, the bit-channels are rated. U3 to U8 bit-channels are utilized for information bits as they have lesser raw BER than U2 and U1 [1]. The error correction performance would be satisfactory if the corresponding flash cells of the information bits are more reliable than the adjacent frozen bits.

Figure 4 displays the block diagram for the realization of the polar codes in flash cells. To practically implement these codes in SSD, first k number of information bits have to be encoded with the addition of certain number of CRC bits|m|, and then encoding is done for the total number of N bits as in polar coding, which is to be stored in the SSD. When the data is to be retrieved, the CA-SCL decoding comes into play in which SCL decoding takes place first and then CRC decryption performs

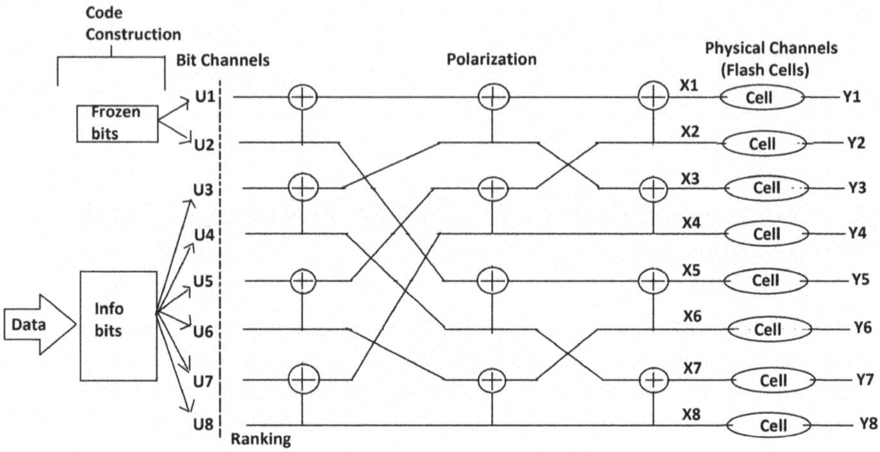

Fig. 3 A pictorial example of polar code (8, 6) implementation in flash cells

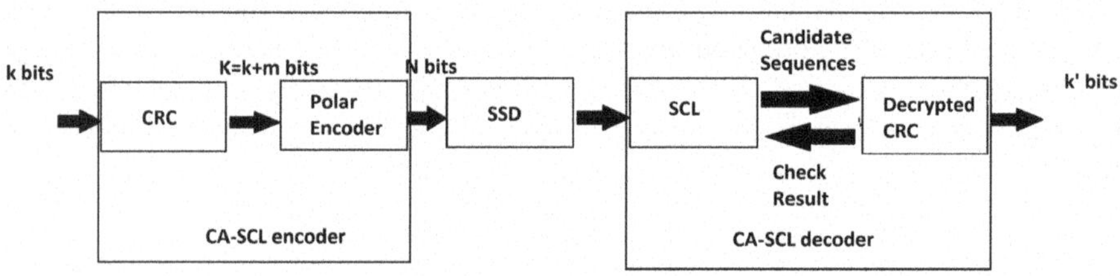

Fig. 4 Block diagram for CRC concatenated polar encoding and list decoding model in SSD

to correct the error bit. This decoding process repeats several times according to the given list size and then the final decoded bits come out with low error probability.

For mathematical modeling of Multilevel Cell (MLC) flash, the voltage distribution given by Intel [18] is generally considered as a Gaussian distribution with probability density function (PDF) as

$$P_0(X) = \frac{1}{\sigma_0\sqrt{2\pi}} e^{\frac{-(x-\mu)^2}{2\sigma_0^2}} \tag{10}$$

where the standard deviation is σ_0 and the expectation is μ. As the distribution of the voltage is given by the Gaussian distribution, thus the implementation of CRC concatenated polar code in SSD by AWGN channel comes under consideration in this work.

4 Result Analysis

The decoding performance of proposed concatenated polar codes is investigated under various varying conditions via simulations over Binary-Input Additive White Gaussian Noise Channels (BI-AWGNCs) as considering flash cell channel conditions and also relative performances are compared with the existing work [10] in this section. The m number of CRC bits are added with the k number of information and all the $K = k + m$ bits are fed into the polar encoders becoming the corresponding code rate $R = K/N$. For all coding schemes, R = 1/2 is taken and CRC generator polynomial are randomly created to identify best error correction performance in terms of BLER.

In Fig. 5, the BLER performance for proposed concatenated polar codes with varying encoded block length N and CRC size are examined but given emphasis to mainly on effect of different list sizes|L| keeping fixed code rate R. The error correction performance of proposed codes increases rapidly with increasing SNR value for a constant CRC size, list size, and encoded bits. Again, for a constant number of encoded bits, CRC size, and SNR value, the performance of correcting

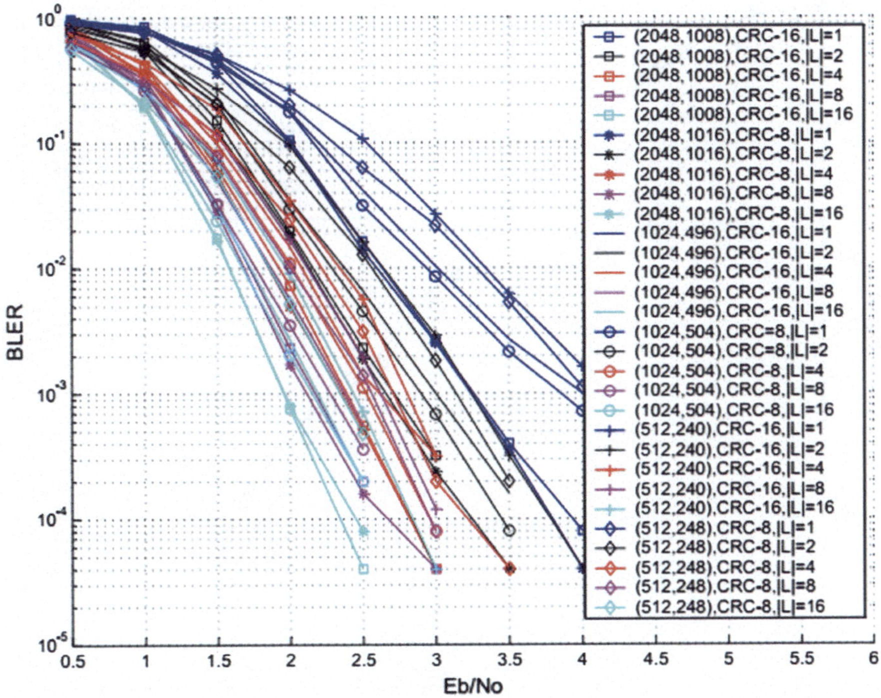

Fig. 5 Plot of BLER performance results for CRC concatenated polar codes with fixed code rate $R = ½$ and varying code length, CRC, and list size

errors increases smoothly as the list size increases greater than 1. From Fig. 5, it can be witnessed that the performance of correcting errors decreases with decreasing |L| or increasing CRC length after a certain limit or decreasing SNR value. Thus, to increase the performance of decoders in SSD, relatively low value of CRC length and comparatively high value of|L| should be used during SCL decoding, though the value of |L| should not be increased above 16 because of excessive rise in latency that gets introduced in high value of |L| during the decoding process. Also, the CRC length is taken as 8 and 16, as low CRC length leads to loss of effective correction of the error bits.

Figure 6 shows basically the graphical representation of Table 2 results where code length N = 1024 and K = 512 with CRC = {8, 16}, list size = {1, 2, 4, 8} and varying code rate R = {0.508, 0.516}, respectively, though L = 16 is not considered since the simulation results for L = 8 and L = 16 are close enough.

From Tables 1 and 2, it can be observed that by keeping fixed code rate as R = 0.5 and varying the information bits with respect to fixed CRC bits would increase the performance of SCL decoder very much, though the information bits cannot be decreased too much because the net bit rate becomes very low compared to gross bit

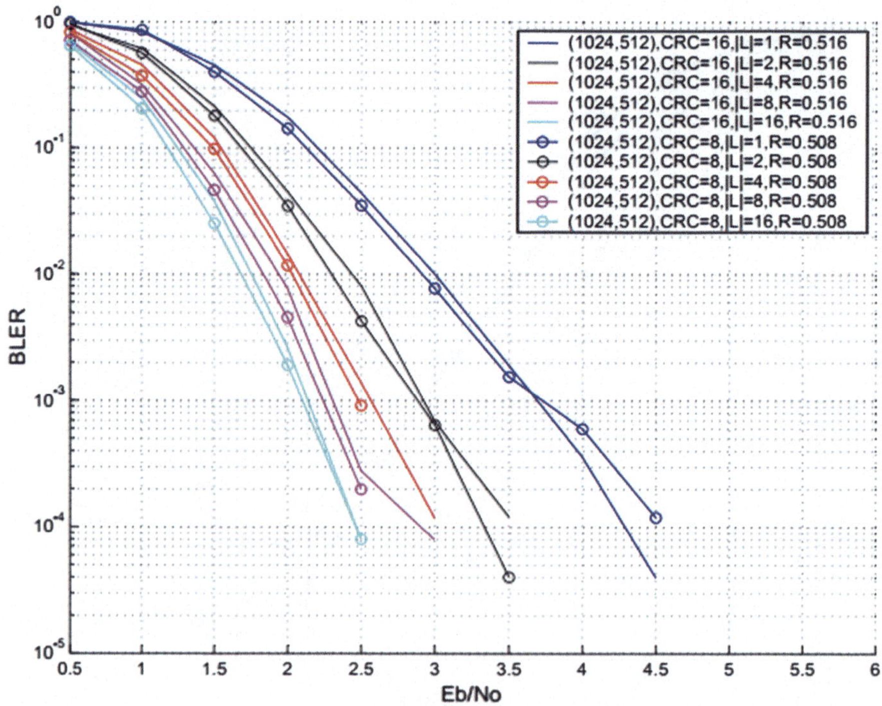

Fig. 6 Plot of BLER performance results for CRC concatenated polar codes with fixed code length N = 1024, CRC = {16, 8}, list size = {1, 2, 4, 8, 16}, and varying code rate

Table 1 Decoding performance of polar codes (with SC decoding) for SSD channel as presented in paper [10]

BLER Performance	E_b/N_o for (1024, 512) with no CRC and R = 0.500	
	Hard decoding	Quantized-soft decoding
10^{-1}	1.72	0.80
10^{-2}	2.92	2.06
10^{-3}	4.12	3.12
10^{-4}	5.10	4.12

Table 2 Decoding performance of CRC concatenated polar codes (with SCL decoding) for SSD channel as proposed in this paper

BLER performance	E_b/N_o for (1024, 512) with CRC = 8 and R = 0.508		
	$\|L\| = 2$	$\|L\| = 4$	$\|L\| = 8$
10^{-1}	1.6	1.5	1.3
10^{-2}	2.3	2.0	1.8
10^{-3}	2.9	2.5	2.25
10^{-4}	3.35	–	–

rate and thus rendering the purpose of SSD of having the capacity of storing large amount of data. Also, if the CRC length is decreased too much, then the performance of the cyclic redundancy check to detect and correct the error bits reduces. Comparing the values of Tables 1 and 2, it is shown in Fig. 7 that by taking N = 1024 and K = 512 for proposed CRC concatenated polar codes with different lists also give much better BLER performance than the decoders (hard and quantized-soft decoder) mentioned in existing work [10] exclusively for NAND flash cells of SSD.

5 Conclusion

In this work, proposed CRC concatenated polar codes with SCL decoding is shown effectively improved error correction ability in flash data storage in SSD. The investigational outcomes confirm that the decoding performance by the proposed method in this paper approaches the preferred error floor behavior for error correction in flash storage devices. As a result, the life span of SSD would be further increased without increasing major hardware cost with the efficient and effective implementation of proposed concatenated polar codes.

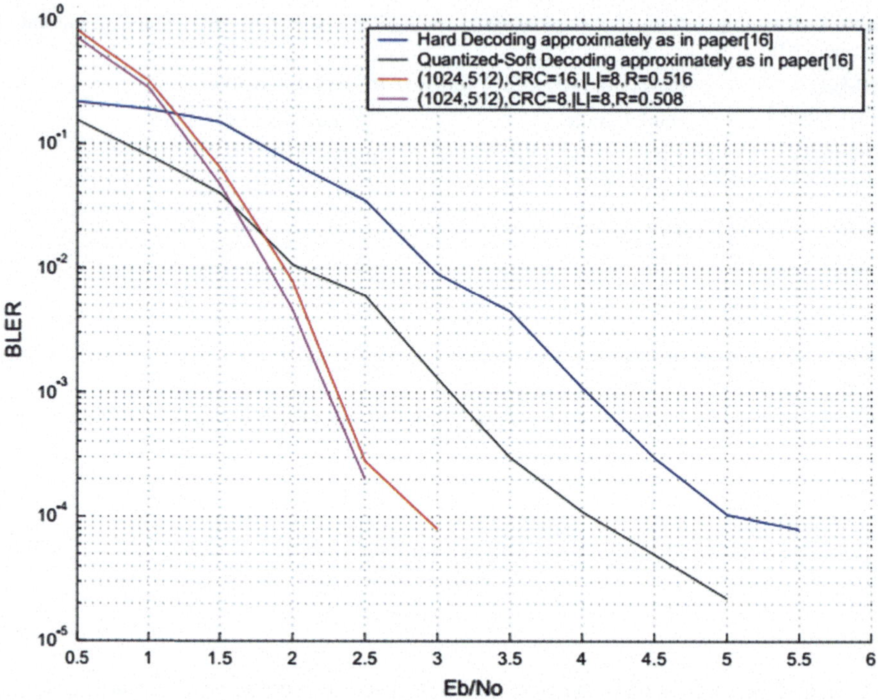

Fig. 7 Plot of BLER performance comparison between the proposed method and existing work (as mentioned in [10])

References

1. Hsu, C.K., Tsao, W.C., Chang, H.Y., Kuo, W.T., Huang, M.Y.: Proactive channel adjustment to improve polar code capability for flash storage devices. In: 55th ACM/ESDA/IEEE Design Automation Conference (DAC), pp. 1–6 (2018)
2. Tal, I., Vardy, A.: How to construct polar codes. IEEE Trans. Inf. Theory **59**(10), 6562–6582 (2013)
3. Hassani, H.S. Urbanke, R.: Universal polar codes. In: IEEE International Symposium on Information Theory, pp. 1451–1455 (2014)
4. Li, Q., Jiang, A., Haratsch, F.E.: Noise modelling and capacity analysis for nand flash memories. In: IEEE International Symposium on Information Theory (2014)
5. Darnell, M.: Error control coding: Fundamentals and applications, IEEE Proc. F-Commun. Radar Signal Process. **132**(68) (1985)
6. Zhang, C., Parhi, K.K.: Low-latency sequential and overlapped architectures for successive cancellation polar decoder. IEEE Trans. Signal Process. **61**(10), 2429–2441 (2013)
7. Arıkan, E.: Channel polarization: a method for constructing capacity achieving codes for symmetric binary-input memoryless channels. IEEE Trans. Inf. Theory **55**(7), 3051–3073 (2009)
8. Cai, Y., Haratsch, F.E., Mutlu, O., Mai, K.: Error patterns in MLC NAND flash memory: measurement, characterization, and analysis. In: Design, Automation and Test in Europe Conference and Exhibition (2012)

9. Niu, K., Chen, K.: CRC-aided decoding of polar codes. IEEE Commun. Lett. **16**(10), 1668–1671 (2012)
10. Song, H., Zhang, C., Zhang, S., You, X.: Polar code-based error correction code scheme for NAND flash memory applications. In: 8th International Conference on Wireless Communications and Signal Processing (WCSP), pp. 1–5 (2016)
11. Chen, K., Niu, K., Lin, J.: A hybrid ARQ scheme based on polar codes. IEEE Commun. Lett. **17**(10), 1996–1999 (2013)
12. Ho, C.C., Liu, P.Y., Chang, H.Y.: Antiwear leveling design for SSDs with hybrid ECC capability. IEEE Trans. VLSI Syst. **25**(2), 488–501 (2017)
13. Goela, N., Korada. B.S, Gastpar, M.: On LP decoding of polar codes. In: Proceedings of the IEEE Inforamtion Theory Workshop (ITW'10), Dublin, Ireland, pp. 1–5 (2010)
14. Mori, R., Tanaka, T.: Performance of polar codes with the construction using density evolution. IEEE Commun. Lett. **13**(7), 519–521 (2009)
15. Li, B., Shen, H., Tse, D.: An adaptive successive cancellation list decoder for polar codes with cyclic redundancy check. IEEE Commun. Lett. **16**(12), 2044–2047 (2012)
16. Murata, T., Ochiai, H.: On design of CRC codes for polar codes with successive cancellation list decoding. In: IEEE International Symposium on Information Theory, Germany, pp. 1868–1872 (2017)
17. Zhang, Q., Liu, A., Pan, X., Pan, K.: CRC code design for list decoding of polar codes. IEEE Commun. Lett. **21**(6), 1229–1232 (2017)
18. Atwood, G., Fazio, A., Mills, D., Reaves, B.: Intel strataflash memory technology overview. Intel Technol. J. (1997)

Measuring Real-Time Road Traffic Queue Length: A Reliable Approach Using Ultrasonic Sensor

Avirup Mandal, Pampa Sadhukhan, Firoj Gaji and Prolay Sharma

Abstract Traffic congestion not only lengthens the travel time of a commuter but also increases the fuel consumption as well as air pollution. Thus, traffic congestion management has drawn significant research interest over the past few years. On the other hand, a reliable estimation of the congestion density via traffic queue length measurement at any segment of the road in real time is highly desirable to control the traffic congestion in the urban areas. Thus in this paper, we focus on how to estimate the congestion density in a reliable way and propose a real-time traffic queue length measuring approach for the signalized intersections using ultrasonic sensor node (USN). The experimental results demonstrate that our proposed approach achieves almost 100% success in vehicle detection if the vehicle queue is formed within a distance of 2.5 m from the position of USN. Moreover, our proposed approach also attains high success rate in the detection of vehicle queue even when USN is placed 2 m away from the vehicle queue and it lies in between two vehicles by having the intervehicle distance falls within the range of 30–80 cm.

Keywords Congestion detection · Congestion density · Traffic queue · Vehicle · Ultrasonic sensor node (USN)

A. Mandal · P. Sadhukhan (✉)
School of Mobile Computing & Communication, Jadavpur University, Kolkata 700032, India
e-mail: pampa.sadhukhan@ieee.org

A. Mandal
e-mail: avirup456@gmail.com

F. Gaji
Department of Computer Science & Engineering, Indian Institute of Technology Kharagpur, Kharagpur 721302, India
e-mail: firojgazi123@gmail.com

P. Sharma
Department of Instrumentation and Electronics Engineering, Jadavpur University, Kolkata 700032, India
e-mail: prolaysharma08@gmail.com

S. Kundu et al. (eds.), *Proceedings of the 2nd International Conference on Communication, Devices and Computing*, Lecture Notes in Electrical Engineering 602, https://doi.org/10.1007/978-981-15-0829-5_38

1 Introduction

Traffic congestion is a major concern to the public transportation systems in each developing country since it not only lengthens the travel time of a commuter but also increases the fuel consumption as well as air pollution. Hence, traffic congestion has a significant negative impact on the overall growth of a country. Thus, various researchers have paid attention in designing traffic congestion control or management system to improve the efficiency of the transportation system over the past few years [1–5]. On the other hand, reliable estimation of the traffic congestion density or measurement of the traffic queue length either at the signalized intersections or on any segment of the road is essential to manage the traffic congestion in a successful way. Although vehicle-to-vehicle (V2V) communication-based traffic congestion monitoring techniques can work at any segment of the road, these techniques cannot provide reliable estimation of the congestion density [6, 7]. This is so, because such techniques require sufficient exchange of the messages carrying localized traffic density information among the vehicles, correlating them followed by dissemination of the final estimate, which in turn creates significant communication overhead among the vehicles. Moreover, deployment of 802.11p compliant network interface card within each vehicle is mandatory to ensure its participation in V2V communication [8].

Various other techniques to measure the road traffic queue length in real time using some wireless technologies such as loop detector, infrared (IR) sensor, acoustic sensor, sensor network, RFID, etc. have been proposed in the literature over the past decade. Among these, loop detector-based system [9] that counts the number of vehicles queued on a segment of road using magnetic loop is required to be installed under the road. Thus, deployment of such system by digging up the existing road infrastructure is a time-consuming and costly operation. The use of IR sensor to estimate the traffic congestion density strictly depends on line-of-sight (LoS) communication between the transmitter–receiver pair placed across the road [10]. RFID-based traffic congestion density estimation has been adopted in [4]. However, such RFID-based technique mandates the installation of a RFID tag in each vehicle and would work perfectly for lane-based traffics so that each vehicle would come within the range of RFID reader deployed at the signalized intersection.

On the other hand, acoustic sensors have been applied by the researchers in [11] to estimate the traffic congestion density. However, such estimation can provide erroneous results due to the lack of uniformity in the sound patterns created by a certain number of vehicles standing in the traffic jam. This is so because the sound pattern created by a set of vehicles standing in the traffic jam depends on nature of the driver. The researchers in [12] have proposed a mechanism to detect the road occupancy by a vehicle based on the variation in RF link quality in case the pair of sensors placed across the road is not in LoS communication due to the presence of some vehicle. Unfortunately, such kind of road occupancy detection by a pair of sensors placed across the road would fail for lane-based traffic if the positions of the sensor pair fall in between two vehicles. Furthermore, the characteristics of the RF link vary over time due to the variation in temperature, humidity, etc.

To address the aforementioned issues in the process of real-time traffic queue length estimation, we present a real-time traffic queue length measuring technique using ultrasonic sensor node (USN) that can detect the presence of vehicle queue on the road in a reliable way. In our earlier work on traffic congestion control system for the signalized intersections [5], we have already presented the traffic density estimation approach via the USN-based traffic density monitoring module (TDMM). But no experimental results have been provided there. Therefore, this paper presents the experimental results of the USN-based traffic queue length measuring technique to demonstrate the reliability of the proposed mechanism. Experimental results show that USN-based detection achieves almost 100% success if the vehicle queue is formed within a distance of 2.5 m from the position of USN. Moreover, such detection achieves 90% or more success even when the USN is located in between two vehicles (as shown in Fig. 2) provided the upper limit of the distance between two vehicles, i.e., intervehicle distance is set to 60 centimeter (cm).

This paper is structured as follows. Section 2 presents our proposed traffic queue length measuring technique. Experimental results are given in Sect. 3. Finally, we conclude and present our goal of future research in Sect. 4.

2 Proposed Traffic Queue Length Measuring Approach

Our proposed USN-based road traffic queue length measuring technique is described in this section. A set of USN-based traffic density monitoring modules (TDMMs) are required to be placed at certain predefined distances (like 10, 20, 30 m, etc.) from the signalized intersections along the sides of the incoming road as shown in Fig. 1 to measure the traffic queue length by our proposed approach. The hardware architecture of traffic density monitoring module (TDMM) that comprises a microcontroller, USN, and Wi-Fi module and also the algorithm employed by a set of TDMMs deployed before the signalized intersection to measure the volume of traffic density or traffic queue length is already described in [5]. The set of TDMMs is required to be deployed on both sides of the road (as shown in Fig. 1) if the road is wider than the maximum detection range of the USN which is illustrated in Fig. 2. Otherwise, placement of the set of TDMMs on only one side of the incoming road is sufficient for reliable detection of vehicle queue on the road. Figure 2 not only shows that USN can detect the presence of vehicle on the road within a certain distance (which is defined as its maximum detection range) but also illustrates the ability of the USN to detect the vehicle queue even if it is placed in between two vehicles.

In our proposed technique, the USN is triggered periodically to emit the sound wave and then it waits to receive the echo of this transmitted signal. Upon receiving the echo of some emitted signal, the attached microcontroller, at first, computes the distance traversed by the signal before it is obstructed and then the presence of some standing vehicle (which acts as an obstacle) on the road is detected if the distance of

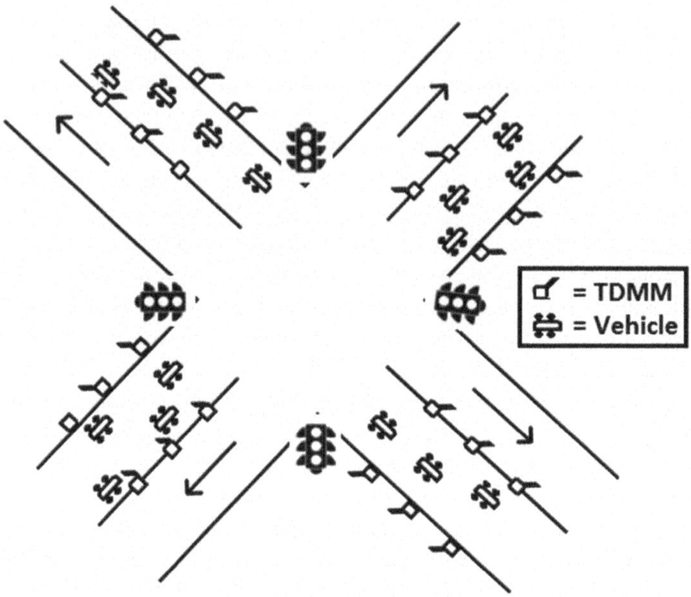

Fig. 1 Placement of traffic density monitoring module (TDMM) for measuring traffic queue length at signalized intersection

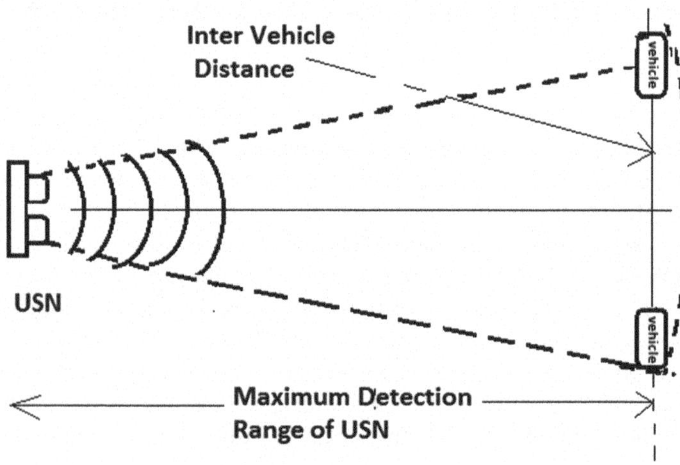

Fig. 2 Illustration of maximum detection range and degree of detection of the USN

the obstacle (denoted as d_o) from the position of USN is less than the width of the road (denoted as d_w). The algorithm adopted by our proposed approach to detect the presence of vehicle queue on the road is presented below.

Algorithm 1: Algorithm for detecting vehicle queue on road

1. Trigger USN to emit sound signal periodically and record transmission time of the signal (T_s).
2. Wait to receive the echo of transmitted signal.
   (a) If signal is echoed back, record reception time (T_r).
   (b) Compute the distance traversed by signal before it is obstructed (d_o) by applying following equation:

$$d_o = C \times \frac{(T_r - T_s)}{2}, \tag{1}$$

   where C is propagation speed of the sound wave.
   (c) Vehicle queue is detected if following condition is satisfied:

$$d_o < d_w,$$

   where d_w is the width of road.

3 Experimental Results and Discussions

This section presents the experimental results to demonstrate the reliability of our proposed USN-based traffic queue length measuring technique. The hardware prototype of the traffic density monitoring module using which this experimentation has been done, as shown in Fig. 3.

The reliability of our proposed approach is demonstrated in terms of the percentage of successful detection of the vehicle on road with respect to different values of the distance between the USN and the vehicle as shown in Fig. 4. On the other hand, it is also demonstrated in terms of the percentage of successful detection of the vehicle queue on the road with respect to the different values of the intervehicle distance as shown in Fig. 5 when the USN is placed 2 m away from the vehicle queue and it lies in between two vehicles as shown in Fig. 2. The graphical results given in Fig. 4 show that the percentage of successful detection remains almost 100% if the vehicle is standing within 2.5 m from the position of the USN and it suddenly drops at 77% when the distance increases to 3 m. Thus, it can be concluded from the graphical results given in Fig. 4 that maximum detection range of the USN is 3 m.

Figure 5, on the other hand, shows that the percentage of successful detection of the vehicle queue on road remains high (80–100%) with respect to the different values of the intervehicle distance in the range of 30–60 cm. However, this detection rate falls to 74% and 67% when the intervehicle distance is increased to 70 cm and 80 cm, respectively, as shown in Fig. 5. In the developing countries like India, the

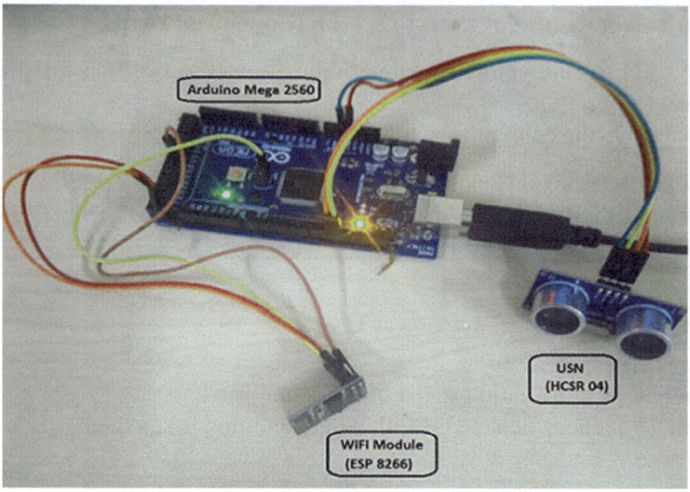

Fig. 3 Hardware prototype of traffic density monitoring module

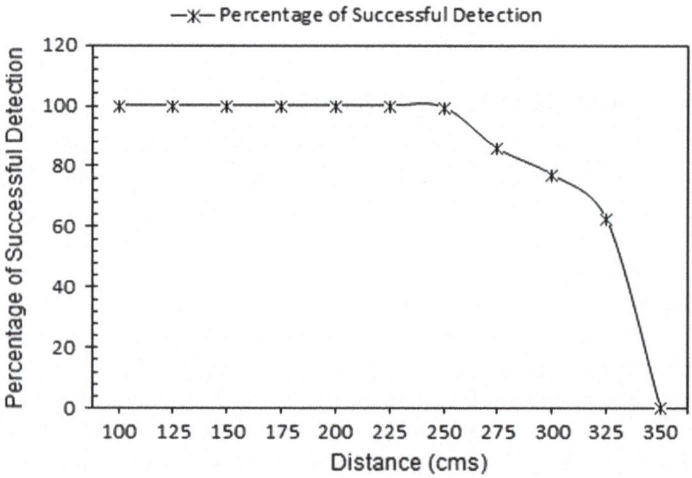

Fig. 4 Percentage of successful detection versus distance between USN and vehicle

vehicles, while standing within a traffic queue created at the signalized intersections, usually maintain a distance in the range of 30–60 cm between them. Thus, achieving 80% or more detection success at an intervehicle distance within the range of 30–60 cm indicates very high reliability of our proposed USN-based approach.

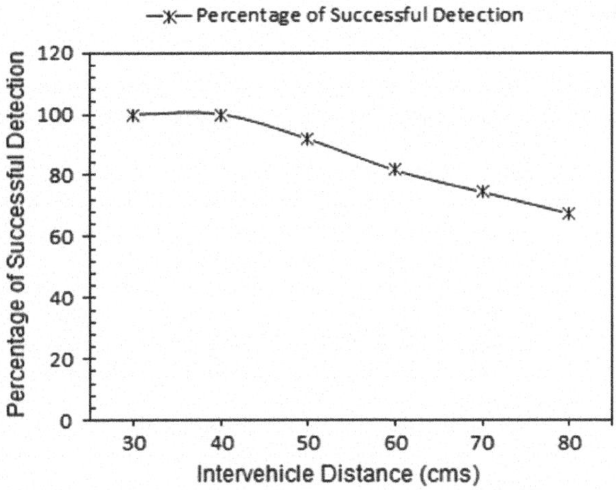

Fig. 5 Percentage of successful detection versus intervehicle distance

4 Conclusion and Future Works

A highly reliable approach based on USN to measure road traffic queue length in real time is presented in this paper. The reliability of the proposed technique is demonstrated in terms of the percentage of successful detection of the vehicle queue on the road with respect to varying distances between the USN and the vehicle as well as the different values of intervehicle distance in case USN is located in between two vehicles on the road. The experimental results demonstrate that our proposed approach achieves almost 100% success in vehicle detection if the distance between the USN and the vehicle queue formed on the road is less than 2.5 m, whereas 80% or even more success in detecting the vehicle queue is obtained even when the USN is placed 2 m away from the vehicle queue and it lies in between two vehicles by having the intervehicle distance falls within the range of 30–60 cm. Moreover, USN-based embedded system (TDMM) used to measure the traffic queue length is also cost-effective.

In our present work, however, we have experimented with the static vehicle. Experimentation with the slowly moving vehicle queue passing through some signalized intersection is required to be done to verify the reliability and efficiency of our proposed approach. We intend to do this in near future.

Acknowledgments The authors gratefully acknowledge the facilities and support provided by the Director and all other staff members of the School of Mobile Computing and Communication, Jadavpur University, a Centre of Excellence set up under the "University with Potential for Excellence" Scheme of the UGC.

References

1. Tubaishat, M., Qi, Q., Shang, Y., Shi, H.: Wireless sensor-based traffic light control. In: IEEE Consumer Communications and Networking Conference (CCNC'08), Las Vegas, NV, pp. 702–706, Jan 2008
2. Mandal, K., Sen, A., Chakraborty, A., Roy, S., Batabyal, S., Bandyopadhyay, S.: Road traffic congestion monitoring and measurement using active RFID and GSM technology. In: Proceedings of 2011 14th International IEEE Conference on Intelligent Transportation Systems (ITSC), Washington, D.C., pp. 1375–1379 (2011)
3. Sarkar, D., et al.: A smart traffic congestion control method. Int. J. Electron. Commun. Technol. **6**(3) (2015)
4. Sundar, R., Hebbar, S., Golla, V.: Implementing intelligent traffic control system for congestion control, ambulance clearance, and stolen vehicle detection. IEEE Sens. J. **15**(2), 1109–1113 (2015)
5. Sadhukhan, P., Gazi, F. :An IoT based intelligent traffic congestion control system for road crossings. In: Proceedings of 2018 International Conference on Communication, Computing & Internet of Things (IC3 IoT 2018), Chennai, India, pp. 371–375, Feb 2018. https://doi.org/10.1109/IC3IoT.2018.8668131
6. Bouza, R., Gozalvez, J.: Traffic congestion detection in large-scale scenarios using vehicle-to-vehicle communications. J. Netw. Comput. Appl. **36**, 1295–1307 (2013)
7. Yuan, Q., et al.: A traffic congestion detection and information dissemination scheme for urban expressways using vehicular networks. Transp. Res. Part C **47**, 114–127 (2014). Elsevier
8. Bhoi, S.K., Khilar, P.M.: Vehicular communication: a survey. IET Netw. **3**(3), 204–217 (2014)
9. Coifman, B., Cassidy, M.: Vehicle reidentification and travel time measurement on congested freeways. Transp. Res. Part Policy Pract. **36**(10), 899–917 (2002)
10. Promila Sinhmar, A.: Intelligent traffic light and density control using ir sensors and micro-controller. Int. J. Adv. Technol. Eng. Res. **2**(2), 30–35 (2012)
11. Sen, R., Siriah, P., Raman, B. :Roadsoundsense: acoustic sensing based road congestion monitoring in developing regions. In: Proceedings of 2011 8th Annual IEEE Communications Society Conference on Sensor, Mesh and Ad Hoc Communications and Networks, Salt Lake City, UT, pp. 125–133 (2011)
12. Sen, R., et al.: Kyun queue: a sensor network system to monitor road traffic queues. In: Proceedings of the 10th ACM Conference on Embedded Network Sensor Systems, Toronto, ON, Canada, pp. 127–140, Nov 2012

Biometric-Based Three-Tier Microservice Architecture for Mitigating the Fraudulent Behaviour

S. Prayla Shyry

Abstract Though monolithic systems have tremendous advantages, it suffers from its own drawbacks like independency and debugging time. So the paradigm has shifted to build the applications with microservice architecture. The microservice is a part of the service-oriented architecture that decomposes the monolithic architecture into different services. All the divided services perform heterogeneous activity, and they are interfaced with proper application programming interface (API). Microservice is highly resilient; hence, when one service fails, it does not have an impact on other services. Traditional systems use monolithic architecture which fails in their design pattern and dependency. In this paper, to avoid duplicity in passport, a novel biometric-based monolithic architecture is developed for authentication system. All the physiological and behavioural biometrics of a person are stored in the repository and compared during the authentication phase. All these services are created as microservice and interfaced with proper APIs which improves parallel processing.

Keywords Application programming interface · Biometric · Microservice

1 Introduction

Large Internet companies like Amazon, Netflix and Gilt need their applications to be upgraded frequently since their competition may hurt their business if they have no innovations. In microservice architecture, a large computer application is decomposed as heterogeneous autonomous services. Each and every microservice adopts a well-suited data model for its need. Microservices interact among each other by well-defined interfaces (API). API gateway is the single entry point for the system. The requests raised by the clients pass through the API gateway that in turn ping the requested microservice. API gateway provides all the clients with specific API that is used to reduce the interaction time between the client and application. Unlike

S. Prayla Shyry (✉)
School of Computing, Department of Computer Science and Engineering, Sathyabama Institute of Science and Technology, Chennai, India
e-mail: suja200165@gmail.com

S. Kundu et al. (eds.), *Proceedings of the 2nd International Conference on Communication, Devices and Computing*, Lecture Notes in Electrical Engineering 602, https://doi.org/10.1007/978-981-15-0829-5_39

microservice architecture, in monolithic architecture, when one service needs to be updated, the other services get disturbed which takes a lot of time. Also, if one service in the application fails, the whole set of service goes down. Each microservice is developed by different technologies and tested.

2 Related Work

Villamizar et al. [1] evaluated monolithic and microservices architecture pattern to display web applications in the cloud. They designed an application to generate and query the payment plans for loans delivered by an institution to their customers and have this application with both monolithic and microservice architectures. They framed two services, generate payment plan and repay existing plan. First service takes 300 ms and the latter 30 ms. Same language is used in all the services of monolithic application. But in microservice application, different technology stacks are used. Each of these services exposes its functionality to a private microservice through API. The request is forwarded from API gateway to end users. They evaluated the results using a Jmeter and compared the average response time. Results showed that microservice architecture has the ability to address a very large complex application as a set of heterogeneous small services.

Gan and Delimitrou [2] have tried to imply the impact of microservices on system bottlenecks and datacenter server design. They have implemented ten microservices on movie renting and video streaming. They have calculated per-microservice breakdown for the movie streaming service and analysed the latency. Their results showed that despite the increase in network traffic, the individual services experience high scalability, elasticity, modularity and abstraction.

3 Methodology

Fraudulent behaviours can be mitigated by biometric-based microservice architecture. In this paper, passport authentication system has two modes, enrolment mode and verification mode. In the enrolment mode, client registers the credentials and the same is stored in a repository. In the verification mode, each time the client checks in for authentication and presents biometric specifications that are matched with the registered credentials in the repository. If a match is found, then the client is authenticated or else further legal procedures will be taken up (Fig. 1).

Credentials are entered during passport application process. Apart from the Aadhar and PAN of a person, the physiological and behavioural biometrics are scanned by a video camera without the cooperation or awareness of the applicant. The data is then processed and stored in the repository or database. After the application process, when a client needs to check-in, the real and present biometrics are scanned

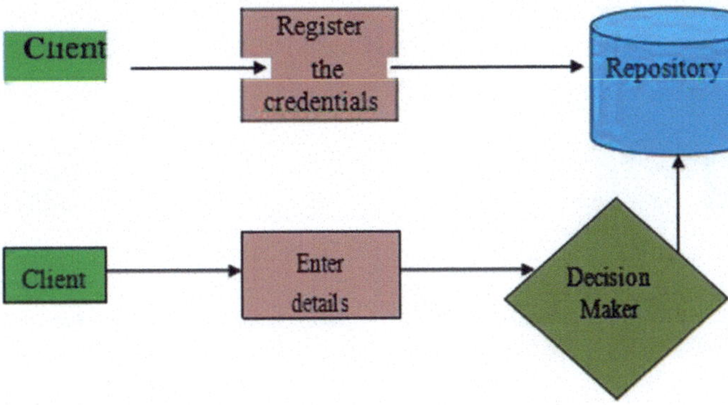

Fig. 1 Overview of authentication system

and matched with the database. The credentials to be registered are gait gestures, keyboard dynamics, mouse dynamics and fingerprint.

1. Gait gestures

Both the physical (Size of the feet and body weight) and behavioural biometrics (human walk) are captured and recorded by a video camera. Segmentation and motion detection methods are used to detect the walking persons, and the personality is segmented from the background. Feature extraction methodologies are used to extract robust and discriminative feature, and a mathematical model is created. The similarity measures are evaluated by comparing gait features and the stored features in gait database.

2. Keystroke dynamics

A person's typing pattern (number of keys used within a particular time and frequency of keys used), the speed of typing on a keyboard and the rhythm are analysed by this biometric. Literature show clear evidence that the elapsed time between (key up and down) releasing a key and pressing the following key or intercharacter timing can give 99% accurate identification of the person who is typing. Individuals are differentiated in the way they involve in the keyboard, evaluating these parameters.

3. Mouse dynamics

Based on the mouse movement from one point to another, the features of the curve are studied. The curve features follow a probability distribution and each curve feature differs with respect to the individual. Also, elapsed time between the mouse key down and up is calculated which also differs from person to person.

4. Fingerprint

In today's world, fingerprint is used in cyber forensics to detect the crimes, and it is the process of comparing two instances of friction ridge skin. No two fingers can have the same features and the fingerprint verification methodologies are used to determine

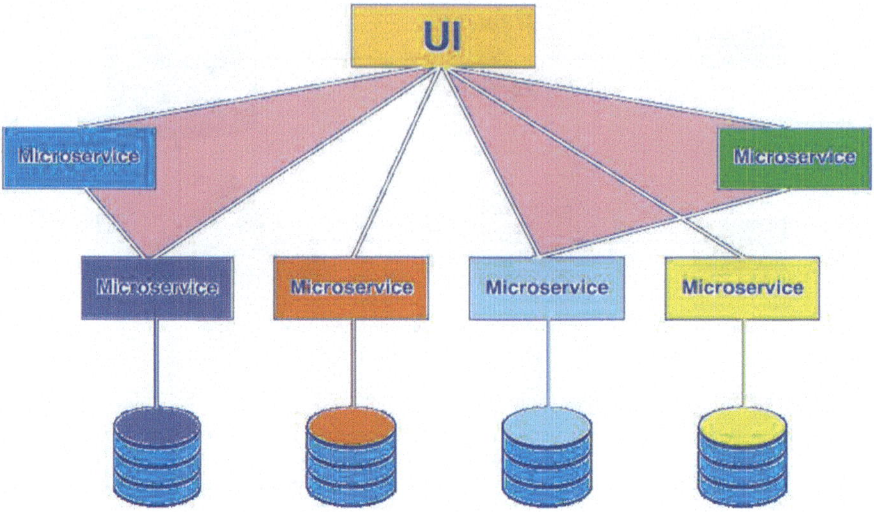

Fig. 2 Microservice architecture

the impressions and decide whether two friction ridge impressions are likely to have originated from the same finger. For all the above, to speed up the process and to attain accuracy, all these services are created as independent heterogeneous microservices. The average response time of an individual to access and finish the check-in procedure is reduced, compared to monolithic architecture (Figs. 2 and 3).

The details of the individual are captured through video camera microservice and using the feature extraction method, the features are extracted and stored in the repository service. During verification process, the present details are matched with the database. The match service acts as a decision-maker and will compare the detail. If match found, then check-in service will run automatically or else legal service will be invoked. This microservice can be developed by Spring Boot developed modern Java-based enterprise applications, and it manages the rest APIs (Figs. 4 and 5).

4 Conclusion

The aggregate throughput of the organisation is improved since microservice fosters the accomplishment of the project. Since all the services are independent, technology or software updation of any single microservice will not affect the other services. Also, selecting the appropriate and optimal technologies for a specific service is quite easy because of its loosely coupled fashion. Passport fraud is stealing the identity of a deceased person to use their passport to facilitate illegal crimes. Every day, world is facing many duplicate or fake passports of individuals trying to indulge in terrorism, bankruptcy and many more illegal issues. This microservice architecture will provide

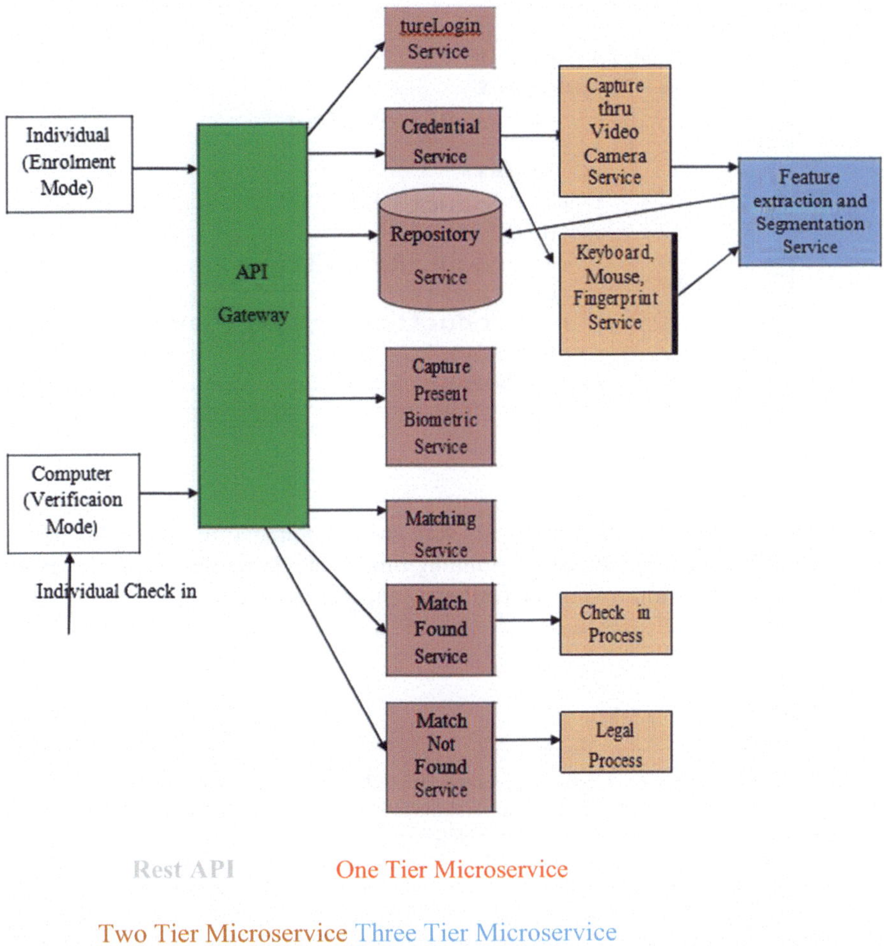

Fig. 3 Microservice architecture for passport authentication system

Fig. 4 Sample code for spring framework

```
<dependency>

<groupId>org.springframework.boot</groupID>

</dependency>
```

```
com
+- tutorialspoint
     +- myproject
          +- Application.java
          |
          +- model
          | +- Product.java
          +- dao
          | +- ProductRepository.java
          +- controller
          | +- ProductController.java
          +- service
          | +- ProductService.java
```

Fig. 5 General layout of Spring Boot application

quick response time and find the frauds very easily compared to monolithic architecture. Waiting time is reduced and running time is increased for the microservice architecture, thereby improving the resiliency and average throughput.

References

1. Villamizar, M., Garcés, O., Castro, H., Verano, M., Salamanca, L., Casallas, R., Gil, S.: Evaluating the monolithic and the microservice architecture pattern to deploy web applications in the cloud. 583–590. IEEE (2015). 978-1-4673-9464-2/15/
2. Gan, Y., Delimitrou, C.: The architectural implications of cloud micro-services. 1–4 (2018)
3. Villamizar, M., Garces, O., Ochoa, L., Castro, H., Salamanca, L., Verano, M., Casallas, R., Gil, S., Valencia, C., Zambrano, A., Lang, M.: Infrastructure cost comparison of running web applications in the cloud using AWS lambda and monolithic and microservice architectures. In: 2016 16th IEEE/ACM International Symposium on Cluster, Cloud, and Grid Computing, pp. 179–182
4. Nginx: Adopting microservices at netflix: lessons for architectural design (2015)
5. Buyya, R.: Cloud computing: the next revolution in information technology. In: 2010 1st International Conference on in Parallel Distributed and Grid Computing (PDGC), pp. 2–3, Oct 2010
6. Lewis, J., Fowler, M.: Microservices (2014)
7. Decomposing twitter: adventures in service-oriented architecture. In: 2014 Microservices Workshop: Why, What, and How to Get There. http://www.slideshare.net/adriancockcroft/microservices-workshop-craft-conference
8. Courjon, D., Sarayeddine, K., Spajer, M.: Scanning tunneling optical microscopy. Opt. Commun. **71**(1-2), 23–28 (1989)

Anticipation of Actuation Properties of IPMC for Soft Robotic Gripper

Srijan Bhattacharya, Pranjal Tiwary, Adil Shayaque, Bikash Bepari and Subhasis Bhaumik

Abstract This paper primarily focuses on mathematical models for anticipation of actuation force and displacement for IPMC actuator which was utilized to actuate a PDMS-based microgripper. To achieve higher actuation force and displacement, higher voltage (4.5 V) and lower frequency (0.001 Hz) values are to be ensured. Moreover, for higher actuation force KCl dopant and for actuation displacement DW work, doping solutions are to be used. In order to select proper material for compliant gripper, a systematic approach taken by adopting MCDM techniques like TOPSIS, MOORA, COPRAS, ELECTRE-II and Polydimethylsiloxane (PDMS) was found to be the best material. ANSYS-14 software was utilized to adjudge the compatibility of actuation force exerted by IPMC and the functionality of PDMS gripper prior to fabrication. A microgripper has been fabricated which is capable to grasp small objects of 1.247 gms.

Keywords Regression · IPMC · Material selection · PDMS gripper

1 Introduction

For the past three decades, the Electroactive Polymers (EAPs) are considered to be used as an actuator by the scientific community due to its compliance, less weight-to-volume ratio, ease of machining and the response under electrical stimuli [1]. The EPAs are widely used as replacement to the conventional actuators and sensors which when activated consumes less energy [2]. EAPs are categorized into two classes: (a) electronic EAPs and (b) ionic EAPs, which depend on their activation characteristics. Electronic EAP calls for high activation voltage and subsequently

S. Bhattacharya
RCC Institute of Information Technology, Kolkata 700015, West Bengal, India

P. Tiwary · A. Shayaque · B. Bepari (✉)
Haldia Institute of Technology, Haldia 721657, West Bengal, India
e-mail: bikashbepari@gmail.com

S. Bhaumik
IIEST, Shibpur, Howrah 711103, West Bengal, India

S. Kundu et al. (eds.), *Proceedings of the 2nd International Conference on Communication, Devices and Computing*, Lecture Notes in Electrical Engineering 602, https://doi.org/10.1007/978-981-15-0829-5_40

reveals higher mechanical actuation energy. Contrastingly, the functionality of ionic EAPs depends on diffusion of ions between two electrodes [3]. Moreover, ionic EPAs show better electromechanical strain when subjected to electrical incentive and resembles the behaviour of biological muscle which can provide articulating power and displacement and thus coined as synthetic muscle [4].

Ionic Polymer Metal Composite (IPMC) belongs to ionic EAP is a three-member sandwich type of structure with an ionomer coated with electrode material. The activation of an IPMC strip depends upon the functional transverse voltage along the thickness and is mainly subjected to ion redistribution [5]. Figure 1 depicts the actuation mechanism of IPMC. When the IPMC actuators are activated, they do provide actuation power to flex and the tip gets deflected [1] and these two attributes are more advisable from functionality perspective [6]. Literature survey reveals that increased functional voltage augments both the tip displacements as intrinsic force. Moreover, the span and area moment of inertia of the strip have enormous consequence on the activating strength and tip deflection. The actuating strength is directly proportional to width, exponential of the thickness and reciprocal to the length as revealed by [7]. In addition, modulus of elasticity of the IPMC is dependent on the thickness and applied frequency as found by previous researchers [1, 8], respectively. Bhattacharya et al. [9] reported a systematic description during design of anthropomorphic microgripper for varied beam length and width of IPMC. Nemat-Nasser et al. [10] addressed the working of IPMCs pertaining to actuation strength (force) and tip deflection when doped with Li^+, K^+ and Na^+ solutions.

Figure 2 represents the inevitable and prevailing elements of IPMC-assisted microgripper design strategies. Herein, IPMC being one of the mostly used EAPs has been considered to act either individually or in conjugation with PDMS-based compliant gripper. Eventually, the reason for selecting PDMS as a compliant robotic

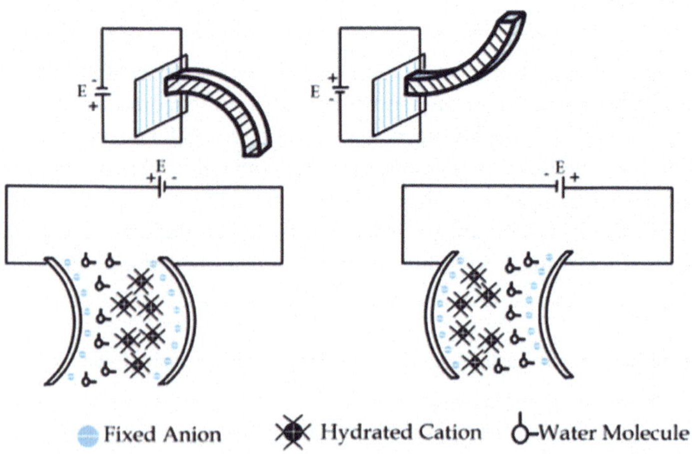

Fig. 1 IPMC actuation mechanisms

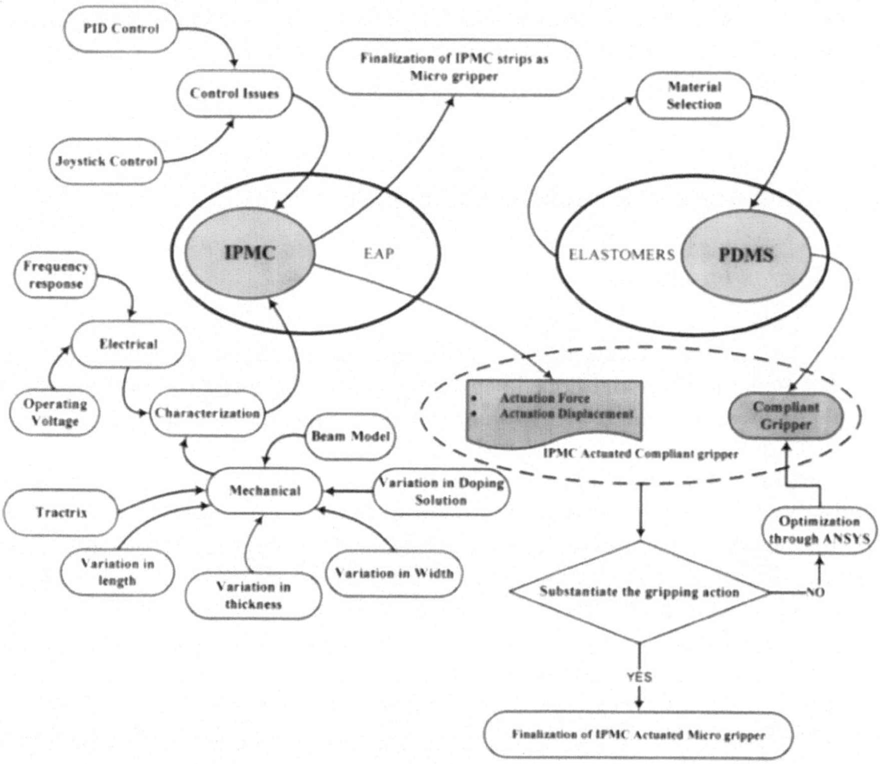

Fig. 2 Issues and aspects of IPMC-assisted microgripper design

material comes into the forefront which must be substantiated by methodological approach. Characterization of IPMC is one of the prima-facie concerns which corroborate both mechanical and electrical aspects. Mechanical aspect includes tractrix (locating the tip position with certainty), equivalent beam model, variation in physical attributes (length, breadth and width) and obviously anticipating the behaviour in doping solution. The electrical counterpart of characterization includes the response of IPMC when it is subjected to different frequency and voltage levels. The control issue entails the mathematical modelling so as to fetch the upshots by tweaking the different gains pertinent to PID control. Nevertheless, if IPMC is harnessed with PDMS microgripper, compatibility should be adjudged so that the actuating force of IPMC should be sufficient enough to actuate the PDMS gripper. In this regard, optimization of the shape and size may be accomplished to enable IPMC to serve as an actuator. The objectives of the present investigation are (a) to characterize and anticipate the behaviour of IPMC as an actuator under process parameters like actuating voltage, frequency and doping solution to act individually or harnessed with PDMS; (b) to correlate the responses (actuating force and actuating displacement) of IPMC with the process parameters through regression analysis; (c) to suitably adjudge the selection of compliant material (PDMS) in commensuration with the responses of IPMC; and (d) to optimize the compliant PDMS gripper through the actuating force of IPMC.

2 Anticipation of Actuation Behaviour

2.1 Experimentation

For the present study, three strips of IPMC were taken with a dimension of 30 × 5 × 0.2 (units in mm) and doped separately within (a) distilled water, (b) NaCl solution with 1.5 (N) and (c) KCl solutions with 1.5 (N) for at least six hours prior to the carrying out of tests. In order to measure the displacement, laser sensor was used. Digital laboratory balance was chosen to a resolution of 0.001 gmf. DC dual changeable power supply was used to regulate the voltage ranging from 0 to 5 V. Tektronix-made single-channel function generator was engaged to provide triggered pulses to IPMC. All the factors (dopants, frequencies and voltages) were laid at three levels. The responses were considered as (i) free end force and (ii) free end displacement of the IPMC cantilever [6]. As all the three governing factors have three levels, therefore a full factorial design of experiments was laid down. Experiments were conducted and employing Grey Relational Analysis (GRA) the suitability was analysed [6]. Table 1 shows the responses in terms of Grey Relational Grade (GRG) which is nothing but a collective way of representing the goodness or badness of an experiment within a specified layout. The recommended mean value of the voltage and frequency was 3 V and 0.01 Hz, respectively, and upper and lower deviations were chosen accordingly.

Table 1 shows that experiment-24 is having higher value of GRG followed by experiment-3. It is noteworthy that experiment-24 reveals higher value of displacement, whereas experiment-3 reveals high value of actuating force and in both cases the voltage (4.5 V) and frequency (0.001 Hz) values are the same. Therefore, depending upon the doping solution the IPMC strips are showing different characteristics for the prediction of activating parameters, and simultaneous expectation in terms of both force and displacement is not possible to obtain because of this trade-off situation. Moreover, a systematic mathematical guideline should be established so that one can tune the factors so that the desired output in terms of actuation force and/or actuation displacement can be ascertained. In this regard, one can obviously go for statistical regression analysis for representing the desired responses in terms of parameters through mathematical equations. Bhattacharya et al. [6] did not give any proper mathematical guideline for the prediction of actuating force and actuating displacement separately for KCL and DW. Keeping this lacuna in consideration, the authors extended an attempt to proceed forward for the regression analysis to obtain mathematical equations for prediction of the actuating force and displacement.

Table 1 Experimental layout and the responses [6]

Expt. No.	Factors			Original sequence		GRG
	Doping solution	Frequency (Hz)	Voltage (V)	Force (mN)	Displacement (mm)	
1	**KCl 1.5 (N)**	0.1	1.5	1.96	0.36	0.3967
2		0.01	3	3.04	1.08	0.4687
3		**0.001**	**4.5**	**4.80**	1.41	**0.6892**
4		0.1	1.5	0.50	0.81	0.3549
5		0.01	3	1.13	1.98	0.3971
6		0.001	4.5	1.31	2.55	0.4174
7		0.1	1.5	0.29	0.54	0.3449
8		0.01	3	0.69	1.89	0.3817
9		0.001	4.5	1.18	3.24	0.4327
10	NaCl 1.5 (N)	0.1	1.5	0.10	0.12	0.3333
11		0.01	3	0.18	0.29	0.3380
12		0.001	4.5	0.22	0.99	0.3511
13		0.1	1.5	0.46	0.54	0.3490
14		0.01	3	0.76	2.56	0.4003
15		0.001	4.5	0.99	4.49	0.4739
16		0.1	1.5	0.69	0.56	0.3557
17		0.01	3	0.93	2.98	0.4169
18		0.001	4.5	1.18	4.32	0.4725
19	**Distilled water**	0.1	1.5	0.30	0.96	0.3526
20		0.01	3	0.48	2.70	0.3963
21		0.001	4.5	0.99	4.80	0.4887
22		0.1	1.5	0.44	1.32	0.3630
23		0.01	3	0.88	1.89	0.3874
24		**0.001**	**4.5**	1.77	**7.20**	**0.7189**
25		0.1	1.5	0.53	0.90	0.3574
26		0.01	3	1.13	2.70	0.4152
27		0.001	4.5	2.27	6.00	0.6142

2.2 *Regression Analysis*

In this section, the authors have synthesized the empirical mathematical model through regression analysis. The objective was to exhibit the responses (free end strength and deflection) in terms of independent variables (Voltage and frequency) for the two dopants, namely, KCl and DW. Regression equations need to be made separately for the two doping solutions by employing Eqs. 1 and 2.

Table 2 Regression study for strength with KCl dopant

Forecaster	Coef	SE Coef	T	P
Invariable	−0.7907	0.4472	−1.77	0.127
v	0.5000	0.1317	3.80	0.009
f	25.576	3.609	7.09	0.000

Notes S = 0.4839, R-Sq = 91.5%, R-Sq (adj) = 88.7%

Table 3 Analysis of variance for strength with KCl dopant

Resource	DF	SS	MS	F	P
Regression	2	15.1372	7.5686	32.32	0.001
Residual error	6	1.4050	0.2342		
Total	8	16.5422			

$$F_a = g(f, v) + \varepsilon_{Fa} \tag{1}$$

$$D_a = g(f, v) + \varepsilon_{Da} \tag{2}$$

2.2.1 Regression for Actuating Force

Statistical application software MINITAB (V-13.1) was used to obtain the correlation for strength (force) for KCl dopant in terms of potential difference and frequency, which is as follows:

$$\mathrm{F}_{\mathrm{a_{KCl}}} = -0.791 + 0.500\mathrm{V} + 25.6\mathrm{f} \tag{3}$$

The upshots of regression and ANOVA for free end strength of actuation for KCl have been provided in Tables 2 and 3, respectively.

3-D wire-mesh plot for force when the IPMC was doped in KCl solution has been depicted in Fig. 3, and the corresponding contour plot has been shown in Fig. 4 which shows voltage and frequency combination to predict desired force.

2.2.2 Regression for Actuating Displacement

The regression equation for tip deflection as a function of potential difference and frequency for distilled water doping has been shown in Eq. 4.

$$D_{a_{DW}} = -0.917 + 1.33\mathrm{V} - 5.96\mathrm{f} \tag{4}$$

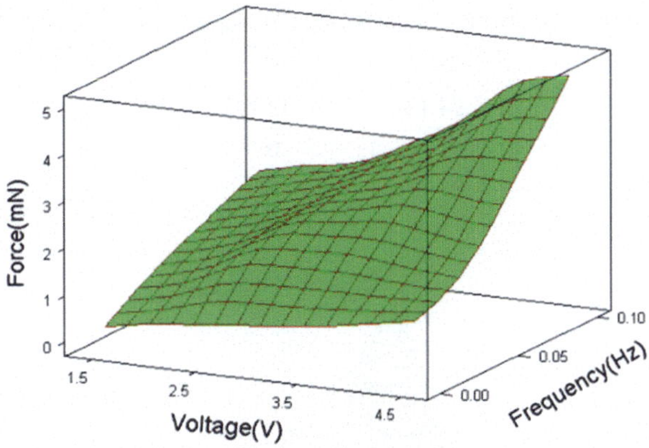

Fig. 3 Wire mesh plot of force for KCl doping solution

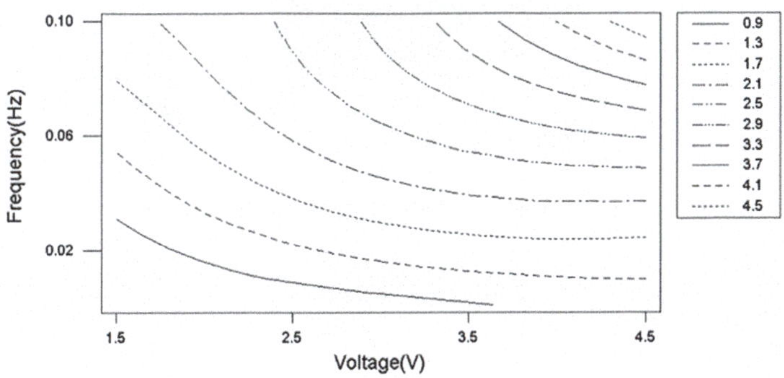

Fig. 4 Contour plot of force for KCl doping solution

The outcome of regression study and ANOVA for displacement is given Tables 4 and 5, respectively. The regression coefficient seem satisfactory. 3-D surface plot for displacement when the IPMC was doped in distilled water and the corresponding contour plot have been depicted in Figs. 5 and 6, respectively.

Table 4 Regression analysis for displacement during DW dopant

Forecaster	Coef	SE Coef	T	P
Invariable	−0.9174	0.6972	−1.32	0.236
v	1.3273	0.2053	6.46	0.001
f	−5.963	5.626	−1.06	0.330

Notes S = 0.7544, R-Sq = 87.7%, R-Sq (adj) = 83.6%

Table 5 Study of variance for displacement during DW dopant

Resource	DF	SS	MS	F	P
Regression	2	24.424	12.212	21.46	0.002
Lasting error	6	3.415	0.569		
Sum	8	27.839			

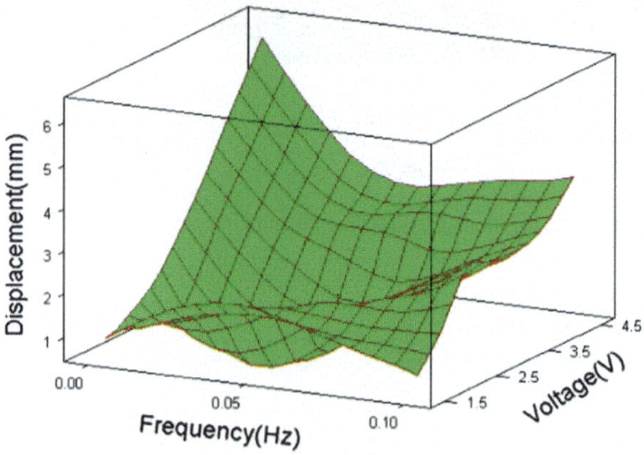

Fig. 5 Wire mesh plot of displacement for IPMC doped in distilled water

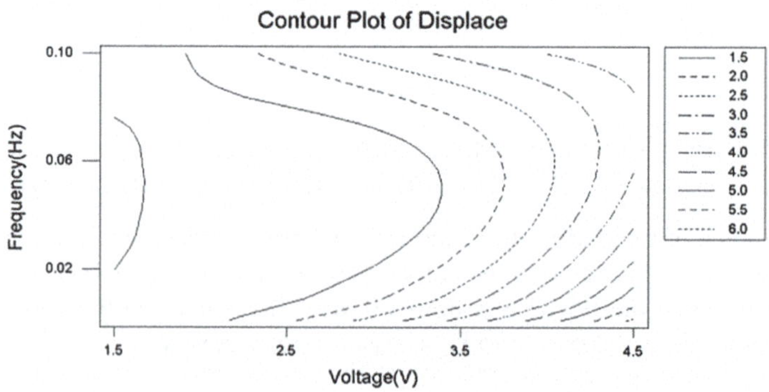

Fig. 6 Contour plot of displacement for IPMC doped in distilled water

3 Selection of Material for Compliant Robotic Gripper

Compliant mechanism is a variety of elastic mechanism that achieves its moveability from relative deformation of its members. When the load is applied, flexible links deform and the joints deflect consequently. Moreover, the kinematic performance is essential for the propagation of the force from one port to the other. Hence, it is evident that, from the plethora of available compliant materials (elastomers), choosing or selecting a particular compliant material subjected to a set of requirements seems to be a daunting task.

Over the past couple of decades, multiple criteria decision-making (MCDM) has pioneered itself as an imperative as well as inevitable tool for decision support systems which fundamentally necessitates multiple decision alternatives, to adjudge their suitability subjected to a finite set of attributes or criteria [11].

Herein, Multicriteria Decision-Making (MCDM) models like TOPSIS [12], MOORA [13], COPRAS [14] and ELECTRE-II [15] were adopted to find out the relative suitability of compliant materials like Ethylene Propylene Diene Monomer (EPDM), Ethylene Vinyl Acetate (EVA), Ethylene Propylene Monomer (EPM), Polydimethylsiloxane (PDMS), Polyurethane (PU), Ethylene Propylene Terpolymer (EPT) and Polyvinylidene Fluoride (PDVF) based on adjudging criteria like hardness, density, elongation at break, tensile stress and cost. The competing material with their criteria-wise attributes has been shown in Table 6.

After fetching the decision matrix and solving the same through the abovementioned four techniques, the comparison of the competing materials has been illustrated in Fig. 7. The figure depicts that PDMS is the most preferred out of the seven competing materials.

Table 6 Decision matrix for selection of compliant materials

Material	Hardness (Shore)	Tensile strength (MPa)	Elongation at break (%)	Density (gm/cc)	Cost ($/kg)
EPDM	65	13.5	400	1.01	1.585
EPM	62.5	14	350	0.86	1.629
EVA	80	19	550	0.945	1.275
PDMS	70	13	65	0.965	0.25
PT	62.5	38.5	575	1.25	1.125
EPT	60	12.5	545	0.85	2.69
PDVF	64.5	43	200	1.75	0.2025

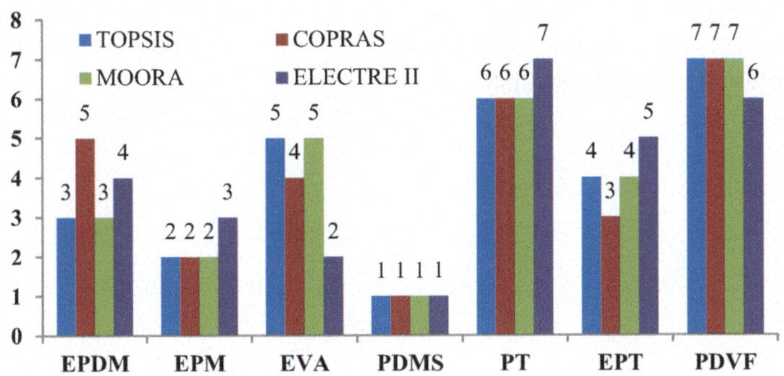

Fig. 7 Comparison of rank orders by different methods

4 Design and Optimization of Compliant Robotic Gripper

Inspired from the previous design of microgripper by Maheshwari et al. [16], the authors fabricated a miniature compliant gripper actuated by IPMC actuators and simulation in ANSYS for gradual movement has been depicted in Fig. 8a–d. Figure 9 shows that the compliant gripper is getting the actuation force from the IPMC strips at the input port and as the gripper is having two inverted four-bar linkages with pivots turn the jaws inward. It has been observed that the compliant gripper is capable to hold an object of 1.247 gms with confidence. This thickness of the PDMS gripper has been finalized as 1 mm through FEM analysis with the aid of ANSYS.

5 Conclusion

The present investigation was primarily aimed to (a) characterize the IPMC for action strength and deflection, (b) characterize tip strength and deflection for actuation in relation with process variable through regression, (c) select acquiescent material for microgripper and (d) optimize the soft PDMS gripper through ANSYS. In congruence with the outcome of the experimentations and subsequent study, the following conclusions are prepared.

i. The mathematical models for anticipation of actuation force as well as actuation displacement have been established and the R-square values ranging in-between 87 and 91% which is satisfactory.
ii. To achieve higher actuation force and actuation displacement higher voltage (4.5 V) and lower frequency (0.001 Hz), values are to be confirmed. But, for higher actuation force KCl dopant and for actuation displacement DW work the best, respectively.
iii. For selection of compliant gripper material, MCDM techniques like TOPSIS, MOORA, COPRAS, and ELECTRE-II were adopted and polydimethylsiloxane (PDMS) was found to be the best material and a microgripper was fabricated which is capable to grasp small object of 1.247gms. Prior to fabrication, ANSYS was used to check the compatibility of actuation force exerted by IPMC and the deformation of PDMS gripper.

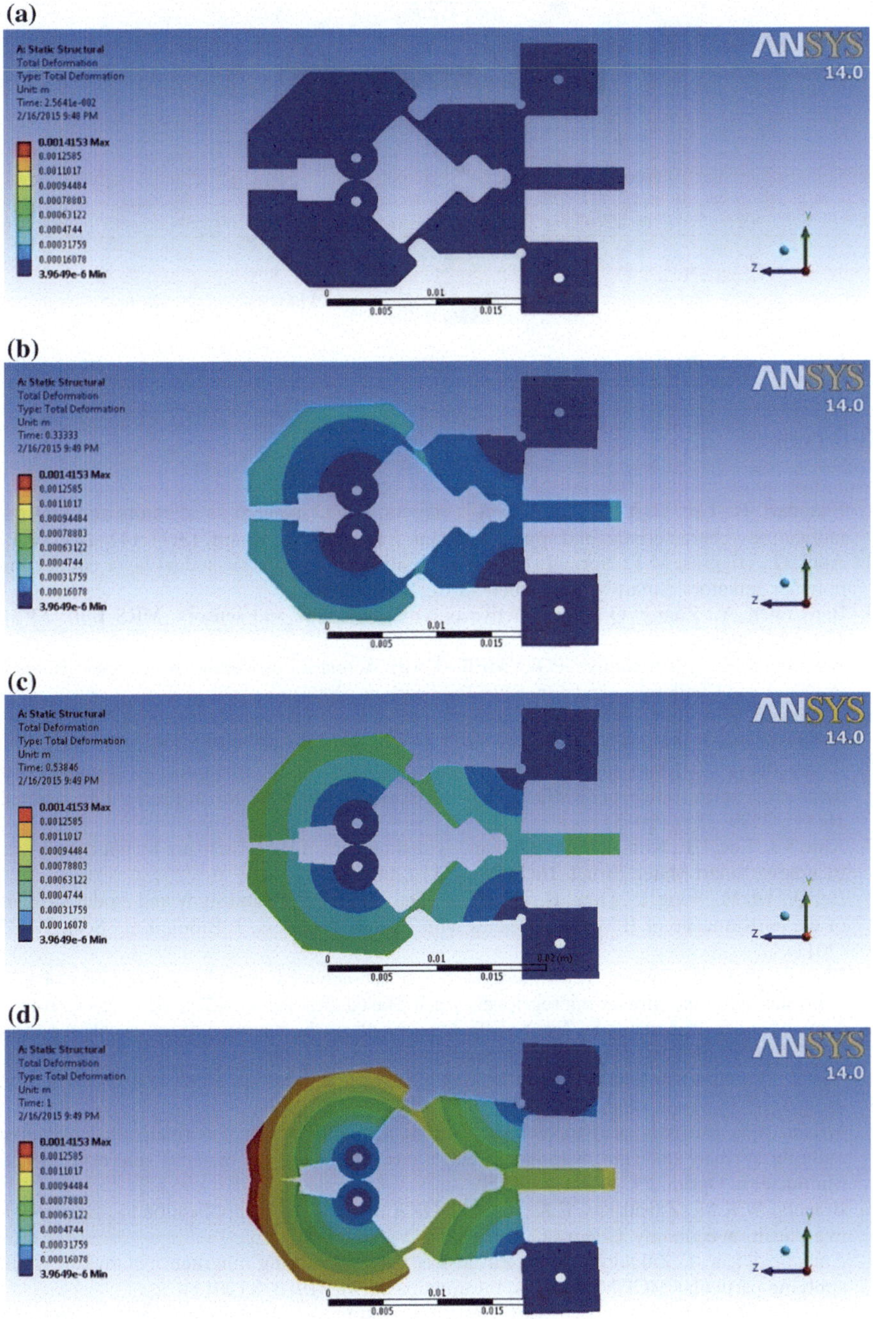

Fig. 8 Working mechanism of the compliant robotic gripper

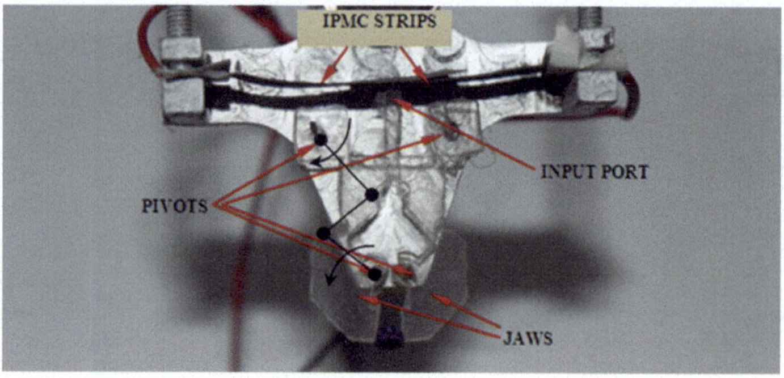

Fig. 9 The compliant microgripper grasping small objects

References

1. Bhandari, B., Lee, G.-Y., Ahn, S.-H.: A review on IPMC material as actuators and sensors: fabrications, characteristics and applications. Int. J. Precis. Eng. Manuf. **13**(1), 141–163 (2012)
2. Alici, G., Higgins, M.J.: Normal stiffness calibration of microfabricated tri-layer conducting polymer actuators. Smart Mater. Struct. **18**(6), 1–10 (2009)
3. Bar-Cohen, Y., Zhang, Q.: Electroactive polymer actuators and sensors. MRS Bull. **33**(3), 173–181 (2008)
4. Bar-Cohen, Y.: Electroactive POLYMER (EAP) Actuators as Artificial Muscles: Reality, Potential and Challenges, 2nd edn. SPIE Publications (2004)
5. Lughmani, W.A., Jho, J.Y., Lee, J.Y., Rhee, K.: Modeling of bending behavior of IPMC beams using concentrated ion boundary layer. Int. J. Precis. Eng. Manuf. **10**(5), 131–139 (2009)
6. Bhattacharya, S., Chattaraj, R., Das, M., Patra, A., Bepari, B., Bhaumik, S.: Simultaneous parametric optimization of IPMC actuator for compliant gripper. Int. J. Precis. Eng. Manuf. **16**(11), 2289–2297 (2015)
7. Kim, S.J., Lee, I.T., Kim, Y.H.: Performance enhancement of IPMC actuator by plasma surface treatment. Smart Mater. Struct. **16**(1), N6–N11 (2007)
8. He, Q., Yu, M., Song, L., Ding, H., Zhang, X., Dai, Z.: Experimental study and model analysis of the performance of IPMC membranes with various thickness. J. Bionic Eng. **8**(1), 77–85 (2011)
9. Bhattacharya, S., Bepari, B., Bhaumik, S.: IPMC-actuated compliant mechanism-based multifunctional multifinger microgripper. Mech. Based Des. Struct. **42**(3), 312–325 (2014)
10. Nemat-Nasser, S., Zamani, S., Tor, Y.: Effect of solvents on the chemical and physical properties of ionic polymer-metal composites. J. Appl. Phy. **99**, 1–17 (2006)
11. Yue, Z.: A method for group decision-making based on determining weights of decision makers using TOPSIS. Appl. Math. Model **35**, 1926–1936 (2011)
12. Milani, A.S., Shanian, A., Madoliat, R., Nemes, J.A.: The effect of normalization norms in multiple attribute decision making models: a case study in gear material selection. Struct. Multidiscipl. Optim. **29**, 312–318 (2005)
13. Brauers, W.K.M., Zavadskas, E.K.: The MOORA method and its application to privatization in a transition economy. Control Cybern. **35**, 445–469 (2006)
14. Antucheviciene, J., Zakarevicius, A., Zavadskas, E.K.: Measuring congruence of ranking result applying particular MCDM methods. Informatica **22**(3), 319–338 (2011)
15. Pang, J., Zhang, G., Chen, G.: ELECTRE I decision model of reliability design scheme for computer numerical control machine. J. Softw. **6**(5), 894–900 (2011)
16. Maheshwari, N., Reddy, A.N., Sahu D.K., Ananthasuresh, G.K.: Miniature compliant grippers with force-fensing. In: Proceedings of 14th National Conference on Machines and Mechanisms, NIT Durgapur, India, pp. 431–439 (2009)

Feature Extraction and Classification of MIT-BIH Arrhythmia Database

Sumanta Kuila, Namrata Dhanda and Subhankar Joardar

Abstract The MIT-BIH arrhythmia database is publicly available dataset which provides standard investigation material for the detection of heart arrhythmia. Since 1980, it is used for purpose of fundamental research and medical device development on cardiac rhythm and related diseases. The objective to build the database is to create an automated arrhythmia detectors that read the diversity of the signal and based on that automated cardiac diagnosis can be done. The different complexities of the electrocardiogram like the variety of the waveform of the heartbeat and corresponding cardiac beat, and the mystifying power of the artifacts and noise, combine to construct study of the signal tricky. So automation of the recording of the Electrocardiograms (ECG) signal is obvious and different publicly available databases are there which contain the recorded ECG signal for future medical use. MIT-BIH arrhythmia database is mostly used for medical and research purpose of different heart arrhythmia detections and analyses. This database tries to provide a fully automated environment to provide exact information for the detection of ventricular arrhythmias.

Keywords Arrhythmia · AHA · AAMI · Train data · Test data · Linear discriminate

S. Kuila (✉) · S. Joardar
Haldia Institute of Technology, Haldia, India
e-mail: sumanta.kuila@gmail.com

S. Joardar
e-mail: subhankarranchi@yahoo.co.in

N. Dhanda
Amity University, Noida, Uttar Pradesh, India
e-mail: ndhanda@lko.amity.edu

S. Kundu et al. (eds.), *Proceedings of the 2nd International Conference on Communication, Devices and Computing*, Lecture Notes in Electrical Engineering 602, https://doi.org/10.1007/978-981-15-0829-5_41

1 Introduction

The initiative to develop the database started at 1975 at Beth Israel Hospital of Boston collaboration with Massachusetts Institute of Technology (MIT) and the process continues till 1980. It also worked with the popular database of the American Heart Association (AHA) and in 1980 it is completed and the distribution is also started and today it covers more than 500 sites globally. Since then, it is considered as the most reliable standard test material for fundamental research of arrhythmia detectors [1]. The database is distributed with 9-track ½ in. digital tape which has 800 and 1600 bpi; it has also FM analog tape of quarter-inch IRIG format. Later in the month of August 1989, a CD-ROM version of the database is published. This database contains 48 recordings each of which has ½ h excerpts of double-channel recordings of ambulatory ECG data. From the time period between 1975 and 1979 in the BIH laboratory, 47 different subjects are studied to generate the database. At Boston's BIH hospital, initially a set of 4000 samples is recorded, each contains ambulatory ECG recordings of duration of 24 h from mixed population of patients of different ages. From the samples, 23 recordings were chosen as normal and 25 records are chosen such that it shows clinically proven arrhythmias. The recording is done at 360 samples/second per channel (11-bit resolution) and it has a range of 10 mV. These records are thoroughly checked by expert cardiologists, and based on the approvals of the specialist physicians the database is produced [2].

2 Data Recording

The data recording is an important part to create the database. Through electrocardiograms, data is generated in the form of analog signal. The data is converted into digital form through digitization and through annotation the complete database is generated.

2.1 Heartbeat Category

Data acquisition is an important phase of this process. Using several hardware and software tools, the ECG data is collected and processed to create the MIT-BIH database. The study of the components of the MIT-BIH heartbeat signal is important for that reason [3] (Fig. 1).

MIT-BIH heartbeat signal is categorized into five different heartbeat classes based on successive processing. In the above diagram, the graphical representations of the signal are given. The class "N" denotes the beat which implements the sinus node, the class S implements the beats of supraventricular ectopic (SVEBs), and ventricular ectopic beats (VEBs) denoted by V. After fusing VEBs and normal beats, the resultant beat creates the F class. Unknown beats and the paced beats create the class Q [4].

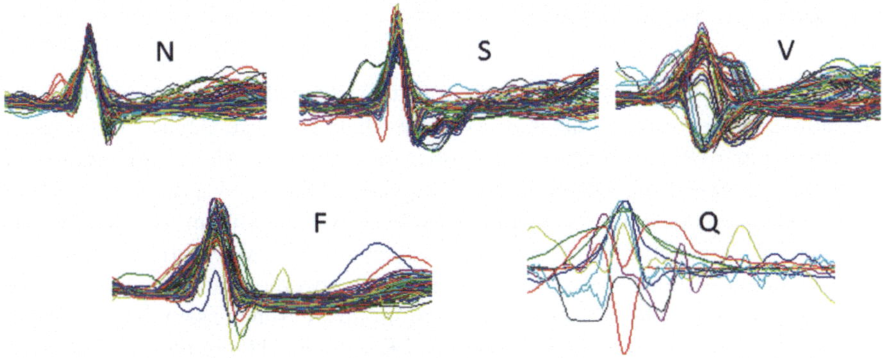

Fig. 1 Group of MIT-BIH heartbeat signal [13]

2.2 Data Acquisition

Data acquisition technique has several steps which relates to the recording device and the quality of the signal generated. The samples are generated by measuring the electrical potential difference of body surface of a certain patient. Figure 2 shows several steps of ECG data acquisition and the methodologies used for it. The sample results are the projections which are generated by the voltage difference of the human body. Initially, the records are generated through waveforms and later through sampling it turns to sample sequences which is an ordered and collective set of samples. Numerous observations are done for the measuring of the potential difference between two points to capture many views of the cardiac activity. The observation series denotes the multiple observations of the recording with a certain time frame [5].

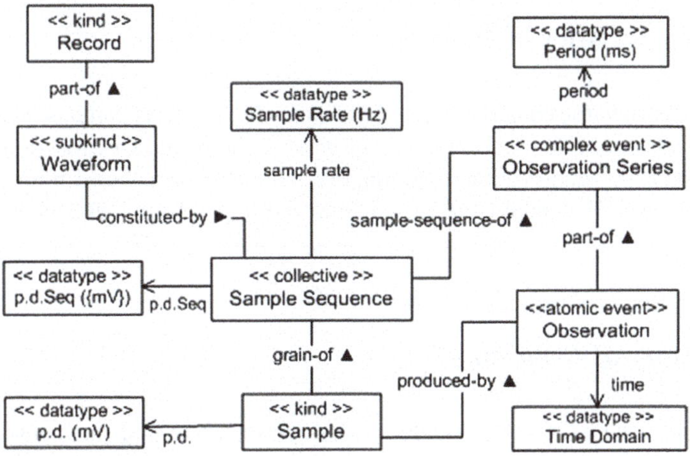

Fig. 2 ECG data acquisition mechanism [5]

2.3 Internal Mechanism

The acquired data is processed internally for ECG signal classification and analysis. The cardiac pulse creates forms in every cycle where cohesive electrophysiological incident is recorded and mapped for further data analysis. The depolarization of ventricles is mapped my QRS complex, the depolarization of atria is mapped by P wave, and the connection between two is done by RR segment [6]. The records generated by elementary forms are different in nature. The waveform generated by the ECG signal is nonelementary; it maintains the specific time boundaries of the recording session for physician's analysis and for this reason the samples are similar in nature. The combination of QRS complex, PR segment, TP segment, ST segment, T wave, and P wave compiles the cardiac cycle and from there the necessary data is collected where QRS complex is the key as it composes more than one wave. Sometimes due to non-visibility of the signal in few cycles, few elementary forms are not found where Q wave is missing and this leads some error to analyze the collected data [7].

2.4 Data Repository

The MIT-BIH arrhythmia database is available in PhysioNet and as a whole the database is the part of repository of PhysioBank. The database contains ECG records for research and commercial purpose. The AHA/MIT-BIH creates limited succession of integer samples which is generated by digitizing nonstop experiential function at permanent sampling frequency. Each of the MIT-BIH database records is 30 min duration, and the data is annotated thoroughly [8].

3 Characteristics of Data

The properties and the characteristics of the data which reflects patients' heart condition are discussed in this section. Evaluation of methods of data collection methods from old days to modern times, the quality judgment of test and train data with compared to AAMI standard and basic feature extraction from that data is discussed here.

3.1 History of Recording

Recent times, the ECGs are widely used as noninvasive and inexpensive means for the observation of heart condition and detection of heart diseases. In the year

1961, Norman J. Holter first introduced the system for nonstop recording of ECG in ambulatory with different subjects over the period 24 h in general. From mid-1970s, the Holter recording technique become the standard procedure to record the electrical activity of cardiac movement. It also records the irregularities of cardiac rhythm (which is popularly known as arrhythmias) by which automated methods are designed for the detection of heart arrhythmias [9]. Different research groups in industry and academia are showing interest to implement the automated arrhythmia detection system. Before 1980, the concerned people working in research on arrhythmia are used to generate data on their own but later MIT-BIH database become popular for this purpose. Study on ECG rhythms reveals that it has very large variability and contains variety of waveform morphologies, and the MIT-BIH database perfectly reflects these characteristics in its recording which covers the subjects as well as individuals over a specific period of time [10]. During the time period of 1960s and later, due to unavailability of the standard database, it was very difficult to analyze and implement arrhythmia algorithms for automation. So before the availability of the standard database, the performance of these algorithms depends upon the locally available database which lacks the variety and variability [11].

3.2 Comparing with AAMI

The MIT-BIH database checked with Association for the Advancement of Medical Instrumentation (AAMI) standards for heartbeat classifications where results of reporting and testing performance of heart rhythm ST segment maintain a standard measurement [12]. Table 1 describes how the MIT-BIH arrhythmia database maintains AAMI standard.

The table data describes the details of classification of heartbeat of AAMI standard (followed by AHA) and following implementation of MIT-BIH arrhythmia database. The database is divided into two different sets of data with separate characteristics, i.e., training and testing sets.

3.3 Train Dataset

The training dataset is generated from the data repository of the MIT-BIH database to process different feature extraction algorithms. The files are picked in by analyzing the waveforms as normal, RBBB (right bundle branch block), LBBB (left bundle branch block), atrial flutter, placed beats, and PVC. It also labels different classifications of the entire database where PVC beats are annotated as "V" and normal beats as "N" [14]. Study reveals that the number of files that generates the excellent results is in a ratio that number of beats exercised or the training data purpose. It is around 2% of the total beats, which is used for the training data. More percentage of

Table 1 Heartbeats described in the MIT-BIH database: group maintained AAMI standard [13]

AAMI class	MIT-BIH original class	Type of beat
Normal (N)	N L R e j	Normal beat Left bundle branch block beat R Right bundle branch block beat Atrial escape beat Nodal (junctional) escape beat
Supraventricular ectopic beat (S)	A a J S	Atrial premature beat Aberrated atrial premature beat Nodal (junctional) premature beat Supraventricular premature beat
Ventricular ectopic beat (V)	V E	Premature ventricular contraction Ventricular escape beat
Fusion beat (F)	F	Fusion of ventricular and normal beat
Unknown beat (Q)	/ f Q	Paced beat Fusion of paced and normal beat Unclassifiable beat

training data may cause over-fitted and make obstruction of feature extraction especially upon PVC beats. Different training algorithms like Fletcher, Shanno update, Broyden, and Goldfarb are there for feature extraction where the iterative methods are used to determine the weight matrix and also minimize the percentage of error [15].

3.4 Test Dataset

The test data determines the performance based on classification from which robustness and accuracy are measured. Generally, the performance of the classifier decreases as the volume of the data and its related components increases [16]. Table 2 describes which dataset belongs to test data and which one is train data according to Chazal et al. [13].

Table 2 Records of train and test datasets [13]

Set	Records
Training	101, 106, 108, 109, 112, 114, 115, 116, 118, 119, 122, 124, 201, 203, 205, 207, 208, 209, 215, 220, 223, 230
Test	100, 103, 105, 11, 113, 117, 121, 123, 200, 202, 210,212, 213, 214, 219, 221, 222, 228, 231, 232, 233, 234

4 Feature Extraction

Feature extraction is the key process of analyzing the data and from there decision-making can be done. For every heartbeat, features relate to the interval of fiducial points. Information generated from the heartbeat is analyzed, and different various characteristics are considered as different features [17].

4.1 Basic Operation

The basic operation starts once the ECG leads collect the ECG signal and for preprocessing the analog signal passes through high-frequency filter for farther annotation. The data is then classified into several characteristics of the heartbeat. The features of the heartbeat are categorized into five different parts and processed for further classification [18]. The feature space is portioned by linear discriminants into several classes by utilizing a set of hyperplanes. The classifier model uses several parameters which obtain the training data using several classification algorithms. The feature extraction for the wave P, QRS, and T requires two ECG signals and for RR interval it requires single beat [19]. Figure 3 describes the processing of ECG signal and from there the process of classification and feature extraction is done. In the initial stage, the ECG signal (ECG lead) is collected. The next phase is preprocessing stage where the signal is analyzed by baseline high-frequency filter where it is divided with separate beats according to heart segmentation. The heart segmentation is done by detecting the fiducial points of the cardiac beats. After segmentation, different

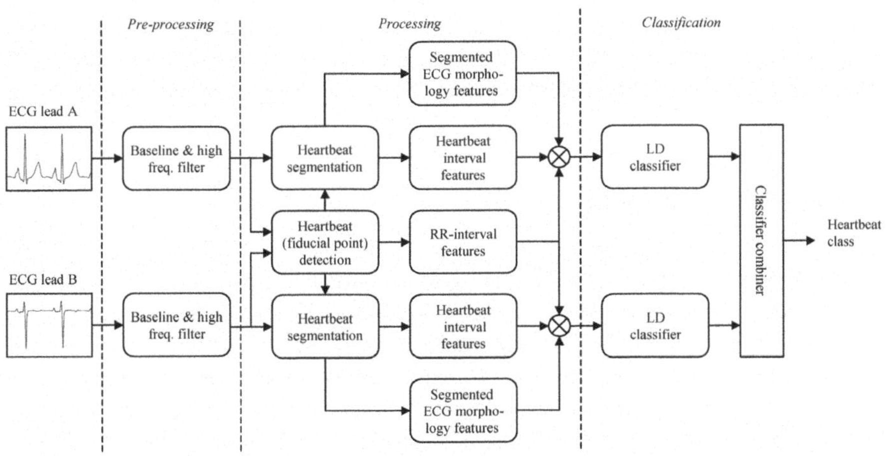

Fig. 3 Diagram to represent the automated processing and classification of ECG signal [20]

features like RR interval, heartbeat interval, and segmented ECG morphology features are generated. The features are then extracted by using linear discriminate (LD) classifier to make it specific heartbeat class [20].

4.2 Feature Selection Process

The feature selection process describes the entire process by which using the appropriate algorithms the final performance evaluation can be done. Now summarize the entire feature selection process. First phase describes the validation and train dataset with several parameters of the appropriate algorithms. Second phase describes the comparative study between different models used and based on the performance and characteristics which database model is selected for future process. In the final phase, the top performance model is picked and activated for final performance with test dataset. The selected model uses the database and produces the final result from the evaluation of concluding performance [21].

5 Process of Digitization

The ECG records collected in analog form needs to be digitized for feature extraction and classification. The analog signals were digitized with the sampling rate of 360 Hz. The experts classified data in 15 classes related to different types of arrhythmias. For digitization, the computers were specially designed and it uses analog-to-digital converter (ADC) with the tape driver controller. The standard digitization rate is followed, i.e., 360 samples/second/channel. Sample digital notch filters are used to eliminate the additional noise of the interface. DC300 digital cartridge has four tracks with the capacity around 400 kB/tracks that were used for storage purpose [22, 23].

6 Annotation of Data

Simple QRS detector (slope sensitive) is used to annotate the data. In the background, the software (usually written in assembly language) is there which generates the signals for the annotation. In the traditional recording, every half-hour long tape was used to generate two equal 150 Ft. (46 m) length of ECG chart recording. From the recorded chart, the expert cardiologists worked to delete false detections and the abnormal beats are deleted. To make the beats synchronous, the discrepancy of the data was properly observed and necessary corrections were made using the annotated editor. In whole process of MIT-BIH arrhythmia database, using software all of the generated annotations were reviewed and test them for the consistency of the database. At the first phase of creation of MIT-BIH arrhythmia database, around 110,000 annotations were generated and checked to complete the database [24].

7 Database Usage

The database is tested first before introducing it for the standard test set for arrhythmia detection. Starting from 1984, the database was checked beat by beat and later episode by episode with the references and the algorithms. The database later found very useful to detect ventricular arrhythmia and maintain the AAMI standards. The annotated files of MIT-BIH database support various machine learning algorithms for the automation of arrhythmia detection. The measurement of the performance is resolved by comparing the reference annotated files with the annotated files generated by the database, and this is an automated process controlled by comparison software stated by AAMI standards. The test results are verified by arrhythmia analyzers and various regulatory agencies using materials like comparison software, test data, and hardware devices for recording of ECG annotated files [25].

8 Limitations of the Database

MIT-BIH database is publicly available test dataset and it is widely used for the research of arrhythmia detection. The database has several limitations which should be carefully handled for better performance. For the study of heart rate variability, the differences in recording speed and the playback speed should be carefully observed since the recording of heart irregularities should be properly measured. At the time of recording of two signals, the low tape speed hampers the exact recording of ECG signal which results in the creation of erroneous database. The database is not fully equipped, and annotation editors are there to prepare the final database after frequent corrections and make the database inconsistent [26]. The annotations are also not unanimous, as total of 33 beats from 6 records (out of 48 records) remain unclassified as in that cases cardiologists–annotators could not take any decision on that particular beat types. As the QRS complex is unable to detect properly, loss of signal/noise reflects on seven episodes where these episodes have short span of time (around 10 s) [27].

9 Conclusion

The MIT-BIH database is publicly available around more than 25 years, and the performance statistics is quite impressive once it is widely understood among group of researchers. For standard test of arrhythmia, the database supports various feature extraction and classification algorithms which trends the process for automated ventricular arrhythmia detection. This well-characterized dataset supports many industry and academic researchers' commercially as well academic purpose without any investment as it is well-circulated publicly available database. In comparison with other available databases for arrhythmia detection like European ST standard, AHA database, the MIT-BIH database has more impact especially on the field of academy research, for its variation with longer recording and various temporal patterns with different subjects.

References

1. Moody, G.B., Mark, R.G.: The MIT-BIH Arrhythmia Database on CD-ROM and Software for Use with it. Massachusetts Institute of Technology, Cambridge, MA, USA, IEEE, pp. 185–188 (1991)
2. https://www.physionet.org/physiobank/database/mitdb
3. Sibushri, G.: Analysis of ECG signals for arrhythmia using MATLAB. Int. J. Innov. Res. Comput. Commun. Eng. **2**(1), 1093–1099 (2014)
4. Ince, T., Kiranyaz, S., Gabbouj, M.: A generic and robust system for automated patient-specific classification of ECG signals. IEEE Trans. Biomed. Eng. **56**(5), 1415–1426 (2009)
5. B. Goncalves, G. Guizzardi, Pereira Filho, J.G.: Using an ECG reference ontology for semantic interoperability of ECG data. J. Biomed. Inf. 126–136 (2010)
6. Hamilton, P.S., Tompkins, W.J.: Quantitative investigation of QRS detection rules using the MIT/BIH arrhythmia database. IEEE Trans. Biomed. Eng. **33**(12), 1157–1165 (1986)
7. Martínez, J.P., Almeida, R., Olmos, S., Rocha, A.P., Lag, P.: A wavelet-based ecg delineator: evaluation on standard databases. IEEE Trans. Biomed. Eng. **51**(4), 570–581 (2004)
8. Moody, G.B., Mark, R.G., Goldberger, A.L.: PhysioNet: A Research Resource for Studies of Complex Physiologic and Biomedical Signals. MIT Room E25-505A, Cambridge, MA 02139 USA
9. Holter, N.J.: New method for heart studies. In: 4th International Conference on Medical Electronics, New York, SCIENCE, vol. 134, pp. 1214–1220 (1961)
10. Balasubramaniam, D., Nedumaran, D.: Implementation of ECG signal processing and analysis techniques in digital signal processor based system. In: International Workshop on Medical Measurements and Applications, Cetraro, Italy, pp. 60–63(2009)
11. Walinjkar, A., Woods, J.: ECG classification and prognostic approach towards personalized healthcare. In: Wearable Systems for Real-time ECG Classification and Prognosis
12. American National Standard.: Testing and reporting performance results of cardiac rhythm and ST segment measurement algorithms. ANSI/AAMI EC57 (2012)
13. de Albuquerque, V.H.C., Nunes, T.M., Pereira, D.R.: Robust automated cardiac arrhythmia detection in ECG beat signals. Neural Comput. Appl. 2–22 (2018)
14. Inan, O.T., Giovangrandi, L., Kovacs, G.T.: Robust neural-network-based classification of premature ventricular contractions using wavelet transform and timing interval features. IEEE Trans. Biomed. Eng. **53**(12), 2507–2515 (2006)

15. Hu, Y.H., Palreddy, S., Tompkins, W.J.: A patient-adaptable ECG beat classifier using a mixture of experts approach. IEEE Trans. Biomed. Eng. **44**(9), 891–900 (1997)
16. Israela, S.A., Irvine, J.M., Cheng, A., Wiederhold, M.D., Wiederhold, B.K.: ECGto identify individuals. Pattern Recogn. **38**, 133–142 (2005)
17. de Chazal, P., Reilly, R.B.: A patient-adapting heartbeat classifier using ECG morphology and heartbeat interval features. IEEE Trans. Biomed. Eng. **53**(12), 2535–2543 (2006)
18. de Chazal, P., Reilly, R.B.: Automatic classification of ECG beats using waveform shape and heart beat interval features. In: Digital Signal Processing Research Group, Department of Electronic and Electrical Engineering, University College Dublin, Republic of Ireland
19. Saritha, C., Sukanya, V., Murthy, Y.N.: ECG signal analysis using wavelet transforms. Bulg. J. Phys. **35**, 68–77 (2008)
20. de Chazal, P., O'Dwyer, M., Reilly, R.B.: Automatic classification of heartbeats using ECG morphology and heartbeat interval features. IEEE Trans. Biomed. Eng. **51**(7), 1196–1206 (2004)
21. Llamedo, M., Martinez, J.P.: Heartbeat classification using feature selection driven by database generalization criteria. IEEE Trans. Biomed. Eng. **58**(3), 616–625 (2011)
22. Patel, S., Datar, A.: ECG data compression using wavelet transform. Int. J. Eng. Trends Technol. (IJETT) **10**(3), 770–776 (2014)
23. Kao, T., Hwang, L.J., Lin, Y.H., Lin, T.H., Hsiao, C.H.: Computer analysis of the electrocardiograms from ECG paper recordings. In: 23rd Annual EMBS International Conference, Istanbul, Turkey, pp. 3232–3234 (2001)
24. Stockbridge, N., Brown, B.D.: Annotated ECG waveform data at FDA. J. Electro Cardiol. **37**, 63–68 (2004)
25. Jenkins, J.M.: Automated electrocardiography and arrhythmia monitoring. Progress Cardiovasc. Dis. **25**(5), 367–408 (1963)
26. Hilton, M.L.: Wavelet and wavelet packet compression of electrocardiograms. IEEE Trans. Biomed. Eng. **44**(5), 394–402 (1997)
27. Silvia Priscila, S., Hemalatha, M.: Diagnosis of heart disease with particle bee-neural network. Biomed. Res. 40–46 (2018). www.biomedres.info

Adaptive Neuro-Fuzzy Approach for Forecasting of Solar Power Generation

Dola Sinha

Abstract This paper highlights the prediction of power generation from a solar panel with the help of the ANFIS. Solar irradiance and atmospheric temperature in a place are not even throughout the year. It depends on geographical position (i.e., longitude and latitude) and the environmental situation of the place. Power generation from a PV panel is highly subjective by solar irradiance and temperature. Pollution, fog, smoke dust cloud, etc. may also affect these two. So, changes in environmental conditions may affect the prediction of power generation from solar PV panel. In this paper, power generation is predicted for different irradiances and temperatures with the help of the ANFIS. ANFIS results are validated with the real-time experimental value which gives an improvement in predictive accuracy. This also signifies the ANFIS capability to estimate power in a different situation which could be helpful for maximum power point tracking.

Keywords Solar irradiance · Fuzzy logic controller · ANFIS · Solar power prediction

1 Introduction

The electrical energy consumption rate is surprisingly increasing now, and it becomes challenging to fulfil the total demand, as the natural sources of fossil fuels are limited and decreasing from nature. So, researchers are paying more concentration to find ways of efficiently utilizing the nonconventional resources. Solar energy is a very significant energy source and most sustainable choice of power generation among the nonconventional and renewable energy families.

Though the efficiency of solar cell is low, the panels can be installed on the rooftop of the house [1] or large area which can produce sufficient power at low cost as sunlight can be achieved free and abundantly from nature.

D. Sinha (✉)
Department of Electrical Engineering, Dr. B. C. Roy Engineering College, Durgapur, India
e-mail: dola.sinha@gmail.com

S. Kundu et al. (eds.), *Proceedings of the 2nd International Conference on Communication, Devices and Computing*, Lecture Notes in Electrical Engineering 602, https://doi.org/10.1007/978-981-15-0829-5_42

Solar cells are basically a p-n junction diode. When sunlight (photon) falls upon the cell, electron–hole pairs are generated, and current starts flowing through it. The transmitted light is absorbed within the semiconductor material, and it uses this energy to excite free electrons from a low energy label to high energy label. Standalone photovoltaic systems are the best solutions in the locality where grid-connected system is impossible to set up or in many small electrical energy demand applications such as communication systems, water pumping, and low power [2]. The efficiency of the cell is less, so it is more beneficial to operate the solar panel at its maximum power point (MPP). MPP of the system depends on the power (voltage and current) generation from the solar panel which in turn depend on temperature and solar irradiation. So, it is necessary to predict the power generation from the solar panel at the different atmospheric conditions to operate the system in its maximum power point. Different strategies are implemented to predict the power generation like fuzzy logic [3, 4] and neural network [5]. Solar radiation is used in [6, 7] to obtain a reference power. In this paper, ANFIS-based power prediction model is designed. In the next section, analytical calculation procedure is mentioned to find nominal operating cell temperature, solar irradiance, and annual solar energy output of the photovoltaic system. In Sect. 3, ANFIS is designed in Matlab/Simulink to obtain a generation of solar power. The power generated from prototype solar PV panel is validated with the results.

2 Mathematical Modeling

2.1 Operating Cell Temperature

The operating cell temperature (expressed in Eq. (1a, 1b)) depends on nominal operating cell temperature and certain environmental conditions like (a) irradiance on the cell surface, (b) air temperature, and (c) wind velocity.

$$T_{\text{cell}} = T_{\text{air}} + \frac{\text{NOCT} - 20}{80} \times S \tag{1a}$$

where S = irradiance in W/m^2.

T_{air} is air temperature, and the nominal operating cell temperature (NOCT) is the temperature obtained by the open-circuited cell in the solar panel at predefined environmental condition stated below:

1. Irradiance on cell surface = 800 W/m^2,
2. Air temperature = 20 °C, and
3. Wind velocity = 1 m/s.

The solar panel temperature will be lower than this when wind velocity is high, but higher under still conditions. The relationship between wind speed and solar panel temperature as used by Kurtz et al. [8] is shown in Eq. 1b:

$$T_{\text{cell}} = T_{\text{air}} + S \times \exp(-3.473 - 0.0594 \times \text{Wind Speed}) \tag{1b}$$

2.2 *Calculation of Solar Irradiance*

Solar irradiance is the energy received from the Sun in the form of electromagnetic radiation per unit area. The current generation of the solar cell mainly depends on solar irradiance, and the relationship is mentioned in Eq. (2).

$$I_{\text{Gen}} = I_{\text{Gen}}\max \times \frac{S_{\text{op}}}{S_{\max}} \tag{2}$$

where I_{Gen} is generated current at S_{op} irradiance, and I_{Genmax} is a current generation at a maximum irradiance of $S_{\max}$, i.e., 1000 W/m^2. The average daily irradiation for each month is broken down into hourly using the ratio r_t and expressed in Eq. (3). The diffused component of the total irradiation for each hour is also shown in Eq. (4).

$$r_t = \frac{\text{hourly total irradiance}}{\text{daily total irradiance}} \tag{3}$$

$$r_d = \frac{\text{hourly diffused irradiance}}{\text{daily total diffused irradiance}} \tag{4}$$

2.3 *Calculation of Annual Electrical Energy Generation from Solar System*

The global formula to estimate the electrical energy generation from the solar photovoltaic system is expressed in Eq. (5):

$$E = A \times r \times H \times PR \tag{5}$$

where

E Electrical Energy (kWh),
A Total solar panel Area (m^2),
r Solar panel efficiency (%) $= \frac{\text{Electrical power of one solar panel}}{\text{The area of that panel}}$
H Annual average solar radiation on tilted panels (shadings not included), and
PR Performance ratio, coefficient for losses (range between 0.5 and 0.9, default value = 0.75).

3 Results and Discussions

ANFIS model is developed using a Sugeno-type fuzzy controller and multilayer feedforward neural network.

3.1 Design of Fuzzy Logic Controller

The fuzzy logic controller has two inputs, irradiance and temperature, and power is considered as output. Input 1 is irradiance and its range varies from 100 to 1000 W/m^2 in a step of 100 W/m^2 and consists of eight membership functions. The structure is shown in Fig. 1. Temperature is Input 2, and it varies from 10 to 30 °C in a step of 5 °C. It consists of six membership functions, and the structure is shown in Fig. 2. The output power has ranged from 1 to 420 mW. The fuzzy rules have been given in Table 1.

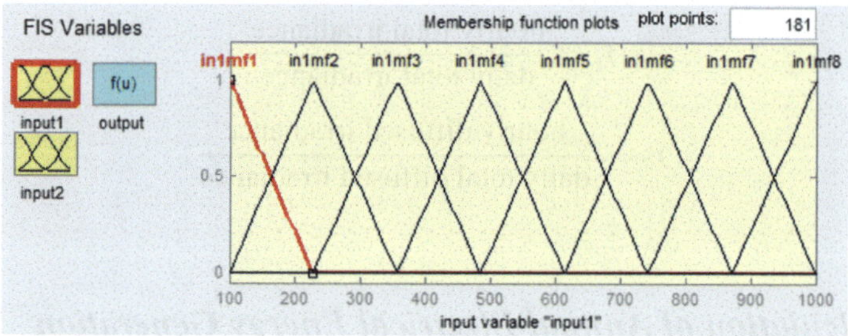

Fig. 1 Membership function of input 1

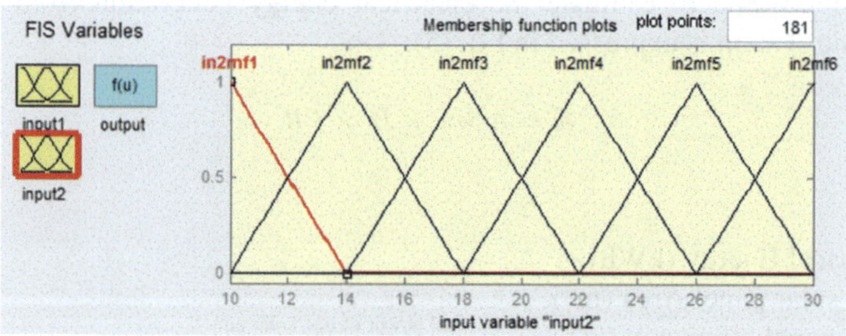

Fig. 2 Membership function of input 2

Table 1 Fuzzy rules

Rules	Input 1 (irradiance)	Input 2 (temperature)	Output (power)
1	1mf1	2mf1	1mf1
2	1mf1	2mf2	1mf2
3	1mf1	2mf3	1mf3
4	1mf1	2mf4	1mf4
5	1mf1	2mf5	1mf5
6	1mf1	2mf6	1mf6
7	1mf2	2mf1	1mf7
8	1mf2	2mf2	1mf8
9	1mf2	2mf3	1mf9
10	1mf2	2mf4	1mf10
11	1mf2	2mf5	1mf11
12	1mf2	2mf6	1mf12
13	1mf3	2mf1	1mf13
14	1mf3	2mf2	1mf14
15	1mf3	2mf3	1mf15
16	1mf3	2mf4	1mf16
17	1mf3	2mf5	1mf17
18	1mf3	2mf6	1mf18
19	1mf4	2mf1	1mf19
20	1mf4	2mf2	1mf20
21	1mf4	2mf3	1mf21
22	1mf4	2mf4	1mf22
23	1mf4	2mf5	1mf23
24	1mf4	2mf6	1mf24
25	1mf5	2mf1	1mf25
26	1mf5	2mf2	1mf26
27	1mf5	2mf3	1mf27
28	1mf5	2mf4	1mf28
29	1mf5	2mf5	1mf29
30	1mf5	2mf6	1mf30
31	1mf6	2mf1	1mf31
32	1mf6	2mf2	1mf32
33	1mf6	2mf3	1mf33
34	1mf6	2mf4	1mf34
35	1mf6	2mf5	1mf35
36	1mf6	2mf6	1mf36

(continued)

Table 1 (continued)

Rules	Input 1 (irradiance)	Input 2 (temperature)	Output (power)
37	1mf7	2mf1	1mf37
38	1mf7	2mf2	1mf38
39	1mf7	2mf3	1mf39
40	1mf7	2mf4	1mf40
41	1mf7	2mf5	1mf41
42	1mf7	2mf6	1mf42
43	1mf8	2mf1	1mf43
44	1mf8	2mf2	1mf44
45	1mf8	2mf3	1mf45
46	1mf8	2mf4	1mf46
47	1mf8	2mf5	1mf47
48	1mf8	2mf6	1mf48
49	1mf9	2mf1	1mf49
50	1mf9	2mf2	1mf50
51	1mf9	2mf3	1mf51
52	1mf9	2mf4	1mf52
53	1mf9	2mf5	1mf53
54	1mf9	2mf6	1mf54
55	1mf10	2mf1	1mf55
56	1mf10	2mf2	1mf56
57	1mf10	2mf3	1mf57
58	1mf10	2mf4	1mf58
59	1mf10	2mf5	1mf59
60	1mf10	2mf6	1mf60

3.2 *Development of ANFIS Model*

It is a multilayer feedforward neural network. The ANFIS structure is shown in Fig. 3. The model is trained by training data, and after training process, the model was tested with the checking data (shown in Figs. 3 and 4). The ANFIS decision surface view is shown in Fig. 4 for the solar power estimation. The overall structure of ANFIS model is shown in Fig. 5 based on fuzzy logic rule base shown in Table 2 and Figs. 6, 7. This system is then compared with the real time, i.e., experimental value of power. The two inputs, power through ANFIS and experimental power through the entire dataset used in this system, are given in Table 1.

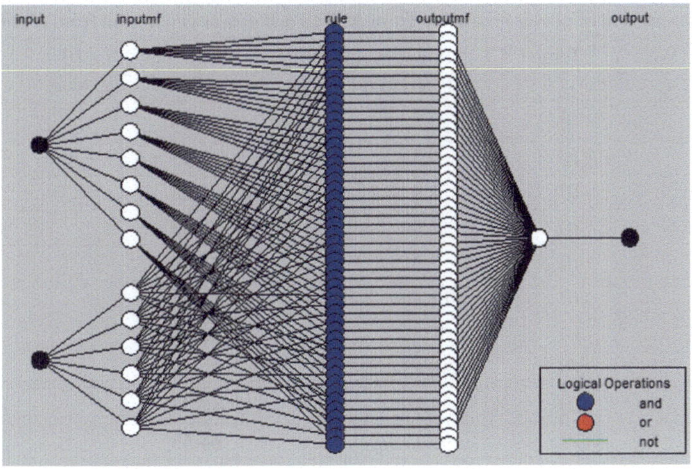

Fig. 3 Structure of the ANFIS model

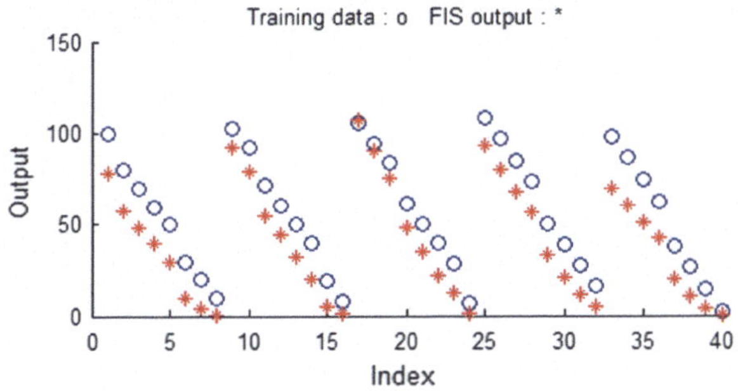

Fig. 4 Trained data

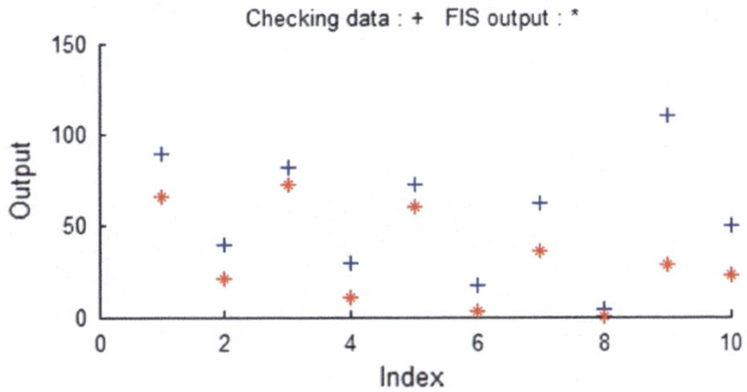

Fig. 5 Validation result

Table 2 Comparison of power prediction from ANFIS and power generation from real-time model

Irradiance (W/m^2)	Temp (°C)	Power from ANFIS (×10 mW)	Expt power (×10 mW)	% Error
100	10	0.1	0.18	−0.467
200	10	0.34	0.29	0.162
300	10	1.9	1.28	0.475
400	10	4.56	3.69	0.235
500	10	7.21	8.29	−0.131
600	10	13.89	14.93	−0.070
700	10	21.51	22.92	−0.062
800	10	32.45	33.59	−0.034
900	10	44.66	47.23	−0.054
1000	10	65.12	64.55	0.009
100	15	0.15	0.11	0.345
200	15	0.299	0.225	0.326
300	15	1.57	1.018	0.541
400	15	2.85	2.91	−0.021
500	15	5.55	6.52	−0.149
600	15	9.56	11.78	−0.189
700	15	17.15	18.27	−0.061
800	15	25.76	27.04	−0.047
900	15	37.65	38.37	−0.019
1000	15	51.22	52.97	−0.033
100	20	0.00075	0.00057	0.305
200	20	0.22	0.122	0.848
300	20	0.6765	0.62	0.085
400	20	1.99	1.83	0.085
500	20	4.2557	4.19	0.014
600	20	7.95	7.44	0.069
700	20	9.687	11.03	−0.122
800	20	15.66	15.55	0.007
900	20	21.5	21.12	0.018
1000	20	27.9	27.70	0.007
100	25	0.0065	0.0052	0.229
200	25	0.09	0.075	0.199
300	25	0.456	0.416	0.094
400	25	1.5	1.23078	0.219
500	25	2.97	2.78	0.068

(continued)

Table 2 (continued)

Irradiance (W/m^2)	Temp (°C)	Power from ANFIS (×10 mW)	Expt power (×10 mW)	% Error
600	25	5.65	5.46	0.035
700	25	9.91	9.66	0.026
800	25	14.77	15.92	−0.073
900	25	24.679	24.70	−0.001
1000	25	36.9	36.67	0.006
100	30	0.027	0.011	1.383
200	30	0.022	0.019	0.112
300	30	0.21	0.197	0.065
400	30	0.68	0.638	0.065
500	30	1.56	1.47	0.059
600	30	3.15	3.64	−0.135
700	30	8.1569	8.37	−0.026
800	30	15.9	15.68	0.014
900	30	26.77	26.26	0.019
1000	30	41.91	40.88	0.025

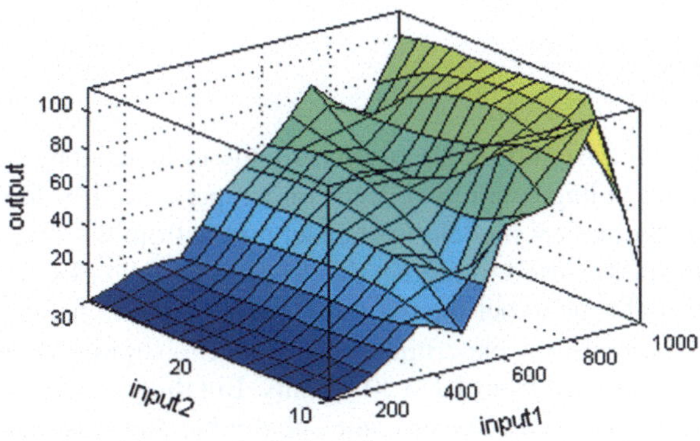

Fig. 6 Surface view of ANFIS model

3.3 Model Validation

The system aims to demonstrate the precision of the proposed model on solar power prediction. This is ensured by a comparison made between the proposed ANFIS model and experimental model. The performance analysis of this model is determined by the number of epochs reached. The result presented in Table 2 indicates that the ANFIS model has the best capability for estimating global solar power. Figure 7 shows the real-time experimental model of solar panel rated as 6 V, 0.5 W.

Fig. 7 Real-time experimental model of solar panel

% Error is calculated as

$$\%\text{Error} = \frac{\text{Power from ANFIS} - \text{Power from Experiment}}{\text{Power from Experiment}} \times 100 \quad (6)$$

4 Conclusion

The current generation from the solar panel is highly influenced by solar irradiance, and the open-circuit voltage mainly depends on temperature. It is highly difficult to form the exact mathematical model to get exact power from the solar panel. Hence, in this paper, the authors tried to predict the power using ANFIS, and the model is validated by the real-time experimental solar panel. The irradiance is varied by a solar simulator. The real-time experiment is done over a stipulated time period, and so the temperature is varied from 10 to 30 °C only. But this model can be utilized for higher temperature also. The % error comes negligible and less than 0.5%, so this model can be utilized in association with MPPT controller also.

References

1. Sick, F., Erge, T.: Photovoltaic in building. Int. Energy Agency (1996)
2. Kottas, L., Boutalis S., Karlis, D.: New maximum power point tracker for PV arrays using fuzzy controller in close cooperation with fuzzy cognitive networks. IEEE Trans. Energy Convers. **21**(3) (2006)
3. Senjyu, T., Uezato, K., Maximum power point tracker using fuzzy control for photovoltaic arrays. In: Proceedings of the IEEE International Conference on Industrial Technology (1994), pp. 143–147

4. Patcharaprakiti, N., Premrudeepreechacharn, S.: Maximum Power Point Tracking Using Adaptive Fuzzy Logic Control for Grid-Connected Photovoltaic System, pp. 372–376. IEEE (2002)
5. Hillowala, R.M., Sharaf, A.M.: A rule-based fuzzy logic controller for a PWM inverter in photovoltaic energy conversion scheme. In: Proceedings of the IEEE Industry Application Society Annual Meeting, Oct 1992, pp. 762–769
6. Lasnier, F., Ang, T.G.: Photovoltaic Engineering Handbook. IOP Publishing Ltd. (1990)
7. Jang, J.S.R.: ANFIS: Adaptive-network-based fuzzy inference systems. IEEE Trans. Syst. Man Cybern. **23**(3), 665–685 (1993)
8. Kurtz, S., Whitfield, K., Miller, D., Joyce, J., Wohlgemuth, J., Kempe, M., et al.: Evaluation of high-temperature exposure of rack mounted photovoltaic modules. In: 34th IEEE Photovoltaic Specialists Conference (PVSC), pp. 2399–2404 (2009)

A Recurrent Neural Network-Based Approach to Automatic Language Identification from Speech

Himadri Mukherjee, Ankita Dhar, Sk. Md. Obaidullah, K. C. Santosh, Santanu Phadikar and Kaushik Roy

Abstract The task of automatically identifying the used language from speech signals is known as automatic language identification. It is very much important prior to speech recognition in multilingual scenarios where speakers use more than a single language in course of communication. In this paper, a recurrent neural network (RNN)-based system with long short-term memory (LSTM) along with handcrafted line spectral frequency-based features is proposed for language identification. Experiments were performed on as many as 21908 clips (more than 30 h of data) from the top three spoken languages of the world, namely, English, Chinese, and Spanish, and a highest average accuracy of 95.22% has been obtained.

Keywords Language identification · Recurrent neural network · Long short-term memory · Line spectral frequency

---

K. C. Santosh, Senior Member, IEEE.

H. Mukherjee (✉) · A. Dhar · K. Roy
Department of Computer Science, West Bengal State University, Kolkata, India
e-mail: himadrim027@gmail.com

A. Dhar
e-mail: ankita.ankie@gmail.com

K. Roy
e-mail: kaushik.mrg@gmail.com

Sk. Md. Obaidullah
Department of Computer Science & Engineering, Aliah University, Kolkata, India
e-mail: sk.obaidullah@gmail.com

K. C. Santosh
Department of Computer Science, The University of South Dakota, Brookings, SD, USA
e-mail: santosh.kc@ieee.org

S. Phadikar
Deptartment of Computer Science & Engineering, Maulana Abul Kalam Azad University of Technology, Kolkata, India
e-mail: sphadikar@yahoo.com

S. Kundu et al. (eds.), *Proceedings of the 2nd International Conference on Communication, Devices and Computing*, Lecture Notes in Electrical Engineering 602, https://doi.org/10.1007/978-981-15-0829-5_43

1 Introduction

Speech-based applications have developed significantly ever since the first attempt of speech recognition by Dudley in 1930 [1]. Numerous speech recognition-enabled systems have simplified the lives of the rustics in multifarious ways. Such systems are now commercially available in certain languages like English, Spanish, etc. Speech recognition is itself a challenging task which complicates even more for multilingual countries where people often use a mixture of languages while talking. Language identification prior to speech recognition is very much necessary in such scenarios. Language identifiers can aid toward speech recognition by first identifying the different languages of the uttered words or phrases and then invoking the language-specific speech recognizers.

RNN was used by Srivastava et al. [2] for designing language-specific phonotactic information. Their adopted method merged statistical and RNN-based technique outperformed the classic DNN-based systems for more than 176 languages with respect to mean F1 score. Tang et al. [3] extracted phonetic features from Babel and AP16-OLR dataset for identifying languages using DNN architecture along with RNN. Their proposed approach outperformed standard acoustic neural models and also produced better results compared to the standard i-vector based systems for short and noisy audio. Watanabe et al. [4] performed various experiments by combining CNNs and bidirectional LSTM along with RNN architecture for language identification. They reported average character error rates of 16.6 and 21.4 for 7 and 10 languages, respectively. Zissman et al. [5] presented four different models that include Gaussian- and phoneme-based approaches. They obtained accuracies of 94.5 and 79.2% for 2 and 10 languages, respectively, on OGI multilingual telephonic speech corpus.

Gupta et al. [6] worked on IIIT-H dataset for identifying Malayalam, Marathi, Telugu, Bangla, Tamil, and Hindi languages using LPC and MFCC features along with random forest and SVM. They achieved highest accuracy of 92.6% for random forest by combining both the features. A system was proposed by He et al. [7] for classifying Uyghur and Kazakh from audio of short texts consisting of less than 14 words each. They obtained precision of 96.5% and 94.6%, respectively, using heuristic-based features combined with maximum entropy-based classifier. Jin et al. [8] explored a senone-based approach on NIST LRE 2009 dataset for identifying 23 languages. Their methodology outperformed the state-of-the-art methods based on DNN structure and i-vectors by adhering to the senones for classification and the formation of i-vector. A system for differentiating seven Indic languages, namely, Hindi, Bangla, Urdu, Telugu, Manipuri, Punjabi, and Assamese using both prosodic and phonotactic features, was proposed by Madhu et al. [9]. They trained an artificial neural network with two hours of data from each of the languages and obtained accuracies of 68% and 72%, respectively, for the prosodic and phonotactic information-based systems.

Here, a system is proposed to identify language from speech. Experiments were performed with English, Chinese, and Spanish which are the top three languages in

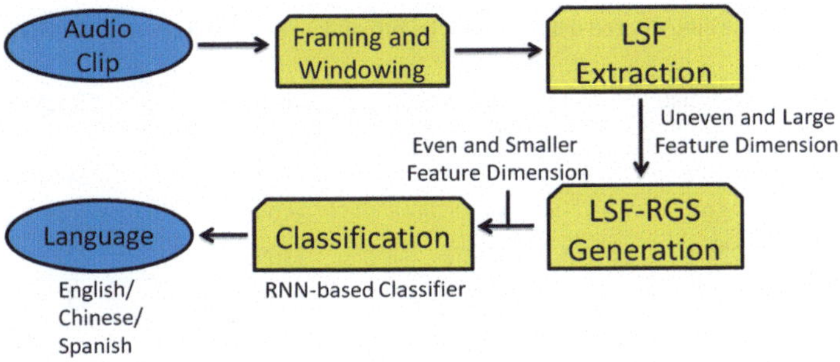

Fig. 1 Block diagram of the proposed system

the world based on number of speakers [10]. The speech signals were characterized with handcrafted line spectral frequency-based features followed by RNN-based classification. A graphical overview of the system is presented in Fig. 1.

In the rest of the paper, the dataset detail is presented in Sect. 2 followed by the proposed methodology in Sect. 3. The results are presented in Sect. 4, and the conclusion is drawn in Sect. 5.

2 Dataset

Data is one of the most important aspects of an experiment. The quality of data plays a pivotal role in the final result. During data collection, it is extremely important to ensure that the dataset portrays real-life characteristics as far as possible which aids in the development of robust systems. In the current experiment, the dataset comprises the top three spoken languages in the world (English, Chinese, and Spanish) [10]. The approximate number of speakers for each of the three along with the number of countries where they are primarily spoken [10] is presented in Table 1.

The dataset was put together with the aid of clips of interviews, talk shows, and news from YouTube [11]. The extracted audio clips were stored in .wav uncompressed format at a bit rate of 1411 kbps. It was observed that in many clips, there was background music, header and trailer music, short pauses, and data from multiple

Table 1 Approximate number of speakers with the number of primarily used countries for the three languages

Language	Approximate number of speakers (millions)	Number of countries mainly spoken in
Chinese	1299	38
Spanish	442	31
English	378	118

Table 2 Details of the engendered datasets with length of data per language in (HH:MM:SS) format

Dataset (Clip length in seconds)	English (10:08:18)	Chinese (10:16:05)	Spanish (10:12:53)	Total (30:37:16)
D1 (5)	7244	7389	7275	21908
D2 (10)	3598	3692	3597	10887
D3 (15)	2384	2460	2369	7213
D4 (20)	1777	1843	1753	5373

environments within a single clip. These factors lead to high acoustic variations, thereby making our dataset versatile. Moreover, it was also observed that the speakers often used keywords from different languages while talking. These attributes were not removed from the dataset in order to uphold the real-world characteristics. The collected data was used to generate four datasets (D1–D4) having clips of lengths 5, 10, 15, and 20 s, respectively, in accordance with [12]. The number of clips in each of the datasets for the three languages along with the total number of clips is presented in Table 2.

3 Proposed Method

3.1 Framing and Windowing

An audio clip consists of multiple frequency components. These components show high deviations when a clip is analyzed as a whole which poses difficulties. In order to cope up with this problem, an audio signal is partitioned into smaller segments. Each of such segments is known as a frame within which the spectral characteristics tend to be quasistationary. In the present experiment, the audio clips were framed into 256 sample point wide frames in overlapping mode with an overlap factor of 100 points due to its utility as presented in [12]. The number of obtained frames (Fr) of size (Sz) with an overlap of (O) points from a signal having (Ns) sample points is presented as follows:

$$Fr = \left\lceil \frac{Ns - Sz}{O} + 1 \right\rceil. \tag{1}$$

Jitters are often observed in the frames post division of the original signal. These components interfere with the frequency-based analysis in the form of unwanted peaks in the results. This is known as spectral leakage. To handle this issue, a windowing function is multiplied with the frames. In the present experiment, we choose hamming window for this purpose due to its utility as illustrated in [12].

3.2 Line Spectral Frequency Extraction

Linear predictive coefficients are efficiently represented by means of a technique known as line spectral frequency [13]. It is unique in nature and also ensures better and higher interpolation property along with effective quantization. In this technique, the cth-order linear predictor coefficient (L(z)) is broken down into two parts which are symmetric and antisymmetric to the original (L(z)) (represented as L1(z) and L2(z), respectively). The cth-order predictor is mathematically presented below:

$$L(z) = 1 + \sum_{i=1}^{c} l_i z^{-i}. \quad (2)$$

The obtained expressions of L1 and L2 in terms of L(z) are presented below. It is the roots of these equations which constitutes the line spectral frequency.

$$L1(z) = L(z) + z^{-(c+1)} L(z^{-1}) and \quad (3)$$

$$L2(z) = L(z) - z^{-(c+1)} L(z^{-1}). \quad (4)$$

The cth-order predictor can be expressed in terms of the polynomials L1 and L2 as shown below:

$$L(z) = \frac{1}{2}(L1(z) + L2(z)). \quad (5)$$

3.3 LSF-Ratio Grade Statistics (LSF-RGS) Computation

We had extracted 5-, 10-, 15-, and 20 [12]-dimensional LSF values for each frame of the audio clips. In real world, audio clips occur in disparate sizes which leads to an uneven number of frames and ultimately to features of disparate dimension. It was also observed that an audio clip of only 1 second produced as many as 440 frames in accordance with Eq. (1). When only 5-dimensional LSFs were extracted (leading to 5 bands), a feature of 2200 dimensions was obtained. In order to deal with this unevenness and large feature dimension, the ratio of distribution of energy values in every band was computed. Along with this, the bands were also graded based on total energy content and this band sequence was also used to form LSF-RGS along with the mean and standard deviation values of each band. Thus when 5-dimensional LSFs were extracted for a clip of 5 s, a feature of 20 dimensions (5 ratio values +5 grades +5 means +5 standard deviations) was obtained as compared to the 2200-dimensional features. Similarly for the 10-, 15-, and 20-dimensional LSFs, LSF-RGS of 40, 60, and 80 dimensions was obtained. The trend of the feature values for the 40-dimensional features of D1 (best result, detailed in later section) is presented in Fig. 2.

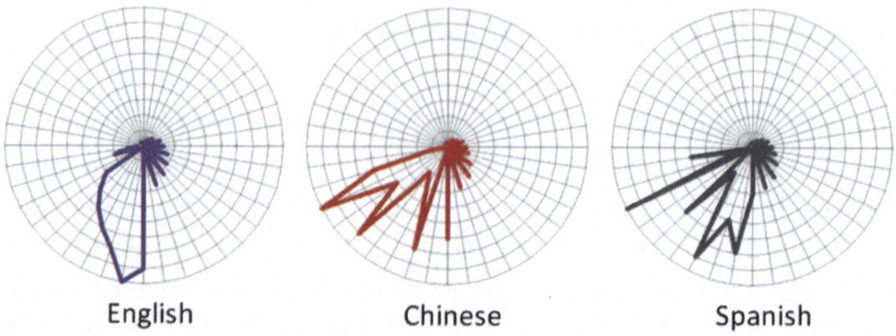

Fig. 2 Trend of the 40-dimensional feature values for English, Chinese, and Spanish

3.4 RNN-based Classification

Recurrent Neural Network (RNN) [3] is a specific neural network used particularly for classification of sequences. It differs from the standard neural networks on the basis of its capability of preserving the states [14] which helps in sequential data classification. It is a generalization of DNNs which can cover longer sequences [15]. The present state S_n is presented as under where S_{n-1} is the previous state, X_n is the input at n*th* state, and F_w is a recursive function.

$$S_n = F_w(S_{n-1}, X_n). \tag{6}$$

Each of the weights is multiplied with respective weights, and the recursive function is a *tanh* function which is presented below:

$$S_n = tanh(W_S S_{n-1} + W_X X_n). \tag{7}$$

The final output Y_n is obtained by introducing a weight W_y to S_n as shown below:

$$Y_n = W_y * S_n. \tag{8}$$

Long Short-Term Memory-RNN (LSTM-RNN) is the solution of the vanishing gradient problem encounters by simple RNNs [15]. It has the ability of solving complex as well as extended problems in which RNN faces difficulties [16]. A single LSTM block consists of long-term memory often termed as the cell state. It involves three gates: forget gate, input gate, and output gate. The input gate (i_n) determines the values associated in determination of the new state shown in Eq. (9) where W_i is the associated weight.

$$i_n = \sigma(W_i S_{n-1} + W_i X_n) \tag{9}$$

The forget gate determines the values required to be discarded or forgotten from the previous state while learning the present state values. It is mathematically illustrated below where W_f is the associated weight:

$$f_n = \sigma(W_f S_{n-1} + W_f X_n). \tag{10}$$

The output gate determines the values from the present state that will be used for determining the next state. It is provided below where W_o is the associated weight.

$$o_n = \sigma(W_o S_{n-1} + W_o X_n). \tag{11}$$

The intermediate cell state c_n can be represented as below:

$$c_n = tanh(W_c S_{n-1} + W_c X_n), \tag{12}$$

which produces cell state C_n as

$$C_n = (i_n * c_n) + (f_n * c_{n-1}). \tag{13}$$

The new state h_n thus obtained is

$$h_n = o_n * tanh(C_n). \tag{14}$$

In the present work, embedding layer was used and the matrix was initialized with a value of 300 followed by a 100-dimensional LSTM layer. The output of this layer was carried forward to a 100-dimensional layer having a ReLU activation function. Now the output of this layer was passed to the second 50-dimensional dense layers with the same activation function. Finally, the values of this layer were provided to the final three-dimensional dense layer having softmax activation. The dimensionality of the layers as well as the number of layers and activation functions were set based on trial. Categorical cross-entropy loss function and an adam-based optimizer were used in the experiments combined with fivefold cross-validation and a maximum of 100 training iterations.

4 Results and Discussion

Each of the datasets was subjected to the RNN-based classifier whose results are presented in Table 3. It can be observed from the Table that the best result was obtained for D1 with 40-dimensional features. This also points toward the systems' capability of working with clips of short durations which is often required in real-world scenario.

The confusion matrix for the best result is presented in Table 4.

It can be seen from the table that English–Chinese was the highest confused pair. 3.42% of the total instances were confused amid Chinese and English. Chinese–Spanish constituted the next highest confused pair leading to 2.20% misclassification over the total set of instances. One probable reason for these misclassifications is the utterance of keywords from different languages. It was observed that in many

Table 3 Obtained accuracies using the different feature dimensions for the engendered datasets

Dataset	Feature Dimension			
	20	40	60	80
D1	91.13	**95.22**	92.42	92.30
D2	93.44	89.52	92.96	92.68
D3	92.07	88.52	93.86	92.53
D4	83.79	88.18	93.95	92.70

Table 4 Confusion matrix for D1 with 40-dimensional features (best result)

	English	Chinese	Spanish
English	6943	266	35
Chinese	484	6762	143
Spanish	1	119	7155

Table 5 Performance of different popular classifiers on D1 with 40-dimensional features (best result)

Classifier	Accuracy (%)
Naïve Bayes	76.18
BayesNet	84.40
RBF Network	78.27
Logistic	87.87
MLP classifier	89.39
RNN	**95.22**

clips, speakers used names of places, brands, people, etc. which are common to all languages, thereby leading to ambiguity. It was also observed that clips from different languages had similar background music at times which further led to the confusions. We had also subjected the same dataset to 10-fold cross-validation but obtained a lesser accuracy of 90.04%. A few other popular classifiers in pattern recognition problems from WEKA [17] in the thick of Naïve Bayes, BayesNet, RBF network, logistic, and MLP classifier were also applied to the same whose results are presented in Table 5.

We had also carried out Friedman test [18] on the best performing dataset for checking statistical significance. The dataset was divided into five parts (N) which were subjected to six classifiers (k) each. The obtained accuracies for the subparts of D2 for the six classifiers along with their ranks based on performance are presented in Table 6.

The Friedman statistic (χ_F^2) was computed with the aid of Table 6 in accordance with Eq. (15), where R_j represents the mean rank of the *jth* classifier.

Table 6 Obtained ranks and accuracies for parts of D1

Classifiers		Parts of the dataset					Mean rank
		1	2	3	4	5	
Bayes Net	A	96.1	92.19	92.85	88.15	97.4	2.0
	R	(1.0)	(3.0)	(2.0)	(3.0)	(1.0)	
Naive Bayes	A	86.69	78.36	79.36	76.51	81.68	6.0
	R	(6.0)	(6.0)	(6.0)	(6.0)	(6.0)	
Logistic	A	92.67	91.76	89.77	92.72	95.56	2.8
	R	(3.0)	(4.0)	(4.0)	(1.0)	(2.0)	
MLP Classifier	A	92.69	92.44	89.84	92.53	95.28	2.4
	R	(2.0)	(2.0)	(3.0)	(2.0)	(3.0)	
RBF Network	A	90.8	88.26	82.69	82.47	83.25	5.0
	R	(5.0)	(5.0)	(5.0)	(5.0)	(5.0)	
RNN	A	91.9	95.11	93.54	87.15	94.78	2.8
	R	(4.0)	(1.0)	(1.0)	(4.0)	(4.0)	

$$\chi_F^2 = \frac{12N}{k(k+1)} \left[\sum_j R_j^2 - \frac{k(k+1)^2}{4} \right]. \tag{15}$$

The standard value of (χ_F^2) for the afore-mentioned values of k and N was found to be 15.086 at a significance level of 0.01. We obtained a value of 18.49, thereby rejecting the null hypothesis.

5 Conclusion

In this paper, a RNN-based approach is presented for language identification. Experiments were performed with the top three spoken languages in the world, namely, Chinese, Spanish, and English. The system was tested with more than 30 h of YouTube data, and encouraging results were obtained. In future, we will experiment with a larger dataset encompassing more languages. We also plan to use other acoustic features and machine learning techniques coupled with different preprocessing methodologies to obtain further improvement of the system performance. Finally, we also plan to use deep learning-based approaches in the feature extraction stage as well and observe the performance of the same.

References

1. Dudley, H.: The vocoder. Bell. Labs. Rec. **17**, 122–126 (1939)
2. Srivastava, B.M.L., Vydana, H., Vuppala, A. K., Shrivastava, M.: Significance of neural phonotactic models for large-scale spoken language identification. In: International Joint Conference on Neural Networks (IJCNN), pp. 2144–2151 (2017)
3. Tang, Z., Wang, D., Chen, Y., Li, L., Abel, A.: Phonetic temporal neural model for language identification. IEEE/ACM Trans. Audio, Speech, Lang. Process. **26**(1), 134–144 (2018)
4. Watanabe, S., Hori, T., Hershey, J.R.: Language independent end-to-end architecture for joint language identification and speech recognition. In: Automatic Speech Recognition and Understanding Workshop (ASRU), pp. 265–271. IEEE (2017)
5. Zissman, M.A., Singer, E.: Automatic language identification of telephone speech messages using phoneme recognition and n-gram modeling. In: ICASSP, pp. 305–308. IEEE (1994)
6. Gupta, M., Bharti, S.S., Agarwal, S.: Implicit language identification system based on random forest and support vector machine for speech. In: 2017 4th International Conference on Power, Control & Embedded Systems (ICPCES), pp. 1–6. IEEE (2017)
7. He, J., Zhang, Z., Zhao, X., Li, P., Yan, Y.: Similar language identification for uyghur and kazakh on short spoken texts. In: 2016 8th International Conference on Intelligent Human-Machine Systems and Cybernetics (IHMSC), vol. 2, pp. 496–499. IEEE (2016)
8. Jin, M., Song, Y., McLoughlin, I., Dai, L.R., Jin, M., Song, Y., McLoughlin, I., Dai, L.R.: LID-senones and their statistics for language identification. IEEE/ACM Trans. Audio, Speech Lang. Process. (TASLP), **26**(1), 171–183 (2018)
9. Madhu, C., George, A., Mary, L.: Automatic language identification for seven Indian languages using higher level features. In: 2017 IEEE International Conference on Signal Processing, Informatics, Communication and Energy Systems (SPICES), pp. 1–6. IEEE (2017)
10. https://www.ethnologue.com/statistics/size Visited on 10.11.2018
11. https://www.youtube.com Visited on 10.11.2018
12. Mukherjee, H., Obaidullah, S.M., Santosh, K.C., Phadikar, S., Roy, K.: Line spectral frequency-based features and extreme learning machine for voice activity detection from audio signal. Int. J. Speech Technol. 1–8 (2018)
13. Paliwal, K.K.: On the use of line spectral frequency parameters for speech recognition. Digit. Signal Process. **2**(2), 80–87 (1992)
14. Lipton, Z.C., Berkowitz, J., Elkan, C.: A critical review of recurrent neural networks for sequence learning (2015). arXiv:1506.00019
15. Li, J., Mohamed, A., Zweig, G., Gong, Y.: LSTM time and frequency recurrence for automatic speech recognition. In: 2015 IEEE Workshop on Automatic Speech Recognition and Understanding (ASRU), pp. 187–191. IEEE (2015)
16. Hochreiter, S., Schmidhuber, J.: Long short-term memory. Neural Comput. **9**(8), 1735–1780 (1997)
17. Hall, M., Frank, E., Holmes, G., Pfahringer, B., Reutemann, P., Witten, I.H.: The WEKA data mining software: an update. SIGKDD Explor. **11**(1), 10–18 (2009)
18. Demšar, J.: Statistical comparisons of classifiers over multiple data sets. J. Mach. Learn. Res. **7**, 1–30 (2006)

On the Determination of Graphic Integer Sequence from Graph Integrity

Debajit Sensarma and Samar Sen Sarma

Abstract Role of networks in our daily lives is very important. Effectiveness of network decreases with the breaking down of some nodes or links. So, less vulnerable communication network is required for greater stability. Vulnerability is the measure of the resistance of the network after failure of communication links. In this article, a graph has been considered for modeling a network and integrity as the measure of vulnerability, and the aim is to construct a graphic integer sequence of a graph or network with maximum number of edges from a given vertex order or number of vertices and graph integrity in linear time. Experiments show some possible graphs with maximum number of edges which obeys the given vertex order and integrity.

Keywords Vulnerability · Graphic integer sequence · Integrity · Tenacity · Rupture degree · Scattering number

1 Introduction

A communication network contains nodes or processors and communication links. Link cuts, software errors, node in interruption, hardware, and also transmission failures at various points can cause service interruption for a long time and as consequence effectiveness may be lost. These events are termed as the vulnerability of communication networks [1]. Two parameters are important for analyzing vulnerability of a communication network and they are (i) the number of nonfunctioning nodes and (ii) the remaining group size within which mutual communication can occur.

D. Sensarma (✉)
Department of Computer Science, Vivekananda Mission Mahavidyalaya,
Vidyasagar University, Midnapore, India
e-mail: debajit.sensarma2008@gmail.com

S. S. Sarma
Department of Computer Science & Engineering, University of Calcutta, Kolkata, India
e-mail: sssarma2001@yahoo.com

S. Kundu et al. (eds.), *Proceedings of the 2nd International Conference on Communication, Devices and Computing*, Lecture Notes in Electrical Engineering 602, https://doi.org/10.1007/978-981-15-0829-5_44

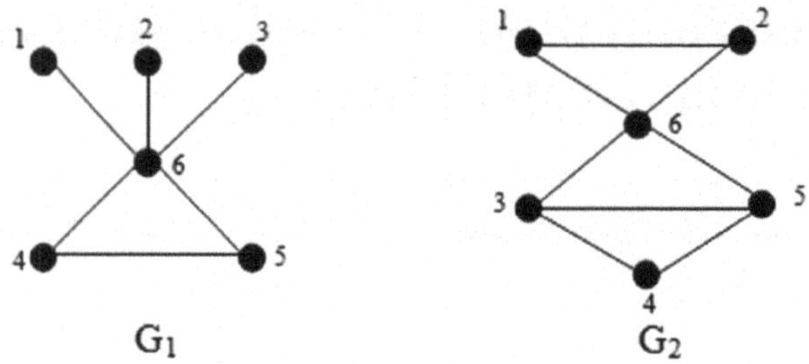

Fig. 1 Connectivity and integrity comparison of graph G_1 and G_2

> In an adversarial relationship, it would be desirable for an opponent network to be such that the two quantities can be made to be simultaneously small [2].

Many graph-theoretic parameters are there that can be used to measure the vulnerability of communication networks which include connectivity, tenacity, integrity, toughness, binding number, and more [1]. This paper considers integrity as the measure of vulnerability as it is a useful measure of vulnerability with respect to other parameters [3]. The concept of integrity of a graph was introduced in [4]. According to [2]

> Integrity of a simple connected graph G is defined by $I(G) = \min\{|S| + m(G - S)\}$, where S and $m(G - S)$ denote the subset of vertices V and the order of the largest component of $G - S$, respectively.

In Fig. 1, a small comparison between connectivity [5, 6] and integrity as the measure of vulnerability has been given. It can be seen that integrity gives good results. The connectivity of both graphs G1 and G2 is 1 but the integrity of G1 and G2 is 3 and 4, respectively. So G2 is less vulnerable than G1.

An integer sequence is an ordered list of integers [7, 8]. There are various integer sequences. This article considers integer sequences which are graphic. Graphic integer sequence [9, 10] is a universal representation structure of a graph. It is an inherent characteristic of any graph. If the labeling is added with a graphic integer sequence, a graph can be uniquely drawn. So, this is a universal superset of all other graph representations [11]. As at least one graph can be drawn from the given graphic integer sequence, the proposed algorithm has a very important application in constructing less vulnerable network

Authors of the article are interested in the following scenario: for any given two integers n and I such that $2 \leq I \leq n$, how a network can be constructed, where n is the number of vertices and I is the integrity.

The article is organized as follows: Sect. 2 illustrates the preliminary ideas behind the proposed technique. Section 3 of this paper contains some known results related to graph integrity. Section 4 contains some related works with vulnerability measurement. In Sect. 5, the proposed algorithm has been described and in Sect. 6, an algorithm is illustrated with an example. Section 7 explains some results. Concluding remarks are given in Sect. 8.

2 Preliminaries

This section contains some terms and theorems which are acting as the basis of the proposed algorithm. For various graph-theoretic terminologies, readers may refer to [5, 6, 12].

Definition 1 *"A sequence* $x = d_1, d_2, d_3, \ldots, d_n$ *of nonnegative integers is said to be graphic sequence if there exists a graph G whose vertices have degree* d_i *[*$1 \le i \le n$*] and G is called the realization of x"* [13].

Theorem 1 *"For a given graph G and an integer k, the problem of deciding whether the integrity of G is at most k is NP-Complete, even for planar graphs"* [14].

Theorem 2 *Graph G(V, I(G)) has the maximum number of edges. i.e.,*

$$\frac{(I(G) - 1)\,(I(G) - 2)}{2} + (V - I + 1)(I - 1)$$

Proof Let I set of the graph be $S \subseteq V$.

So, $I(G) = |S| + m(G - S)$

Let the components of $G - S$ be $V_1, V_2, \ldots, V_C$,

So, $G - S = G_1 \cup G_2 \cup \cdots \cup G_C$ and $I(G) = |S| + m(G - S), \sum_{i=1}^{C} V - |S|$.

Now, let $|S| = t$.

To achieve the maximum number of edges of a graph G, the following conditions must be held.

(a) G(S) is the complete graph of G.
(b) $G_i[\forall i = 1, \ldots, C]$ are complete.
(c) All vertices in S must be adjacent to all the vertices of $G_i[\forall i = 1, \ldots, C]$.

Let $f(V_1, V_2, \ldots, V_C, t)$ denote the number of edges of the graph G, then

$$\begin{aligned} f(V_1, V_2, \ldots, V_C, t) &= \sum_{i=1}^{C} \binom{V_i}{2} + \binom{t}{2} + t\sum_{i=1}^{C} V_i \\ &= 1/2(\sum_{i=1}^{C} V_i^2 - V_i) + \binom{t}{2} + t\sum_{i=1}^{C} V_i \\ &= 1/2\sum_{i=1}^{C} V_i^2 - 1/2\sum_{i=1}^{C} V_i + \binom{t}{2} + t\sum_{i=1}^{C} V_i \\ &= 1/2(\sum_{i=1}^{C} V_i^2) + (t - 1/2)\sum_{i=1}^{C} V_i + \binom{t}{2} - \sum_{1<i<j<C} V_i V_j \end{aligned} \quad (1)$$

So, to get the maximum value of $f(V_1, V_2, \ldots, V_C, t)$, $\sum_{1<i<j<C} V_i V_j$ must be minimum and $1 \le V_i \le V - t - (C - 1)$ [i = 1, …, C].

The problem becomes

$$\text{Minimize } h(N) = \sum_{1<i<j<C} V_i V_j \text{ such that } \begin{cases} 1 \le V_i \le V - t - (C-1) \\ i = 1, 2, \ldots, C \\ \sum_{i=1}^{C} V_i = V - t \\ V_i \in Z \\ Z = positive\ integers \\ N = (V_1, \ldots, V_C) \end{cases}$$

Suppose, $N^0 = (V_1{}^0, \ldots V_C{}^0)$ is an arbitrary solution of the above nonlinear programming. Let V_j^0 be the first number larger than 1, i.e., $N^0 = (1, 1, \ldots, 1, V_j{}^0, \ldots, V_C{}^0)$. Then construct new solution $N' = (1, 1, \ldots, 1, V_{j+1}^0, \ldots, V_C^0)$. Then, $h(N') \le h(N^0)$.

Continuing like this, an optimal solution can be achieved, i.e., $N' = (1, 1, \ldots, V - t - C + 1)$, where $V_C = V - t - C + 1$.

So, $f(V_1, V_2, \ldots, V_C, t)$ is minimized.

Now, substituting the values in Eq. (1)

$$f(V_1, V_2, \ldots, V - t - C + 1, t) = \binom{V - t - C + 1}{2} + \binom{t}{2} + t(V - t) \quad (2)$$

As Eq. (2) must be minimized

$$\therefore \quad V - t - C + 1 = 1$$
$$\therefore \quad V - t - C = 0 \quad (3)$$

Finally, after substituting $t = I(G) - 1$ in Eq. (2)

$$F(1, 1, \ldots, t) = \frac{(I(G) - 1)(I(G) - 2)}{2} + (V - I + 1)(I - 1)$$

■

3 Known Results on Graph integrity

Theorem 3 [2] *For any graph G when I(G) is the integrity of G, then*

(a) $I(G) \le \alpha(G) + 1$,
(b) $I(G) \ge \delta(G) + 1$ [where $\delta(G)$ = Minimum vertex degree of G],

(c) I (G) ≥ χ (G) [where χ (G) = Chromatic number of G], and
(d) I (G) = κ (G) +1 if and only if κ (G) = α (G) [where κ (G) = Connectivity of G].

Theorem 4 [2] *The integrity of*

(a) The Complete graph K_n is n [where n is the number of vertices],
(b) The Null graph-K_n is 1,
(c) The P_n is $\lceil 2\sqrt{n+1} \rceil - 2$,
(d) The Cycle C_n is $\lceil 2\sqrt{n} \rceil - 1$, and
(e) The Complete Bipartite Graph $K_{m,p}$ is 1 + min {m, p}.

4 Related Works

There are many graph-theoretic parameters for measuring the vulnerability of networks. This section contains some works related to that work. First, in [15, 16], authors used connectivity as the measure of vulnerability. Next, in [17], author discussed the relationship between safe number and integrity of graph and in [18], authors studied the bounds and relationships among scattering number, integrity, and tenacity which are better than other parameters for measuring stability of networks. Besides this, authors of [19–23] studied integrity of fuzzy graphs, hub integrity, relationship among integrity and component order connectivity, domination integrity and edge, and hub integrity of the graph, respectively. Next, in [14, 24], authors used tenacity as the vulnerability measure. Scattering number and rupture degree have been discussed in [25–30]. The authors of [3, 31–33] proved that computing the vulnerability parameters such as integrity and scattering number of a graph is NP-hard in general.

5 Proposed Algorithm

In this algorithm, a graphic integer sequence is constructed from the given vertex order and integrity. Also, a possible graph has been produced that obeys the graphic integer sequence. The details of graph construction from the given graphic integer sequence can be found in [9]. The time complexity of the algorithm proposed in this article is O(n) where n is the number of vertices. This is the main advantage of the proposed algorithm.

Algorithm:

Input: Integrity (I(G)) and number of vertices n of the graph G.

Output: Sequence of nonnegative integers $d_1, d_2, \ldots, d_n$, number of vertices (n).

Begin

Step 1: If I(G) =n, then construct graphic integer sequence x= (d_1 = n-1, d_2 = n-1,…, d_n =n-1) and goto step 4.

Else goto step 2.

Step 2: If 2 ≤ I(G) ≤ n-1 then

i) Construct a graphic integer sequence x= (d_1 = I(G)-2, d_2 = I(G)-2, …, $d_{I(G)-1}$ =I(G)-2).

ii) Connect other (n-I(G)+1) vertices with the graphic integer sequence x and goto step 3, else goto step 5.

Step 3: Construct the graphic integer sequence x_F = (d_1= I(G)-2 + (n-I(G) +1) = n-1, d_2= n-1, …, $d_{I(G)-1}$ = n-1, $d_{I(G)}$ = I(G)-1, $d_{I(G)+1}$ = I(G) -1,…,d_n= I(G) -1) and goto step 4.

Step 4: Construct the graph from the graphic integer sequenced produced from step 3 using algorithm depicted in [6].

Step 5: End

6 Illustration with an Example

The examples only contain graphic integer sequence construction. The details of graph construction from the given graphic integer sequence can be found in [9].

Example 6.1. Consider a graph G with number of vertices n = |V| = 5 and integrity I(G) = 5.

According to step 1 of the proposed algorithm, the graphic integer sequence will be x = (4, 4, 4, 4, 4) with the maximum edges = (4 × 3)/2 + 1 × 4 = 10 (from Theorem 2). The graph is shown in Fig. 2.

Example 6.2. Consider a graph G with number of vertices n = |V| = 5 and integrity I(G) = 3.

According to step 1 of the proposed algorithm, the graphic integer sequence will be

x = (1, 1) with the maximum edges = (4 × 3)/2 + 1 × 4 = 10 (from Theorem 2).

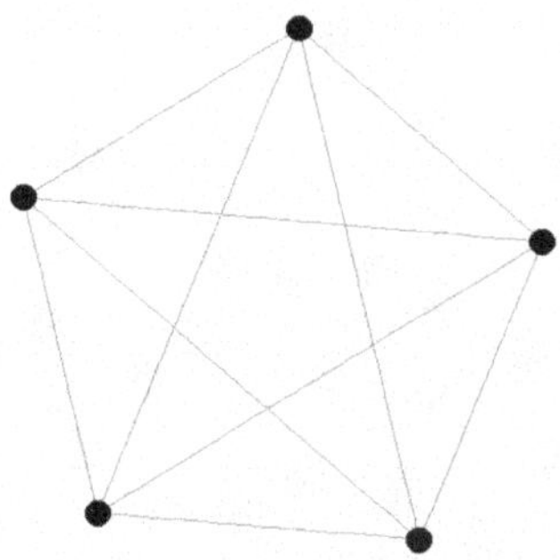

Fig. 2 Graph with five vertices and graphic integer sequence = (4, 4, 4, 4, 4)

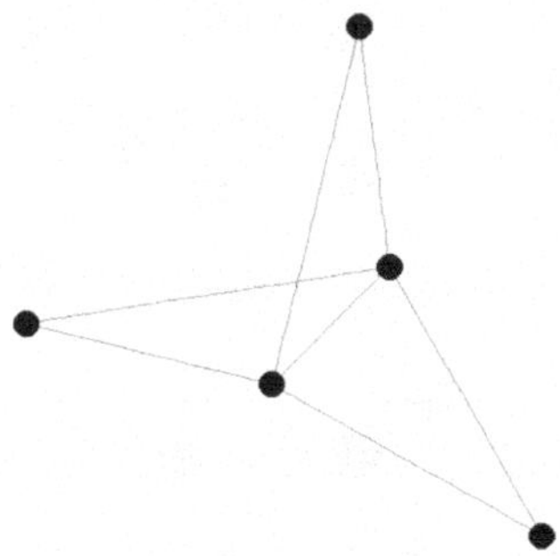

Fig. 3 Graph with five vertices and graphic integer sequence = (4, 4, 2, 2, 2)

$x_F = (4, 4, 2, 2, 2)$
Maximum edges = $(2 \times 1)/2 + 3 \times 2 = 7$. The graph is shown in Fig. 3.

7 Results

The algorithm proposed in the article has been coded in C, compiled using Borland C++ 5.0 compiler in Win32 mode and in Intel® AtomTM 1.60 Hz Processor with 1.0 GB DDR2 RAM. Figures 4, 5, 6, and 7 show some possible graphs with given vertex order and integrity. Table 1 shows the execution time in milliseconds and the maximum number of edges produced by the algorithm where the number of vertices or vertex order and the graph integrity is given.

All graphs from the graphic integer sequence produced by the algorithm were generated with software R [34].

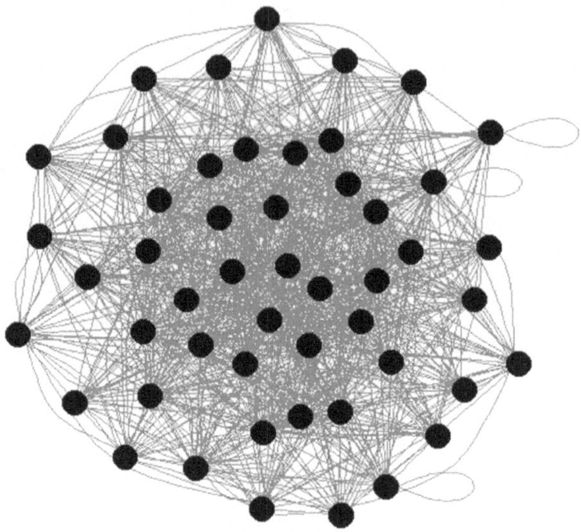

Fig. 4 One possible graph with vertex order = 50, integrity = 30

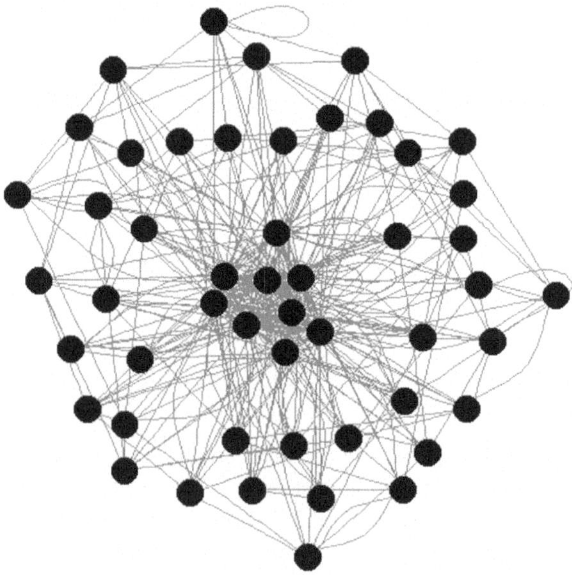

Fig. 5 One possible graph with vertex order = 50, integrity = 10

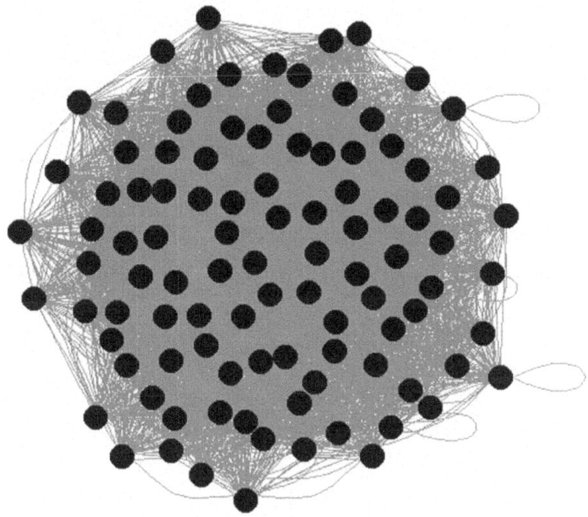

Fig. 6 One possible graph with vertex order = 100, integrity = 70

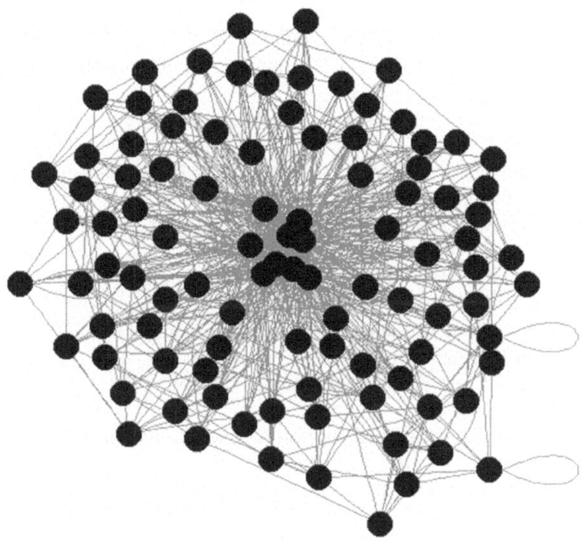

Fig. 7 One possible graph with vertex order = 100, integrity = 10

8 Conclusion

This article combines universal graph representation, i.e., graphic integer sequence with the integrity to measure the network vulnerability. The authors present and prove a formula to construct a network with the maximum number of edges where vertex

Table 1 Execution time required and the maximum number of edges produced by the proposed algorithm

Number of vertices	Graph integrity	Maximum number of edges	Time in millisecond
500	300	104,650	31
1000	900	494,450	85
5000	1000	4,495,500	533
10,000	7000	45,493,500	1187
15,000	10,000	99,990,000	1803
20,000	16,000	191,988,000	2460

order and integrity are given. Though integrity is a useful parameter for vulnerability measure and has been studied extensively, there are many unsolved problems. In future, author will try to cope with those problems.

References

1. Ye, Q.: On vulnerability of power and total graphs. WSEAS Trans. Math. **11**, 1028–1038 (2012)
2. Bagga, K.S., Beineke, L.W., Goddard, W.D., Lipman, M.J., Pippert, R.E.: A survey of integrity. Discret. Appl. Math. **37**, 13–28 (1992)
3. Moazzami, D.: Vulnerability in graphs—a comparative survey. J. Combin. Math. Combin Comput. **30**, 23–32 (1999)
4. Barefoot, C.A., Entringer, R., Swart, H.: Vulnerability in graphs—a comparative survey. J. Combin. Math. Combin. Comput. **1**(38), 13–22 (1987)
5. Deo, N.: Graph Theory with Applications to Engineering and Computer Science. Courier Dover Publications (2017)
6. Hakimi, S.L.: On realizability of a set of integers as degrees of the vertices of a linear graph. I. J. Soc. Ind. Appl. Math. **10**(3), 496–506
7. Sloane, N.J.A.: A Handbook of Integer Sequences. Academic Press (2014)
8. Sloane, N.J.: The on-line encyclopedia of integer sequences (2003)
9. Knuth, D.E.: The Art of Computer Programming. Combinatorial Algorithms, Part 1, vol. 4A (2011)
10. Bondy, J.A., Murty, U.S.R.: Graph theory with applications, vol. 290. Macmillan, London (1976)
11. Basuli, K.: Graphic integer sequence and its applications. Ph.D. Thesis, University of Calcutta (2017)
12. Chartrand, G., Zhang, P.: A First Course in Graph Theory. Courier Corporation (2013)
13. Basuli, K.: Role of degree sequence in determination of maximal clique of a graph. J. Glob. Res. Comput. Sci. **1**(2) (2010)
14. Cozzens, M.B., Moazzami, D., Stueckle, S.: The tenacity of the Harary graphs. J. Combin. Math. Combin. Comput. **16**, 33–56 (1994)
15. Doty, L.L.: Extremal connectivity and vulnerability in graphs. Networks **19**(1), 73–78 (1989)
16. Jun, W.U., Barahona, M., Yue-Jin, T., Hong-Zhong, D.: Natural connectivity of complex networks. Chin. Phys. Lett. **27**(7), 078902 (2010)
17. Fujita, S., Furuya, M.: Safe number and integrity of graphs. Discr. Appl. Math. (2018)
18. Li, Y., Shi, Y., Gu, X.: Spectrum bounds for the scattering number, integrity, tenacity of regular graphs. Future Gen. Comput. Syst. **83**, 450–453 (2018)

19. Saravanan, M., Sujatha, R., Sundareswaran, R., Sahoo, S., Pal, M.: Concept of integrity and its value of fuzzy graphs. J. Intell. Fuzzy Syst. (Preprint) 1–11 (2018)
20. Drange, P.G., Dregi, M., van't Hof, P.: On the computational complexity of vertex integrity and component order connectivity. Algorithmica **76**(4), 1181–1202 (2016)
21. Mahde, S.S., Mathad, V., Sahal, A.M.: Hub-integrity of graphs. Bull. Int. Math. Virtual Inst. **5**, 57–64 (2015)
22. Vaidya, S.K., Kothari, N.J.: Some new results on domination integrity of graphs. Open J. Discret. Math. **2**(03), 96 (2012)
23. Mahde, S.S., Mathad, V.: Edge hub-integrity of graphs
24. Moazzami, D.: Towards a measure of vulnerability, tenacity of a Graph. J. Algorithms Comput. **48**(1), 149–153 (2016)
25. Aslan, E.: A measure of graphs vulnerability: edge scattering number. Bull. Soc. Math. Banja Luka **4**, 53–60 (2014)
26. Bacak-Turan, G., Şenoğlu, M.Ü., Altundağ, F.N.: Neighbor rupture degree of some middle graphs. Süleyman Demirel Üniversitesi Fen Bilimleri Enstitüsü Dergisi **22**(1), 75–80 (2018)
27. Aytac, A.: The common-neighbourhood of a graph. Boletim da Sociedade Paranaense de Matemática **35**(1), 23–32 (2017)
28. Bacak-Turan, G., Demirtekin, E.: Neighbor rupture degree of gear graphs. CBU J. Sci. **12**(2), 319–323 (2017)
29. Aslan, E.: Weak-rupture degree of graphs. Int. J. Found. Comput. Sci. **27**(06), 725–738 (2016)
30. Aslan, E., Bacak-Turan, G.: Mean rupture degree of graphs. Univ. Politeh. Buch. Sci. Bull. Ser. Appl. Math. Phys. **78**(1), 233–242 (2016)
31. Clark, L.H., Entringer, R.C., Fellows, M.R.: Computational complexity of integrity. J. Combin. Math. Combin. Comput **2**, 179–191 (1987)
32. Li, F., Li, X.: Computing the rupture degrees of graphs. In: Proceedings of 7th International Symposium on Parallel Architectures, Algorithms and Networks, 2004, pp. 368–373. IEEE (2004)
33. Zhang, S., Li, X., Han, X.: Computing the scattering number of graphs. Int. J. Comput. Math. **79**(2), 179–187 (2002)
34. https://cran.r-project.org/manuals.html

An Automatic Classification of Magnetic Resonance Brain Images Using Machine Learning Techniques

R. Murugan

Abstract In medical image analysis, an automatic Magnetic Resonance (MR) brain image classification is very important. This research paper's objective is to classify the abnormality of MR brain image. This paper presents a Discrete Wavelet Packet to separate highlights from images, followed by using principal component analysis to decrease the measurement of highlights which were submitted to a Kernel support vector machine(KSVM). The procedure of K fold stratified cross-validation was utilized to upgrade the speculation of KVSM. The proposed method has been tested by 160 Magnetic Resonance brain images, both normal and abnormal which are collected from the Harvard Medical School website; it achieves better accuracy of classification as 99.38%. The performance of this proposed method has been evaluated with four different kernels. The comparison has been done by various states of the art methods. From the experimental data, our method was effective and rapid. This classification method can help doctors to diagnose the patients.

Keywords Magnetic Resonance images · Principle component analysis · Discrete wavelet transform · Kernel support vector machine

1 Introduction

Magnetic Resonance (MR) Imaging is a method that uses magnetic field and radio frequencies to view the internal organs of the human body [1]. It provides high-quality pictures because of which it will be easy to identify and characterize the diseases. So MRI plays a vital role in early detection and diagnosis. MRI is the best imaging modality to diagnose the brain tumor. It displays the slices of the brain which can be combined to form the 3D picture of the tumor [2]. A powerful method for feature extraction of MRI brain image is wavelet transform because it

R. Murugan (✉)
Department of Electronics and Communication Engineering, National Institute of Technology Silchar, Silchar 788010, Assam, India
e-mail: murugan.rmn@gmail.com

S. Kundu et al. (eds.), *Proceedings of the 2nd International Conference on Communication, Devices and Computing*, Lecture Notes in Electrical Engineering 602, https://doi.org/10.1007/978-981-15-0829-5_45

is suitable for time–frequency analysis [3]. It is used for de-noising and finding the discontinuities, irregularities, break point location, etc. Although many advantages are there, the main shortcoming of this method is intensive computation and storage. To overcome this issue, principal component analysis (PCA) is used to reduce the space of dimensionality and data analyzing the cost of computation by removing redundant data. However, it is not up to the mark in classification. In classification problems, SVM plays a vital role; it is one of the supervised machine learning algorithms. The biggest limitation of this method is the selection of kernel, speed, and it will not handle the multiclass. K-Nearest Neighbor algorithm is used for classification because of its easy interpretation and less computation time. It is cost effective and requires full training datasets [4].

As MRI is connected with a titanic information vault, manual interpretation of each picture winds up unimaginable, clearing course for automated mechanical assemblies' headway [5]. In completely modified procedures, the tumor shapes are obtained with no human intercession and are outstandingly charming on a fundamental level. Exactly when the sort of tumor does not fit the division exhibit as picked up from the readiness dataset, the division result may not be right. Specialists [6] propose a technique that naturally fragments cerebrum tumors using Markov irregular field show, in light of super-voxels and terms that get the power probabilities and edge prompts. Another exploration [7] shows a division for gliomas incorporating their subareas meanwhile isolating between edema, necrotic focus, and dynamic cells. This remarkable methodology relied upon decision timberlands. The approval was done on two instances of review III tumors and 38 instances of review IV tumors acquired from multichannel MR pictures. These programmed yields were appropriate for the estimation of tumor volume and supported intuitive treatment arranging. An examination going for irregular glioblastoma was proposed [8] to improve productivity and exactness of treatment response evaluation naturally.

A changed characterization of K-Nearest Neighbor (kNN) method was used by MR pictures from 13 patients to overview the alterations in tumor volume. This system was then differentiated and manual volumetric estimations and Macdonald's criteria. Despite all yields with infiltrative tumors having obscure edges, this method was appropriate. A high association existed between the manual tumor volume estimations and the modified strategy ($r = 0.96$) yet their match with Macdonald's criteria was only 68% in spite of the way that these yields were affirmed using Magnetic Resonance Spectroscopy (MRS) and moreover by a neuroradiologist. Researcher [9] portrayed gliomas from Susceptibility Contrast (DSC) imaging, to check whether tumor volume is learned using FCM bundling gave unclear symptomatic productivity as the manual significance of tumor volumes. It took around 4 min to make a twofold glioma volume for each patient by this methodology and 10 min physically. At the moment that the low quality and high-audit glioma volumes were considered, the customized system demonstrated more unmistakable affectability than manual technique: 83% for below-average gliomas and 69% for high-survey gliomas. In spite of the way that DSC imaging shows believability in the depiction of gliomas before the restorative system, these methods are kept to driving investigation arranged associations as it were. Automatic tumor division using multi-terrible data examination

[8], Artificial Neural Networks (ANN) [5, 10], Support Vector Machine classifier [11–14] and learning-based FCM gathering [9] strategies are promising procedures. The upsides of performing tumor division normally are viability in time, the nonattendance between and intra-spectator changeability and tumor depiction using solid benchmarks.

2 Methodology

The proposed strategy comprises the accompanying three phases, (1) Preprocessing (feature extraction and feature reduction), (2) Training the kernel SVM, (3) Load MRI brain images to be trained kernel SVM and predict the output. The proposed work block diagram is shown in Fig. 1.

2.1 Preprocessing

Feature extraction The 2D biorthogonal wavelet packet is connected to the information 2D MR brain images. A biorthogonal wavelet is a place the related wavelet change is invertible yet not by any stretch of the orthonormal functions. Designing biorthogonal wavelets permits a larger number of degrees of flexibility than orthogonal wavelets. The extra level of flexibility is the likelihood to develop symmetric wavelet capacities. Numerous filtering applications require channels with symmetrical coefficients to accomplish a direct stage. None of the symmetrical wavelet frameworks with the exception of Haar are having symmetrical coefficients. Yet, Haar is excessively deficient for some handy applications. Bi-symmetrical wavelet framework can be intended to have this property. That is our inspiration for outlining such wavelet framework. In filter analysis and synthesis, nonzero coefficients are not the same.

$$\Phi(t) = \sum_{k=0}^{N-1} h(k)\sqrt{2}\phi(2t - k) \tag{1}$$

and its translates form the scaling function of primal basis and let the corresponding space be ϕ

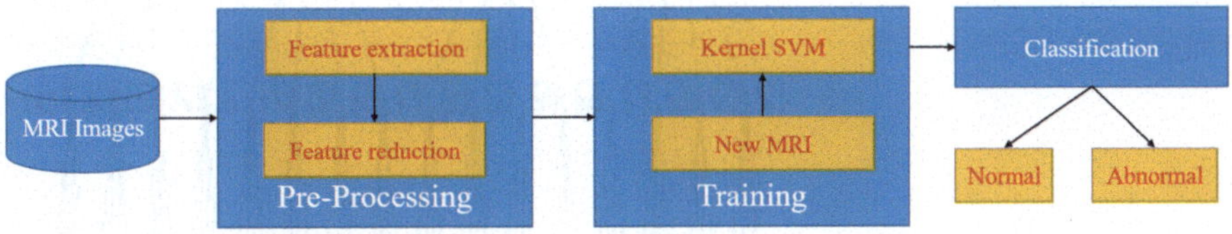

Fig. 1 Proposed methodology

$$\bar{\Phi}(t) = \sum_{k=0}^{N-1} \bar{h}(k)\sqrt{2}\bar{\phi}(2t - k) \quad (2)$$

and translates form the dual scaling function basis and corresponding space be ϕ. This means that $\phi(t)$ and its translates are not orthogonal among themselves but orthogonal to translates of three-level wavelet packet decomposition (shown in Figs. 2 and 3) are utilized to reduce the size of the total input image. The coefficients of wavelet packet represent the approximation coefficient of level 3 which is $32 \times 32 = 1024$ s.

This method has three effects: it orthogonalizes the parts of the info vectors so is uncorrelated with one another, it orchestrates the resulting symmetrical portions so that those with the greatest assortment begin things out, and it takes out those sections contributing the base to the assortment in the dataset. It should be seen that the information vectors be standardized to have zero mean and solidarity difference before performing PCA. Standardization is a standard strategy. Insights regarding PCA could be found in Ref. [15].

Kernal SVM The presentation of SVM is a historic point in machine learning. The advantages of SVMs incorporate high exactness, rich numerical tractability, and direct geometric elucidation [16]. As of late, numerous enhanced SVMs have developed quickly, among which the piece SVMs are the most famous and viable. Part SVMs have the accompanying points of interest [17]: (1) Work extremely well by and by and have been strikingly effective in such differing fields as characteristic dialect arrangement, bioinformatics, and PC vision; (2) Have a couple of tunable parameters; and (3) Preparing frequently includes arched quadratic enhancement [18]. Consequently, arrangements are worldwide and typically one of a kind, hence maintain a strategic distance from the union to nearby minima displayed by other measurable learning frameworks, for example, neural systems. Conventional SVMs built a hyperplane to arrange information, so they cannot manage characterization

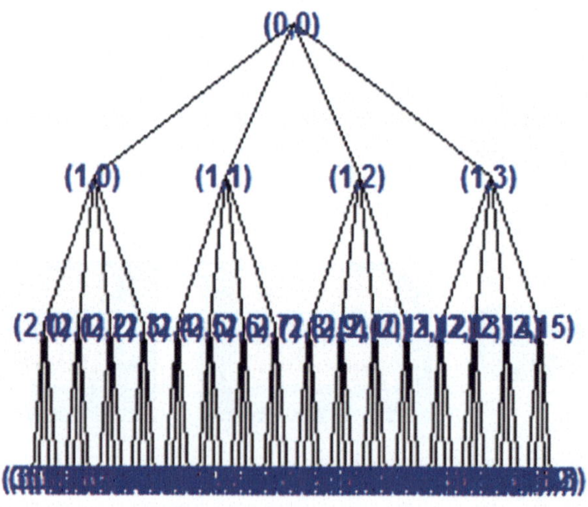

Fig. 2 A decomposition tree of three-level wavelet

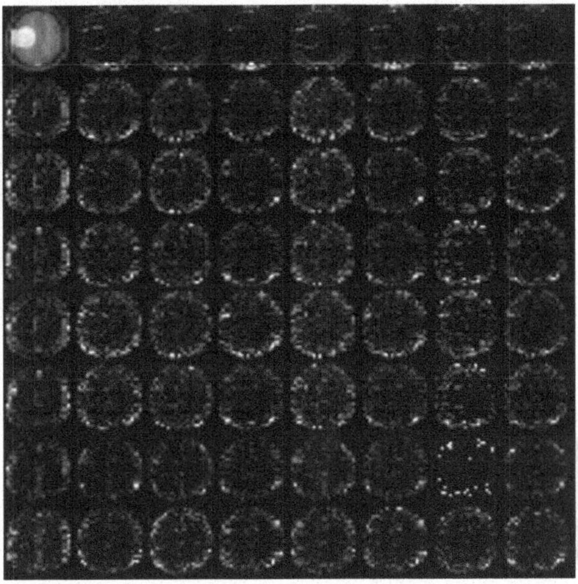

Fig. 3 2D Biorthogonal wavelet packet schematic diagram

issue of which the distinctive kinds of information situated at various sides of a hypersurface, the portion system is connected to SVMs [19]. The subsequent calculation is formally comparable, aside from that each dab item is supplanted by a nonlinear piece work. The kernel is represented to the transform $\phi(xi)$ by the equation, $k(xi, xj) = \phi(xi)\phi(xj)$.

3 Results and Discussion

3.1 Dataset

The datasets consist of T2-weighted MR brain images of 256×256 in-plane resolution, downloaded from the Harvard Medical School website, OASIS dataset, and ADNI dataset. The abnormal MR brain images of the dataset comprise the accompanying sicknesses: glioma, meningioma, Alzheimer's malady, Alzheimer's infection in addition to visual agnosia, Pick's ailment, sarcoma, and Huntington's ailment. Arbitrarily, there are 20 images for each sort of cerebrum that was chosen for testing.

3.2 Results

The proposed method was implemented in MATLAB (R2018b) where runs needed on average 0.0448 s for each image on a PC with Dell Alienware, i7 6th gen, GTX

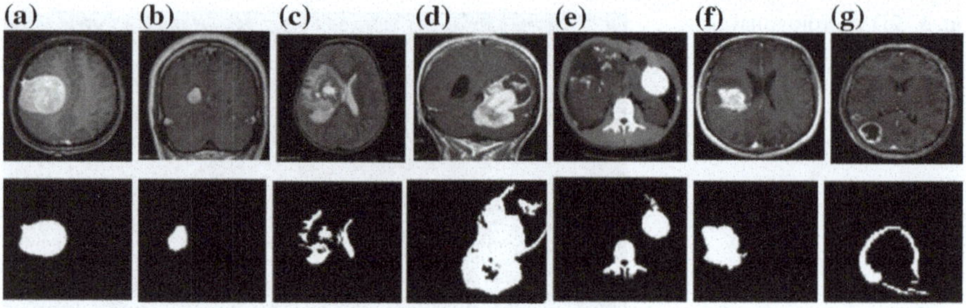

Fig. 4 MRI brain image samples and its segmentation: **a** glioma; **b** meningioma; **c** Alzheimer's disease; **d** Alzheimer's disease with visual agnosia; **e** Pick's disease; **f** sarcoma; **g** Huntington's disease

Fig. 5 Variances against no. of principle components (x axis is log scale)

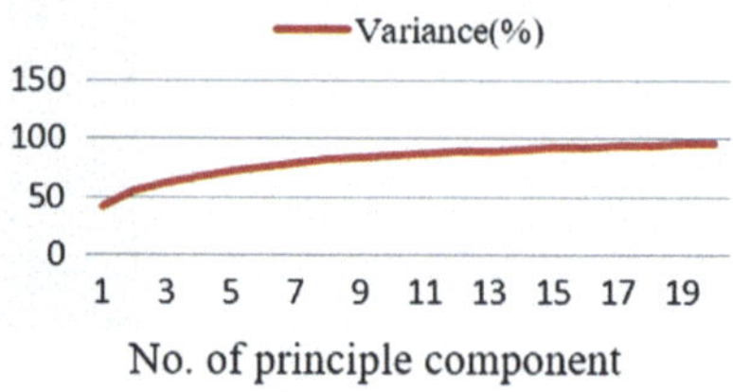

1080 GPU, 64 GB DDR4 RAM, 64-bit OS. The sample segmentation is shown in Fig. 4.

3.3 *Performance Analysis*

The variance versus the number of principle components from 1 to 20 is shown in Fig. 5. It demonstrates that only 19 principle components which are only 1.86% of the original features, could preserve 95.4% of total variance (shown in Fig. 5). The classification exactness correlation of six distinct calculations for a similar MRI dataset and the same number of pictures are shown in Table 1. The proposed technique has better precision with other cutting-edge strategies.

3.4 *Validation*

Since the classifier is set up by a given dataset, it may achieve high portrayal precision only for this readiness dataset, not yet for different independent datasets. To dodge this overfitting, we need to facilitate cross approval into our technique. Cross approval will not grow the last portrayal precision; nonetheless, it will make the classifier trustworthy and can be summed up to other free datasets.

Table 1 Comparison with other state-of-the-art methods

Algorithms	Classification accuracy (%)
Discrete Wavelet Transform+SOM [12]	94
Discrete Wavelet Transform+SVM with linear kernel [12]	96
Discrete Wavelet Transform+SVM with RBF-based kernel [12]	98
Discrete Wavelet Transform+PCA+ANN [10]	97
Discrete Wavelet Transform+PCA+KNN [10]	98
Discrete Wavelet Transform+PCA+ACPSO+FNN [25]	98.75
Discrete Wavelet Transform+PCA+KSVM [Proposed]	99.38

Cross-approval systems include three sorts: Random sub-assessing, K-overlay cross-endorsement, and Overlook one endorsement. The K-cover cross-endorsement is associated with account of its properties as direct, straightforward, and using all data for getting ready and endorsement. The part is to make a K-overlay bundle of the whole dataset, repeat K times to use K-1 folds for getting ready and a left cover for endorsement, finally make the screw-up rates of K tests ordinary. The schematic graph of 5-overlay cross approval has been shown in Fig. 6.

The K-overlay cross approval is connected due to its properties as fundamental, basic, and using all data for preparing and approval. The segment is to make a K-overlay parcel of the whole dataset, repeat K times to use K-1 folds for getting ready and a left wrinkle for endorsement, in conclusion, make the botch rates of K tests ordinary. The schematic outline of 5-overlap cross approval is shown in Fig. 6.

The K folds can be completely indiscriminately allotted; in any case, a couple of folds may have an extremely one of a kind courses from various folds. Thusly, stratified K-crease cross approval was used, where each cover has right around a comparative class appointments [20]. Another test is to choose the number of folds. In case K is set excessively huge, the inclination of the authentic mix-up rate estimator will be close to nothing, yet the difference in the estimator will be far reaching and the figuring will be dreary. Of course, if K is set pretty much nothing, the figuring time will decrease, the distinction of the estimator will be close to nothing, anyway, the inclination of the estimator will be broad [21]. In this examination, we have

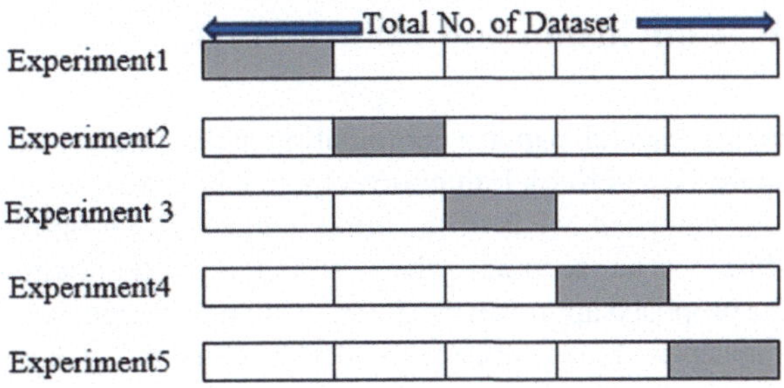

Fig. 6 A fivefold cross-validation

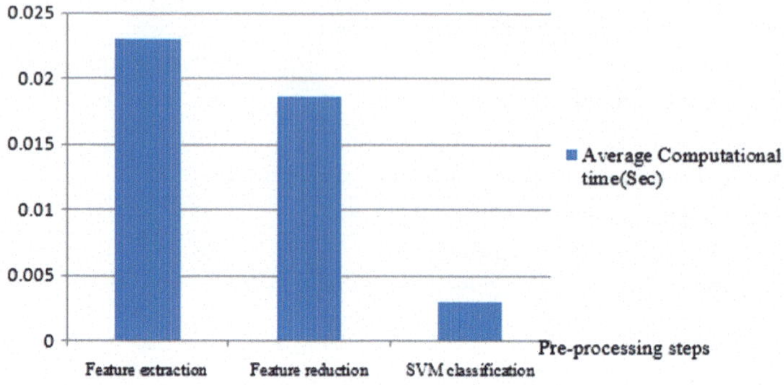

Fig. 7 Computation times at different stages

tentatively chosen K as 5 through the experimentation method, which suggests that, we accept parameter K varying from 3 to 10 with extending advance as 1, and after that we set up the SVM by each regard. Finally, we select the perfect K regard contrasting with the most significant request precision [22].

3.5 Timing Analysis

Computation time is another important factor to evaluate the classifier. The time for SVM training was not considered since the parameters of the SVM keep unchanged after training. The 160 images were tested by a classifier, recorded relating calculation time, registered the normal esteem, portrayed devoured time of various stages as shown Fig. 7. For each 256×256 image, the averaged computation time on feature extraction, feature reduction, and SVM classification is 0.023 s, 0.0187 s, and 0.0031 s, respectively. The feature extraction stage is the most time consuming as 0.023 s. The feature reduction costs 0.0187 s. The SVM order costs the minimum time just 0.0031 s. The aggregate calculation time for each 256×256 size picture is around 0.04 s, which is sufficiently fast for continuous analysis.

4 Conclusion and Future Work

In this research paper, a combination of discrete wavelet packet, principal component analysis, and Kernel support vector machine method have proposed to classify normal and abnormal MRIs of the brain images. The proposed method has achieved better classification accuracy when compared to other state-of-the-art methods. In future, the proposed approach can be tested to other large, benchmark, and publicly available datasets and also enhance the performance using deep learning.

References

1. Mohsen, H., El-Dahshan, E.S.A., El-Horbaty, E.S.M., Salem, A.B.M.: Classification using deep learning neural networks for brain tumors. Future Comput. Inf. J. **3**(1), 68–71 (2018)
2. Sergio, P., Raphael, M., Richard, M., Roland, W., Victor, A., Carlos, A.S., Mauricio, R.: Enhancing interpretability of automatically extracted machine learning features: application to a RBM-Random Forest system on brain lesion segmentation. Med. Image Anal. **44**(1), 228–244 (2018)
3. Geethu Mohan, M.: Monica Subashini: MRI based medical image analysis: Survey on brain tumor grade classification. Biomed. Signal Process. Control. **39**, 139–161 (2018)
4. Deepak, R., Ratnakar, D., Banshidhar, M., Shuihua W.: Combining extreme learning machine with modified sine cosine algorithm for detection of pathological brain. Comput. Electr. Eng. **68**, 366–380 (2018)
5. El-Dahshan, E.S.A., Mohsen, H.M., Revett, K., Salem, A.B.M.: Computer-aided diagnosis of human brain tumor through MRI: a survey and a new algorithm. Expert Syst. Appl. **41**(11), 5526–5545 (2014)
6. Menze, B., Reyes, M., Jakab, A., Gerstner, E., Kirby, J., Kalpathy-Cramer, J., Farahani, K.: NCI-MICCAI challenge on multimodal brain tumor segmentation. In: Proceedings of NCI-MICCAI BRATS (2013)
7. Zikic, D., Glocker, B., Konukoglu, E., Criminisi, A., Demiralp, C., Shotton, J., Thomas, O.M., Das, T., Jena, R., Price, S.J.: Decision forests for tissue-specic segmentation of high-grade gliomas in multi-channel MR. Med. Image Comput. Comput. Interv. **15**, 369–376 (2012)
8. Liberman, G., Louzoun, Y., Aizenstein, O., Blumenthal, D.T., Bokstein, F., Palmon, M., Corn, B.W., Bashat, D.B.: Automatic multi-modal MR tissue classication for the assessment of response to bevacizumab in patients with glioblastoma. Eur. J. Radiol. **82**, 87–94 (2013)
9. Emblem, K.E., Nedregaard, B., Hald, J.K., Nome, T., Due-Tonnessen, P., Bjornerud, A.: Automatic glioma characterization from dynamic susceptibility contrast imaging: brain tumor segmentation using knowledge-based fuzzy clustering. J. Magn. Reson. Imaging **30**, 1–10 (2009)
10. Havaei, M., Davy, A., Warde-Farley, D., Biard, A., Courville, A., Bengio, Y., Pal, C., Jodoin, P.M., Larochelle, H.: Brain tumor segmentation with deep neural networks. Med. Image Anal. **35**, 18–31 (2017)
11. Tanoori, B., Azimifar, Z., Shakibafar, A., Katebi, S.: Brain volumetry: an active contour model-based segmentation followed by SVM-based classication. Comput. Biol. Med. **41**, 619–632 (2011)
12. Sachdeva, J., Kumar, V., Gupta, I., Khandelwal, N., Ahuja, C.K.: A package-SFERCB-segmentation, feature extraction, reduction and classication analysis by both SVM and ANN for brain tumors. Appl. Soft Comput. J. **47**, 151–167 (2016)
13. Luts, J., Heerschap, A., Suykens, J.A.K., Van Huffel, S.: A combined MRI and MRSI based multiclass system for brain tumour recognition using LS-SVMs with class probabilities and feature selection. Artif. Intell. Med. **40**, 87–102 (2007)
14. Sachdeva, J., Kumar, V., Gupta, I., Khandelwal, N., Ahuja, C.K.: Multiclass brain tumor classication using GA-SVM. Dev. Syst. Eng. **97**, 182–187 (2011)
15. Zhang, Y., Wu, L., Wei, G.: A new classifier for polarimetric SAR images. Prog. Electromagn. Res. **94**, 83–104 (2009)
16. Martiskainen, P.: Cow behaviour pattern recognition using a three-dimensional accelerometer and support vector machines. Appl. Anim. Behav. Sci. **119**(2), 32–38 (2009)
17. Bermejo, S., Monegal, B., Cabestany, J.: Fish age categorization from otolith images using multi-class support vector machines. Fish. Res. **84**(2), 247–253 (2007)
18. Muniz, A.M.S.: Comparison among probabilistic neural network, support vector machine and logistic regression for evaluating the eect of subthalamic stimulation in Parkinson disease on ground reaction force during gait. J. Biomech. **43**(4), 720–726 (2010)
19. Acevedo-Rodrguez, J.: Computational load reduction in decision functions using support vector machines. Signal Process. **89**(10), 2066–2071 (2009)
20. Deris, A.M., Zain, A.M., Sallehuddin, R.: Overview of support vector machine in modeling machining performances. Proc. Eng. **24**, 308–312 (2011)

21. Chaplot, S., Patnaik, L.M., Jagannathan, N.R.: Classification of magnetic resonance brain images using wavelets as input to support vector machine and neural network. Biomed. Signal Process. Control. **1**(1), 86–92 (2006)
22. Zhang, Y., Wang, S., Wu, L.: A novel method for magnetic resonance brain image classification based on adaptive chaotic PSO. Prog. Electromagn. Res. **109**, 325–343 (2010)

Raspberry PI 3B+ Based Smart Remote Health Monitoring System Using IoT Platform

Samik Basu, Mahasweta Ghosh and Soma Barman (Mandal)

Abstract A portable smart health monitoring system with remote access to the physician is very essential for senior citizens and rural healthcare personnel. This can be achieved by a low-power, accurate, compact, cost-effective, user-friendly system which is capable of measuring the patient's vital parameters. These data are displayed locally at the patient-end and also regularly sent to the physician-end using the IoT platform. The proposed system uses noninvasive sensors to monitor the vital parameters like heart rate, SpO_2, and body temperature accurately. Raspberry Pi 3B+ minicomputer is used to integrate, process the sensors' data, and remotely communicate to the hospital(s) or the physician using MathWorks cloud and applet. The reliability of the uploaded vital parameter(s) is also verified with the locally displayed data. The use of Webhook applet to notify the physician about the health condition of the patient makes the proposed system novel.

Keywords Smart remote health monitoring system · Raspberry Pi 3B+ · MAX30100 · DS18B20 · IoT · ThingSpeak · Webhooks · IFTTT · Email

1 Introduction

Deaths in India due to cardiovascular diseases increased from 1.3 million in 1990 to 2.8 million in 2016, i.e., by 34.3%. The death rate from cardiovascular diseases in India in 2016 among adults aged 70 years or older was 2777 per 100,000, i.e., 46.6%. It is also found that mortality rates from ischaemic heart disease in populations aged 30–69 increased rapidly in rural India and surpassed those in urban areas between the

S. Basu (✉) · M. Ghosh · S. Barman (Mandal)
Institute of Radio Physics and Electronics, University of Calcutta, 92, A.P.C. Road, Kolkata 700009, India
e-mail: samikbasu2010@gmail.com

M. Ghosh
e-mail: mahasweta94g@gmail.com

S. Barman (Mandal)
e-mail: barmanmandal@gmail.com

S. Kundu et al. (eds.), *Proceedings of the 2nd International Conference on Communication, Devices and Computing*, Lecture Notes in Electrical Engineering 602, https://doi.org/10.1007/978-981-15-0829-5_46

years 2000 and 2015 [1]. So, in today's lifestyle, a smart remote health monitoring system is of absolute necessity in our homes as well as in rural healthcare facilities.

Several studies and researches have been done in the field of technology-based healthcare services. Kumar and Rajasekaran [2] in their paper propose the use of an IR transmitter and receiver for heart rate measurement and a thermistor for temperature measurement. Kodali et al. [3] in their proposed IoT based healthcare system use an LM35 temperature sensor to measure the patient's body temperature and upload the data to the cloud using a Zigbee connectivity. Thermistors and analog temperature sensors like LM35 are very useful for measuring the ambient or surface temperature but are not suitable for the measurement of core body temperature. Jain et al. [4] use ATmega 32 to monitor heart rate, body temperature, etc., and serially communicate continuously with external peripherals using the RS232 interface along with the GSM module. This continuous monitoring may be essential for chronically ill patients but will require huge storage facilities. Kaur and Jasuja [5] propose a health monitoring system using Raspberry Pi 2 involving the IBM Bluemix cloud service. However, ThingSpeak cloud is free, good for monitoring purpose and graphical visualizations compared to some other clouds [6].

IoT (*Internet of Things*) can be defined as the network of interconnected sensor-equipped electronic devices and servers that exchange data and connect with each other [7]. IoT plays a significant role to magnify the medical information system. The main objective of IoT in healthcare system is to achieve quality health services at a minimum cost. Our system measures three important vital parameters: heart rate, blood oxygen saturation, and body temperature accurately using MAX30100 and DS18B20 sensors. The processor Raspberry Pi 3B+ processes the sensed parameters. The vital patient physiological parameter (PPP) data is displayed locally and is also transmitted over the IoT platform to the ThingSpeak cloud. The corresponding physician gets notified regarding the patient's health daily via email and these vital data can also be downloaded regularly to the hospital's server. The main objectives of our system design are low power consumption, portability, cost-effectiveness, and accuracy.

The novelty of our system design is manifold as follows:

(i) DS18B20 has better accuracy as it can measure core body temperature.
(ii) MAX30100 integrates sensors for measurement of two vital parameters—heart rate and SpO_2. So, separate sensors are not required.
(iii) Use of Wi-fi connectivity which is cheap and secure compared to Zigbee connectivity in today's world, portable routers or mobile hotspots can help to connect our device to the Internet in almost every location.
(iv) Use of Webhook applet to notify the physician by email is a new approach in the healthcare system.
(v) Indicator module alerts the patient regarding their present health condition.
(vi) Push-button switch for selecting the desired sensor for measuring the patient's vital parameters.

The paper is organized as follows: Sect. 2 describes our system architecture. Section 3 presents the methodology of our entire health monitoring system. In Sect. 4, the results (graphical representation) of the vital parameters (as displayed by the system and as uploaded to the cloud) are shown along with the physician's analysis at the remote end. Finally, the paper is concluded in Sect. 5.

2 System Architecture

An overall block diagram of our proposed smart remote health monitoring system is shown in Fig. 1. The proposed system is divided into five main parts—Patient Body Parameter Sensing Unit, Embedded Processing Unit, Local Display Unit, Indicator Unit, and Remote Monitoring Unit.

2.1 Patient Physiological Parameter (PPP) Sensing Unit

In our proposed system, two integrated sensor modules are used to sense three vital parameters of the patient: one is *DS18B20* which is used to measure the patient's body temperature and another is *MAX30100* which measures both blood oxygen saturation (SpO_2) and heart rate. A push-button switch is used to select the sensor module which can measure the patient's desired vital body parameter.

DS18B20. It is a probe-ended digital thermometer sensor that measures the core body temperature accurately [8].

MAX30100. It is an integrated pulse oximeter and heart rate sensor which combines two LEDs (a red LED and an IR LED), a photodetector, optimized optics, and

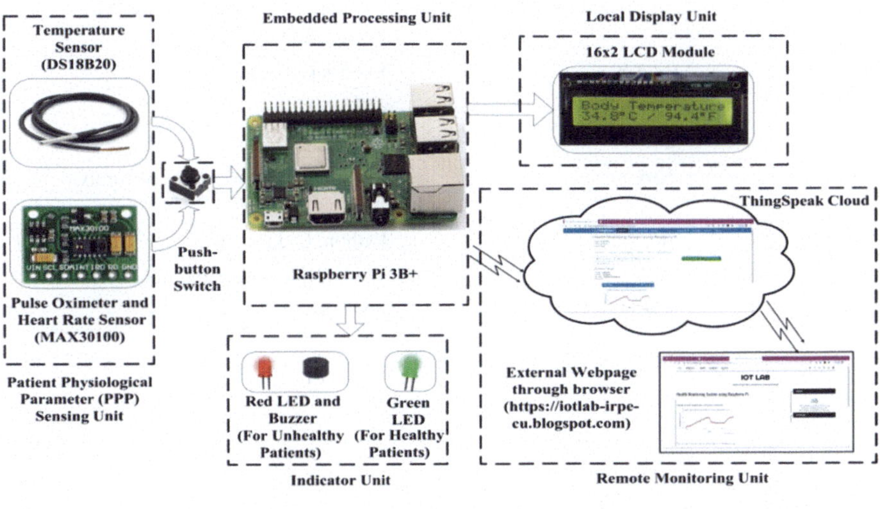

Fig. 1 Block diagram representation of the proposed smart remote health monitoring system

low-noise analog signal processing [9]. MAX30100 uses the *reflective photoplethysmography (PPG)* technique to measure blood oxygen saturation (SpO_2) level and heart rate [10].

2.2 Embedded Processing Unit

The embedded processing unit mainly comprises the *Raspberry Pi 3 model B+*, which uses the open-source *Raspbian OS*. Raspberry Pi 3B+ is a minicomputer with inbuilt modules which is suitable for integrating various application-oriented systems [11]. In our system design, we use Raspberry Pi 3B+ because of its several advantages compared to a popular microcontroller Arduino UNO as listed in Table 1 [12]. The codes for the remote health monitoring system are written in *Python 3 IDE* (Integrated Development Environment) [13].

2.3 Local Display Unit

The display unit consists of a 16-pin *16X2 LCD* (Liquid Crystal Display) module compatible with the Hitachi HD44780 driver. In our proposed system, the LCD is used in the 4-bit mode (data pins D4–D7).

Table 1 Comparative study of Raspberry Pi Model 3B+ with Arduino UNO [11, 12]

Features	Arduino UNO	Raspberry Pi 3B+
Processor	16 MHz AVR ATmega 328P	1.4 GHz 64-bit quad-core ARM Cortex-A53
Storage	32 KB	External SD card
RAM	2 KB	1 GB
I/O pins	20	17
OS	N/A	Raspbian, other varieties of Linux, Android
Languages	Arduino	Python, C, Java, etc.
Integrated modules	N/A	Wi-Fi, Bluetooth, Ethernet, etc.
Power supply	5 V USB or DC jack	5 V USB

2.4 Indicator Unit

The indicator unit is used to give the patient an idea of whether his/her body vital parameters are within normal limits or not. This unit consists of a *5 V HYDZ piezo-electric buzzer* and *two LEDs*: one *green* and another *red*. When the sensing parameters are outside the range of the typical values, the Raspberry Pi turns "ON" the buzzer and the red LED starts blinking, showing the sign of an unhealthy patient. When these vital parameters are within the normal range, only the green LED starts blinking, indicating the good health of the patient.

2.5 Remote Health Monitoring Unit

The remote health monitoring unit uses the IoT platform to establish communication. In order to remotely monitor the patient's vital parameters, the ThingSpeak API is used to regularly upload and store these data. The uploaded data can be observed both on the ThingSpeak API and on an external Webpage (https://iotlab-irpe-cu.blogspot.com, accessed through a web browser) [14]. The patient or his physician finally gets to view the data through a *web browser* of any connected device like a computer or smartphone across the world. Email notifications are sent using the Webhooks service on IFTTT [15]. Finally, the stored data from the ThingSpeak cloud is exported to a remote server (namely, the hospital's server) for detailed analysis. Inbuilt Wi-Fi module is used for Internet connectivity.

3 Methodology

The entire experimental setup for our proposed system is shown in Fig. 2.

The working principles of the proposed system are illustrated into a number of steps as follows:

1. Turn ON the system. All modules are automatically initialized.
2. Press the push-button to measure the vital parameters of the patient. For default OFF condition, body temperature is measured and for ON condition, heart rate and SpO_2 are measured.
3. Update the data measured by the sensor system after every 15 s.
4. Display the measured data locally on the LCD. Body temperature is displayed in °F and °C both, heart rate in BPM (beats per minute) and SpO_2 in %.

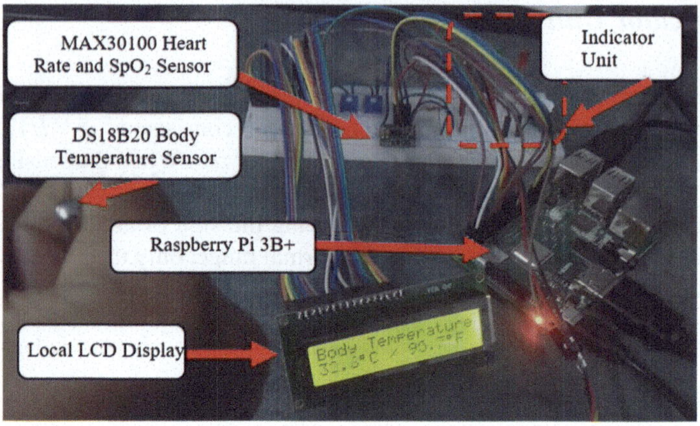

Fig. 2 Experimental setup of the smart health monitoring system

[For Heart rate, use the formula:

$$H_{\text{rate}} = (j_C - j_P) \times 4. \tag{1}$$

where j_C = Current instant's heart rate data and j_P = Past instant's heart rate data]

5. The indicator module will be turned ON whenever vital data are not within the typical range.
6. Check whether the internet connectivity is available or not. If the connection is available, upload the measured data to the ThingSpeak cloud's private channel using a proper API key, otherwise, display "Connection Failure!".
7. The graphical representation of these vital parameters is displayed on both ThingSpeak cloud's channel (Health Monitoring System using Raspberry Pi) or our website (https://iotlab-irpe-cu.blogspot.com).
8. The last uploaded data and their health analysis are sent to the corresponding physician automatically using Webhook services at the day's end every day.
9. Every fortnight, these data are downloaded at the hospital's remote server in ".csv" format.

The flowchart of this entire system operation is given in Fig. 3.

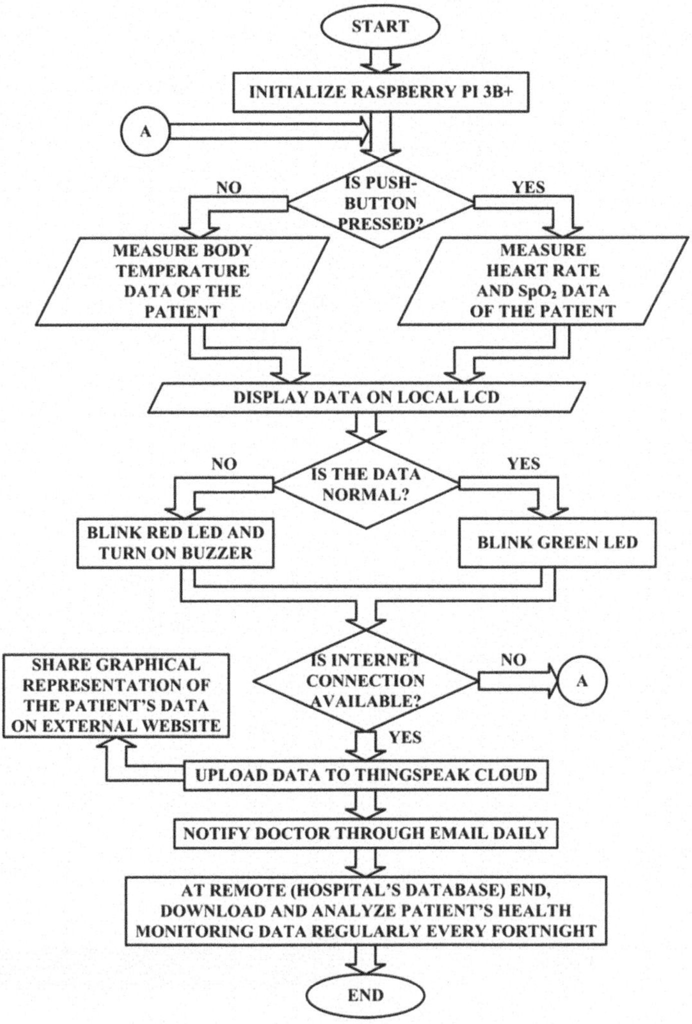

Fig. 3 Flowchart of the entire operation of the smart health monitoring system

4 Results

Table 2 tabulates the vital parameters as displayed locally compared to the uploaded data during various times of the day (4 July 2018). In Fig. 4, the variation of body temperature (in degree Fahrenheit) [Field 1] of a single patient with time is shown—(a) For the temperature data as displayed on the LCD and (b) For the temperature data uploaded and stored in the ThingSpeak cloud. Similarly, in Figs. 5 and 6, the graphical variation of heart rate (in bpm) [Field 2] and blood oxygen saturation SpO_2 (in %) [Field 3] with time is shown, respectively.

Table 2 Vital parameters as displayed locally and as uploaded at different time instants

Vital parameter	Time instant	Displayed locally on LCD	Uploaded to ThingSpeak cloud
Body Temperature	11:17 am	96.80 °F	96.80 °F
	11:49 am	96.35 °F	96.35 °F
	12:31 pm	96.01 °F	96.01 °F
	01:16 pm	96.01 °F	96.01 °F
	01:55 pm	96.57 °F	96.57 °F
	02:36 pm	96.24 °F	96.24 °F
	03:11 pm	96.24 °F	96.24 °F
Heart rate	11:20 am	88 bpm	88 bpm
	11:52 am	80 bpm	80 bpm
	12:34 pm	72 bpm	72 bpm
	01:19 pm	72 bpm	72 bpm
	01:22 pm	80 bpm	80 bpm
	01:57 pm	64 bpm	64 bpm
	02:40 pm	72 bpm	72 bpm
	03:13 pm	68 bpm	68 bpm
SpO_2	11:21 am	97%	97%
	11:23 am	97%	97%
	12:34 pm	96%	96%
	01:23 pm	97%	97%
	01:58 pm	97%	97%
	02:41 pm	96%	96%
	03:14 pm	96%	96%

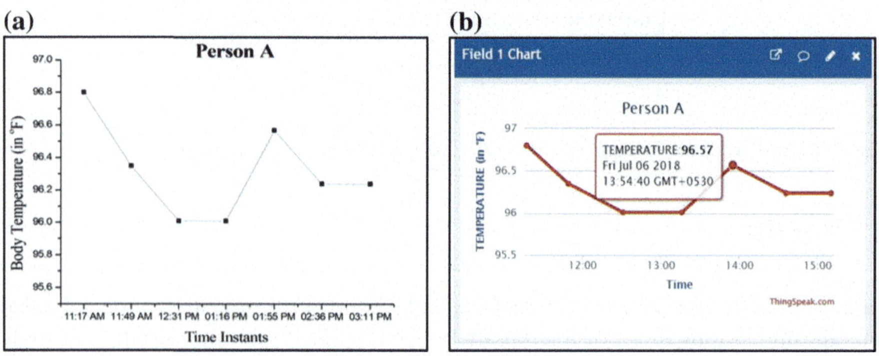

Fig. 4 Body temperature (in °F) of Person A versus time plot. **a** With data as displayed on the LCD module. **b** With data uploaded on the ThingSpeak cloud

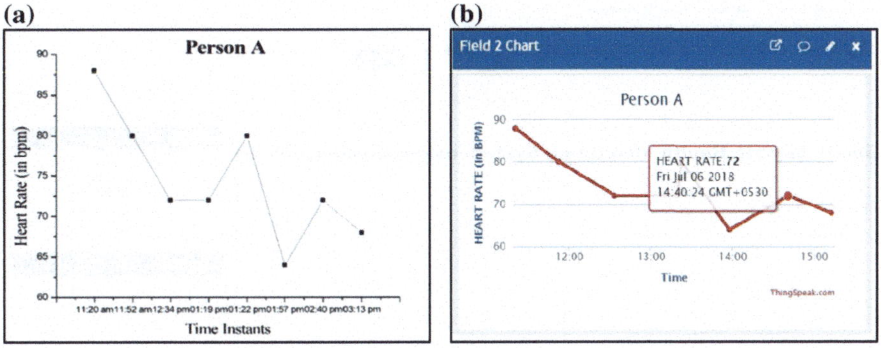

Fig. 5 Heart Rate (in bpm) of Person A versus time plot. **a** With data as displayed on the LCD module. **b** With data uploaded on the ThingSpeak cloud

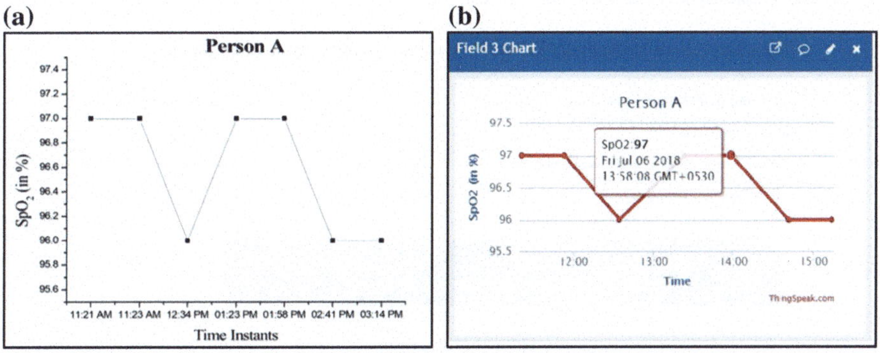

Fig. 6 Blood Oxygen Saturation SpO_2 (in %) of Person A versus time plot. **a** With data as displayed on the LCD module. **b** With data uploaded on the ThingSpeak cloud

Figures 4, 5, and 6 as well as Table 2 show the consistency of the health parameter values between the data at the local LCD display and the IoT communicated data at the remote end. Thus, we can conclude that our system reliably transfers the measured vital parameters to the remote end using the IoT platform. Figure 7 shows the graphs plotted on the ThingSpeak private channel as viewed from an external web page.

Figure 8 shows the email notification as obtained by the physician at the day's end. The alarm indications, the general analysis of these vital parameters, and doctor's consultation decisions are based on Table 3 [16–19].

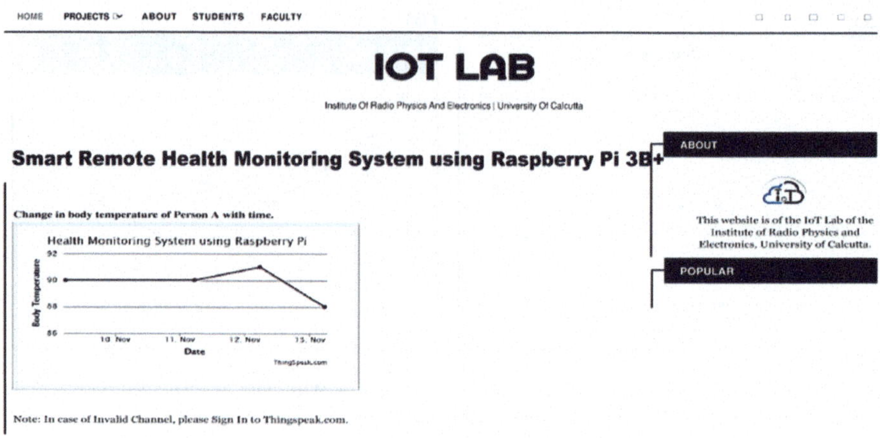

Fig. 7 The graphs as is seen when viewed from an external web page

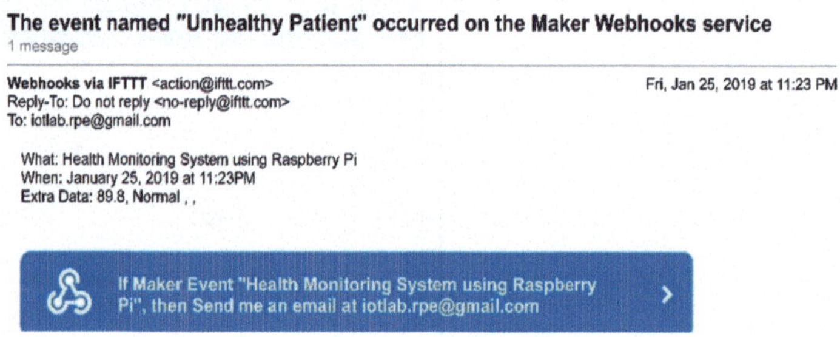

Fig. 8 An email notification as received by the physician

Table 3 Vital parameters and their significance in analysis and alarm indications

Vital parameter	Value	Analysis	Indication	Remarks
Body Temperature	<95.5 °F	Hypothermia	Red LED + buzzer	Consult physician
	97.8–99.0 °F	Normal	Green LED	Vitals normal
	>100.9 °F	Fever/Hyperthermia	Red LED + buzzer	Consult physician
	>104.0 °F	Hyperpyrexia	Red LED + buzzer	Consult physician
Heart rate	<60 bpm	Bradycardia	Red LED + buzzer	Consult physician
	60–100 bpm	Normal	Green LED	Vitals normal
	>100 bpm	Tachycardia	Red LED + buzzer	Consult physician
SpO_2	>95.0%	Normal	Green LED	Vitals normal
	<90.0%	Hypoxemia	Red LED + buzzer	Consult physician

5 Conclusion and Future Scope

This paper presents a low-power, cost-effective, high-accuracy, compact, portable, remote health monitoring system. This system will allow the physicians to monitor vital signs remotely and suggest a visit only in case of an emergency. Healthcare personnel in remote rural regions can also use this system. In the proposed system, the use of the Webhook applet services for sending email notification to the remote doctor's end is a new approach in healthcare applications.

This system can be improved in the future by including more sensors to measure other health parameters like blood pressure, ECG, respiration pattern, EEG, etc., to make it a complete smart remote health monitoring system. A solar panel can reduce our system's dependence on electricity and thus used in remote rural areas. Our proposed telehealth system may be collaborated with other apps, to provide telemedicine and ambulance services too. Also, this patient-friendly system can play a major role in building India's own medical database which can be accessed (with restrictions) by physicians to help in the setting up of a *digital integrated healthcare service in India* [20]. This will reduce the hospitals' burden of regular health checkup.

Acknowledgements The authors would like to thank UGC UPE II, "Modern Biology Group B: Signal Processing Group", the University of Calcutta for providing research facilities. The authors are also thankful to West Bengal Higher Education, Science and Technology and Biotechnology (Science and Technology) funded project "Cytomorphic CMOS Circuit Modeling and Ultra-Low Power Design of P53 Protein Pathway for Synthetic Biology Applications" for partial support of infrastructure. The authors of the paper are thankful to all the student volunteers and the faculty members of the Institute of Radio Physics and Electronics, University of Calcutta for their active support to conduct our research work. There is no conflict of interest.

References

1. India State-Level disease burden initiative CVD Collaborators: The changing patterns of cardiovascular diseases and their risk factors in the states of India: the Global Burden of Disease Study 1990–2016. Lancet Glob. Health. **6**, e1339–e1351 (2018)
2. Kumar, R., Rajasekaran, M.P.: An IoT based patient monitoring system using Raspberry Pi. In: International Conference on Computing Technologies and Intelligent Data Engineering, pp. 1–4 (2016)
3. Kodali, R.K., Swamy, G., Lakshmi, B.: An Implementation of IoT for Healthcare. In: IEEE Recent Advances in Intelligent Computational Systems, pp. 411–416 (2015)
4. Jain, N.P., Jain, P.N., Agarkar, T.P.: An embedded, GSM based, multiparameter, realtime patient monitoring system and control—an implementation for ICU patients. In: World Congress on Information and Communication Technologies, pp. 987–992 (2012)
5. Kaur, A., Jasuja, A.: Health monitoring based on IoT using RASPBERRY PI. In: International Conference on Computing, Communication and Automation, pp. 1335–1340 (2017)
6. Ray, P.P.: A survey of IoT cloud platforms. Futur. Comput. Inform. J. **1**(1–2), 35–46 (2016)
7. Atzori, L., Iera, A., Morabito, G.: The Internet of Things: a survey. Comput. Netw. **54**, 2787–2805 (2010)
8. Maxim Integrated: DS18B20 programmable resolution 1-wire digital thermometer, Rev 4. https://datasheets.maximintegrated.com/en/ds/DS18B20.pdf (2015)

9. Maxim Integrated: MAX30100 pulse oximeter and heart-rate sensor IC for wearable health, Rev 0. https://datasheets.maximintegrated.com/en/ds/MAX30100.pdf (2014)
10. Hu, S., Zheng, J., Chouliaras, V., Summers, R.: Feasibility of imaging PPG. In: International Conference on BioMedical Engineering and Informatics, pp. 72–75 (2008)
11. Raspberry Pi Foundation: Raspberry Pi documentation. https://www.raspberrypi.org/documentation (2018). Accessed 07 Nov 2018
12. Arduino: Arduino Uno Rev 3. https://store.arduino.cc/usa/arduino-uno-rev3 (2018). Accessed 07 Nov 2018
13. Python: Python 3.7.1 documentation. https://docs.python.org/3 (2018). Accessed 08 Nov 2018
14. ThingSpeak: IoT analytics. https://thingspeak.com (2018). Accessed 13 Nov 2018
15. IFTTT: Webhooks. https://ifttt.com/maker_webhooks (2018). Accessed 13 Nov 2018
16. URMC: Health Encyclopedia vital signs. https://www.urmc.rochester.edu/encyclopedia/content.aspx?ContentTypeID=85&ContentID=P00866 (2018). Accessed 11 Nov 2018
17. Mayo Clinic: Bradycardia. https://www.mayoclinic.org/diseases-conditions/bradycardia/symptoms-causes/syc-20355474 (2018). Accessed 11 Nov 2018
18. Mayo Clinic: Tachycardia. https://www.mayoclinic.org/diseases-conditions/tachycardia/symptoms-causes/syc-20355127 (2018). Accessed 11 Nov 2018
19. Mayo Clinic: Hypoxemia. https://www.mayoclinic.org/symptoms/hypoxemia/basics/definition/sym-20050930 (2018). Accessed 11 Nov 2018
20. Dimitrov, D.V.: Medical Internet of Things and big data in Healthcare (Review article). Healthc. Inform. Res. **22**(3), 156–163 (2016)

Study of Evolution by Searching *Alu* Pattern from Primate Transposon

Rachita Ghoshhajra, Sanghamitra Chatterjee and Soma Barman Mandal

Abstract All living beings have Deoxyribonucleic Acid (DNA) as the blueprint of their characteristics. Among noncoding part of DNA sequences, the analysis of transposon is gaining interest in recent years as the mystery of human evolution is hidden inside Transposon or Transposable Element (TE). Scientists around the world are trying to develop methods for automatic TE identification to uncover the mystery of the evolution process. Researchers mainly focused on sample-based TE classification but identification of TE classes using computational tools is yet to be explored. *Alu,* one of the major TE classes, is believed to be mostly responsible for evolution of eukaryotes especially primates. In this paper, the authors address the evolution process of primates by automated searching of *Alu* pattern from TE sequences. The authors proposed "AP Suche" algorithm to identify *Alu* pattern from TE sequence. Results indicate the existence of *Alu* pattern more in human chromosome compared to other primates.

Keywords Evolution · Transposon · SINE · Alu · Primates

1 Introduction

DNA is the building block of all living organisms. DNA nucleobase contains four chemical bases: Adenine (A), Cytosine (C), Thymine (T), and Guanine (G). A always pairs up with T, and C always with G, forming units called base pairs (BP). Human

---

R. Ghoshhajra
MCKV Institute of Engineering, Liluah, Howrah, India
e-mail: rghhajra@gmail.com

S. Chatterjee (✉)
Camellia Institute of Technology, Kolkata, India
e-mail: sangha3030@gmail.com

S. Barman Mandal
Institute of Radio Physics & Electronics, University of Calcutta, Kolkata, India
e-mail: barmanmandal@gmail.com

S. Kundu et al. (eds.), *Proceedings of the 2nd International Conference on Communication, Devices and Computing*, Lecture Notes in Electrical Engineering 602, https://doi.org/10.1007/978-981-15-0829-5_47

genome contains approximately 3 billion base pairs that reside in 23 pairs of chromosomes. A segment of DNA called genes, wrapped in chromosomes, passes information from parents to offspring. The sequential arrangements of base pairs are different in various species. DNA sequences are mainly categorized into two parts: Coding and Noncoding regions. The coding region is responsible for synthesis of Amino Acids, i.e., proteins [1]. The region of DNA that does not code for a protein is called "Noncoding" region. Over the years, "Noncoding" DNAs (NCDNA) have emerged with new insights. These regions are responsible for genetic diversity, genetic disorders, and genetic diseases in humans. The detailed classification of NCDNA proposed by Shanmugam in 2017 [2] shows that Trans Regulatory Elements are one of the major classes of NCDNA and among its subclasses, transposon or transposable element (TE) play a decisive role in evolution. In 2016, Arensburger et al. [3] proposed that "TEs are discrete segments of DNA capable of moving within a host genome from one chromosome or plasmid location to another, or between hosts by using parasitic vectors that they use for lateral transfers." TEs consist of two major classes: Class I or Retrotransposons and Class II or DNA transposons. The behavior of these TEs depends on where it lands in the genome. Landing inside a gene may result in a mutation. The human genome contains roughly of 50% TE whereas it is 90% in the plant genome. Figure 1 depicts the classification of DNA. Non-LTR retrotransposons include Long Interspersed Nuclear Element (LINE), Short Interspersed Nuclear Element (SINE), *Alu*, and SVA elements, collectively accounting for 33% of the human genome [2].

Since *Alu* plays a significant role in evolution, authors in this paper have concentrated mainly on the *Alu* class of transposon. SINE is an important indicator to identify *Alu* pattern. The structure of *Alu* and SINE and their roles are described in Sect. 2. Section 3 describes the "AP Suche" algorithm developed for finding SINE

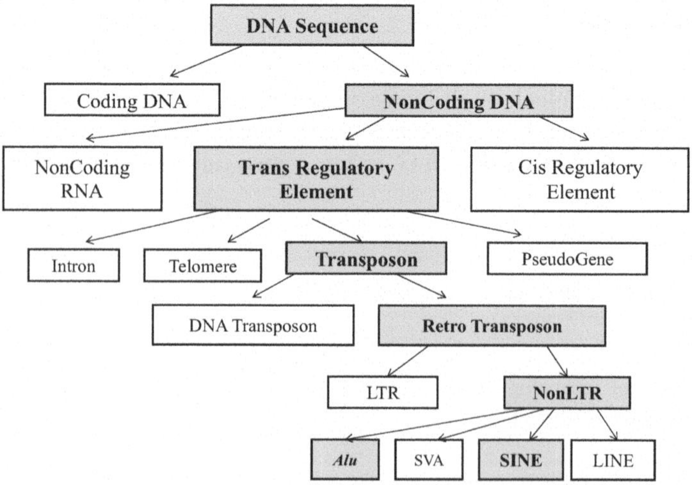

Fig. 1 Classification of DNA

and *Alu* sequence inside transposon. Results and discussions are presented under Sect. 4. Based on the response of the "AP Suche" algorithm, concluding remarks have been drawn in Sect. 5.

2 Structure and Role of *Alu* and SINE

Non-LTR retrotransposon class includes Long Interspersed Nuclear Element (LINE), Short Interspersed Nuclear Element (SINE), *Alu,* and SVA element. These elements collectively account for about one-third of the human genome. The structure of SINE (Fig. 2) as proposed by Bowen in 2002 [4] shows that SINEs are repetitive structures with a set of consecutive A's found interspersed in a DNA sequence.

Alu element is a short stretch of DNA originally characterized by the action of the *Arthrobacter luteus (Alu)* restriction endonuclease. Among all transposable elements, *Alu* is the most abundant and huge no. of copies are dispersed throughout the eukaryotic genome [5, 6]. The proper structure and function of *Alu* were unknown prior to 1997 as NCDNA was treated as "junk" at that time. Mighel et al. [7] reported that there is a chance of large number of *Alu* patterns present in NCDNA which function adaptively or mal-adaptively. Researchers started to recognize that *Alu* is one of the key elements in the study of human evolution [8–10]. The human genome modified by *Alu* element is also responsible for human diseases through insertion or recombination between them [11].

Several scientists are working rigorously to find the exact structure of *Alu* [10, 12]. Lee et al. [10] proposed the typical structure of *Alu* as shown in Fig. 3. However, the detailed structure of the left and right monomer is not specified in his paper. The special sequence present between the left and right monomer of the *Alu* structure

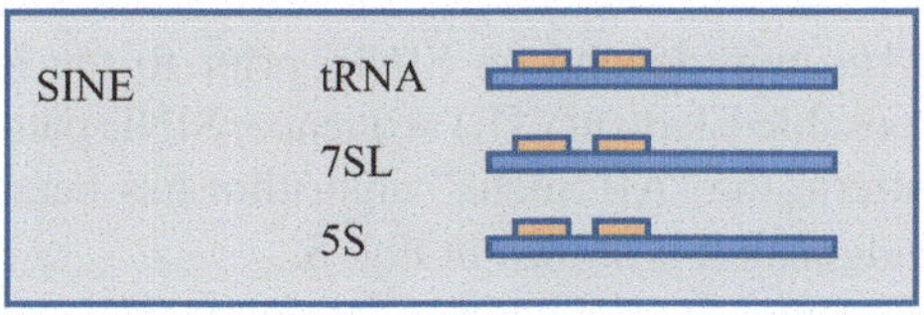

Fig. 2 Structure of SINE sequence

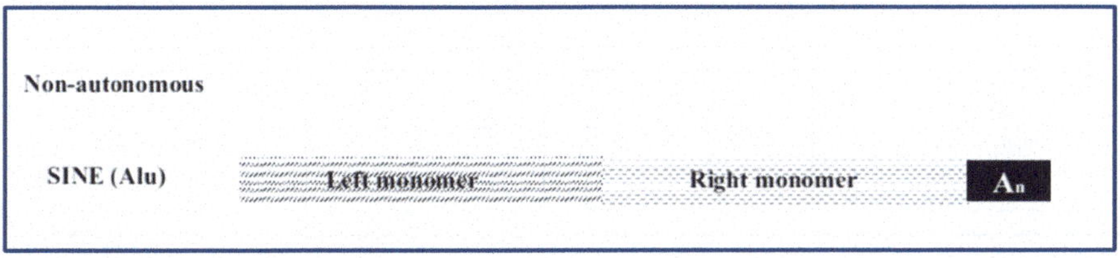

Fig. 3 Schematic structure of *Alu* [10]

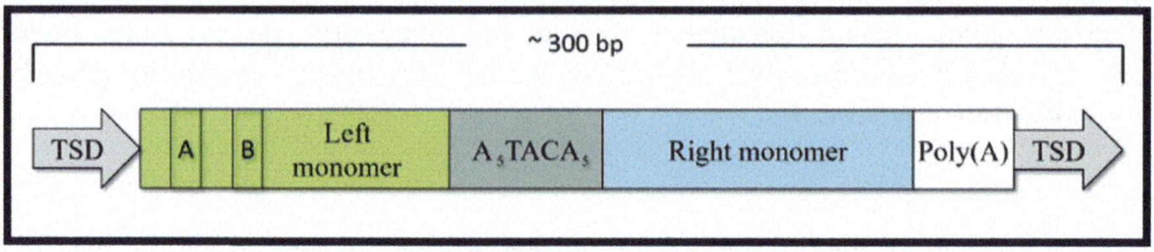

Fig. 4 Structural schema of *Alu* element [11]

is depicted in Fig. 4 [11]. Length of *Alu* sequence is 300 bp, as mentioned in the literature [5, 11].

The Alu element is transcribed into a messenger RNA by RNA polymerase and then converted into a double-stranded DNA molecule by reverse transcriptase. The new double-stranded DNA molecule is then inserted into a new location in the genome [5]. Due to this reason, *Alu* may be classified as SINE [13]. *Alu* are nonautonomous like all SINEs. *Alu* elements frequently duplicate when they jump and scientists estimate that the human genome acquires one new *Alu* insert in approximately every 200 births.

Scientists working on genetic ancestry consider some key features of *Alu* elements like (i) probability of two different *Alu* insertions at the same place is practically zero and (ii) most *Alu* elements cannot be removed after insertion [6]. These features signify that two DNA sequences who share the same *Alu* sequence may have a common ancestor. Thus, the behavior of *Alu* sequences plays a key role in the study of human evolutionary relationships with chimpanzee and gorillas [14, 15].

3 Methodology

As the existence of *Alu* may depend on SINE, prior to the identification of *Alu* sequence from Transposable Element (TE) sequence, SINE pattern is also searched. For searching *Alu* pattern, the "AP Suche" algorithm has been developed and the block diagram of the algorithm is shown in Fig. 5.

Preprocessing or Filtering (Step1) searches for a typical "consecutive A" sequence "LongA" in the search field, i.e., transposon sequence considered. In SINE, the length of "LongA" should be greater than or equal to 6. So, in this algorithm, a sliding

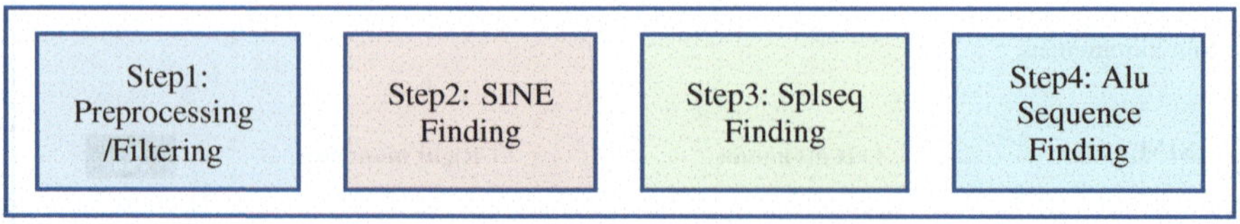

Fig. 5 Block diagram of "AP Suche" algorithm

window "w" of length 6 is moved along the input sequence. If "LongA" is present, the transposon sequence is to be processed further. If it is absent, rest of the processing steps are unnecessary. The complexity of this preprocessing step is O(N). Though complexity is high in case of a long sequence, this step is important as it reduces the chance of false search in Step2 or Step3.

The presence of the SINE sequence in transposon is found through Step2. The existence of SINE signifies the presence of at least two sets of "consecutive A" (A6) of length more than 6. Here, EOS is End of Sequence.

"Splseq" acts as an indicator for the presence of the *Alu* sequence. "Splseq" is A_5TACA_5 or similar sequence that may exist in the transposon sequence. Step3 searches for the presence of "Splseq" on the sequence ("seq") which passes Step1 or Step2. Once the required "Splseq" is found in Step3, the sequence is passed to Step4, otherwise, the processing stops. This process reduces the total time required by the algorithm when it is applied to a big TE sequence.

Step4 of the algorithm identifies the presence of the *Alu* sequence. With reference to Fig. 4, Green part, i.e., A + B + Left monomer is denoted by *SecI* and Blue part, i.e., right monomer is denoted by *SecII* in this algorithm. As mentioned earlier, total length of *Alu* sequence is taken to be 300 bp.

Detailed steps of the "AP Suche" algorithm are mentioned below:

'AP Suche' Algorithm

Step1: Preprocessing or Filtering

Step1.1: Read Transposon sequence
Step1.2: Remove unwanted spaces and store in 'Seq'
Step1.3: Find longest consecutive sequence of 'A' (LongA) and its position
Step1.4: if LongA >=6,
Go to Step2 and Step3
Step1.5: else
print "Small 'LongA' sequence: Processing Stops",
Stop.

Step2: SINE Finding

Step2.1: Read the Filtered Transposon sequence ('Seq') obtained in Step1
Step2.2: Repeat steps 2.3 to 2.5 until EOS reached
Step2.3: Find 1st loc_A6
Step2.4: Find 2nd loc_A6 after 1st occurrence.
Step2.5: if both exist,
print "SINE present, May continue Step3"
Step2.6: Print "SINE absent, May continue Step3"

Step3: 'Splseq' Finding
Step3.1: Read the Filtered Transposon sequence ('Seq') obtained in Step1
Step3.2: Repeat step 3.3 and 3.5 until EOS reached
Step3.3: Search for 'SplSeq' using sliding window of size 'w'
Step3.4: If 'SplSeq' found go to Step4
Step3.5: else Print "SplSeq not found",
Stop.

Step4: Alu sequence Finding
Step4.1: Find location of consecutive A chain 'ConA' where
(length_LongA>=length_ConA) and (Loc_ConA > Loc_Splseq)
Step4.2: If 'Loc_ConA' not found Stop
Else continue steps 4.3 to 4.5 until EOS reached
Step4.3: 'SecII' = 'Loc_ConA' - 'Loc_SplSeq'.
Step4.4: Find 'SecI' using
Alu_Length = Length_SecI +Length_Splseq + Length_SecII +
Length_ConA = 300bp
Step4.5: Alu_Seq = 'SecI'+ 'SplSeq'+ 'SecII' + 'ConA'
Step4.6: Stop

4 Result and Discussion

"AP Suche" algorithm is applied on different datasets available from the NCBI website https://www.ncbi.nlm.nih.gov [16]. Databases are collected for Human, Chimpanzee, Pygmy Chimpanzee, Rhesus Monkey, Crab Eating Macaque, Gelada, and Olive Baboon. Also Human Chromosome1 available in different organs of the human body is used as input. The algorithm is implemented using Python language and run on "Intel Core i5 @ 2.3 GHz" processor with 6 GB RAM. Datasets of varying size are stored in .txt files.

DataSet1 consists of different sequences of Chromosome1 available in different organs of humans. The response of the algorithm for this dataset is shown in Table 1.

From Table 1, it is observed that human Chromosome1 contains SINE sequence in all cases whereas the *Alu* sequence is not present always. Also, length of "ConA" is almost the same for this dataset.

The algorithm applied on DataSet2 consisting of different human chromosome sequences and outputs is displayed in Table 2.

From Table 2, it is observed that all human chromosomes do not contain SINE sequence and the chromosomes where SINE is absent, *Alu* sequence is also absent. From Tables 1 and 2, it is clearly visible that

(i) presence of *Alu* confirms the presence of SINE in any transposon.
(ii) presence of SINE does not confirm the presence of *Alu* in any unknown transposon sequence. It is evident from Table 2 row 4, that in spite of the presence

Table 1 Presence of SINE and *Alu* sequence in human Chromosome1

CH_No	Gene ID	SINE	*Alu*	Length SecI (bp)	Length SecII (bp)	ConA
CH1_1	84944	*Present*	*Present*	131	149	A6
CH1_2	9204	*Present*	*Present*	131	149	A6
CH1_3	1160	*Present*	*Present*	131	149	A6
CH1_4	100381270	*Present*	**Absent**	–	–	–
CH1_5	79605	*Present*	*Present*	131	149	A6
CH1_6	113802	*Present*	*Present*	142	137	A7
CH1_7	267002	*Present*	**Absent**	–	–	–

Table 2 Presence of SINE and *Alu* sequence in different human chromosomes

CH_No	NCBI ref sequence	SINE	*Alu*	Length SecI (bp)	Length SecII (bp)	ConA
HS_CH2	NC_000002.12	**Absent**	**Absent**	–	–	–
HS_CH4	NC_000004.12	*Present*	*Present*	131	149	A6
HS_CH5	NC_000005.10	*Present*	**Absent**	–	–	–
HS_CH8	NC_000008.11	**Absent**	**Absent**	–	–	–
HS_CH11	NC_000011.10	**Absent**	**Absent**	–	–	–

of SINE in HS_CH5, *Alu* sequence is absent. Figure 6 depicts the presence and absence of *Alu* and SINE sequences in each of the human chromosomes considered.

DataSet3 consists of different chromosomes of Chimpanzee and Pygmy Chimpanzee. After applying the "AP Suche" algorithm on this dataset, results are tabulated in Table 3.

Three significant observations in Table 3 are as follows:

(i) presence of SINE does not confirm the presence of *Alu*.
(ii) among the dataset taken, *Alu* is present only in chromosome 6 of both chimpanzees.
(iii) all chromosomes do not contain the *Alu* sequence.

DataSet4 consists of Chromosome1 of different primate species like Chimpanzee (CH1_C), Crab Eating Macaque (CH1_CM), Gelada (CH1_GL), Olive Baboon(CH1_OB), and Rhesus Monkey (CH1_RM). The "AP Suche" algorithm is applied to this dataset and the result obtained is shown in Table 4.

Observations from Table 4 are

(i) presence of SINE does not confirm the presence of *Alu* for this set also.
(ii) *Alu* is not present in Chromosome1 of Gelada or Olive Baboon, but present in Rhesus Monkey and Crab-Eating Macaque.

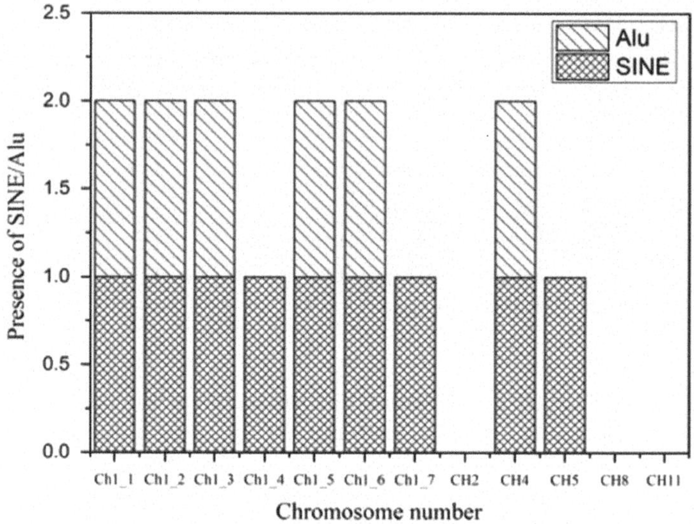

Fig. 6 Chart to represent the presence or absence of SINE and *Alu* sequence in human chromosomes

Table 3 Presence of SINE and *Alu* sequences in Chimpanzee and Pygmy Chimpanzee chromosomes

CH_No	NCBI ref sequence	SINE	*Alu*	Length SecI (bp)	Length SecII (bp)	ConA
C_CH1	NC_036879.1	**Absent**	**Absent**	–	–	–
C_CH5	NC_036884.1	*Present*	**Absent**	–	–	–
C_CH6	NC_036885.1	*Present*	*Present*	131	149	A6
PyCH1	NC_027868.1	**Absent**	**Absent**	–	–	–
PyCH5	NC_027873.1	**Absent**	**Absent**	–	–	–
PyCH6	NC_027874.1	*Present*	*Present*	131	149	A6

Table 4 Presence of SINE and *Alu* sequence in chromosomes of different species

CH_No	NCBI ref sequence	SINE	*Alu*	Length SecI (bp)	Length SecII (bp)	ConA
CH1_C	NC_036879.1	*Present*	**Absent**	–	–	–
CH1_CM	NC_022272.1	*Present*	*Present*	129	150	A7
CH1_GL	NC_037668.1	*Present*	**Absent**	–	–	–
CH1_OB	NC_018152.2	*Present*	**Absent**	–	–	–
CH1_RM	NC_027893.1	*Present*	*Present*	129	150	A7

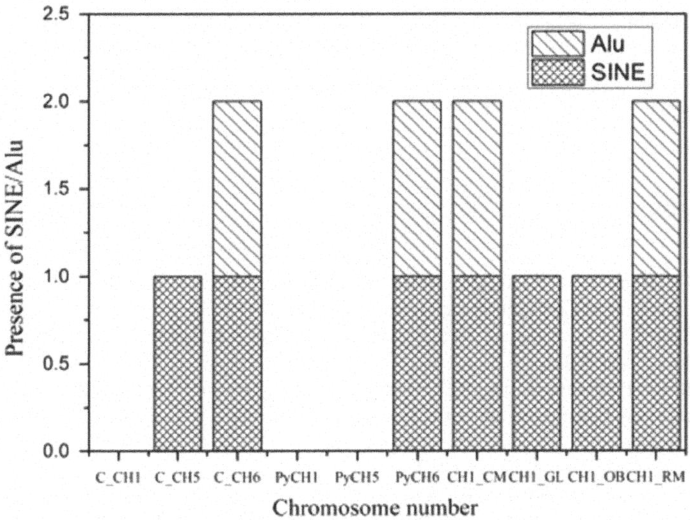

Fig. 7 Chart to represent the presence or absence of SINE and *Alu* sequence in primate chromosomes

(iii) it is interesting to note that "ConA" sequences present in Crab-Eating Macaque and Rhesus Monkey consists of seven consecutive A's (A7) whereas it is mostly A6 in human and Chimpanzee chromosomes. The presence of SINE and *Alu* in primate is shown in Fig. 7.

Thus the "AP Suche" algorithm can be used to find the presence of both SINE and *Alu* sequences from any DNA sequence. It is observed from the results that the possibility of finding the *Alu* sequence in human chromosomes is much higher than chromosomes of any other primates.

5 Conclusion

The signature of *Alu* is significant in the evolution process due to its presence in primates only. The proposed "AP Suche" algorithm is trying to locate the presence of the *Alu* sequence inside any DNA sequence. In the implementation of this algorithm, a static "Splseq", the most significant part of *Alu*, is taken. By dynamically varying the "Splseq", better results are expected to be obtained. In future, dynamic programs may be incorporated to modify this algorithm for searching presence of multiple *Alu* sequences in any chromosome. This may play an important role in the study of the evolution process. Hence, proper identification methods are required to fully characterize *Alu* and its presence in primates. This work is only a stepping stone in the path of long journey toward characterizing human evolution.

Acknowledgements The authors thank West Bengal Higher Education Science and Tech funded Project Cytomorphic Circuit Modeling of P53 protein pathway for synthetic biology applications for technical support.

References

1. Lin, T.-Y., Shah A.H.: Stochastic finite automata for the translation of DNA to protein. In: IEEE International Conference on Big Data (2014)
2. Shanmugam, A., Nagarajan, A., Pramanayagam, S.: Non-coding DNA—a brief review. J. Appl. Biol. Biotechnol. **5**(05), 42–47 (2017)
3. Arensburger, P., Piegu, B., Bigot, Y.: The future of transposable element annotation and their classification in the light of functional genomics. Mob. Genet. Elem. **6**(6), (2016)
4. Bowen, N.J., Jordan, I.K.: Transposable elements and the evolution of eukaryotic complexity. Curr. Issues Mol. Biol. **4**, 65–76 (2002)
5. Dagan, T., Sorek, R., Sharon, E., Ast, G., Graur, D.: AluGene: a database of Alu elements incorporated within protein-coding genes. Nucleic Acids Res. **32**, (2004). https://doi.org/10.1093/nar/gkh132
6. Pray, L.: Functions and utility of Alu jumping genes. Nature **1**(1), 93 (2008)
7. Mighell, A.J., Markham, A.F., Robinson, P.A.: Alu sequences. FEBS Lett. **417**, 1–5 (1997)
8. Hormozdiari, F., et al.: Alu repeat discovery and characterization within human genomes. Genome Res. **21**, 840–849 (2011)
9. Dridi, S.: Alu Mobile Elements: From Junk DNA to Genomic Gems, 11 pp. Hindawi Publishing Corporation Scientifica (2012). http://dx.doi.org/10.6064/2012/545328
10. Lee, H.-E., Ayarpadikanna, S., Kim, H.-S.: Role of transposable elements in genomic rearrangement, evolution, gene regulation and epigenetics in primates. Genes Genet. Syst. **90**, 245–257 (2015)
11. Kim, S., Cho, C.-S., Han, K., Lee, J.: Structural variation of *Alu* element and human disease. Genom. Inf. **14**(3), 70–77 (2016)
12. David, M., Mustafa, M., Brudno, M.: Detecting Alu insertions from high-throughput sequencing data. Nucleic Acids Res. **41**(17), (2013). https://doi.org/10.1093/nar/gkt612
13. Deininger, P.: Alu elements: know the SINEs. Genome Biol. **12**, 236 (2011). http://genomebiology.com/2011/12/12/236
14. Xing, J., Witherspoon, D.J., Ray, D.A., Batzer, M.A., Jorde, L.B.: Mobile DNA elements in primate and human evolution. Phys. Anthropol. **50**, 2–19 (2007)
15. Trizzino, M., Park, Y.S., Holsbach-Beltrame, M., Aracena, K., Mika, K., Caliskan, M., Perry, G.H., Lynch, V.J., Brown, C.D.: Transposable elements are the primary source of novelty in primate gene regulation. http://www.genome.org/cgi/doi/10.1101/gr.218149.116
16. https://www.ncbi.nlm.nih.gov

Identity Management in Internet of Things: A Software-Defined Networking Approach

Kazi Masum Sadique, Rahim Rahmani and Paul Johannesson

Abstract Internet of things (IoT) network needs to accommodate billions of connected smart objects (things). Connected IoT objects should be able to communicate with each other. Objects should be able to travel between different networks regardless of their locations, network providers, and manufacturers. To allow transparent movement of IoT objects, it is very important to have a trust relationship between these objects. To establish trusted relationship, unique identity is one of the key properties for any IoT object. It will not be possible for the devices to freely move within the networks if we do not have a common identity solution. The use of software-defined networking (SDN) approach in IoT is increasing these days, due to its flexibility and easy adaptability with any network. In this paper, we have presented an SDN-based identity management system which will possibly solve the unique identity problem and increase trust in heterogeneous IoT network.

Keywords Internet of Things · Identity management · Security · Trust · Software defined network · Security of IoT devices · IoT security management

1 Introduction

The quality of human life is improving every day with the vast implementation of smart Internet of Things (IoT) objects in everything. The application of IoT technology includes smart health, smart cities, smart farming, smart energy, smart home, smart vehicle, smart waste management, and many more [1]. IoT technology introduces networks of anything. To achieve network of anything, heterogeneous and

K. M. Sadique (✉) · R. Rahmani · P. Johannesson
Department of Computer and Systems Sciences, Stockholm University, Borgarfjordsgatan 8, 164 07 Kista, Sweden
e-mail: sadique@dsv.su.se

R. Rahmani
e-mail: rahim@dsv.su.se

P. Johannesson
e-mail: paul@dsv.su.se

S. Kundu et al. (eds.), *Proceedings of the 2nd International Conference on Communication, Devices and Computing*, Lecture Notes in Electrical Engineering 602, https://doi.org/10.1007/978-981-15-0829-5_48

interconnection capabilities between smart objects are crucial. It means that the IoT objects should be able to be connected or disconnect to any network that it could belong to, regardless of time, location, and its hardware manufacturer [2]. One of the common use cases of IoT objects/nodes is to act as sensors or actuators. These sensors and actuators are typically communicating with a local gateway. The gateways connect to a centralized server to share date produced from the end nodes. The centralized servers are located on the cloud. A centralized control from a cloud platform for globally distributed IoT nodes is becoming difficult with respect to security when billions of nodes need to transfer data to and from a centralized server. Real-time authentication and authorization of these smart nodes from a centralized server is also challenging [3]. In many critical situations, waiting for response from a centralized cloud server could be very harmful [4]. These lead to the concept of decentralized management of IoT nodes instead of controlling from a cloud server.

2 Background and Motivations

Mobility capabilities of wireless sensor nodes open new opportunities and use cases for smart healthcare, smart agriculture, cattle tracking, smart traffic management, vehicle location tracking, battlefield surveillance, and so on [16]. Management of billions of dynamic smart nodes from a central server is not the best solution for the future IoT. There are also security issues such as node authentication, integrity, and confidentiality of data, key management, and identity management which need to be addressed. Previous research shows that a public/private key pair based approach is appropriate for resource constraint sensor nodes [17]. We have discussed authentication and identity management below before describing our proposed solution.

2.1 Authentication

Authentication is the process of ensuring a valid entity within any system. It is a kind of service that is concerned with assuring that relation within premises is authentic. Authentication is the first step to perform any computing operation to go to the next step. To be granted access to the system, usually, the entity needs to provide some kind of secret or piece of information known by the privileged entity. Identification is crucial for authentication [5].

2.2 Identity Management

Previous researches show that identity management is one of the issues that need further research [5]. Future IoT infrastructure will combine billions of smart objects with flexible movement. All these smart objects carry context information that can allow further track within the network. To allow flexible movement of smart objects in the IoT networks, the unique identity of devices is very crucial. Also to ensure authentication and authorization, devices need unique identity.

Considering the future IoT infrastructure scenario and above-described security issues, we found that a unique identity management solution is essential for future IoT paradigms. A question can be arrived at—why do we need a distributed solution and not a centralized control? The answers are

- Billions of smart nodes need to be able to connect to the network dynamically without any human assistance.
- Without any difficulty, future IoT nodes should be able to move between different gateways regardless of the location.
- Single point of failure at a centralized solution will possibly interrupt the system, even if we keep redundant standby servers.
- Management of the identity of billions of nodes from a centralized server located at cloud is not easy and not an efficient solution.
- Time-critical decision-making is not possible using a centralized solution.

3 Related Work

There are many recent papers that addressed the software-defined networking (SDN)-based IoT security model and architectures; but as per our knowledge, none of those have proposed a decentralized identity management architecture. Mike et al. have proposed SDN-IoT architecture with network function virtualization (NFV) implementation to address challenges in IoT. The paper also addressed business aspects of using SDN- and NFV-based IoT architecture [6]. Oliver et al. pointed out that a combination of SDN and OpenFlow-based software switches affect the performance of IoT devices. The proposed SDN model was designed to secure both wired and wireless network communication including Ad Hoc network objects like mobile phones, sensors, tablets, or household apparels [7]. Sahoo et al. highlighted the security challenges in IoT and proposed an SDN-based network architecture. In a table, the authors have also presented several security issues such as unauthorized access, data leakage, data modification, malicious application, denial of service, and configuration issue for different layers of SDN. The table represents that control layer and data interface are most vulnerable for an SDN based solution. There are two major reasons that are found out throughout this research are single-point failure and coordination among multiple controllers. To prevent attacks, grid of security was introduced even though a large portion of issues still needs to take care of [8]. Ding et al. discussed that fast networking comes with lots of opportunities as well as unwanted security issues that lead to a security breach. So, their focus in the paper is that it possible to overcome wireless network security issues using SDN. The key elements of the authors' solution framework include local agents, central controller, and security layer for wireless mobile networks. According to the authors' observation, there is a lack of concern for SDN security in the wireless and mobile domains. The OpenFlow API and its northbound interfaces already provide a good foundation for innovative solutions on top of the existing SDN framework, especially for security elements [9].

4 Distributed SDN Based Identity Management

In our previous work [10], we suggested the architecture for future IoT networks where we described identity management based on context information. In our proposed architecture, we have considered location and time information to produce unique identity of any smart object in IoT network. We have designed a location-based identity provider server for IoT networks. There are a few basic assumptions: we have assumed that the smart objects are the end nodes that are involved in different specific tasks as per the requirement of the application. The nodes are always transmitting or receive data via an edge device (a gateway). We did not consider the distributed software-defined networking approach. The magic of SDN-based solution is layered separation of control and data plane. This makes SDN solution efficient and easily maintainable [11]. We realized that our previously proposed solution needs extension. This paper is a significantly modified presentation of our previous work. Figure 1 presents the network architecture of our previously proposed context-based identity solution. Identity management is done using identity provider server. Every node and gateway with a specific location gets their unique identity from the identity provider server. When a node moves to a new location, it needs to confirm its own identity before performing any kind of data transfer to and from a gateway. The identity provider server at the new location confirms the identity of the new node with the identity server at the previous location.

A distributed identity management architecture shown is in Fig. 1; it can be an example scenario for future 5G networks where all the edge devices connect to the cloud storage via a 5G network and the unique identities of those devices are assigned by the software-defined identity management system. The problem with the above-proposed solution is single point of failure and also communication between different identity servers can be time-consuming and we also found that with a traditional presentation of identity server with routers and switches, it is not a robust solution. Based on our analysis, we have modified our previously proposed architecture with distributed software-defined network solution. In our proposed solution, in Fig. 2, we have suggested distributed solution with sync capabilities. Local nodes and gateways share parameters with the local SDN server to get the identity. The SDN device

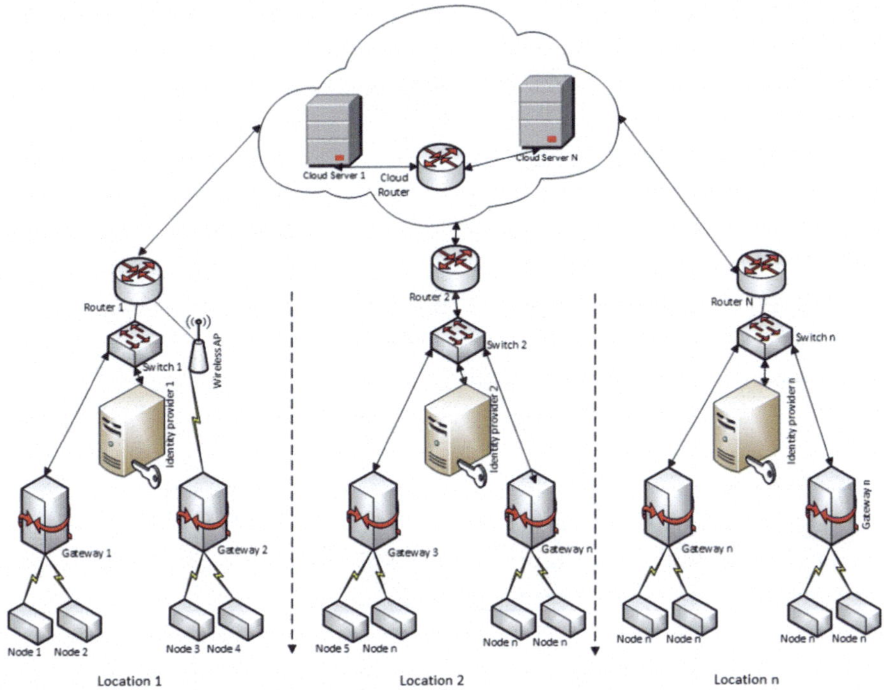

Fig. 1 Architecture of location-based identity management system [10]

verifies the request and assigns a unique identity to the node or gateway. In Fig. 2, we have considered nodes and gateways as nodes only. To ensure reliability, the local SDN device syncs its identity database with the Global SDN device. All Global SDN devices also sync their databases.

To understand the operation of our designed identity provider, we have to consider a few scenarios. An example scenario for context-based moving entities is a vehicle network. Moving vehicles needs to collaborate and share information between each other which we can define as vehicle's cooperation as a cluster [13]. In a multi-vehicle infrastructure, nodes can make decisions based on context at two levels in response to (a) the unavailability of communications paths and (b) application-level events and context changes [14]. Finally, we require interworking between the multi-vehicles, which in essence is a peer-to-peer IoT infrastructure [15] and a cloud-based IoT platform hosted in a data center (Fig. 3).
Scenario 1: A newly connected smart node needs to transmit data to the gateway. The join to a gateway is confirmed in a few steps. We have described the steps below.

- Node makes a RequestToJoin to the nearest gateway.
- Gateway replies with a RequestToGetIdentity from the local SDN device.
- Gateway sends information about the RequestToJoin from a node to the local SDN device.

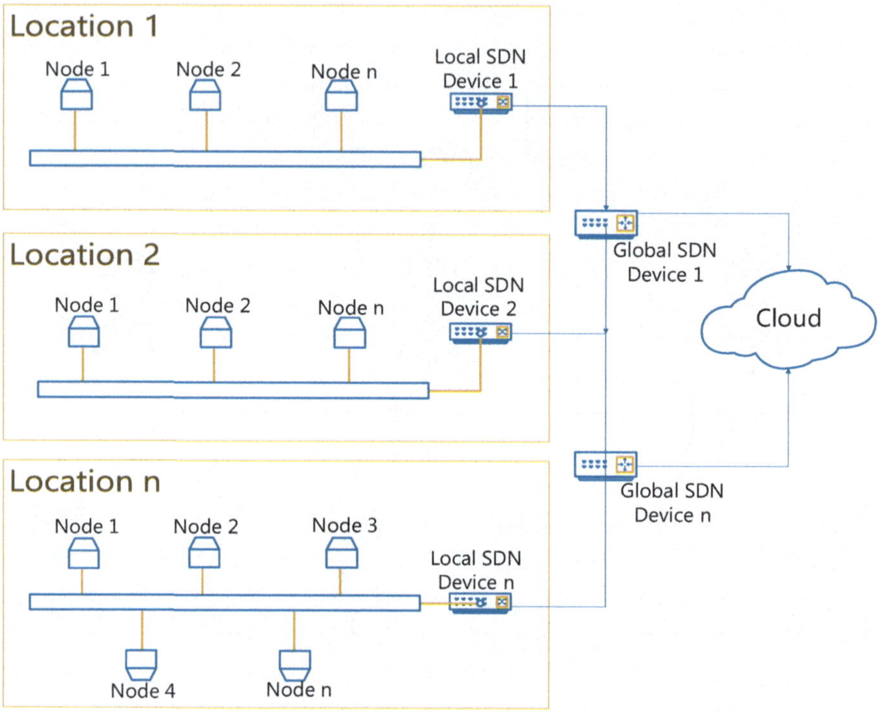

Fig. 2 Architecture of distributed SDN based identity management system

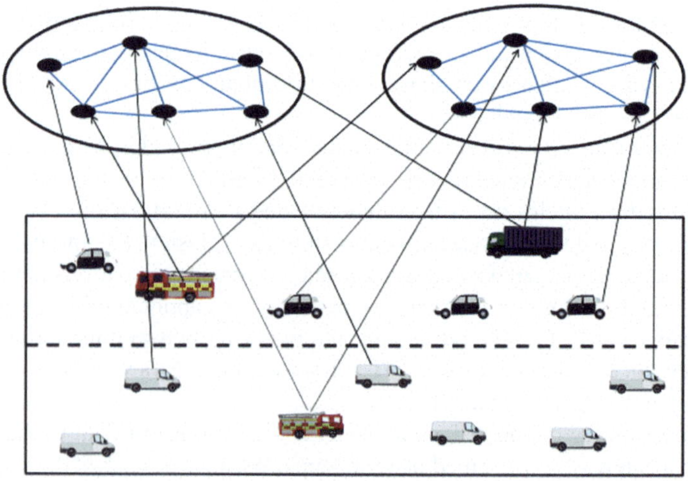

Fig. 3 Example scenario of a context-aware vehicular network diagram [12]

- Node makes RequestToGetIdentity to the local SDN device.
- Local SDN device verifies the requests received from the node and gateway and assigns a unique identity to the node and shares IdentityAssigned with the node and the gateway.

Messages with context information for the above scenario:

RequestToJoin: A newly installed smart object sends a request to join to the nearest gateway. The object includes own hardware address, current location, and own public key in a packet.
RequstToGetIdentity: When the gateway gets a request to join, it checks if the object trying to join has a unique identity information or not. If no identity information is found at the gateway database, it sends a RequstToGetIdentity message with the identity and public key of the local SDN device.
IdentityAssigned: When a unique identity is assigned by a local SDN device, it shares this information to the respective gateway and the requesting node.
RequestToSync: This message is sent when a local SDN device shares an update of the unique identity table with the global SDN device.

Scenario 2: A node is moved to a new location and needs to join a new gateway. The join to a gateway is confirmed in a few steps. We have described the steps below.

- Node makes a RequestToJoin to the gateway at the new location with the unique identity received from the local SDN device at the previous location.
- Gateway checks if there is any identity information that exists for this node in its database, if not the gateway checks for the identity of the node in the local SDN device.
- If local SDN device cannot find an entity at its local database, it communicates with the SDN device located at the previous location of the device.
- Based on the reply from the SDN device located at the previous location, the local SDN server replies to the gateway about the identity of the node.
- Gateway confirms or denies the RequestToJoin request from the node based on the identity verification result from the local SDN device.

We have proposed the following algorithm for the SDN-based identity management device when it assigns a unique identity to an end node:

```
program IdentityAssignment (RequstToGetIdentity)
  {Assuming all the end nodes share their location
  information and information of the gateway they wants
  to join};
  const  ownLocationInformation;
  var    endNodeLocationInformation,
  gatewayLocationInformation, uniqueIdentityForDevice;
  begin
      if requestForIdentityInformation from a node is
  same as requestToAssignIdentity from a gateway are
  same then
      SDN device sends IdentityAssigned message to an
  end node;
  else
      decline reply for RequstToGetIdentity is send;
  end.
```

The below algorithm is for the synchronization of identity information between local and global identity devices.

```
program SinkIdentityTable (RequestToSync)
  {Assuming that the local SDN device pushes updates
  about change to the global SDN device};
    const  ownLocationInformation,
  connectedGlobalSDNDeviceLocationINformation;
  begin
      if new update in local identity table
  then
      share identity table update with the connected
  SDN devices;
  end.
```

In Table 1, we have presented different variable parameters for different devices which will be connected in the proposed architecture to share and update identity information of different devices.

5 Discussion and Future Work

We have proposed a fault-tolerant aggregation algorithm based on distributed software-defined networking. This software-defined identity management solution will have the following benefits:

- The proposed architecture has the possibility to solve the unique identity problems of billions of IoT devices.
- The problem of a single point of failure will be reduced by using this distributed software-defined architecture.
- The system should be easily manageable and configurable.

Table 1 Parameters for the proposed algorithms

Device	Parameter
End node	Own current location information, own previous location information, own key pair (public and private), own identity information, current gateway identity information, current gateway public key, previous gateway identity information, and software-defined identity provider's identity and public key
Gateway device	Own location information, own key pair (public and private), own identity information, connected nodes identity information public key and location information, and software-defined identity provider's identity and public key
Local software defined identity provider	Own location information, own key pair (public and private), own identity information, identity information of end nodes and gateways within a certain location, and identity information of the global software-defined identity provider
Global software-defined identity provider	Own location information, own key pair (public and private), own identity information, identity information of end nodes and gateways within different locations (synced by the local SDN device), and identity information of the local software-defined identity provider devices

- Controller update and the upgrade process will be very easy without any interruption of the authentication process.

Our fault-tolerant approach is proactive and is based on a novel design of a future architecture for authentication and identity management of IoT nodes and gateways. Although the proposed approach distributes trust and noise for different privacy among the meters, we showed that it can gracefully tolerate a single node fail due to communication failures.

References

1. Talari, S., Shafie-Khah, M., Siano, P., Loia, V., Tommasetti, A., Catalão, J.: A review of smart cities based on the internet of things concept. Energies **10**(4), 421 (2017)
2. Patel, K.K., Patel, S.M.: Internet of things-IOT: definition, characteristics, architecture, enabling technologies, application & future challenges. Int. J. Eng. Sci. Comput. (2016)
3. Yu, W., Liang, F., He, X., Hatcher, W.G., Lu, C., Lin, J., Yang, X.: A survey on the edge computing for the internet of things. IEEE Access **6**, 6900–6919 (2018)
4. Lin, J., Yu, W., Yang, X., Yang, Q., Fu, X., Zhao, W.: A real-time en-route route guidance decision scheme for transportation-based cyberphysical systems. IEEE Trans. Veh. Technol. **66**(3), 2551–2566 (2017)

5. Mahmoud, R., Yousuf, T., Aloul, F., Zualkernan, I.: Internet of things (IoT) security: current status, challenges and prospective measures. In: 2015 10th International Conference for Internet Technology and Secured Transactions (ICITST), pp. 336–341 (2015)
6. Ojo, M., Adami, D., Giordano, S.: A SDN-IoT architecture with NFV implementation. In: 2016 IEEE Globecom Workshops (GC Wkshps), pp. 1–6 (2016)
7. Olivier, F., Carlos, G., Florent, N.: New security architecture for IoT network. Procedia Comput. Sci. **52**, 1028–1033 (2015)
8. Sahoo, K.S., Sahoo, B., Panda, A.: A secured SDN framework for IoT. In: 2015 International Conference on Man and Machine Interfacing (MAMI) (2015)
9. Ding, A.Y., Crowcroft, J., Tarkoma, S., Flinck, H.: Software defined networking for security enhancement in wireless mobile networks. Comput. Netw. **66**, 94–101 (2014)
10. Sadique, K.M., Rahmani, R., Johannesson, P.: Trust in internet of things : an architecture for the future IoT network. In: International Conference on Innovation in Engineering and Technology (ICIET) (2018)
11. Jammal, M., Singh, T., Shami, A., Asal, R., Li, Y.: Software defined networking: state of the art and research challenges. Comput. Netw. **72**, 74–98 (2014)
12. Kanter, T., Rahmani, R., Li, Y., Xiao, B.: Vehicular network enabling large-scale and real-time immersive participation. In: International Conference on Internet of Vehicles, pp. 66–75 (2014)
13. Rahmani, R., Rahman, H., Kanter, R.: On performance of logical-clustering of flow-sensors (2014). arXiv preprint arXiv:1401.7436
14. Rahmani, R., Rahman, S.M.H., Kanter, T.: Context based logical clustering of flow sensors exploiting HyperFlow and hierarchical DHTs. RNIS: Res. Notes Inf. Serv. Sci. **14**, 721–728 (2013)
15. Rahmani, R., Kanter, T.: Layering the internet-of-things with multicasting in flow-sensors for internet-of-services. Int. J. Multimedia Ubiquit. Eng. **10**(12), 37–52 (2015)
16. Khan, S.U., Pastrone, C., Lavagno, L., Spirito, A.M.: An energy and memory-efficient key management scheme for mobile heterogeneous sensor networks. In: 2011 6th International Conference on Risks and Security of Internet and Systems (CRiSIS). IEEE (2011)
17. Seo, S.-H., Won, J., Sultana, S., Bertino, E.: Effective key management in dynamic wireless sensor networks. IEEE Trans. Inf. Forensics Secur. **10**(2), 371–383 (2015)

Peltier Module-Based Water Generation and Waste Heat Management System

Trisha Patra, Samik Basu and Soma Barman (Mandal)

Abstract Human civilization, from the very beginning of its existence, has relied on natural resources. With time, the resources are depleting and energy generation from such resources is reaching toward crisis period. In order to cope up with such situation, an alternative technology needs to be discovered that can utilize waste energy and generate energy in useable form. This paper is addressing such problems and presenting an alternate technology of Peltier effect as a solution to energy scarcity. In the proposed system the cold side of the Peltier module generates water from the moist air. Additionally, waste heat from the hot side of the module is utilized by our system to generate low-power regulated voltage using thermistor sensor. This regulated voltage is used to control the speed of 3 V motor to run a fan. The Peltier module directly converts waste heat energy into electrical energy without utilizing natural resources and without harming environment. Hence, this system is to be considered as a green system.

Keywords Peltier module · Peltier effect · Condensation · Waste heat energy · Thermistor · Voltage generation · Green system

1 Introduction

Day by day the human civilization is attaining new heights. With the growth in population, demand for resources is also growing. The main problem occurs when enough resources are not available to meet the needs of increasing demands. As a result, fuel price is increasing which leads to further discomfort for people. In recent

T. Patra · S. Basu (✉) · S. Barman (Mandal)
Institute of Radio Physics & Electronics, University of Calcutta, 92 APC Road, Kolkata 700009, India
e-mail: samikbasu2010@gmail.com

T. Patra
e-mail: trishapatra92@gmail.com

S. Barman (Mandal)
e-mail: barmanmandal@gmail.com

S. Kundu et al. (eds.), *Proceedings of the 2nd International Conference on Communication, Devices and Computing*, Lecture Notes in Electrical Engineering 602, https://doi.org/10.1007/978-981-15-0829-5_49

years, green and renewable energies have found utmost importance in technological fields. Renewable resources are the need of the hour, resulting in pollution-free and cost-friendly alternatives. In countries like India, reusing the waste resources, like heat and water, can help to overcome the water shortage issues in dry areas and also solve the electricity problems. Peltier modules are very popular for green technologies [1] because of its small and lightweight structure and immovable parts. These modules are based on the principle of Peltier effect.

With the increase in demand for renewable sources of energy, Peltier modules are gaining importance. Many noticeable works have been done in the recent past utilizing the Peltier effect. Due to its cooling capacity, the modules are utilized in building eco-friendly refrigerator [2]. Allwin Jose et al. proposed an air conditioner based on Peltier module [3]. Utilizing both the hot and cold sides of the module, portable food warmer and cooler were proposed by Pavan Attavane et al. [4]. Since the module can perform Seebeck effect also, works on generating electricity from heat and maintaining a temperature difference has been done by Prisilla A. J. Stecanella et al. [5, 6]. Sagar Venkateshwar Nemani et al. proposed a system based on generating electrical energy from exhaust gases of automobiles [7]. The cold side has been used for experimental investigations on generating portable freshwater by Joshi et al. [8]. The figure of merit, thermal conductivity, and electrical conductivity of Bi_2Te_3 and $CoSb_3$ based modules has been compared as a function of temperature by Andreas Larsson et al. [9, 10]. Kumar G. B. Arjun et al. proposed a Peltier module-based refrigerator powered by solar panel [11, 12].

Since Peltier module shows its potentiality in renewable energy generation, the authors in this paper proposed a green system using Peltier module. The module functions in twofold. First, the cold side of the module is used for water generation from atmospheric moisture. Second, the waste heat from the hot side of the module is used for voltage generation. The water generated from the cold side can further be processed for household and laboratory purposes. The collected water can further be used to measure atmospheric pollutants. The hot face of the module generates regulated power supply of the range 3.2–6.25 V using thermistor sensor with some electronic circuit arrangement. The regulated voltage is tested to control the speed of a 3 V motor along with a fan. It can also be used to run low-power devices, for example, Arduino, MSP430, and electronics vehicles.

The paper consists of five sections. In Sect. 2, the system architecture of the system is briefly explained. The experimental setup and methodology is presented in Sect. 3. In Sect. 4, the discussion is presented based on the results obtained. Finally, in Sect. 5, the paper is concluded along with future scope.

2 System Architecture

The general block diagram of the proposed system is given in Fig. 1. The system is divided into two main units—water generation unit and power generation unit.

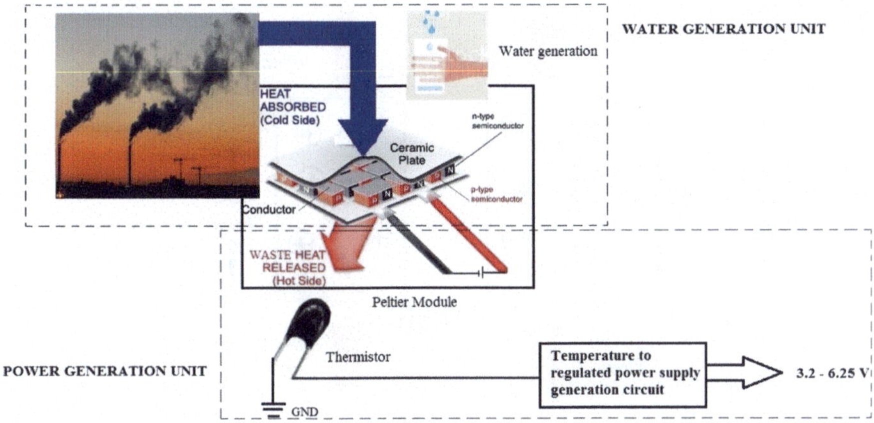

Fig. 1 General block diagram of the proposed system

2.1 Water Generation Unit

Peltier module. A Peltier module is a collection of array of Bismuth Telluride p- and n-type semiconductor cubes, sandwiched between two ceramic plates. The semiconductors are electrically connected in series and thermally connected in parallel. A pair of p- and n-type semiconductor cubes is called a couple. A thermocouple produces low voltage and high current. Thus, in order to obtain a significantly high voltage, a number of thermocouples are connected.

The module works on the basic principle of Peltier effect, which states that when current is made to flow at the junction of two dissimilar semiconductors, one junction gain heat (cold side), and the junction releases heat(hot side) [13]. The Peltier coefficient, π_{AB}, between two conductors, A and B, is equal to the ratio of the rate Q of heating or cooling at each junction to the electric current I, is represented in the following form:

$$\pi_{\mathrm{AB}} = Q/I. \tag{1}$$

The basic reason being jumping of electron to higher energy state (when current flows from p- to n-side), it gain energy and makes the surface cold. In the next cycle (n- to p-side), the electron jumps to lower energy level and releases energy, hence, making the surface hot. Figure 2 shows the current and heat flow between the couples of a module. The hot and cold sides are interchangeable based on the polarities of supply voltage [14]. In this system, TEC1-12706 module has been used. It consists of 127 couples and provides maximum temperature difference of 66 °C at 25 °C [15]. The main advantages of using Peltier module, in addition to being eco-friendly, are no moving parts, no noise, no vibration, compact, and no Freon refrigerants.

Atmospheric air. When the atmospheric air comes in contact with the cold side of the Peltier module, due to condensation the moisture in the air settles down on the

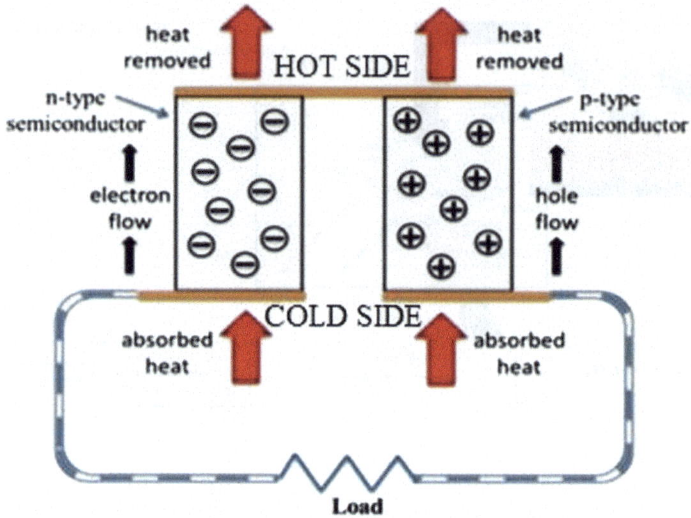

Fig. 2 Schematic of current and heat flow

cold surface. As a result, water droplets are formed. These water droplets are collected and can be utilized for various fields like laboratory purposes. The collected water can be a good source to quantify the atmospheric pollution level. The amount of water condensed on the module depends on the amount of moisture in the area. So, if the system is installed in coastal areas, the amount of water derived will be maximum.

2.2 Power Generation Unit

Thermistor sensor. The hot side of the Peltier module is kept in contact with an NTC thermistor sensor, which is thereby connected to a temperature to voltage converter circuit. The main function of the thermistor is to utilize the waste heat from the hot side of the module. It senses the heat and reduces the resistance to the circuit and a regulated voltage is obtained. After neglecting the self-heating of the thermistor, its resistance at temperature T K is related to its resistance at temperature T_0 K by the relation of the form

$$R_T = R_{T_0} e^{\mathrm{B}(1/T - 1/T_0)}, \tag{2}$$

where B is the material constant of thermistor.

3 Experimental Setup and Methodology

Experimental setup of our proposed system is shown in Fig. 3. The Peltier module is connected to a 9 V battery. As soon as the setup is connected, the module instantly shows the temperature difference, i.e., one face becomes hot and the opposite face becomes cold. The Peltier module is connected to an external heat sink with fan on the cold face. This increases the surface area for cooling operation and also increases the rate of condensation. As the moist air comes in contact with the cold side, the air starts cooling and its temperature reaches to the dew point and condensation begins. Due to gravity, the water droplets start flowing downward in the container.

For the second part of the system, the waste heat from hot face of the module is sensed by thermistor sensor and converted into voltage by temperature to voltage converter circuit shown in Fig. 4 [16]. The converter circuit consists of opamp and zener diode for maintaining regulated voltage at the output. The thermistor is basically a negative temperature coefficient (NTC) type thermistor of 10 KΩ rating. As the name suggests, it shows a negative slope in temperature versus resistance plot. With increasing temperature, the resistance of the thermistor decreases, as a result of which, the output voltage from the opamp increases [17]. However due to the opamp, the voltage is regulated after the temperature increases. The output voltage is measured across output pin of opamp and ground. It is measured using multimeter. The output-regulated voltage is used to regulate the speed of motor to run a fan.

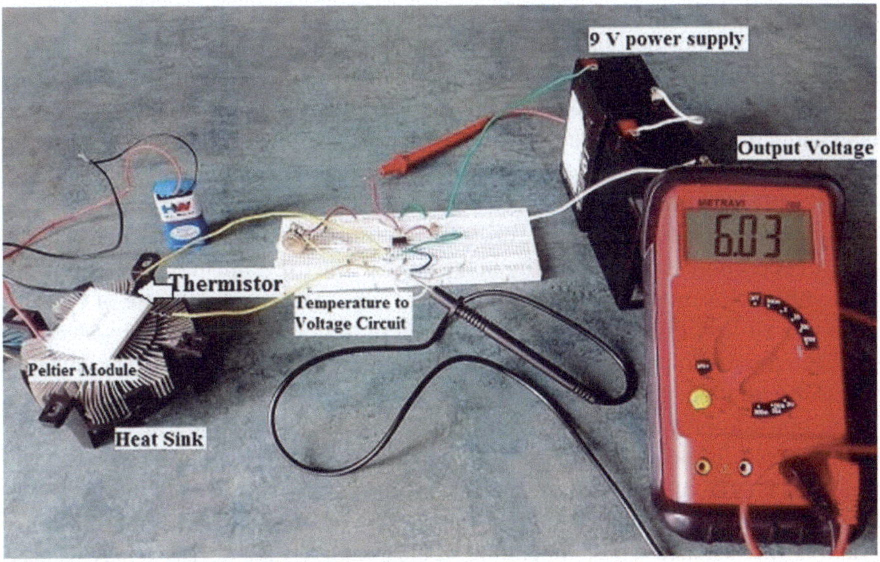

Fig. 3 Experimental setup of the proposed system

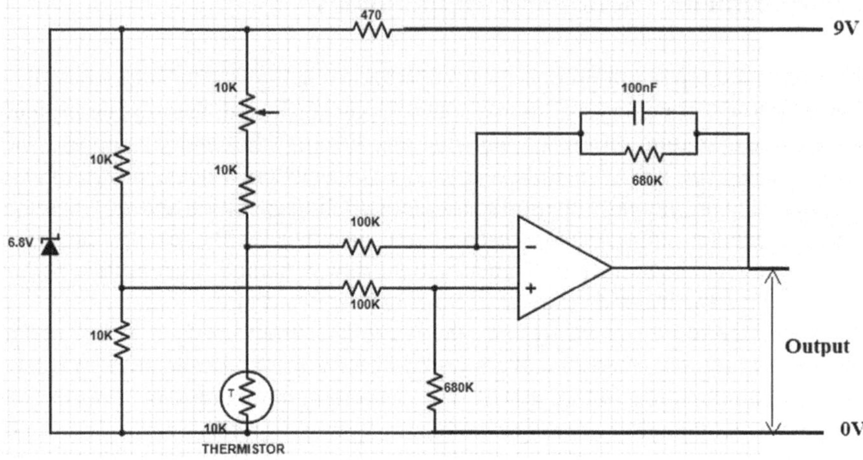

Fig. 4 Temperature to voltage converter circuit

4 Results and Discussion

After connecting Peltier module to a 6 V battery, the temperature difference on both faces of the module is instantly observed. The change in temperature values of both sides is very rapid and changes in few seconds. The cold side temperature goes as low as 21.1 °C and the hot side temperature goes up to 39.38 °C. The module's efficiency as a heating system is more prominent than as a cooling system. This leads to a slight increase in temperature on cold side. This happens because of the heat conduction from hot side through the modules. In Fig. 5, the graph shows the variation of temperature of two faces of the Peltier module with time.

The cold side of the module, when comes in contact with atmospheric moisture, undergoes condensation and water droplets are formed. The rate of condensation is directly proportional to the humidity in the air [5]. The experiment was performed in the month of November and the humidity was 50%. As a result, the rate of condensation is a bit low. Figure 6 shows graphical representation of collected water with time. The amount of water collected at the end of 143 min is 0.5 ml. The quantity increases with time.

The hot side of the module, when it comes in contact with the thermistor sensor connected in electronic circuitry, produces a regulated voltage [18, 19]. The thermistor is NTC type, which results in decrease in its resistance with increase in temperature. As a result, the output voltage through opamp increases. Figure 7 shows variation of resistance with temperature. It clearly shows the decrease in resistance and increase in temperature. The variation of voltage at the output of opamp with increase in temperature is shown in Fig. 8. With the decrease in resistance, voltage is increasing.

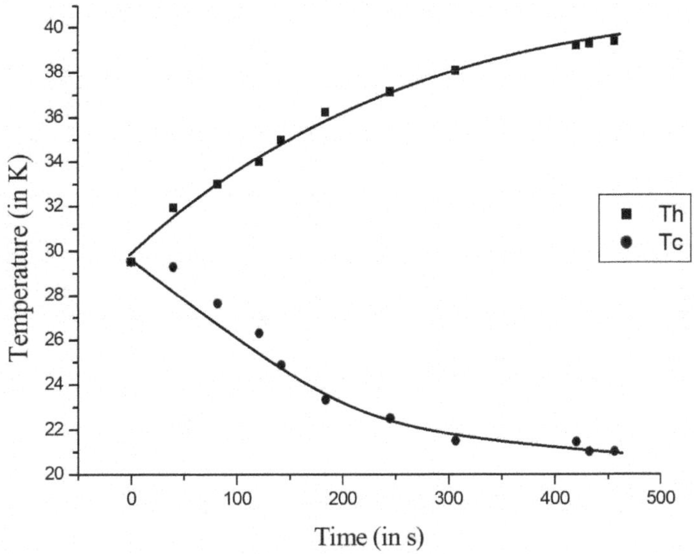

Fig. 5 Variation of the temperature of two sides of Peltier module with time, where *Th* and *Tc* are hot side temperature and cold side temperature, respectively

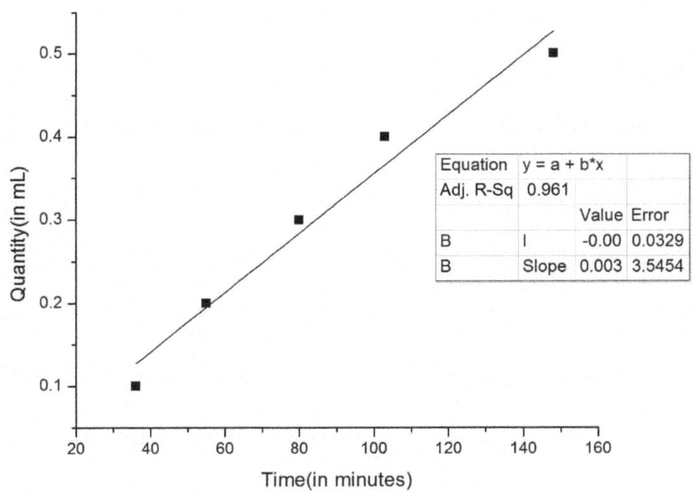

Fig. 6 Experimental result: amount of water collected versus time plot

This output voltage, is thereby, utilized to control speed of 3 V motor and a fan, as shown in Fig. 9. The fan runs for approximately 8 min and the speed of the motor is controlled using 10 KΩ potentiometer. This voltage can also be used to run other low-power devices, for example, Arduino and MSP430.

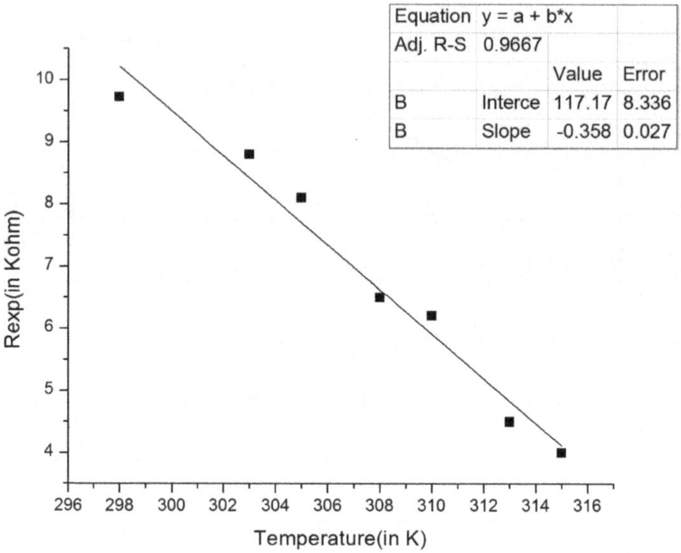

Fig. 7 Variation of resistance with temperature

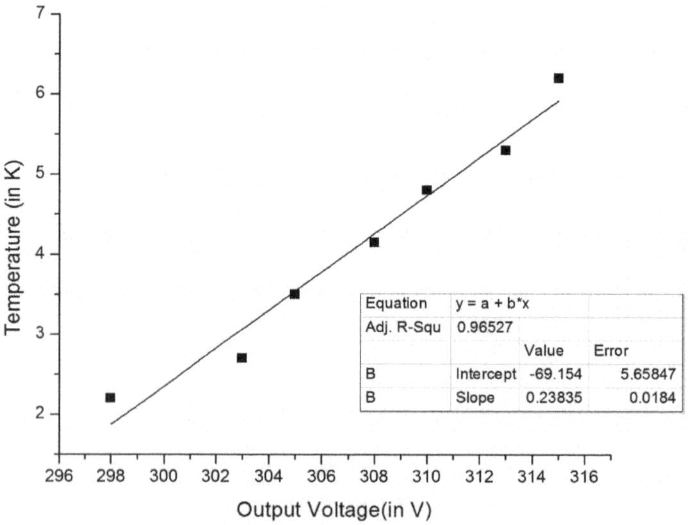

Fig. 8 Variation of output voltage with temperature

5 Conclusion and Future Work

The proposed system is an energy conservation system that generates water and also utilizes waste heat from hot side of the Peltier module for voltage generation [20].

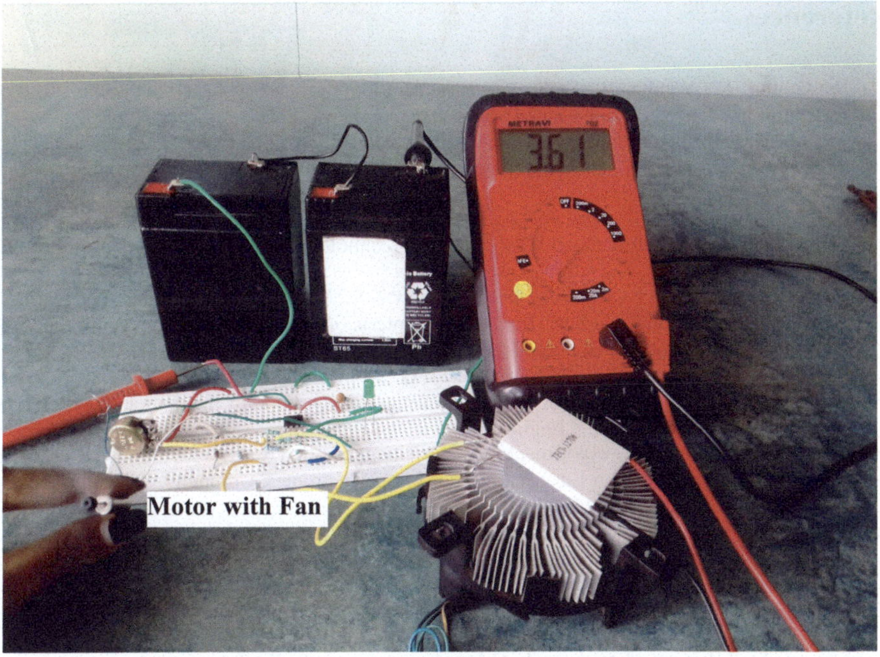

Fig. 9 The running motor with fan utilizing regulated output voltage

The temperature to voltage converter circuit is powered through two 6 V batteries collectively providing 9 V supply. The system will be potential for sufficient quantity of water generation if it is installed in high humidity areas like coastal regions. This water, on further processing, can be made consumable and for other commercial uses. The water droplets which are generated from moist air also be used to measure atmospheric pollution contents. The chemicals dissolved in the air can be tested using these water droplets.

The regulated voltage is tested here to run a motor and fan with variable speed. It can be used for operating low-power systems. This voltage can further be utilized to power the Peltier module itself. The system can also be powered using solar panels, making it a greener and more eco-friendly option.

Acknowledgements The authors would like to thank UGC UPE II, "Modern Biology Group B: Signal Processing Group", University of Calcutta for providing research facilities. The authors are also thankful to West Bengal Higher Education, Sci & Tech and Biotech (Sci & Tech.) funded project "Cytomorphic Circuit Modeling of P53 Protein Pathway for Synthetic Biology Applications" for partial support of infrastructure.

References

1. Prashantha, K., Wango, S.: Smart power generation from waste heat by thermo electric generator. Int. J. Mech. Prod. Eng., 45–49 (2016)
2. Chaudhari, V., Kulkarni, M., Sakpal, S., Ubale, A. Sangale, A.: Eco-friendly refrigerator using peltier device. In: International Conference on Communication and Signal Processing, pp. 817–819 (2018)
3. Jose, A., D'Souza, A., Dandekar, S., Karamchandani, J., Kulkarni, P.: Air conditioner using Peltier module. In: International Conference of Technologies for Sustainable Development, pp. 1–4 (2015)
4. Attavane, P., Arjun, G.B., Radhakrishna, R., Jadav, S.R.: Solar powered portable food warmer and cooler based on Peltier effect. In: 2nd IEEE International Conference On recent Trends in Electronics Information & Communication Technology, pp. 1975–1978 (2017)
5. Steanella, P.A.J., Faria, M.A.A., Domingues, E.G., Gomes, P.H.G., Calixto, W.P., Alves, A.J.: Electricity generation using thermoelectric generator TEG. In: 15th International conference on Environment and Electrical Engineering
6. Ouitrakul, S.: Preliminary experiment for electricity generation using peltier modules. In: 11th International Conference on Electrical, Engineering/Electronics, Computer, Telecommunications and Information Technology (2014)
7. Nemani, S., Shahi, D., De, M.: Effective generation of electrical energy gases from exhaust gases in automobiles. In: 7th Power India International Conference
8. Joshi, P., Joshi, S., Kothari, H.A., Mahajan, D., Chaudhari, B., Sant, K.D.: Experimental Investigation on a portable fresh water generator using a thermoelectric cooler. In: International Conference on Recent Advancement in Air Conditioning and Refrigeration
9. Larson, A., Tollefsen, T.A., Lovvik, O.M., Aasmundtveit, K.E.: Thermoelectric module for high temperature application. In: 16th IEEE ITHERM Conference, pp. 719–725 (2017)
10. Chalub, A.M.P., Ferreira da Rocha, P.H., Vilcanqui, O.A.C.: Modeling and parameter estimation of thermoelectric modules (TEM). In: CHILEAN Conference on Electrical, Electronics Engineering, Information and Communication Technologies, pp. 95–99 (2015)
11. Arjun, K.G.B., Pruthviraj, B.G., Chethan, K.Y.K., Rashmi, P.: Design and implementaton of peltier based solar powered portable refrigeration unit. In: 2nd IEEE International Conference on Recent Trends in Electronics Information & Communication Technology (RTEICT), pp. 1971–1974 (2017)
12. Morales-Acevedo, A., Gastelo-Roque, J.A.: Design of a photovoltaic system using thermoelectric Peltier cooling for vaccines refrigeration. IEEE MIT URTC (2017)
13. https://en.m.wikipedia.org/wiki/Thermoeletric-cooling. Accessed on 24 Oct 2018
14. Thakkar, M.: A report on "Peltier cooling module". https://www.researchgate.net/publication/292976771. Accessed on 10 Nov 2018
15. Hebei I.I. (Shanghai) Co., Ltd.: Thermoelectric cooler TEC1-12706. http://Peltiermodules.com/petier.datasheet/TEC1-12706.pdf
16. https://electronicsarea.com/temperature-to-voltage-converter-using-thermistor-pcb/. Accessed on 10 Oct 2018
17. https://en.m.wikipedia.org/wiki/Thermistor. Accessed on 01 Oct 2018
18. Skala, A., Waradzyn, Z.: Investigation and determination of efficiency of the waste heat recovery system using peltier modules. In: Conference on Electrotechnology: Processes, Models, Control and Computer Science (2018)
19. Prantor, T.T., Hasan, M., Hossain, C.A.: Generation of electricity using concentrated solar power and thermo-electric module. In: IEEE Region 10 Conference, pp. 1278–1283 (2016)
20. Kumar, S., Gupta, A., Yadav, G., Singh, H.P.: Peltier module for refrigeration and heating using embedded system. In: International Conference on Recent Developments in Control, Automation and Power Engineering (RDCAPE), pp. 314–319 (2015)

Modern Smart Classroom-Based Touch Technology Using Digital Image Processing

Sayan Tripathi, Jhilam Jana, Jagannath Samanta and Suvam Barui

Abstract New Technology of cooperative learning requires an upgrade from standard classrooms to interactive classrooms. A smart classroom is a modern technology that amalgamates information with novel teaching and other new technologies. The smart classroom should be developed in such a manner that it can bring up active participation and interaction among learners. This provides an important element that allows teachers to help them for interaction with each other. The smart classroom concept appeared with the evolution of the internet-based learning such as e-learning, and many other distance education systems. In this paper, we have proposed the smart touch technology by employing the concept of infrared image processing. Our proposed technology requires an IR modulated camera, a smart pen, and a projector. The smart pen which consists of infrared light emitting diode (IR LED) fitted with a circuit of two batteries and a relay switch. We have developed new software for camera to detect single stroke of the switch. The proposed design is surface independent and very cost-effective compared to traditional smart classroom technology.

Keywords Smart classroom · E-learning · Infrared LED · Smart touch technology

S. Tripathi (✉) · J. Jana
Dept. of ETCE, Jadavpur University, Jadavpur, Kolkata 700032, India
e-mail: tripathysayan@gmail.com

J. Jana
e-mail: jhilamjana2014@gmail.com

J. Samanta · S. Barui
Haldia Institute of Technology, Haldia, India
e-mail: jagannath19060@gmail.com

S. Barui
e-mail: shuvamb@yahoo.com

S. Kundu et al. (eds.), *Proceedings of the 2nd International Conference on Communication, Devices and Computing*, Lecture Notes in Electrical Engineering 602, https://doi.org/10.1007/978-981-15-0829-5_50

1 Introduction

In today's world, people employ the smart technologies to reduce their workload and to save their valuable time. Many new innovations have been invented in the recent past which directly lead us to the generation who are depleting our resources. In earlier days, teachers have employed markers and pens to teach on whiteboard. After invention of the smart classroom [1, 2], every classroom has been equipped with a projector and a desktop on the teachers desk [3, 4]. The shortcomings that occurred in the existing technology have been overcome and in this process, the system has turned out to be more interactive and efficient. The method or technique incorporated to teach, plays a very vital role in an individuals life.

A study was conducted on adaptable spaces and their impact on learning which recognized four key factors that affect the learning process: basic human need, teaching, learning, and engagement [2]. When classroom furniture is easily moved to allow for comfort and practicality students' learning experience was elevated with increased sitting comfort, being able to clearly understand the professor, and view materials. Besides students being better serviced by redesigned and modifiable classrooms, educators also reported the benefits of increased lighting, better access to internet connections, improved ability to hear students and having more whiteboard space. From the mentioned facts it is clear, what significance the way of learning or interacting can have on a person. Hence to increase the range and reach, of this setup we need more active participation and interaction, which can be achieved by handling the pointing device from a device by multiple users so that they can use the setup simultaneously. It can be estimated that the operable range of the device can be increased, so that it can be operated from any corner of a presentation or a classroom.

In the proposed infrared image processing technology, we need an infrared (IR) modulated camera, a smart pen, and a projector. In this paper, we deal with a smart pen which consists of infrared light emitting diode (IR LED) fitted with a circuit of two batteries and a relay switch. The smart pen has a IR filter which we use to point on the whiteboard where the display of the desktop is displayed through a projector. The camera detects no IR light from the IR LED in the smart pen when the circuits are closed, but as we point something on the whiteboard by switching on the circuit the camera detects IR light and the camera being connected to the desktop detects a single stroke of the switch which is equivalent to a single click of the mouse [5]. We can use the smart pen from a specific distance in the range of 8–45 m and the function of the touch screen can be performed without touching it. This technology can be operated on any surface which makes it cost-effective, portable and is easy to operate. It is a tool that will enable us to make the already interactive classroom more interactive with less expenses and is solely designed to benefit the whole scenario of teaching inside a classroom.

This paper is organized for remainder portion as follows. Section 2 provides the related work. Section 3 presents the proposed smart touch technology. Section 4 provides the operations of this proposed design. Section 5 presents advantages and limitations. Section 6 contains the application. Section 7 presents the conclusion.

2 Related Works

Classrooms and meetings are being made more interactive by a setup which includes smart board, a pointing device (mouse or light pen), a projector. Though this setup meets all the requirements for the proper functioning of a class or meeting, it costs a hefty amount and cannot be afforded by many organizations under budget constraints.

A smart board is a touch-controlled screen that enables a user to project a computer screen on a larger scale. It works on the basis of a touch screen that works with the help of a computer and projector [6]. The projector projects the desktop image onto the smart board, which acts as a monitor and an input device simultaneously [3]. When the board is pressed with a finger or a pen, a contact point is registered and as the smart board is designed such as some pressure on the board's surface is sufficient to perform pointing functions on the board [7].

Projection of an image is done by a projector through a transparent lens or directly with the help of lasers [8]. In the next section, we will be discussing the proposed technology.

3 Proposed Smart Touch Technology

A basic touch screen technology is depicted but we need not to touch it. Touch screen is an electronic visual display which can be operated using one or more fingers. A smart board uses a resistance of touch screen which consists of two layers and the point of intersection of the layer points where the screen has been touched. We can use the smart pen from a specific distance in the range of 8–45 m and all the functions of the touch screen can be performed easily without touching it. The efficiency of the device is limited as we need to work more on it, efficiency meaning the distance from which it operates, as the device is cost-efficient. The efficiency increases as we increase the power of the IR LED. To implement this technology, we need an IR modulated camera, a smart pen, and a projector. Here, we deal with a smart pen which consists of IR LED fitted with a circuit of two batteries and a relay switch.

3.1 Block Diagram

Figure 1 represents the block diagram of smart touch technology. The projector is connected to the computer which projects the screen of the computer at the wall or board. The camera points at the board and the image that is captured by the IR camera is received by the computer and then processed which consecutively simulates touch events. The resolution of the screen in which the projector is projecting and the resolution of the image the camera is capturing are different. Thus, there has to be a map of the computer screen inside the captured camera images. A matrix is made

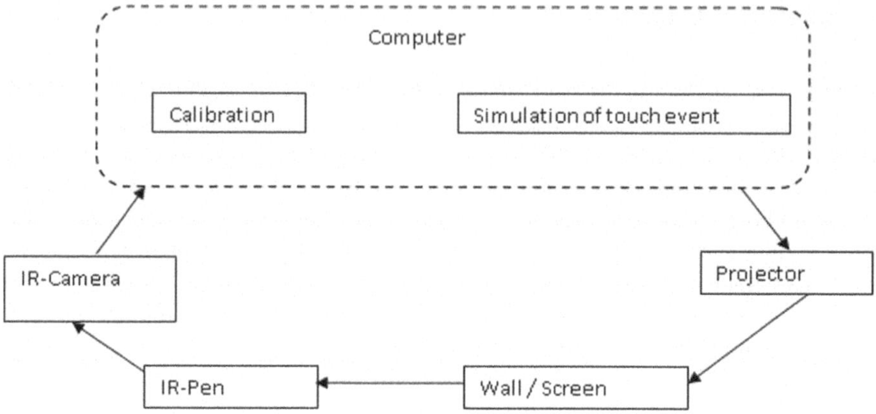

Fig. 1 Block diagram of smart touch technology

Fig. 2 Working flow of proposed smart touch technology

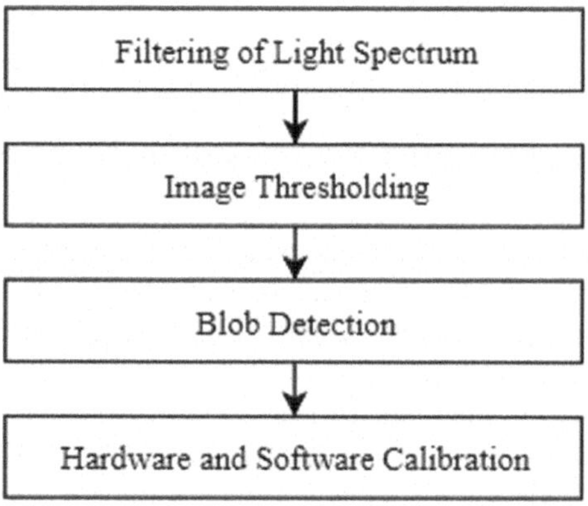

from different points on the captured image to find the corresponding screen pixel values. The method of creating the matrix is called calibration of the device setup. The user is asked to touch nine points showed in the projected screen and corresponding captured image pixels are noted and taking those as reference the mapping matrix is created and the hardware is calibrated with the software.

The working flow of proposed smart touch technology is described in Fig. 2. Filtering of light spectrum is a very important step as the infrared spectrum is debarred from flowing which is a main step of this setup. The captured image of the IR camera is then put through segmentation so that the original image is simplified and easy to calibrate. Blob detection method is incorporated to understand the image better and hence which will facilitate in mapping. Resolution of the projected screen differs from the resolution of the image captured by the IR camera so the hardware and software have to be calibrated.

3.2 *Component Required*

In this technology, we need a IR modulated camera, a smart pen, and a projector. So in this work, we deal with a smart pen which consists of IR LED fitted with a circuit of two batteries and a relay switch.

3.2.1 Near IR Sensor Modulated Webcam and Its Modification

Near-infrared (NIR) is the part of the electromagnetic spectrum directly adjacent to the visible range; as it cannot get in touch with the human eye. NIR optimized cameras have applications that utilize this specific wavelength range [9, 10]. We have made a slight modification in the normal web camera. A normal web camera captures the normal vision i.e. passes the visible light and blocks the IR. The modification we have made is by adding a negative/camera's film as a filter now the camera can capture the IR. Hence, the IR LED can be tracked. We have hence reversed the process of a normal camera.

3.2.2 Smart Pen

Figure 3 represents the prototype for the smart pen. A smart pen is a newly designed device that is made of from a basic circuit fitted with IR LED at its tip, two batteries which give the required power and a basic relay switch attached to it for control. A very important component of this device is the projector which will be discussed in the following.

3.2.3 Projector

Projectors project image through a transparent lens or through lasers. Video projectors are used extensively nowadays and are replacements for slide projectors. Lasers and LEDs are being used in modern-day projectors to project images.

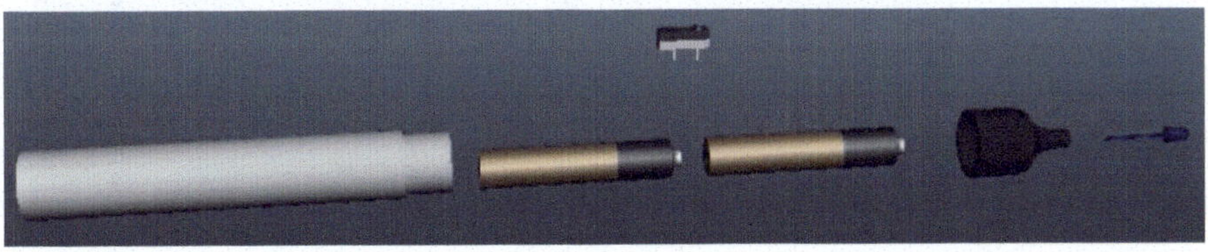

Fig. 3 Prototype for the smart pen

4 Operation

The whole operation is successfully achieved by the syncing of both the hardware and software which are described in the following.

4.1 Software and Hardware Calibration

The projector connected to the computer is projecting the screen of the computer at the wall or board. The camera is pointing at the board. The image that is captured by the IR camera is received by the computer and then processed which consecutively simulates touch events.

4.2 Mapping of the Two Screens

In Fig. 4 the mapping of the two screens is discussed. The resolution of the screen in which the projector is projecting and the resolution of the image the camera is capturing are different. Thus, there has to be a map of the computer screen inside the captured camera images. A matrix is made from different points on the captured image to find the corresponding screen pixel values. The method of creating the matrix is called calibration of the device setup. The user is asked to touch nine points showed in the projected screen and corresponding captured image pixels are noted and taking those as reference the mapping matrix is created and the hardware is calibrated with the software.

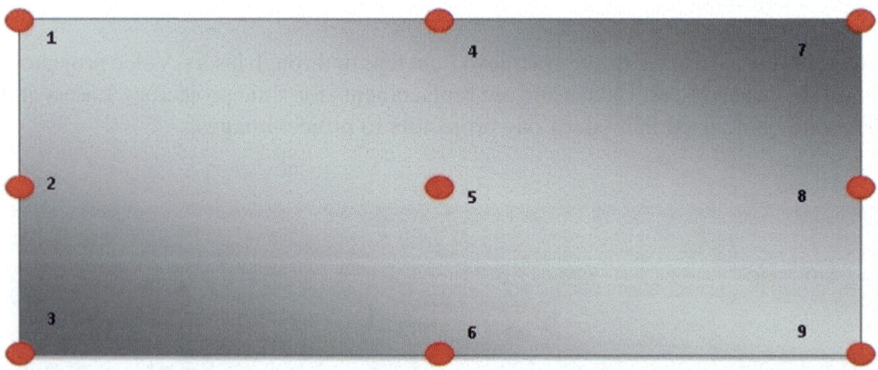

Fig. 4 Screen mapping

Fig. 5 Live operation of proposed technology

4.3 *Live Operation*

Figure 5 shows the live operation of this proposed technology. The projector projects on a surface and the image captured by the IR Camera is processed which helps in simulation of the touch events.

4.4 *Getting Coordinate*

The user points the stylus toward the board and the IR camera captures the IR led of the pen. The image captured by the camera is then threshold and the position of the touchpoint is detected by a blob detection algorithm. After getting the pixel coordinate of the touchpoint, the corresponding screen position is found out from the mapping matrix generated by the calibration process and the screen coordinate is found.

4.5 *Simulating Touch Event*

After detection, on the detected screen point, a touch simulation algorithm is applied and according to the user gesture, touch operation is used, i.e., left-click, right-click.

5 Advantages and Limitations

5.1 Advantages

(a) The most significant advantage of this technology is that it can be operated on any kind of surface as it does not come in contact of it rather it maps the coordinates of the screen and sends a signal, so as to where the pixels have to be highlighted.
(b) It is very cost-effective as when we use a smart board setup, we require multiple components especially a smart board which costs around 50,000–75,000 INR, if we remove that component, with the help of our technology the environment can continue to be interactive even in the absence of the smart board. Hence, hugely diminishing the overall cost of the setup.
(c) It overthrows the traditional method of learning and helps to create an environment which has a more practical approach, which in turn makes it a good method of learning and interaction.
(d) It is a wireless device which runs on a portable source of power, making it portable and feasible to carry and use anywhere.
(e) It runs on a very simple circuit and if damaged can be repaired very easily on a low-cost budget.
(f) It is user-friendly and does not any prior knowledge or training for operation, any person with hands can use this device.

5.2 Limitations

(a) It runs on a limited power source and will run out if batteries are not replaced after extensive use, i.e., only after the batteries have totally drained.
(b) It has a limited range of operations, i.e., it can be operated within a specific area.

6 Application

Smart boards are used in teaching, training, meetings, presentations, etc. A smart board not only makes the learning session more interactive but also enables the user and his counterparts to actively participate in it. The users may modify access in live time and make the surrounding more fun and interactive. This smart board technology offers an environment that allows students to collaborate actively. The smart board allows users to work with huge amount of data, it offers an environment that demands collaboration actively among people, the output produced is creative and interactive.

7 Conclusion

In this work, we have introduced a smart touch technology to develop the modern smart classroom communication systems. The proposed technology can significantly lower the cost of operating a smart classroom with a more interactive environment. Thus, any organization, school or college, can use this as per their need. This device connects the bridge between digital world and the reality. Using very minimum amount of hardware we have successfully implemented a large screen touch monitor. Smart classes use all interactive elements like videos and presentations and portray the visual methods of teaching which makes teaching interesting for students who are already struggling with the traditional method of teaching in a classroom. This kind of interaction is both captured by eye and perceived by the brain and young students can easily relate to them as the audio-visual room targets the senses of students and it helps the students to capture the information fast and in an effective way. It also saves a lot of time as the time taken to draw or demonstrate is minimized in this process.

References

1. Nishantha, G.G.D., Pishva, D., Hayashida, Y.: Smart classrooms: architectural requirements and deployment issues. In: Industrial and Information Systems, 2008. ICIIS 2008. IEEE Region 10 and the Third international Conference on (pp. 1–6). IEEE (2008)
2. Dufresne, R.J., Gerace, W.J., Leonard, W.J., Mestre, J.P., Wenk, L.: Classtalk: a classroom communication system for active learning. J. Comput. Higher Educ. **7**(2), 3–47 (1996)
3. Bouslama, F., Kalota, F.: Creating smart classrooms to benefit from innovative technologies and learning space design. In: Current Trends in Information Technology (CTIT), 2013 International Conference on (pp. 102–106). IEEE (2013)
4. Singh, S.P., Kumar, A., Singh, A., Jain, K.: Smart and intelligent next generation classrooms over cloud. In: MOOCs, Innovation and Technology in Education (MITE), 2015 IEEE 3rd International Conference on (pp. 273–277). IEEE (2015)
5. Swaroop C.H.: A Byte of Python. Revision 0.10, 14 Nov 2003
6. Mun, S.H., Abdullah, A.H.: A review of the use of smart boards in education. In: Engineering Education (ICEED), 2016 IEEE 8th International Conference on (pp. 120–125). IEEE (2016)
7. Lee, K., Kim, Y., Wang, Y., Kim, S., Baek, K.H.: A remote pointing device by using modulated IR signals with a 2-D striped pattern. IEEE Trans. Consum. Electron. **59**(3), 699–704 (2013)
8. Webb, D.C., Morozov, A.K., Ensign, T.H.: A new approach to low frequency wide-band projector design. In: OCEANS'02 MTS/IEEE, vol. **4**, pp. 2342–2349. IEEE (2002)
9. Vollmer, M., Mllmann, K.P., Shaw, J.A.: The optics and physics of near infrared imaging. In: Education and Training in Optics and Photonics (p. TPE09). Optical Society of America (2015)
10. Mangold, K., Shaw, J.A., Vollmer, M.: The physics of near-infrared photography. Eur. J. Phys. **34**(6), S51 (2013)

An Experimental Analysis of Machine Learning Classification Algorithms on Biomedical Data

Himansu Das, Bighnaraj Naik and H. S. Behera

Abstract Enormous growth of data in biomedical engineering domain has posed a big challenge for data analysis and processing. This massive volume of biomedical data leverage leads to a great challenge. To address this issue, this paper compares the performance measures of various classification algorithms of machine learning on biomedical data. The various classification algorithms such as Decision Tree (DT), K-Nearest Neighbors (KNN), Naive Bayes (NB), Linear Support Vector Machine (L-SVM), Radial Basis Function Support Vector Machine (RBF-SVM), Polynomial Support Vector Machine (P-SVM), Random Forest (RF), and Adaboost are used for classification process to measure the classification accuracy with 14 number of biomedical datasets. It is observed that the nature of dataset has very strong impact on the performance of the classifiers. The performance of all the aforesaid classification algorithms are analyzed on 14 biomedical datasets and observed that the Adaboost classifier outperforms than rest of the classifiers.

Keywords Classification · Machine learning · Biomedical data · Decision tree · KNN · Naive bayes · SVM · Random forest and adaboost

---

H. Das (✉) · H. S. Behera
Department of Information Technology, Veer Surendra Sai University of Technology, Burla, Sambalpur 768018, Odisha, India
e-mail: das.himansu2007@gmail.com

H. S. Behera
e-mail: hsbehera_india@yahoo.com

B. Naik
Department of Computer Application, Veer Surendra Sai University of Technology, Burla, Sambalpur 768018, Odisha, India
e-mail: mailtobnaik@gmail.com

S. Kundu et al. (eds.), *Proceedings of the 2nd International Conference on Communication, Devices and Computing*, Lecture Notes in Electrical Engineering 602, https://doi.org/10.1007/978-981-15-0829-5_51

1 Introduction

Currently, large number of machine learning algorithms are most frequently used in biomedical engineering to validate clinically diagnostic techniques. These algorithms learn the key properties of the training dataset and correctly predict the disease diagnosis in biomedical applications [1–3]. With the increasing amount of complexity in real-life problems it is very important for machine learning algorithms to satisfy large number of requirements in terms of accuracy, precision, speed, and learning ability. So it is essential to design models that can be fine-tuned and customized for very specific requirements are being used in a very large extent in industries, research, and other academic purposes. These models are generally used in disease diagnoses such as brain disorder, cardiac disease, breast cancer, Alzheimer, thyroid disorder, leukaemia, and hypertension. It is also extensively used in biomedical datasets for classification and recognition. So, these models can be fine-tuned and customized for very specific requirements that are being used in a very large extent in industries, research, and other academic purposes. Analysis of biomedical data for disease classification and prediction using efficient machine learning algorithm is a challenging task due to its noisy and irrelevant features. These biomedical datasets usually contain large features to describe the patient records. Most of these biomedical datasets contain large set of features, noisy information, and complex patterns (mixed data types contains both continuous and discrete features). It is essential to analyze these high dimensional and multivariate biomedical datasets efficiently and accurately using various machine learning algorithms with the following objectives. First, which machine learning algorithm is most suitable for biomedical datasets. Second, how the performance of various machine learning algorithms has an impact on biomedical datasets. Selecting a correct machine learning algorithm is really tedious task in machine learning. So, different factors need to be considered during the development process of machine learning algorithm such as number of features need to be considered that are significantly contributing in the decision-making process of classification [4–8] or prediction of result, dividing the dataset into appropriate training and testing datasets based on the nature of its data, selecting the correct method of analysis and validation of classifiers. The classifier is useful only when it selects the correct and appropriate features that contain significant amount of medically relevant information. These reduced attributes speed up the classification process and also accuracy. The remainder section of this paper is organized as follows. Section 2 presents the classification procedure and various classification algorithms in machine learning, Sect. 3 analyzes the experimental results of various classification algorithms with 14 biomedical datasets, Sect. 4 describes the statistical significance of results, Sect. 5 discusses the various steps involved in classification process and its open issues, and Sect. 6 concludes the paper.

2 Classification Modeling Methodology

Classification is one of the primary tasks in data mining and pattern recognition that extracts knowledge from real-world problems. It is a process of constructing a model that accurately predicts the target class from the data into different class levels. This section describes the classification procedure and various machine learning algorithms as follows.

2.1 *Classification Procedure*

The detailed step-by-step procedure that is adapted during classification process is described in Fig. 1. In the first stage data is pre-processed in which data cleaning, data selection, and data transformation are performed. In data cleaning, noisy data, and redundant data are removed. The data cleaning step also handles the missing values by replacing the mean value for that attribute based on statistical analysis. In data selection, statistical correlation analysis is used to discard redundant attributes so that only the relevant attributes are taken from the given dataset. In data transformation, the data is normalized and transformed all the values of attributes in the range of -1.0 to $+1.0$. In the second stage, the original dataset is divided into two subsets called training data and testing data. In the training phase, the training dataset is applied to the model to train the model for classification. In the testing phase, the

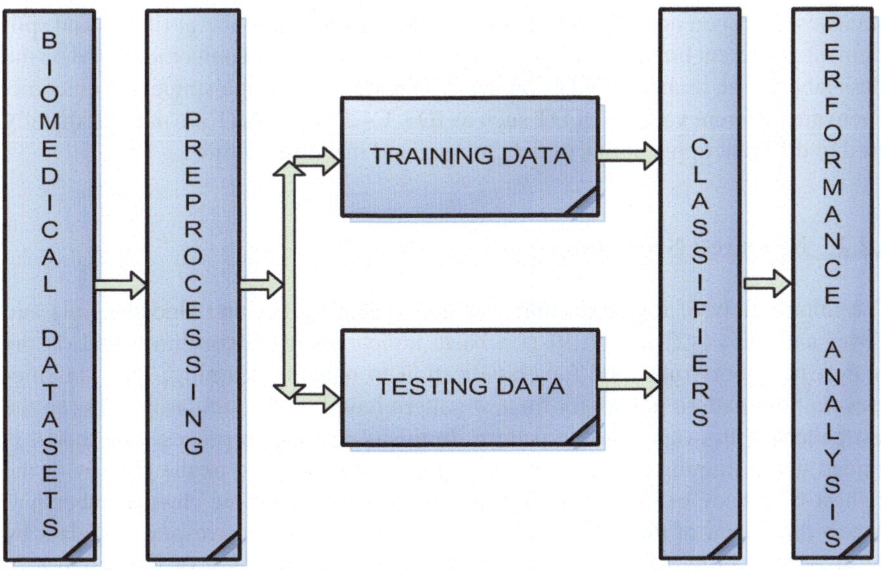

Fig. 1 Classification procedure

model is used to test the testing data for estimating the model performance of the classification algorithms. Finally, the performance of the model is computed based on different performance measures such as accuracy, precision, recall, F-measure, and support.

2.2 *Classification Algorithms*

Classification algorithms are machine learning models used to discriminate between the different objects based on certain principles. In this section, various classification algorithms such as DT, KNN, NB, SVM, RF, and Adaboost are used in the classification process.

2.2.1 Decision Tree

DT [9, 10] is a supervised classification and regression algorithm that can be modeled in the form of tree structure commonly used in data mining and machine learning. It can be represented in two ways such as classification tree in which target variables are discrete in nature and regression tree in which the target variables are continuous data. It is a tree representation in the decision-making process in which the decision nodes continuously expand the nodes until the stopping criteria meet. It recursively selects the best attribute to splits the data continuously based on certain parameters. The decision tree having three components such as decision nodes, leaf nodes, and branches. Decision nodes generally conduct the test on particular attributes and split them into different branches. The leaf nodes represent the final outcomes based on the class labels. The branches of DT on decision nodes represent a single test outcome. There are different variants of DT such as ID3, C4.5, and CART are most frequently used in different applications of data mining and machine learning.

2.2.2 K-Nearest Neighbors

The initial study of this algorithm was undertaken by Fix and Hodeges [11] and Cover and Hart [12]. KNN [9] is a basic nonparametric algorithm based on the nearest neighbors rule used for classification in machine learning. This classifier uses training pattern to classify the test pattern based on the similarity between the test pattern with every training pattern. In this classification process, the model is trained using training data and the testing can be performed as the class with the highest frequency of k-most similar instances that produce the class membership. A new data point of the testing pattern is classified into its corresponding class by the majority voting of its neighbors based on the computation of the distance to the nearest training pattern. The selection of suitable value of k is really crucial task that influences the performance of the classifier [13] (classification accuracy).

Generally, the value of k can considered as a small odd and even number for even and odd number of class problem, respectively, to break the ties. The larger values of k reduce the effect of noisy points in the datasets.

2.2.3 Naive Bayes

NB [9, 14] classifier belongs to the family of probabilistic classifier based on the principle of Bayes' Theorem commonly used in machine learning. This classifier works based on the assumption that the prior probabilities of the patterns are known and the posterior probabilities are assigns to the class labels. The outcome of this classification process is based on the maximum posterior probability. This classifier uses Bayes' Theorem for conversion of prior probability to posterior probability using maximum likelihood values to predict the class labels. It is suitable to handle the problems with high dimensions of independent features. This classifier results the outcome with the highest probability or with the lowest probability error.

2.2.4 Support Vector Machine

SVM [15] is a supervised machine learning algorithm used for classification and regression analysis. This algorithm finds an optimal hyperplane that maximizes the margin between the data points of both the classes in n-dimensional space. These hyperplanes are generally decision boundaries that can distinctly classify the data points. The dimensions of the hyperplane can be considered based on the number of features available in the dataset. Support vectors are those data points that are closer to the hyperplane and it can also influence the position and orientation of the hyperplane. The distance between the hyperplane and nearest data points from either class is known as margin. The margin of the classifier can be maximized using these support vectors. The objective of SVM algorithm is to choose an optimal hyperplane that maximizes the margin between the data points.

2.2.5 Random Forest

RF [16, 17] is an ensemble of randomized decision tree classifiers used for classification and regression kind of problems in machine learning. Random forest consists of many classification trees. The class labels of testing dataset are measured based on the voting of each classification tree. The outcome of the classifier depends on the class labels that have the maximum voting by the classification trees. This algorithm has some advantages such as high classification accuracy, nonparametric in nature, effectively handles large datasets, capability to determine the importance of variable, and also robust to handle the outliers.

2.2.6 Adaboost

Boosting algorithms are ensemble of techniques that creates a strong classifier from a number of weak classifiers to improve the performance. This can be achieved by building a model from the training data and then it creates another model that corrects the error from the first model. The models are continuously added until the training set perfectly predicts or maximum models are generated. The final prediction of this classifier is based on the majority voting of the individual classifiers. Adaboost (Adaptive Boosting) [18, 19] is proposed by Freund and Schapire in 1996. It is an iterative ensemble of multiple classifiers used to boost the performance of machine learning classifiers for binary classification problems. Generally, Adaboost classifier is sensitive to outliers and noisy data.

3 Experimental Results and Discussion

This section investigates the performance of different classification algorithms such as DT, KNN, NB, L-SVM, RBF-SVM, P-SVM, RF, and Adaboost on 14 different biomedical datasets and found that Adaboost classification algorithm is superior than the rest of the classification algorithms based on average rank method. These biomedical datasets [20] are collected from UCI machine learning repository and split into training data (75%) and testing data (25%). The details information about each biomedical datasets such as number of samples, number of features excluding class-level attributes, and the number of classes are available which are presented in Table 1. The implementation of these classification algorithms is done using Python and the classification accuracy is presented in Table 2. The other performance measures such as precession, recall, F-measure, and support of all the classification algorithms for 14 biomedical datasets are presented in Tables 3 and 4. From the classification accuracy, it is noticed that the variance of classification accuracy of various classifiers for a particular dataset is significantly low as compared to the variance of classification accuracy of a particular classifier with different datasets. It implies that the nature of dataset has very strong impact on the performance of the classifiers (i.e., classification accuracy). Apart from that, the performance of the classifiers also affect significantly in the classification accuracy with the percentage of presence of missing values, and also minor impact due to the imbalance ratio of the datasets.

4 Statistical Analysis

The statistical analysis is used to analyze and validate the performance of various algorithms in terms of classification accuracy or error. It also helps to analyze the nature of data and its significance. One-way ANOVA test is a parametric statistical

Table 1 Biomedical dataset information

Dataset name	Number of samples	Number of features	Number of classes
Pima Indian diabetes	768	8	2
Breast Cancer Wisconsin (diagnosis)	569	30	2
Breast Cancer Wisconsin	699	9	2
Mammographic mass	961	5	2
Thyroid	215	5	3
Blood Transfusion Service Center	748	4	2
Dermatology	366	33	6
Heart statlog	270	13	2
Lung cancer	32	56	3
SPECTF heart	267	44	2
E. coli	336	7	8
Haberman	306	3	2
Liver	345	6	2
Hepatitis	155	19	2

test [21] used for comparison of three or more groups mean for statistical significance of the datasets. It uses null and alternative hypothesis. The acceptance of null hypothesis indicates that there is no significant different among the performance of the classification algorithms. The acceptance of alternative hypothesis indicates that at least anyone classifier is different from rest of the classification algorithms. But one-way ANOVA test does not reveal which classifier is different from which classifier. In our experiment, one-way ANOVA test rejects the null hypothesis and alternatively accepts the alternative hypothesis. The one-way ANOVA test has been carried out in Duncan's multiple test range with 95% confidence interval and 5% significant level using SPSS (Version 16.0) and the result has been presented in Table 5.

5 Discussion and Open Issues

This section discusses the various factors that affect the performance of the classification algorithms. These various factors need to be considered and addressed while applying the classification algorithms in various domains. The first step of analysis is preprocessing in which data cleaning, noisy data removal, outlier detection, missing value imputation, and transformation of features are performed. In our work, we have used Min–Max data normalization, KNN data imputation in preprocessing step. The issue in this step is that mean value data imputation techniques may not always be

Table 2 Comparison of classification accuracy of DT, KNN, NB, L-SVM, RBF-SVM, P-SVM and RF

Datasets/classifiers	Accuracy (%)							
	DT	KNN	NB	L-SVM	RBF-SVM	P- SVM	RF	Adaboost
Pima Indian diabetes data set	69.605 (7)	80.2 (1)	76.56 (4)	80.2 (1)	77.6 (3)	75 (5)	72.39 (6)	79.16 (2)
Breast Cancer Wisconsin (diagnosis)	93.972 (5)	95.1 (4)	91.6 (6)	97.2 (2)	96.5 (3)	91.6 (6)	97.2 (2)	97.9 (1)
Breast Cancer Wisconsin	93.92 (6)	97.71 (1)	95.42 (4)	96 (3)	95.42 (4)	94.28 (5)	96.57 (2)	94.28 (5)
Mammographic mass	73.29 (6)	79.8 (2)	79.32 (3)	81.73 (1)	79.8 (2)	77.88 (4)	75.1 (5)	81.73 (1)
Thyroid	96.92 (3)	98.14 (2)	100 (1)	98.14 (2)	100 (1)	96.29 (4)	96.92 (3)	–
Blood Transfusion Service Center	73.55 (2)	71.12 (5)	72.19 (4)	70.58 (6)	70.58 (6)	71.12 (5)	75.55 (1)	72.72 (3)
Dermatology	96.75 (2)	95.55 (3)	82.22 (7)	93.33 (5)	94.44 (4)	91.11 (6)	99.07 (1)	–
Heart statlog	70.98 (5)	77.94 (3)	80.88 (2)	80.88 (2)	77.94 (3)	83.82 (1)	83.82 (1)	75 (4)
Lung cancer	66.67 (1)	57.14 (2)	42.85 (3)	42.85 (3)	57.14 (2)	42.85 (3)	57.14 (2)	–
SPECTF heart	86.19 (3)	73.86 (7)	75 (6)	78.4 (5)	85.22 (4)	72.72 (8)	88.63 (1)	87.5 (2)
E. coli	75.24 (5)	79.76 (3)	77.38 (4)	80.95 (2)	82.14 (1)	75 (6)	79.76 (3)	–
Haberman	68.47 (1)	61.03 (3)	59.74 (4)	59.74 (4)	57.14 (5)	59.74 (4)	63.63 (2)	61.03 (3)
Liver	60.09(4)	57.47 (5)	54.02 (6)	65.51 (3)	66.66 (2)	48.27 (7)	70.11 (1)	66.66 (2)
Hepatitis	79.78 (3)	79.48 (3)	69.23 (5)	87.17 (1)	79.48 (3)	76.92 (4)	82.05 (2)	82.05 (2)
Average rank	4.41 (6)	3.67 (5)	4.91 (7)	3.34 (3)	3.58 (4)	5.67 (8)	2.67 (2)	2.5 (1)

Table 3 Comparison of precision, recall, F-measure, and support of DT, KNN, and NB

Datasets/classifiers	DT				KNN			
	Precision	Recall	F-measure	Support	Precision	Recall	F-measure	Support
Pima Indian diabetes	0.691	0.697	0.692	231	0.8	0.8	0.8	192
Breast Cancer Wisconsin (diagnosis)	0.941	0.941	0.938	171	0.95	0.95	0.95	143
Breast Cancer Wisconsin	0.938	0.938	0.938	210	0.98	0.98	0.98	175
Mammographic mass	0.735	0.73	0.735	249	0.8	0.8	0.8	208
Thyroid	0.97	0.97	0.97	65	0.98	0.98	0.98	54
Blood Tansfusion Service Center	0.705	0.735	0.705	225	0.67	0.71	0.65	187
Dermatology	0.97	0.965	0.965	108	0.96	0.96	0.96	90
Heart statlog	0.715	0.71	0.705	81	0.78	0.78	0.78	68
Lung cancer	0.77	0.8	0.775	47	0.79	0.79	0.76	39
SPECTF heart	0.89	0.865	0.865	105	0.78	0.74	0.75	88
E. coli	0.805	0.75	0.77	101	0.77	0.8	0.78	84
Haberman	0.7	0.68	0.69	92	0.59	0.61	0.56	77
Liver	0.6	0.605	0.595	104	0.57	0.57	0.57	87
Hepatitis	0.735	0.73	0.735	249	0.8	0.8	0.8	208

(continued)

Table 3 (continued)

Datasets/classifiers	NB				Adaboost			
	Precision	Recall	F-measure	Support	Precision	Recall	F-measure	Support
Pima Indian diabetes	0.76	0.77	0.76	192	0.79	0.79	0.79	192
Breast Cancer Wisconsin (diagnosis)	0.92	0.92	0.92	143	0.98	0.98	0.98	143
Breast Cancer Wisconsin	0.96	0.96	0.96	175	0.94	0.94	0.94	175
Mammographic mass	0.79	0.79	0.79	208	0.82	0.82	0.82	208
Thyroid	1	1	1	54	–	–	–	–
Blood Tansfusion Service Center	0.69	0.72	0.68	187	0.73	0.73	0.65	187
Dermatology	0.8	0.82	0.78	90	–	–	–	–
Heart statlog	0.81	0.81	0.81	68	0.75	0.75	0.75	68
Lung cancer	0.86	0.69	0.71	39	–	–	–	–
SPECTF heart	0.82	0.75	0.76	88	0.88	0.88	0.88	88
E. coli	0.71	0.77	0.74	84	–	–	–	–
Haberman	0.56	0.6	0.5	77	0.59	0.61	0.55	77
Liver	0.55	0.54	0.53	87	0.67	0.67	0.67	87
Hepatitis	0.79	0.79	0.79	208	0.82	0.82	0.82	39

Table 4 Comparison of precision, recall, F-measure, and support of L-SVM, RBF-SVM, P-SVM, and RF

Datasets/classifiers	L-SVM				RBF-SVM			
	Precision	Recall	F-measure	Support	Precision	Recall	F-measure	Support
Pima Indian diabetes	0.8	0.8	0.8	192	0.77	0.78	0.77	192
Breast Cancer Wisconsin (diagnosis)	0.97	0.97	0.97	143	0.96	0.97	0.96	143
Breast Cancer Wisconsin	0.96	0.96	0.96	175	0.95	0.95	0.95	175
Mammographic mass	0.83	0.82	0.81	208	0.8	0.8	0.8	208
Thyroid	0.98	0.98	0.98	54	1	1	1	54
Blood Transfusion Service Center	0.5	0.71	0.58	187	0.5	0.71	0.58	187
Dermatology	0.94	0.93	0.93	90	0.95	0.94	0.94	90
Heart statlog	0.81	0.81	0.81	68	0.78	0.78	0.78	68
Lung cancer	0.89	0.87	0.86	39	0.79	0.79	0.76	39
SPECTF heart	0.77	0.78	0.78	88	0.85	0.85	0.85	88
E. coli	0.76	0.81	0.77	84	0.76	0.82	0.78	84
Haberman	0.36	0.6	0.45	77	0.45	0.57	0.45	77
Liver	0.67	0.66	0.64	87	0.69	0.67	0.65	87
Hepatitis	0.83	0.82	0.81	208	0.8	0.8	0.8	208

(continued)

Table 4 (continued)

Datasets/classifiers	P-SVM				RF			
	Precision	Recall	F-measure	Support	Precision	Recall	F-Measure	Support
Pima Indian diabetes	0.76	0.75	0.71	192	0.71	0.72	0.71	192
Breast Cancer Wisconsin (diagnosis)	0.93	0.92	0.91	143	0.97	0.97	0.97	143
Breast Cancer Wisconsin	0.94	0.94	0.94	175	0.97	0.97	0.97	175
Mammographic mass	0.79	0.78	0.78	208	0.76	0.75	0.75	249
Thyroid	0.96	0.96	0.96	54	0.97	0.97	0.97	65
Blood Transfusion Service Center	0.8	0.71	0.6	187	0.73	0.76	0.73	225
Dermatology	0.94	0.91	0.91	90	0.99	0.99	0.99	108
Heart statlog	0.84	0.84	0.84	68	0.84	0.84	0.84	68
Lung cancer	0.75	0.77	0.72	39	0.33	0.57	0.42	7
SPECTF heart	0.53	0.73	0.61	88	0.89	0.89	0.89	88
E. coli	0.7	0.75	0.7	84	0.78	0.8	0.78	84
Haberman	0.36	0.6	0.45	77	0.63	0.64	0.60	77
Liver	0.41	0.48	0.39	87	0.7	0.7	0.7	87
Hepatitis	0.79	0.78	0.78	208	0.82	0.82	0.82	39

Table 5 ANOVA test with descriptive statistics of eight classifiers on 14 dataset

Classifiers	Datasets	Mean	Std. deviation	Std. error	95% confidence interval for mean		Minimum	Maximum
					Lower bound	Upper bound		
DT	14	78.9591	12.35614	3.30232	71.8248	86.0933	60.09	96.92
KNN	14	78.8786	14.15887	3.78412	70.7035	87.0537	57.14	98.14
NB	14	75.4579	15.63011	4.17732	66.4333	84.4824	42.85	100
L-SVM	14	79.4771	15.69597	4.19493	70.4146	88.5397	42.85	98.14
RBF-SVM	14	80.0043	13.79111	3.68583	72.0415	87.967	57.14	100
P-SVM	14	75.4714	16.28584	4.35257	66.0683	84.8746	42.85	96.29
RF	14	81.2814	13.24558	3.54003	73.6337	88.9292	57.14	99.07
Adaboost	14	75.5957	13.73724	3.67143	67.6641	83.5274	44.44	97.9
Total	112	78.1407	14.12429	1.33462	75.496	80.7853	42.85	100

suitable for every application. In the second step, dataset is split into training data and testing data. Training data with class label is used to train the model in order to predict the outcome of the model on test data. In this step, overfitting and underfitting issue may arise during the training of the model. Overfitting means model is trained so well that it will closely fit into the training data. This situation may arise when model is too complex that contains large number of features with small number of instances. In contrast, underfitting does not fit the model to the training data. It is probably due to very simple model. In the third step, various machine learning classification algorithms are used for classification. Each classification algorithms has its own pros and cons. Some of these issues are addressed as follows. In DT, any changes in the input features such as eliminating redundant features can lead to significant changes in structure of the DT. Apart from this some more issues may be considered such as handling continuous data, selection of decision nodes, handling missing values, and handling features with differing costs. In KNN, selection of suitable value of 'k' is really critical task that influences the performance of the classifier. Generally, the value of k can considered as a small odd and even number for even and odd number of class problem, respectively, to break the ties. NB can suitable to handle missing values and imbalance data. It has also some issues such as handling incomplete data, continuous data, and independent features. The challenges of SVM are handling unbalanced data, multi-class problems, and large datasets. In RF, determination of forest size is a challenging task. Generally, more number of trees are ensembles to improve the accuracy of the model which makes the model slower. Adaboost classifier is suitable for two-class problems but it cannot handle multi-class problems. In the last step, performance measures and compared for various classification algorithms.

6 Conclusion

This research work compared the performance of various classification algorithms such as DT, KNN, NB, SVM, RF, and Adaboost that are most successfully used in various applications of data mining and machine learning. The working principle of these classification algorithms is described in details and also compute different performance measures such as accuracy, precession, recall, F-measure, and support for analysis. This work is used in 14 biomedical datasets for analysis of performance and found Adaboost classification algorithms outperforms than the rest of the classification algorithms. This analysis also reveals that the overall performance of all the classification algorithms differ based on the nature of datasets. The trends of data in the dataset have strong impact on the performance of classification algorithms.

References

1. Tanwani, A.K., Farooq, M.: The role of biomedical dataset in classification. In: Conference on Artificial Intelligence in Medicine in Europe (pp. 370–374). Springer, Berlin, Heidelberg (2009)
2. Tanwani, A.K., Afridi, J., Shafiq, M.Z., Farooq, M.: Guidelines to select machine learning scheme for classification of biomedical datasets. In: European Conference on Evolutionary Computation, Machine Learning and Data Mining in Bioinformatics (pp. 128–139). Springer, Berlin, Heidelberg (2009)
3. Pradhan, C., Das, H., Naik, B., Dey, N.: Handbook of Research on Information Security in Biomedical Signal Processing (pp. 1–414). IGI Global, Hershey, PA (2018). https://doi.org/10.4018/978-1-5225-5152-2
4. Pattnaik, P.K., Rautaray, S.S., Das, H., Nayak, J.: Progress in Computing, Analytics and Networking. Springer (2018)
5. Das, H., Naik, B., Behera, H.S.: Classification of diabetes mellitus disease (DMD): a data mining (DM) approach. In: Progress in Computing, Analytics and Networking (pp. 539–549). Springer, Singapore (2018)
6. Sahani, R., Rout, C., Badajena, J.C., Jena, A.K., Das, H.: Classification of intrusion detection using data mining techniques. In: Progress in Computing, Analytics and Networking (pp. 753–764). Springer, Singapore (2018)
7. Das, H., Jena, A.K., Nayak, J., Naik, B., Behera, H.S.: A novel PSO based back propagation learning-MLP (PSO-BP-MLP) for classification. In: Computational Intelligence in Data Mining, vol. 2 (pp. 461–471). Springer, New Delhi (2015)
8. Dey, N., Ashour, A.S., Kalia, H., Goswami, R., Das, H.: Histopathological Image Analysis in Medical Decision Making, pp. 1–340. IGI Global, Hershey, PA (2019). https://doi.org/10.4018/978-1-5225-6316-7
9. Murty, M.N., Devi, V.S.: Pattern Recognition: An Algorithmic Approach. Springer Science & Business Media (2011)
10. Quinlan, J.R.: Induction of decision trees. Mach Learn **1**(1), 81–106 (1986)
11. Fix, E., Hodges Jr, J.L.: Discriminatory Analysis-Nonparametric Discrimination: Consistency Properties. California Univ Berkeley (1951)
12. Cover, T., Hart, P.: Nearest neighbor pattern classification. IEEE Trans. Inf. Theory **13**(1), 21–27 (1967)
13. Hall, P., Park, B.U., Samworth, R.J.: Choice of neighbor order in nearest-neighbor classification. Ann. Stat. **36**(5), 2135–2152 (2008)

14. Rish, I.: An empirical study of the naive Bayes classifier. In: IJCAI 2001 workshop on empirical methods in artificial intelligence, vol. 3, no. 22, pp. 41–46. IBM, New York (2001)
15. Cortes, C., Vapnik, V.: Support-vector networks. Mach Learn **20**(3), 273–297 (1995)
16. Ho, T.K.: Random decision forests. In Document analysis and recognition. In: Proceedings of the Third International Conference on (vol. 1, pp. 278–282). IEEE (1995)
17. Barandiaran, I.: The random subspace method for constructing decision forests. IEEE Trans. Pattern Anal. Mach. Intell. **20**(8), (1998)
18. Freund, Y., Schapire, R.E.: A decision-theoretic generalization of on-line learning and an application to boosting. J. Comput. Syst. Sci. **55**(1), 119–139 (1997)
19. Freund, Y., Schapire, R., Abe, N.: A short introduction to boosting. J. Japan. Soc. Artif. Intell. **14**(771–780), 1612 (1999)
20. Blake, C.: UCI repository of machine learning databases. http://www.ics.uci.edu/~ mlearn/MLRepository.html (1998)
21. Fisher, R.A.: Statistical Methods and Scientific Inference, 2nd edn. Hafner Publishing Co., New York (1956)

A Noble Approach Toward Security Implementation in Cloud Virtualization Framework

Tanmoy Mukherjee, Sudipta Sahana and Debabrata Sarddar

Abstract Virtual machine infrastructure has been one of the most powerful technologies over the years. In the mid of its growing popularity in the market, there lies a fear in using this technology. Various survey papers discussed the security issues concerning this technology, for instance, the attack at the Virtualization level, man-in-the-middle attacks, denial of service (DoS). In our paper, we discussed various security issues which can occur when we use an application in a virtualized environment. Those issues can be related to the physical machine, in which the virtualized environment is being set up. Attackers can also gain access to the hypervisor which acts as an interface between the virtual machines and the host machine. Security threats can also come up when we login into an application in a virtual machine. Those issues can be related to the login credentials, the data which is to be transferred to the destination system and last but by no means the least, the vulnerabilities of the virtual machines, through which an attacker can gain access to the hypervisor, which is the heart of the virtualized environment. Hence, it is imperative that we do some survey on how we can implement the measures by which we can minimize these security threats and counter the same.

Keywords Virtual machine · Hypervisor · Network spoofing · Filtering · Secure file transfer protocol (SFTP)

T. Mukherjee · S. Sahana (✉)
Department of CSE, JIS College of Engineering, Kalyani, Nadia, India
e-mail: ss.jisce@gmail.com

T. Mukherjee
e-mail: tanmoymukherjee29@gmail.com

D. Sarddar
Department of CSE, University of Kalyani, Kalyani, Nadia, India
e-mail: dsarddar1@gmail.com

S. Kundu et al. (eds.), *Proceedings of the 2nd International Conference on Communication, Devices and Computing*, Lecture Notes in Electrical Engineering 602, https://doi.org/10.1007/978-981-15-0829-5_52

1 Introduction

Virtualization forms an important part of cloud computing. The role of Virtualization [1, 2] in cloud computing is immense as this layer maintains the virtual servers. It also hides the physical characteristics/properties of computing resources. Although the role of Virtualization has gained immense popularity over the years in the information technology (IT) market and other business organizations, the risks associated with this technology are many. Here, we analyze some of the threats that might come when we run an application in a virtualized environment.

In an organization, there can be many security issues that can come up when we run any application in a virtual machine server. There may be attacks on the host hardware system, the virtual machine monitor which acts as an interface between the virtual machines located on the virtual server and the host hardware system. An attacker can also gain access to any of the virtual machines, steal the login credentials, for logging into the application where the data, which is to be transmitted to the destination system, is being stored. The data which is to be transferred also needs to be highly secured and free from all sorts of vulnerability.

In this paper, we are going to analyze all these security threats and suggest some ways to mitigate these attacks.

2 Related Work

Joyce Beryl Princess et al. [3] surveys various security threats during live migration and the methods to mitigate those attacks. It narrates formal methods for live migration of virtual machines from source physical server to destination physical server which are—Pre-Copy Algorithm and Post-Copy Algorithm. It discusses various security threats during live migration of virtual machines which are—Man-in-the-middle attack, DoS Attack, Overflow Attack, Replay Attack. It surveys various techniques to mitigate the security attacks during live migration of virtual machines which are—Security Preserving Live Migration, Hypershield, Protection Aegis for Live Migration of VMs (PALM), Comprehensive live migration (COM), Seamless Live Migration, Remote Attestation Approach, Virtual Trusted Platform Module (vTPM) Migration Protocol, VM-vTPM Migration Protocol, Live migration defense framework (LMDF), Framework for secure live migration, Secure Context Migration, Token-Based Migration, Xensploit, RSA algorithm with Secure Socket Layer (SSL) Protocol.

In this paper [4], the author discusses Virtualization and its types, how virtual machines are attacked at Virtualization level and the attacks at this level. It also describes the different methods of Virtualization which are—Operating System-Based Virtualization, Hypervisor-Based Virtualization, and Application-Based Virtualization. The Virtualization concerns are also described here which are—Virtualization and Virtual Machines, Virtual Machine (VM) Sprawl, DoS and VM Escape Attack.

Sahana et al. [5] examine a technique that can reduce the cost of storing data in Cloud using Gossip protocol, which is used to distribute and update information across wireless networks and large-scale database systems. It suggested that using this technique will help the users in reducing time required to search for data. Efficiency is exhibited and a lot more energy is saved through the demonstration of this technique, as suggested in the paper. It also mentions that using this approach will result in achieving the objective of a green cloud computing infrastructure. The algorithms for accumulation of new information at regional data center (RDC) and information retrieval are also discussed in this paper.

In paper [6] the author describes the security techniques for securing a Virtual Cloud Computing Infrastructure (VCCI), Virtual Machine Monitors (VMMs) such as Encryption and Key Management (EKM), Access Control Mechanisms (ACMs), Virtual Trusted Platform Module (vTPM), Virtual Firewall (VF), and Trusted Virtual Domains (TVDs).

Sahana et al. [7] devise the concept of server utilization to regulate the temperature of servers operating in cloud data center infrastructure. The concept of Mean Utilization Factor is introduced here, that allows detecting and regulating the amount of cool air that is to be channeled in and around the servers within a cloud data center to bring down the operating temperature. The problem of increasing efficiency of a data center in a cloud infrastructure is approached by designing a framework that allows effective monitoring of the cooling system in place.

In paper [8] the author discusses an overview of the cloud computing, cloud service models, and typical cloud architecture with respect to Virtualization, Virtualization impact on cloud security, and security threats related to Virtualization. It discusses the essential characteristics of cloud infrastructure which are—On-Demand self-service, Broad Network Access, Resource Pooling, Rapid Elasticity, and Measured Service. It also describes the service models which are—Infrastructure as a Service (IaaS), Platform as a Service (PaaS), and Software as a Service (SaaS). The deployment models—private cloud, community cloud, public cloud, and hybrid cloud are also discussed here. The concept of Virtualization and the reasons for Virtualization which are—sharing, isolation, aggregation, dynamics, and ease of management are also discussed here. The problems caused by Virtualization are also discussed here which are—isolation, hypervisor vulnerabilities, multitenancy, workload complexity, loss of control, network topology, no physical endpoints, single point of access.

Bose et al. [9] introduce a communication network that is aimed at augmenting network performance in situations where a company needs to operate a branch through means other than wired or mobile networks. It utilizes a mix of very small aperture terminal (VSAT) and cable technologies that have been designed to improve efficiency of a cloud data center. It also facilitates access to cloud infrastructure from remote terrestrial regions where wired connectivity is either unavailable or unreliable to the point where sustained data and information exchange is impossible to conduct. It also bolsters security by putting together two forms of network modes, i.e., cable and VSAT. It also discusses the algorithm for data sending from cloud data center. The mode selection procedure is also discussed in this paper.

Sahana et al. [10] introduce an entirely new technique of storing data in cloud networks by file type classification. Combining Redundant Array of Inexpensive Disk (RAID) types with this, it proposes a model that is robust and can withstand instances of hard disk failures. It also discusses a methodology that is embedded in the proposed framework that supports user authentication for data storage and retrieval. The secure user authenticated cloud RAID model shields user data from prying eyes and even from unauthorized attempts by the cloud service provider. It also aims at presenting secure storage management that should be able to handle assorted file types using several RAID implementations. It also uses the cryptographic technique for document files. The algorithms for data storing, encryption and decryption are also discussed in this paper.

Tayab et al. [11] discuss the common exploits of security uses in the virtualized environment and focuses on the security threats from the attacker's perspective. It discusses the major areas of virtualized model environment and also addresses the security concerns. It also provides a solution for secure Virtualization in IT infrastructure and to protect intercommunication of virtual machines. It discusses the Virtualization methods which are—Operating system-based Virtualization, Hypervisor-based Virtualization. It also discusses the proposed security architecture which is the Virtual Firewall architecture and its implementation in mitigating the security risks.

Divyambika et al. [12] discuss the migration of the virtual machines in the cloud environment and thus avoid threats and vulnerabilities. It discusses the issues in threats and vulnerabilities—Man-in-the-middle attack, the system architecture—security in live migration, process migration, memory migration. It also discusses the methods to implement all those techniques to mitigate the security risks which are—Elliptic Curve Cryptography, XEN Hypervisor—Regular and Live Migration, Multipass Algorithm.

Farhad Ahamad et al. [13] propose a conceptual model to develop security profiles for virtual machines by designing a fuzzy logic-based mechanism to amalgamate all the threat parameters namely inner vulnerability, intrusion behavior analysis and trusted membership for each of the virtual machine. Hence, the existing research gap is filled by proposing a design to construct security profiles for virtual machines.

In data centers, all the security policies are commonly being applied on the applications irrespective of their category or security threats that it faces. Phalguna Krishna et al. [14] developed to segregate the applications as per the type of threats (by adapting detection mechanisms) being faced. This approach is optimized where the latency associated with applying security policies can be efficiently reduced.

Garima Bhardwaj et al. [15] introduces the concept of Virtualization and present a debate over various pros and cons of using virtual machines. It discusses a variety of compelling benefits posed by Virtualization and also gives an overview of different threats that make some enterprises hesitant to adopt this technology. It illustrates the virtual machine architecture. It also discusses the potential benefits of Virtualization which are—Abstraction, Isolation, Snapshots, Deployment, Reduced downtime and administration cost, and External monitoring. It also discusses the challenges offered through Virtualization which are—Attacking Hypervisor, Guest-to-guest attack, DoS, Mobility.

James Poore et al. [16] explore how Virtual Machine Introspection (VMI) improves traditional digital forensics to overcome its downfalls when used to investigate virtual machines, especially during a live analysis of the machine. It gives the concept of Digital Forensics, Live and Dead Analysis, Issues with Digital Forensics in Virtualization, Implementing VMI with Traditional Digital Forensics.

Amani S. Ibrahim et al. [17] explore the security problem of the cloud platform virtual infrastructure identifying the existing security threats and the complexities of this virtual infrastructure. It also discusses the existing security approaches to secure the cloud virtual infrastructure and their drawbacks. It also proposes some key research challenges of implementing new Virtualization-aware security solutions that can provide the pre-emptive protection for complex and ever dynamic cloud virtual infrastructure.

Jinpeng Wei et al. [18] discuss the new risks that face administrators and users (both image publishers and image retrievers) of a cloud's image repository. It proposes an image management system that controls access to images, tracks the provenance of images, and provides users and administrators with efficient image filters and scanners that detect and repair security violations, to address those risks.

3 Methodology

This paper is a literature survey that analyzes the various security issues when we try to login into an application in a virtual machine environment. In a virtualized environment, there may be several server physical machine infrastructures. There may be several virtual machines that are running in a single server. Different virtual machines run using different operating systems.

Now, let's take a scenario where several virtual machines are running in a single server. Suppose we want to login into an application in one of the virtual machines. There may be several security issues that may come up while doing this operation. First, an attacker can **attack the host operating system and can gain access to the main physical machine infrastructure**, which includes the host hardware and the physical resources that serve all the virtual machines lying in the same network server. Second, an attacker can **gain control of the virtual machine monitor (VMM) or hypervisor**, and this can have a catastrophic effect on the entire network as VMM, which is the heart of the virtualized environment, is a single point of failure. Assaulters can also implement **Trojan or any malicious software** to gain control of the virtual machines, as a vulnerable virtual machine becomes an easy entry point for attackers, who could potentially access the hypervisor, other VMs, and the host operating system (OS). There can be problems of **user id and password sniffing**, wherein login details can be intercepted or stolen by capturing the network traffic using a sniffer. Using a sniffer application, an attacker can analyze the network and gain information to eventually cause the network to crash or to become corrupted or read the communications happening across the network. Security issues can also come up when the data is getting transmitted from the sender to the receiver which can lead to **network spoofing**, wherein packets of data are created or manipulated using a false Internet Protocol (IP) address for the purpose of hiding the identity of the sender or impersonating another computer system. We need to diagnose these security issues and suggest ways to counter these security issues.

The physical machine or the host machine should have a **highly secured username and password**. The most important way to ensure this is to use an alphanumeric password which will be really difficult for hackers to crack. There can be many other ways. For instance, we can perform software security updates on a timely basis. We need to enable firewall. We need to install powerful antivirus and anti-spyware software.

The key to securing the hypervisor or virtual machine monitor (VMM) is to use the **virtual machine introspection (VMI) technique**. The other way to secure VMM is to **monitor events within the virtual environment** so that any unusual behavior can be flagged early. Through the VMI technique, we can also monitor the state of application software, guest operating systems, and virtual machines running on a physical server.

Securing virtual machines can be done by a number of ways. For instance, we need to keep our operating system and applications current for both virtual and host machines. We can isolate the virtual machines by installing firewall. Antivirus programs should be installed on the virtual machines and they should be updated on a timely basis. There should be a strong encryption between the host machine and the virtual machines.

Packet filtering is one defense mechanism against network spoofing attacks. Here, the outside attacker can be prevented from spoofing the address of the machine where we want to login. This can be done by a gateway to a network that performs **ingress filtering**. Through this method, the packets are blocked from outside the network with a source address inside the network. The gateway also performs the **egress filtering** on outgoing packets. Here, the packets can be blocked from inside the network with a source address that is not inside. This way an attacker within the network can be prevented from performing filtering from launching IP spoofing attacks against external machines.

Encryption of login details and keeping away from applications that are using insecure protocols like basic HTTP authentication, File Transfer Protocol (FTP), and Telnet can help in countering the problems of user id and password sniffing. **Secure protocols** such as **HyperText Transfer Protocol Secure (HTTPS)**, **Secure File Transfer Protocol (SFTP)**, and **Secure Shell (SSH)** should be preferred. In case there is necessity for using any insecure protocol in any application, all the data transmission should be encrypted.

The **Algorithm** to counter the entire security issue can be detailed in the following steps:

i. Switch on the host machine or the physical machine
ii. At first, the user signs into the operating system (OS) of the host machine using highly secured username and alphanumeric password.
iii. The user ensures the security of the virtual machine monitor (VMM) by employing the virtual machine introspection (VMI) technique and monitoring the events in a virtual environment so that any unusual event can be detected early.
iv. The user ensures that the virtual machines are secured by making use of updated antivirus programs and updating the operating system and the applications for the host machine and the virtual machines. He also needs to maintain strong encryption between the host machine and the virtual machines.
v. The user switches on any one of the virtual machines on the network server and opens the application in which he needs to work. He needs to open the application in a secured manner by making use of secure protocols such as https instead of using insecure protocols such as http.
vi. He needs to ensure that his login credentials are highly secured by making use of alphanumeric passwords which may be difficult to crack and deploying other encryption algorithms.
vii. Once he logs in into the Home Page of the application using his login credentials, he is ready to transfer the data to the receiver. But here, the user has to ensure that he is a valid user and not a hacker who is intending to corrupt and steal the secret data that is to be sent to the receiver. This can be done in two ways, one of which is given in the next step.
viii. The egress filtering technique can be employed which ensures that the user, if he happens to be an attacker, is not able to launch spoofing attacks against the machine where the receiver is waiting to get the data from the sender.
ix. Next, while the data is getting transmitted from the sender to the receiver, there can be a chance that the data on its way to transmission can be modified or tampered by an attacker, which is called the man-in-the-middle attack. To counter the same, we need to properly encrypt the data using the concepts of private-key and public-key cryptography, which are popularly known as symmetric key and asymmetric key cryptography.
x. Next, the ingress filtering technique can be employed to ensure that the incoming data to the receiver is from that particular network where the source system (sender) is present. This way, the user will not be able to hide the identity.
xi. Last, but not the least, once the data is received by the destination machine (receiver), the encrypted data has to be decrypted by employing the concept of asymmetric key cryptographic technique or public-key cryptography.
xii. The receiver received the plain data sent by the sender.
xiii. Once, the above steps are done, the entire security system can be assured to be risk-free.

The **Flowchart** for the algorithm is shown in Fig. 1.

The flowchart (Fig. 1) shows a host machine, which needs to be switched on. Then it needs to be logged in using a secured user id and password. Then the user needs to

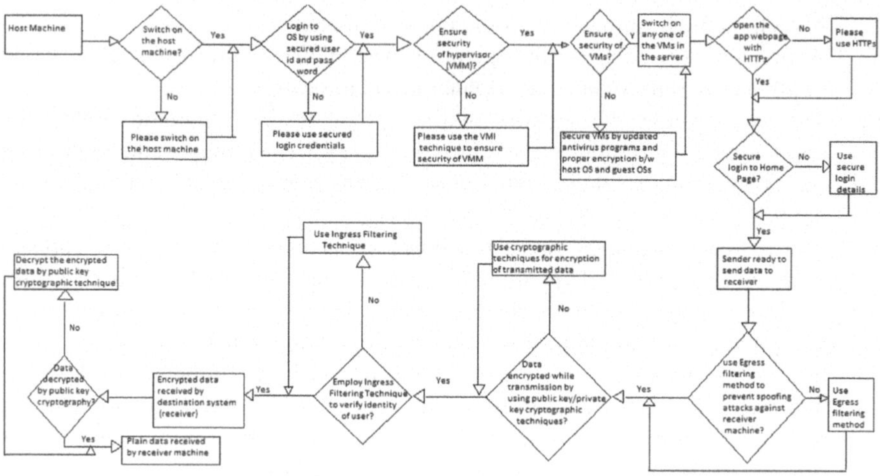

Fig. 1 Flowchart for our algorithm

ensure that the hypervisor is fully protected using the VMI technique. Then virtual machines must be made secure by updated antivirus programs and proper encryption between the operating system of the host physical system and the operating systems of the guest virtual machines. After ensuring that the virtual machines are fully secured, the user needs to login to any of the virtual machines and open the application webpage using secured sites such as https. Then the user needs to login to home page using secured login credentials which will give him access to data, which needs to be sent to the receiver. Once the user is ready to send data, the egress filtering technique has to be applied to prevent spoofing attacks against receiver machine. Then the data, before it is sent to the receiver, needs to be properly encrypted using cryptographic techniques. Then the ingress filtering technique needs to be applied to verify the identity of the user who is sending the data. Once the encrypted data reaches the destination system, the receiver decrypts the data using public-key cryptography. Finally, the plain text or data is received by the destination system.

The position of the virtual server in cloud environment is shown in Fig. 2. The foolproof security system which follows the algorithm and the flowchart (Fig. 1) is shown in Fig. 3.

4 Result Analysis

In our work, we have made use of the virtual machine introspection (VMI) technique to secure the virtual machine monitor (VMM). Unlike the traditional models which does not provide enough efficiency with increasing number of nodes in the network in security process, the VMI technique possesses several advantages, for instance, it does not affect the operation of hypervisor and real hardware resources. It is

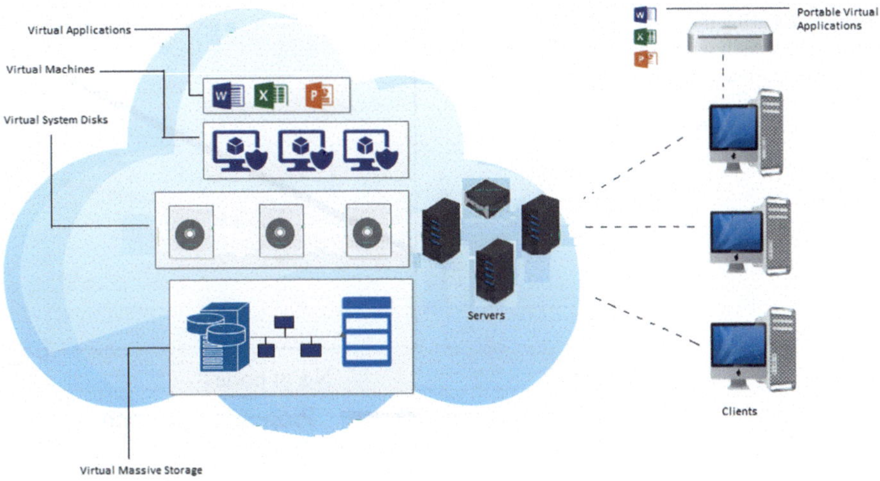

Fig. 2 The position of virtual server in cloud environment

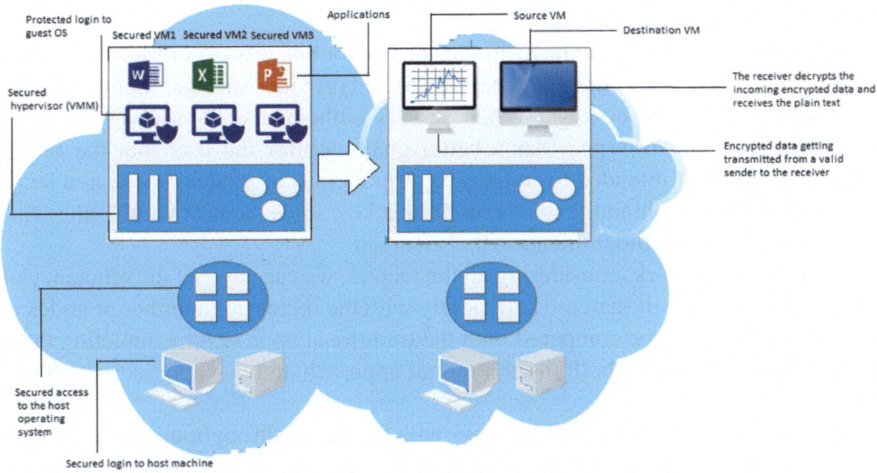

Fig. 3 The secured running of the application in a virtualized environment

independent of any exclusive feature of the hypervisor architecture, and hence, it is applicable to any type of hypervisors. The modifications to guest operating systems (OS) are not required. It does not slow down the performance of the hypervisor, the guest VMs, and the host hardware machine.

Also, unlike most of the other proposed works, we have made use of the encryption technique in our proposed work. It ensures complete data protection. Also, data encryption software allows us to ensure that all data, across any device, are completely encrypted with the same protections in place that we would find in data stored on

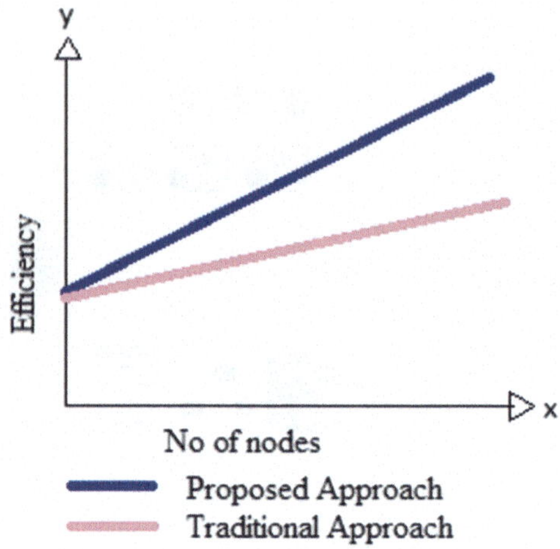

Fig. 4 Comparing the proposed approach with the traditional approach showing the efficiency will increase with increasing number of nodes

a desktop computer. An effective encryption technique is highly useful in terms of files sharing or uploading to cloud systems as our files will remain totally intact even during the transmission process.

The public-key cryptographic technique has major benefits over most of the other techniques as it is more secured than passwords and provides stronger identity checking through secret private-keys. When servers authenticate themselves to users, public-key authentication provides a better guarantee for the user that the server is the server the user intended to connect to. A malicious user cannot pose as a legitimate server without obtaining that server's host key, since the user would otherwise be warned that the host identification had changed.

In our proposed work, considering all the factors, we can say that the efficiency of our security system will increase sufficiently with the increasing number of nodes in the network. This can be compared with the traditional approach by depicting them in the form of graphs for both the traditional approach and our approach as shown in Fig. 4.

Also, the degree of node will increase with increased throughput. The graphical representation for both the traditional approach and our proposed approach is shown in Fig. 5.

5 Conclusion

Cloud security and Virtualization security go hand in hand, as Virtualization remains the heart of cloud computing. Hence, it is of utmost importance to secure the virtualized environment. In our paper, we discussed most of the possible security threats through which the virtualized environment can get affected. We mentioned the attacks by taking an example of an application, through which data will be sent from the

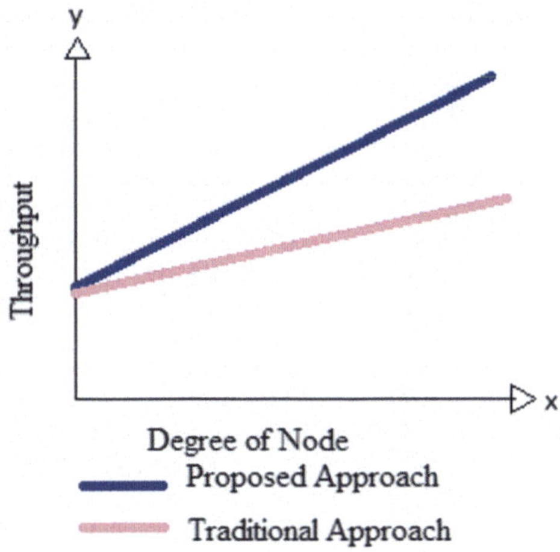

Fig. 5 Comparing the proposed approach with the traditional approach showing that the degree of node will increase with increased throughput

source VM to the destination VM in the virtual server. While suggesting the threats, we also considered the total architecture of the virtual machine wherein the hypervisor, being the heart of the virtual machine and the interface between the host machine and the virtual machines, remains the target point of the attackers. Accordingly, we devised some methods and techniques, which according to our opinion, can help mitigate those threats as far as possible. If we can guarantee a risk-free virtualized environment, it will lay the platform for a secure cloud environment. In the future, we are planning to implement a virtualized setup, wherein all the suggested measures for protecting the virtualized environment will be put to practical purpose.

References

1. Loganayagi, B., Sujatha, S.: Creating virtual platform for cloud computing. In: The Proceedings of IEEE International Conference on Computational Intelligence and Computing Research (ICCIC 2010), pp. 1–4, 28–29 Dec 2010
2. Sun, D., Chang, G., Guo, Q., Wang, C., Wang, X.: A dependability model to enhance security of cloud environment using system-level virtualization techniques. In: The Proceedings of First International Conference on Pervasive Computing, Signal Processing and Applications (PCSPA), pp. 305–310 (2010)
3. Princess, J.P., Paulraj, G.J.L., Jebadurai, I.J.: Methods to mitigate attacks during live migration of virtual machines—a survey. Int. J. Pure Appl. Math. **118**(20), 3663–3670 (2018)
4. Rekha, G.S.: A study on virtualization and virtual machines. Int. J. Eng. Sci. Invention (IJESI). **7**(5), 51–55 (2018)
5. Sahana, S., Bose, R., Sarddar, D.: An enhanced search optimization protocol based on gossip protocol for the cloud. Int. J. Appl. Eng. Res. **12**(19), 8436–8442 (2017). ISSN 0973–4562
6. Althobaiti, A.F.S.: Analyzing security threats to virtual machines monitor in cloud computing environment. J. Inf. Secur. **8**, 1–7 (2017)
7. Sahana, S., Bose, R., Sarddar, D.: Server utilization based smart temperature monitoring system for cloud data center. In: The Proceedings of International Conference on Industry Interactive

Innovations in Science, Engineering and Technology. Lecture Notes in Networks and Systems 11. Published on 24–25 Oct 2016. https://doi.org/10.1007/978-981-10-3953-9_30
8. Prashanthi, M.: Analysis of security issues in virtualization cloud computing. Int. J. Comput. Sci. Mobile Comput. IJCSMC **5**(8), 274–281 (2016)
9. Bose, R., Sahana, S., Sarddar, D.: An adaptive cloud communication network using VSAT with enhanced security implementation. In: The Proceedings of International Conference on Intelligent Computing and Communication 2016 (ICIC^2 2016) on 18th and 19th Feb 2016. https://doi.org/10.1007/978-981-10-2035-3_13
10. Sahana, S., Bose, R., Sarddar, D.: Harnessing RAID mechanism for enhancement of Data Storage and Security on Cloud. Braz. J. Sci. Technol. **3**, 12 (2016). https://doi.org/10.1186/s40552-016-0024-5
11. Tayab, A., Talib, W., Fuzail, M.: Security challenges for virtualization in cloud. University of Engineering and Technology (UET) Taxila, Pakistan, Tech. J. **20**(3), (2015)
12. Divyambika, R., Umamakeswari, A.: Protection of virtual machines during live migration in cloud environment. Indian J. Sci. Technol. **8**(S9), 333–339 (2015)
13. Ahamed, F., Shahrestani, S., Javadi, B.: Developing security profile for virtual machines to ensure secured consolidation: conceptual model. In: The Proceedings of the 13th Australasian Symposium on Parallel and Distributed Computing (AusPDC 2015), Sydney, Australia, 27–30 Jan. 2015
14. Phalguna Krishna, E.S., Sandhya, E., Ganesh Karthik, M.: Managing DDoS attacks on virtual machines by segregated policy management. Glob. J. Comput. Sci. Technol. E Netw. Web Secur. **14**(6), 2014
15. Bhardwaj, G., Sharma, P., Sharma, G.: Virtualization as an opportunity and a threat. Int. J. Adv. Res. Comput. Sci. Softw. Eng. **4**(2), 2014
16. Poore, J., Flores, J.C., Atkison, T.: Evolution of digital forensics in virtualization by using virtual machine introspection. ACMSE'13, 4–6 April 2013, Savannah, GA, USA. ACM (2013). 978-1-4503-1901-0/13/04
17. Ibrahim, A.S., Hamlyn-Harris, J., Grundy, J.: Emerging security challenges of cloud virtual infrastructure. In: The Proceedings of APSEC 2010 Cloud Workshop, Sydney, Australia, 30 Nov 2010
18. Wei, J., Zhang, X., Ammons, G., Bala, V., Ning, P.: Managing security of virtual machine images in a cloud environment. CCSW'09, Chicago, Illinois, USA. ACM, 13 Nov 2009. 978-1-60558-784-4/09/11

Fuzzy Logic Systems for Transmission Control Protocol

Hardik K. Molia and Amit D. Kothari

Abstract TCP-Transmission Control Protocol acts as a logical vehicle between two processes running on two different end devices. Standard TCP was proposed for wired networks where the primary cause of packet loss is congestion. When the same TCP is used with wireless networks, a significant amount of performance degradation can be found. The main issue is TCP's default consideration of any loss as a cause of congestion, ignoring the possibilities of channel issues or route failures. Researchers have proposed TCP variants for loss differentiation and transmission rate management according to the network status. TCP often gets incomplete and inaccurate network states for which intelligent interpretation is required. Fuzzy logic systems could be useful to analyze network state more accurately and intelligently. This paper discusses various fuzzy logic system-based TCP variants for loss differentiation, transmission rate control, and early detection of congestion conceptually. This paper compares these variants and discusses research directions.

Keywords TCP · Losses · Fuzzy logic system · Random early detection

1 Introduction

TCP-Transmission Control Protocol ensures connection-oriented and reliable process to process communication as a logical vehicle between two end devices [1]. TCP was initially introduced for wired networks where the primary cause of packets loss is network congestion. TCP's congestion control mechanism maintains the transmission rate according to the network congestion. When the same TCP is used

H. K. Molia (✉) · A. D. Kothari
Gujarat Technological University—Ahmedabad, Gujarat, India
e-mail: hardik.molia@gmail.com

A. D. Kothari
e-mail: amitdkothari@gmail.com

H. K. Molia
Government Engineering College—Rajkot, Gujarat, India

S. Kundu et al. (eds.), *Proceedings of the 2nd International Conference on Communication, Devices and Computing*, Lecture Notes in Electrical Engineering 602, https://doi.org/10.1007/978-981-15-0829-5_53

with wireless networks, a significant amount of end-to-end performance degradation can be found. One of the main reasons is the default consideration of any loss as a cause of congestion. Wireless networks may suffer from non-congestion losses due to other issues like channel issues, route failures, etc. The performance affects when the transmission rate is reduced unnecessarily by misinterpreting a non-congestion loss as a congestion loss [2–4]. It has been seen that the differentiation of loss solely depends upon TCP's awareness about network status. Being a transport layer protocol, TCP always gets incomplete and inaccurate network states for which intelligent interpretation is required. Artificial intelligence techniques could be used for TCP to differentiate losses and to act accurately.

A FLS-Fuzzy Logic System processes incomplete and inaccurate inputs and provides acceptable outputs. It also resembles human reasoning for decision-making. Fuzzy logic system-based TCP may deal with uncertain and inaccurate network state easily. Conventional logic accepts precise inputs and produces definite outputs such as Yes or No. Fuzzy logic resembles human reasoning to include all intermediate possible values between Yes and No. A FLS has four components which are as follows:

- Fuzzification module transforms crisp inputs into fuzzy values.
- Knowledge Base has rules in the form of If.Else.
- Inference Engine selects a specific If.Else rule based on fuzzy values.
- Defuzzification converts output of inference engine from fuzzy values to crisp values.

This paper explores various fuzzy logic system-based TCP variants. Section 2 discusses Fuzzy Logic System-based Loss Handling. Section 3 discusses Fuzzy Logic System-based Transmission Rate. Section 4 discusses Fuzzy Logic System-based RED Techniques. Section 5 compares all these variants and concludes with future directions.

2 Fuzzy Logic System-Based Loss Handling

2.1 FEDM

FEDM-Fuzzy-based Error Detection Mechanism [5] uses fuzzy logic controller to differentiate congestion losses and random channel losses (losses due to wireless errors). FEDM has IED-Improved Error Detector module to output possible reason of a loss. IED output can be any of the three flags: C-Congestion, U-Uncertain, and B-Bit Error (wireless errors). Error Recovery Mechanism accepts output of IED and acts accordingly. RTO-Retransmission Time Out with Congestion or Uncertain (C flag is set or U flag is set) retransmits and reduces transmission rate. RTO timeout with Bit Error (B flag is set) retransmits without reduction of transmission rate. Along with the FEDM's error recovery, fast retransmission phase may detect and handle errors in its own way (on receiving three duplicate ACKs-Acknowledgements).

FEDM mechanism is activated on receiving an ACK. It monitors end-to-end number of hops and variations in RTT-Round Trip Time values. An increase of RTT can be of due to two reasons: Congestion or increase in number of hops. NH-Number of Hops is counted using TTL-Time to Live field of IP header. RR-RTT Rate module identifies n number of increases of RTT where each increase is larger than of α. In [5], it has been mentioned that satisfactory results could be found when n is set to 2, α is set to 20% and sampling duration between two RTT values to 100 ms. NH information is used to set IED status to Congestion immediately, when there is no increase in number of hops but significant increase in RTT is found. The input variables are t-RTT mean and Δ_t-RTT Variance.

$$t = \frac{1}{n}\sum_{i=1}^{n} t_i \tag{1}$$

$$\Delta_t = \frac{1}{n}\sum_{i=1}^{n}(t_i - t)^2 \tag{2}$$

There are three fuzzy sets for t and Δ_t: small, medium, and large. The output of fuzzy engine is described with three singleton sets corresponding to Congestion, Bit Error, and Uncertain status. Fuzzy engine is designed with Gaussian membership functions. The set of fuzzy rules are shown in Table 1 [5].

The simulation has been performed in NS-2 simulator. The performance of FEDM is compared with TCP Reno. A satisfactory performance improvement has been found.

Table 1 Fuzzy rules—FEDM [5]

Input		Output
t	Δ_t	Status
Small	Small	Bit error
Small	Medium	Bit error
Small	Large	Bit error
Medium	Small	Congestion
Medium	Medium	Uncertain
Medium	Large	Bit error
Large	Small	Congestion
Large	Medium	Congestion
Large	Large	Congestion

Table 2 Fuzzy rules—TCP with FLS [6]

Input		Output
Δ RTT	ΔRTO_{RTT}	Δ RTO
Negative big	Very high	Negative big
Negative medium	High	Negative medium
Negative small	Medium high	Negative small
Zero	Medium	Zero
Positive small	Medium low	Positive small
Positive medium	Low	Positive medium
Positive big	Very low	Positive big

2.2 TCP with FLS

TCP with Fuzzy Logic Strategy [6] estimates value of RTO-Retransmission Time Out timer. Being a reliable protocol, TCP retransmits a packet if not acknowledged within a time interval (RTO timer). Along with retransmission, it reduces the transmission rate and increases RTO value as a part of its congestion control mechanism. Smaller RTO value may cause unnecessary activation of congestion control which may lead to underutilization of the available bandwidth. Larger RTO value may cause pauses in transmission which may lead to underutilization of the available bandwidth in case of absence of congestion. Conventional algorithms of RTO estimation are based on statistical analysis of RTT-Round Trip Time values. These algorithms are not capable to estimate RTO precisely in a highly dynamic network (Complex traffic, Dynamic topology, etc.) due to frequent nonlinear variations in RTT values. In such cases, RTO value must need to be set accurately. Fuzzy logic system is designed to represent nonlinear relationship between RTT value and RTO value.

Input variables are ΔRTT-Variance between Current RTT and Previous RTT, and ΔRTO_RTT-Difference between Current RTO and Current RTT. Output variable is ΔRTO-Difference between Next RTO and Current RTO. It is easy to calculate RTO_{Next} as

$$RTO_{\text{Next}} = RTO_{\text{Current}} + \Delta RTO. \tag{3}$$

Each of the input and output variable is defined with seven fuzzy sets. Fuzzy sets for Δ RTT and Δ RTO are Negative Big, Negative Medium, Negative Small, Zero, Positive Small, Positive Medium, and Positive Big. Fuzzy sets for ΔRTO_{RTT} are Very High, High, Medium High, Medium, Medium Low, Low and Very Low. Membership functions of Δ RTT and Δ RTO are symmetric while asymmetric for ΔRTO_{RTT}. The set of fuzzy rules are shown in Table 2 [6].

MATLAB and NS2 are used to simulate fuzzy logic controller and TCP Reno-based network. It has been seen that RTO estimated with FLS is smaller and sensitive to network dynamics as compared with Karn's algorithm estimation. Satisfactory performance improvement is also found.

3 Fuzzy Logic System-Based Transmission Rate

3.1 *Fuzzy TCP*

Fuzzy TCP [7] estimates value of $Cwnd$-Congestion Window (TCP Parameter to control transmission rate) using a fuzzy logic controller. It uses current value of $Cwnd$, $SSThresh$-Slow Start Threshold and RTT values to estimate the next value of $Cwnd$. TCP's Slow Start and Congestion Avoidance phases show exponential and linear growth of transmission rate, respectively. Slow Start increases $Cwnd$ slowly in beginning and too fast in ending. On a packet loss, after halving $Cwnd$, TCP switches to Congestion Avoidance. At the time of transition from Slow Start to Congestion Avoidance packet loss may occur. Fuzzy TCP tries to smooth out this transition of $Cwnd$ [7].

Fuzzy TCP has two input parameters: SC and dRTT. SC parameter is a combination of $Cwnd$ and $SSThresh$. It defines how much $Cwnd$ can still grow under Slow Start phase. dRTT defines the difference between subsequent RTTs. SC and $dRTT$ can be calculated as below.

$$SC = \frac{ssthresh - cwnd}{ssthresh} \tag{4}$$

$$dRTT = \frac{RTT_{\text{New}} - RTT_{\text{Prev}}}{0.5 * (RTT_{\text{New}} + RTT_{Prev})} \tag{5}$$

When $Cwnd$ is very small, SC becomes close to 1 indicating $Cwnd$ should be increased. When $Cwnd$ is very high, SC becomes close to 0 or negative indicating $Cwnd$ should not be increased. When dRTT is positive, RTT_{New} is higher than RTT_{Prev} indicating increase in traffic. When dRTT is negative, RTT_{New} is lower than RTT_{Prev} indicating decrease in traffic. The output of fuzzy logic controller is $\Delta Cwnd$—Amount of change in current value of $Cwnd$. The fuzzy sets for SC are: Low, Medium and High. The fuzzy sets for $dRTT$ are Decrease, Zero, and Increase. The fuzzy sets for $\Delta Cwnd$ are: Very small, Small, Moderate, More than moderate, Large, and Very large. Trapezoidal membership functions are used. The set of fuzzy rules is shown in Table 3 [7].

The simulation has been performed in NS-2 simulator. Fuzzy TCP is implemented with various TCP variants keeping other phases intact. The performance of Fuzzy TCP is compared with Tahoe, Reno, NewReno, and SACK. A satisfactory performance improvement in terms of packet loss rate and efficient use of bottleneck's bandwidth is found. Moreover this approach is a TCP sender-side modification which is easy to implement.

Table 3 Fuzzy rules—Fuzzy TCP [7]

Input		Output
SC	dRTT	$\Delta Cwnd$
Low	Zero	Very small
Low	Decreasing	Small
Medium	Increasing	Very small
Medium	Zero	Moderate
Medium	Decreasing	More than moderate
High	Increasing	Small
High	Zero	Large
High	Decreasing	Very large

Table 4 Fuzzy rules—Fuzzy-based transport protocol [8]

Input		Output
Variable	Value	*f*
D	Very low	Extreme large
D	Low	Large
D	Medium	Medium
D	High	Small
D	Very high	Extreme small
SP	Very low	Extreme large
SP	Low	Large
SP	Medium	Medium
SP	High	Small
SP	Very high	Extreme small

3.2 Fuzzy-Based Transport Protocol

Improved Transport Protocol [8] is proposed to improve the performance of transport layer in MANETs—Mobile Adhoc Networks. ITP is a cross-layer approach where suitable data rate is decided using the information collected by MAC layer. ITP starts with a probe packet which is sent to the network for exploration of a route. Every intermediate node sums queuing delay and transmission delay and forwards with a probe packet. This information is used to find perceived data rate. Perceived data rate and signal power values are used to derive output parameter f.

Fuzzy controller has two input parameters: D-perceived data rate and SP-Signal Power. The fuzzy sets for D and SP are very low, low, medium, high and very high. The fuzzy sets for the output parameter f are extreme small, small, medium, large, and extreme large. Triangular membership functions are used. The set of fuzzy rules are shown in Table 4 [8].

TCP receiver calculates D_{avg}-next average data rate using D_{cur}-current values of data rate and D_{avg} average data rate as below

$$D_{avg} = (1 - f) * D_{avg} + f * D_{cur} \tag{6}$$

TCP receiver sends two types of feedback information to the TCP sender. Rate feedback sends Davg and Reliability feedback uses SACK-Selective Acknowledgement to send information about lost packets. TCP sender receives average data rate with a feedback packet. TCP sender increases data rate if received average data rate is higher than current data rate and decreases data rate if received average data rate is lower than current data rate.

The simulation has been performed in NS-2 simulator. Performance of ITP is compared with TCP NewReno and ATCP. Satisfactory performance improvement in terms of throughput and time spent in slow start phase is found.

3.3 FL-TCP

FL-TCP [9] is a layered approach, sender-side modification to estimate value of $Cwnd$. It is an enhancement of TCP Vegas. TCP Vegas uses the expected throughput and actual throughput to set value of $Cwnd$ using thresholds α and β [10]. FL-TCP uses fuzzy logic controller instead of thresholds. The input parameters are E—Expected throughput and A—Actual throughput and output parameter are Cw—Change in Cwnd. The seven fuzzy sets for E and A are very low, low, little low, medium, little high, high and very high. The six fuzzy sets for Cw are decrease very large, decrease large, decrease medium, decrease small, no change and increase. triangular membership functions are used with limit of maximum throughput. The range for Cw is [−2.0 to 0.005]. Some of the important fuzzy rules are shown in Table 5 [9].

Table 5 Fuzzy rules—FL-TCP [9]

Input		Output
E	A	Cw
Very low	Very low	Increase
Low	Very Low	No change
Medium	Low	Decrease small
Little high	Little	Decrease medium
High	Low	Decrease large
Very high	Low	Decrease very large

The simulation has been performed in NS-2 simulator. The performance of FL-TCP is compared with TCP Vegas, TCP Reno, and TCP NewReno. Satisfactory performance improvement in terms of throughput and packet delay is found.

4 Fuzzy Logic System-Based RED Techniques

4.1 RED

Active routers adopt RED-Random Early Detection-based methods using AQM-Active Queue Management. A RED enabled router predicts possibility of future congestion based on the status of buffer queue (how much it is full). To avoid congestion, packets are probabilistically dropped to avoid severe congestion due to buffer overflow. RED method defines Min. and Max. thresholds for queue length. The calculation of dropping probability is a function of queue length. On receiving a packet, if current queue length exceeds Min threshold then packet is dropped with dropping probability. Once queue length exceeds Max threshold, dropping probability becomes 1 which drops every packet. In another RED variant, GRED-Gentle RED, three thresholds are introduced: Min., Max., Double Max. Packets are dropped with low probability when queue length is between Min. and Max. but with high probability when it is between Max. and Double Max. In some variations like RED-ECN [11], packets are probabilistically marked to inform sender about congestion. Some RED-based solutions use average queue length as a parameter. Average queue length-based analysis works on long-term basis where instantaneous (current) queue length is not considered for decision-making. The detailed algorithm and calculations of average queue length and dropping (or marking) probability are discussed in [11]. Though RED methods outperform as compared to drop-tail policy, they suffer from linearity and parameterization problems. These methods are not suitable for dynamic networks where the number of sources is changed rapidly. As the network traffic is nonlinear and uncertain most of the time, fuzzy logic system-based RED solutions are proposed.

4.2 AFRED

AFRED-Adaptive Fuzzy-based RED [12] shows how fuzzy logic controller can be used to avoid parameterization problem of RED. AFRED is adaptive in terms of dynamic adjustment of fuzzy rules. AFRED has two control modules: FCM-Fuzzy Controller Module and AAM-Adaptive Adjust Mechanism. FCM calculates PDP-Packet Drop (or mark) Probability with instantaneous (current) QL-Queue Length as an input. AAM uses PDR-Packet Drop Ratio to set fuzzy rules dynamically. On arrival of a packet, FCM calculates PDP based on QL as input and fuzzy rules set by AAM. AAM periodically measures PDR. A large PDR is indication of overload so the PDP should set to a larger value. For a small PDR, PDP should set to a smaller value.

The total queue length is divided into N levels. As a part of fuzzification, input value of QL is fuzzified into any of these N levels based on its value. For example, $L(i)$ refers to the ith level of queue, $L(1)$ and $L(N)$ refers to the minimal and maximal fuzzified values, respectively. The fuzzy value represented by $L(i)$ is $FQ(L(i))$ where $FQ(L(i))$ is smaller than $FQ(L(i+1))$. A queue length level defines the level of congestion. For ith level, if i is larger, high congestion can be considered and if i is smaller, low congestion can be considered. Corresponding to each of the queue levels, a probability is defined. For example, $D(i)$ refers to the drop probability for the ith level whose fuzzy value is $FP(D(i))$. Trapezoidal membership functions are defined for each of the queue length levels. Single line functions are defined for each of the drop probabilities. The simple fuzzy rule is, If QL is $L(i)$ then PDP is $D(i)$.

A too small value of PDP may invite congestion while a too large value of PDP may invite underutilization of bandwidth. PDP value should be modified based on the current utilization of the link. There are two possibilities as given below.

1. If link is underutilized or PDP is set to too large value, If QL is $L(i)$ then PDP should set to $D(j)$ where $i > j$.
2. If link is over utilized or PDP is set to too small value, If QL is $L(i)$ then PDP should set to $D(j)$ where $j > i$.

But it is difficult to find out value of j. AAM adjusts the value of $D(i)$ based on the current PDR. The detail arithmetic of calculation for PDP is given in [12].

The simulation has been performed in NS-2 simulator. The performance of AFRED is compared with RED and PI methods. A satisfactory performance improvement in terms of goodput and stable queue length is found.

4.3 FLRED

The main goal of FLRED [13] is to improve TCP's performance in terms of reducing packet delay, packet loss, and problems which arise due to TCP global synchronization and TCP starvation. FLRED's Fuzzy logic controller is designed with two input variables: AQL-Active Queue Length and Ds-Speculated Delay. AQL is calculated using low pass filter to avoid drastic effect on its value due to all of sudden traffic burst or due to a temporary congestion. The Delay referred here is the time for which a packet waits in buffer queue before transmitted. Ds is calculated based on arrival speculation and departure speculation calculated from arrival rate and departure rate, respectively. The detail arithmetic of calculation for AQL and Ds is given in [13].

As a part of fuzzification, crisp values of AQL and Ds are converted into fuzzy values. The fuzzy sets of AQL are null, trivial, normal, and large. The fuzzy set of Ds are zero, low, moderate, high. The fuzzy set of output Dp are zero, trivial, average, long. trapezoidal membership functions which are used. Sixteen rules are defined between every possible combination of fuzzy value of AQL and Ds for output Dp.

The simulation has been performed with discrete-time queue model within a network. The performance of FLRED is compared with RED and ERED in different scenarios according to the level of congestion. It has been seen that FLRED detects congestion earlier in case of heavy congestion. Even if packet drop rate is higher than of RED and ERED, overall performance in terms of packet loss and packet delay is improved. Performances of RED and ERED are better in case of light congestion. Performance of FLRED is better in case of no congestion.

4.4 Other Schemes

Fuzzy RED [14] was proposed for efficient congestion control in Integrated Service Networks. The base idea is to calculate dropping probability based on current queue length and rate of change of queue length. A Fuzzy Controller for AQM [15] was proposed with queue length and variance of queue length-based input values. The input variables are queue length error (difference between expected queue length and actual queue length) and error variation. The output is the change in dropping probability. A Fuzzy logic-based RED solution [16] has been proposed with average queue length and packet drop ratio as input variables and dropping probability as output variable. This solution tried to reduce the number of packet drops as compared with RED. GREDFL-Gentle RED with Fuzzy Logic [17] uses average queue length and packet delay as input variables and dropping probability as output variable.

5 Conclusion

Fuzzy logic system-based TCP variants are discussed for loss differentiation, adjustment of transmission rate, and early detection of congestion. Wireless networks may suffer from transmission issues which cause packet loss. Wireless networks like MANETs-Mobile Adhoc Network may suffer from route failures which cause packet loss too. The primary focus of using fuzzy logic system-based TCP variants is to improve end-to-end performance by accurate interpretation of network state information.Various Fuzzy logic system-based TCP variants are compared in Table 6.

Fuzzy logic systems are mainly introduced to control value of $Cwnd$-Congestion Window to set transmission rate. Further research work could be done toward setting following parameters using fuzzy logic systems.

- Maximum congestion window size
- Receiver's advertised window
- Slow start threshold

Table 6 Comparision—TCP with fuzzy logic systems

Sr	TCP variant	Features	Limitations
1	**TCP-FEDM** [5] Fuzzy error detection mechanism	Simple way to differentiate congestion loss and channel loss. Input variables are easy to calculate	Losses are differentiated on timeout only. Static thresholds for RTT variation estimation are not suitable for MANETs
2	**TCP-FLS** [6] TCP with Fuzzy Logic Strategy	RTO value is estimated using fuzzy logic. Nonlinear relationship between RTT and RTO is defined	Only RTO value estimation is performed using fuzzy logic system. More performance can be improved if other related parameters like number of duplicate ACKs to trigger fast retransmission, number of packets to retransmit could be defined using fuzzy logic
3	**Fuzzy TCP** [7]	Simple to implement. The purpose of reducing packet drop at transition of slow start to congestion avoidance is fulfilled	Other related parameters could be estimated based on the input parameters
4	**Fuzzy-based transport protocol** [8]	Estimation of data rate is accurate. Reliability is assured with two ways	Complex due to cross-layer design. Not suitable for MANETs with high mobility due to frequent change in intermediate nodes
5	**FL-TCP** [9] Fuzzy logic-based TCP	No threshold values are used for changing the value of Cwnd	Membership functions are defined with fixed ranges for throughput and change in Cwnd. These ranges may not be suitable under all scenarios
6	**Fuzzy RED techniques** [12–17]	These methods do not suffer from parameterization problem of traditional RED techniques. Some techniques are adaptive to change their fuzzy rules to be suitable for dynamic networks	As these methods need to be implemented at intermediate nodes only, Not suitable for highly mobile MANETs where intermediate nodes may change frequently

- Congestion window increase factor in slow start
- Congestion window increase factor in congestion avoidance
- Number of duplicate acknowledgements to trigger fast retransmission
- Number of retransmitted packets per fast retransmission
- Congestion window reduction factor on time out
- Congestion window reduction factor on fast retransmission
- Slow start threshold reduction factor on time out
- Slow start threshold reduction Factor on fast retransmission
- Retransmission time out timer increase factor for TCP's exponential backoff algorithm
- Acknowledgement delay timer
- Maximum number of packets to delay before sending a cumulative acknowledgement.

References

1. Forouzan, B.: TCP/IP Protocol Suite. McGraw-Hill (2009)
2. Mast, N., Owens, T.J.: A survey of performance enhancement of transmission control protocol (TCP) in wireless adhoc networks. EURASIP J. Wirel. Commun. Network. (Springer) (2011)
3. Al-Jubari, A. M., Othman, M., Ali, B. M., Hamid, N.A.W.A.: TCP performance in multi-hop wireless ad hoc networks challenges and solution. EURASIP J. Wirel. Commun. Netw. (Springer) (2011)
4. Tsaoussidis, V., Matta, I.: Open issues on TCP for mobile computing. J. Wirel. Commun. Mobile Comput. **2**, 3–20 (2002)
5. de Oliveira, R., Braun, T.: A delay-based approach using fuzzy logic to improve TCP error detection in ad hoc networks. IEEE Wirel. Commun. Netw. Conf. **3**, 1666–1671 (2004)
6. Zhang, Z., Li, Z., Suthaharan, S.: Fuzzy logic strategy of prognosticating TCP's timeout and retransmission. In: Computational Intelligence for Modelling and Prediction, vol. 2, pp. 309–320. Springer, Berlin (2005)
7. Nejad, H.V., Yaghmaee, M.H., Tabatabaee, H.: Fuzzy TCP: optimizing TCP congestion control. In: IEEE Asia-Pacific Conference on Communications, vol. 1, pp. 1–5 (2006)
8. Wang, N.-C., Huang, Y.-F., Liu, W.-L.: A fuzzy-based transport protocol for mobile ad hoc networks. In: IEEE International Conference on Sensor Networks, Ubiquitous, and Trustworthy Computing, vol. 1, pp. 320–325 (2008)
9. Ibrahim, K., Tabash, M., Beg, S., Ahmad, N.: FL-TCP: a fuzzy logic based scheme to improve TCP performance in MANETs. In: IEEE International Conference on Multimedia, Signal Processing and Communication Technologies, vol. 1, pp. 188–191 (2011)
10. Brakmo, L.S., OMalley, S.W., Peterson, L.L.: TCP vegas: new techniques for congestion detection and avoidance. In: SIGCOMM '94 the Conference on Communications Architectures, Protocols and Applications, pp. 24–35 (1994)
11. Ramakrishnan, K., Floyd, S., Black, D.: The addition of explicit congestion notification (ECN) to IP. RFC **3168** (2001)
12. Wang, C., Li, B., Sohraby, K., Peng, Y.: AFRED: an adaptive fuzzy-based control algorithm for active queue management. In: 28th Annual IEEE International Conference on Local Computer Networks, pp. 12–20 (2003)
13. Abualhaj, M.M., Abu-Shareha, A.A., Al-Tahrawi, M.M.: FLRED: an efficient fuzzy logic based network congestion control method. In: Neural Computing and Applications, pp. 1–11. Springer, Berlin (2016)

14. Loukas, R., Kohler, S., Andreas, P., Tran-Gia, P.: Fuzzy RED: congestion control for TCP/IP diff-serv. In: 10th IEEE Mediterranean Electrotechnical Conference. Information Technology and Electrotechnology for the Mediterranean Countries, vol. 1, pp. 19–22 (2000)
15. Fengyuan, Ren, Yong, Ren, Xiuming, Shan: Design of a fuzzy controller for active queue management. Comput. Commun. **25**(9), 874–883 (2002)
16. Abdel-jaber, H., Mahafzah, M., Thabtah, F., Woodward, M.: Fuzzy logic controller of random early detection based on average queue length and packet loss rate. In: International Symposium on Performance Evaluation of Computer and Telecommunication Systems, pp. 428–432 (2008)
17. Baklizi, M., Abdel-Jaber, H., Abu-Shareha, A.A., Abualhaj, M.M., Ramadass, S.: Fuzzy logic controller of gentle random early detection based on average queue length and delay rate. Int. J. Fuzzy Syst. **16**(1) (2014)

Improved Brain Tumor Detection by Region Competition Method

Tejal Kothari and Ankita Pramanik

Abstract Brain tumor segmentation is still a manual process requiring pathological aid. This is a tedious and time-consuming process. Improved segmentation can be achieved by computer-aided diagnosis. To improve the accuracy of the detection of tumor, a multi-region segmentation based on the region competition method is proposed in this work. The image with constant or varying intensity can be segmented without knowing the number of regions a priori, by this method. This work uses region competition method for brain tumor segmentation. The segmentation is followed by feature extraction for providing additional information about the tumor, namely, shape, texture, etc. The proposed segmentation method is carried out on three-dimensional images from BRATS-2012 (25 images) dataset on T2 modality of magnetic resonance imaging. On an average, ~88% of dice value is achieved against ~67%, ~72.9% and ~73.57% of dice values obtained in the existing work that uses simple linear iterative clustering version 0, fuzzy c-means, and statistical region merging algorithm, respectively.

Keywords Segmentation · Brain tumor · MRI images · Region competition method · Dice values

1 Introduction

The estimated number of people living with cancer is around 2.5 million, with over seven lakh new cases being registered every year. One of the deadly manifestations of the disease is the brain tumor. An uncontrolled, abnormal growth of cells in the brain is referred to as brain tumor, which can be further classified as benign or malignant. Again, the tumor in the glial cells (supporting cells of the brain) is called glioma.

T. Kothari · A. Pramanik (✉)
Department of Electronics and Telecommunication Engineering, IIEST, Shibpur, West Bengal, India
e-mail: ankita@telecom.iiests.ac.in

T. Kothari
e-mail: kothari.tejal17@gmail.com

S. Kundu et al. (eds.), *Proceedings of the 2nd International Conference on Communication, Devices and Computing*, Lecture Notes in Electrical Engineering 602, https://doi.org/10.1007/978-981-15-0829-5_54

This glioma can be low or high grade [1]. The present work focuses on high-grade gliomas (HGGs). Low-grade gliomas are often referred to as benign and HGGs as malignant. According to a study, 33% of all brain tumors and central nervous system (CNS) tumors, and 80% of all malignant brain tumors are accounted by gliomas [2]. In India, it ranges from 5 to 10 per one lakh population with a malignancy of 2% [3]. Early detection of gliomas is very crucial to contain the outbreak of the disease.

The dominantly used detection method for gliomas is the biopsy. However, this technique is physically invasive, time-consuming, tedious, and is prone to errors. Brain tumors being fatal requires a fast and accurate detection method. Medical image analysis through computer-aided diagnosis (CAD) methods, an active research area, is quite effective for quick and early detection of the disease [4]. CAD helps in achieving maximum possible accuracy. This method aids in pathologists' decision [5]. Computed tomography (CT), positron emission tomography (PET), magnetic resonance imaging (MRI) [6], etc., are few imaging modalities employed for tumor detection. Apart from possessing good resolution and contrast, MRI helps in early detection of abnormal changes in tissues and organs with detailed description of tumor. A large amount of research effort has been concentrated on MRI brain tumor detection of different modalities. Most of the existing research works require prior knowledge of images as in [7–9]. Also, medical images require the enhancement process [10, 11] to increase the contrast. This highlights the need for developing a quick and efficient segmentation method for brain tumor. The above-mentioned problems are dealt with in the proposed work.

The organization of the rest of the paper is as follows: The literature survey along with the scope and contributions of the proposed work are presented in Sect. 2. Section 3 briefly discusses the tools and techniques used in this work. The proposed methodology is presented in Sect. 4 and the results and discussions associated with the presented methodology are given in Sect. 5. Finally, the conclusions and future scope of works are discussed in Sect. 6.

2 Literature Review

Some of the existing works incorporate different color spaces, other than red, green, and blue (RGB) color model, namely, hue, saturation, value (HSV); hue, saturation, intensity (HIS), etc., to have effective segmentation algorithm. The work in [12] is based on transformation of the RGB image to the HSV color model. Then, the watershed algorithm and Canny edge detection are applied to the image to extract the tumor.

Preprocessing is often an initial step in tumor detection which improves the overall segmentation process. The work in [10] uses sharpening and median filters for the preprocessing of MRI brain images and the resulting image is then enhanced by histogram equalization. This is followed by segmentation through thresholding. After morphological operations to reveal underlying features, the tumor region is obtained by using the technique of image subtraction.

The work in [11], preprocessing is done using Sobel operator and then the enhancement of the images is done by histogram equalization. Thresholding followed by segmentation using morphological operators is then done. Features like area and the perimeter of the tumor are calculated for further analysis.

Few of the recent works combine principal component analysis (PCA) and k-means clustering method for segmentation [7]. Feature dimensionality reduction is achieved through PCA. The k-means clustering helps in the extraction of tumor but with a lesser time as compared to other clustering techniques.

Noises present in the image can hamper the performance of different algorithms. So, noise removal becomes a priority as used in many existing works. The work in [13] uses the median filter and bilateral filter to remove the noise from the MRI images and the best image out of both is selected based on performance evaluation. Then, segmentation is done using watershed followed by connected component labeling (CCL).

In [14], the statistical region merging algorithm is proposed. It is a faster algorithm that groups the region based on particular merging criteria, for example, size, and range of color. It can cope up with noise and can process atypical data.

The work in [15] uses marker controlled watershed algorithm proposed by Meyer. The segmentation of image is done using a software named Ilastik by combining different features like color, orientation, edge, etc. Dice [16] and Tanimoto [17] coefficients are used for parametric evaluation.

To improve the segmentation process, in recent times, super-pixels are being widely used [8]. The work in [8] outlines about simple linear iterative clustering version 0 (SLICO) segmentation for brain tumor CT images. Then region merging is used to group the clustered images. Super-pixel based approach gives better segmentation of the images [8]. However, SLICO over-segments the images. Also, it becomes harder to operate on different images. The segmentation procedure is seriously affected by the noise present. So, the segmentation algorithm should inherently include a noise cleaning step.

For better segmentation, some works use the fuzzy c-means (FCM) method [9]. In [9], T1-contrast MRI modality images are used which are first N3 corrected, then, anisotropic diffusion filtering and unsharp masking are applied for preprocessing of images. Segmentation and labeling of the images are done by FCM. The unsharp masking method may amplify noise. Also, the "c" clusters must be initialized in FCM which is a slow process.

Thus, from the above discussions, it can be inferred that choice of segmentation is a key factor in the accuracy of the image analysis.

2.1 Scope and Contributions of the Present Work

Various segmentation techniques, each with its own pros and cons are used in the existing methods. Thus, developing efficient segmentation algorithms is essential. Brain tumor detection is difficult through conventional methods like pixel/edge/region-based algorithm. It has been seen that super-pixel based methods yield efficient segmentation of the test image. However, the images are often over segmented. The problem of over-segmentation encountered [8] is tackled here by proposing region competition (RC) segmentation method. RC method has been previously used in extracting agricultural fields from aerial images, segmentation of white matter (WM), gray matter (GM), etc. The research gap is seen in existing work that uses RC method. This work uses RC for HGG tumor detection.

The contributions of the present work can be summarized as follows:

(i) Three-dimensional brain tumor images are segmented directly.
(ii) The presented algorithm works well on either of the constant or varying intensity images. Thus, contrast enhancement methods are not required.
(iii) To aid analysis before operating on the tumor, various features like area and eccentricity are extracted for each of the images.
(iv) The proposed method being a three-step process is simple and easy to use.
(v) The efficacy of the presented algorithm is demonstrated by rigorous simulations.

3 Existing Tools and Techniques

The present work uses RC method for the segmentation of the brain tumor. Different features are extracted from the binarized image for giving better information about the tumor. These existing tools and techniques are discussed below.

3.1 Region Competition Method

RC [18] is a multi-region image segmentation method which can work for both two- and three-dimensional images. Through this method, the image with constant or varying intensity can be segmented without a prior knowledge of the number of regions. In the segmented image, the different regions are marked with different colors. This algorithm is based on minimization of general Bayes or minimum description length (MDL) principle. The assumption made in this method is that the foreground region has connected set of pixels.

The contour of image is represented by particles placed on the pixel grid. The particles mark the outlines of different regions in the image. For a region in the image, the particles are present at the outline of that region. Here, face connectivity is assumed. Since three-dimensional images are used, a six-connected pixel neighborhood is considered for foreground region. Segmentation is done in iterations with every particle searching for the set of adjacent pixels it can possibly move to. Then, the difference in energy of all potential moves is calculated with the minimum energy move being executed. Iteration continues until no further improvement is seen in the movement of the particles. As the adjacent region particles compete with common boundary pixels for ownership, this algorithm is named as RC.

3.2 Area

The tumor area is calculated from 256 × 256 (JPEG) sized binarized image which consists of only two values, either black (0) or white (1). The mathematical representation of the binary image is summation of total number of white and black pixels. The unit of white tumor area is millimeter square (mm^2).

3.3 Dice Similarity Coefficient

A metric used for comparing the similarity of two images, i.e., the final segmented binary image with the ground truth image is dice similarity coefficient. The formula used in this paper is given by

$$\text{Dice} = 2 \times (|R \cap S|)/(|R| + |S|) \tag{1}$$

where R denotes the tumor region obtained through RC segmentation and S refers to the ground truth.

3.4 Major Axis Length

The length of the major axis of an ellipse, returned as a scalar value, is referred to as major axis length. It is the longest diameter value.

3.5 Minor Axis Length

The scalar length of the minor axis of an ellipse is called minor axis length. It is the shortest diameter value.

3.6 Eccentricity

The ratio of the distance between the foci and the major axis length of an ellipse is eccentricity. Its value lies between 0 and 1. It can help to determine the shape of the tumor.

3.7 Convex Area

The number of pixels in a convex image in scalar is convex area. Convex image refers to the image which has all the foreground pixels within the convex hull.

3.8 Equiv-Diameter

It is the circle's diameter and has the area the same as the region. It is a scalar value supported for only two-dimensional label matrices.

3.9 Extent

The ratio of pixels in the tumor region extracted to the pixels in the total bounding box gives the extent value. It is also scalar.

3.10 Solidity

It is computed as the ratio of extracted tumor area to the convex area. Area/Convex Area is the mathematical representation of the scalar value solidity.

3.11 Filled Area

The number of pixels in the filled image (the image filled with holes) is the filled area. It is a scalar value.

3.12 Euler Number

The number of objects in the region minus the number of holes in those objects gives the scalar Euler number.

3.13 Orientation

The orientation is the angle between the horizontal axes and the major axis, ranging from −90 to +90.

4 Proposed Methodology

To minimize human interaction and thus the errors, most of the current research is targeted at semiautomatic segmentation of brain tumors on two-dimensional images [19]. To improve the chances of tumor detection for diagnosis, the present work focuses on three-dimensional MRI images of the tumor. MRI has different modalities. The present work uses T2 modality of MRI. This modality is chosen as they provide higher contrast and clearer vision compared to others.

The proposed tumor detection algorithm is highly effective yet very lucid. The algorithm uses three main steps for three-dimensional MRI image detection. The block diagram of the proposed method is given in Fig. 1. The input to the proposed system is an MRI image of brain taken from the BRATS 2012 dataset [20]. The various steps of the algorithm are described below.

First, the three-dimensional images are segmented using RC method. Through RC method, different objects in an image are characterized as individual regions. Using this technique, the brain image can be segmented into multiple regions. Every particle searches for the set of adjacent pixels it can possibly move to. Then, the difference in energy of all potential moves is calculated with the minimum energy move being executed. Segmentation is done in iterations. Iteration continues until no further improvement is seen in the movement of the particles. The adjacent region particles compete with common boundary pixels for ownership.

In the next step, the segmented image is converted into the binary image using Otsu method as it is the simplest, fastest, and standard method for automatic threshold selection. In the last step of this algorithm, various features from the binary image like area of the extracted tumor, eccentricity, convex area, perimeter, etc., are calculated using "regionprops" function of MATLAB. These extracted features help in shape and texture like coarseness, regularity, etc., analysis of the tumor. This can help the doctor analyze the tumor efficiently and can aid her to decide on the size of the region that is to be operated.

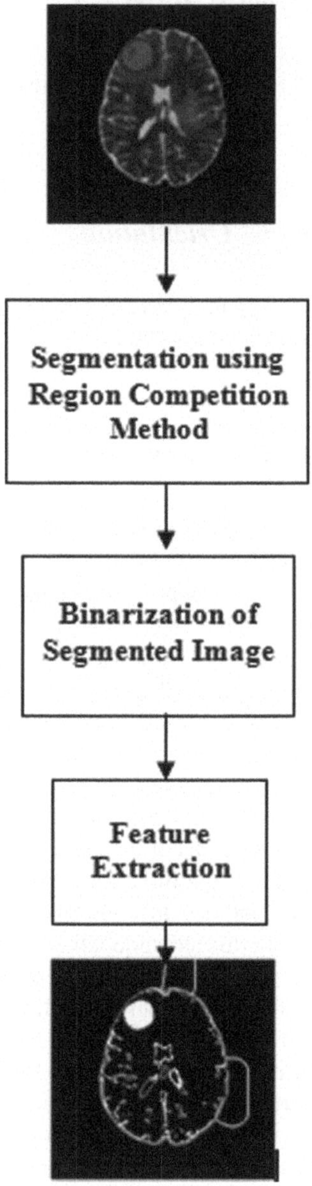

Fig. 1 Block diagram of the proposed method

RC method, being a multi-region segmentation method is faster and accurate than the conventional methods like pixel/edge/region-based algorithm as it doesn't require the knowledge of number of regions the image is to be segmented a priori. Also, it is capable of tackling over-segmentation problem. Since RC can work well with images having constant or varying intensities, so no preprocessing step is required in the present work.

The proposed method is effective and simple to use. The segmented image is compared to the ground truth image and the dice similarity coefficient is calculated for each image. The experimental results show that the features of tumor extracted and that of the ground truth do not differ much. The next section presents the performances of the proposed algorithm.

5 Results and Discussions

The experimental output of the proposed methodology is discussed here. The experiment is carried on 25 images from the BRATS dataset [20]. However, due to the constraint of space, output of various stages for only two test images is shown in Figs. 2 and 3, with the extracted features in Tables 1 and 2, respectively. The efficacy of the proposed algorithm is demonstrated by comparing with other existing works and also with the ground truth. Ground truth is the tumor area manually mapped by the radiologist.

The dice value for the image given in Fig. 2 is 94.89 and that given in Fig. 3 is 94.60%. The efficacy of the proposed algorithm is also demonstrated here. The comparison of the proposed method with the existing works is shown in Table 3. The average dice score of all the images is ~88%, with the highest value being 94.89% and the lowest being 79.38%. The average dice similarity coefficient obtained by the works in [8, 9, 14] is ~67%, ~72.9%, and ~73.57%, respectively against the dice similarity coefficient of ~88% with the proposed method. Thus, it can be seen from Table 3 that the proposed algorithm gives improved results.

Various segmentation techniques, each with its own pros and cons are used in the existing methods. Brain tumor detection is difficult through conventional methods like pixel/edge/region-based algorithm (Statistical region merging [14], Fuzzy c-means [9]). It has been seen that super-pixel based methods yield efficient segmentation of the test image. However, the images are often over segmented. The problem of over-segmentation encountered [8] is tackled here by proposing RC segmentation method which gives improved results. The work in [21] uses fuzzy region competition for segmenting various images including liver vessel, but no quantitative analysis is done. In [22], the region competition method is used for segmentation of textures.

Apart from the dice similarity coefficient, the other parameters used for quantitative analysis in this paper are the error value and error percentage calculated for each image. The error value represents the variation between the area of the segmented tumor region and the area of the ground truth tumor region. Error percentage is the

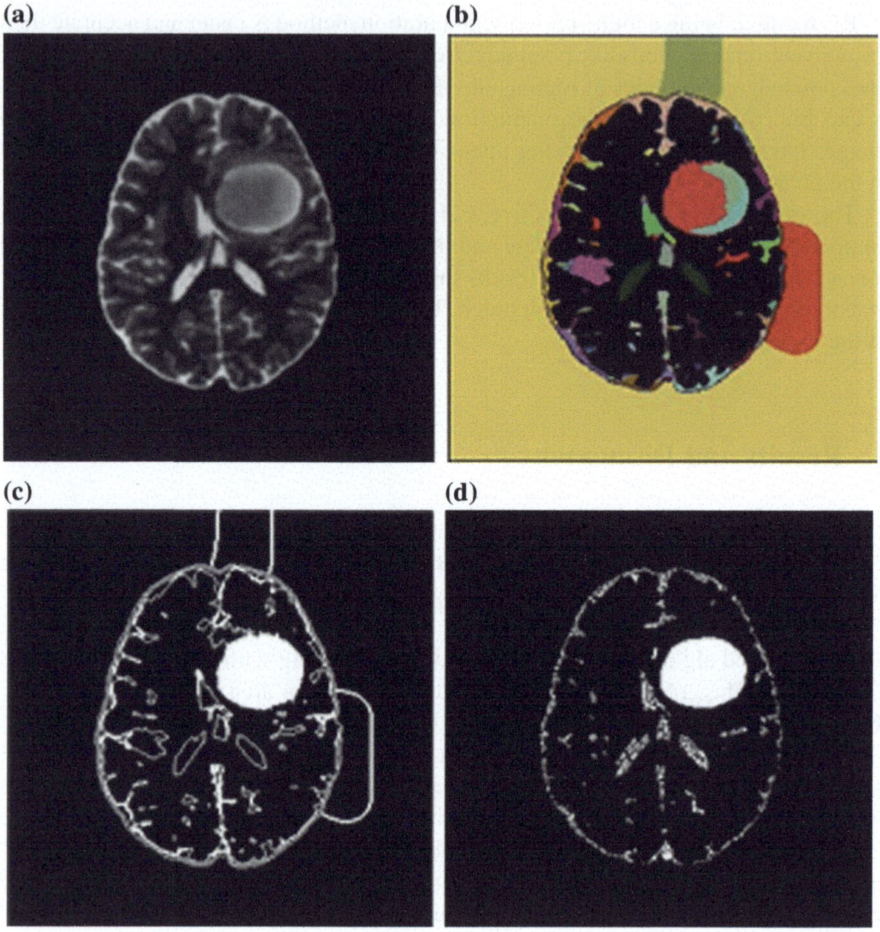

Fig. 2 For patient 1 **a** original image, **b** RC segmented image, **c** binary image, **d** ground truth image

ratio of error value to ground truth. The error percentage for each image comes out to be the lowest with an average of ~0.107724. The error value and error percentage for the above two images are shown in Table 4. Thus, it can be seen that the tumor area obtained with the proposed algorithm almost matches that given in the ground truth.

6 Conclusions and Future Works

In this work, a semiautomatic method of segmentation is proposed. The conversion of image from two dimensions to three dimensions can cause information losses

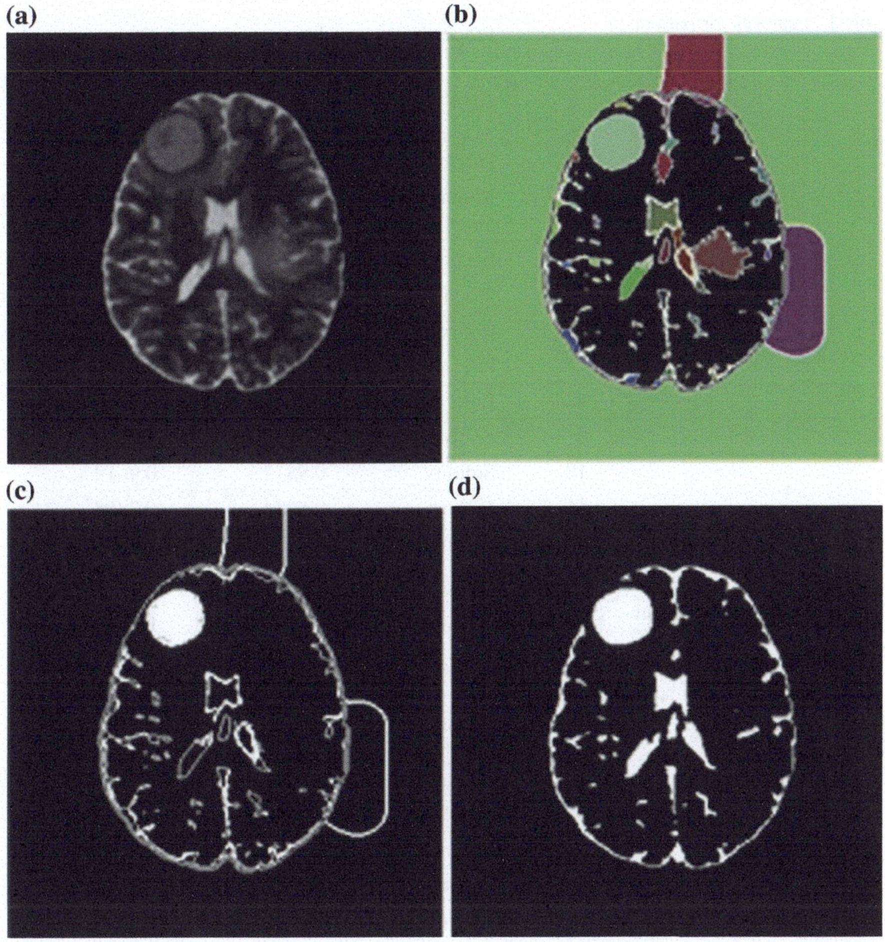

Fig. 3 For patient 2 **a** original image, **b** RC segmented image, **c** binary image, **d** ground truth image

in images. To deal with this problem, the three-dimensional images are directly segmented and binarized for further analysis. The presented algorithm is simple to use and is much efficient compared to the existing works. A higher dice similarity index of ~88% is achieved by the proposed algorithm as against ~73.57% obtained by statistical region merging method. The area of extracted tumor is close to that of the ground truth image. Also, very low values of error percentage between the extracted tumor and the ground truth with an average of ~0.107724 shows that the extracted tumor is closer to the ground truth.

The future scope of work includes achieving even better dice similarity coefficient. Preprocessing of images by different techniques of three-dimensional images and comparison of the same can be done as an extension, to have more accuracy. Testing the proposed methodology on the other modalities of brain MRI, namely T1, T1c,

Table 1 Features for patient 1

Sr. No.	Properties	For ground truth	For segmented image
1.	Area	1723	1841
2.	Major axis length	51.8438	52.1899
3.	Minor axis length	42.4030	45.0978
4.	Eccentricity	0.5754	0.5033
5.	Orientation	7.3922	4.4610
6.	Convex area	1756	1914
7.	Filled area	1723	1841
8.	Euler number	1	1
9.	Equiv-diameter	46.8379	48.4152
10.	Solidity	0.9812	0.9619
11.	Extent	0.8044	0.7868
12.	Perimeter	146.0510	161.0370

Table 2 Features for patient 2

Sr. No.	Properties	For ground truth	For segmented image
1.	Area	938	890
2.	Major axis length	36.2254	35.0136
3.	Minor axis length	33.0890	32.5971
4.	Eccentricity	0.4070	0.3651
5.	Orientation	24.1842	33.5278
6.	Convex area	955	926
7.	Filled area	938	896
8.	Euler number	1	−5
9.	Equiv-diameter	34.5586	33.6628
10.	Solidity	0.9822	0.9611
11.	Extent	0.8121	0.7479
12.	Perimeter	106.3380	107.1400

Table 3 Comparison with existing papers

Sl. No.	Paper	Dice similarity coefficient (%)
1.	[8]	67
2.	[9]	72.9
3.	[14]	73.57
4.	Proposed method	88

Table 4 Error value and error percentage

Sr. No.	Ground truth area	Segmented tumor area	Error value	Error percentage
1.	1723	1841	118	0.0684
2.	938	890	48	0.0511

FLAIR, etc., or on the combination of these modalities can be undertaken. The segmentation of brain tumor into edema, active tumor, and necrotic part can also be carried out as future work.

References

1. Goodenberger, M.K.L., Jenkins, R.B.: Genetics of adult glioma. Cancer Genet. **205**(12), 613–621 (2012)
2. Johns Hopkins Medicine Health Library: http://www.hopkinsmedicine.org/healthlibrary. Accessed 19 Apr 2018
3. Dasgupta, A., Gupta, T., Jalali, R.: Indian data on central nervous tumors: a summary of published work. South Asian J. Cancer **5**(3), 147–153 (2016)
4. Jamaludin, A., Kadir, T., Zisserman, A.: Automatic modic changes classification in spinal MRI. In: International Workshop on Computational Methods and Clinical Applications for Spine Imaging, pp. 14–26. Springer, Munich, Germany (2016)
5. El-Dahshan, E.-S.A., Mohsen, H.M., Revett, K., Salem, A.-B.M.: Computer-aided diagnosis of human brain tumor through MRI: a survey and a new algorithm. Expert Syst. Appl. **41**(11), 5526–5545 (2014)
6. Kharrat, A., Benamrane, N., Messaud Mohamed B., Abid, M.: Detection of brain tumor in medical images. In: 3rd International Conference on Signals, Circuits and Systems (SCS), pp. 1–6. IEEE, Medenine, Tunisia (2009)
7. Katkar, J., Baraskar, T., Mankar, V.R.: A novel approach for medical image segmentation using PCA and K-means clustering. In: International Conference on Applied and Theoretical Computing and Communication Technology (ICATCCT), pp. 430–435. IEEE, Davangere, India (2015)
8. Wang, X., Ma, P., Zhao, J.: Brain tumor CT image segmentation based on SLIC0 superpixels. In: International Congress on Image and Signal Processing, BioMedical Engineering and Informatics (CISP-BMEI), pp. 427–431. IEEE, Datong, China (2016)
9. Sehgal, A., Goel, S., Mangipudi, P., Mehra, A., Tyagi, D.: Automatic brain tumor segmentation and extraction in MR images. In: Conference on Advances in Signal Processing (CASP), pp. 104–107. IEEE, Pune, India (2016)
10. Natarajan, P., Krishnan, N., Kenkre, N.S., Nancy, S., Singh, B.P.: Tumor detection using threshold operation in MRI brain images. In: IEEE International Conference on Computational Intelligence & Computing Research (ICCIC), pp. 1–4. IEEE, Coimbatore, India (2012)
11. Murthy, T.S.D., Sadashivappa, G.: Brain tumor segmentation using thresholding, morphological operations and extraction of features of tumor. In: International Conference on Advances in Electronics, Computers and Communications (ICAECC), pp. 1–6. IEEE, Bangalore, India (2014)
12. Maiti, I., Chakraborty, M.: A new method for brain tumor segmentation based on watershed and edge detection algorithms in HSV colour model. In: National Conference on Computing and Communication Systems (NCCCS), pp. 1–6. IEEE, Durgapur, India (2012)
13. Dhage, P., Phegade, M.R., Shah, S.K.: Watershed segmentation brain tumor detection. In: International Conference on Pervasive Computing (ICPC), pp. 1–5. IEEE, Pune, India (2015)

14. Nock, R., Nielsen, F.: Statistical region merging. IEEE Trans. Pattern Anal. Mach. Intell. **26**(11), 1452–1458 (2004)
15. Benson, C.C., Lajish, V.L., Rajamani, K.: Brain tumor extraction from MRI brain images using marker based watershed algorithm. In: International Conference on Advances in Computing, Communications and Informatics (ICACCI), pp. 318–323. IEEE, Kochi, India (2015)
16. Sørensen, T.: A method of establishing groups of equal amplitude in plant sociology based on similarity of species and its application to analyses of the vegetation on Danish commons. Biol. Skr. **5**, 1–34 (1948)
17. Rogers, D.J., Tanimoto, T.T.: A computer program for classifying plants. Science **132**(3434), 1115–1118 (1960)
18. Cardinale, J., Paul, G., Sbalzarini, I.F.: Discrete region competition for unknown numbers of connected regions. IEEE Trans. Image Process. **21**(8), 3531–3545 (2012)
19. Gordillo, N., Montseny, E., Sobrevilla, P.: State of the art survey on MRI brain tumor segmentation. Magn. Reson. Imaging **31**(8), 1426–1438 (2013)
20. Menze, B.H.: The multimodal brain tumor image segmentation benchmark (BRATS). IEEE Trans. Med. Imaging **34**(10), 1993–2024 (2015)
21. Mory, B., Roberto A.: Fuzzy region competition: a convex two-phase segmentation framework. In: International Conference on Scale Space and Variational Methods in Computer Vision, pp. 214–226. Springer, Berlin, Heidelberg (2007)
22. Xu, Q., Yang, J., Ding, S.: Texture segmentation using LBP embedded region competition. In: Electronic Letters on Computer Vision and Image Analysis, pp. 41–47. China (2005)

Binary and Continuous Feature Engineering Analysis on Twitter Data Stream for Classification of Spam Messages

Cinu C. Kiliroor and C. Valliyammai

Abstract Twitter, the social network evolving faster and regular usage by millions of people. Spam is defined as unwanted content that appears on online social networks (OSN) sites that are driven by several goals such as to spread advertisements, generate sales, viruses, phishing, or simply just to compromise system reputation. Twitter has gained a lot of popularity throughout the world, it continues to be a very interesting target for spammers and malicious users for spreading spam messages. Several machine learning algorithms exist to detect the spam messages in the OSN but still it is a challenging issue. Not only the significant classifiers used are essential but also the selection of relevant features is also very crucial. Binary and Continuous Feature Engineering (BCFE) analysis is proposed in this paper for better feature selection. The features are evaluated using significance strategies and the best features are selected for effective spam classification.

Keywords Online social network · Feature engineering · Feature analysis · Kernel density estimate

1 Introduction

Twitter is a social media platform that plays a dual role in social networking and microblogging. Users of Twitter web application communicate with one another by posting short texts called tweets, and they acquire the latest information from other users' tweets by following them if they find their posts interesting. Studies show that Twitter has become an important mechanism for users to keep up with friends as well as the latest popular topics reaching more than 1 billion users, 255 million of which are active monthly, sending out 500 million tweets per day and averaging 208 followers per user. Spam refers to a variety of prohibited behaviors that violates the Twitter rules. Spam can be generally defined as unsolicited, redundant actions that

C. C. Kiliroor (✉) · C. Valliyammai
Department of Computer Technology, Madras Institute of Technology, Anna University, Chennai, India
e-mail: cinuck191@gmail.com

S. Kundu et al. (eds.), *Proceedings of the 2nd International Conference on Communication, Devices and Computing*, Lecture Notes in Electrical Engineering 602, https://doi.org/10.1007/978-981-15-0829-5_55

will create a negative impact on users. Spam on Twitter is driven by several goals such as to spread advertisements, generate sales, viruses, phishing, or simply just to compromise system reputation.

Features can be of two major types based on the dataset. Basic features which can collect straight from the datasets with no additional manipulation. Derived features are usually obtained from feature engineering, where features are extracted from existing data attributes. Feature engineering process infers the useful information from the datasets with respect to the data analysis for the designing of accurate models and it makes a huge impact for effective predictions when it compares with the basic classification model. Selection of useful and relevant features from the existing features happens in feature selection method. Feature selection is used for simplification of models, shorter training times, avoiding the curse of dimensionality, enhanced generalization by reducing overfitting. Redundant or irrelevant features are removed without much loss of information happens in feature selection process. Redundant and irrelevant are two distinct notions, since one relevant feature may be redundant in the presence of another relevant feature with which it is strongly correlated. Researchers focused on various classification algorithms for spam message classification in social networks. To make the spam message classification more effective, the consideration of feature selection is also important in addition to the selection and designing of a classifier. The proposed model performs a feature engineering process to find relevant features which help in the modeling of an effective spam classifier. Section 2 of the paper includes the literature survey. Section 3 includes the proposed work. Section 4 includes the data analysis and results and the last section includes the conclusion part.

2 Prior Arts

The user behavior based spam classification considers various features such as user-generated content [5], social graph connections, and user profile activities. User can be authenticated using captcha. Mutual confidence is established using Social Graph-User conversations [6]. Collected data is used to analyze and generate classifier model. Usage of social interaction graph yields better results. Finding difference between inactive users and malicious users is difficult [9]. Feature selection based spam filtering is done by reviewing posts, comments, and studying their features. Some of the optimized algorithms used here are simulated annealing, particle swarm optimization [11], ant colony optimization. To separate spam comments, hybrid model is used. Hashtags, reply, likes, and comments are used for evaluating the features [7]. Genetic algorithm is applied for evaluation criteria based feature selection [1]. Hig- dimensional vector space is handled for the classification of short text in [3]. Conversion of original format of tweets to a particular format for the classification purpose is difficult. Incorporating the Twitter rules for the sentiment analysis of tweets is carried out in [4]. A hybrid approach to detect spammers in social network is proposed in [8].

Boosted logistic regression is used for the detection of spam in Twitter [2]. Restricting the features leads to improper classification. Usage of the existing features will not lead to good performance. The methods such as decision tree, particle swarm optimization, and genetic algorithm are combined and proposed a hybrid model for classification of spam messages which is presented in [10]. Finding the smaller subset of the attributes for spam classification in social networks with the help of a rough set theory based approach is proposed in [13]. The ranking method for feature selection can be used only for discrete or categorical attributes and it needs to be improved for generalizing the algorithm for applying across different types of datasets. Analysis of cross-validation and hold-out validation for different datasets happened in [12]. Classification of unwanted messages in a semi-supervised way is performed in [14]. Deep learning based unwanted message detection in Twitter datasets is proposed in [15]. All existing research works for spam message classification in social networks are concentrated on either combining different classifiers or combining different features related to various contexts. There is also a need for methods to identify and extract useful features which will lead to the design of effective classifier model and also will reduce the time complexity. In this regard, the proposed model presents an outline for such a solution in the following section.

3 The Proposed Method

Binary and continuous feature extraction happened in the proposed model for effective spam message classification in social networks. BCFE approach creates a new set of binary and continuous features from the existing features. A deep data analysis is conducted for every estimation feature in the dataset. The continuous features are analyzed using Kernel Density (KDE) and the binary features are analyzed using count bar-plot to estimate the distribution of features in the dataset. The workflow diagram for the proposed model is shown in Fig. 1.

The unwanted and null data are removed in the preprocessing stage and the balanced dataset with respective fields is given as input for feature engineering task. In feature engineering phase, BCFE is carried out where a new set of binary and continuous features are extracted using existing fields. The features are analyzed and tested for significance and the best feature vectors are used for building supervised machine learning model. BCFE feature extraction is shown in Algorithm 1.

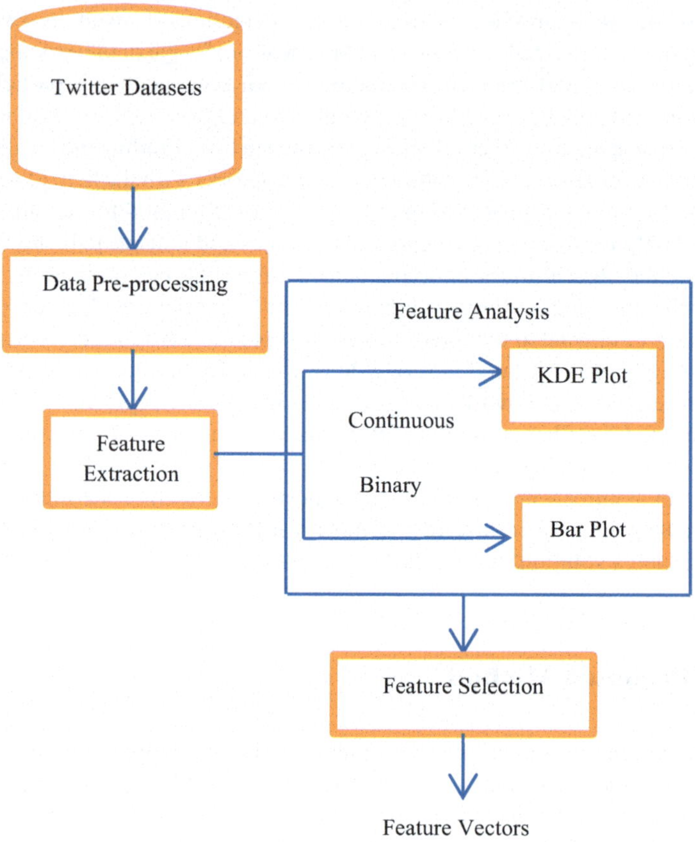

Fig. 1 Workflow diagram

Algorithm 1: BCFE

Input: **D**, collected Dataset
Output: **Lbc**, *list of binary and continuous features*
Begin
1. Initialize **F** ← [ ], **Lb** ← [ ], **Lc** ← [ ] , **Lbc** ← [ ]
2. **F** ← extract_f(D)
3. **for each** Fi in F do
4. Lb ← extract_bf (Fi)
5. Lc ← extract_cf(Fi)
6. **end for**
7. **Lbc** ← con(Lb , Lc)
8. **return** Lbc

End

4 Exploratory Data Analysis

Data mining process is done for extracting the essential data from a huge stream of Twitter. Class balancing technique is applied to make a balanced dataset. The main contribution of the work focuses on the feature engineering task in the process of identifying spam messages. The features vectors generated are tested for validation using Sample T-test shown in Eq. (1) and the features with considerable significance are selected for designing model. In order to make features more meaningful, new binary and continuous values are created from features. The BCFE features considered are shown in Table 1.

$$\mathbf{t} = \frac{\bar{x} - \mu}{\sqrt{\frac{s^2}{n}}} \tag{1}$$

where $\bar{x}$ is the sample mean, s^2 is the sample variance, n is the sample size, μ is the specified population mean, and t is the test statistic value. The features that are selected based on their significance are used to build the model for spam classification.

Exploratory Data Analysis is conducted for every feature in the dataset. For continuous attributes, the distribution of the attribute is analyzed using KDE, to visually identify any significant differences in the distribution across each type of tweet is shown in Figs. 2, 3, 4, and 5. Similarly, for categorical attributes, bar plots of the counts are plotted to visually identify any significant differences in the distribution across each type of tweet are shown in Figs. 6, 7, 8, 9, 10, 11, 12, 13, 14, and 15.

Table 1 BCFE features

Binary features	**Description**
has_hashtags	Whether a tweet has hashtag or not
has_media	Whether a tweet has media content or not
has_urls	Whether a tweet has url or not
has_favorite_count	Whether a tweet has favorite count or not
has_place	Whether a tweet has location or not
has_retweet_count	Whether a tweet has retweet count or not
has_user_name	Whether user name is present or not
has_user-description	Whether user description is present or not
is _retweet	Whether a tweet is a retweet
is_user_verified	Whether the user is verified or not
Continuous features	**Description**
Digits	Number of digits present in a tweet
Cap	Number of capital words in a tweet
length	Length of a tweet
f_ratio	Followers: followings

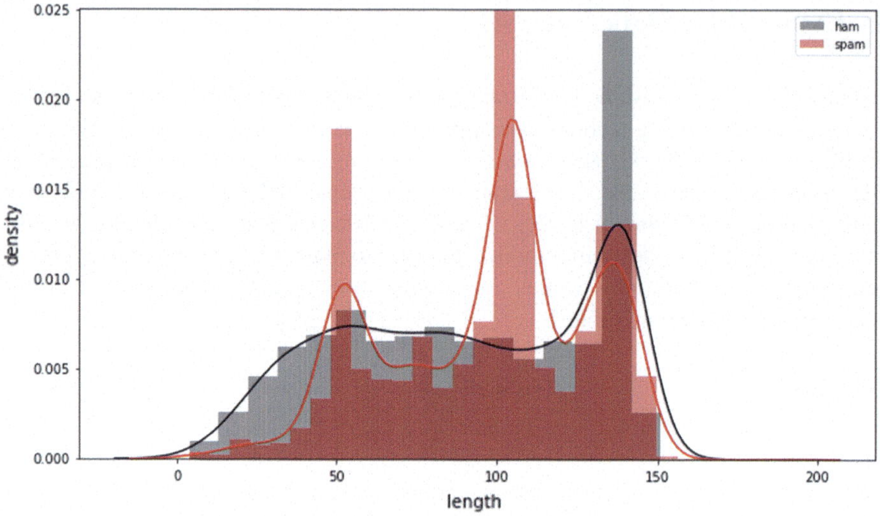

Fig. 2 KDE plot for length of each tweet

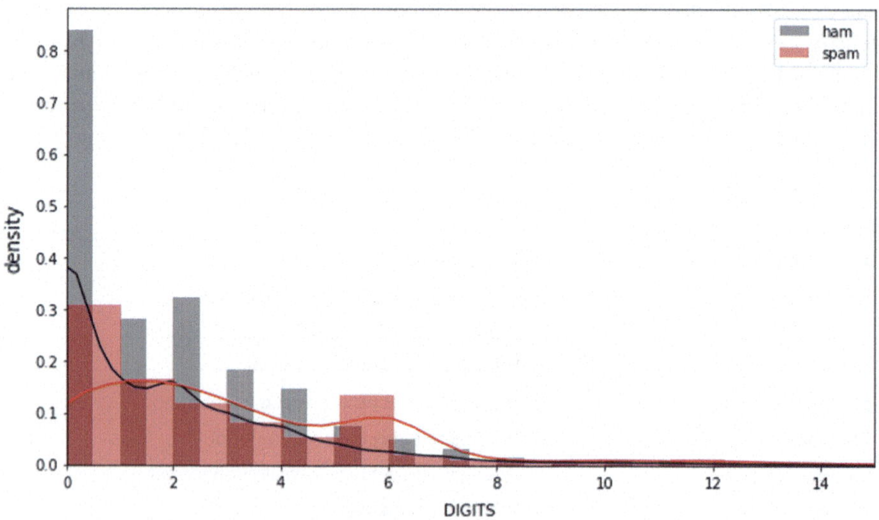

Fig. 3 KDE plot for number of digits in each tweet

The analysis of Twitter datasets based on various features in a visualized way clearly shows which extracted features are more important for designing a spam message classification model. The knowledge inferred from the graphical visualization based on the Twitter datasets is very significant. The probability distribution of spam or ham messages in the datasets vividly depends on the features. So, selection of features is very important in addition to the selection of machine learning based

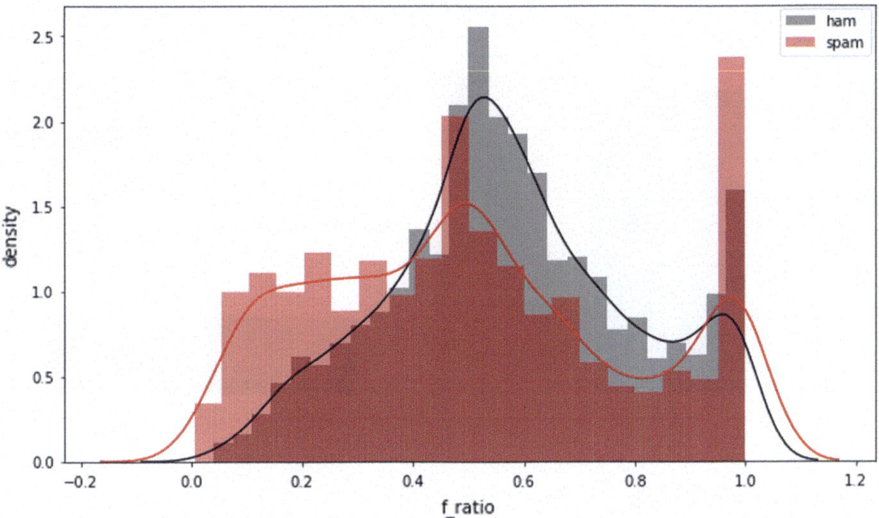

Fig. 4 KDE plot for f_ratio of each user

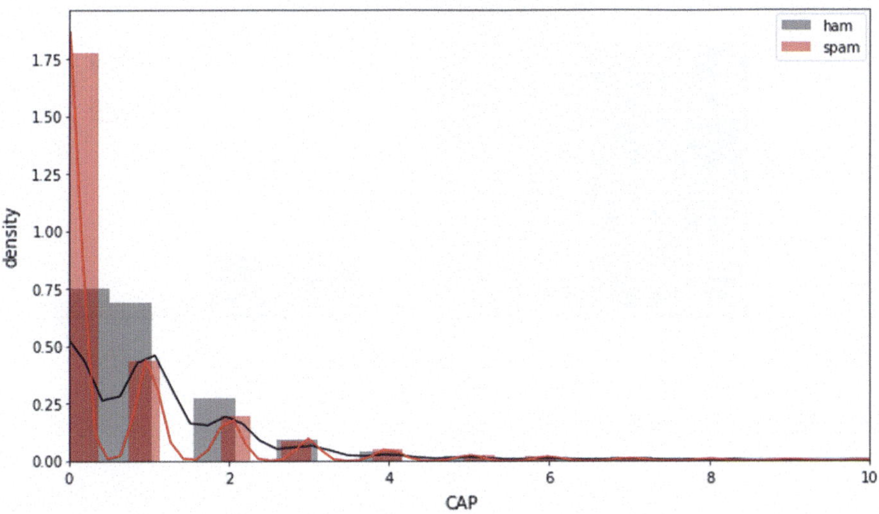

Fig. 5 KDE plot for number of capitalized words in each tweet

classification models. Sample T-test is conducted for the validation of the input features to find the most relevant set of features—a subset of features, which would significantly aid the prediction of the tweet label.

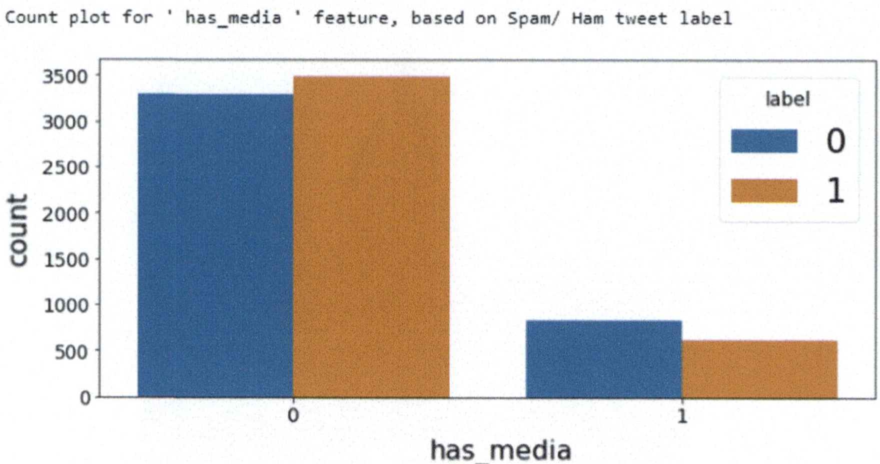

Fig. 6 barplot for has_media

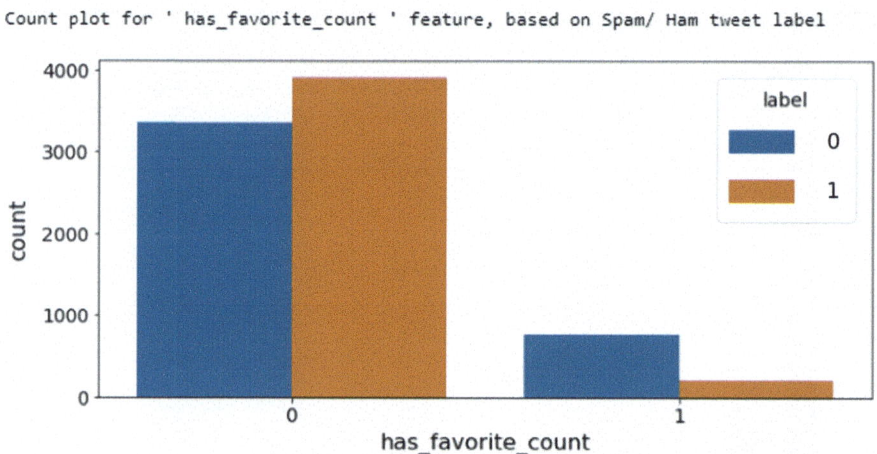

Fig. 7 barplot for has_favorite_count

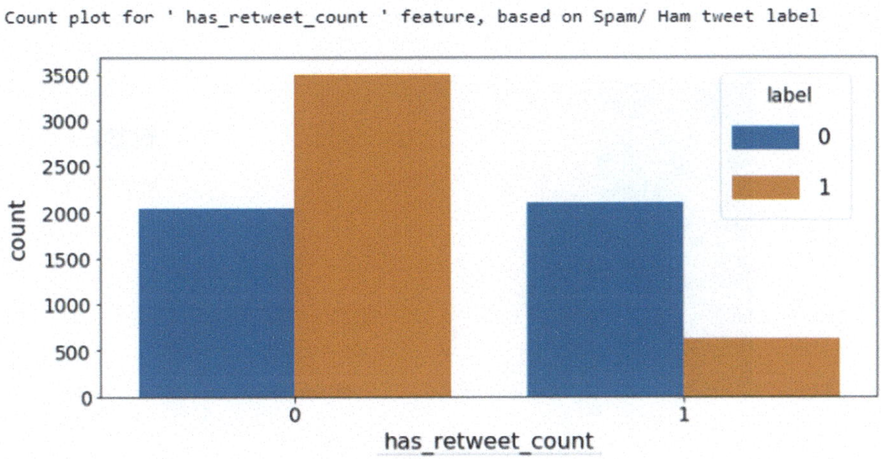

Fig. 8 barplot for has_retweet_count

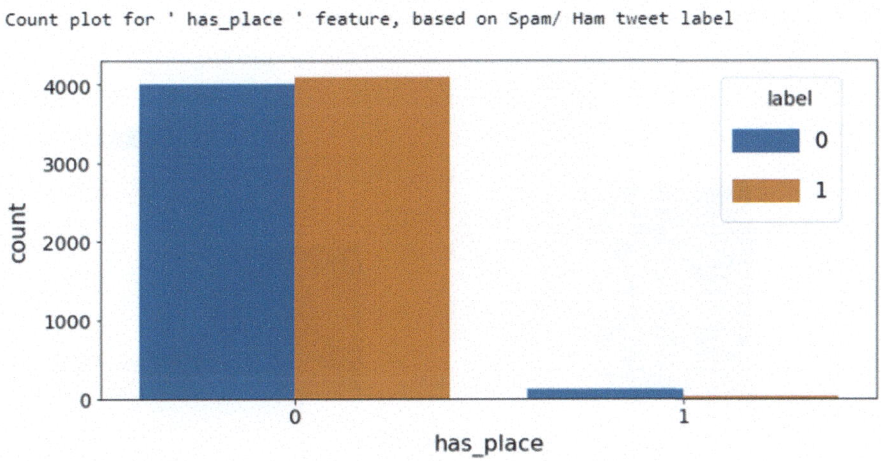

Fig. 9 barplot for has_place

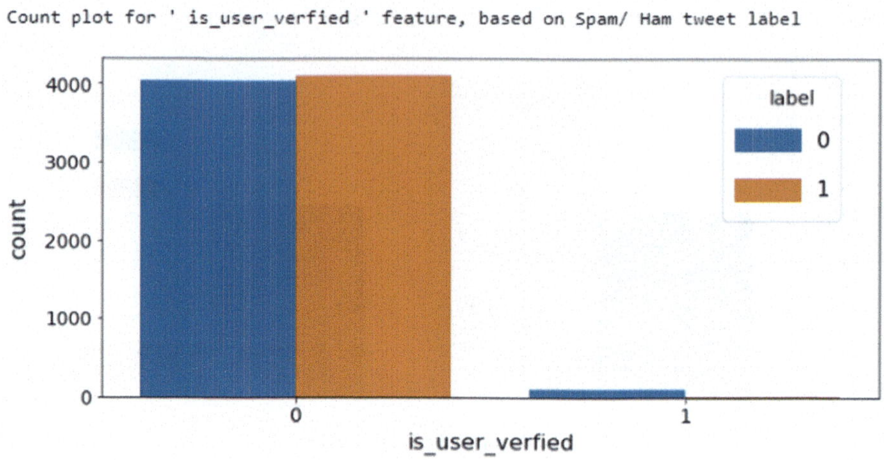

Fig. 10 barplot for is_user_verified

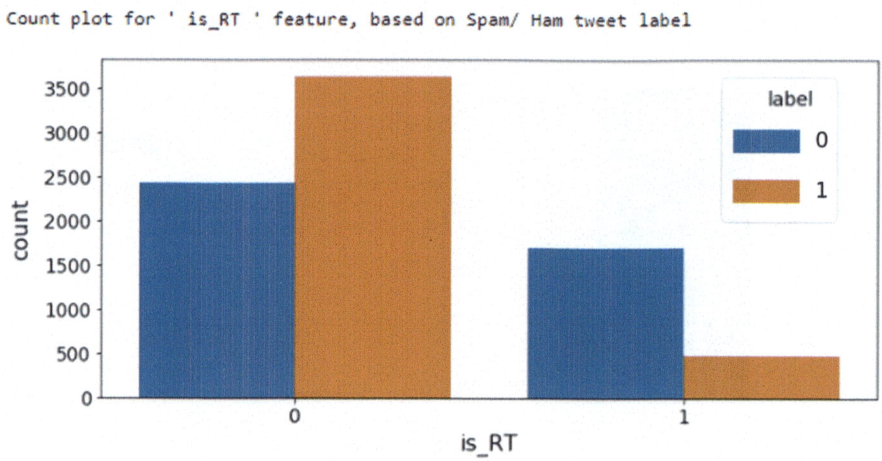

Fig. 11 barplot for is_RT

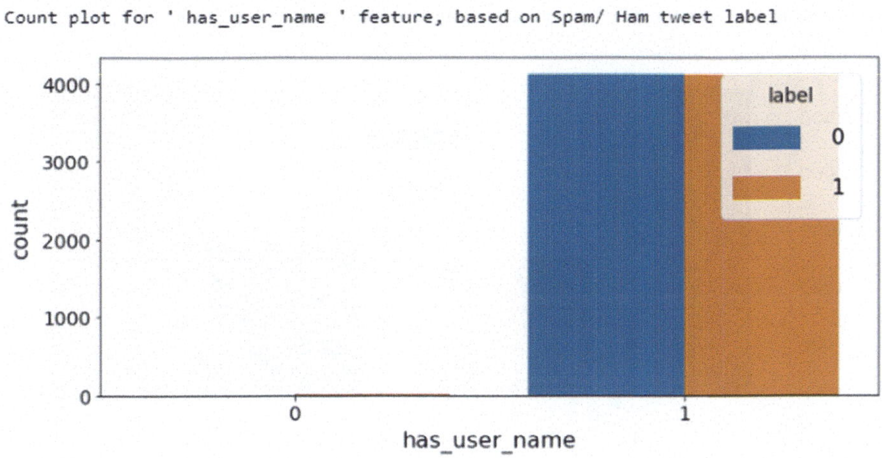

Fig. 12 barplot for has_user_name

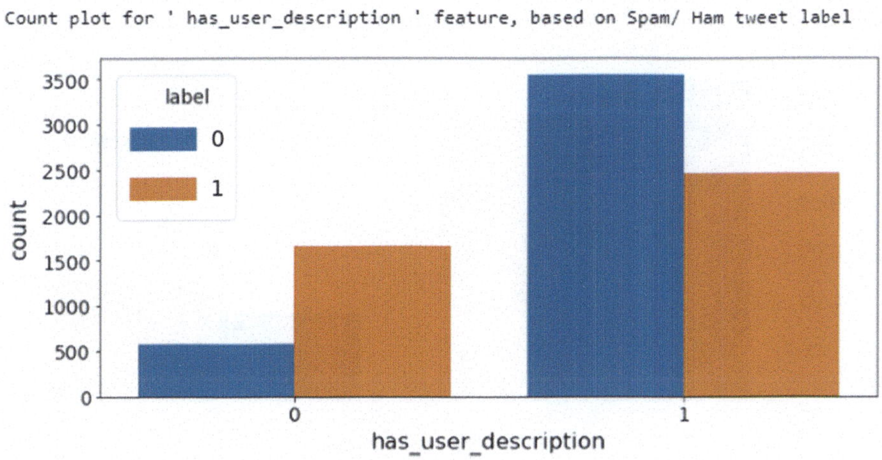

Fig. 13 barplot for has_user_description

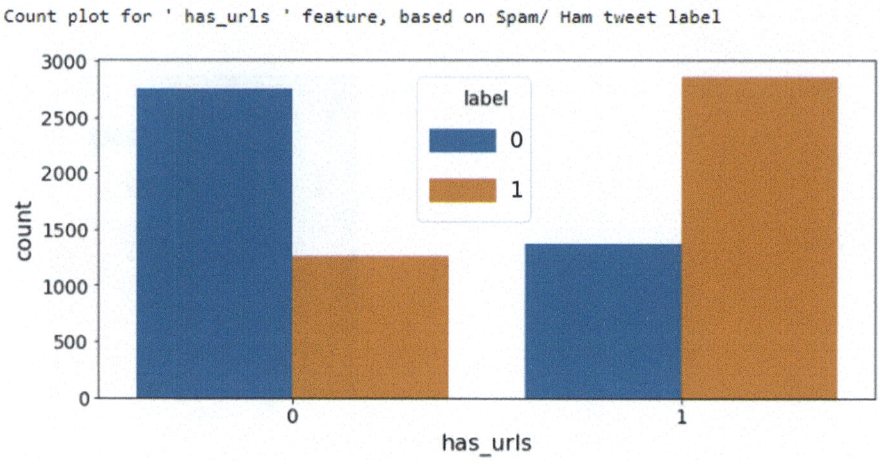

Fig. 14 barplot for has_urls

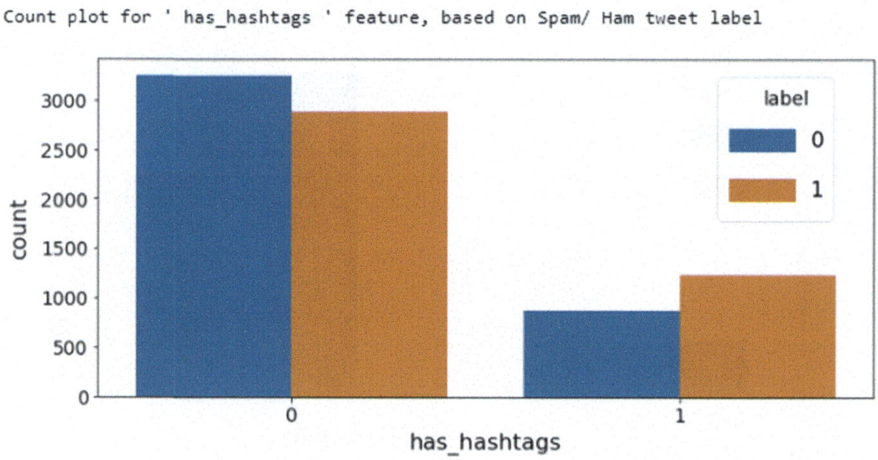

Fig. 15 barplot for has_hashtags

5 Conclusion and Future Work

The proposed work aimed to analyze the inherent nature of tweets based on the features which can model a robust classifier that could help in classifying a tweet is spam or ham. Class balancing technique is applied to make a balanced dataset. Data mining process is done for extracting the essential data from a huge stream of Twitter. The main contribution of the work focuses on the BCFE method in the process of identifying spam messages. KDE and BAR-PLOT are used to understand the features and their important correlation with a tweet being spam or ham. Feature Selection like STT is used for significance test and list of features of high importance is derived. The derived features can used to build classification model. Future work envisions to build a spam classifier model in a hybrid way in addition to the BCFE method.

Acknowledgements This work is financially supported under grants provided by the Visvesvaraya Ph.D. Scheme for Electronics and IT.

References

1. Bing, X., Mengjie, Z., Will, N., Browne, Xin, Y.: A survey on evolutionary computation approaches to features selection. IEEE Trans. Evol. Comput. **20**, 606–626 (2016)
2. Guanjun, L., Nan, S., Surya, N., Jun, Z., Yang, X., Houcine, H.: Statistical twitter spam detection demystified: performance, stability and scalability. IEEE Access Big Data Anal. Internet Things Cyber-Phys. Syst. **5**, 11142–11154 (2017)
3. Issa, A., Gan, K.H.: Term weighting scheme for short-text classification: Twitter corpuses. Int. J. Neural Comput. Appl. (2018) (Springer)
4. Jaishree, R., Allen, S., Irudayaraj, A., Tzacheva, A.A: Action rules for sentiment analysis on Twitter data using spark. In: IEEE International Conference on Data Mining Workshops, pp. 51–60 (2017)
5. Lalitha, L.A., Hulipalled, V.R., Venugopal, K.R.: Spamming the mainstream: a survey on trending Twitter spam detection techniques. In: IEEE International Conference on Smart Technology for Smart Nation, pp. 444–448 (2017)
6. Malik, M., Muhammad, A., Iqbal, M.A., Muhammad, A.I.: A hybrid approach for spam detection for Twitter. In: IEEE 14th International Bhurban Conference on Applied Sciences & Technology, 466–471 (2017)
7. Mohammad, K.S., Firoozeh, K.: A feature selection approach to detect spam in the Facebook social network. Arab. J. Sci. Eng. **43**, 949–958 (2018)
8. Mohd, F., Muhammad, A.: A hybrid approach for detecting automated spammers in Twitter. IEEE Trans. Inf. Forensics Secur. **13**, 2707–2719 (2018)
9. Muhammad, A.M., Shamim, H., Majed, A., Mizanur, S.M.R., Atif, A.: Leveraging analysis of user behavior to identify malicious activities in large-scale social networks. IEEE Trans. Ind. Inform. **14**, 799–813 (2018)
10. Senthil, N., Usha, G.D.: Detecting streaming of Twitter spam using hybrid method. Wirel. Pers. Commun. **103**, pp. 1353–1374 (2018) (Springer)
11. Samina, K., Tehmina, K., Shamila, N.: A survey of feature selection and feature extraction techniques in machine learning. Sci. Inf. Conf., pp. 372–378 (2014)

12. Sanjay, Y., Sanyam, S.: Analysis of k-fold cross-validation over hold-out validation on colossal datasets for quality classification. In: IEEE 6th International Conference on Advanced Computing, pp. 78–83 (2016)
13. Soumi, D., Sujata, G., Ratnadeep, D., Asit, K.D., Saptarshi, G.: Attribute selection for improving spam classification in online social networks: a rough set theory based approach. In: Social Network Analysis and Mining, vol. 8, pp. 1–16. Springer, Berlin (2018)
14. Surendra, S., Aixin, S.: Semi-supervised spam detection in Twitter stream. IEEE Trans. Comput. Soc. Syst. **5**, 169–175 (2018)
15. Tingmin, W., Shigang, L., Jun, Z., Yang, X.: Twitter spam detection based on deep learning. In: Australasian Computer Science Week Multiconference, pp. 1–8 (2017)

Visual Computing for Blast and Brown Spot Disease Detection in Rice Leaves

Nilanjana Guchait, Ishita Bhakta, Santanu Phadikar and Koushik Majumder

Abstract Automatic diagnosis of rice diseases is a growing field of research in E-Agriculture. It increases the productivity with a sustainable environment. Disease detection at primary stage is a tedious job for the farmer. They identify the diseases from experience and apply pesticides at fixed amount which further affects the environment. Sometimes personal biasness of a farmer like fatigue and eyesight may cause wrong identification of the disease. To overcome these challenges, an automated computer-aided system is developed to detect rice diseases. As a case study, the brown spot and blast rice diseases are considered. Images are analyzed to extract texture, shape, color, and intensity distribution features to evaluate the disease pattern. At first, leaf and stem of the rice plants are identified to detect the infected region. Then, healthy and infected leaves are identified and at last, diseases are detected from the infected leaves with 81.81% accuracy in IBK Classifier.

Keywords Rice leaf disease classification · Rice blast disease · Brown spot disease · IBK classifier

1 Introduction

Rice is the major crop of India. It provides 21% of global human per capita energy and 15% of global human per capita protein [1]. The scientific name of rice is Oryza sativa [2]. It needs massive amount of rain during its growing season. The field needs to be dried up before harvesting. Hence, India's hot and humid climate is favorable for rice cultivation. As a result, rice becomes dominating staple food in West Bengal as well as India. In West Bengal, 95% farmers cultivate rice in small and marginal scale which contributes more than 15% of total rice production in India [3]. Nowadays, a huge loss is found in the rice yield due to various types of rice plant diseases. In India per year, 37.02% loss is found due to weed, 27.9% loss due to insect, and 15.6%

N. Guchait · I. Bhakta (✉) · S. Phadikar · K. Majumder
Department of Computer Science and Engineering, Maulana Abul Kalam Azad University of Technology, BF 142, Sector 1, Salt Lake City, Kolkata 700064, West Bengal, India
e-mail: ishita.official@gmail.com

S. Kundu et al. (eds.), *Proceedings of the 2nd International Conference on Communication, Devices and Computing*, Lecture Notes in Electrical Engineering 602, https://doi.org/10.1007/978-981-15-0829-5_56

loss due to diseases [4]. Due to these huge losses over 50 years, a significant number of populations suffer from hunger and malnutrition [5]. It raises a question about the food security of these rising population in India [6]. Hence, rice plant diseases are required to be prevented by accurate identification as early as possible. Manual rice disease detection faces various challenges like lack of expertise, working conditions, expensive monitoring, etc. Researchers invent new automated technologies for accurate and reliable detection and classification of rice plant diseases. Automated disease detection and classification system can easily overcome the challenges in manual identification. Automation introduces precision in farm management practice. Visual computing with image processing technology brings this automation in existence. The major steps involved in an automated disease diagnosis system based on visual computing are data collection, preprocessing, segmentation, feature extraction, and classification.

In the literature, image processing technologies are widely used for plant disease detection. In [7], the author discussed the importance of color, texture, and shape feature in analyzing the disease pattern for different types of crops. In [8], the author pointed out the morphological changes in rice leaves due to blast and brown spot diseases through histogram analysis. They classified diseases with different classifiers like Bayes' and SVM Classifiers. The authors in [9] highlighted the early chili plant disease detection methods by inspecting the leaf features. In [10], the authors described various types of segmentation algorithms as well as feature extraction methods for detecting the diseased leaf of a plant. In [11], the author described features extraction method of diseased area through the pattern recognition technology of healthy and infected leaf.

After reviewing all these literature, a common shortcoming has been observed during the image acquisition procedure. Due to improper acquisition method, different kinds of image noises appeared in the sample, which prevent to achieve the desired accuracy for the automated system. Most of the paper shows that data has been acquired in a nonstandard manner (directly from field, damaging the plant). In this paper, we have used a standard box for data collection to reduce the noises and developed an automated system for identifying the rice blast and brown spot leaf diseases. The images are collected with this standardized system. After collecting the images, noises are removed at the preprocessing phase. After preprocessing, segmentation has been done. The features are extracted from the segmented images to quantify the changes due to diseases. After feature extraction, classification has been done in three stages. At first stage, rice leaf and stem are classified to detect the region of disease. Then, healthy and infected leaves are classified. At last stage, Blast and Brown Spot diseases are classified. Five classifiers are compared to find the best classification model for these three stages.

Section 2 highlights the proposed work, Sect. 3 gives supportive results and describes each result, and Sect. 4 concludes the paper.

2 Proposed Method

Automated rice disease detection is an important research topic as it reduces manual labor, saves time, and brings accuracy in the diagnosis procedure. In this paper, blast and brown spot diseases are detected automatically based on the symptoms that appear on the rice leaves. The system follows six steps in a pipeline. In the first step, normal and infected images of the rice plants are collected from the field. The collected images are preprocessed to remove noise and eliminate the background.

Figure 1 represents the flow chart of the proposed method. At first, the collected images are read in MATLAB 2014b software. Then, these images are converted to HSV color model to get the image intensity and color information value separately. Then image hole-filling algorithm is applied to define the edge of the leaves clearly. After that, median filter is applied to remove sharp noises. The area to be cropped is calculated to eliminate the background. After all these preprocessing steps, segmentation of the images is done using Otsu method. After segmentation, the features were extracted for classification. Classification has been done in three stages. At first, leaf and stem are detected from collected images to detect the region of interest in a rice plant. Then infected and healthy leaves are detected. At last stage, Blast and Brown spot diseases are detected.

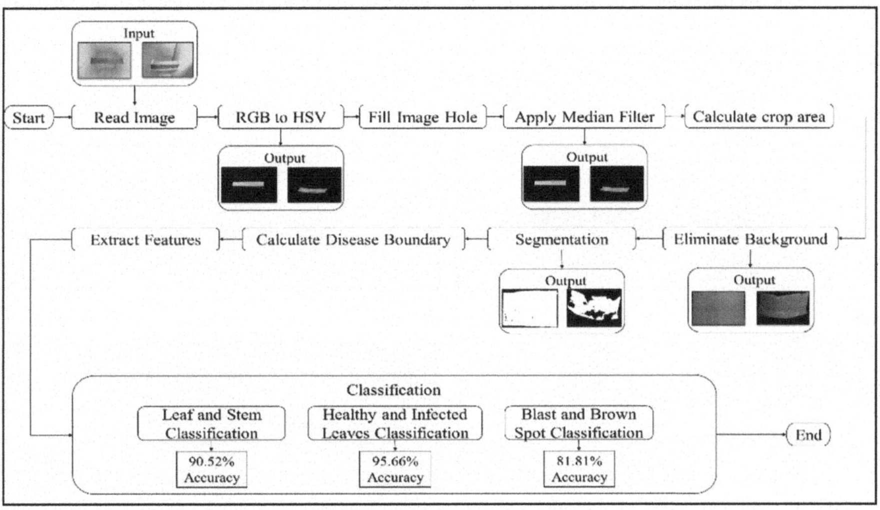

Fig. 1 Work flow chart of the proposed method

2.1 Image Acquisition

The first step of an automated rice disease detection system is image acquisition. The images of normal and infected rice leaves are collected from different parts of West Bengal like Singur, Hooghly, and Midnapur. The challenges that can be faced during the data collection from real-world field are manual error during image capture like angle, shivering of hand, distance, aperture resolution, etc. To overcome these challenges, an image acquisition toolbox (Fig. 2) has been prepared for data collection in uniform environment. The images were captured using CANON IXUS155 digital camera in this box. Total 541 images were collected. It consists of three types of rice diseases—blast, brown spot, and sheath rot. Figure 3a, b is the sample image of healthy leaf and healthy stem, Fig. 3c, d is the sample image of infected leaves with Brown Spot and Blast, respectively.

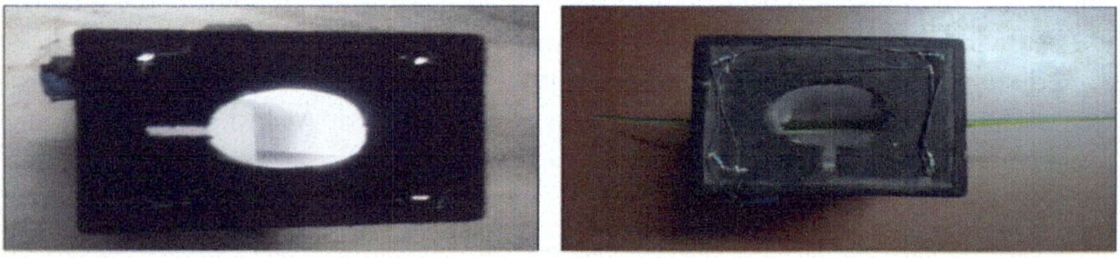

Fig. 2 Snapshot of the device used for data collection

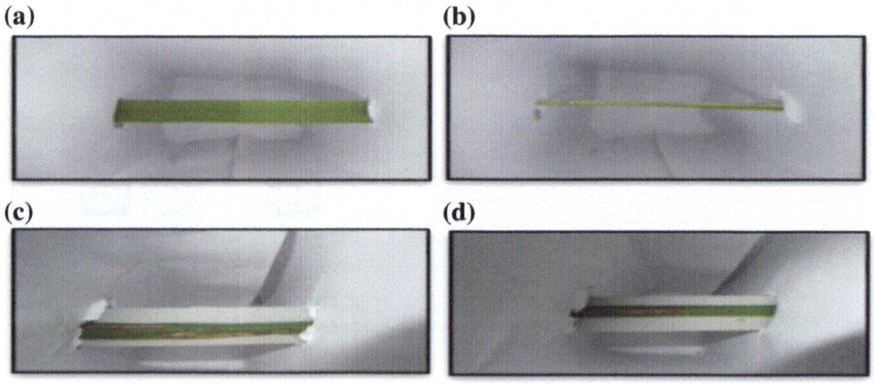

Fig. 3 Acquired images **a** healthy leaf, **b** healthy stem, **c** leaf infected by brown spot, **d** leaf infected by blast

2.2 Preprocessing

Image preprocessing is required to remove the noise in acquired images. In this study, three methods are applied for preprocessing. This step takes acquired images as input and outputs preprocessed image by RGB to HSV conversion, noise reduction, and background elimination.

RGB to HSV Model Conversion
HSV describes the image using color, vibrancy, and brightness. Hue is generally related to the wavelength of light and intensity shows the amplitude of light. Saturation represents the vibrancy of the color. The value represents the brightness of the color. The first step of converting RGB to HSV model [12] is to divide the R, G, B values by 255 to change the range from [0.. 255] to 0..1. The rest of the process has been done on saturation channel image as it gives better results.

Noise Reduction
Sometimes acquired images lack in contrast and brightness because of the limitations of imaging subsystems and illumination conditions while capturing image. Hence, acquired images may have different types of noises. In this paper, a hole-filling method is used. It is based on morphological reconstruction algorithm. Another noise removing method, the median filter is used in this paper. It is particularly effective at removing "salt and pepper" type noise. It is also known as impulse noise.

Background Elimination
After noise reduction images are fed forward for background elimination. The objective of this step is to remove the background part of the sample data. In this paper, an algorithm has been implemented to find all nonzero elements of the filtered image for automatic crop to detect the region of interest. Then, eliminate rest of the part and cropped the image into the leaf and stem area only.

The result of all these preprocessing steps is shown in Fig. 4 corresponding to Fig. 3.

2.3 Image Segmentation

Image segmentation is the process of partitioning a digital image into multiple segments of those with sets of similar type pixels. The goal of image segmentation is to simplify and change the representation of an image into something that is more meaningful and easier to analyze. Once the acquired images are preprocessed, the most commonly used Otsu's method [13] is applied to preprocessed images for segmentation. The result of segmentation is shown in Fig. 5.

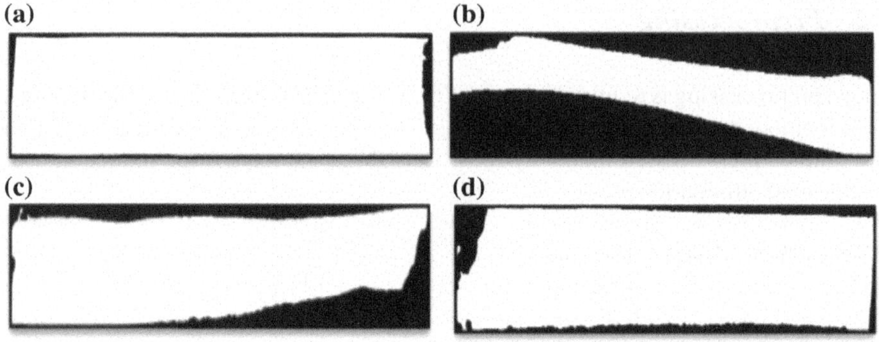

Fig. 4 Images after background elimination corresponding to Fig. 3: **a** healthy leaf, **b** healthy stem, **c** leaf infected by brown spot, **d** leaf infected by blast

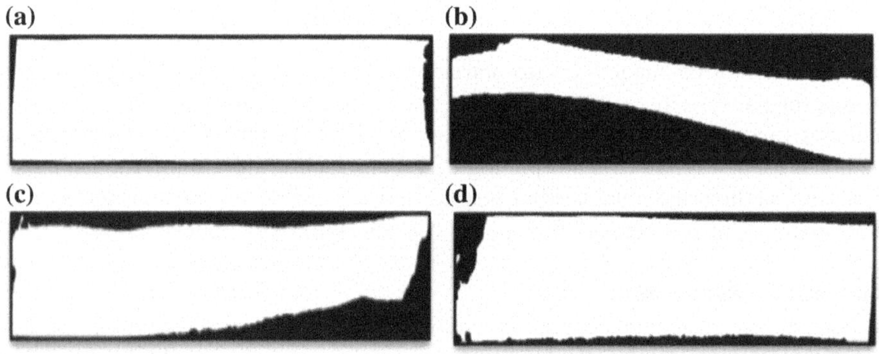

Fig. 5 Segmented image: **a** healthy leaf, **b** healthy stem, **c** leaf infected by brown spot, **d** leaf infected by blast

2.4 *Feature Extraction*

A total of 59 features are extracted in this paper. These are grouped into four main categories—texture, color, shape, and intensity distribution.

Texture Feature: The texture describes the basic pattern of the surface in a particular direction. It describes the physical composition of a surface which can be distinctly classified. We have extracted 16 texture features from gray-level co-occurrence matrix (GLCM) including—Contrast, Correlation, Energy, and Homogeneity in four different angles-0°, 45°, 90°, and 135°. The gray co-matrix function creates a gray-level co-occurrence matrix (GLCM) [8] by calculating how often a pixel with the intensity (gray-level) value occurs in a specific spatial relationship to a pixel with the value j.

Color Feature: Color features are extracted from both RGB and HSV color space. The variation of color of an image can be quantified by five main features—Mean, Mode, Median, Standard Deviation, and Variance. Mean indicates the average brightness of an image [14]. Higher value of mean indicates brighter images and lower value of mean indicates darker images. Mode calculates most frequent values in array. Median separates high-intensity pixels from low-intensity pixels. Standard Deviation and Variance measure the contrast of an image [15]. High value of standard deviation and variance means high-contrast images and low value means low-contrast images.

Shape Feature: Generally, expert identifies diseased region and disease spot with color but shape is required for objective evaluation of disease regions. In this paper, a total of eight shape descriptors [16]—area, filled area, eccentricity, width, height, extent, major, and minor axis length, are extracted from segmented region to identify rice leaf diseases automatically. Area is used to quantify disease region. Extent calculates the ratio of pixels within disease region and total image. The filled area gives the number of pixels in the disease region and all wholes included that region to make it continuous. Eccentricity defines how much an image is elliptical. Major and Minor axis length define the length of the axis that passes through the centroid of the elliptical region. The height is the maximum distance between any two points on the perimeter of the particle parallel to the major axis. Likewise, the width is the maximum distance between any two points on the perimeter of the particle parallel to the minor axis.

Intensity Distribution: Intensity refers to the numerical value of a pixel. An intensity distribution of an image is a data matrix whose values represent intensities within some range. It is difficult to describe the spatial arrangements [17] of the leaves and stem of rice plants in words. The arrangements are not regular, but there is some quality of the arrangement that would make one indication that there is a noticeable arrangement in sample leaf and stem image. Figure 6 shows the intensity distribution of Leaf and Stem. From this figure, it is clear that the distribution is different in both cases. The leaf region gives straight-line base intensity distribution and the stem region gives radial shape base intensity distribution.

2.5 *Classification*

After collecting all the 59 features for a total of 541 images, a data set is prepared with these features values. This data set is fed into the classification tool WEKA [18]. Classification of this automated computing system has been done in three phases—First phase classifies leaf and stem of the rice plant to identify the region of occurrence of the disease, at second phase, healthy and infected leaves are classified and at last phase, Blast and Brown spot diseases are classified from infected leaves.

The automatic leaf and stem classification is applied to the 299 sample images of leaf and 242 sample images of stem. Leaf and stem classification has been done based on intensity distribution, texture, color, and shape features. For Healthy and infected leaf classification, a feature set of 299 images is prepared. For Blast and Brown

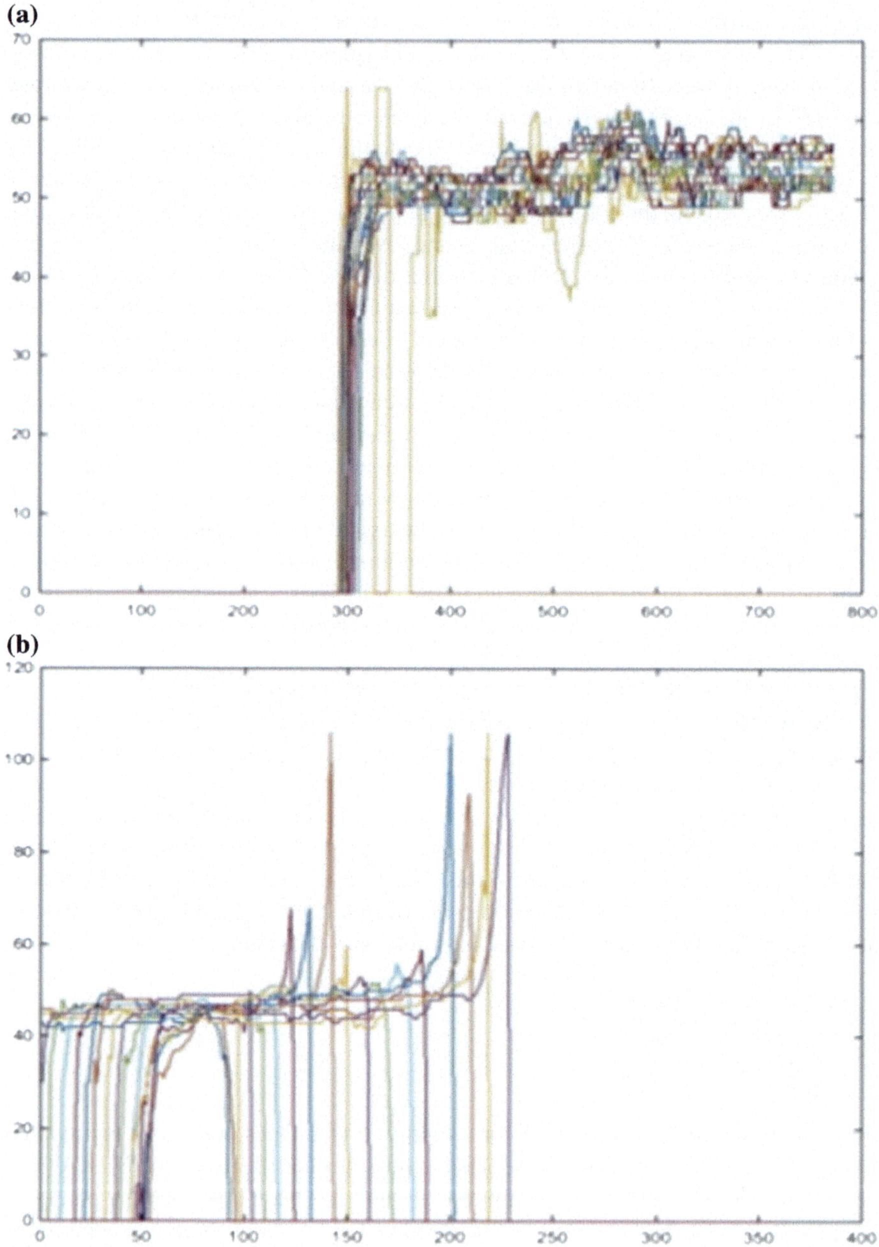

Fig. 6 Intensity distribution of Fig. 3a, b. **a** Intensity distribution of leaf, **b** intensity distribution of stem

spot disease classification, a feature set of 151 images is prepared. The automated system classified the healthy leaf according to its difference in texture, color, shape features from infected leaf. The infected leaf has an unhealthy region and the test results showed some specific certain features of those unhealthy regions. For the Blast disease, the lesion is diamond shape and for the Brown spot, the lesion is circular shape. As a result, the value of major axis length, minor axis length, width, height, extent, eccentricity, mean, and mode is different for diamond and circular shape lesion.

Five classifiers Random Committee, Random Forest, Random Tree, Bagging, and IBK are compared to choose best classification model for this three-layer classification. At first layer, rice leaf and stem classification has been done in these five classifiers. Then healthy and infected leaves are classified and at last, Blast and Brown Spot classification has been done. The result of this comparison is given in the following section.

3 Result and Discussion

The classification has been done in WEKA with 10-fold cross-validation. Five classifiers are compared with respect to given accuracy. A meta classifiers Random Committee gives 90.52% accuracy for the leaf and stem classification. Random Tree classifier provides best result in healthy and infected leaf classification with accuracy 95.66%. K-nearest neighbor based classifier IBK gives the best accuracy of 81.81% for Blast and Brown Spot disease classification. The best classifiers with accuracy and the decision class are tabulated in Table 1.

In Figs. 7, 8, and 9, the accuracy of different classifiers are compared for leaf and stem region classification, healthy and infected leaf classification, and Blast and Brown Spot classification, respectively.

Table 1 Best classifier for each decision class with their accuracy

Classifier	Decision class	Accuracy (%)
RandomCommittee	Leaf and stem	90.52
RandomTree	Healthy and infected leaf	95.66
IBK	Blast and brown spot	81.81

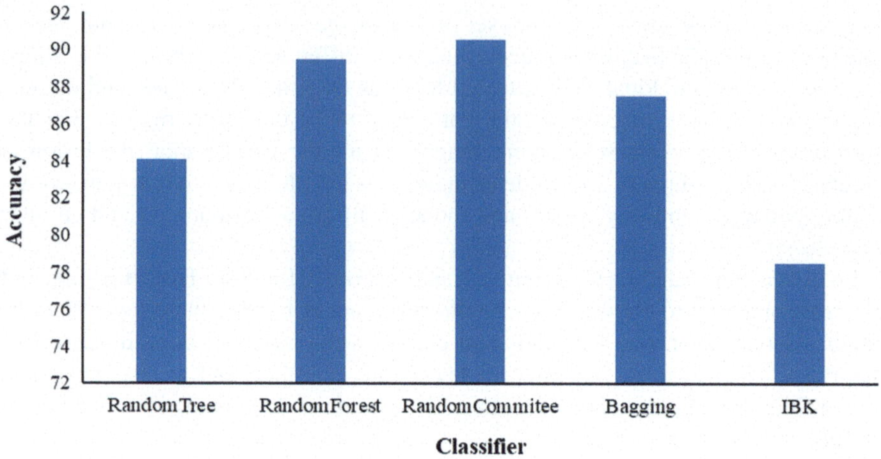

Fig. 7 Comparison of five classifiers for leaf and stem classification

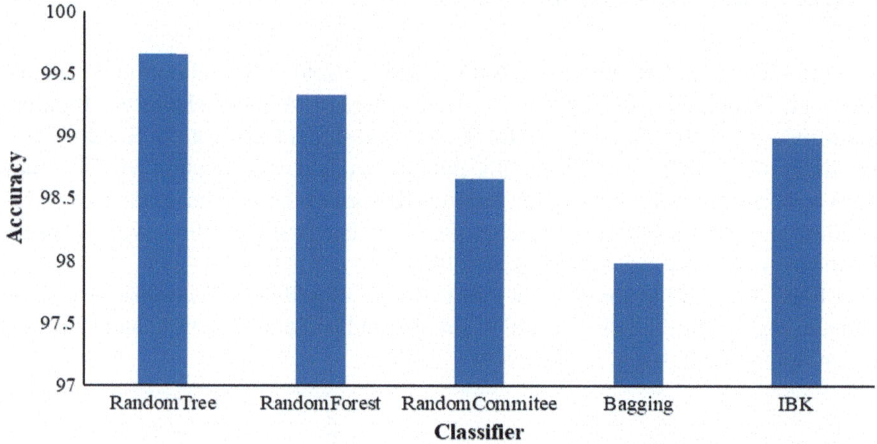

Fig. 8 Comparison of five classifiers for healthy and infected leaf classification

4 Conclusion

An automated system for classification of rice leaf diseases has been developed in this paper using image processing technology based on different types of disease symptoms. After collecting the sample of rice plants, images are arranged according to their classes identified by the expert. After that, different preprocessing methods are applied to improve the quality of the images. These preprocessed images are segmented using Otsu method to find out the diseased region. After segmentation, 59 features are extracted to classify leaf and stem diseases. The accuracy of classification for rice plant leaf diseases is 81.81%. The main advantages of this work are—(1) A

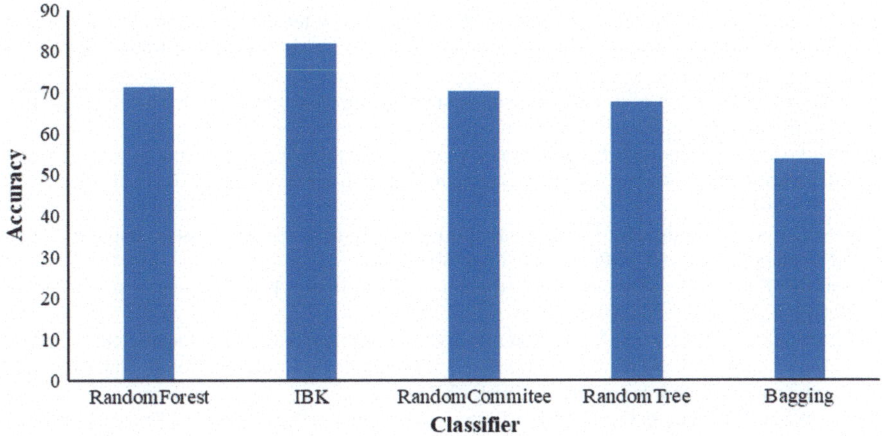

Fig. 9 Comparison of five classifiers for leaf disease blast and brown spot classification

standard data acquisition system has been built up to acquire data in homogeneous manners without damaging the plant. (2) A complete automated system for detecting rice plant diseases. Our research work concentrates only two rice plant diseases. Future works may include more types of rice plant diseases, detecting the early stages of a particular disease and can provide various solutions to the problem at that early stage.

References

1. Chose, L.M., Ghatge, B., Subramanyan, V.: Rice in India. Indian Council of Agricultural Reserarch, New Delhi (1956)
2. Datta, D.: Principles and Practices of Rice Production. International Rice Research Institute (1981)
3. Rice as a crop—Ricepedia: http://ricepedia.org/rice-as-a-crop (2018)
4. Mondal, D., et al.: Yield loss assessment of rice (Oryza Sativa L.) due to different biotic stresses under system of rice intensification (SRI) (2017)
5. Yellareddygari, S.K.R., et al.: Rice sheath blight: a review of disease and pathogen management approaches. J. Plant Pathol. Microbiol. **5**(4), 1 (2014)
6. Tilman, D., et al.: Agricultural sustainability and intensive production practices. Nature **418**(6898), 671–677 (2002)
7. Song, Y., et al.: Image feature extraction of crop disease. In: IEEE Symposium Electrical & Electronics Engineering (EEESYM) (2012)
8. Phadikar, S., Sil, J., Das, A.K.: Classification of rice leaf diseases based on morphological changes. Int. J. Inf. Electron. Eng. **2**(3), 460 (2012)
9. Zulkifli, H., et al.: Feasibility study on plant chili disease detection using image processing techniques (2012)
10. Khirade, S.D., Patil, A.B.: Plant disease detection using image processing. In: Computing Communication Control and Automation (ICCUBEA). IEEE (2015)
11. Phadikar, S., Sil, J.: Rice disease identification using pattern recognition techniques. In: 11th International Conference on Computer and Information Technology (ICCIT). IEEE (2008)

12. Al Bashish, D., Braik, M., Bani-Ahmad, S.: A framework for detection and classification of plant leaf and stem diseases. In: International Conference on Signal and Image Processing (ICSIP). IEEE (2010)
13. Vala, H.J., Baxi, A.: A review on Otsu image segmentation algorithm. Int. J. Adv. Res. Comput. Eng. Technol. **2**(2), 387–389 (2013)
14. Wicaksono, Y., Wahono R.S., Suhartono, V.: Color and texture feature extraction using gabor filter-local binary patterns for image segmentation with fuzzy C-means. J. Intell. Syst. **1**(1), 15–21 (2015)
15. Deole, P.A., Longadge, R.: Content based image retrieval using color feature extraction with KNN classification. IJCSMC **3**(5), 1274–1280 (2014)
16. Andaló, F.A., et al.: Shape feature extraction and description based on tensor scale. Pattern Recognit. **43**(1), 26–36 (2010)
17. Phadikar, S., Goswami, J.: Vegetation indices based segmentation for automatic classification of brown spot and blast diseases of rice. In: 3rd International Conference on Recent Advances in Information Technology (RAIT). IEEE (2016)
18. Bouckaert, R.R., Frank, E., Hall, M., Kirkby, R., Reutemann, P., Seewald, A., Scuse, D.: WEKA Manual for Version 3-7-8. Hamilton, New Zealand (2013)

Review of Atmospheric Precursors for Earthquake Prediction Model

Tilak Mukherjee and Dalia Nandi (Das)

Abstract This paper analyzes the signature effects of earthquake (EQ) events on the lower atmospheric levels mainly in the troposphere region. Significant changes and anomalies in temperature, relative humidity, and radio refractive index profiles are observed in most of these seismic events prior to the impending event, thereby providing a precursor method to predict the occurrence of EQ in the faulty zones. Temperature, relative humidity, and radio refractive index anomalies using electromagnetic signals have been considered mainly in the paper. A better understanding of the precursory signals involving radio wave propagation based on lower atmospheric conditions associated with such EQ devastating event is discussed in the paper.

Keywords Signature effect · EQ · Anomaly · Precursor · Radio refractive index

1 Introduction

Earthquakes (EQs) are considered to be one of the worst natural disasters leading to immense irreparable loss to life as well as the property of an individual and to the society in totality. It is a sort of seismic activity where sudden energy is released in the earth's crust in the form of seismic waves. These waves travel through the interior of the earth leaving the focus like body waves or may spread from the epicenter like surface waves. Tectonic shifts in the lithosphere, volcanic and surface effects are the prominent factors behind such a hazard. This seismic activity is changed according to the frequency, type, and magnitude of the event. In the past years, several techniques have been employed to specify the time, location, and magnitude of the EQ within stated limits so as to reach a prediction model with certain reliability using various

T. Mukherjee (✉)
Department of Electronics & Communication Engineering, Haldia Institute of Technology, Haldia, West Bengal, India
e-mail: mukherjeetilak@gmail.com

D. Nandi (Das)
Department of Electronics & Communication Engineering, Indian Institute of Information Technology Kalyani, Kalyani, West Bengal, India

S. Kundu et al. (eds.), *Proceedings of the 2nd International Conference on Communication, Devices and Computing*, Lecture Notes in Electrical Engineering 602, https://doi.org/10.1007/978-981-15-0829-5_57

remote-sensing methods [1]. The growing need is to strengthen the forecast using various proxies termed as EQ precursors.

Some commonly used earthquake precursor techniques are

(i) Sensing of tectonic plate movement.
(ii) Electro-sensitivity experiments.
(iii) Extraction of seismic signal or reverse tracing of precursor.
(iv) Radio wave techniques using atmospheric studies using satellite data or Radiosonde data.

Various other precursor methods have been studied leading to the prediction of a seismic event in the faulty zones. However, the prediction of EQs is of great challenge, despite modern technology and high-resolution data. Recognizable precursors along with proper detection of anomalies are not precise enough to predict the occurrence of the EQ event [1, 2]. Understanding this fundamental link between the nature of precursor and the EQ event will definitely contribute to hazard reduction. In the present paper, characteristic deviation from the expected values or anomalies is observed for important parameters mainly temperature profile, relative humidity, and radio refractive index of the lower atmosphere just few days prior to the event and its repercussion.

2 Signature Effect on Troposphere

The troposphere is the lowest layer of our atmosphere starting from ground level, that extends upward to about 10–12 kms above sea level. Human beings live in the troposphere, and nearly all weather changes like fog, clouds, dew, frost, hailstorm, thunder, and lightning effects occur in this lower layer. Air pressure drops and temperatures get colder, as height increases at the rate of 6.5 °C/km. This is the lapse rate. This layer is of much interest to communication and radio engineers.

Consistent efforts are being made to investigate a possible correlation between atmospheric parameter observation and EQ events, which would have an impact on our further understanding of the physics of EQ and the phenomena related to their energy release [3].

Three different physical parameters are studied characterizing the state of the atmosphere during the periods associated with the major earthquake events, having a magnitude greater than 5.5 (M > 5.5).

They are mainly

1. Temperature profile.
2. Humidity profile.
3. Radio refractive index (or Radio Refractivity) and Infrared(IR) emissions. This sudden deviation of atmospheric parameters associated with the EQ event is commonly termed as the "Signature Effect" of the seismic activity on lower atmospheric profile [3, 4].

3 Effect on Temperature Profile

EQ events are accompanied by strong gravity waves which interfere with the normal temperature profile of the troposphere. A few days before the occurrence of any seismic activity in a particular fault zone, the tropopause level temperature rises. The temperature profile variation can be observed using a Radiosonde or online archived GPS data extended over a period of time prior to the EQ event, as the case may be.

Apart from this, the gravity waves have the tendency to move upward in the lower atmosphere levels thereby causing lower tropospheric temperature profile to deviate unexpectedly. This temperature variation might indicate about the impending EQ event. As radon gas is emitted from underground, the increase of the temperature is marked immediately on the ground surface region. The temperature difference between faults area and far from the faults leads to horizontal air movements, air mixture, and air temperature rise over all around the earthquake preparation zone, thus leading to thermal anomaly [3–6]. Close examination of the precursor anomalies as observed from remote-sensing measurements for the EQ preparatory zones indicates the strong air ionization effect on the thermal balance of the boundary layer of the atmosphere.

Figure 1 shows seismic moment on day 0 (on x-axis) with the inverted triangle symbol and dashed lines indicate the parameter peculiarity 6 days before the EQ event interpreted as precursor phenomenon. Colima EQ (M = 7.6) that occurred on January 21, 2003 (latitude and longitude positions of 19.22° N and 103.7° W) has been observed [6].

Quite similar observation is studied with Nepal Gorkha EQ (latitude 28.1° N and longitude 103.7° W). The devastating event took place on April 25, 2015 with magnitude M = 7.8. During their occurrences, abrupt increase in greenhouse gases like CO_2, CH_4, H_2, etc., and enhancement of radon emanations were found leading to huge latent heat release. The data collected from climate data online (http://

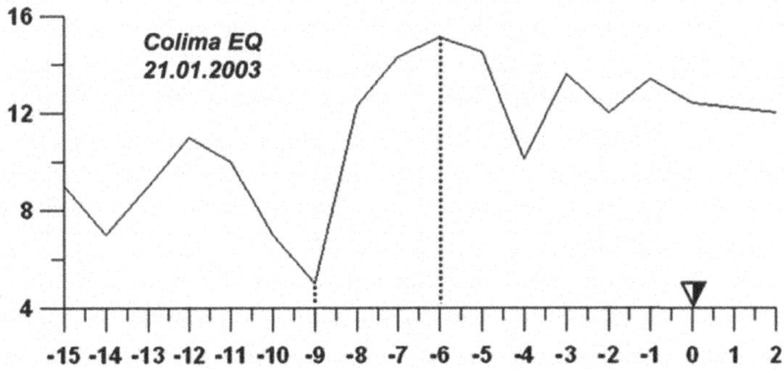

Fig. 1 Variation of air temperature in degree Centigrade (T max to T min on y-axis) registered close to the seismic zone for Colima EQ. X-axis marks the day variation with respect to the EQ event [6]

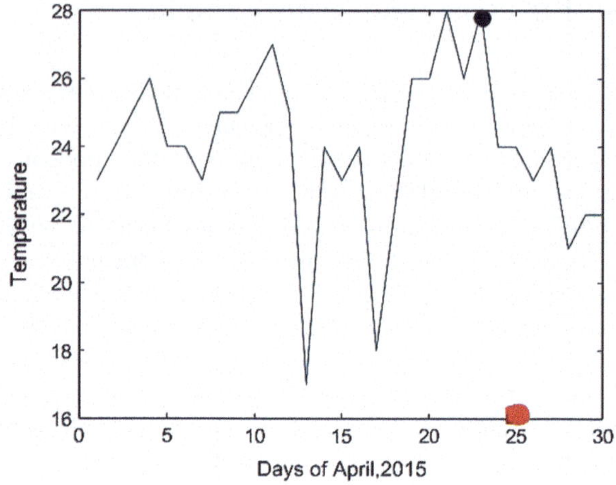

Fig. 2 Variation of average air temperature (degree Centigrade) for Kathmandu Gorkha EQ for the month of April 2015. Red dotted mark indicates the occurrence of the EQ event. Black dot indicates temperature rise a few days before the event (anomaly)

www.timeanddate.com/weather) during April in the year 2015 for Kathmandu region shows rise in air temperature just few days prior to the EQ event.

From the above Fig. 2, it is observed that 4–5 days before the EQ event there is a sharp increase in air temperature profile, which could possibly be one important precursor information indicating the catastrophe [8, 9].

4 Effect on Humidity Profile

As already discussed, the possible EQ zones release huge amounts of radon causing an increase in the ionization intensity of the lower atmospheric levels. The water vapor molecules adhere to these ions leading to the formation of light and heavy ions or aerosols. Water vapor attachment to these ions decreases the free water vapor content in air, and so relative humidity decreases accordingly. The relative humidity is inversely proportional to the water holding capacity of the air, and when the air is dry, it decreases. In this entire process, a large amount of latent heat is released leading to thermal anomalies in the lower atmosphere. Consequently, change in the relative humidity parameter is another important precursor anomaly associated with the seismic event. This could possibly be indicated as another signature effect of the alarming seismic event [3, 6].

From Table 1, it is observed that the temperature is maximum on April 21, 2015. The data is collected from climate data online (http://www.timeanddate.com/weather) for April 2015. Humidity level is low few days just before the impending EQ event, i.e., on April 23, 2015, indicating as prominent precursor anomaly for the Nepal zone EQ. Such deviation behavior in the atmospheric parameters related to the EQ event is of immense importance to data science researchers in prediction modeling.

Table 1 Climate profile (April 2015) for Nepal EQ

Value	Temperature profile	Humidity profile
High	28 °C (April 21, 2015)	100% (April 4, 2015)
Low	10 °C (April 5, 2015)	21% (April 23, 2015)
Average value for the month	18 °C	71%

5 Effect on Radio Refractive Index and IR Emission

The troposphere causes the bending of line of sight (LOS) path of electromagnetic (EM) waves. The effective limit to the path of a signal that propagates through the troposphere is determined by the Earth's curvature, known as the LOS path or popularly radio horizon. However, this horizon does not remain constant, and the major influence on LOS propagation comes from changes in the refractive index (n). The humidity of the troposphere modifies this parameter, but the variations are very small so this is designated by a factor $N = (n-1) \times 10^6$, called as radio refractive index (RRI) or radio refractivity. Its gradient with respect to height is denoted by (dN/dh). The atmospheric radio refractive index modifies and thereby decreases the gradient of radio refractive index parameter (dN/dh) few days prior to the EQ event.

When (dN/dh) decreases rapidly, then atmospheric ducts are formed thus allowing the wave to reach beyond the LOS limit. Atmospheric stratification and permittivity changes cause the EM waves to bend further. Modification in atmospheric refractive index followed by drop in relative humidity levels above the epicenter region few days prior to the EQ event could possibly be another signature effect of the alarming seismic event. Temperature rise along with drop in relative humidity often accompany it [5–7]. Additionally, increased ionization intensity of the lower atmosphere gives rise to large number of aerosols which are both positively and negatively charged. Electric field is generated by charge separation in the atmosphere that further tends to accelerate the charged particles. This electron energy is adequate to cause excitation and emission of quanta of energy in the Infrared(IR) region, which is due to the excitation of CH_4, CO, CO_2, N_2O, and O_3 molecules, occurring at an altitude of around 15–20 km in the lower atmosphere levels during EQ preparation times [8–12].

However, atmospheric parameters that have been considered in this review paper are mainly temperature, humidity profile, and radio refractive index variation pertaining to lower atmospheric region only that are closely associated with the occurrence of the EQ event.

6 Conclusion and Future Scope

This present paper mainly takes into account atmospheric parameters like air temperature, humidity profile, and changes in radio refractive index for the consideration of EQ preparation zones and their sudden deviation that form precursor anomalies for seismic zones. Significant changes are observed in the temperature, humidity, and refractivity gradient profiles prior to the seismic activity or EQ event. But proper identification of anomalies should be done based on long-term regional analysis to prevent subjectivity bias. Statistical evidence for a correlation is lacking as the pattern of such precursors may vary from one EQ to another. The probability of occurrence has to be determined with parameters having adequate precision, reliability, and accuracy so as to minimize the aftermath of such vulnerable havoc.

There can also be strong nonhomogeneity as well as temporal variation in the atmospheric parameters. Local seasonal changes, instrumentation errors, or background noise if any must be filtered out prior to the proper identification of the precursors. Attempts are to establish a better prediction model with enhanced accuracy of prediction using machine learning techniques and relate the variability of atmospheric parameters as future scope of the research work in the field of data science analysis and remote sensing.

References

1. Huang, F.Q., Jian, C.L., Tang, Y., Xu, G.M., Deng, Z.H., Chi, G.C.: Response changes of some wells in the mainland subsurface fluid monitoring network of China, due to the September 21, 1999, Ms 7.6 Chi-Chi earthquake. Tectonophysics **390**, 217e234 (2004)
2. Huang, F., et al.: Study of earthquake precursors in China: last 50 years, Geod. Geodyn. J. **8**, 1–12 (2017)
3. Pulinets, S.A., Boyarchuk, K.A.: Ionospheric Precursors of Earthquakes, p. 315. Springer, Berlin, Germany (2004)
4. Geller, R.J., Jackson, D.D., Kagan, Y.Y., Mulargia, F.: Earthquakes cannot be predicted. Science **275**, 1616–1617 (1997)
5. Yonaiguchi, N., Ida, Y., Hayakawa, M., Masuda, S.: On the statistical correlation of over horizon VHF signals with meteorological radio ducting and seismicity. J. Atmos. Solar-Terr. Phys. **69**, 661–674 (2007)
6. Pulinets, S.A., et al.: Physical nature of thermal anomalies observed before strong earthquakes. Phys. Chem. Earth **31**, 143–153 (2006)
7. Devi, M., Barbara, A.K., Ruzhin, Y.Y., Depueva, A.H.: Beyond the horizon propagation of VHF signals: atmospheric features and earthquake beyond the horizon propagation of VHF signals, atmospheric features and earthquake. Electron. J. (Russia) **39**, 1331–1340 (2007)
8. Dey, S., Singh, R.P.: Surface latent heat flux as earthquake precursors. Nat. Hazards Earth Syst. Sci. 749–755 (2003)
9. Hazra, P., Dey, S., Paul, S., Guha, G., Ghosh, A.: Thermal anomalies around the time of Nepal earthquakes M7.8 April 25,2015 and M7.3 May 12, 2015. Int. J. Geotech. Earthq. Eng. **8**(1), 58–73 (2017)

10. Evison, F., Rhoades, D.: Model of long-term seismogenesis. Ann. Geophys. **44**(1) (2001)
11. Liperovsky, V.A., et al.: Electric field and infrared radiation in the troposphere before earthquakes. Nat. Hazards Earth Syst. Sci. **11**, 3125–3133 (2011)
12. Medha, A., Devi, M., Goswami, H., Barbara, A.K.: Atmospheric turbulences over Guwahati and their association with tropospheric dynamics. Indian J. Radio Space Phys. **44**, 35–44 (2015)

Analysis of a Data Acquisition System for a Compact Electronic Nose

Shambo Roy Chowdhury, Amol P. Bhondekar, Sudeshna Bagchi and Vinod Karar

Abstract Electronic nose systems have been recently in use for various applications ranging from environmental monitoring to biomedical applications. This paper introduces a compact electronic nose system integrating conducting polymer sensors with data acquisition, data processing and storage unit. The data acquisition system presented here is equipped with a graphical user interface (GUI) to control the acquisition parameters as well as perform analysis on obtained data. The real-time data acquisition system is capable of collecting data from an 8X8 resistive sensor array as per the parameters passed by the GUI. Finally, the paper is concluded with results and discussion obtained using the proposed electronic system. The result shows with proper classification algorithm, the system is capable of identifying various amine and alcohol groups.

Keywords Electronic nose · Signal processing · Data acquisition · Data processing

---

S. R. Chowdhury (✉) · A. P. Bhondekar · S. Bagchi · V. Karar
CSIR-Central Scientific and Instrument Organisation, Chandigarh 160030, India
e-mail: shamboroychowdhury@gmail.com

A. P. Bhondekar
e-mail: amolbhondekar@csio.res.in

S. Bagchi
e-mail: sudeshna_bagchi@csio.res.in

V. Karar
e-mail: vinodkarar@csio.res.in

Academy of Scientific and Innovative Research, New Delhi, India

S. R. Chowdhury
Manipal University, Jaipur, India

S. Kundu et al. (eds.), *Proceedings of the 2nd International Conference on Communication, Devices and Computing*, Lecture Notes in Electrical Engineering 602, https://doi.org/10.1007/978-981-15-0829-5_58

1 Introduction

The perception of odour is used for various purposes such as finding food, avoiding danger or choosing mate [1, 2]. It is always an essential sense whether one is on air, land or under water. However, even after several decades of research, the science to smell still lacks objectivity. This is owing to the fact that the perception of odours depends on multiple aspects such as the relative frequency of exposure, ecological impacts or the domicile of the observer. This subjectivity has drawn the interest of researchers to build a causal link between the various physico-chemical properties of the odorants and their perceptions. One such attempt involves building an electronic analogue of the biological olfactory system, commonly known as electronic nose. An electronic nose qualitatively analyses sample of odorants to classify them based on their chemical compositions [3].

Major research for electronic nose systems is generally segregated to inspection of different aspects of the system such as the sensing techniques [4–7] or the gas identification algorithms and applications [8–12]. When research concentrates on developing sensor, the acquisition technique generally includes specially calibrated commercial instruments to measure resistances and control of power supply [4, 13–15]. These ensure well-calibrated measures and are suited for first proof of concept. But for portable devices, we require data acquisition arrangement which is cheap and compact, trading off with accuracy and reliability. The system as a whole requires an optimised integration of data between the sensing techniques and the classification algorithm. Gas sensors based on semiconductor metal oxides and conducting polymers as active layers have been extensively reported in literature [16–21]. The conductivity of these sensors changes based on the binding of the vapour molecules to the polymers [7, 14]. These changes are to be captured and processed by a data acquisition system to improve sensitivity and remove noise. The data acquisition system does the work of olfactory neurons and the olfactory glomeruli which connect the receptor cell to the olfactory region of the brain. This data acquisition system similarly collects the spatio-temporal signal of the chemical gas sensor and conveys the message to the particular classifier. Detecting these resistances involves multiple amplification stages and other signal processing arrangements for improving SNR of the sensor data. The challenge in this part involves high-frequency sampling of elements in sensor array, which as a consequence, increases ADC conversion errors, and also induces pseudo capacitance [22, 23]. Morsi et al. [24] reported in their work, an electronic nose setup by acquiring data from an array of FIGARO metal oxide sensors with the help of microcontrollers. The data is analysed after transferring data to a PC. Work with similar objectives have been published with conducting polymers [6, 18, 25, 26] and metal oxide semiconductors [27, 28]. Fifty-five chemically diverse conducting polymer sensing elements were used by Ramesh Kumar et al. [6], to test the odour identifying capability of their electronic nose. The hardware for the systems had been designed as an integrated system within a chip with sensor array and data processing embedded in it.

Most of these works isolate the data acquisition portion and the analysis unit. In this work, we present a compact electronic nose system capable of differentiating between various VOCs.

2 Electronic Nose

To build a compact portable electronic nose system, we selected the type of sensor array and the hardware components for complete data acquisition and analysis. This section would describe the arrangement made for data acquisition and some basic analysis of the hardware arrangement. The sensor elements and the array arrangements are discussed in the first subsection of the chapter. This is followed by a subsection with a brief description of the chamber design for sensor array. The third subsection will describe the electronic setup for data sampling and recording. The above three subsections complete the basic building of a portable electronic nose. The experimental result with the enose setup is shown and discussed in subsection four and five, respectively. The experimental result would help to analyse the basic properties of the proposed electronic nose, such as sensitivity, repeatability, etc. The device can be utilised for electrical characterisation for resistive gas sensors and sensor arrays.

2.1 The Sensor Array

A sensor array of 64 sensor elements has been designed, in an 8×8 array. The concentric circular sensor elements have a diameter of 1 mm and been placed at centre to centre distance of 2 mm. The image of the sensor array (before coating of sensor material) is shown in Fig. 1a. The two concentric circles are of conducting material (copper for our experiments) and act as two electrodes. This is a standard practice for using conducting polymers as gas sensor arrays where the sensor is coated on a substrate in-between electrodes. As a result, the electrodes act as the connection for the variable conductivity gas sensors.

2.2 The Chamber

A plain hollow chamber was designed for the sensor array, with diagonally opposite inlet and outlet for gas (Fig. 1b). This intuitionally should provide an optimal coverage of the gas within the chamber for all sensor elements as the gas need to cover the maximum volume inside the chamber. The material has been chosen to be aluminium considering low cost and no porosity. The design was optimised based on various design parameters, manufacturing constraints and simulations in Ansys. Aluminium

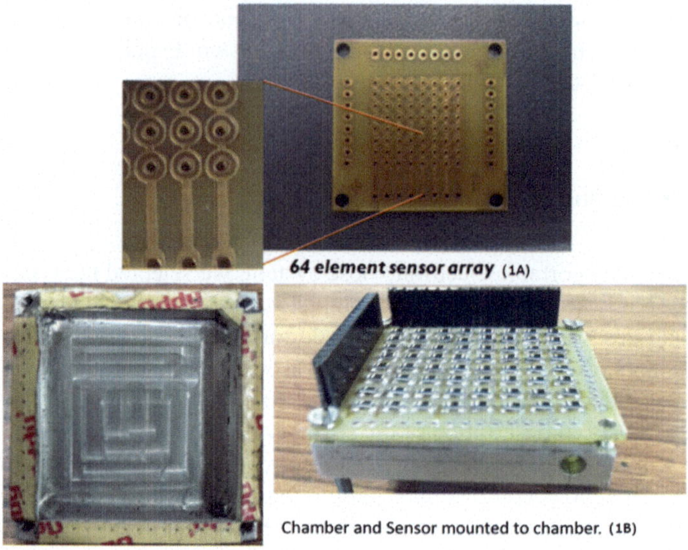

Fig. 1 The sensor array and the chamber prototype

is a popular choice for sensor chamber of enose system as it is non-reactive to most gases and is non-porous as well as cheap compared to other options. Further machining is easier for aluminium compared to plastic substitutes and various design of the chamber was fabricated in-house for experimentation.

2.3 The Hardware

The data acquisition system for an 8 × 8 sensor array was built with ADCs microcontroller and analogue multiplexers. The silicon lab C8051f350 microcontroller was chosen for controlling the real-time data acquisition from the gas sensor array. The i/o pins of the controller are used to control the multiplexers and valves. The multiplexer used are two 8:1 multiplexer, used to activate individual row and columns of the sensor arrays, respectively. The UART capability of the controller is used to communicate with a Raspberry pi (with 9600 baud rate) for data storage, data analysis and user interfacing. For data conversion, the microcontroller provides a 24 bit fully differential ADC. The basic block diagram of the data acquisition arrangement is shown in Fig. 2. The integration of 8051 microcontroller with arm base raspberry pi enables the system for real-time data acquisition along with implementing data analysis algorithms and user interface. The microcontroller is programmed to accept user input from the raspberry pi to adjust the gas sampling cycle. The accepted variable would control the sampling loop as well as generate the control signal for the valves. The sampling and data communication are sequentially programmed following the

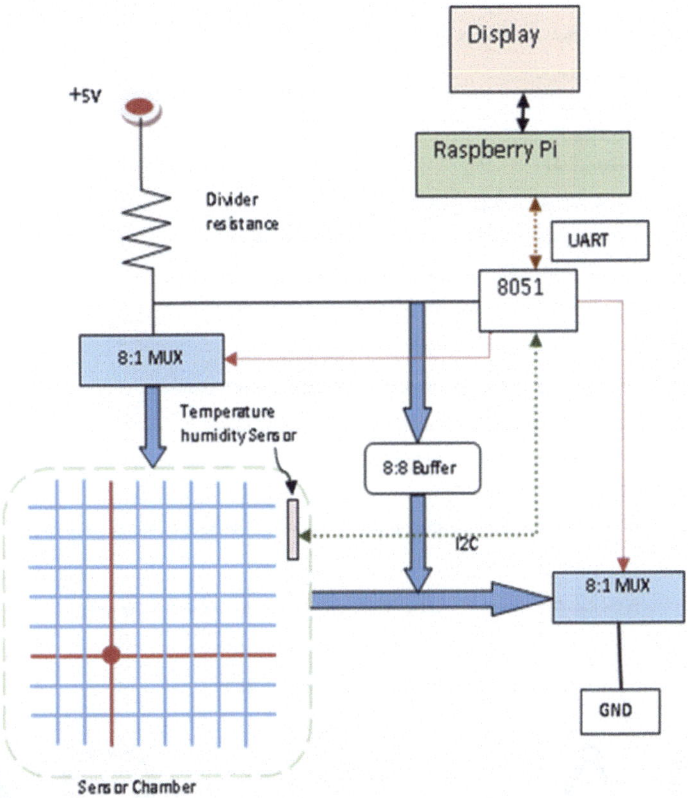

Fig. 2 Block diagram of the complete data acquisition arrangement

algorithm shown in Fig. 3. The sampled data is transferred to raspberry pi for further data processing. With 64 gas sensor elements and each element generating 24 bit of data (because of 24 bit ADC), a total cycle of data acquisition contains (30 × 64) 1920 bits. This could allow us for (9600/1920) five samples of each element per second, considering maximum overhead of the constituting programme in the microcontroller. But the execution cycle of the codes also consumes a considerable amount of time and needs to be synchronised with the sampling and the UART. The UART and the ADC conversion work on individual interrupt modes and the algorithm manages the interrupts to avoid any interrupt overlaps.

The sensor values are sent in hexadecimal values which are later converted to decimal in raspberry pi. A multithreaded graphical user interface was created in Raspberry Pi for user interactions, backend communication management with 8051 and data processing. The multithreading helps in separating the communication module to run independently of actions in the GUI, which in turn protects the communication module being interrupted by any error in GUI or data saving and vice versa. Screenshots of the GUI explaining the steps or flow of instructions are shown in Fig. 4

- Receive the parameters from raspberry pi user interface
- Arrange variables accordingly
  - A= time for buffer phase
  - B= time for input phase
  - C= time for washing phase
- Set the timer as per A and open pathway for buffer gas
  - Timer interrupt is used to control the valves as it enables parallel threads to operate
- Multiplexer inputs are serially controlled with column wise scanning of each element
  - For each column selected, row multiplexer scans each element of the selected column
- The data for each sensor is stored in linear array
- Once scanning is done for one batch of 64 sensors, the batch data is serially transmitted to raspberry pi
- The data is passed as text in format of sensor no. followed by the sensor value
  - For example the value for sensor 12 will be sent as "S12 0234"
  - 8 character are sent out for each sensor value
- On the timer thread, as time A is over the path for buffer air is closed and the route for sample gas is opened for time B
- The data flow is continued for time A and B
- As time B is over the data flow is stopped and the sample gas path is exchanged with buffer gas

Fig. 3 Algorithm for data acquisition by 8051 microcontroller

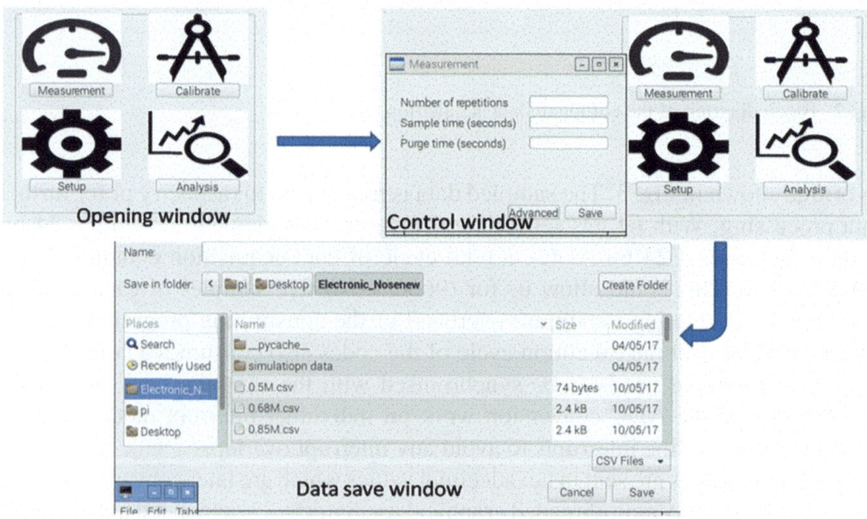

Fig. 4 Different windows for parameter setting and data storage in the GUI of the data acquisition system

- The GUI is initiated and prompts for appropriate input from users.
- User enters the series of inputs and reaches the save window for the data file
- As save button is clicked, the serial communication with 8051 unit starts in another thread
- The experiment parameters are sent to the microcontroller to set the control output
- The serial communication port waits for data from the microcontroller 8051
- The data from the 8051, for a particular experiment starts with a 'start' text from the microcontroller
- The serial data look for sensor number to mark the beginning of a sampling instance.
- The data are stored in 2-D array with 64 column for 64 sensors
- Each row has sampled data for each sensing element at a given time instant
- As per the sensor number the accompanied sensor value is stored in respective column of the current row
- For a particular column the latest value is appended below to the last value of the column
- The communication is closed with 'END' text received from the microcontroller 8051.
- As the communication is ended the array is returned back to the GUI thread and is saved as CSV file

Fig. 5 Algorithm for the data acquisition user interface

and the backend algorithm is defined by the flow chart in Fig. 5. The raspberry pi is equipped with a 7-inch capacitive touch display for user interface.

To obtain a machine-readable format of the instantaneous resistance of the sensors, a voltage divider circuit is used between the voltage supply and the ground with a fixed resistance and the sensing element. The voltage is measured in-between the sensor and the ground and the sensor would be directly proportional to the sensor resistance. In ideal case, the relationship defined between the output voltage (Vo) and the sensor resistance is given by Eq. 1.

$$Vo = Vcc.\frac{R_{\text{sensor}}}{R_{\text{sensor}} + R_{\text{bias}}} \tag{1}$$

In practice, because of a very high resistance value of the sensor elements, the voltage divider circuit will experience various loading effects such as input of ADC, connection resistances, and microcontroller pin resistances. Due to these reasons, the measured voltage has deviated from the voltage calculated from Eq. 1. To deal with this condition, authors have reportedly used relative voltage such as change in voltage with respect to base voltage. This is obtained by controlling the airflow and measure voltage at different stages of gas flow cycle in the chamber. The sequence adopted for our process is defined in the next subsection.

2.4 *Gas Cycles*

The system goes through different stages of gas flow in the chamber. The first stage involves passing of a standard gas such as zero air or sometimes more conveniently ambient air. During this phase, the sensors would be exposed to the standard gases and would provide the base value of the sensor for the experiment. The next phase involves the input of the gas or VOC to be tested into the chamber and is maintained at constant flow rate. During this time, the sensors react to the stimulus gas and generate the corresponding responses. After some time, the flow of sample gas is stopped and again the base gas is purged through the chamber. These three stages are being referred to as the gas cycle for the electronic nose. The gas pressure inside the chamber is also an essential parameter to be controlled for repeatability of the system.

3 Experimental Setup

To achieve proper stability and repeatability in the system, experiments were performed in various arrangements of micro-pumps and micro-valves. Finally, to achieve optimised immunity from the ambient effects, gas was bubbled in the chamber using arrangement shown in Fig. 6. This arrangement utilises the same flow of base air to bubble out VOCs out of sample vial. The advantage of the arrangement is the ability

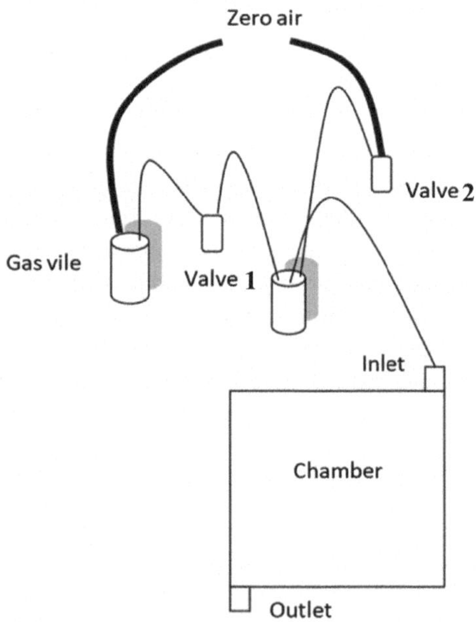

Fig. 6 Optimised gas delivery arrangement with two valves and one pump

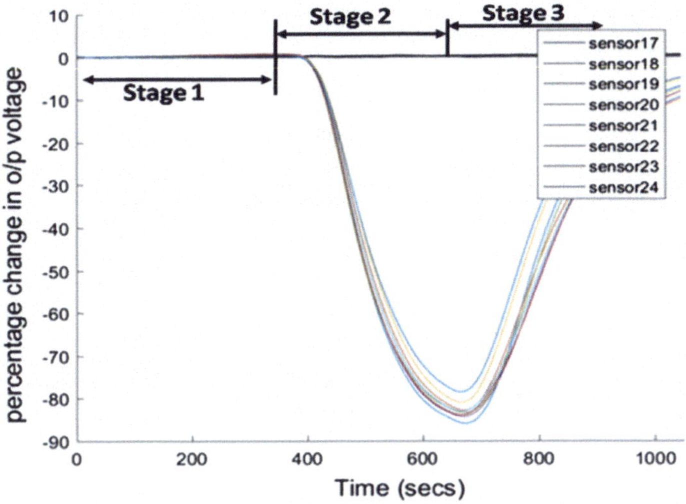

Fig. 7 Output recorded for Blank and Ethanol with various stages of the gas cycle

to eliminate the flow rate difference between the sample channel and the zero air channel. The channels are selected by alternating two valves connected in series to each channel as shown in Fig. 6.

Valve 1 stays open during **stages 1** and **3** of the gas cycle and allows zero air to pass on to chamber, while valve 2 blocks any supply of VOCs in the chamber. In **stage 2**, valve 2 is opened and valve 1 closed. During this time, the zero air pushes out the sample gas from the vial (the process is commonly known as bubbling) and passes on to the chamber via valve 2 and mixing vial. Valve 1 blocks the flow of direct zero air into the chamber. Since the gas is delivered by bubbling in both the case, the flow rate inside the chamber is maintained in all three stages of gas cycle. The different stages of the gas cycles are approximately marked in Fig. 7. The zero air is supplied via a mass flow controller (MFC) with constant pressure and temperature. The sensors are very sensitive to surrounding and process environments. Since the experiments are conducted at room temperature and the carrier gas comes from a controlled environment, the temperature for a set of experiments is assumed to be constant.

4 Result and Discussion

For the purpose of experimentation, four different gases were used in different concentrations. The gas with their concentration is listed in Table 1, along with a sample number. Henceforth the samples will be referred by their number as in Table 1.

Table 1 List of VOCs used for experimentation with their concentrations

Sample no.	VOC type	Concentration (parts of volume)
S1	Ammonia (NH_3)	1/100
S2	Ammonia (NH_3)	1/250
S3	Ammonia (NH_3)	1/500
S4	Tri-ethyleAmine	1/100
S5	Tri-ethyleAmine	1/250
S6	Tri-ethyleAmine	1/500
S7	Ethanol	1/20
S8	Ethanol	1/100
S9	Ethanol	1/500
S10	Methanol	1/300

Each sample was experimented for multiple times, to validate the results statistically. To validate the arrangement, we first check the capability of the system to differentiate between base air and VOCs. Figure 7 display the results of the same set of random sensor responses to base gas and Ethanol. We can clearly see that the sensor gives no response for base gas, whereas there is clear change in response for ethanol. This establishes the fact that the system is immune to ambient effects such as moisture and air pressure. This confirms the proper functionality of the acquisition system along with the delivery system. The system was tested with all the gases mentioned in Table 1.

To have a better understanding of the time-domain data, we calculate a few features based on the sensor response. Few time-domain features for each sensor were extracted from data set

- Peak value: the maximum peak that has been obtained for a particular experiment
- Rise time: the time required for the measured value to reach 90% percent of the peak value from 10% of the peak value
- Fall time: the time required for the measured value to reach 10% percent of the peak value from 90% of the peak value

Figure 8a, b shows the rise time and peak value of 14 different sensors for various VOC configurations. The above results show that sensors behave differently to different scenarios, but are not linearly correlated. For this, researchers opt for higher dimensional feature space for classification of complex gases. For further analysis of sensor behaviours, peak values for each sensor element were used to identify the most sensitive sensor elements to each gas type. Figure 9 shows the histogram plot of occurrence of each element in top 10 sensing elements for different gases in the experiment. The primary sensing element reduces the probability of accounting the elements showing sensitivity due to crosstalk. The histogram is plotted as columns with maximum sensitivity. Figure 9a has the histogram for tri-ethyle amine for five repetitions. As per the figure, the sensors in column 1 (sensor 1–8), column 3 (sensor

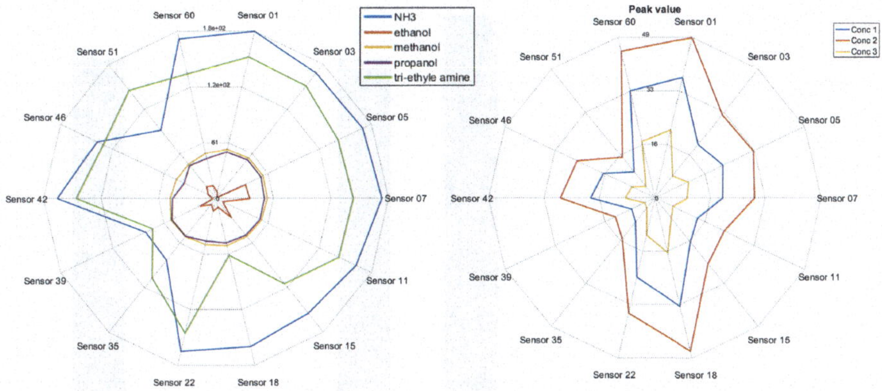

(a) Rise time (b) Peak Value

Fig. 8 Rise time of different VOCs are shown in **a** and **b** shows the difference in peak value for different concentration of Methanol

17 –24) and column 8 (sensor 57–63) were most sensitive to the Tri-ethyl amine. Figure 9b which represents the result for ethanol samples shows the response was pretty much distributed over all the columns with few exceptions. The hysteresis plot for ammonia is shown in Fig. 9c, and column 1, 3 and 8 turns out to be most sensitive. These results show that ammonia and tri-ethyl amine shares the group of sensitive sensors and differs from the response pattern of ethanol. This pattern supports the fact that both ammonia and tri-ethyl amine-containing NH groups bind with similar kinds of sensors, whereas alcohol follows different patterns. These results are subjective to types of conducting polymer used and their characteristics. The results would also depend upon the gas distribution and cycle timing. The CP sensors are also sensitive to temperature and while comparing the results of different VOCs, the temperature of the gas also plays a critical role in the value obtained. Thus for better analysis, the results must be calibrated with the temperature component.

5 Conclusion

The electronic nose proposed here is a compact device for capturing and analysing volatile organic compounds. As discussed earlier, the scope of applications for such a device is vast. The paper presents results which support the device's capability to differentiate between different types of VOCs. The device has isolated real-time data acquisition with analysis software, which adds on to the reliability of the system. The advantage of the proposed model is its capability to analyse the obtained data. This feature enhances its usability and portability. The design has been optimised

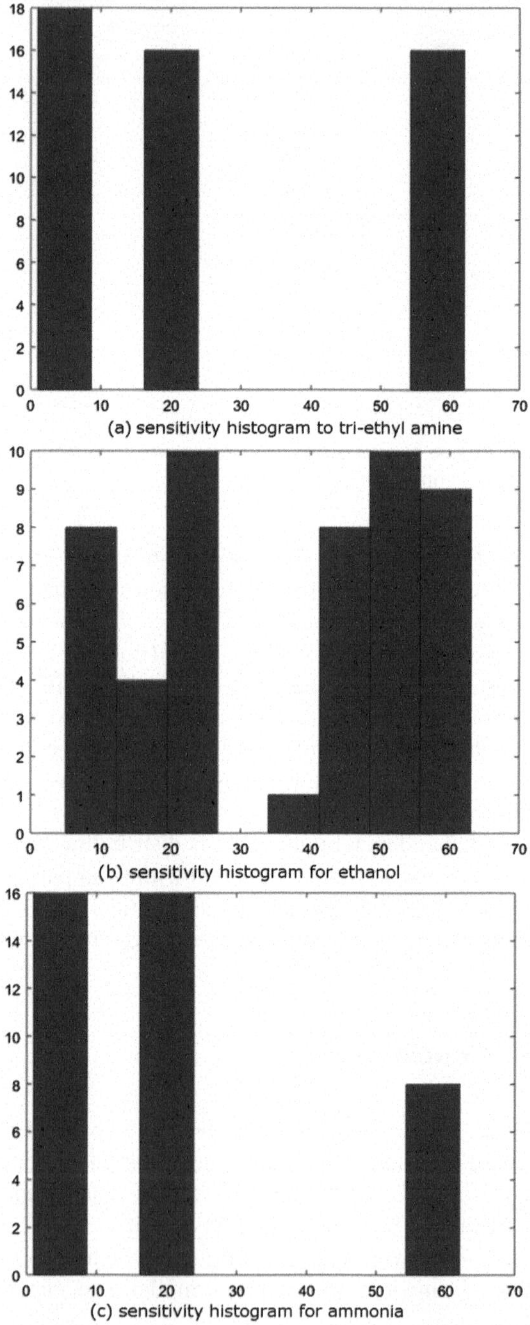

Fig. 9 Histogram of sensors reacting to respective VOCs

for improved repeatability and sensitivity by controlling airflow and compact chamber design. The device can be equipped with proper classification and calibration algorithms in future work as per any need of application in view.

References

1. Gardner, J.W., Bartlett, P.N.: A brief history of electronic noses. Sens. Actuators B **19**, 18–19 (1994)
2. Goldstein, E., Brockmole, J.: Sensation and Perception (2016)
3. Persaud, K., Dodd, G.: Analysis of discrimination mechanisms in the mammalian olfactory system using a model nose. Nature **299**(5881), 352–355 (1982)
4. Xie, C., Xiao, L., Hu, M., Bai, Z., Xia, X., Zeng, D.: Fabrication and formaldehyde gas-sensing property of ZnO-MnO2 coplanar gas sensor arrays. Sens. Actuators B Chem. **145**(1), 457–463 (2010)
5. Sadek, A.Z., Wlodarski, W., Kalantar-Zadeh, K., Baker, C., Kaner, R.B.: Doped and dedoped polyaniline nanofiber based conductometric hydrogen gas sensors. Sens. Actuators A Phys. **139**(1–2 Special issue), 53–57 (2007)
6. Ramesh Kumar, M., Ryman, S., Tareq, O., Buchanan, D.A., Freund, M.S.: Chemical diversity in electrochemically deposited conducting polymer-based sensor arrays. Sens. Actuators B Chem. **202**, 600–608 (2014)
7. Liu, X., Cheng, S., Liu, H., Hu, S., Zhang, D., Ning, H.: A survey on gas sensing technology. Sensors **12**(7), 9635–9665 (2012)
8. Benrekia, F., Attari, M., Bouhedda, M.: Gas sensors characterization and multilayer perceptron (MLP) hardware implementation for gas identification using a field programmable gate array (FPGA). Sens. (Switz.) **13**(3), 2967–2985 (2013)
9. Wongchoosuk, C., Lutz, M., Kerdcharoen, T.: Detection and classification of human body odor using an electronic nose. Sensors **9**(9), 7234–7249 (2009)
10. Lozano, J., Santos, J.P., Horrillo, M.C.: Classification of white wine aromas with an electronic nose. Talanta **67**(3), 610–616 (2005)
11. Jana, A.: Classification of Aromatic and Non-Aromatic Rice using Electronic Nose and Artificial Neural Network, pp. 291–294 (2011)
12. Santos, J.P., Lozano, J.: Real time detection of beer defects with a hand held electronic nose. In: Proceedings of 2015 10th Spanish Conference on Electron Devices, CDE 2015, pp. 1–4 (2015)
13. Prajapati, C.S., Soman, R., Rudraswamy, S.B., Nayak, M., Bhat, N.: Single chip gas sensor array for air quality monitoring. J. Microelectromech. Syst. **26**(2), 433–439 (2017)
14. Bai, H., Shi, G.: Gas sensors based on conducting polymers. Sensors **7**(3), 267–307 (2007)
15. Ha, S.C., et al.: Integrated and microheater embedded gas sensor array based on the polymer composites dispensed in micromachined wells. Sens. Actuators B Chem. **105**(2), 549–555 (2005)
16. Fine, G.F., Cavanagh, L.M., Afonja, A., Binions, R.: Metal oxide semi-conductor gas sensors in environmental monitoring. Sensors **10**(6), 5469–5502 (2010)
17. Barsan, N., Koziej, D., Weimar, U.: Metal oxide-based gas sensor research: How to?. Sens. Actuators B: Chem. **121**(1),18–35 (2007)
18. Janata, J., Josowicz, M.: Conducting polymers in electronic chemical sensors. Nat. Mater. **2**(1), 19–24 (2003)
19. Guadarrama, A., Rodríguez-Méndez, M.L., De Saja, J.A., Ríos, J.L., Olías, J.M.: Array of sensors based on conducting polymers for the quality control of the aroma of the virgin olive oil. Sens. Actuators B Chem. **69**(3), 276–282 (2000)

20. Sharma, S., Hussain, S., Singh, S., Islam, S.S.: MWCNT-conducting polymer composite based ammonia gas sensors: A new approach for complete recovery process. Sens. Actuators B Chem. **194**, 213–219 (2014)
21. Wang, Y., Hu, Y., Wang, D., Yu, K., Wang, L.: The analysis of volatile organic compounds biomarkers for lung cancer in exhaled breath, tissues and cell lines. Cancer Biomark. **11**, 3233 (2012)
22. Agarwal, A., Sampath, H., Yelamanchili, V., Vemuri, R.: Accurate estimation of parasitic capacitances in analog circuits. In: Proceedings Design, Automation and Test in Europe Conference and Exhibition, pp. 1364–1365
23. Chowdhury, S. R., Bhondekar, A. P., Kumar, R., Bagchi, S., Kaur, R., Karar, V.: Circuit arrangement to suppress crosstalk in chemo-resistive sensor arrays. IET Sci. Measur. Tech. 12(8), 1039–1046 (2018)
24. Morsi, I.: Electronic nose system and artificial intelligent techniques for gases identification. In: Data Storage, pp. 175–200 (2010)
25. Chueh, H.T., Hatfield, J.V.: A real-time data acquisition system for a hand-held electronic nose (H2EN). Sens. Actuators B Chem. **83**(1–3), 262–269 (2002)
26. Tang, K.-T., Chiu, S.-W., Pan, C.-H., Hsieh, H.-Y., Liang, Y.-S., Liu, S.-C.: Development of a portable electronic nose system for the detection and classification of fruity odors. Sensors **10**(10), 9179–9193 (2010)
27. El Barbri, N., Llobet, E., El Bari, N., Correig, X., Bouchikhi, B.: Electronic nose based on metal oxide semiconductor sensors as an alternative technique for the spoilage classification of red meat. Sensors **8**(1), 142–156 (2008)
28. Frey, U., Graf, M., Taschini, S., Kirstein, K.U., Hierlemann, A.: Digital systems architecture to accommodate wide range resistance changes of metal-oxide sensors. In: Proceedings of IEEE Sensors, pp. 593–595 (2008)

On the Implementation and Performance Evaluation of Security Algorithms for Healthcare

Sohail Saif and Suparna Biswas

Abstract With rapid and successful advancement in communication technology and semiconductor devices Internet of Things based smart healthcare is not a concept, but reality. Patients' vital information sensed by sensing devices is transmitted through heterogeneous communication links to fog/cloud layer for processing, storage, and decision-making or knowledge extraction. Sensitive health data is vulnerable to various security attacks and threats of different levels and impact while getting sensed, are saved or getting transmitted. Off-the-shelf security techniques such as encryption, trust can be applied for privacy and security of data but application-specific demands such as latency/end to end delay in data transmission, additional overhead in terms of computation time, etc., must be taken care of, especially in time-critical healthcare application. This work implements encryption-decryption algorithms and hashing techniques on ECG signals of varying size for secure transmission through communication layer to cloud layer for performance evaluation based on computation time (as in healthcare maximum permissible end to end delay is 250 ms) to ensure Confidentiality, Integrity, and Authentication.

Keywords Healthcare · ECG · Encryption · Decryption · Hashing etc.

1 Introduction

In recent times researchers are getting attracted towards the developments of smart applications such as smart healthcare, smart transportation, smart surveillance, smart energy systems [1], etc. These smart systems and applications are inevitable to build a smart city. And smart city is now not a concept, but reality with the success of the Internet of Things, Machine Learning, Big Data, Artificial Intelligence, Security, and

S. Saif (✉) · S. Biswas
Department of Computer Science & Engineering, Maulana Abul Kalam Azad University of Technology, Kolkata, WB, India
e-mail: sohailsaif7@gmail.com

S. Biswas
e-mail: mailtosuparna@gmail.com

S. Kundu et al. (eds.), *Proceedings of the 2nd International Conference on Communication, Devices and Computing*, Lecture Notes in Electrical Engineering 602, https://doi.org/10.1007/978-981-15-0829-5_59

Trust. Smart applications have several layers such as sensing layer, communication layer, processing and storage layer, and control and feedback layer [2]. In sensing layer, heterogeneous sensing devices are to sense data of varying sizes and at varying rates. Data type also may vary such as some data may be simple and numeric, some may be complex and dynamic signals. Biological sensing devices like temperature, pressure, blood pressure, sugar or pulse rate sense numeric vital signals, value of the signals may vary time to time and person to person. Biological sensors such as ECG, EMG, EEG, etc., sense complex dynamic signals. Accelerometer, gyroscope, etc., sense vital data related to person's movement, etc. These sensors are physically unprotected mostly wearable by the human body and due to unawareness of the user, legitimate sensors may be replaced by illegitimate nodes [3] or may come under control of adversary which may become inside attacker, difficult to predict and detect. This attack probability is at sensing layer. The next layer, i.e., the communication layer is comprised of heterogeneous communication links such as wired or wireless. Various link types have different transmission bandwidth and varying rate of transmission. WiFi, Bluetooth, Zigbee, WiMax, Cellular or Adhoc communication links suffer from several security attacks and threats such as man-in-the-middle attack, sinkhole attack, wormhole attack, replay attack, routing attack, denial-of-service attack, etc. [4]. Overall, attacks may lead to information leakage, modification, impersonation, unauthorized data access, non-repudiation, nonavailability, etc. So, data at transmission through communication layer must be protected and secured. Vital data goes ultimately to cloud for storage, further analysis and information extraction for knowledge building. Though for data processing at cloud in any system, mostly the underlying security system of cloud is trusted with, but there is security breach which must be additionally tackled. For example, at cloud source of data received and its integrity to be verified, data may be saved in encrypted format at storage to prevent unauthorized access. Now to be build up smart systems, knowledge or information extraction at processing layer will be input to the control layer. This control layer will send feedback to sensing layer if any adjustment, change or modification to be done in physical system. Security of this controlled feedback is utmost important as if this information is wrongly modified, physical adjustment will be based on this and ultimately the purpose of the whole smart system will be jeopardized. So, Internet of things based smart system must be secure against several attacks and threats that are inevitable in such systems due to its architecture comprising of several layers where each layer having heterogeneity, resource limitations, energy challenges, etc. Following Fig. 1 is the three-layer architecture for IoT based secure healthcare.

The rest of the work is organized as follows: Sect. 2 describes literature survey, Sect. 3 illustrates motivation and unique contribution of proposed work, Sect. 4 presents experimental results and whole work is concluded in Sect. 5.

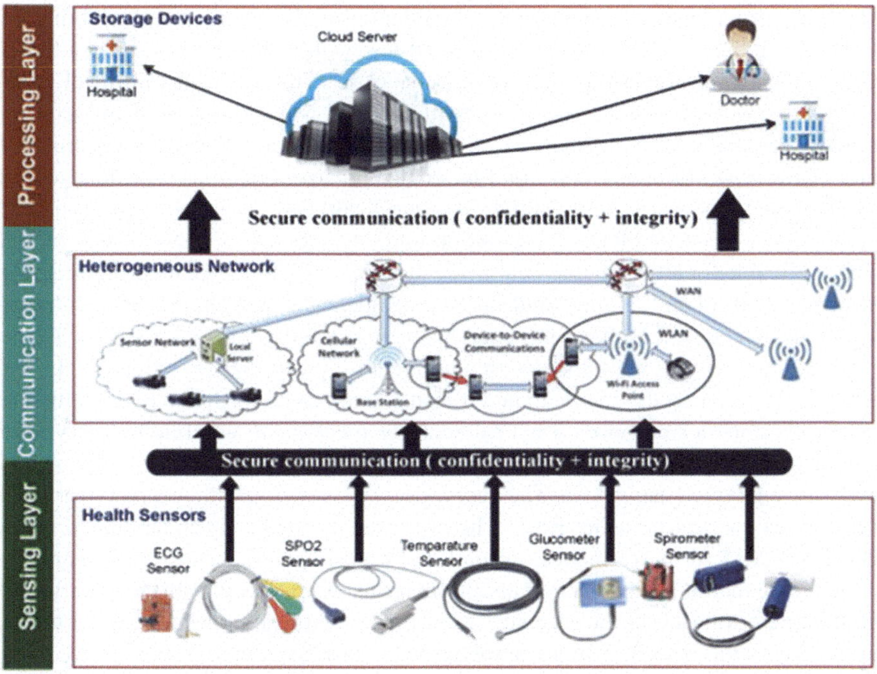

Fig. 1 Three-layer secure architecture

2 Literature Survey

In [5], the authors consider several cryptography techniques such as DES, 3DES, AES, RSA, and blowfish techniques for cost-performance trade-off analysis. This implementation based performance evaluation is purposefully done to select application-specific appropriate cryptography techniques. They have used both text and image files as input to cryptography algorithms. Based on the specific parameters, specific algorithm outperforms the other, e.g., in terms of encryption and decryption time, blowfish performs best, RSA performs worst for same file size. Other performance parameters based on which performance evaluation has been done are memory used, avalanche effect, entropy, etc. In [6], authors have investigated security attacks and threats in IoT based healthcare system. They have highlighted a password guessing attack that can cause confidentiality loss at deice level, i.e., sensing layer. To protect system against such attack, they have used personal information for password building. Several password creation mechanisms have been discussed and then a password strength evaluation technique has been proposed. The implementation method and results rank selected passwords based on the difficulty in password guessing following different methods. In [7], authors has proposed an IoT and cloud based e-health system also they have exploited various vulnerabilities at different levels in IoT Devices and cloud. They have defined security requirements of e-health system in terms of three components: confidentiality, integrity, and availability. A new secure IoT and cloud-based architecture for e-healthcare are proposed to address device or hardware-level security, network-level security, and software-level security. But this work does not provide any implementation or simulation of any security technique or attack for performance evaluation purposes. In [8], an extensive study of state-of-the artworks on IoT and cloud-based healthcare models has been done to find out their strengths, weaknesses, suitability for e-healthcare along with exploration of security and privacy challenges. This work uniquely tries to propose a generic model to be applicable for all IoT based healthcare system. Also they have identified each of the components of an IoT and cloud-based e-healthcare system and security vulnerabilities at each point. Emphasis has been given on suitability of LPWAN for IoT. The proposed model is not implemented or simulated for performance evaluation. In [9], authors propose a secure medical data transmission model for IoT based healthcare system. The model proposes a hybrid security mechanism combining encryption scheme using AES and RSA techniques and steganography. Input data both text and image are encrypted using AES. The decryption key is shared with receiver over open channels by encrypting using RSA. The proposed model has been simulated using MATLAB 2015 for performance evaluation. Proposed work ensures confidentiality of the image maintains the desired quality in terms of imperceptibility, capacity, and minimal deterioration in the received stego-image. Table 1 shows the recent works security techniques for IoT based applications in last few years.

2.1 Threat Model

Attacks mainly considered here are active and passive attacks as these attacks change the context of vital privacy requirements of sensitive health data, respectively. The passive attacks considered here are: (i) Ciphertext only attack (COA) (ii) Known plain text attack (KPA) (iii) Chosen plain text attack (CPA) (iv) Brute Force attack (BFA) (v) packet sniffing, (vi) Eavesdropping, etc. These kind of attacks mainly observes the data which are being transmitted in different levels such as at sensing layer, communication layer or at processing layer.

Active attacks such as man-in-the-middle attack, rushing attack are considered for the probability of alteration of sensitive health information. Proposed work implements and analyzes performance of security techniques that can protect a system against these attacks as it may prove to be fatal causing patients' life.

3 Contribution of the Work

Following are the contribution of this proposed work:

Table 1 Comparative literature survey

Authors, year	Security aspects					Implementation
	Confidentiality	Authentication	Integrity	Access control	Trust	
Yin et al. [10], 2017	✓	✓	×	✓	×	×
Al Hamid et al. [11], 2017	✓	✓	×	×	×	✓
Alrawis et al. [12], 2017	✓	×	×	×	×	✓
Diro et al. [13], 2018	✓	×	×	×	×	✓
Zheng et al. [14], 2017	✓	✓	×	×	×	✓
Abdallah et al. [15], 2017	✓	×	✓	✓	✓	×
Chen et al. [16], 2016	✓	✓	✓	×	×	✓
Kao et al. [17], 2013	✓	✓	×	×	×	×
Shen et al. [18], 2017	✓	×	✓	×	×	✓
Kim et al. [19], 2017	×	✓	×	✓	✓	×

- Through this work, we have experimentally analyzed the time required for encryption and decryption for medical data, also we have analyzed the time required for the generation of hash and its verification.
- Different sizes of ECG Datasets are used for the experiments. ECG-ID database has been collected from Physiobank Database [20].
- Well-known symmetric algorithms AES, DES, 3-DES, Blowfish, and CAST are used for encryption which ensures confidentiality.
- ECB, CBC, and CTR modes are used for performance comparison.
- MD5, SHA-1, SHA-256, SHA-384, and SHA-512 are used in Hash Based message authentication technique for ensuring integrity of the data.
- All the above mentioned symmetric algorithms and hashing algorithms are implemented in PHP and simulated in cloud server.
- We have computed and compared the time required for using the above mentioned encryption and hashing algorithms.

4 Proposed Work and Result Analysis

Health data of the human body is sensitive in nature and it's vulnerable to various security attacks when transferred through an insecure channel. The adversary can apply passive attacks like Cipher text only attack (COA), Known plain text attack (KPA), Chosen plain text attack (CPA), and Brute Force attack (BFA) if the encryption algorithm is not strong enough. There are some other common active attacks on health data like man-in-the-middle, eavesdropping where adversary can alter the data, to prevent this type of attack we need some strong Hash Based Message Authentication Code (HMAC) techniques. In this work, we have implemented well known popular symmetric encryption and hashing algorithms in PHP, we have taken ECG data of human body as input message in the algorithms. Phpcrypt library is used for the implementation. It is an open-source library file that contains all symmetric cryptographic algorithms, this library can be used for key generation, encryption, decryption, hashing, and hash verification. For computation of time delay we have used different ECG data of different duration, if duration is increased then data size will be increased, for experiments we have used ECG for 10 s, 30 s, 1 min, 3 min, and 5 min. Since in medical application there is a need for continuous monitoring of health parameters like ECG, so we have chosen varying duration of ECG data. These datasets are primarily stored in local PC. For simulation we have designed a PHP based web application, this application will run on cloud hosting. First the web application reads the ECG data from local PC in plain text format, then it encrypts the data using randomly generated key and encrypted data is decrypted into plain text. We have recorded the total time for each algorithm for varying duration of ECG data. Data sizes for the ECG duration of 10 s, 30 s, 1 min, 3 min, and 5 min are 30 kb, 87 kb, 172 kb, 518 kb, and 988 kb, respectively. End to end time is the only parameter based on which the performance comparison is done because, in healthcare, maximum permissible end to end delay in data transmission is 250 ms [21]. For integrity verification Hash Based Message Authentication Code (HMAC) is used, here random key is generated and then well-known hashing algorithms are used to generate hash of the ECG data. Similarly, varying size of ECG data is used as input here, we have recorded the time required for each algorithm and compared with each other.

4.1 Simulation Setup

Here we have used a cloud server on pay-as-you go basis, hardware and software specifications of that cloud server is shown in Table 2. Encryption and HMAC algorithms settings are shown in Table 3 and Table 4 respectively.

Table 2 Simulation environment of cloud server

Description	Value/name
Server location	Chicago, Illinois, United States (US)
Server name	Intel E3-1220LV2
CPU	2.50 GHZ Quad core
RAM	4 GB
Storage	1000 GB HDD in RAID 1
Bandwidth	5 TB/month
Apache version	2.4.27
PHP version	7.2.3
MYSQL version	5.7.19
Architecture	X84_64
OS	Linux

Table 3 Encryption algorithm settings

Algorithm	Block size (bits)	Key size (bits)
AES	128	128
DES	64	64
3-DES	64	192
BLOWFISHS	64	64
CAST	64	128

Table 4 HMAC algorithm settings

Algorithm	Digest size (bits)	Key size (bits)
MD5	128	128
SHA-1	160	128
SHA-256	256	128
SHA-384	384	128
SHA-512	512	128

4.2 Simulation Result

We have simulated the encryption and decryption process in four phases with varying data sizes. In the first phase, all five cryptographic encryption algorithms are executed in ECB (Electronic codebook) mode, results are shown in Table 5 and graphically plotted in Fig. 2. We can see the total time required by AES, DES, 3-DES, BLOWFISH, and CAST lies between 1.14–27.31 ms, 2.94–74.53 ms, 10.9–280.81 ms,

Table 5 Time required (ms) in ECB mode

Data size (KB)	AES	DES	3-DES	BLOWFISH	CAST
30	1.14	2.94	10.90	0.19	0.39
87	3.04	8.28	29.03	0.47	1.15
172	5.98	14.55	52.83	1.16	2.35
518	15.29	37.45	148.77	4.41	9.15
988	27.31	74.53	280.81	8.45	16.29

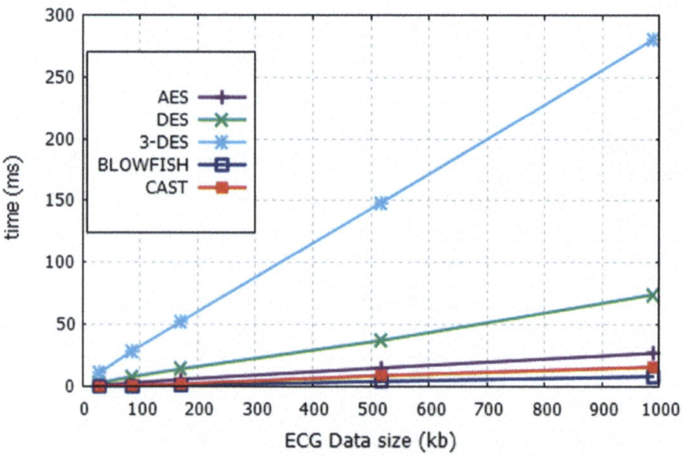

Fig. 2 Total time required by different symmetric algorithms in ECB mode

0.19–8.45 ms, and 0.39–16.29 ms, respectively, for ECB mode. Similarly in the second phase, all the above-mentioned algorithms are executed in Cipher Block chaining (CBC) mode. Figure 3 reflects the required time that lies in between 1.16–32.36 ms, 2.95–75.33 ms, 10.92–290.14 ms, 0.20–9.45 ms, and 0.40–17.29 ms in CBC mode. Here we can see that the required time is more than ECB mode because the processing time of CBC is higher due of its chaining nature, average time difference between two modes is 1.23 ms, which is relatively low. In the third phase we have chosen Counter (CTR) mode, the required time for encryption and decryption of data is graphically plotted in Fig. 4 and detail results are shown in Fig. 5, here we can

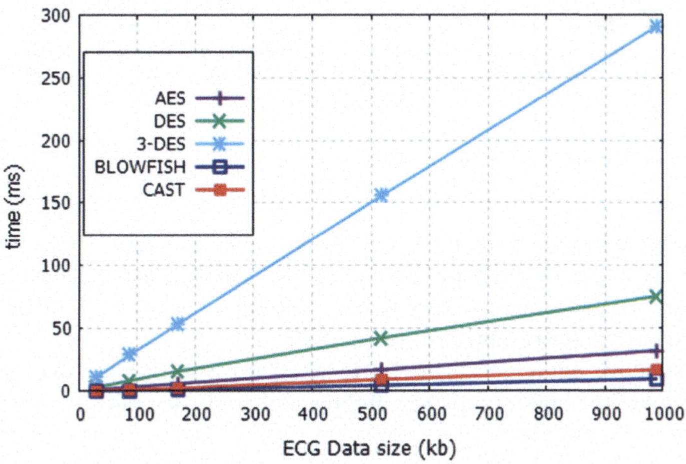

Fig. 3 Total time required by different symmetric algorithms in CBC mode

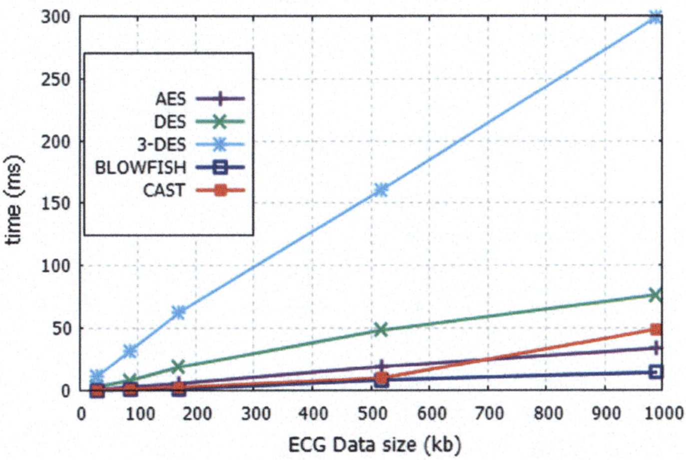

Fig. 4 Total time required by different symmetric algorithms in CTR mode

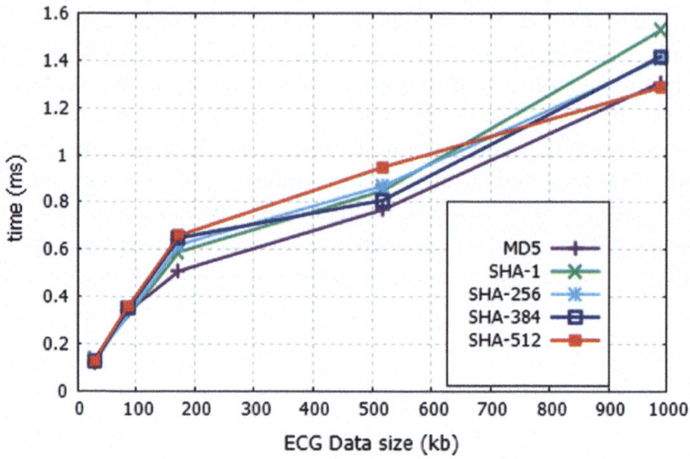

Fig. 5 Total time required by different hashing algorithms

see the total time lies in between 1.18–34.06 ms, 2.96–76.53 ms, 10.99–298.81 ms, 0.25–14.45 ms, and 0.45–49.29 ms. Because of the complex design of the CTR mode, this mode takes more time than other two modes, but security of this mode is higher than compared to other two modes (Tables 6 and 7).

From the above figures, we can clearly see that Blowfish takes the lowest time and 3-DES takes the highest time for all three cipher block modes.

Table 6 Time required (ms) in CBC mode

Data size (KB)	AES	DES	3-DES	BLOWFISH	CAST
30	1.16	2.95	10.92	0.20	0.40
87	3.09	8.29	29.10	0.48	1.18
172	6.08	16.19	53.83	1.22	2.37
518	17.29	42.45	155.77	4.49	9.18
988	32.36	75.33	290.14	9.45	17.29

Table 7 Time required (ms) in CTR mode

Data size (KB)	AES	DES	3-DES	BLOWFISH	CAST
30	1.18	2.96	10.99	0.25	0.45
87	3.20	8.47	32.41	0.55	1.59
172	5.92	19.02	62.83	1.26	2.74
518	19.29	48.25	160.77	8.49	10.15
988	34.06	76.53	298.81	14.45	49.29

Table 8 Time required (ms) in by different hashing algorithms

Data size (KB)	MD5	SHA-1	SHA-2 56	SHA-384	SHA-512
30	0.12	0.14	0.13	0.13	0.13
87	0.34	0.40	0.36	0.35	0.36
172	0.51	0.59	0.62	0.65	0.66
518	0.77	0.85	0.87	0.81	0.95
988	1.31	1.53	1.41	1.42	1.29

In the last phase of the experiment we have executed different hashing algorithms, here we have also used varying sizes of ECG data as input. Figure 5 shows the total time required using MD5, SHA-1, SHA-256, SHA-384, and SHA-512 lies between 0.12–1.31 ms, 0.14–1.53 ms, 0.13–1.41 ms, 0.13–1.42 ms, and 0.13–1.29 ms, respectively. From Table 8 we can see that MD5 takes the lowest time than other techniques.

So, the above performance evaluation gives us direction towards transmission of block of vital health signal (ECG signal is taken here) ensuring confidentiality using BLOWFISH and integrity using MD5. Implementation results show that BLOWFISH takes lowest time to encrypt and decrypt same size input in comparison to AES, DES, 3-DES, and CAST algorithms. Among MD5, SHA-1, SHA-256, SHA-384, and SHA-512 techniques for hash generation, MD5 takes the lowest time.

5 Conclusion

In this work, an implementation based performance evaluation of different encryption-decryption techniques, AES, DES, 3-DES, BLOWFISH, and CAST and hashing algorithms, MD5, SHA-1, SHA-256, SHA-384, and SHA-512 has been done. Here healthcare is considered as a smart application of IoT and ECG data from standard data base is used as input health data to all the algorithms. Experimental results states that BLOWFISH and MD5 outperforms the other encryption and Hash algorithms in terms of computation time. Unlike many other works, this paper does performance evaluation implementation of several security algorithms at cloud instead of just analytical comparison to give direction to choose a more appropriate algorithm while confidentiality and integrity of input data are important in a real time application like healthcare.

References

1. Zanella, A., Bui, N., Castellani, A., Vangelista, L., Zorzi, M.: Internet of things for smart cities. IEEE Internet Things J. **1**(1), 22–32 (2014)
2. Zhang, K., Ni, J., Yang, K., Liang, X., Ren, J., Shen, X.S.: Security and privacy in smart city applications: challenges and solutions. IEEE Commun. Mag. **55**(1), 122–129 (2017)
3. Singh, M.M., Singh, A., Mandal, J.K.: Preventing node replication attack in static wireless sensor netwroks. In: Proceedings of 3rd International Conference on Reliability, Infocom Technologies and Optimization, Noida, pp. 1–5 (2014)
4. Saif, S., Gupta, R., Biswas, S.: Implementation of cloud-assisted secure data transmission in WBAN for healthcare monitoring. In: Bhattacharyya, S., Chaki, N., Konar, D., Chakraborty, U., Singh, C. (eds.) Advanced Computational and Communication Paradigms. Advances in Intelligent Systems and Computing, vol. 706. Springer, Singapore (2018)
5. Patil, P., Narayankar, P., Narayan, D., Meena, S.: A comprehensive evaluation of cryptographic algorithms: DES, 3DES, AES, RSA and Blowfish. Procedia Comput. Sci. **78**, 617–624 (2016)
6. He, D., Ye, R., Chan, S., Guizani, M., Xu, Y.: Privacy in the internet of things for smart healthcare. IEEE Commun. Mag. (2018)
7. Ben Ida, I., Jemai, A., Loukil, A.: A survey on security of IoT in the context of eHealth and clouds. In: IEEE 2016 11th International Design and Test Symposium (IDT) (2016)
8. Baker, S.B., Xiang, W., Atkins, I.: Internet of things for smart healthcare: technologies, challenges, and opportunities. IEEE Access **5**, 26521–26544 (2017). Electronic ISSN: 2169-3536
9. Elhoseny, M., Ramírez-González, G., Abu-Elnasr, O.M., Shawkat, S.A., Arunkumar, N., Farouk, A.: Secure medical data transmission model for IoT-based healthcare systems. IEEE Access **6**, 20596–20608 (2018)
10. Yin, L., Zhang, A., Ye, X., Wang, L.: Security-aware attribute-based access control for fog-based eldercare system. In: 2017 3rd IEEE International Conference on Computer and Communications (ICCC), Chengdu, pp. 2680–2684 (2017)
11. Al Hamid, H.A., Rahman, S.M.M., Hossain, M.S., Almogren, A., Alamri, A.: A security model for preserving the privacy of medical big data in a healthcare cloud using a fog computing facility with pairing-based cryptography. IEEE Access **5**, 22313–22328 (2017)
12. Alrawais, A., Alhothaily, A., Hu, C., Xing, X., Cheng, X.: An attribute-based encryption scheme to secure fog communications. IEEE Access **5**, 9131–9138 (2017)
13. Diro, A.A., Chilamkurti, N., Nam, Y.: Analysis of lightweight encryption scheme for fog-to-things communication. IEEE Access **6**, 26820–26830 (2018)
14. Zheng, Q., Wang, X., Khurram Khan, M., Zhang, W., Gupta, B.B., Guo, W.: A lightweight authenticated encryption scheme based on chaotic SCML for railway cloud service. IEEE Access **6**, 711–722 (2018)
15. Abdallah, E.G., Zulkernine, M., Gu, Y.X., Liem, C.: TRUST-CAP: a trust model for cloud-based applications. In: 2017 IEEE 41st Annual Computer Software and Applications Conference (COMPSAC), Turin, pp. 584–589 (2017)
16. Chen, R., Mu, Y., Yang, G., Gao, F., Huang, X., Wang, X., Wang, Y.: Server-aided public key encryption with keyword search. IEEE Trans. Inf. Forensics Secur. **11**(12), 2833–2842 (2016)
17. Kao, Y.W., Huang, K.Y., Gu, H.Z., Yuan, S.M.: uCloud: a user-centric key management scheme for cloud data protection. IET Inf. Secur. **7**(2), 144–154 (2013)
18. Shen, J., Liu, D., He, D., Huang, X., Xiang, Y.: Algebraic signatures-based data integrity auditing for efficient data dynamics in cloud computing. IEEE Trans. Sustain. Comput. (2017)
19. Kim, H., Lee, E.A.: Authentication and authorization for the internet of things. IT Professional **19**(5), 27–33 (2017)
20. Link: https://physionet.org/physiobank/database/ecgiddb/
21. Movassaghi, S., Abolhasan, M., Lipman, J., Smith, D., Jamalipour, A.: Wireless body area networks: a survey. IEEE Commun. Surv. Tutor. **16**(3), 1658–1686 (2014)

Smart Garbage Monitoring and Alert System Using IoT

Jayeeta Das, Ankita Pramanik and Susanta Kumar Parui

Abstract With the overcrowding of cities, the clearance of garbage has become the most challenging task for civic bodies. Thus, a well-planned garbage monitoring system is essential to keep the environment clean. Keeping in mind the above scenario, this paper proposes a smart garbage monitoring and alert system using the concept based on the Internet of Things (IoT) to maintain cleanliness and hygiene in any locality. The system, proposed here can monitor the solid or liquid waste level in garbage bin through sensors attached to it. An Android application is also developed for accessing information such as coordinates, temperature, humidity, and methane gas value of the bin. Through monitoring of these values, the clearance of garbage can be done on a need basis. As the proposed system, is an internet-based system, Wi-Fi connection is essential for its implementation. The developed system uses IFTTT protocol to send notification in a time-efficient fashion.

Keywords IoT · Garbage monitoring · Ultrasonic sensor · Gas sensor · Wi-Fi

1 Introduction

Internet of Things (IoT) is the latest revolution of the internet which helps to communicate among billions of internet-connected objects through some agreed protocols [1]. It can identify, locate, track, monitor, and manage the various real-world objects effectively. IoT has wide applications in varied domains like healthcare, smart cities, smart hospitals; smart environment, industrial and agricultural monitoring [2]. The perception of the world is changing with the rapid increase of IoT connected devices

J. Das
Tata Consultancy Services, Kolkata, West Bengal, India
e-mail: jayeetadasat1993@gmail.com

A. Pramanik (✉) · S. K. Parui
Indian Institute of Engineering Science and Technology, Shibpur, West Bengal, India
e-mail: ankita@telecom.iiests.ac.in

S. K. Parui
e-mail: susanta_p@telecom.iiests.ac.in

S. Kundu et al. (eds.), *Proceedings of the 2nd International Conference on Communication, Devices and Computing*, Lecture Notes in Electrical Engineering 602, https://doi.org/10.1007/978-981-15-0829-5_60

[3]. The paper [4] provides brief overview of latest technologies and sensors used in different applications of IoT. Some of the sensors used in IoT are proximity sensors, infra-red (IR) sensor, gas sensor, temperature sensor, accelerometer sensor, etc. IoT can be efficiently integrated into various applications to bring improvement in the quality of life.

Nowadays the general awareness towards the improvement of cleanliness level is increasing in our country. The government has also initiated various movements to increase the cleanliness of the surroundings. But the problem of garbage disposal is an important challenge for civic bodies. Overflowing, stinky bins lead to health hazards. IoT can be efficiently used to develop a proper waste monitoring system that can be placed in homes, schools, colleges, offices, shopping malls, and roads so that the vision of maintaining smart cities [5] and forming better environment for living becomes true. It brings new direction to the traditional wireless waste management system. It takes a step towards "Swachh Bharat Abhiyaan" for cleanliness [6].

For the smart city vision, the present work proposes an IoT based smart garbage monitoring and alert system which will notify the cleaning authority to empty the bin on time. It uses the features of the Arduino microcontroller and Wi-Fi module, DHT11, Ultrasonic, and MQ4 Methane (CH4) gas sensors. Garbage bins located in different areas of a city can be constantly monitored by placing distance sensors on top of each bin, thus giving us readings of the garbage level accordingly. When a dustbin reaches maximum capacity, the garbage monitoring authorities are informed to take immediate action. Moreover, the android app can be used to check the status of the bin anytime. Additionally temperature and humidity sensor and gas sensors are also attached to our system for monitoring atmospheric conditions and detection of harmful gas emission.

IFTTT (If This Then That) is a free web-based service to create applets. Applets are chains of conditional statements. Changes within web services such as Gmail, Twitter, Facebook, etc., trigger applets [7]. Triggers are the "this" part of an applet and Actions are the "that" part of an applet that results from the input of the trigger. Within next few years majority of IoT devices will be connected to IFTTT.

The rest of the paper is organized as follows: Sect. 2 gives a brief literature survey. Section 3 provides the proposed architecture along with the details of the hardware and the developed software application. Section 4 provides the results and discussions part and Sect. 5 gives the conclusion of the project.

2 Literature Survey

This section surveys the related previous works on garbage monitoring and waste management systems. IoT based garbage management system has become a significant and growing concept to establish a clean and green smart city. The work in [8] proposes a system for garbage clearance which is able to send an alert message to the municipal authority with proper verification when garbage is filled. This system uses an ultrasonic sensor attached to Arduino UNO to monitor the garbage level. The drivers confirm the cleanliness of the dustbins with the aid of RFID Tag. Users get notification in their android application based mobiles.

A wireless dustbin monitoring and alert system using Arduino is proposed in [9]. It can send an alert message to the server about the filling of the garbage or the emission of harmful gases. Each dustbin is connected with GPS which helps to monitor the location and status of the bins.

In [10] the authors propose IOT based smart garbage monitoring and clearance system using sensor systems to check the level of garbage in the dustbins. The system uses GSM/GPRS to know the position of the dustbins. A microcontroller is used to maintain an interface between GSM/GPRS and the sensor system. An android application to monitor the dustbin at different locations is also developed.

The paper [11] introduces a smart solid waste management method to overcome the growing problem of solid waste disposal and maintain a green environment. The system consists of renesas microcontroller, force sensor, passive infra-red sensor, ultrasonic sensor, LCD, GPRS, GPS, Amazon cloud-based web server, and mobile application. The web server updates the status of the garbage bins. PIR sensor is able to detect human being and opens the lid automatically. The system can be used for dry solid waste and wet waste separately. Daily clearance and proper disposal are also necessary to maintain hygiene and cleanliness.

The work in [12] implemented a smart dustbin using AVR family microcontroller and Wi-Fi module for clean and safe public space. Both authorities and common users can access the uploaded data on the clouds to monitor the garbage bins in real time. The system is user-friendly and cost-effective in terms of keeping the surroundings clean and safe.

2.1 *Contribution of the Paper*

Most of the existing works have made use of Arduino Nano or Uno or Raspberry pie along with ESP8266 Wi-Fi module. The usage of these components increases the system cost. The present work proposes a smart garbage monitoring system at a lower cost. The contributions of the present work can be briefly summarized as follows:

- The proposed system ESP8266 NodeMCU board with inbuilt Arduino microcontroller features. Thus, a single board makes the proposed system cheaper, lightweight, and reliable compared to the existing works [8–10].
- An android application is developed to help in monitoring the system easily, situated at any place, over the internet in real time.
- IFTTT (If This Then That) protocol is used here for notification purposes when the dustbin is full. Thus, small applets that can be easily triggered by the changes are created.
- Internet facility is used here for notification purpose rather than the usage of heavy GSM modules [13].

- The proposed system has an additional feature of weather monitoring with its temperature, humidity and gas sensors.
- The system detects harmful gases when it reaches the threshold value.

Overall the proposed system is very effective for usage in residential, commercial, industrial or municipality places.

3 Proposed Architecture

The proposed system includes various modules like NodeMCU ESP8266 Wi-Fi module, DHT-11 sensor, HC-SR04 Ultrasonic sensor, and MQ4—Methane natural gas sensor. The open-source IoT platform NodeMCU is a self-contained Wi-Fi with the features of microcontroller. DHT11 provides calibrated output of temperature and humidity. Dry and wet waste can be differentiated depending upon the output temperature and humidity value. It strengthens the monitoring purpose of the developed system. The distance between sensor and the garbage level is measured by ultrasonic sensor. MQ4 gas sensor is able to detect the traces of methane (CH_4) and natural gas in the sensitivity range from 300 to 10000 ppm. Any other gas sensor like MQ2 smoke gas detector, MQ5 natural gas detector, MQ135 air quality detector, MQ-7 carbon monoxide gas detector, etc., can be used as per the requirements. The Block diagram is represented below in Fig. 1.

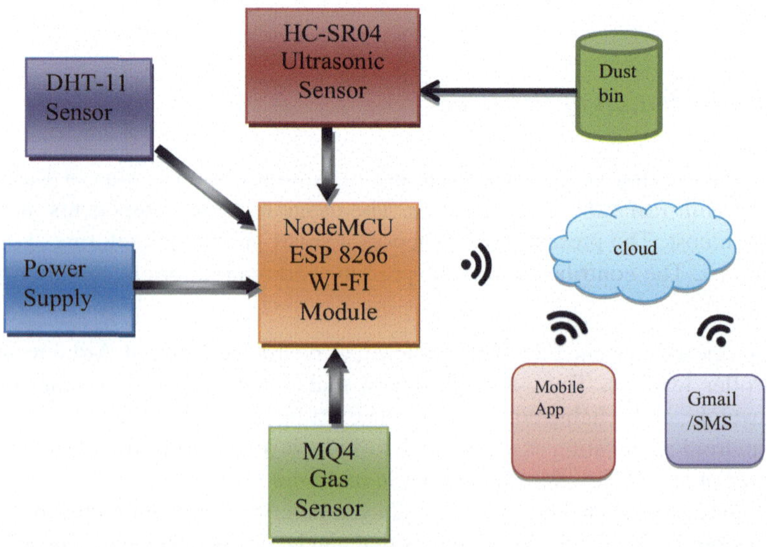

Fig. 1 Block diagram of the proposed system

Fig. 2 IFTTT with NodeMCU Gmail and SMS notification

An Android application is developed to access real time information. The application is to be logged in with the username and password. This prevents any unauthorized use of sensitive information. When the garbage bin is full, SMS or email is sent to the garbage monitoring system authority by the NodeMCU using the IFTTT protocol. Figure 2 shows the integration of IFTTT with NodeMCU so that an SMS or Email can be sent to the garbage monitoring system authority when the garbage bin is full. A user account is needed on IFTTT.com for connecting the maker channel within IFTTT [14]. We downloaded the IFTTT app on our android device. A key is generated after creating the applet which is needed in the program code to run the applet when the given condition is fulfilled. C++ programming is used on ESP8266 NodeMCU to implement the system.

The android phone needs to be connected to the internet. Arduino is programmed based on two levels, i.e., the dustbin is empty and dustbin is full. For the purpose of demonstration, garbage levels less than or equal to 6 cm is considered as empty bin otherwise the bin is considered full. Figure 3 shows the flowchart of the proposed system. At first NodeMCU reads the sensor value and stores data in the database. After the user is logged in, he/she can view the status of the dustbin on main screen of the android application.

Figure 4 shows the implemented system. Initially, the garbage bin is empty or partially filled. The user sees the message “Dustbin is Empty” on the main screen of their app. The information regarding the bin such as temperature, humidity, distance, gas sensor value, latitude, and longitude are updated on the database. In the developed app, the temperature, humidity, distance, gas sensor value, latitude, and longitude value of the dustbin are displayed. After that the dustbin is started to be filled and when it comes below the specified threshold value the message “Dustbin is Full” is displayed on the main screen of the Android application as well as the server and an Email or SMS will be sent to the garbage controlling authority so that the dustbin may be cleared properly in time. After clearance of the dustbin the status is again updated in the server and in Android application and the process repeats.

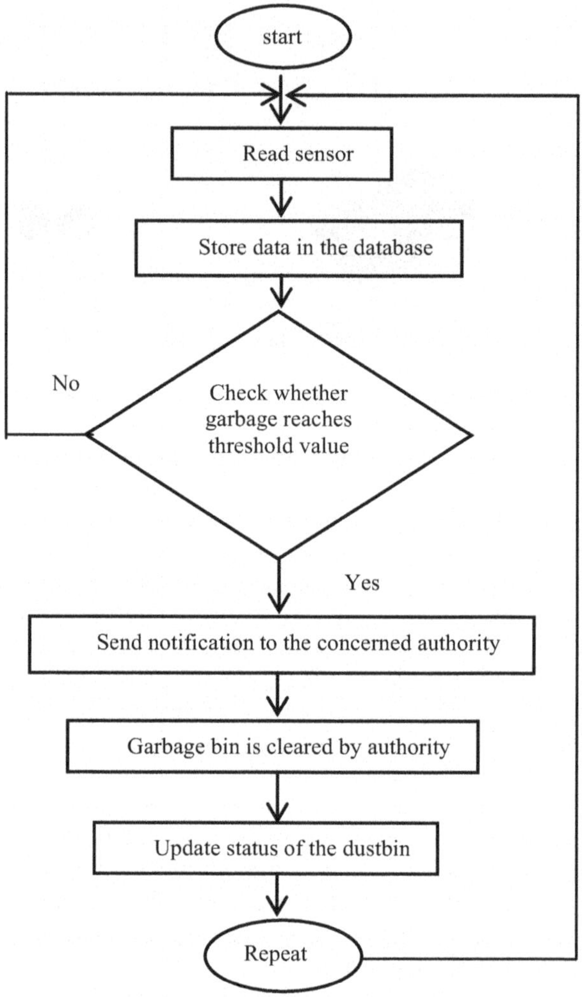

Fig. 3 Flowchart of the proposed system

4 Results and Discussions

Results that are shown in the Android app and web pages can also be shown on serial monitor whenever the NodeMCU is plugged in with mobile or laptop or any other display device. Figure 5 shows a snapshot of sample output which shows the status of the developed system on serial monitor.

Test runs have been performed after the final system has been made which completed 30 full cycles of loading and unloading of garbage at three different places, i.e., in air-conditioned room, at room temperature, and at open ground. Various possible combinations have been tested to assess performance of the proposed system. As shown in Table 1 the garbage bin is empty first. Then it was started to be filled

Fig. 4 Dustbin

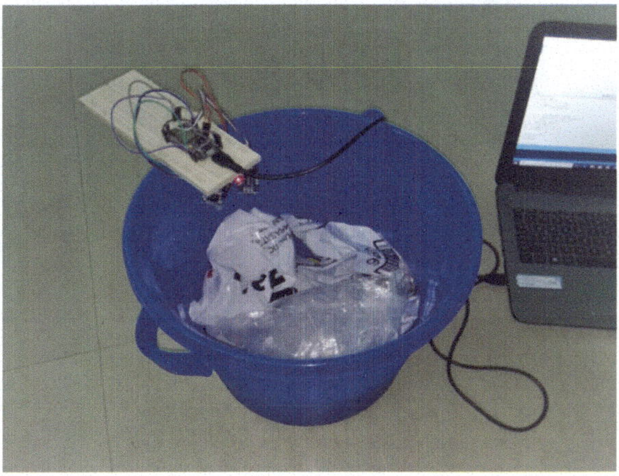

for further experimentation. The bin was unloaded again and the same process was repeated at the rest of the two places. The readings taken at those two places are shown in Tables 2 and 3. The tables provide sample outputs of garbage distance, temperature, humidity, gas value, latitude and longitude for a particular time instant. Temperature and humidity change as the experimental setup was moved from the AC room to normal room and then to the open ground. Tables 1, 2 and 3 show that the proposed system works correctly as it shows real time values of the sensors attached to it.

For further processing of the data, the android phone application is logged in by entering proper User Name and Password. This feature is added to data security. After the user is logged in, the information of the bins can be viewed on the main screen. Figure 6 shows screenshot of the login screen of the Android application. After logging in, the information displayed on the main screen is shown in Figs. 7, 8 and 9 for the three different places mentioned above. Figures 7, 8 and 9 show the snapshot of sample outputs for different time instants. The sample output gives a feel of the working of the developed app. It is to be noted that the app also provides real time data. Google firebase cloud server is used here for real time monitoring purposes. Google firebase is a development platform that provides database to the users. Figure 10 shows the snapshot of the webpage.

To analyze the performance of the developed system, the collection of real time data for five consecutive days from 29.09.2018 to 03.10.2018 has been taken. Data has been taken between 9 a.m. to 9 p.m. of each day. After that the average values of the experimental data of the consecutive five days are calculated. Only the final average table is shown in Table 4.

In this system when the garbage bin is filled with waste beyond the threshold value then an Email notification will be sent to the concerned garbage monitoring authority or user to alert them about the situation using IFTTT protocol via the internet. Alternatively, an SMS can also be sent to the registered authority mobile number. The snapshot of the Email and SMS is shown in Fig. 11.

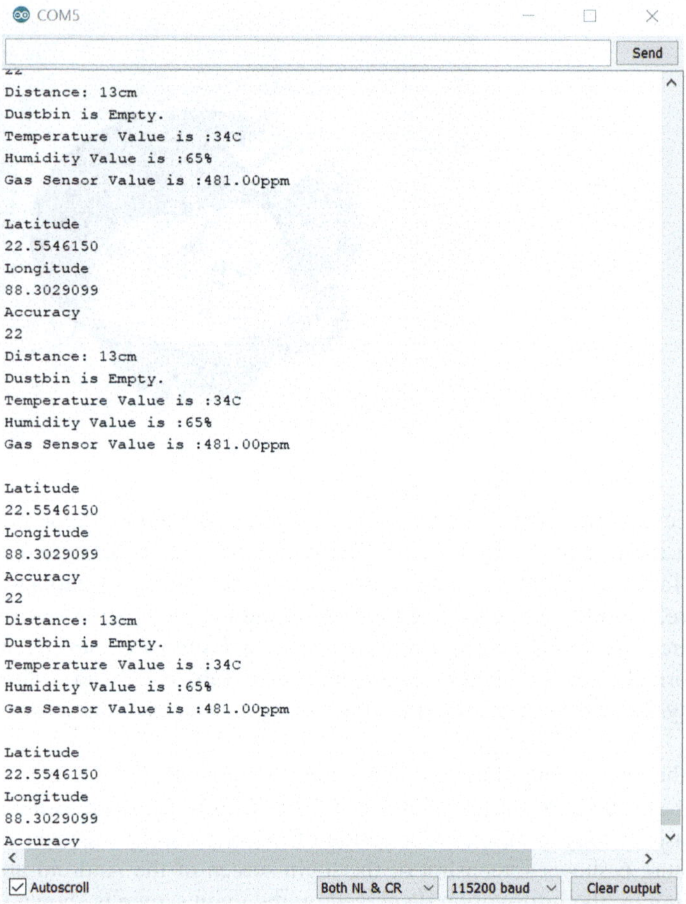

Fig. 5 Serial monitor output

Table 1 Experimental data from several tests run at AC room

Test no	Temperature (°C)	Humidity (%)	Distance (cm)	Gas sensor value (ppm)	Latitude (°N)	Longitude (°E)	Status of the bin
1	28	57	21	684	22.55498	88.30676	Empty
2	28	56	17	673	22.55498	88.30676	Empty
3	28	57	12	654	22.55498	88.30676	Empty
4	29	57	7	640	22.55498	88.30676	Empty
5	29	57	5	624	22.55498	88.30676	Full

Table 2 Experimental data from several tests run indoors at room temperature

Test no	Temperature (°C)	Humidity (%)	Distance (cm)	Gas sensor value (ppm)	Latitude (°N)	Longitude (°E)	Status of the bin
1	32	79	15	616	22.55508	88.30675	Empty
2	32	79	9	607	22.55508	88.30675	Empty
3	32	78	7	595	22.55508	88.30675	Empty
4	32	78	6	598	22.55508	88.30675	Full
5	32	78	5	601	22.55508	88.30675	Full

Table 3 Experimental data from several tests run at open ground

Test No	Temperature (°C)	Humidity (%)	Distance (cm)	Gas sensor value (ppm)	Latitude (°N)	Longitude (°E)	Status of the bin
1	39	58	15	598	22.498	88.30651	Empty
2	38	59	12	595	22.498	88.30651	Empty
3	37	66	8	601	22.498	88.30651	Empty
4	38	58	4	591	22.498	88.30651	Full
5	39	58	2	587	22.498	88.30651	Full

Fig. 6 Login screen of android application

Fig. 7 Main screen display of mobile app when the garbage bin is in ac room

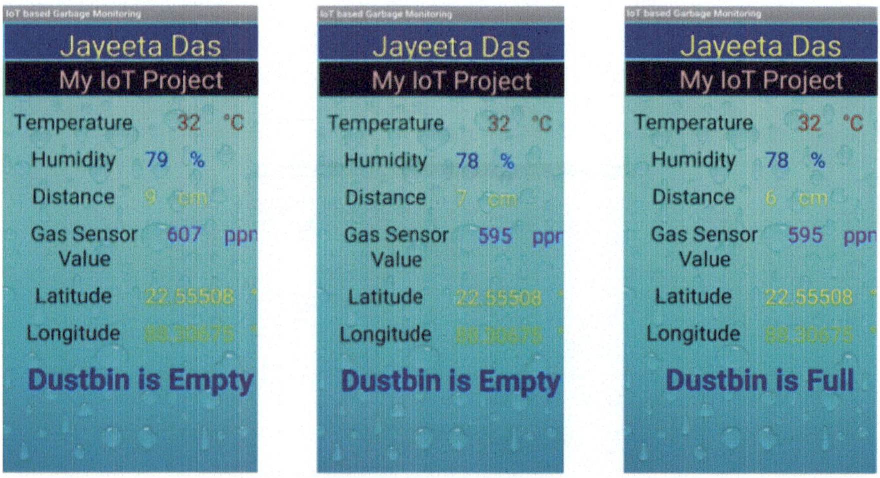

Fig. 8 Main screen display of mobile app when the garbage bin is at normal temperature

The system is useful for both indoor and outdoor usage. It has some additional features of sensing environmental temperature and humidity also. The features of the developed board are compared with the existing works. The comparison of the performances is presented in Table 5. It can be seen that the developed system is much better than the existing systems in terms of form, various sensor used, ease of working, etc.

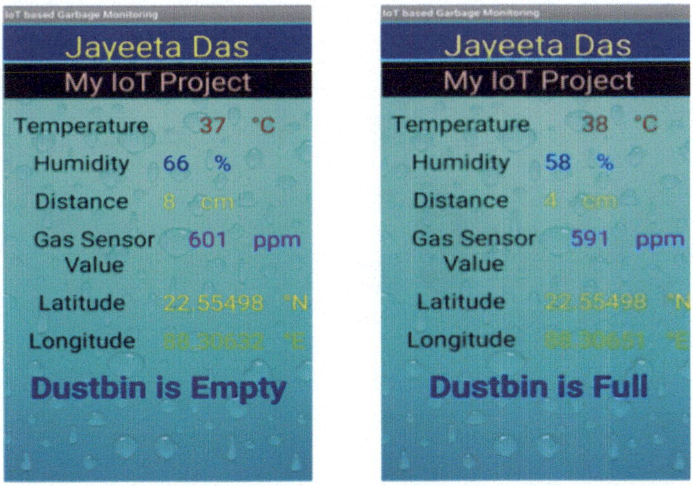

Fig. 9 Main screen display of mobile app when the dustbin is on open ground

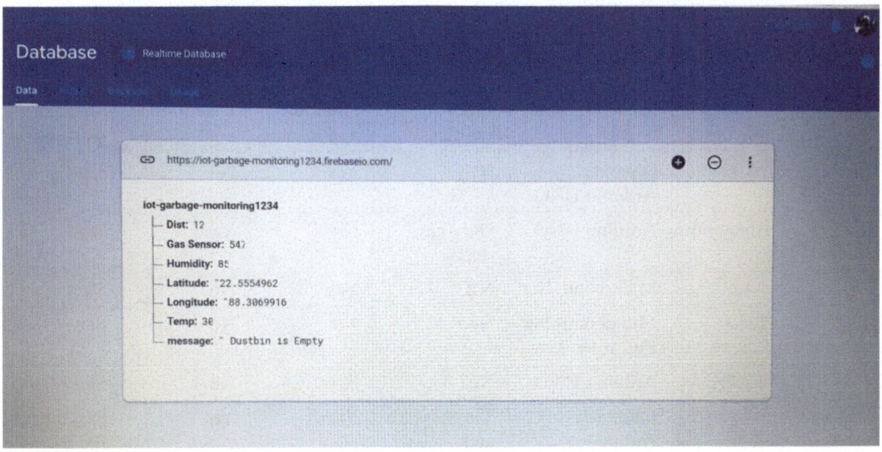

Fig. 10 Webpage to remotely monitor the status of the dustbin

Table 4 Average experimental data of the consecutive five days test

Time	09:00	10:30	12:00	13:30	15:00	16:30	18:00	21:00
Temperature (°C)	29.2	30.8	32	32.6	32.6	32.2	31.8	32.8
Humidity (%)	70.2	70.4	67.8	64	62.8	64	66.2	72.4
Distance (cm)	24	20.4	15.4	12.4	9	6.4	5.2	3.6
Gas sensor (ppm)	560.4	530	495.6	494.8	499.4	492	490.8	479.6
Status of the bin	Empty	Empty	Empty	Empty	Empty	Full	Full	Full

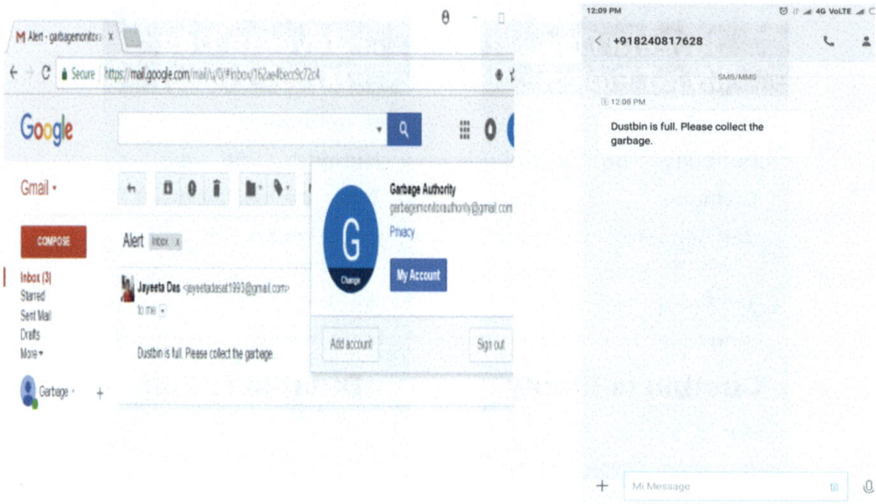

Fig. 11 Snapshot of Gmail and SMS notification to the user authority

Table 5 Comparison with existing works

Works	Wireless module	Microcontroller board	Other sensors	Android application	Email/SMS notification	Speed of notification
[8]	Wi-Fi	Separate Arduino UNO	NIL	Yes	No	Slow
[9]	Bluetooth	Arduino UNO	Only gas sensor	No	Yes	Slow
[10]	Wi-Fi	Microcontroller	NIL	Yes	No	Slow
[13]	TSRF	Microcontroller PIC 16F8 77A	No	No	No	Slow
[12]	Wi-Fi	Arduino UNO	No	No	No	Slow
Proposed	Wi-Fi	Integrated Arduino + Wi-Fi	Temperature and humidity sensor, gas sensor	Yes	Yes	Fast due to IFTTT protocol

5 Conclusions

The developed system finds multiple usages. The presence of any harmful gas can be detected through the developed board. If the gas sensor value exceeds the threshold value, the government authorities can be notified directly, i.e., without any human intervention. Arduino based sensors are used here to reduce the cost and also make the system efficient and portable. Here only 3 sensors have been connected. But a maximum of 13 sensors can be connected with the developed system. Email or an

SMS can be sent directly to the garbage collection center when the dustbin is full, by the developed app. The system can also be used at home. Compared to the existing works, the use of IFTTT protocol in the developed system can send notifications quickly and efficiently. The use of ESP 8266 NodeMCU brings down the cost and complexity of the system.

There are some minor limitations on which future research can be done. Google firebase cloud server is used for real time continuous monitoring and only the anomalous data are saved for notification purposes and for further analysis. If the good data is needed to be saved then, a large storage or customized database is required. The android software can be modified and a dedicated web server can be developed to store large data chunk for a long time for future use. Through this dedicated server various dustbins can be connected through a common network and routing among various garbage bins could be done in near future. Thus, the system can be made ready for global use in a smart city.

References

1. Shanzhi, C., Hui, X., Dake, L., Bo, H., Hucheng, W.: A vision of IoT: applications, challenges and opportunities with china perspective. IEEE Internet Things J. **1**(1), 349–359 (2014)
2. Sindhura, V., Ramya, P., Yelisetti, S., Manne, S.: A platform approach for unlocking the value of IoT (internet of things). In: Second International Conference on Electrical, Computer and Communication Technologies (ICECCT), Coimbatore, pp. 1–7 (2017)
3. Stankovic, A.: Research directions for the internet of things. IEEE Internet Things J. **1**(1), 3–9 (2014)
4. Bhuvaneswari, V., Porkodi, R.: The internet of things (IoT) applications and communication enabling technology standards: an overview. In: International Conference on Intelligent Computing Applications. IEEE (2014)
5. Medvedev, A., Fedchenkov, P., Arkady Zaslavsky, A.: Waste management as an IoT enabled service in smart cities. In: Conference on Smart Spaces, pp. 104–115. Springer, Cham (2015)
6. Thakker, S., Narayanamoorthi, R.: Smart and wireless waste management. In: 2015 International Conference on Innovations in Information, Embedded and Communication Systems (ICIIECS), Coimbatore, pp. 1–4 (2015)
7. Corno, F., Russis, L.D., Roffarello, A.M.: A semantic web approach to simplifying trigger-action programming in the IoT. Computer **50**(11), 18–24 (2017)
8. Kumar, N.S., Vuayalakshmi, B., Prarthana, R.J., Shankar, A.: IOT based smart garbage alert system using Arduino UNO. In: IEEE Region 10 Conference (TENCON), Singapore, pp. 1028–1034 (2016)
9. Reddy, P.S.N., Naik, R.N., Kumar, A.A., Kishor, S.N.: Wireless dust bin monitoring and alert system using Arduino. In: Second International Conference on Electrical, Computer and Communication Technologies (ICECCT), Coimbatore, pp. 1–5 (2017)
10. Kumar, S.V., Kumaran, T.S., Kumar, A.K., Mathapati, M.: Smart garbage monitoring and clearance system using internet of things. In: IEEE International Conference on Smart Technologies and Management for Computing, Communication, Controls, Energy and Materials (ICSTM), Chennai, pp. 184–189 (2017)
11. Jain, A., Bagherwal, R.: Design and implementation of a smart solid waste monitoring and collection system based on internet of things. In: 8th International Conference on Computing, Communication and Networking Technologies (ICCCNT), Delhi, pp. 1–5 (2017)

12. Neetha, S.S., Vaishnavi, V., Bedhi, V.: Smart bin—an "Internet of Things" approach to clean and safe public space. In: International Conference on I-SMAC (IoT in Social, Mobile, Analytics and Cloud) (I-SMAC), Palladam, pp. 652–657 (2017)
13. Nirde, K., Mulay, P.S., Chaskar, U.M.: IoT based solid waste management system for smart city. In: International Conference on Intelligent Computing and Control Systems (ICICCS), Madurai, pp. 666–669 (2017)
14. Kauling, D., Mahmoud, Q.H.: Sensorian hub: an IFTTT-based platform for collecting and processing sensor data. In: 14th IEEE Annual Consumer Communications and Networking Conference (CCNC), Las Vegas, NV, pp. 504–509 (2017)

Automated Real-Time Parking Management for Smart Cities

Pampa Sadhukhan and Arijit Talukdar

Abstract The proper management of available parking spaces to enable a driver to locate the parking slot quickly during busiest hours of the day in the urban areas is a major concern to the city authorities. So, a significant amount of research works have been devoted to the development of smart parking systems over the past few years. Most of these have paid attention to the detection of free parking slot within the parking areas along with guiding the driver to reach the parking area. Very few works have paid attention to the real-time management of all parking spaces available within a city. By keeping track of the occupancy status of all the parking spaces available within a city, it is possible to aid not only the driver to locate his/her preferable parking spot quickly but also the city authorities in generating revenues by proper utilization of the parking spaces. Thus, this paper presents an automated real-time parking management system for smart cities to address the above mentioned issues. The proposed parking management system provides online reservation facility for the parking spot in a secured manner via an android application that is to be installed on the driver's phone and also integrate an electronic payment gateway to enable automatic collection of the parking charges by the city authorities. The experimental results and various screenshots of the mobile client application provided in this paper validate the effectiveness of the proposed system.

Keywords Parking · Reservation · Real-time · System

1 Introduction

The proper management of the available parking spaces in the urban areas is a necessity to assist the drivers in locating the parking spot quickly during the busiest hours of the day as well as to enable the city authorities to earn the revenue by

---

P. Sadhukhan (✉) · A. Talukdar
School of Mobile Computing & Communication, Jadavpur University,
Kolkata 700032, India
e-mail: pampa.sadhukhan@ieee.org

A. Talukdar
e-mail: arijittalukdar2010@gmail.com

S. Kundu et al. (eds.), *Proceedings of the 2nd International Conference on Communication, Devices and Computing*, Lecture Notes in Electrical Engineering 602, https://doi.org/10.1007/978-981-15-0829-5_61

proper utilization of the parking areas. So, smart parking management is treated as an important aspect of the smart city services [1] in today's world. Thus, the development of smart parking system (SPS) for metropolitan areas to address the abovementioned issues have become an important field of research. Although a lot of research efforts have been contributed in designing various SPSs over the past few years, most of them focus on the detection of free parking slot within a parking area by using some wireless technology like radio frequency identification (RFID), sensor node, sensor network, etc. [2–7]. Very few works concentrate on the automated management of the parking spaces which is a necessity to keep track of available free parking slots within the parking area as well as to enable the driver to reserve some parking spot of his/her choice. Longer the search for a suitable parking slot, more increases not only the fuel consumption, but also the traffic congestion as well as the air pollution.

On the other hand, none of the automated parking management systems proposed in literature untill now has been designed with the aim of managing all parking areas available throughout the city in real-time. Moreover, such systems should ensure the automatic collection of the parking charges based on the duration of parking slot's occupancy by some vehicle. To address the abovementioned issues, the framework of an IoT based E-parking system for smart cities has been proposed in the earlier work [8]. But no experimental results have been provided to validate the proposed work. Thus, this paper not only presents the detailed design of the proposed automated real-time parking management system but also the experimental results to demonstrate the effectiveness of the proposed system. The automated parking management system (APMS) proposed in this paper aids the driver to find out and reserve a parking spot of his/her choice in real-time in a secured way via some control server which is responsible for the management of all parking spaces available throughout the city as well as the city authorities in the automatic collection of the parking charges. Various screenshots depicting the graphical user interface (GUI) of the android based mobile application that enables the authenticated drivers to locate the nearest parking spot and then reserve it have also been provided in this paper. This paper has been structured in the following way. In Sect. 2, a detailed review of existing smart parking management systems has been provided. Section 3 presents the design and implementation details of the proposed automated real-time parking management system. In Sect. 4, some experimental results as well as the prototype of the mobile application used by the drivers to locate and reserve the parking spot are provided in order to validate the proposed system and to demonstrate its effectiveness in managing all parking spaces throughout the city in real-time. Finally, the concluding remarks are given in Sect. 5.

2 Related Work

This section briefly reviews various parking management systems proposed in literature over the past few years. An IoT based smart parking system that also integrates cloud that has been proposed in [9]. The proposed system can monitor the availability status of each parking space through an IoT module deployed at the parking space and communicate it to the cloud. Moreover, this system provides a mobile application for the end user in order to check the parking space's availability and then reserve a parking slot accordingly. A smart parking system for the commercial stretch in cities abbreviated as SPSCSC has been proposed in [10]. The proposed SPSCSC focuses on reducing the searching time for the suitable parking slots and unnecessary travel for that purpose by deploying sensors and image detectors within each parking area. In [11], the researchers have proposed the design of a parking guidance system that consists of parking slot sensors, sink node, and the parking manager. The parking slot sensors communicate with the sink node via the ZigBee protocol to transmit real-time parking information to the sink node. Upon receiving the status of all parking lots from the sink node, the parking manager will display the information and position of available parking spaces. However, none of the above-mentioned SPSs cannot provide citywide parking management facility as well as an automatic collection of parking charges.

The authors in [12], have presented a Smart Parking System which is based on the integration of Radio Frequency Identification (RFID) and Wireless Sensor Network (WSN) technologies to detect the occupancy state of the parking spaces. The proposed system also employs a customized software application to direct the drivers to the nearest vacant parking spot. Such an application also provides the users the option of paying the parking fees through some near field communication (NFC) based e-wallet. Moreover, it can alert the traffic cops about the improper use of a reserved space or expiration of the purchased time via an Android mobile application. The researchers in [13] have proposed an architecture based on cyber-physical system (CPS) for car parking management. In this system, the server finds out available parking slots based on the current location of the car provided by the driver through some smartphone app and sends this information along with the route directions to the driver. A context-aware parking space locating mechanism which has been implemented using network simulator-3 has been proposed in [14]. This proposed system provides the driver with the shortest route to the nearest parking space by applying a particle swarm optimization algorithm. Moreover, this system also enables the vehicle to gather knowledge from its environment and also to perceive the surroundings via the microsensors and wireless sensors mounted on the vehicle. An automatic car parking and controlling system proposed in [15], has been designed by a small programmable logic controller (PLC). In this system, Infrared sensors (IR) were deployed at the entrance and departure gates to keep track of the availability status of parking lots within the parking space and to ensure proper management of the parking spaces. On the other hand, a vision-based smart parking framework to aid the drivers to locate the free parking slot within a parking space and then reserve it has been proposed in [16]. This system has segmented the parking space into several blocks by applying calibration and then those blocks have been classified to determine how many freely available parking slots are there. The architecture and design of the Arduino based car parking system have been proposed in [17]. This proposed system ensures the security of a vehicle within the parking space by restricting the entry of unauthorized users. This system also helps to park the vehicle in multi-floor parking spaces. A real-time parking reservation service that employs a mixed-integer programming model in order to allocate the parking slots and then make the schedule of drivers' travel plan accordingly, has been proposed in [18].

3 Proposed Automated Real-Time Parking Management (ARPM) System

This section presents the proposed automated real-time parking management system. The proposed system consists of three modules, which are local parking management system (LPMS), central parking management system (CPMS), and parking reservation module (PRM) as shown in Fig. 1. The detailed design along with the functional specification of each of these modules are provided below.

3.1 *Local Parking Management System (LPMS)*

This module is deployed onto the local parking management server available within each parking facility area. It is responsible for monitoring the whole parking facility area. For that purpose, it maintains a record for each parking slot to keep track of its status (which can be empty, reserved or occupied) by communicating with the parking meter (PM) deployed at each parking slot. The detailed architecture of PM, its functional specification as well as the architecture of LPMS is already provided in

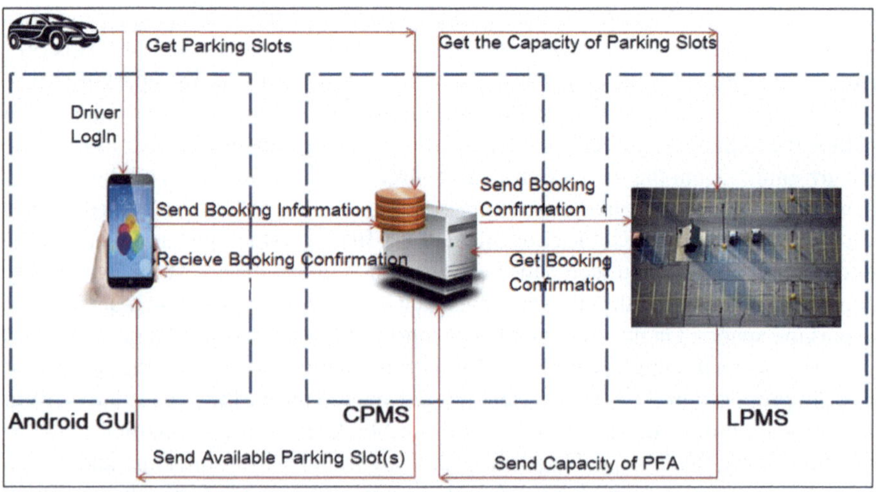

Fig. 1 Architecture of proposed automated real-time parking management system

the earlier work [8]. On the other hand, LPMS has an uninterrupted connection with the CPMS to receive reservation requests for some parking slots as well as to notify CPMS of the number of free parking slots available within the associated parking facility area in a periodic way. This module accomplishes the following tasks.

- It receives the reservation request for some parking slot from CPMS and also notify CPMS of the number of free parking slots available within its associated parking area.
- Upon reserving some parking slot, it sends reservation command to the associated PM.
- Upon receiving the image of a vehicle's license plate from some PM, it attempts to extract the license plate number from it, failing which it immediately sends the signal of improper parking to the corresponding PM.
- If a vehicle's exit is reported by some PM, it can calculate the parking charges and provide this information to CPMS.

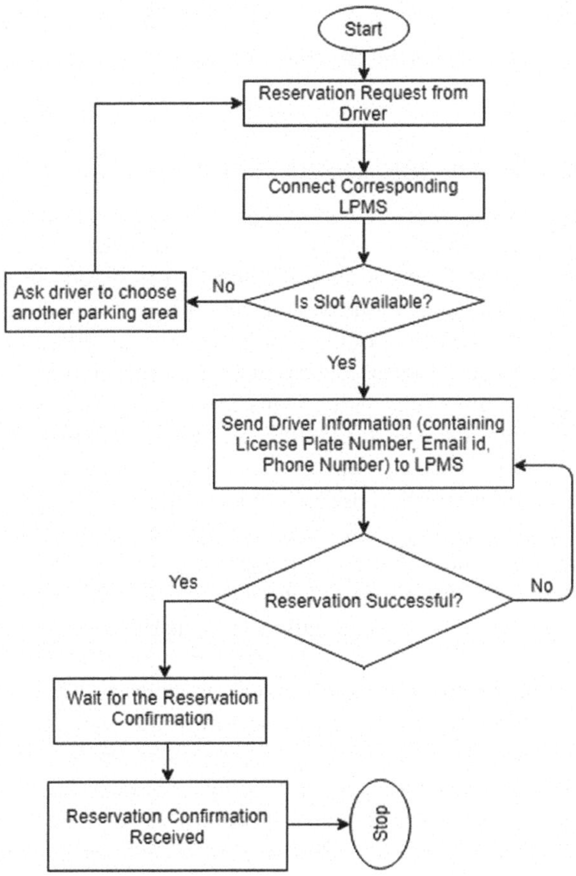

Fig. 2 Work flow diagram of central parking management system

3.2 Central Parking Management System (CPMS)

This module is deployed on the central parking management (CPM) server which is responsible for monitoring all parking facility areas available throughout the city. The CPM server is a high-end server with global IP in order to facilitate the parking reservation module running on the user's handheld device to communicate with it over the Internet. The detailed software specification of this module is already presented in the previous work. Figure 2 shows the workflow diagram of this module.

3.3 Parking Reservation Module (PRM)

This is a GUI based mobile application that would run on the driver's handheld device as mentioned in our previous work [8]. It is developed using android programming. This module enables the driver to reserve as well as locate some parking slot based on his/her choice and also assists the city authorities to receive the online payments from the drivers as already described in our earlier work [8]. Figure 3 depicts the workflow of this proposed client application.

4 Results and Discussion

The performances of the proposed system are evaluated in terms of the response time of the CPMS server upon receiving a reservation request; average booking time of a new user and an already registered user; the waiting time of real-time booking as well as advance booking with respect to the different times of the day. The graphs are plotted based on data provided by the mobile client application deployed and running at various android smartphones at different times of the day.

Figure 4 shows that CPMS takes longer duration to respond during the peak hours and office time as compared to the same in the morning and night, which is a very usual scenario because of higher number of reservation requests received during the peak hours and office time. Figure 5, on the other hand, compares the average waiting time to reserve some parking slot by a new user and an already registered user at different times of the day. The graphical results provided in Fig. 5 depicts that average booking duration of the parking slot is high around 10 a.m. and 6 p.m. due to significantly high reservation requests and it becomes lower in the morning and night. Since the parking management system presented in this paper provides two types of parking slot reservation facilities, i.e., real-time booking as well as advance booking, henceforth Fig. 6 provides the comparison between the waiting time for these two sorts of bookings. Figure 6 shows that the real-time booking that reserves the parking slot at the current time, always take a longer duration than that of advanced booking. This happens because the probability of getting a parking slot

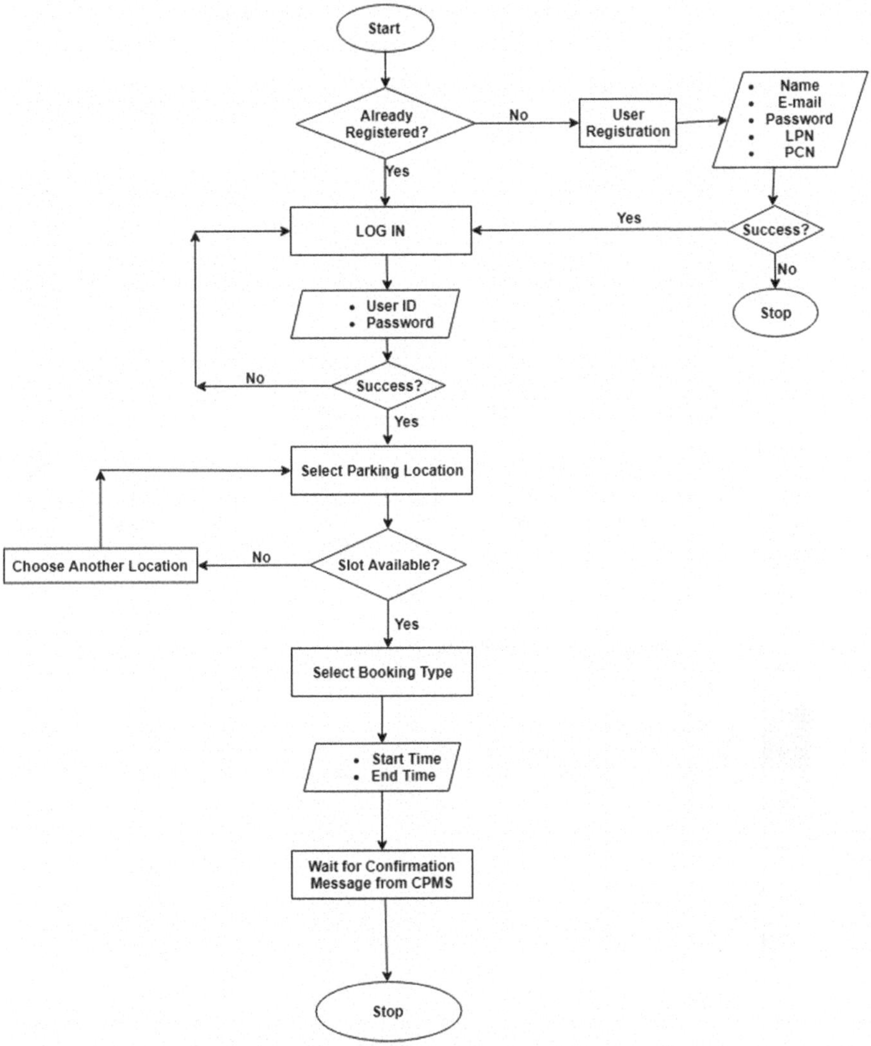

Fig. 3 Work flow diagram of parking reservation module

is more via advance booking compared to real-time booking which make on spot reservation at the current time.

Various screenshots of the parking reservation module, i.e., mobile client application while running on the user's android based smart phone, are displayed by Figs. 7, 8, 9, and 10. Figure 7 shows the GUI of user registration and user login with CPMS before searching for suitable parking spaces. On the other hand, GUI of the android based client application upon locating nearby parking spaces and then trying to reserve some suitable parking slot either via real-time booking or via advance

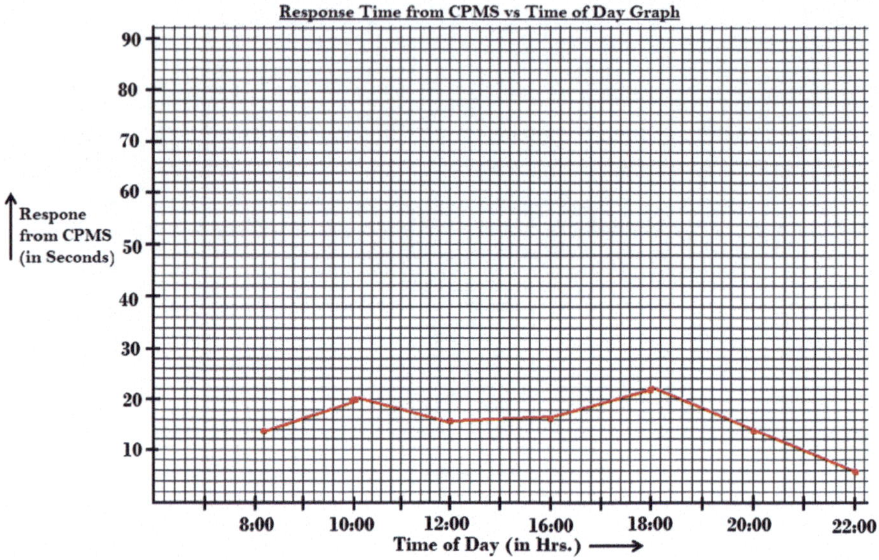

Fig. 4 Response time of central parking management server versus time of day

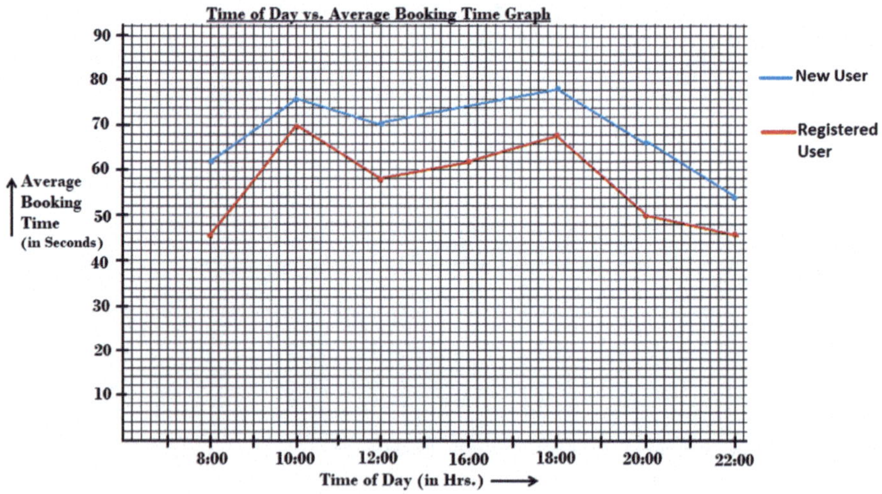

Fig. 5 Average booking time versus time of day

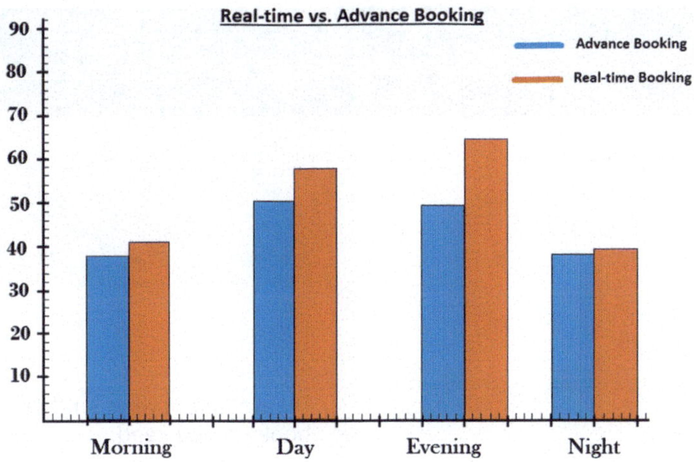

Fig. 6 Comparison of real-time booking and advance booking versus time of day

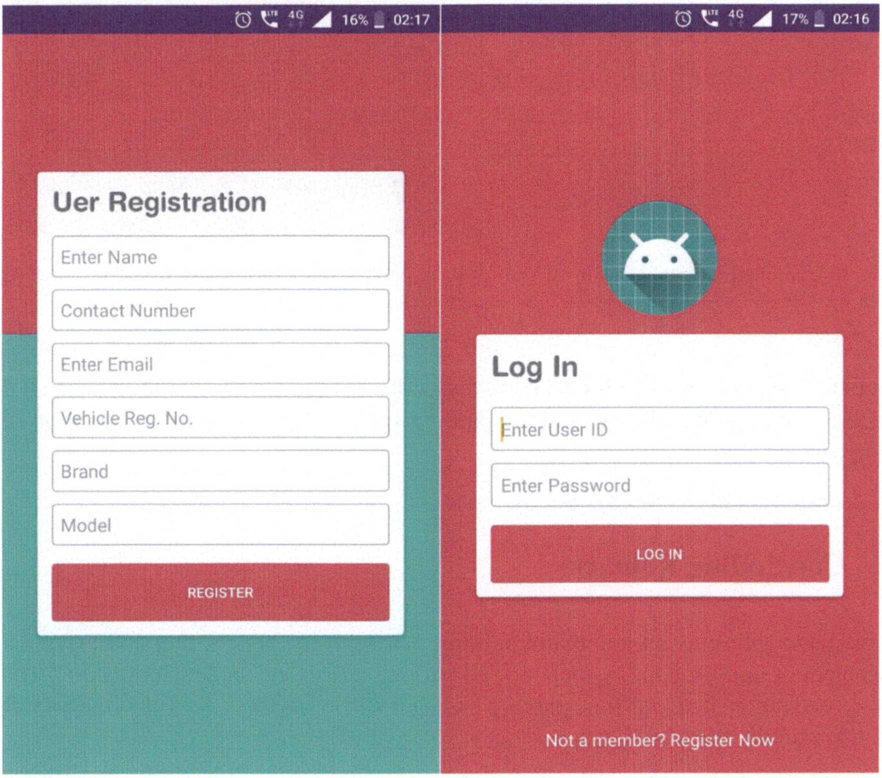

Fig. 7 GUI of user registration and user login

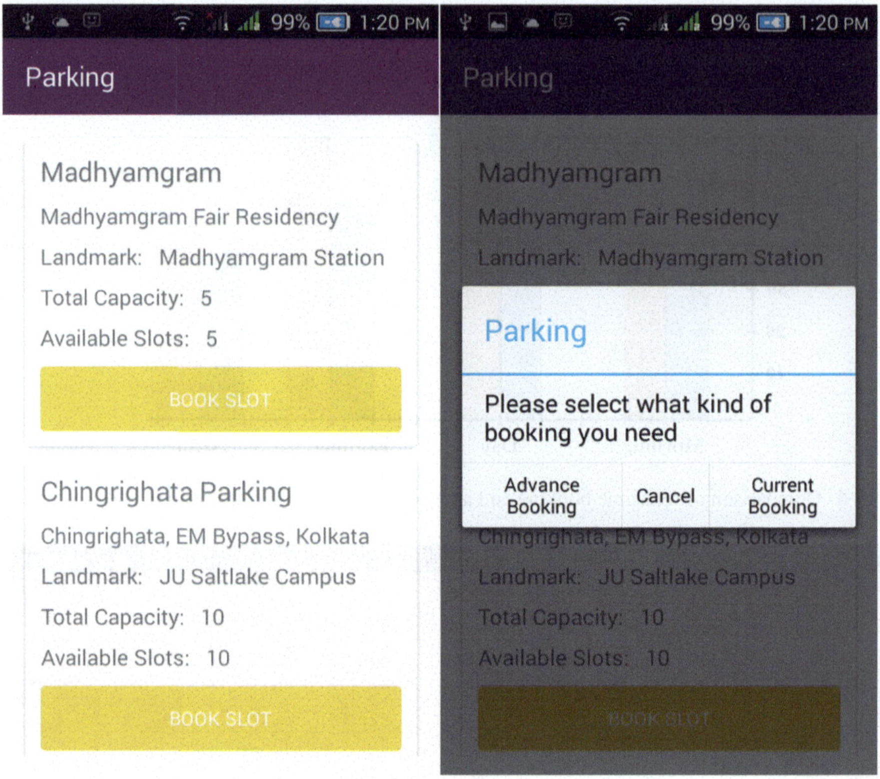

Fig. 8 GUI of the android client upon locating some parking spaces and trying to reserve a parking slot

booking, are depicted in Figs. 8 and 9, respectively. Figure 10 finally shows the GUI of available parking slots within the parking spaces after successful reservation of one parking slot.

5 Concluding Remarks

This paper presents an automatic parking management system that addresses the problem of real-time management of all parking spaces available within a city. The proposed automated real-time parking management system has the following important features.

- It aids the driver to locate his/her preferable parking spot quickly.
- It can detect improper parking of a vehicle at some parking slot.
- It provides an online reservation facility of the parking spot in a secured manner via an android application.

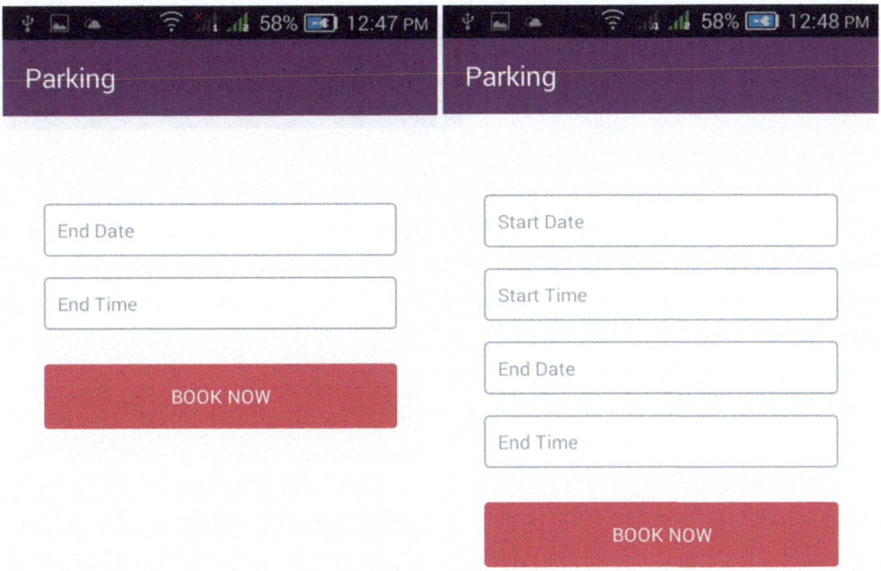

Fig. 9 GUI of real-time booking and advance booking

– It also ensures an automatic collection of the parking charges by the city authorities via an electronic payment gateway.

The experimental results and various screenshots provided in this paper validate the effectiveness of the proposed system. Moreover, the parking reservation module proposed in this paper is much simpler and more cost-effective compared to those that rely on Google Maps and GPS technology.

Acknowledgements The authors gratefully acknowledge the facilities and support provided by the Director and all other staff members of the School of Mobile Computing and Communication, Jadavpur University, a Center of Excellence set up under the "University with potential for Excellence" Scheme of the UGC.

Fig. 10 GUI of an android client after successful reservation of a parking slot

References

1. Sadhukhan, P.: An IoT based framework for smart city services. In: Proceedings of 2018 International Conference on Communication, Computing & Internet of Things (IC3IoT 2018), Chennai, India, pp. 376–379, Feb 2018. https://doi.org/10.1109/IC3IoT.2018.8668103
2. Cho, W., et al.: Robust parking occupancy monitoring system using random forests. In: Proceedings of 2018 International Conference on Electronics, Information, and Communication (ICEIC), Honolulu, HI, pp. 1–4 (2018)
3. Baroffio, L., Bondi, L., Cesana, M., Redondi, A.E., Tagliasacchi, M.: A visual sensor network for parking lot occupancy detection in Smart Cities. In: Proceedings of 2015 IEEE 2nd World Forum on Internet of Things (WF-IoT), Milan, pp. 745–750 (2015)
4. Fabian, T.: An algorithm for parking lot occupation detection. In: Proceedings of 2008 7th Computer Information Systems and Industrial Management Applications, Ostrava, pp. 165–170 (2008)
5. Suhr, J.K., Jung, H.G.: Sensor fusion-based vacant parking slot detection and tracking. IEEE Trans. Intell. Transp. Syst. **15**(1), 21–36 (2014)
6. Ng, C., Cheong, S., Haji, E., Yap, W.: Mobile outdoor parking space detection application. In Proceedings of 2017 IEEE 8th Control and System Graduate Research Colloquium (ICSGRC), Shah Alam, pp. 81–86 (2017)

7. Delibaltov, D., Wu, W., Loce, R.P., Bernal, E.A.: Parking lot occupancy determination from lamp-post camera images. In: Proceedings of 16th International IEEE Conference on Intelligent Transportation Systems (ITSC 2013), The Hague, pp. 2387–2392 (2013)
8. Sadhukhan, P.: An IoT-based E-parking system for smart cities. In: Proceedings of 2017 International Conference on Advances in Computing, Communications and Informatics (ICACCI), Udupi, pp. 1062–1066 (2017)
9. Khanna, A., Anand, R.: IoT based smart parking system. In: Proceedings of 2016 International Conference on Internet of Things and Applications (IOTA), Pune, pp. 266–270 (2016)
10. Kanteti, D., Srikar, D.V.S., Ramesh, T.K.: Smart parking system for commercial stretch in cities. In: Proceedings of 2017 International Conference on Communication and Signal Processing (ICCSP), Chennai, pp. 1285–1289 (2017)
11. Cai, W., Zhang, D., Pan, D.: Implementation of smart Parking Guidance System based on parking lots sensors networks. In: Proceedings of 2015 IEEE 16th International Conference on Communication Technology (ICCT), Hangzhou, pp. 419–424 (2015)
12. Mainetti, L., Marasovic, I., Patrono, L., Solic, P., Stefanizzi, M.L., Vergallo, R.: A Novel IoT-aware Smart Parking System based on the integration of RFID and WSN technologies. Int. J. RF Technol.: Res. Appl. **7**(4), 175–199 (2016)
13. Rehman, S., Gruhn, V.: Recommended architecture for car parking management system based on cyber-physical system. In: Proceedings of 2017 International Conference on Engineering & MIS (ICEMIS), Monastir, pp. 1–6 (2017)
14. Gupta, A., Sharma, V., Ruparam, N.K., Jain, S., Alhammad, A., Ripon, M.A.: Integrating pervasive computing, infostations and swarm intelligence to design intelligent context-aware parking-space location mechanism. In: Proceedings of 2014 International Conference on Advances in Computing, Communications and Informatics (ICACCI), New Delhi, pp. 1381–1387 (2014)
15. Mamun, M.A.A., et al.: Automatic car parking and controlling system using programmable logic controller (PLC). Int. J. Appl. Eng. Res. **10**, 69–75 (2015)
16. Bibi, N., Majid, M., Dawood, H., Guo, P.: Automatic parking space detection system. In: Proceedings of 2017 2nd International Conference on Multimedia and Image Processing (ICMIP), Wuhan, pp. 11–15 (2017)
17. Chaudhary, H., Bansal, P., Valarmathi, B.: Advanced CAR parking system using Arduino. In: Proceedings of 2017 4th International Conference on Advanced Computing and Communication Systems (ICACCS), Coimbatore, pp. 1–5 (2017)
18. He, H., Zhang, Z., Yan, P.: A real-time reservation service for smart parking system. In: Proceedings of 2018 15th International Conference on Service Systems and Service Management (ICSSSM), Hangzhou, pp. 1–6 (2018)

Lifetime Enhancement of Supernode Based WSNs with Optimal Size Cluster Formation by Using Gravitational Search Algorithm

Ashish Pandey, Amiya Dey and Arnab Nandi

Abstract In this paper, a Gravitational Search Algorithm (GSA) based multi-objective optimization scheme to design an energy-efficient wireless sensor networks (WSNs) has been presented. An energy management scenario is represented maintaining the Quality of Service (QoS) requirement. It also considered the selection of appropriate clusterheads as well as next hop for a given cluster head. The sensor network is designed with the help of Gravitational Search Algorithm to fulfill all the application-centric requirements along with the energy conservation characteristics. Focus has been made on the formation of uniform size clusters while maintaining the other qualities of the networks. Efficient management of energy is incorporated to guarantee maximum lifetime for WSNs without losing the network requirements. Supernodes are being used to increase the network lifetime and also to maintain the QoS requirements of the network. The impact of population of supernodes on the network lifetime is also estimated. Moreover, the effect of node population density on the network lifetime is demonstrated.

Keywords Gravitational search algorithm · Wireless sensor networks · Cluster head · Routing · Clustering

1 Introduction

WSNs consist of a huge number of tiny sensing, computing as well as communicating devices. These devices are of low cost, require less power, and possess the capability of observing, communicating, and reacting to occurrence of the particular situation.

A. Pandey (✉) · A. Dey · A. Nandi
Department of Electronics and Communication Engineering, National Institute of Technology Silchar, Cachar 788010, Assam, India
e-mail: ashish94023@gmail.com

A. Dey
e-mail: amiyo4u@gmail.com

A. Nandi
e-mail: dr.arnab.nandi@gmail.com

S. Kundu et al. (eds.), *Proceedings of the 2nd International Conference on Communication, Devices and Computing*, Lecture Notes in Electrical Engineering 602, https://doi.org/10.1007/978-981-15-0829-5_62

The work environment of these sensors might be related to any biotic system or some information technology framework. The distinctive use of these sensors comprises collection of data, observation, monitoring the incident, therapeutic applications, etc. Sensors may also be deployed in the field of control and automation. They are usually positioned arbitrarily for the measurement of the numerous quantities as per the requirement. The WSNs nodes are primarily equipped with batteries for power and having energy constraints. It is not always possible to recharge the battery in WSNs which creates the constraint in the communication ranges as well. Another main concern of WSNs is the proper connectivity in the network which is mainly dependent on overall sensor nodes population in individual cluster because individual cluster head can work with some limited number of sensor nodes concurrently. Apart from that connectivity can also be determined by the coverage of the individual sensor node. Researches in wireless sensor networks are focused primarily on attaining the near-optimum along with non-dominated solutions within relatively less computational time [1].

Almost in the last decade, various prominent researches have been carried out to enhance the performance of the wireless sensor networks. A heterogeneous WSN proposed consists of both battery-powered sensor nodes (BPSNs) as well as energy harvesting sensor nodes (EHSNs) for enhancing the lifetime of WSNs [2]. Various routing and clustering schemes have been proposed for enhancing the lifetime of the WSNs [3, 4]. The concept of particle swarm optimization (PSO) and "Enhanced-optimized energy-efficient energy protocol" (E-OEERP),has been proposed and has been used to enhance the lifetime of the networks [5–8].

Work also been done by using genetic algorithm (GA) to enhance the lifetime of wireless sensor network [9–12]. Also, a real-time genetic algorithm has been introduced for finding out a suitable direction for node movement. It also considers to minimize the energy consumption [13]. Also, various data aggregation schemes along with the linear programming framework have been implemented to reduce the energy consumption. Also, various work have been done with the help of optimization to secure the wireless sensor network. Apart from these, some researchers also analyzed the security requirement of the WSNs to save the network from various types of attacks [14]. Some work also is done on the link protocols [15, 16] and it is also done using Gravitational Search Algorithm [17] to increase the lifetime of the WSNs [18] after some modifications and results are compared with ECHSSDA [19] and ABC-based node clustering algorithm [20]. Some recent work is also done for enhancing the lifetime in different scenarios including harvesting node wireless sensor networks [21–24].

GSA-based approach is utilized to overcome the problem of quick drainage of power. In this paper, an effort has been made to form the cluster of uniform size as much as possible. The initial population of individual is generated randomly for attaining broader search space. Then fitness value has been calculated for each individual in the population separately using the clustering as well as routing objective function. The main goal is to achieve optimum solution for the problem of clustering and routing problems for the minimization of the energy expenditure of the network minimized efficiently. After deploying the gravitational search algorithm, the outcome has been equated with the Least Distance Clustering algorithm and GA [25]. The energy consumed by each of the higher energy nodes has been analyzed. The data generated within each of the clusters is also analyzed after accomplishing the energy-aware routing and energy-efficient clustering using the GSA and other algorithms.

2 Brief Introduction of Energy Model

Both the fading channel models are adopted here namely free space as well as multipath fading channel model. Implementation of these models is decided by a threshold distance. If the distance of receiver from the transmitter is smaller to threshold distance, then free space (power loss) fading channel model will be taken otherwise, multipath fading (power loss) channel model is used as in [1, 3].

$$\begin{aligned} E_{\text{Tx}}(l,d) &= E_{\text{Tx-elec}} + E_{\text{Tx-amp}}(l,d) \\ &= lE_{\text{elec}} + \xi_{\text{fs}} d^2 \qquad \text{for } d < d_0 \\ &= lE_{\text{elec}} + \xi_{\text{mp}} d^4 \qquad \text{for } d \geq d_0 \end{aligned} \tag{1}$$

and

$$E_{\text{Rx}}(l) = E_{\text{Rx-elec}}(l) = lE_{\text{elec}} \tag{2}$$

where $E_{\text{Tx}}(l,d)$ the energy is required for the transmission of l bits message over a distance of d meters and $E_{\text{Rx}}(l)$ is energy required for reception of l bits message. Threshold distance is denoted by d_0. $E_{\text{Rx-elec}}(l)$ denotes the energy requirement for receiving l bits message. $E_{\text{Tx-amp}}(l,d)$ denotes amplification energy required for l bits of message. $E_{\text{Tx-elec}}$ (l) and $E_{\text{Rx-elec}}$ (l) both are the amount of energy which is needed by electronic circuitry of the transmitter as well as receiver for the transmission of l bits of message. E_{elec} is the energy needed by electronic circuitry. $E_{\text{Rx}}(l)$. Here, d is the distance of receiver from the transmitter. ξ_{fs} and ξ_{mp} are the energy requirement by the amplifier in free space and multipath, respectively.

3 Brief Introduction of Least Distance Clustering

The least distance clustering algorithm can be used in the network during the cluster formation phase as well as during the routing phase in the WSNs. In this algorithm, the distance of each sensor node to base station (BS) and distance between each supernode and BS is calculated using the distance formula [5, 17]. During the cluster formation phase, each of the supernodes is sorted with respect to their distance from the base station and sensor node chooses the supernode which is less distant from the sink node. In the routing phase, each supernode is sorted again the basis of distance from the base station means less distance implicates less distant from the sink node. So every supernode chooses as the next hop to the supernode which is less distant from the sink node and within its range of communication [5, 17].

4 Brief Introduction to Gravitational Search Algorithm

In the gravitational search algorithm, agents reflect as objects and performance of them is contrary to their masses. Gravitational force work among all these objects and creates an overall movement of all objects. Objects move proximate the objects with heavyweight masses. In other words, all these masses collaborate by the use of direct form of communication with the help of gravitational force. In this algorithm, the heavy masses which already giving good solution to the problem moves slowly compared to lighter masses. This kind of movement provides the guarantee of the better solution. In this algorithm, each mass has their location and three types of masses. The location of each of them resembles the solution to the problem while the inertial and gravitational masses are calculated by the use of objective function of the problem. Each mass denotes the solution and the algorithm moves further by modifying the gravitational as well as inertial masses. As the time passes, it is anticipated that masses are attracted by weightiest mass which represents the optimum solution. The next velocity $v_i^d(t+1)$ of an agent is calculated as given below [17]:

$$v_i^d(t+1) = rand_i \times v_i^d(t) + a_i^d(t) \tag{3}$$

$$x_i^d(t+1) = x_i^d(t) + v_i^d(t+1) \tag{4}$$

where $rand_i$ denotes the random variable between 0 and 1. $a_i^d(t)$ is acceleration at a specific time t. $v_i^d(t)$ is velocity at time t. x_i^d denotes the position of ith agent in the dth dimension. The acceleration can be calculated as [17]

$$a_i^d(t) = \frac{F_i^d(t)}{M_{ii}(t)} \tag{5}$$

The inertial mass of the ith agent is denoted by M_{ii} and $F_i^d(t)$ is the total force on agent ith agent in the dth dimension and it can be given as [17]

$$F_i^d(t) = \sum_{j=1, j\neq i}^{N} rand_j F_{ij}^d(t) \tag{6}$$

where $rand_j$ is a random number in the interval [0, 1]. Here in the above equation, $F_{ij}^d(t)$ is defined as [17]:

$$F_{ij}^d(t) = G(t)\frac{M_{pi}(t) \times M_{aj}(t)}{R_{ij}(t) + \xi}(x_j^d(t) - x_i^d(t)) \tag{7}$$

where active gravitational mass related to j is denoted by M_{aj} and M_{pi} denotes the passive gravitational mass related to i. Gravitational force constant at time instant t is given by $G(t)$. ξ is a tiny constant and $R_{ij}(t)$ is the distance between the i and j. So $F_i^d(t)$ may be expressed as [17]

$$F_i^d(t) = \sum_{j=1, j\neq i}^{N} rand_j \times G(t)\frac{M_{pi}(t) \times M_{aj}(t)}{R_{ij}(t) + \xi}(x_j^d(t) - x_i^d(t)) \tag{8}$$

5 GSA-Based Approach for Lifetime Enhancement of WSNs

Here, no of agents is taken as 30 for the simulation purpose. GSA has been applied in routing as well as in clustering to enhance the lifetime of Wireless Sensor Networks. Fitness function for routing is represented as [1]

$$F_1 = A \times (\mathrm{Max_{Dist}} - \mathrm{Min_{Dist}}) + B \times \mathrm{Max_{Hop}} \tag{9}$$

where A and B are constants between 0 and 1. Such that $A = 1 - B$ and $\mathrm{Max_{Hop}}$ denotes the maximum no of hops in the network. $\mathrm{Max_{Hop}}$ denotes the maximum distance between the clusterheads and $\mathrm{Min_{Dist}}$ denotes the minimum distance between the clusterheads. Here, the goal is minimization of the difference between the maximum and minimum distance as well as minimization total number of hops in routing. So our overall objective is to minimization of the value of fitness function F_1. Also a trade-off exists between the distance between the clusterhead and no of hops in the network.

Fitness function for clustering [1]

$$F_2 = \frac{L_{\mathrm{min}}}{\sum_{i=1}^{N} rms\_avg_{\mathrm{Dist}}} \tag{10}$$

where L_{min} is the minimum life of the sensor node in WSNs and rms_avg_{Dist} is the root mean square average of distance between each sensor node and their own clusterhead. Here, the main goal is the maximization of the lifetime along with the minimization of the average distance between individual sensor node and clusterhead for minimizing the expenditure of energy. So maximization of the objective function F_2 is required. In this paper, after the deployment of the supernodes, all of them are provided a unique id. These ids start from the most distant point from the base station in order to make every successive supernodes have the knowledge of the load at its predecessor supernodes. It helps in the exact calculation of the total no of the incoming packets at any supernodes. The normalization of the routing function has been carried out in order to achieve better accuracy. The optimization has been done using above approach to achieve the optimal sized cluster in the wireless sensor networks that has not been achieved in any of the previous work.

6 Simulation Parameters

Simulations are carried out with the help MATLAB R2013a software. Population of sensor nodes positioned primarily is 300. The population of nodes is changed from 300 to 600 for various cases. Initial population supernodes is fixed at 50. It is assumed that individual sensor nodes have energy of 1 J and supernodes are given initial energy of 10 J. The values of various parameters taken for the simulation study are shown in Table 1 [1, 3, 5].

Figure 1 shows the wireless sensor network WSN#1, with a sink node at center of network. The sensor nodes, supernodes, and base station are represented by "*", "+", and "diamond", respectively.

Table 1 Simulation Parameters

Parameters	Values
Area	500×500 m^2
Sensor nodes	300–600
Communication range of Supernodes and sensor nodes	150 m
Supernodes	50
Energy for sensor nodes	1 J
Initial energy of each supernode	10 J
Number of iterations per round	200
Amplification energy of multipath (ξ_{mp})	0.0013 pJ/bit/m^4
Data aggregation energy (E_{DA})	5 nJ/bit
Amplification energy of free space (ξ_{fs})	10 pJ/bit/m2
Data aggregation energy (E_{DA})	5 nJ/bit
Threshold distance (d_{0})	87.0 m
Energy required by electronic circuit (E_{elec})	50 nJ/bit
Packet size	2048 bits

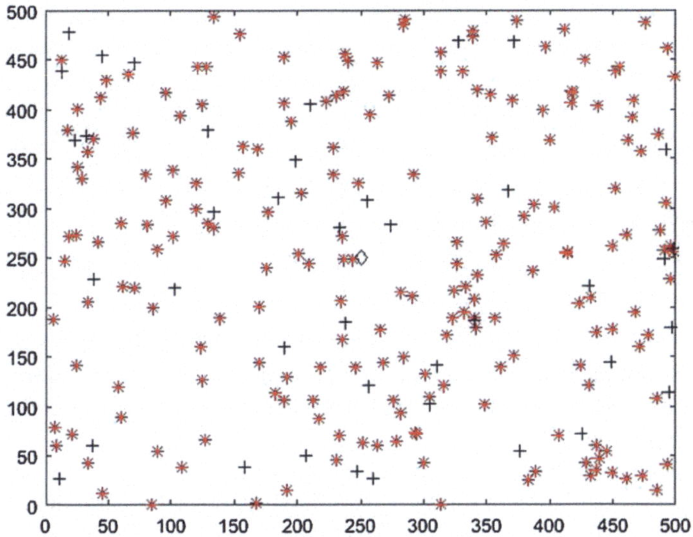

Fig. 1 Wireless sensor network (WSN#1) with base station at center

7 Results and Discussions

Here, two scenarios for two different locations of base station have been considered. The area of sensor network is taken as 500 × 500 m^2. Location of base station of WSN#1 is at centre (250, 250). Location of base station of WSN#2 is at (500, 500), i.e., at the corner of the network. In this experiment, multi hop routing has been considered. In the first study, for WSN#1, the supernode population is fixed at 50 and population of sensor node is changed from 300 to 600. Figure 2a shows the network lifetime for above-mentioned case when total supernodes are 50. Figure 2b shows the lifetime of network for each case of WSN#2. It is shown in Fig. 2a that for

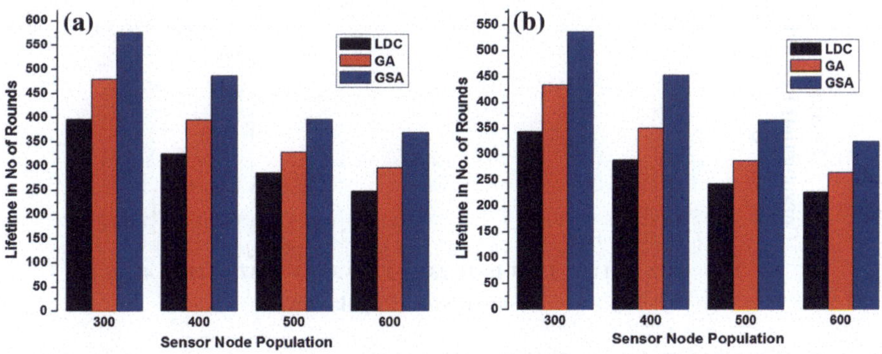

Fig. 2 Lifetime of WSNs with 50 supernodes **a** for WSN#1 and **b** for WSN#2

Table 2 Lifetime of sensor network for variation in sensor nodes population from 300 to 600 while keeping supernode population fixed at 50 with base station at centre for WSN#1 and at the corner for WSN#2

	For WSN#1				For WSN#2			
Sensor node population	300	400	500	600	300	400	500	600
Lifetime with LDC	395	325	284	247	342	288	242	226
Lifetime with GA	479	394	328	295	433	349	287	263
Lifetime with GSA	575	486	395	368	536	452	365	324

WSN#1 the network is increased from 395 to 575 while population of sensor nodes is fixed at 300 nodes and population of supernodes at 50. Also lifetime increases from 325, 284, and 247 to 486, 395, and 368 when number of sensor nodes are 400, 500, and 600 (Table 2).

It is shown in Fig. 2b that for WSN#2 the network lifetime is increased from 342 to 536 while the population of sensor node is fixed at 300 and supernodes at 50. Also, lifetime increases from 288, 242, and 226 to 452, 365, and 324, respectively when number of sensor nodes increases from 400 to 500 and then 600 while keeping the total number of supernodes at 50, results are shown below in Fig. 2b. It is observed that the use of GSA improved the network lifetime significantly for both the locations of base station.

Energy Consumption by each supernodes in the WSNs is shown in Fig. 3. In the experiment when GSA is used, energy consumption increases from 88.545 to 115.894 J while lifetime increases from 433 rounds to 536 rounds in case if GSA is used. In this experiment, it has been found that when Gravitational Search Algorithm is used, then the network lifetime increases significantly. It can be said that discharge of energy is more uniform in the case when Gravitational Search Algorithm is used. Hence, no supernode will die more quickly that will increase the lifetime of network significantly.

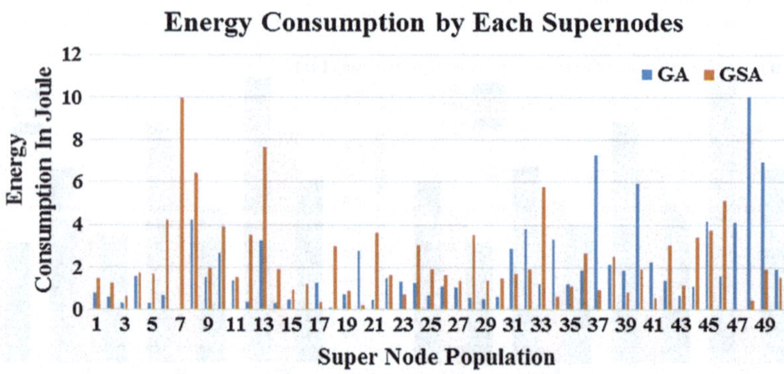

Fig. 3 Distribution of energy consumption in the network

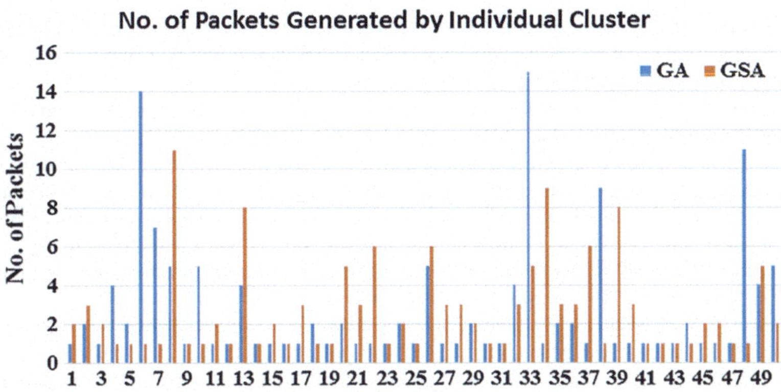

Fig. 4 Packet load at each cluster head

The participation of each supernode at the last round of the experiment in terms of generated packets in individual cluster is shown in Fig. 4. It has been observed that more uniform clusters formation takes place with the help of GSA. It is also helping in the uniform battery drainage hence increasing the lifetime. The need for some load balancing algorithm also demonstrated as only almost around 24% of energy is consumed by the network in its lifetime due to uneven discharge of sensor energy.

8 Conclusions

It has been found with the help of the various simulations and analysis that with the use of Gravitational Search Algorithm, the wireless sensor networks lifetime has increased significantly as compared to least distance clustering and genetic algorithms. The network lifetime decreases if we increase the sensor node population due to increment in the incoming packet at clusterheads in the networks. It is also observed that GSA also works better for the variable location of the sink node. When GSA is used, the drainage of energy is more uniformly distributed which increases the lifetime of WSNs. The performance of the system may be enhanced efficiently with the help of hybrid gravitational search algorithm along with the efficient load balancing algorithms.

References

1. Pandey, A., Rajan, A., Nandi, A., Lifetime enhancement of wireless sensor networks by using sine cosine optimization algorithm. In: 7th International Conference on Reliability, Infocom Technologies and Optimization (Trends and Future Directions) (ICRITO), Aug 2018
2. Zonouz, A.E., Xing, L., Vokkarane, V.M., Sun, Y.: Hybrid wireless sensor networks: a reliability, cost and energy-aware approach. IET Wirel. Sens. Syst. **6**, 42–48 (2016)

3. Heinzelman, W.B., Chandrakasan, A.P., Balakrishnan, H.: An application-specific protocol architecture for wireless microsensor networks. IEEE Trans. Wirel. Commun. **1**, 660–670 (2002)
4. Al-Karaki, J.N., Kamal, A.E.: Routing techniques in wireless sensor networks: a survey. IEEE Werel. Commun. **11**(6), 6–28 (2004)
5. Kuila, P., Jana, P.K.: Energy efficient clustering and routing algorithms for wireless sensor networks: particle swarm optimization approach. Eng. Appl. Artif. Intell. **33**, 127–140 (2014)
6. Rejinaparvin, J., Vasanthanayaki, C.: Particle swarm optimization-based clustering by preventing residual nodes in wireless sensor networks. IEEE Sens. J. **15**, 4264–4274 (2015)
7. Kennedy, J., Eberhart, R.: Particle swarm optimization. In: 1995 IEEE International Conference Neural Networks (ICNN'95), pp. 1942–1948 (1995)
8. Kulkarni, R.V., Kumar, G.: Particle swarm optimization in wireless-sensor networks : a brief survey. IEEE Trans. Syst. Man Cybern. Part C (Appl. Rev.) **41**, 262–267 (2011)
9. Bari, A., Wazed, S., Jaekel, A., Bandyopadhyay, S.: A genetic algorithm based approach for energy efficient routing in two tiered sensor networks. Ad Hoc Netw. **7**, 665–676 (2009)
10. Gupta, S.K., Jana, P.K.: Energy efficient clustering and routing algorithms for wireless sensor networks: GA based approach. Wirel. Pers. Commun. **83**, 2403–2423 (2015)
11. Hu, X.S., Zhang, J.S., Yu, Y., Chung, H.S., Li, Y., Shi, Y., Luo, X.: Hybrid genetic algorithm using a forward encoding scheme for lifetime maximization of wireless sensor networks. IEEE Trans. Evol. Comput. **14**, 766–781 (2010)
12. Yoon, Y., Kim, Y.H.: An efficient genetic algorithm for maximum coverage deployment in wireless sensor networks. IEEE Trans. Cybern. **43**, 1473–1483 (2013)
13. Norouzi, A., Babamir, F.S., Zaim, A.H.: An interactive genetic algorithm for mobile sensor networks. Stud. Inf. Control **22**(2), 213–218 (2013)
14. Kaschel, H., Mardones, J., Quezada, G.: Safety in wireless sensor networks: types of attacks and solutions. Stud. Inf. Control **22**(3), 323–330 (2013)
15. Kaschel, H., Sánchezy, L., Mardones, J.G., Quezada, G.: Design and construction of link quality and localization protocol algorithms at WSN over IEEE802.15.4 physical protocol. Stud. Inf. Control **20**(3), 285–292 (2013)
16. Rashid, I., Nasim, M.A., Akram, A.: Relay based clustering with least distance cluster head selection for wireless sensor networks. In: 15th IEEE International Conference on Communication Technology (ICCT), pp. 545–548 (2013)
17. Rashedi, E., Nezamabadi-pour, H., Saryazdi, S.: GSA: a gravitational search algorithm. Inf. Sci. **179**(13), 2232–2248 (2009)
18. Maraiya, K., Kant, K., Gupta, N.: Efficient cluster head selection scheme for data aggregation in wireless sensor networks. Int. J. Comput. Appl. **23**, 10–18 (2011)
19. Karaboga, D., Okdem, S., Ozturk, C.: Cluster based wireless sensor network routing using artificial bee colony algorithm. Wirel. Netw. **18**, 847–860 (2012)
20. Yazdani, D., Meybodi, M.: A modified gravitational search algorithm and its application. In: IKT2015 7th International Conference on Information and Knowledge Technology (2015)
21. Parvin, J.R., Vasanthanayaki, C.: Gravitational search algorithm based mobile aggregator sink nodes for energy efficient wireless sensor networks. In: 2013 International Conference on Circuits, Power and Computing Technologies, ICCPCT (2013)
22. Gupta, R.K., Pandey, A., Nandi, A.: Lifetime enhancement of WSN using evolutionary clustering and routing algorithms. In: 2018 IEEE International Students' Conference on Electrical, Electronics and Computer Science (2018)
23. Bahbahani, M.S., Alsusa, E.: A cooperative clustering protocol with duty cycling for energy harvesting enabled wireless sensor networks. IEEE Trans. Wirel. Commun. **17**(1), 101–111 (2018)
24. Al-Kiyumi, R.M., Foh, C.H., Vural, S., Chatzimisios, P., Tafazolli, R.: Fuzzy logic-based routing algorithm for lifetime enhancement in heterogeneous wireless sensor networks. IEEE Trans. Green Commun. Netw. **2**(2), 517–532 (2018)
25. Goldberg, D.E.: Genetic Algorithms in Search, Optimization, and Machine Learning. Addison-Wesley (1989)

Attack Detection in Recommender Systems Using Subspace Outlier Detection Algorithm

Partha Sarathi Chakraborty

Abstract Recommender system plays an important role in today's online business. At the time of choosing right product or service from Internet, recommender system helps customers by providing a personalized recommendation. Collaborative filtering algorithms generate recommendations by considering the opinions of the similar-minded customers where similarity between customers is calculated by taking into account the previous purchase information of customers. Recommender systems are vulnerable to shilling attacks where fake customer profiles are inserted to promote or demote a product. In this paper, we use outlier detection algorithm to detect such types of attacks. Due to high dimensionality of the data that recommender systems deal with, we have applied subspace outlier detection algorithm in this context.

Keywords Recommender system · Collaborative filtering · Shilling attack · Subspace outlier detection algorithms

1 Introduction

E-commerce recommender systems provide recommendation to the customers who purchase product or service online for helping them in choosing the right product or service from a large number of alternatives. Recommender systems can be categorized into three groups—Collaborative filtering-based recommender systems [1, 2], content-based recommender systems, and hybrid recommender systems. Collaborative filtering-based recommender systems [1, 2] uses the previous purchase history of the customers and consider the ratings of *k*-nearest neighbors (similar minded customers based on ratings of previous purchases) of the customer for whom the recommendation is to be generated. In another variation of Collaborative filtering (called item-based collaborative filtering) [3], the similarity between items is calculated instead of users in generating predictions.

P. S. Chakraborty (✉)
University Institute of Technology, University of Burdwan, Bardhaman 713104, India
e-mail: Psc755@gmail.com

S. Kundu et al. (eds.), *Proceedings of the 2nd International Conference on Communication, Devices and Computing*, Lecture Notes in Electrical Engineering 602, https://doi.org/10.1007/978-981-15-0829-5_63

Selected Items			*Fillrer Items*			*Items with no ratings*			*Target Item*
i_1	...	I_k	I_{k+1}	...	I_m	I_{m+1}	...	I_p	I_t
r_1	...	r_k	r_{k+1}	...	r_m	*null*	...	*null*	R_t

Fig. 1 Generic attack profile

The paper is organized as follows. Section 1 provides a brief introduction on recommender system. Section 2 discusses about profile-injection attack (also called shilling attack) on recommender systems. Section 3 summarizes the major contributions of the researchers in the direction of attack detection on recommender system. Section 4 discusses subspace outlier detection technique and Sect. 5 elaborates the results obtained after applying subspace outlier detection technique in detecting attacks on recommender systems and finally, the paper is concluded in Sect. 6.

2 Attacks on Recommender System

In recommender systems, fake user profiles are inserted in target of influencing the recommendation provided to customers. This type of attack is called profile-injection attack or shilling attack [4, 5]. A generic attack profile is shown in Fig. 1. Target item is the item in which the attacker tries to promote or demote. In general, the successfulness of the attack depends greatly on the amount of information gathered by the attacker. Random attack, bandwagon attack, reverse bandwagon attack, love/hate attack, and hybrid attack are considered as low-knowledge attacks whereas favorite attack and average attacks are examples of high-knowledge attacks.

In this paper, the random, average, and bandwagon attack models have been taken into consideration for measuring the effectiveness of subspace outlier detection algorithm as a technique for detecting fake profiles (which are considered as outliers in this context). In case of random attack and average attack ratings for selected items in generic profile (shown in Fig. 1.) are missing. The two models differ in the ratings of the filler items. In case of average attack, the filler items contain average ratings for the respective items but in case of random attack, the filler items contain random rating drawn from the distribution of all ratings. In bandwagon attack model, the aim of the attacker is to tie up the target product with the frequently rated items. Keeping this idea in view, the attacker assigns maximum rating value for the target item as well as for some selected items (frequently rated).

3 Related Work

A number of works have been reported in the literature in the direction of identifying attacks on recommender system. Authors of paper [5] used supervised classification techniques for identifying fake profiles. Principal Component Analysis (PCA) based clustering algorithm has been used in attack detection in paper [6]. They assumed that there will be high correlation among the attack profiles. Authors of paper [7] use a clustering based and then an angle-based outlier detection algorithm [8] for identifying attack profiles. In paper [9], authors applied proximity-based (*k*-Nearest Neighbor based) outlier detection algorithm for detecting attacks.

4 Subspace Outlier Detection Algorithm

Subspace outlier detection is based on the idea that, in the high dimension, all the data points seem almost equidistant, if euclidian or similar types of distance measures are used and also some subspaces may be more relevant depending on the data generating process. Rather than dealing with the data objects in full dimension, it will be more effective to search for outlierness in subspaces. In other words, some subspaces of features of a data object may reveal more information regarding the nature of a point whether it is an outlier or not. The steps of the Subspace Outlier Detection (SOD) algorithm [10] that have been applied in this paper are as follows.

For calculating the outlierness of a point, x

1. A meaningful reference set of points, Ref(x) is chosen by Shared Nearest Neighbor (SNN) approach.
2. Based on this reference set, Ref(x), an axis-parallel hyperplane H is generated.
3. Now, distance between the point, x and the hyperplane H is calculated as

$$dist(o, H(S)) = \sqrt{\sum_{i=1}^{d} v_i^S \cdot (o_i, \mu_i^S)^2} \tag{1}$$

where v_i is the subspace defining vector and μ_i^S is the mean of the points of the reference set.

4. Finally, the outlier degree of x, OUTDEGREE(x) is calculated as follows:

$$SOD_{R(p)}(p) = \frac{dist(o, H(R(p)))}{\left\| v^{R(p)} \right\|_1} \tag{2}$$

In the Shared Nearest Neighbor (SNN) approach, the similarly between two points x and y is calculated as the number of common neighbors of x and y, i.e., $sim(x, y) = |N(x) \cap N(y)|$.

Points corresponding to the l-largest similarity values are selected as shared neighbors of x which are, in turn, forms $ref(x)$, the reference set of x.

Once the reference set of a point is known, an axis-parallel subspace hyperplane is chosen such that variance of the points in that subspace is high whereas along the axis to which the hyperplane is parallel, the variance of the points in that subspace is low. If d is the dimensionality of the data points and var($ref(x)$) is the total variance of the points in $ref(x)$ in full dimension, then subspace defining vector v is constructed as follows. For each attribute, Ai in the subspace, if variance of the points in $ref(x)$ in that attribute, vAi exceeds *alpha* times var$(ref(x))/d$ then the corresponding bit position in v contains 1, otherwise, it contains zero. The distance of the data point, x from the hyperplane, Hf corresponding to reference set, $ref(x)$ is calculated using Vi's as weights to individual dimensions, di. Subspace outlier degree of point, x is defined as the distance of the point from Hf normalized by the number of relevant dimensions, i.e., the dimensions for which corresponding bit positions in V is 1.

5 Experiment

MovieLens dataset (movielens.umn.edu) has been used in conducting the experiments which contains 100,000 ratings from 943 users and 1682 movies (items), with each user rated at least 20 items. Experiments have been made using HighDimOut R package [8, 10–12]. In this experiment, three attack models have been chosen, namely, Random attack, Average attack, and Bandwagon attack. For each of these attack models, push, and nuke both attack types have been considered. Attack percentage has been set to 1%. In creating fake user profiles under different attack models percentage of filler items has been set to 10%. Figure 2 shows sensitivity and specificity for different *alpha* (mentioned in algorithm) values. Consulting Fig. 2 value of *alpha* has been chosen as 0.6 for the next experiments.

For different combinations of nearest neighbor and selection values the algorithm has been run (result is shown in Table 1) and (40, 25) pair is chosen as the best combination for the next experiments.

Figures 3, 4, and 5 show the effectiveness of SOD in detecting attacks under Random, Average, and Bandwagon attack model, respectively. For each of these attack models, results for both push and nuke attacks have been shown. In case of detection of attacks following Bandwagon model, the SOD algorithm shows poor

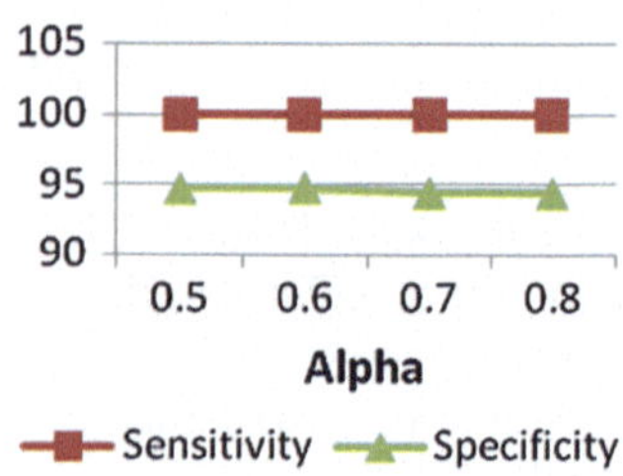

Fig. 2 Sensitivity and specificity of SOD for different Alpha values

Table 1 Sensitivity and Specificity for different *k-nn* and *sel* values

k-nn	sel	Sensitivity	Specificity
30	15	100	94
40	25	100	94.39
50	35	100	94.3
60	55	100	94.3

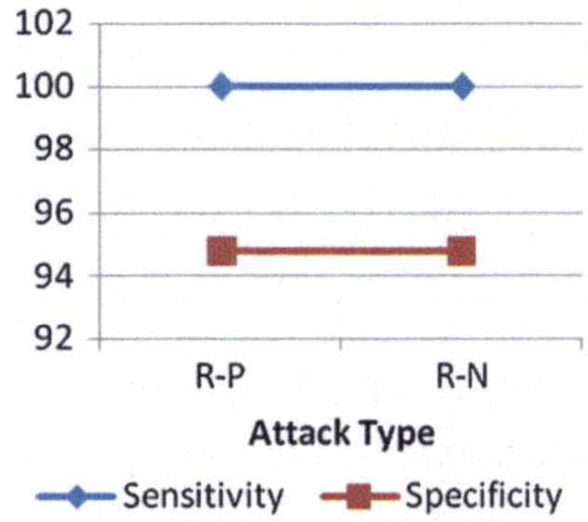

Fig. 3 Sensitivity and specificity of SOD in case of Random push (R-P) and Random nuke (R-N) attack

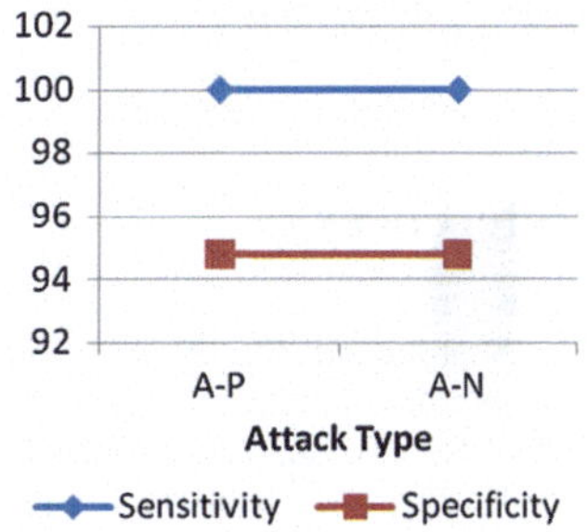

Fig. 4 Sensitivity and specificity of SOD in case of Average push (A-P) and Average nuke (A-N) attack

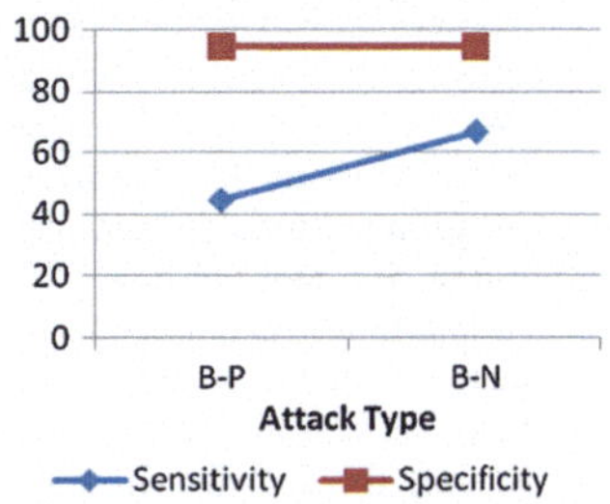

Fig. 5 Sensitivity and specificity of SOD in case of Bandwagon push (B-P) and Bandwagon nuke (B-N) attack

performance compared to other two models, namely, Random attack model and Average attack model. Figure 6 shows the precision of the algorithm for push and nuke attacks under the above mentioned three attack models.

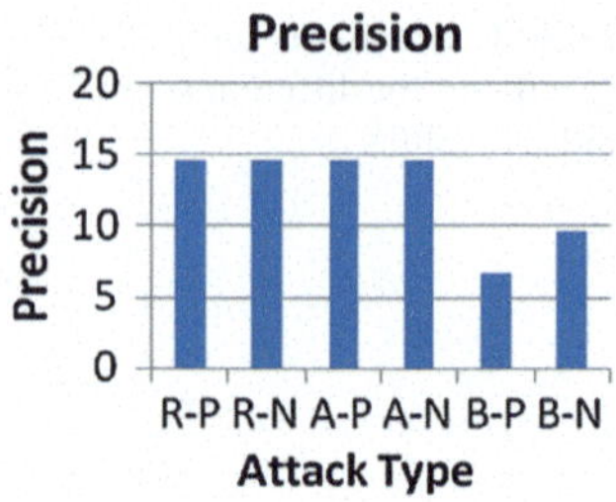

Fig. 6 Precision of SOD in case of Random push (R-P), Random nuke (R-N), Average push (A-P), Average nuke (A-N), Bandwagon push (B-P), and Bandwagon nuke (B-N) attack

6 Conclusion

In this paper, shilling attack on e-commerce recommender systems has been detected using subspace outlier detection technique where fake customer profiles which are injected into the system during attack event are considered as outliers. The experimental results show the effectiveness of the approach. In future work, the target will be to reduce the false positive rate and compare the result with approaches of attack detection in recommender system using other outlier detection techniques for high-dimensional data.

References

1. Shardanand, U., Maes, P.: Social information filtering: algorithms for automating "word of mouth". In: Proceedings of the SIGCHI Conference on Human Factors in Computing Systems (CHI'95). ACM Press/Addison-Wesley Publishing Co., pp. 210–217, New York, NY (1995)
2. Herlocker J., Konstan A., Riedl J.: Explaining collaborative filtering recommendations. In: Proceedings of ACM Conference on Computer Supported Cooperative Work, Philadelphia, PA (2000)
3. Sarwar, B., Karypis, G., Konstan, J., Reidl, J.: Item-based collaborative filtering recommendation algorithms. In: Proceedings of the 10th International Conference on World Wide Web (WWW'01), pp. 285–295. ACM, New York, NY (2001)
4. Lam, S., Riedl, J.: Shilling recommender systems for fun and profit. In: Proceedings of the 13th International WWW Conference, New York, NY (2004)
5. Burke, R., Mobasher, B., Williams, C., Bhaumik, R.: Detecting profile injection attacks in collaborative recommender systems. In: Proceedings of the IEEE Joint Conference on Ecommerce Technology and Enterprise Computing, E-Commerce and E-Services, CEC/EEE 2006, Palo Alto, CA (2006)
6. Mehta, B.: Unsupervised shilling detection for collaborative filtering. Association for the Advancement of Artificial Intelligence (2007). www.aai.org
7. Chakraborty, P.S., Karforma, S.: Detection of profile-injection attacks in recommender systems using outlier analysis. In: Proceedings of International Conference on Computational Intelligence: Modeling Techniques and Applications (CIMTA), Department of Computer Science, The University of Burdwan. Published in Procedia Technology, Elsevier (2013)
8. Kriegel, H., Schubert, M., Zimek, A.: Angle-based outlier detection in highdimensional data. KDD **444–452**, 2008 (2008)
9. Chakraborty, P., Karforma, S.: Effectiveness of proximity-based outlier analysis in detecting profile-injection attacks in E-Commerce Recommender Systems. In: Proceedings of

Second International Conference on Information Systems and Design and Intelligent Applications(INDIA-2015). Published by Springer in series of Advances in Intelligent Systems and Computing (AISC) (2015)
10. Kriegel, H.P., Kroger, H., Schubert, E., Zimek, A.: Outlier detection in axis-parallel subspaces of high dimensional data. In: PAKDD 2009, pp. 831–838 (2009)
11. Kriegel, H., Kroger, P., Schubert, M., Zimek, A.: Interpreting and unifying outlier scores. In: SDM 2011, pp. 13–24 (2011)
12. Lazarevic, A., Kumar, V.: Feature bagging for outlier detection. In: Proceedings of the 11th ACM International Conference on Knowledge Discovery and Data Mining (SIG KDD), Chicago, IL, pp. 157–166 (2005)

Measuring Song Complexity by Statistical Techniques

Sudipta Chakrabarty, Sangeeta Banik, Md. Ruhul Islam and Hiren Kumar Deva Sarma

Abstract The paper proposes a method to determine the complexity of a particular song. A song is one of the units that consists of a set of the music elements, which is sung by human and it comprises structures of different note patterns. In the context of Indian music, songs may vary with their style, genre, complexity, melody, etc. Therefore, song complexity is one of the quality metric factors that determine the character of the song. Hence, determination of song complexity of a particular song is an important task. This work presents an application using some statistical measures that can determine the complexity of a song. The ultimate goal of this work is to explore the complexity part of songs, which is one of the quality metric factors of a song and it also has great impact in the field of musicology.

Keywords Indian Music · Music information retrieval · Song complexity · Wave surfer · Standard deviation · Relative rank

1 Introduction

Song is the most powerful and important factor in Entertainment Industries. It can control all the emotional feelings of human being, such as mood, behavior, different mental set ups like depression, anxiety, insomnia, etc. In the context of Indian Music,

---

S. Chakrabarty (✉)
Department of MCA, Techno India, Salt Lake, Kolkata, West Bengal, India
e-mail: chakrabarty.sudipta@gmail.com

S. Banik
Department of CSE, SNU, DG Block, Newtown, Kolkata, West Bengal, India

Md. Ruhul Islam
Department of CSE, SMIT, Majhitar, Rangpo, East Sikkim, India

H. K. Deva Sarma
Department of IT, SMIT, Majhitar, Rangpo, East Sikkim, India

S. Kundu et al. (eds.), *Proceedings of the 2nd International Conference on Communication, Devices and Computing*, Lecture Notes in Electrical Engineering 602, https://doi.org/10.1007/978-981-15-0829-5_64

song (which is a form of music) can be categorized into different classes like, Classical Song, Folk Song, Modern Song of Low Beats, Modern Song of High Beats, Devotional Song, Ghazal, etc.

Furthermore, song complexity is one of the primary features of any song in Indian music. Some songs have low complexities whereas some others have high complexities. Therefore, measures of the song complexity of a particular song are very useful for singers, music learners, music listeners, and for Music Information Retrieval (MIR). Song complexity depends on different parameters like song lyrics, song and genre pattern, and fundamental frequency ranges of extracted note pattern structures of a particular song. The paper introduces a probabilistic approach for computing the song complexity of a particular song with the help of a few statistical concepts like mean, standard deviation, rank, relative rank, relative non-rank percentage, etc.

2 Related Works

Computational musicology is one of the most promising areas that depend on different concepts of computer science. Indian Music is very complicated, difficult, and large area and maybe, for that reason it is relatively unexplored from computing perspective. The paper [1] proposes a modeling technique and useful tool that formalizes the music composition rules and also increases music analysis speed with the help of Music Petri Nets. It introduces Schoenberg's Rules. The paper [2] introduces an approach that determines the similarity mapping between two songs and this is achieved by the notes and their fundamental frequencies of each note of the two songs and then calculating the Pearson's Correlation of Coefficient. The paper [3] presents a method to generate song list for listening and downloading according to age factor of users online. It is a web-based application that can recommend different songs depending on the listener's choice based on their age group. Songs are selected from the music library.

This paper [4] introduces a model of musical creativity rather than algorithmic music variations with the help of genetic algorithms and the implementation of the model is based on Genom, which is an application software. The paper [5] presents a statistical approach that can find similar song patterns with the help of coefficient of variance. Another paper [6], proposes the time-based raga recommendation system by using Neural Networks. The work presented in [7], is a music recommendation system that classifies different song clusters suitable for different times of a day. Another paper [8], proposed an intelligent mechanism to identify the density of a given music rhythm and complexity of that music rhythm automatically. The work presented in paper [9] illustrates different music research areas and their applications. The paper claims that raga and song can be represented by the Unified Modeling Language [10]. Automatic raga recognition is a method that identifies the notes of a song and then map to Raga Knowledgebase [11]. In [12], genetic algorithm and its operators are applied for creating new offspring music rhythm in producing the quality music rhythm. The basic target of this work is to create a context-aware tool for the field of computational musicology for education. The works presented in papers [13, 14] describe the Petri Nets modeling features and implement the whole song composition comprising vocal and rhythmic cycle structures. The paper [15] proposes a new application of genetic algorithm for automated creation of versatile drum set rhythms.

3 Proposed Work

Complexity is one of the most important quality metrics that is used to measure the overall criticalness of any particular process. Song is also a rendition process and thus, all song rendition processes have some complexities. Therefore, measuring the song complexity of a particular song is very useful for singers, music learners, music listeners, and for Music Information Retrieval (MIR). The workflow of the proposed work has been enlisted below:

Step 1: Take a song.
Step 2: Convert the song in f0 format from the song first, open wave surfer software, and click on file and choose a song, then click on transform and fix sample rate at 22050, channel is Mono and Pitch Method is Average Magnitude Difference Formula(AMDF) for the Conversion process.
Step 3: After that .f0 file is converted to the equivalent excel format (.csv).
Step 4: Now compute the number of occurrences of all the fundamental frequencies of each song.
Step 5: Fix 16 fundamental frequencies which have the highest occurrences respectively, in the fundamental frequency list, from the .csv file of each of the songs.
Step 6: Compute total frequency by the following formula:

$$\text{Total frequency} = \text{Frequency}\ (F) \times \text{Occurrence}\ (O) \tag{1}$$

Step 7:

$$\text{Calculate Mean} = \frac{1}{N}\sum \text{Total Frequency} \tag{2}$$

Here, $N = 16$
Step 8:

$$\text{Calculate Frequency Difference}\ (FD) = \text{Total Frequency} - \text{Mean} \tag{3}$$

Step 9:

$$\text{Calculate Standard Deviation}\ (SD) = \sqrt{\frac{\sum FD^2}{N}}$$

Step 10:

$$\text{Calculate Rank } (R) = \frac{\text{Individual Total Frequency}}{\sum \text{All Individual Total Frequencies}} = \frac{F \times O}{\sum F \times O} \tag{4}$$

Step 11:

$$\text{Calculate Relative Rank } (RR) = \frac{\text{Individual Rank}}{\text{Individual Highest Rank}} \tag{5}$$

Step 12:

$$\text{Compute Relative Non Rank } (RNR)\ \% = 1 - \text{Relative Rank } \% \tag{6}$$

Step 13:

$$\text{Calculate Song Complexity} = \frac{\sum \text{Relative NonRank } (RNR)\%}{\text{Standard Deviation } (SD)} \times 100 \tag{7}$$

4 Result Set Analysis

To achieve the work, there are quite a lot of songs that have been taken into consideration and have been found their complexities. In this paper, we consider three songs—one classical song from Hindi movie, one devotional song from Hindi movie, and one folk song popularly known as Goalparia Lokgeet from the state of Assam in India. Complexity of each of the song has been computed and that is depicted in Table 1, Table 2, and Table 3, respectively. There is a well-known truth that classical music rendition is very critical in comparison to devotional and folk songs. Complexity of classical songs is expected to be higher than that of devotional and folk songs.

The results also reflect the same thing, song 1 is a classical song and its complexity (20.45478) is quiet higher than the complexity (2.1424282) of devotional song (song 2) and the complexity (10.24407684) of folk song (song 3).

Test Case 1:

Song 1: "Laaga Chunri Mein Daag"
Song Type: Hindi Classical Song of movie "Dil Hi to Hai"
Singer: Manna Dey

The complexity of the classical song "Laaga Chunri Mein Daag" is 20.45478; computed results are depicted in Table 1.

Table 1 Measuring complexity of song "Laaga Chunri Mein Daag"

Frequency (F)	Occurrence (O)	Total frequency (F * O)	Mean	Frequency difference (FD)	FD^2	Standard deviation (SD)	Rank (R)	Relative rank (RR)	Relative non rank % (RNR)	Song complexity
110	546	60060	51754.875	8305.125	68975101	7759.9144	0.072529399	0.92304855	99.076951	20.45478
123	529	65067		13312.125	177212672		0.078575931	1	99	
113	525	59325		7570.125	57306793		0.071641802	0.9117525	99.088247	
111	519	57609		5854.125	34270780		0.069569533	0.88537969	99.11462	
119	495	58905		7150.125	51124288		0.071134603	0.90529762	99.094702	
109	461	50249		−1505.875	2267659.5		0.060681482	0.77226551	99.227734	
121	449	54329		2574.125	6626119.5		0.065608554	0.83497011	99.16503	
116	445	51620		−134.875	18191.266		0.062337123	0.7933361	99.206664	
117	438	51246		−508.875	258953.77		0.061885475	0.78758818	99.212412	
114	428	48792		−2962.875	8778628.3		0.058921986	0.74987321	99.250127	
98	426	41748		−10006.875	100137547		0.050415541	0.64161557	99.358384	
136	424	57664		5909.125	34917758		0.069635952	0.88622497	99.113775	
108	405	43740		−8014.875	64238221		0.052821111	0.67223016	99.32777	
126	399	50274		−1480.875	2192990.8		0.060711672	0.77264973	99.22735	
107	388	41516		−10238.875	104834561		0.050135374	0.63805001	99.36195	
106	339	35934		−15820.875	250300086		0.043394463	0.55226152	99.447738	
		$\sum F * O = 828078$							$\sum RNR = 1587.2735$	

Table 2 Measuring complexity of song "Sun Ri Pawan Pawan Purvaiya"

Frequency (F)	Occurrences (O)	Total frequency (F * O)	Mean	Frequency difference (FD)	FD^2	Standard deviation (SD)	Rank (R)	Relative rank (RR)	Relative non rank % (RNR)	Song complexity
208	1544	321152	132086.4375	189065.5625	35745786923	74374.459	0.1519611	1	99.00000002	2.1424282
210	1263	265230		133143.5625	17727208235		0.1255002	0.8258706	99.1741294	
206	1161	239166		107079.5625	11466032705		0.1131674	0.7447128	99.25528723	
204	744	151776		19689.5625	387678871.4		0.0718166	0.4725986	99.52740136	
212	679	143948		11861.5625	140696664.9		0.0681126	0.4482239	99.55177611	
214	536	114704		−17382.4375	302149133.4		0.0542751	0.3571642	99.6428358	
202	520	105040		−27046.4375	731509781.4		0.0497023	0.3270725	99.67292747	
220	494	108680		−23406.4375	547861316.4		0.0514247	0.3384067	99.66159327	
222	493	109446		−22640.4375	512589410.2		0.0517871	0.3407919	99.65920811	
218	447	97446		−34640.4375	1199959910		0.046109	0.3034264	99.69657359	
216	446	96336		−35750.4375	1278093781		0.0455838	0.2999701	99.7000299	
242	370	89540		−42546.4375	1810199344		0.0423681	0.2788088	99.72119122	
183	358	65514		−66572.4375	4431889435		0.0309996	0.2039969	99.79600314	
185	348	64380		−67706.4375	4584161679		0.030463	0.2004658	99.79953418	
225	337	75825		−56261.4375	3165349350		0.0358785	0.2361031	99.76389685	
200	326	65200		−66886.4375	4473795521		0.030851	0.2030191	99.79698087	
		$\sum F * O = 2113383$							$\sum RNR = 1593.419369$	

Table 3 Measuring complexity of song "Bhawaiya"

Frequency (F)	Occurrences (O)	Total frequency (F * O)	Mean	Frequency difference (FD)	FD^2	Standard deviation (SD)	Rank (R)	Relative rank (RR)	Relative non rank % (RNR)	Song complexity
160	625	100000	69156.938	30843.062	951294474	15510.767	0.090374158	1	99	10.24407684
179	527	94333		25176.062	633834098		0.085252655	0.94333	99.05667	
180	517	93060		23903.062	571356373		0.084102191	0.9306	99.0694	
159	485	77115		7958.062	63330750.8		0.069692032	0.77115	99.22885	
177	464	82128		12971.062	168248449		0.074222489	0.82128	99.17872	
162	440	71280		2123.062	4507392.26		0.0644187	0.7128	99.2872	
176	387	68112		−1044.938	1091895.42		0.061555647	0.68112	99.31888	
182	385	70070		913.062	833682.216		0.063325173	0.7007	99.2993	
158	371	58618		−10538.94	111069214		0.052975524	0.58618	99.41382	
175	341	59675		−9481.938	89907148.2		0.053930779	0.59675	99.40325	
157	327	51339		−17817.94	317478915		0.046397189	0.51339	99.48661	
165	325	53625		−15531.94	241241098		0.048463142	0.53625	99.46375	
183	323	59109		−10047.94	100961058		0.053419261	0.59109	99.40891	
164	318	52152		−17004.94	289167916		0.047131931	0.52152	99.47848	
173	305	52765		−16391.94	268695631		0.047685924	0.52765	99.47235	
214	295	63130		−6026.938	36323981.7		0.057053206	0.6313	99.3687	
		$\sum F * O = 1106511$							$\sum RNR = 1588.93489$	

Test Case 2:

Song 2: "Sun Ri Pawan Pawan Purvaiya"
Song Type: Hindi Devotional Song of movie "Anuraag"
Singer: Lata Mangeshkar

Complexity of the devotional song "Sun Ri Pawan Pawan Purvaiya" is 2.1424282; computed results are depicted in Table 2.

Test Case 3:

Song 3: "Bhawaiya"
Song Type: Goalparia Lokgeet
Singer: Krishnamoni Chutiya

The complexity of the Goalparia Lokgeet "Bhawaiya" is 2.1424282; computed results depicted in Table 3.

5 Conclusion

In this work, a useful model has been proposed to determine the song complexity of any particular type of song. The measures of song complexity of a particular song are very useful for singers, music learners, music listeners, and for Music Information Retrieval (MIR) as a whole. Song complexity depends on different parameters like song lyrics, song and genre pattern, and fundamental frequency ranges of extracted note pattern structures of a particular song.

The primary aim of this contribution is to explore the nature of any song in the form of complexity. This contribution may be very much effective for all song composers for choosing the appropriate singer for appropriate song, depending on the song complexity value of a particular song. This work has been implemented by using a few statistical concepts like mean, standard deviation, rank, relative rank, relative non-rank percentage, etc. The findings of this work may be applied as a step toward contributing and assisting song composers or music professionals, in finding song complexity for evaluating the nature of the song rendition process and mapping the appropriate singer.

References

1. Baratè, A., Mauro, A.D., Haus, G., Ludovico, L.A.: Formalizing Schoenberg's fundamentals of musical composition through Petri Nets. In: 15th Sound and Music Computing Conference (SMC2018), Limassol, Cyprus, pp. 254–258, 4–7 July 2018
2. Chakrabarty, S., Ruhul Islam, Md., Deva Sarma, H.K.: An approach towards the modeling of pattern similarity in music using statistical measures. Presented paper in the 5th IEEE International Conference on Parallel, Distributed and Grid Computing (PDGC), in JUIT, Waknaghat, Himachal Pradesh, Dec 2018

3. Chakrabarty, S., Banik, S., Ruhul Islam, Md., Deva Sarma, H.K.: Context aware song recommendation system. Presented Paper in the 3rd National Conference on Communication, Cloud, and Big Data (CCB), in SMIT, Majhitar, East Sikkim, Nov 2018
4. Arutyunova, V., Averkin, A.: Genetic algorithms for music variation on genom platform. In: 9th International Conference on Theory and Application of Soft Computing, Computing with Words and Perception, ICSCCW 2017, Budapest, Hungary, 24–25 Aug 2017. Procedia Comput. Sci. **120**, 317–324 (2017)
5. Chakrabarty, S., Md. Ruhul Islam, De, D.: Modelling of song pattern similarity using coefficient of variance. Int. J. Comput. Sci. Inf. Secur. 388–394 (2017). ISSN 1947-5500
6. Roy, S., Chakrabarty, S., De, D.: Time-based raga recommendation and information retrieval of musical patterns in indian classical music using neural network. IAES Int. J. Artif. Intell. (IJ-AI) 33–48 (2017). ISSN: 2252-8938
7. Chakrabarty, S., Roy, S., De, D.: Handbook of research on intelligent analysis of multimedia information (hardcover) (Chapter 12). In: Siddhartha, B., Hrishikesh, B., Sourav, D., Klepac, G. (eds.) Time-Slot Based Intelligent Music Recommender in Indian Music. IGI Global, USA (2016). ISBN13: 9781522504986, ISBN10: 1522504982
8. Chakrabarty, S., Karmakar, G., Md. Ruhul Islam, De, D.: Raconing of music rhythm density and complexity through mathematical measures. In: Proceedings of the Advanced Computational and Communication Paradigm, LNEE Lecture Note. Springer (2017)
9. Chakrabarty, S., Roy, S., De, D.: A foremost survey on state-of-the-art computational music research. In: Proceedings of the Recent Trends in Computations and Mathematical Analysis in Engineering and Sciences, pp. 16–25. International Science Congress Association (2015)
10. Chakrabarty, S., Roy, S., De, D.: Behavioural modelling of ragas of indian classical music using unified modelling language. In: Proceedings of the 2nd International Conference on Perception and Machine Intelligence, pp. 151–160. ACM (2015)
11. Chakrabarty, S., Roy, S., De, D.: Automatic raga recognition using fundamental frequency range of extracted musical notes. In: The Proceedings of Eight International MultiConference on Image and Signal Processing (ICISP 2014), pp. 337–345, Elsevier (2014)
12. Chakrabarty, S., Roy, S., De, D.: Pervasive diary in music rhythm education: a context-aware learning tool using genetic algorithm. In: The Proceedings of Advanced Computing, Networking and Informatics, vol. 1, pp. 669–677. Springer International Publishing (2014)
13. Roy, S., Chakrabarty, S., Bhakta, P., De, D.: Modelling high performing music computing using Petri Nets. In: The Proceedings of the IEEE International Conference on Control, Instrumentation, Energy and Communication (CIEC 2014) (2014)
14. Roy, S., Chakrabarty, S., De, D.: A framework of musical pattern recognition using Petri Nets. In: The Proceedings of Emerging Trends in Computing and Communication, pp. 245–252. Springer India (2014)
15. Chakrabarty, S., De, D., Quality measure model of music rhythm using genetic algorithm. In: Proceedings of IEEE International Conference on Radar, Communication and Computing (ICRCC), pp. 125–130 (2012)

An IOT-Based System for Classification and Identification of Plastic Waste Using Near Infrared Spectroscopy

Irfan Bhati and Manasija Bhattacharya

Abstract In this paper, an Internet of things (IoT) system for the identification of different types of plastics has been presented using economical near infrared (NIR) spectral sensor. The presented system is highly accurate and economical in distinguishing between various kinds of plastics such as high-density polyethylene (HDPE), polyethylene terephthalate (PET), polystyrene (PS), and polypropylene (PP). Reflectance at four different wavelengths is used to classify and identify these four different plastics. Moreover, multiple machine learning algorithms are applied and compared to improve the accuracy of the system.

1 Introduction

Plastics are one of the most widely and regularly used products in everyday life. They are easy and cheap to manufacture but its prime feature is also its biggest disadvantage. Plastics are hard to be broken down by microbes due to their long hydrocarbon chains and thus wide usage of plastics causes a wide array of pollution-related problems. Therefore, recycling plastics reduce a lot of waste and is crucial for reducing pollution in the environment. There is a wide range of plastic resins that are incompatible with each other [1]. Since different kinds of plastics come mixed in the household waste, sorting plastics according to their constituent resin is mandatory to avoid contamination as such mixing can make the recycled plastic unusable. If the plastic comes presorted then it can heavily reduce the cost of processing and sorting the plastics. There are four major categories of plastic, namely, high-density polyethylene (HDPE), polyethylene terephthalate (PET), polystyrene (PS), and polypropylene (PP). The distinction is done between these four categories. Current methods for identification and sorting of waste plastics include both manual

I. Bhati (✉) · M. Bhattacharya
Haldia Institute of Technology, Haldia, West Bengal, India
e-mail: irfanhb3@gmail.com

M. Bhattacharya
e-mail: manasijait.bhattacharya@gmail.com

S. Kundu et al. (eds.), *Proceedings of the 2nd International Conference on Communication, Devices and Computing*, Lecture Notes in Electrical Engineering 602, https://doi.org/10.1007/978-981-15-0829-5_65

and automated sorting systems. Sorting and separation techniques usually involve finding differences in chemical, optical, or physical properties between the various plastics to be sorted [2]. Manual sorting [3] is still used in many places but requires a trained worker to sort through the plastics and is also slower than automated sorting. On the other hand, automated sorting machines are faster but require expensive machinery to operate. For an average person, it is quite difficult to separate the plastics in the absence of plastic identification label printed on the plastic material. Various methods of automated sorting have been put forward such as X-Ray detector and NIR sensor/camera. Both are quite accurate but X-Rays emit harmful radiation that require expert handling and shielding and are even costlier than NIR camera/sensor [4]. Another approach includes using sound wave signal produced by plastics [5]. In 1995, Scott [6] put forward a method for the identification and separation of PET and PVC resins from each other using NIR [7]. In this paper, we will be proposing a portable and highly economical system to identify a given plastic material using NIR spectroscopy other than a costly non-portable solution [8, 9]. This device will enable anyone to identify the plastic waste before or after disposing of. The device will generate the result directly to the users' phone in real time so that the device can be used in households too to separate their plastic waste before mixing and disposing it into the garbage, thus making it easier to recycle the waste. A dataset containing 100 different readings of the disparate plastic materials was constructed. Various machine learning algorithms were used and compared on the dataset, such as—Linear Regression (LR), K Nearest Neighbors (KNN), Decision Tree Classifier (CART), and Gaussian Naive Bayes (NB). It was found out that Naive Bayes performed the best with an accuracy of 98.75% under normal room lighting condition. Furthermore, an IOT-based system [10] was constructed using the machine learning model to get the data delivered straight to a smartphone or a personal computer.
Naive Bayes equation:

$$P(Y|X) = \frac{P(X|Y) \times P(Y)}{P(X)} \tag{1}$$

where Y is class variable and X is a dependent feature vector (of size n) where

$$X = (x_1, x_2, x_3, \ldots, x_n) \tag{2}$$

2 Materials

A total of 20 different samples of HDPE, PP, PS, and PET were collected from a scrapyard, which was presorted by the manual workers. Some samples had dirt and some were clean, some had different sizes and colors to make the data more versatile. A NIR spectral sensor called the AS7263 was used to collect the spectral reading of the given samples. The AS7263 is a digital six-channel spectrometer for spectral identification in the near IR (NIR) light wavelengths. AS7263 consists of six independent optical filters whose spectral response is defined in the NIR wavelengths from approximately 600–870 nm with full-width half-max (FWHM) of 20 nm. The data from the sensor was fed to a NodeMCU board. NodeMCU is an open-source IoT platform. It includes firmware which runs on the ESP8266 Wi-Fi SoC from Espressif Systems, and hardware which is based on the ESP-12 module.

3 Methods

The AS7263 comes pre-built for detection of 610, 680, 730, 760, 810, and 860 nm of light each with 20 nm of full-width half-max detection. AS7263 comes with an inbuilt bulb (5700 K LED) that was illuminated to take the readings. The NIR sensor was surrounded by a small cardboard casing to stabilize the sensor for reading. The sensor was kept at a distance of 1 cm from every sample with the onboard bulb illuminated while taking the readings. The readings were collected in the form of R(610 nm), S(680 nm), T(730 nm), U(760 nm), V(810 nm), and W(860 nm) floating-point variables along with tempF(temperature in Fahrenheit). The readings were labeled initially as HDPE, PP, PS, and PET for construction of the subsequent machine learning model. The sensor was connected to a NodeMCU board via a wired connection. The collected data was structured and put into a CSV file for constructing a machine learning model. Four different machine learning algorithms, namely, Linear Regression (LR), K Nearest Neighbors (KNN), Decision Tree Classifier (CART), and Gaussian Naive Bayes (NB) were used and compared for the dataset. The following features—S(680 nm), T(730 nm), U(760 nm), V(810 nm), and W(860 nm) were used and R(610 nm) was discarded as it varied a lot according to the color of plastic material and reduced the efficiency of the algorithms. A 98.75% accuracy(NB) was found when wavelengths S, T, U, V, and W were selected as features. The temperature factor is being disregarded in the current research paper. We can look at the interactions between the variables using multivariate scatterplots. This can be helpful to spot structured relationships between input variables (Fig. 1).

The diagonal grouping of some pairs of attributes suggests a high correlation and a predictable relationship (Fig. 2).

4 Results and Discussion

It is found out that the spectral readings at S(680 nm), T(730 nm), U(760 nm), V(810 nm), and W(860 nm) can be used as features and no feature scaling is required as we can see from the highly correlated Scatter plot. A random exert of 40 readings from the data sample of 100 is given below (Table 1):

The following algorithms Linear Regression (LR), K Nearest Neighbors (KNN), Decision Tree Classifier (CART), and Gaussian Naive Bayes (NB) when applied to the collected data give the following results with their corresponding standard deviations—LR: 0.814286 (0.126320), KNN: 0.948214 (0.063613), CART:

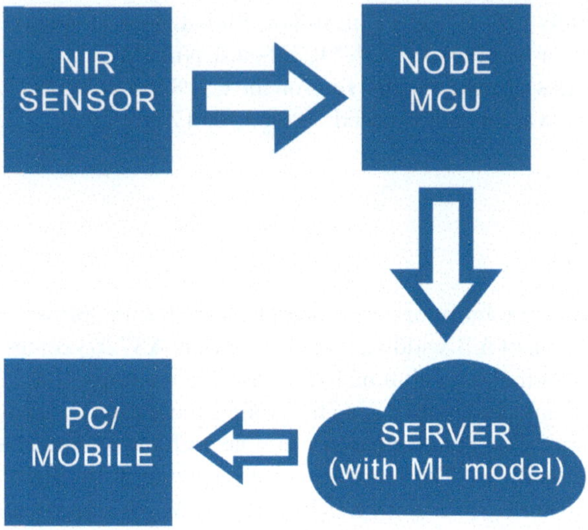

Fig. 1 IOT architecture

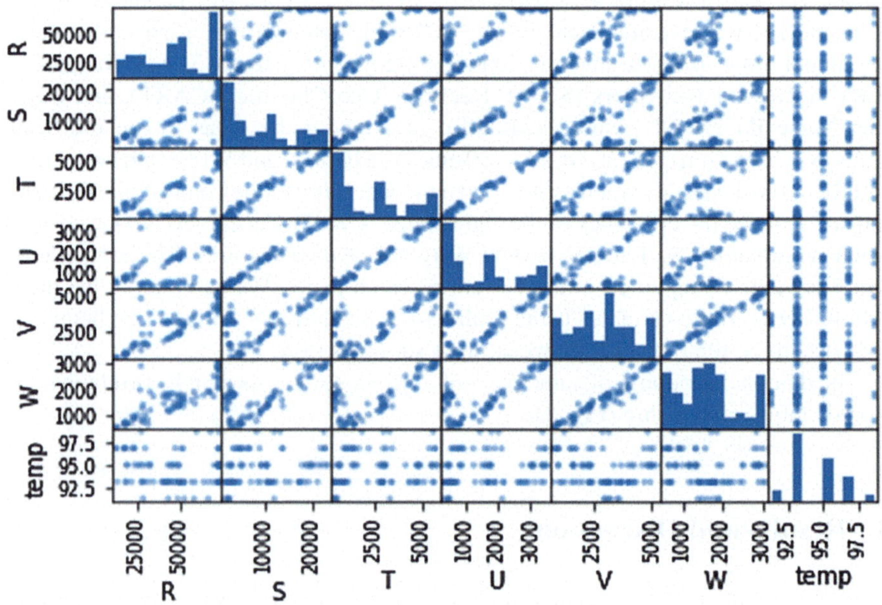

Fig. 2 Scatter plot

Table 1 Spectral readings

S	T	U	V	W	tempF	Class
3115.5	655.46	455.6	3370.34	1963.21	95	pet
3221.84	679.02	469.23	3684.51	1874.19	95	pet
2160.53	447.68	330.99	2326.43	1176.48	95	pet
2138.42	434.83	327.1	2314.5	1257.08	93.2	pet
2672.23	729.36	472.15	3536.37	1500.08	93.2	pet
6394.19	1253.09	930.68	1730.9	890.18	96.8	hdpe
6912.21	1711.48	1248.04	1543	922.66	96.8	hdpe
5516.08	1216.67	941.38	1532.06	927.47	96.8	hdpe
8925.34	2221.29	1600.45	1761.72	1005.66	96.8	hdpe
8287.29	1884.99	1447.61	1263.63	771.09	98.6	hdpe
22443.38	5606.77	3359.58	4809.95	2841.36	98.6	ps
22751.87	5675.32	3399.5	4887.49	2882.26	96.8	ps
21772.68	5407.56	3256.39	4663.8	2780.01	96.8	ps
22600.25	5714.94	3468.62	5047.56	3014.59	96.8	ps
18745.63	5769.57	3462.77	4130.91	2920.76	95	ps
11399.63	2899.24	1930.47	2853.36	1660.07	93.2	pp
10744.73	2926.02	1860.38	2927.92	1605.94	93.2	pp
11782.88	3460.45	2180.66	3486.66	1761.12	93.2	pp
11997.67	3348	2132.96	3621.88	1894.64	91.4	pp
12359.86	3397.26	2167.03	3670.59	1937.95	93.2	pp

0.975000 (0.050000), and NB: 0.987500 (0.037500). It is evident that NB gives an accuracy of 98.75% followed by CART with an accuracy of 97.5%. So, NIR spectroscopy is a highly efficient method for classification of plastics as different plastic materials absorb and reflect light at different intensities according to their chemical properties (Fig. 3).

The resulting Gaussian Naive Bayes model is exported and used to classify a new plastic material into one of the four categories. For an unknown classification, the nodeMCU sends the collected features from the sensor to our given server where the features are analyzed with the help of the Naive Bayes model that we constructed from the collected data. The prediction (classification) is then stored in a database where it can be viewed accordingly. A web app was built which connects the database to a web interface where we can view the data from a smartphone or computer (Table 2).

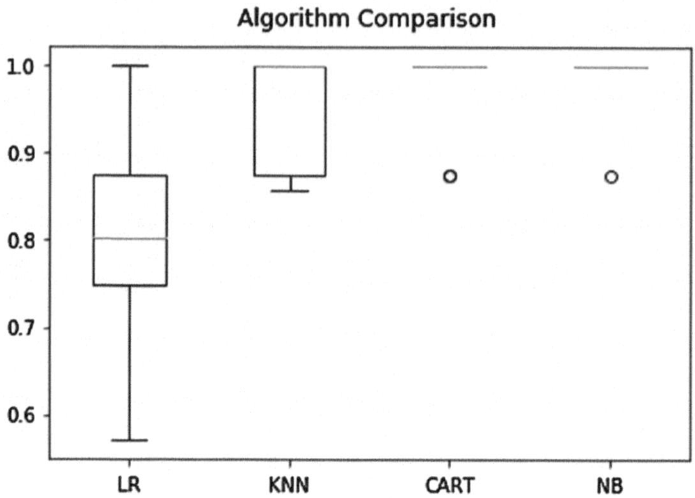

Fig. 3 Whiskers plot

Table 2 Gaussian NB parameters

Type	Precision	Recall	F1-score	Support
hdpe	1.00	0.94	0.97	18
pet	1.00	1.00	1.00	19
pp	0.95	1.00	0.97	18
ps	1.00	1.00	1.00	21
Avg/total	0.99	0.99	0.99	76

5 Conclusion

This way we constructed a cheaper and highly efficient alternative to expensive spectroscopic equipments. The current dataset is small and will be further expanded to broaden our generalization. The model was made under normal room lighting condition and temperature. One important feature that we did not take into consideration was the outside lighting conditions and we plan on including it in our further research works. More importantly, the model created in the study can be used to aid manual sorting of plastic at home or recycling facilities and can also be developed further to complete the whole task automatically. This will ultimately help to reduce pollution and provide better recycling.

References

1. Carvalho, M.T., Ferreira, C., Portela, A., Santos, J.T.: Application of fluidization to separate packaging waste plastics. Waste Manag. **29**, 1138–1143 (2009)
2. Scheirs, J.: Polymer Recycling. Wiley Publication (1998)
3. Wahab, D.A., Abidin, A., Azhari, C.H.: Recycling trends in the plastics manufacturing
4. American Plastics Council: Best Practices in PET Recycling, Automated Sorting Systems NAPCOR (1997)
5. Anuoluwapo, O.A., Jimoh, K.A., Ayilara, O.A.: Evaluation of plastic waste classification systems. Br. J. Math. Comput. Sci. (2016)
6. Scott, D.M.: A two-color near-infrared sensor for sorting recycled plastic waste. Meas. Sci. Technol. **6**, 156–159; and recycling companies in Malaysia. J. Appl. Sci. **7**(7), 1030–1035 (2007)
7. Masoumi, H., Safavi, S.M., Khani, Z.: Identification and classification of plastic resins using near infrared reflectance spectroscopy. Int. J. Mech. Ind. Eng. **6** (2012)
8. Hearn, G.L., Ballard, J.R.: The use of electrostatic techniques for the identification and sorting of waste packaging materials. Resour. Conserv. Recycl. **44**(1), 91–98 (2005)
9. Taylor, B.: Visions of plastic. Recycl. Today **44**(4), 48–56 (2006)
10. Abed, A.A.: Internet of Things (IoT): architecture and design. In: Al-Sadeq International Conference on Multidisciplinary in IT and Communication Science and Applications (AIC-MITCSA) (2016)

Voting Rule Applied to Mobile Service Provider's Survey—An Indian Perspective

Avijit Bose, Prasenjit Das and Sumit Majumdar

Abstract One of the major concepts of voting is alternative. There are numerous examples of alternatives, e.g., several versions of an immigration reform bill, electing the political party, etc. In this paper, we are considering the survey result, which shows the preferences over different mobile service provider's alternatives. Statistical data referred from TelecomLead (TelecomLead—Magazine. https://www.telecomlead.com, [1]) choosing mobile service provider's customer share playing a major role in Indian Telecommunication scenario. We use here Plurality, Copeland, and Borda's rule and tried to find the correlation between MSP's.

Keywords QoS · Voting · Social choice function (SCF) · Plurality · Copeland · Borda rule

1 Introduction

Currently, the Indian market is facing tough competition in the field of Cellular Data Communication. Earlier voice communication played the main role but with new spectrum allocation, voice communication has become insignificant and it is the data and value-added services, which is playing a key role in the revenue generation of MSP's. Rather than to focus on the statistical inference and penetrating the parameters, which lead to the statistical inference result we develop this problem into a social choice problem and apply voting rule to these data sets which helps us to conclude some facts.

A. Bose (✉) · P. Das · S. Majumdar
Computer Science & Engineering, MCKV Institute of Engineering, Howrah, India
e-mail: avijit2691@gmail.com

P. Das
e-mail: mr.das.prasenjit@gmail.com

S. Majumdar
e-mail: sm071985@gmail.com

S. Kundu et al. (eds.), *Proceedings of the 2nd International Conference on Communication, Devices and Computing*, Lecture Notes in Electrical Engineering 602, https://doi.org/10.1007/978-981-15-0829-5_66

1.1 Voting

One of the major concepts of voting is the alternative. Voting may take place in the following scenario as mentioned in [2]. The scenarios may be (a) different amounts to spend on building and (b) several versions of an immigration reform bill. While dealing with voting at first the paper comes across May's rule, which states, "only reasonable voting method is the majority rule." However, it is not true always especially the case where multicandidate voting is to be taken into account. Voting theory defines Social Computing Function (SCF) as "Each voter submits a linear ordering of the alternatives specifying first, second, third choice so there may be a winner or groups of winner as outcome." However, while dealing with voting rule, researchers often come across majority cycle proposed by Condorcet in 1785 which says that collective preference violates what might be expected from an individual. Arrow's theorem plays a crucial role in voting theory, which states that the "relative merits should not be influenced by individual's voter opinion about an irreverent third alternative." According to him having three or more alternatives is a dictatorship. Gibbard–Satterthwaite theorem states, "Every election will have an individual winner." According to him SCF (f) sometimes helps an individual voter to get an incentive and manipulate the voting result. This paper does not consider the above two conditions as an influencing parameter.

1.2 Voting and Social Choice Function: Revisited

Voting generally takes place whenever a group of voters cast ballots that are used as a basis for collective decisions reached through the application of a voting rule. This paper has taken the statistical data from Telecom Lead, a portal [1] that has given the current data set depicting mobile operator's customer share as on April 2018.

1.2.1 Mathematical Notation for a Voting Rule

(1) $N = \{1, 2, 3, \ldots, n\}$ Is finite set of voters
(2) $A =$ finite set of m alternatives with $m \geq 2$
(3) $\geq_i$ A is the ballot cast by voter i in a linear ordering if $a \geq_i b$ and $b \geq_i c$ then $a \geq_i c$ for all $a, b, c \in A$.
(4) It is complete if $a \geq_i b$ or $b \geq_i a$ for all $a \neq b \in A$.
(5) It is reflexive if $a \geq_i a$, for all $a \in A$.
(6) It is antisymmetric if $a \geq_i b$ and $b \geq_i a$ then $a = b$ for $a, b \in A$.
(7) L (A) ➔ set of all linear orderings for given A. The anti-reflexive $x > y$ means $x \geq_i y$ holds and $y \geq_i x$ fails.

A profile $P = (\geq_1, \geq_2, \ldots, \geq_n)$ specifies such a ballot for each voter $i \in N$, $\mathcal{L}(A)^n$ denotes the set of all such profiles for a given n. $\mathcal{L}(A)^{<\infty}$ stands for $U_{n\in N}\ \mathcal{L}\ (A)\ ^n$ (where N denotes the set of all natural numbers).

2 Literature Survey

In [3] there are different parameters, e.g., billing performance, help services, network connectivity and coverage, call drop, and customer satisfaction is covered and they get an average of 77% Customer Satisfaction Index (CSI). In [2] researchers did address voting as a social choice approach where different approaches like plurality, pairwise majority, Copeland, and Borda rules are mentioned. This acts as an inspiration for this paper to implement preferences and alternatives. In [4] authors reveal the idea that different mobile Internet applications can cause deterioration in mobile network performance. QoE is the concept proposed where QoE stands for Quality of Experience, and the paper proposes a framework describing the process of estimating or predicting perceived QoE based on data sets. In [1] Telecom Regulatory Authority of India (TRAI) has released the latest report on mobile operators market share up to April 2018 which reveals Airtel has 27.44% share, Vodafone has 19.74% share, and Reliance Jio has 17.44% share. In [5] authors mention that in the age of digital technology how Reliance is capturing the market through its bundled packages. With the aid of univariate and multivariate statistical analysis, the paper concludes that customers are satisfied with JIO service. In [6] researchers did study an objective to understand the Indian consumer's perception choice in selecting cellular mobile communication service providers. They carried out factor analysis based on the parameters like communication quality, call service, facilities, price, customer care, and service provider's alternatives. The conclusion is that product quality and availability has a significant impact on consumer's perception choice in selecting cellular mobile service provider. In [7] we come to note that by 2020 mobile industry will have 82% contribution to India's GDP. Total 810 million users will be using Smartphone by 2020. Net effective price will be less than $25 and data rate will be less than $1 per GB. All the above papers have influenced the current paper where voting rule indirectly signifies the customer satisfaction parameters.

3 Problem Formulation and Solution

As per the statistical data, this paper has taken into account the customer base of major three providers

(1) Bharti Airtel
(2) Vodafone
(3) Reliance Jio.

Table 1 Depicting the statistical data (*Source* Telecom Lead April 2018)

Alternatives	Voters		
	27.44	19.74	17.44
	Airtel	Vodafone	Reliance
	Vodafone	Airtel	Airtel
	Reliance	Reliance	Vodafone

As per [1] 27.44 million users prefer Airtel to Vodafone and Reliance. 19.74 million prefer Vodafone over Bharti and Reliance. 17.44 million prefer Reliance Jio over Airtel and Vodafone. So from the definition mentioned in Sect. 1.2.1 (1) we have N = {27.44, 19.74, 17.44} and A = {Airtel, Vodafone, Reliance-Jio} which satisfies the condition $m \geq 2$. So now, we can represent the tabular structure.

Case-I—Plurality Rule

Plurality rule states that once the alternatives are arranged in descending order number of votes that put the first alternative in first position depicts the Plurality Value. So from the above definition we can write mathematically

$$P(x) = \{n : n \in N \text{ that puts } x \in A \text{ in the 1st position}\} \quad (1)$$

So directly from Table 1 and from Eq. (1) we get the values of P = {27.44, 19.74, 17.44} which indicates the fact that P(Airtel) = 27.44, P(Vodafone) = 19.74, and P(Reliance) = 17.44. The time complexity is simple as just we have to search the number of columns and index the first element. Therefore, we get an order of O(n) complexity.

Case-II—Pairwise Majority Rule

The pairwise majority rule basically depicts the pairwise difference between number of votes received by a compared to b and number of votes received by b compared to a, mathematically this can be denoted by

$$Net_p(a > b) = |\{j \in N | a >_j b\}| - |\{j \in N | b >_j a\}| \quad (2)$$

So we get γ (Airtel > Vodafone) = 44.88 and γ (Vodafone > Airtel) = 19.74.

So Net_p (Airtel > Vodafone) = [γ (Airtel > Vodafone) − γ (Vodafone > Airtel)] = 25.14, where γ is the intermediate Pairwise majority calculation constant. In this way, we can get Net_p (Vodafone > Reliance) = 29.74 and Net_p (Airtel > Reliance) = 29.74.

So Net_p {A > V, V > R, A > R} = {25.14, 29.74, 29.74}

Thus, it is seen that pairwise majority can only compare customer base index between the two or three MSPs but individual discrete alternative value cannot be calculated. The order of complexity is $O(n^2)$.

Case-III Copeland Score Calculation

Copeland score calculation deals with a rule that voting rule will score candidates according to their win loss record in the pairwise majority case. Defining from above, we get

$$\mathrm{Copeland}(x) = |\{y \in A|x >_{\mu} y\}| - |\{y \in A|y >_{\mu} x\}| \quad (3)$$

where μ indicates the intermediate Copeland constant. For example when we calculate

$$\begin{aligned}\mathrm{Copeland}_{\mathrm{Vodafone}}(\mathrm{Airtel}) &= \mu(\mathrm{Airtel} > \mathrm{Vodafone}) - \mu(\mathrm{Vodafone} > \mathrm{Airtel})\\ &= 2 - 1 = 1\end{aligned}$$

Indicating that in the Table 1 comparing the number of columns we get Airtel is greater than Vodafone two times, whereas Vodafone is greater than Airtel one time. Similarly, we can calculate the value of

$$\begin{aligned}\mathrm{Copeland}_{\mathrm{Reliance}}(\mathrm{Airtel}) &= \mu(\mathrm{Airtel} > \mathrm{Reliance}) - \mu(\mathrm{Reliance} > \mathrm{Airtel})\\ &= 2 - 1\\ &= 1\end{aligned}$$

Thus

$$\begin{aligned}\mathrm{Copeland}(\mathrm{Airtel}) &= \left\{\mathrm{Copeland}_{\mathrm{Vodafone}}(\mathrm{Airtel}) + \mathrm{Copeland}_{\mathrm{Reliance}}(\mathrm{Airtel})\right\}\\ &= \{2\}\end{aligned}$$

Similarly, we get the values of Copeland (Vodafone) = {0} and Copeland (Reliance) = {−2}. So by set theory notation Copeland {Airtel, Vodafone, Reliance} = {2, 0, −2} which is the union of all the three individual Copeland values. This algorithm also has approximately $O(n^2)$ complexity.

Case IV Borda's Rule

This is an interesting calculation. In mathematics $\mathrm{Borda}_p{}^{sym}(x)$, i.e., symmetric Borda operation is defined as

$$\mathrm{Borda}_p{}^{sym}(x) = y \in A \sum \mathrm{Net}_p(x > y) \quad (4)$$

The equation depicted in (4) clearly indicates that calculation methodology is same as that of Copeland. But there is a relationship that exists between $\mathrm{Borda}_p{}^{sym}(x)$, i.e., symmetric Borda^{calculation} and $\mathrm{Borda}_p{}^{sym}(x)$, i.e., Asymmetric Borda calculation. The relationship is as follows

$$\mathrm{Borda}_p{}^{asym}(x) = n + 1/2\ \mathrm{Borda}_p{}^{sym}(x) \quad (5)$$

Table 2 Table constructed for calculation of $\text{Borda}_p^{\text{asym}}(x)$

Alternatives	Weightage value	Voters		
	–	27.44	19.74	17.44
	2	Airtel	Vodafone	Reliance
	1	Vodafone	Airtel	Airtel
	0	Reliance	Reliance	Vodafone

where n is the total number of voters.

So from Eq. (5) we see that if we calculate $\text{Borda}_p^{\text{asym}}(x)$ we can very easily get the values of $\text{Borda}_p^{\text{sym}}(x)$.

While calculating the asymmetric Borda value weightage value, given to the alternatives, which is set in the descending order as mentioned in Table 2. So a weightage value of 0, 1, and 2 is given from lower to upper as mentioned in Table 2.

So from here we calculate the following asymmetric values:

$$\text{Borda}_p^{\text{asym}}(\text{Airtel}) = [2 \times 27.44 + 1 \times 19.74 + 1 \times 17.44] = 92.06$$
$$\text{Borda}_p^{\text{asym}}(\text{Vodafone}) = [1 \times 27.44 + 2 \times 19.74 + 0 \times 17.44] = 66.92$$
$$\text{Borda}_p^{\text{asym}}(\text{Reliance}) = [0 \times 27.44 + 0 \times 19.74 + 2 \times 17.44] = 34.88$$

So thus we see that $\text{Borda}_p^{\text{asym}}$ {Airtel, Vodafone, Reliance} = {92.06, 66.92, 34.88}. This algorithm also has O(n) complexity.

4 Result and Discussion

This paper does not indicate any statistical inference but by applying voting rule we come across certain inference. The graphs indicating the above methodologies are as follows (Figs. 1, 2, 3 and 4).

We come across certain important inferences

1. Plurality rule indicates market share of Airtel still exists with Vodafone second and Reliance Jio in the third position.

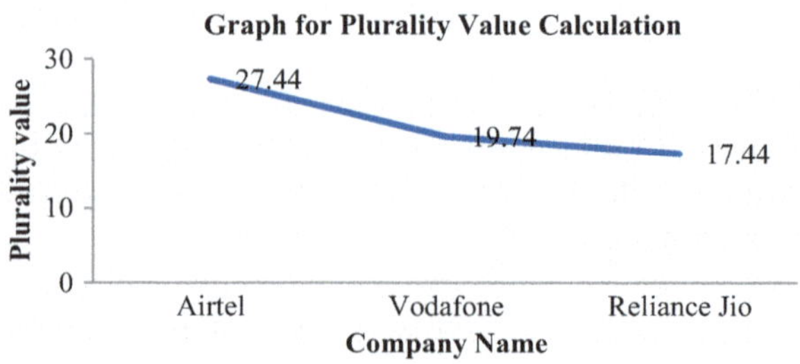

Fig. 1 Graph indicating the Plurality Value Calculation

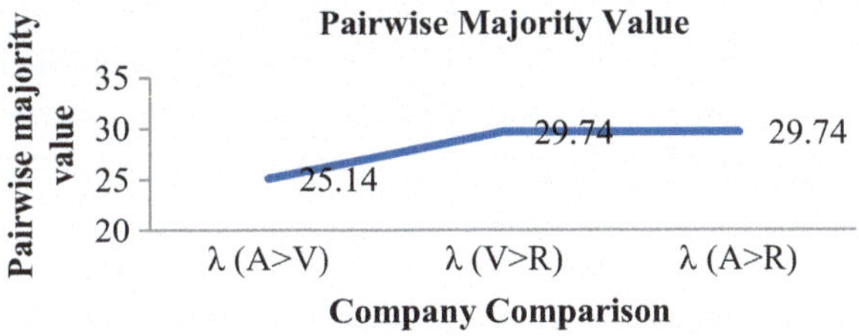

Fig. 2 Graph indicating the Pairwise Majority Value calculation

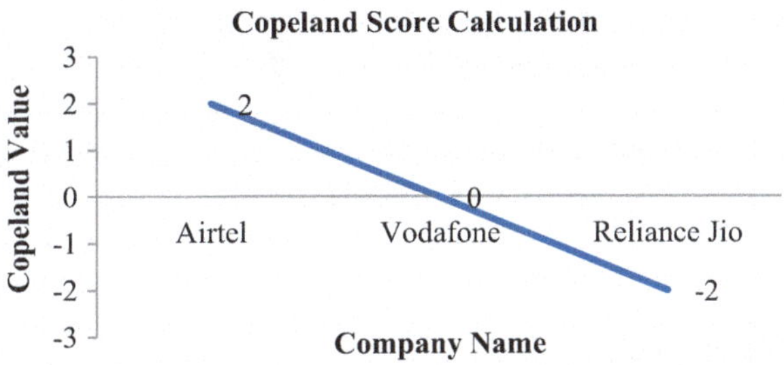

Fig. 3 Graph indicating the Copeland Score Calculation

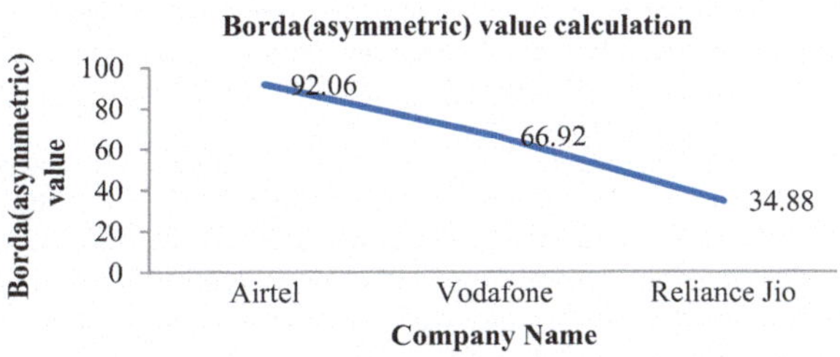

Fig. 4 Graph indicating Borda value of MSP's

2. Pairwise majority rule reveals Airtel > Vodafone > Reliance.
3. Copeland score reveals S = {Airtel, Vodafone, Reliance} = {2, 0, −2} which indicates Reliance is approaching very near to Vodafone.
4. Borda's rule gives {Airtel, Vodafone, Reliance} values of {92.06, 66.92, 34.88}, which is mainly due to Reliance's third preference by customers.

But the above model is stable as per the statistics given in Table 1 is concerned, if Reliance captures another 3 million customers as mentioned in [1] the whole result is likely to change.

5 Conclusion

From Sect. 4 we have come across certain inferences. Copeland's score and Pairwise majority rule clearly indicate that Reliance Jio is approaching very near to Vodafone and Airtel. The above voting rule described in the paper clearly indicates that another

30% shift in the customer base for Reliance is going to tilt the market. The next spectrum allocation is going to see a tough fight between the major MSP's in India and newer schemes to attract customers are going to prevail. Further numerical analysis or models should come out for simultaneous existence of MSP's.

6 Scope and Future Work

In future, there are needs to work on the attributes, which is affecting the customer base shift from one MSP's to another. One of the major parameters is the cost of billing amount. Other parameters may be Quality of Service (QoS), rate of call drop, rate of bandwidth, delay, fault tolerance, etc. Individual parameter's improvability needs to be taken care by MSPs.

References

1. TelecomLead—Magazine. https://www.telecomlead.com
2. Brandit et al.: Handbook of Computational Social Choice. Cambridge University Press
3. Deoskar, A.A., Shodhganga: A Study of Mobile Services from Customer's Perspective. http://hdl.handle.net/10603/1997
4. Wuraola, Ayisat, et al.: Framework for modelling mobile network quality of experience through big data analytics approach. J. Inf. Commun. Technol. (JICT) **17**(1), 79–113 (2018)
5. Patlolla, Ramaya, et al.: Distracting the Indian telecommunication sector: an analytical study on Reliance Jio network. Am. J. Mark. Res. **4**(2), 34–43 (2018)
6. Paulrajan et al.: Service quality & customer preference of cellular mobile service provider. J. Technol. Manag. Innov. **6**(1) (2011)
7. Technology, Media and Telecommunication Predictions 2018—India edition, White paper Deloitte. https://www2.deloitte.com/content/dam/Deloitte/in/Documents/technology-media-telecommunications/in-tmt-predictions-2018-noexp.pdf

Pronominal Anaphora Resolution in Hindi Language Using Number Agreement and Animistic Knowledge

Kavita Tewani

Abstract Anaphora resolution is always been one of the most active research areas in the domain of Natural Language Processing (NLP). Anaphora resolution refers to the act of referring, it is commonly used for pronoun resolution in discourse. It tends to figure out who or what the referent is used for. Much of the work is been carried out and still going on in this field. A number of algorithms has been proposed to solve and to find out the referent. Anaphora resolution approaches rely on domain knowledge and linguistic. The disadvantage of such approach is that it takes lot of effort and hence is laborious. This paper focuses only on pronominal anaphora resolution for Hindi language sentences that are preprocessed by a part-of-speech tagger. Inputs are checked using number agreement and animistic of the noun phrase. The paper defines resolving anaphora on the basis of limited knowledge of the language. A method called transitive anaphora resolution is also proposed, where the antecedent may not be present in the same or the previous sentence.

Keywords Anaphora resolution · Animistic knowledge · Natural language processing · Number agreement · Transitive anaphora resolution number agreement

1 Introduction

Referential pronouns can be anaphoric cataphoric, exophoric, or modified to form of independent references [1]. Anaphora points back to previously defined item (John was happy to see his presents, he liked the watch gifted by his friend). Cataphora is the use of an expression that refers (or co-refers) with later expression in discourse (He was John who met with an accident). Here he refers to John that comes later in the sentence. As given in [1], Exophors refer outside the discourse to entities in the discourse setting (He who digs a pit for others falls in himself).

K. Tewani (✉)
Computer Engineering Department, Institute of Technology, Nirma University, Ahmedabad, Gujarat, India
e-mail: kavitatewani012@gmail.com

S. Kundu et al. (eds.), *Proceedings of the 2nd International Conference on Communication, Devices and Computing*, Lecture Notes in Electrical Engineering 602, https://doi.org/10.1007/978-981-15-0829-5_67

There are different types of anaphora resolution present in a language, [2] lists the types of anaphora resolution in any language. Pronoun resolution is one of the key problems in a language, the way of referring to antecedent changes from one language to another. The algorithm that works for one particular set of language may not work for another. There are number of factors that affect the pronoun resolution in a sentence such as recency, gender agreement, number agreement, and animistic knowledge to name a few [3]. Comparing the gender agreement in English and Hindi language, it may be observed that in English it is comparatively feasible to resolve pronoun on the basis of gender. The [4] also mentions that in English language "he" and "she" can help us to categorize the gender of noun, but in Hindi language "veh" is used for both the genders; however; verb plays a big role to identify the pronominal noun on gender basis. Consider the following statements:

s1: "Khushi is a brilliant student, she is good at maths."
s2: "Jay was too curious to meet his friend that he could not wait for a day."

In both the sentences, one may resolve corresponding pronoun on the basis of gender agreement only and come up with a conclusion that in s1 {she} refers to {Khushi} whereas in s2 {he} refers to Jay.

However, when we change the same set of sentences in Hindi, the gender agreement alone cannot work for resolving pronoun resolution. Consider the sentences in Hindi language:

s3: खुशी एक शानदार छात्र है, वह गणित में अच्छी है।
s4: जय अपने दोस्त से मिलने के लिए उत्सुक था कि वह एक दिन का इंतजार नहीं कर सका।

Here, it can be observed that {वह} in s3 refers to {खुशी}, whereas {वह} in s4 refers to {जय}. So in Hindi language the pronoun resolution cannot be solved just on the basis of gender agreement. The paper proposes a naive approach for pronoun resolution on the basis of number agreement. Consider this example

s5: ख़ुशी और जय ने आम खाए और उन्हें वे पसंद आये।

Khushi and Jay ate mangoes and they liked them.

s6: ख़ुशी और जय बगीचे में गए वे वहाँ फूलों को देख के खुश हो गए।

Khushi and Jay went to the garden, they were happy to see the flowers.

In s5 वे refers to आम, whereas in s6 वे refers to ख़ुशी और जय.

2 Factors to Resolve Anaphora

Anaphora resolution is based on certain factors; however, it may vary from one language to another. As per [5, 6] factors used frequently in the resolution process that include gender and number agreement, c-command constraints, semantic consistency, syntactic parallelism semantic parallelism, salience, proximity, etc.

Whereas as per [3] recency, gender and number agreement as well as animistic knowledge form the base of anaphora resolution. Few of them are explained in detail.

Recency: As per [7] sentence recency salient factor is created for the current sentence. Its scope is all discourse referents introduced by the current sentence. Rich and LuperFoy [8] quote recency as something whose function is to propose referents that have occurred in the recently preceding discourse.

Gender Agreement: It refers to only those antecedents that are referring to particular gender. As in English language {He, Him, and His} refers to masculine, whereas {She and her} refers to feminine.

Rest two of the factors, i.e., number agreement and animistic knowledge are discussed in detail in the next subsections.

2.1 Number Agreement

As per [9], Anaphora is sometimes characterized as the phenomenon whereby the interpretation of an occurrence of one expression depends on the interpretation of an occurrence of an expression has its referent supplied by an occurrence of some other expression in the same or another sentence. Anaphora resolution depends on many factors and number agreement is one of them. The process starts with part-of-speech tagging where the tokens are labeled as per their types. The work mentioned by Dalal et al. [10] claims to provide the accuracy of 89.3% for part-of-speech tagging in Hindi sentences. The part of speech is then checked for singularity and plurality of proper noun(s) as well as a pronominal noun(s). All the tags apart from a proper noun, pronominal nouns, and conjunctions are ignored. As in s6, only proper nouns (ख़ुशी, जय), pronominal noun (वे), and conjunction (और) are considered, where the proper nouns are antecedent and pronominal nouns are anaphora. This helps to focus only on the words which deal with number agreement, although working only with number agreement will not give the perfect result, it has to be combined with other methods of anaphora resolution in order to resolve which antecedent it refers to. Encountering number agreement in a sentence is frequent irrespective of any language and hence it is important to resolve anaphora that is focusing on number agreement.

2.2 Animistic Knowledge

Animistic knowledge is another factor that helps to clutter the task of pronominal anaphora resolution in a sentence. It helps to filter out inanimate noun from the sentence when the anaphor is referring to animistic noun only. The knowledge of animistic and inanimate nouns are required before resolving such problems. After closely observing the statement s5 it can be noted that there are two anaphoras in the sentence: {उन्हें, वे}. The anaphora उन्हें is referring to {ख़ुशी, जय}, whereas the second anaphora {वे} refers to {आम}.

The number agreement will not be enough to solve the problem of anaphora, as both of them refer to the plural noun. The sentence says that "Khushi and Jay ate mangoes and they liked them". It can be observed with the translating statement that the verb liking is an action that can be performed by an animistic noun. The same way in Hindi sentence, although both the anaphora are plural but one of the anaphora is performing the action of liking, i.e., {उन्हें} which directly refers to the animistic nouns {ख़ुशी, जय}.

3 Proposed Work

The proposed work is solely focusing on two of the factors named: number agreement and animistic knowledge. Also, antecedent, in which anaphora is to be resolved is restricted to two previous sentences. In some of the cases, anaphora may not be present in the current sentence or previous sentence but somewhere before that. This sort of anaphora is named as transitive anaphora. The following is the sentences that are taken from a paragraph of Hindi story from the link [11]:

कुछ ही देर में बहुत सी मखियाँ उस ज़मीन पर गिरे हुए शहद पर आकर बैठ गयी। मीठा मीठा शहद उन्हें बड़ा अच्छा लगा। वे जल्दी जल्दी उसे चाटने लगी।

Let us mark the sentences as s7 s8 and s9, respectively.

s7: कुछ ही देर में बहुत सी मखियाँ उस ज़मीन पर गिरे हुए शहद पर आकर बैठ गयी।

In a while, many flies came on the honey that fell on the ground.

s8: मीठा मीठा शहद उन्हें बड़ा अच्छा लगा।

They liked sweet honey.

s9: वे जल्दी जल्दी उसे चाटने लगी।

They started licking it quickly.

While considering the above set of sentences, one can find the anaphora based on the following set:

In s7: {उस}
In s8: {उन्हें}
In s9: {वे, उसे}

In the process, we are interested to find out the anaphora for the pronominal noun {मखियाँ}, where the antecedent is plural and alive, as the focus is only on number agreement and animistic knowledge with respect to a noun. The following search process is proposed to look for proper antecedent from the given set of sentences.

1. After applying the number agreement and animistic knowledge, we can filter out that anaphora {उस} in set s7 is irrelevant so is {उसे} in set s9 as they are referring to singular noun but we are interested in plural one. It narrows down our search for the possible pronoun and hence increases the search process. Here, it reduced the search set by 50%.

2. Now we are left with two anaphoras, one in the set s8 and another in set s9. After applying back search, it may be noted that {वे} in statement s9 do not have any valid antecedent in the same sentence. The search is to be now brought to statement s8. Here the only pronoun is {शहद}, which is not valid for the current anaphora on the basic animistic, as it is not an alive entity. So we are going to look for the anaphora in statement s8 that may be valid for the anaphora in statement s9.
3. As per number agreement and animistic knowledge, the anaphora in statement s9: {वे} is getting a match with anaphora in statement s8: {उन्हें}. The search now starts with looking for antecedent of {उन्हें} in statement s8.
4. Again there is no match for {उन्हें} in the same sentence, we start looking for valid antecedent in statement s7. The only valid antecedent for the anaphora is {मखियाँ} and hence as per the transitive anaphora, this antecedent gets valid for two anaphoras {उन्हें} in s8 and {वे} in s9.

If the antecedent is not present even in the previous two sentences, then it may be looked into previous sentences to find out the proper antecedent.

The proposed work is checked in three Hindi short stories from the link [12]:

title of the story: पक्षी की परेशानी (story 1)
title of the story: सारे रिश्ते टूट गए (story 2)
title of the story: हाथी क्यों हारा (story 3).

The result of these three Hindi stories are mentioned in the table below. The table is created to give details on the number of lines in the story and hence define the search space. The short stories up to lines 22 are considered to fetch the result. As per the proposed work, we are interested in pronominal anaphora only and their corresponding antecedent.

The table shows the result of applying the proposed steps in the three stories. However, the ambiguity arises when the same anaphora may refer to the animistic noun as well as a non-animistic noun. If we talk about Hindi Language, then {वह, उस} are valid anaphora for both the kinds of the noun (animistic as well as non-animistic).

These ambiguities can be resolved with a good knowledge of the subject. While considering the number agreement on the pronominal noun, then in the Hindi Language, {वे , वो , उन्हें} are used as anaphoras that refer to plural nouns as well as in the honorific context (Table 1).

The next table lists the anaphoras that are used in the same three sets of stories and also gives the count of transitive anaphoras encountered in the stories (Table 2).

It can be observed that in the case of pronominal anaphora, the frequency of transitive anaphora is high in the given set of stories. The search for transitive anaphora has also been applied to another set of stories taken from the link [13], and it has been found that in most of the cases, the frequency of pronominal anaphora is high.

Table 1 Results of anaphora resolution on number agreement and animistic knowledge

Story	Number of lines	Number of pronominal anaphora	Number of anaphora resolved on the basis of number agreement and animistic knowledge	Percentage of anaphora resolved
Story 1	11	13	11	84.6
Story 2	22	27	22	81.4
Story 3	5	7	6	85.7

Table 2 Pronominal anaphora used in the stories and count of transitive anaphora

Story	Number of lines	Pronominal Anaphora found along with their frequency	Number of transitive anaphora
Story 1	11	वह-3 उस-1 उसके-1 उतने-1 उन-1 उसकी-1 उसे-2 उसका-1 अपने-2	10
Story 2	22	वह-3 उसे-5 उसकी-1 उसके-5 तुम्हे-2 उसने-2 उस-1 तुम-3 अपने-2 वो-1	17
		अपनी-1 तुम्हारे-1	
Story 3	5	उसे-2 तुम्हारे-1 ये-1 इन्हें-1 यह-2 उस-1 वह-1	3

4 Issues and Challenges

Resolving Anaphora requires deep knowledge of the subject. Especially when the data to be processed is in native language. The following are some of the issues and challenges for Indian Language:

(1) Gender Agreement: Gender agreement has been briefed in the earlier section of this paper. One of the challenges with gender agreement is that in languages like Hindi, the pronominal anaphora that refers to masculine as well feminine is the same. Unlike the English language which has a dedicated set of anaphora on the basis of language.

(2) Preprocessed Data: The pronominal anaphora resolution needs the sentences that are correctly framed as per the language grammar. Also, the part-of-speech tagging done on the sentences should be 100% accurate to find the corresponding antecedent from the set of sentences.

(3) Language Knowledge: Irrespective of any techniques that are applied to resolve the anaphora, the language knowledge is required. However, some articles like [14, 15] propose and quote knowledge-poor approaches to solve anaphora, where only specific domain knowledge is required. Still, some sort of knowledge about the language is required to find out the correct pronoun, without having any knowledge about the language it is difficult to get appropriate results.

(4) Translating to Native Language: It has been observed that while translating the sentences from one language to another, the sentence formation changes and hence required to be resentenced in order to make it language compatible. Consider a sentence in English: "what a great task done by him!" after translating this sentence to Hindi the convertor the sentence as "उसके द्वारा क्या किया गया एक महान काम!". The sentence does not give the proper translation and hence has to be reconsidered.

(5) Tools to work with Native Language: Although for highly acceptable languages like Hindi, many of the translators give an accurate answer. The problem arises when the sentences (especially phrases) are translated to some inferior language, the translator gives the wrong sentence. Let us see one of the sentences that have been translated from English to Hindi, English to Gujarati, and English to Marathi.

    English: What goes around comes around.
    Hindi: जैसा काम करोगे वैसा ही फल मिलेगा
    Gujarati: આસપાસ શું આવે છે આસપાસ આવે છે
    Marathi: काय सुमारे सुमारे येतो

The phrase has been correctly translated to the Hindi Language. But in rest of the languages, the translation has a different meaning from the original one. And hence working on the direct data given by the translator is not an adequate way.

(6) Use of Possessive Reflexives: Possessive reflexives in the Hindi Language, in contrast to the English, are not inflected with the gender and number of possessor [16], but that of possession. Examples of such reflexives are {अपना, अपने, खुदका}.

5 Related Work

Anaphora resolution is not a new topic in the field of Natural Language Processing (NLP). Much of the work is been done or proposed in different languages in order to find the proper antecedent from the given sentences. The conventional method of anaphora resolution is been merged with certain methodologies and algorithms to reach the results. In this section, the related work of anaphora resolution in the Hindi Language is listed along with the techniques used to solve the resolution problem.

(1) A machine learning approach has been applied for the classification of indirect anaphora, the anaphora that does not have an explicit referent in discourse, in Hindi corpus [17].
(2) The work of anaphora resolution based on dependency structures using rule-based module is done for the Hindi Language [16].
(3) A generic anaphora resolution engine using shallow parse text has been proposed in [18] for various Indian Languages.
(4) The paper [19] focuses on the use and interpretation of third person pronouns for anaphoric resolution.
(5) The method proposed in [20] assigns a score to the antecedent on the basis of gender and number agreement after preprocessing the data by a part-of-speech tagger. The antecedent with higher score is then assigned to the anaphora. This technique used is robust and knowledge-poor approach to reach the final solution.
(6) The paper [21] applied Centering principles in Tamil Language.

6 Future Work

The future work focuses on resolving the honorific pronouns where the literature knowledge is limited and to come up with robust approach for resolving such pronoun in Hindi Language. Also, the challenges mentioned in the paper shows that much of the work is yet to be done in inferior native languages, as no proper translators are available when it comes to some less known/used languages. The sentence formation of language is different from other and hence, makes this task more challenging. The work of anaphora resolution can be merged with other applications like translation, text summarization, dialogue processing, etc.

7 Conclusion

The paper discussed the problem of anaphora resolution and narrowed down the search of anaphora on two of the most important features of resolution especially with respect to the Hindi language-number agreement and animistic knowledge. The paper proposes the methodology to resolve anaphora that may not be present in the same or previous statement and hence the term transitive anaphora resolution is given to such form of a problem to resolve pronominal anaphora.

The test has been conducted on three sets of stories where the pronominal anaphora is resolved purely on the basis of number agreement and animistic knowledge. The results show that it is quite sensible to focus on a few of the important features of resolution instead of dealing with all of them. This will help to concentrate on the targeted antecedent only as per the requirement of a problem and hence reduce the targeted noun which ultimately decreases the time for the search.

Different issues and challenges for resolving anaphora in native languages are also discussed in brief.

References

1. Byron, D.K.: The uncommon denominator: a proposal for consistent reporting of pronoun resolution results. Comput. Linguist. **27**(4), 569–577 (2001)
2. Ahir, P., et al.: Anaphora resolution for hindi language
3. Lakhmani, P., Singh, S.: Anaphora resolution in hindi language. Int. J. Inf. Comput. Technol. **3**, 609–616 (2013)
4. Dutta, K., Prakash, N., Kaushik, S.: Resolving pronominal anaphora in hindi using hobbs algorithm. Web J. Formal Comput. Cognit. Linguist. **1**(10), 5607–5611 (2008)
5. Mitkov, R.: Factors in anaphora resolution: they are not the only things that matter: a case study based on two different approaches. In: Proceedings of a Workshop on Operational Factors in Practical, Robust Anaphora Resolution for Unrestricted Texts. Association for Computational Linguistics (1997)
6. Mitkov, R.: Anaphora Resolution. Routledge (2014)
7. Lappin, S., Leass, H.J.: An algorithm for pronominal anaphora resolution. Comput. Linguist. **20**(4), 535–561 (1994)
8. Rich, E., LuperFoy, S.: An architecture for anaphora resolution. In: Proceedings of the Second Conference on Applied Natural Language Processing. Association for Computational Linguistics (1988)
9. Dakwale, P.: Anaphora resolution in Hindi. Diss. Ph.D. thesis, International Institute of Information Technology Hyderabad (2014)
10. Dalal, A., et al.: Hindi part-of-speech tagging and chunking: a maximum entropy approach. Proc. NLPAI Mach. Learn. Compet. (2006)
11. http://www.indif.com/kids/hindi_stories/hindi_short_stories_08.aspx
12. http://badhtechalo.com/hindi-stories-with-moral/
13. http://indif.com/kids/hindi_stories/hindi_short_stories_01.aspx
14. Mohana, R.: Anaphora resolution in hindi: issues and directions. Indian J. Sci. Technol. **9**(32) (2016)
15. Sikdar, U.K.: Ekbal, A., Saha, S.: A generalized framework for anaphora resolution in Indian languages. Knowl. Based Syst. **109**, 147–159 (2016)

16. Dakwale, P., Mujadia, V., Sharma, D.M.: A hybrid approach for anaphora resolution in hindi. In: Proceedings of the Sixth International Joint Conference on Natural Language Processing (2013)
17. Dutta, K., Kaushik, S., Prakash, N.: Machine learning approach for the classification of demonstrative pronouns for indirect anaphora in hindi news items. Prague Bull. Math. Linguist. **95**, 33–50 (2011)
18. Devi, S.L., Ram, V.S., Rao, P.R.K.: A generic anaphora resolution engine for Indian languages. In: Proceedings of COLING 2014, the 25th International Conference on Computational Linguistics: Technical Papers (2014)
19. Prasad, R., Strube, M.: Discourse salience and pronoun resolution in Hindi. Penn Work. Pap. Linguist. **6**(3), 189–208 (2000)
20. Mitkov, R.: Robust pronoun resolution with limited knowledge. In: Proceedings of the 36th Annual Meeting of the Association for Computational Linguistics and 17th International Conference on Computational Linguistics, vol. 2. Association for Computational Linguistics (1998)
21. Jagan, B., et al.: Anaphora resolution in tamil using universal networking language. In: IICAI (2011)

In Silico Design and Homology Modeling of Helicase C-Terminal Domain of Nonstructural Protein NS3 of West Nile Virus (Strain NY-99)

Shamrat Kumar Paul, Abul Bashar Ripon Khalipha, Lutful Kabir, Pranta Ray, Masuma Akter Mina and Hossain Mohammad Masum

Abstract West Nile virus (WNV) is a mosquito-transmitted single-stranded (ss) RNA flavivirus which causes West Nile infection and human disease of variable severity. Most of the people infected with WNV either have only minor signs or symptoms such as mild headache and fever or do not develop any signs or symptoms. Sometimes WNV causes a life-threatening illness as well as inflammation of the spinal cord or brain in infected patients. The genome of WNV is single-stranded RNA (ssRNA) that contains the characteristics of positive polarity (mRNA sense). The strains of WNV are classified into two groups, such as lineage (1 and 2). The basis of grouping these strains based on the substitutions and deletions of signature amino acid in the envelope protein sequence. Among them, Lineage 1 WNV strains are responsible for developing the West Nile infection in humans. The genome of WNV consists of a single open reading frame (ORF) and it produced ten mature viral proteins which are classified into structural and nonstructural. Nonstructural protein 2B (NS2B) and Serine Protease (NS3) are two of nonstructural proteins of them. NS2B-NS3 has a major role in proteolytic processing of these nonstructural proteins. NS3 has several enzymatic activities like NTPase and RNA helicase. Among them,the helicase activities is one of them. There is evidence that Helicase C-terminal domain of West Nile Virus has a major role in the unwinding of dsRNA. But Helicase C-terminal domain is not yet explored, so our purpose is to investigate the physicochemical, structural, and functional features of Helicase C-terminal domain. Molecular modeling of the unexplored Helicase C-terminal domain was generated by using Phyre2 and Swiss Model. The prediction of active ligand binding sites is generated by using PredictProtein server. This results that Helicase C-terminal

S. K. Paul · L. Kabir · M. A. Mina
Department of Biochemistry and Molecular Biology, Bangabandhu Sheikh Mujibur Rahman Science and Technology University, Gopalganj, Bangladesh

A. B. R. Khalipha (✉) · P. Ray
Department of Pharmacy, Bangabandhu Sheikh Mujibur Rahman Science and Technology University, Gopalganj, Bangladesh
e-mail: khalipha1982@gmail.com

H. M. Masum
Institut National de La Recherche Agronomique, Paris, France

S. Kundu et al. (eds.), *Proceedings of the 2nd International Conference on Communication, Devices and Computing*, Lecture Notes in Electrical Engineering 602, https://doi.org/10.1007/978-981-15-0829-5_68

domain protein is stable and its nature is acidic which thermostability is high and has better hydrophilic property. By the putative transferase and catalytic functional activity, 14 binding sites were predicted. In Homology Modeling, it is revealed that 14 binding sites are predicted as ligand binding sites. This investigation decoded the role of this unexplored Helicase C-terminal domain of West Nile virus (WNV), and so it can pave the way for enriching our knowledge for pathogenesis and medication of WNV infection.

Keywords West Nile Virus · Helicase C-terminal domain · Open reading frame · Homology modeling · Ligand binding site

1 Introduction

In the middle of 1999, this mosquito-borne zoonosis was first identified in the United States. More than 39,000 reported clinical cases, of which 17,463 were presented with neuroinvasive disease, 22,094 with West Nile fever, and 1668 with a dangerous condition of patients caused by it [1]. Mosquitoes obtain this virus by biting the infected birds and this virus is transmitted to a human when that infected mosquitoes bite on human [2]. Characteristically the WNV genome is a single-stranded RNA of positive polarity mRNA sense. At the 5′ end there is a type 1 cap structure but at the 3′ end, it terminates with CUOH because compared with other viruses Flavivirus genomes are the only mammalian plus-strand RNA virus genomes which do not contain any 3′ Poly A tail [3]. The length of the WNV genome is 11,029 nt and it has only a single open reading frame (ORF) of 10,301 nt. In its single ORF, there is structural and nonstructural part [4]. From these parts first, it produces a single polyprotein and by proteolytic processing of viral serine protease (NS2B-NS3) and various cellular proteases that polyprotein turns into ten mature viral protein [5]. From the structural part which is situated at the 5′ portion it produces three viral structural proteins, capsid (C), membrane (prM/M), and envelope (E) and the rest seven nonstructural proteins (NS1, NS2A, NS2B, NS3, NS4A, NS4B, and NS5) are encoded from the 3′ nonstructural portion [6]. Recently, the molecular study by Brinton, M. A. [7] suggested that the NS2B and NS3 combinedly act as serine protease which has a great role in proteolytic processing of that single polyprotein. Furthermore, it uptakes energy from NTP hydrolysis for RNA unwinding by Helicase activity. The literature review reveals that in the NS3 protein, it has Helicase C-terminal Domain which has a crucial role in this process [8]. However, the structure of this Helicase C-terminal Domain is not reported yet. Also, the detailed physicochemical characterization and putative structure with ligand binding active sites are not elucidated, so we proposed an in silico 3-D structure prediction of Helicase C-terminal Domain using homology modeling.

2 Methodology

2.1 *Retrieval of the Target Amino Acid Sequence of Helicase C-Terminal Domain*

The amino acid sequence of Helicase C-terminal domain of NS3 (Genome Polyprotein-GP1) was obtained from UniProtKB with the ID Q9Q6P4. Due to the unavailability of 3-D structure in PDB, modeling of this unexplored domain was undertaken utilizing 166 aa long sequence of Helicase C-terminal domain of West Nile virus.

2.2 *Physicochemical Characterization*

Physicochemical properties of the retrieved sequence were determined using two web-based servers. ProtParam tool of Expasy (http://us.expasy.org/tools/protparam.html) employed for the prediction of amino acid composition, instability and aliphatic indices, extinction coefficients, and grand average of hydropathicity (GRAVY) [9]. Theoretical isoelectrical point (pI) was calculated using Sequence Manipulation Suite (SMS) Version 2 (http://www.bioinformatics.org/sms2/protein_iep.html).

2.3 *Secondary Structure Prediction of Helicase C-Terminal Domain Protein*

The self-optimized prediction method with alignment (SOPMA; https://npsa-prabi.ibcp.fr/NPSA/npsa_sopma.htm) software [10] and (PSIPRED program; http://bioinf.cs.ucl.ac.uk/psipred) was used to predict the secondary structure of Helicase C-terminal Domain protein (target). Disorder prediction was performed using DISOPRED tool (Predict Protein server; https://predictprotein.org) used to predict secondary structure [11].

2.4 *Homology Modeling and Validation of Helicase C-Terminal Domain*

There is no experimentally deduced 3D structure available for Helicase C-terminal domain protein in protein data bank (PDB); therefore, homology modeling of the protein of Helicase C-terminal domain was done using two programs (Swiss Model and Phyre2; http://www.sbg.bio.ic.ac.uk/phyre2) [12, 13]. Secondary structure has also been predicted using Phyre2. 3D model of Helicase C-terminal domain generated from Swiss-Model and Phyre2 was compared and only the most suitable 3D model was selected for final validation. The final modeled structure was validated using Ramachandran plot analysis (PROCHECK; http://nihserver.mbi.ucla.edu/SAVES) for stereochemical property.

3 Result and Discussion

The amino acid sequence of Helicase C-terminal domain was retrieved in FASTA format and used as query sequence for determination of physicochemical parameters. The instability index of 37.18 (<40) indicated the stable nature of Helicase C-terminal domain [30]. The protein is acidic in nature (pI 6.23, 5.82*) with molecular weight of the 18.66 kDa. High extinction coefficient values (28,670) indicate the presence of Cys, Trp, and Tyr residues [14]. Higher aliphatic index values (70.57) of the query protein suggested as a positive factor for increased thermos-stability for a wide temperature range [15]. Hydrophilic nature of the protein and the possibility of better interaction with water [16] were indicated by the lower grand average of hydropathicity GRAVY indices value (−0.319) as shown in Table 1. The default parameters (similarity threshold: 8; window width: 17) were considered by SOPMA for the secondary structure prediction with >70% prediction accuracy. Utilizing 511 proteins (sub-database) and 15 aligned proteins, SOPMA predicted 40.96% of residues as random coils in comparison to Alpha helix (25.90%), extended strand (24.10%) and Beta turn (9.04%) as shown in Table 2. PSIPRED showing the higher confidence of prediction of helix, strand and coil (Fig. 1). Secondary structure prediction by PredictProtein employing neural network system, provides the prediction accuracy of more than 72%. 22.9% helix conformation (α; π; 3_10-helix), 44.6% loop (L)

Table 1 Physicochemical parameters computed using Expasy's ProtParam and SMS tool

Physicochemical parameters	Values
No. of amino acid (aa)	166
Molecular weight (MW)	18658.94
Theoretical isoelectric point (pI)	6.23, 5.82*
Aliphatic index	70.57
Instability index	37.18
Extinction coefficient (All Cys form Cysteine)	28,670
Extinction coefficient (All Cys reduced)	28,420
Total no. of negatively charged residues (Asp + Glu)	17
Total no. of positively charged residues (Arg + Lys)	15
GRAVY (Grand average of hydropathicity)	−0.319

*pI determined by SMS Version2

Table 2 Secondary structure elements prediction by SOPMA

Secondary structure elements	Values (%)
Alpha helix (Hh)	25.90
310 helix (Gg)	0.00
Pi helix (Ii)	0.00
Beta bridge (Bb)	0.00
Extended strand (Ee)	24.10
Beta turn (Tt)	9.04
Bend region (Ss)	0.00
Random coil (Cc)	40.96
Ambiguous states (?)	0.00

followed by 32.5% beta strand (E = extended strand in beta sheet conformation) was predicted in Helicase C-terminal domain. Intrinsic disorder profile was computed using DISOPRED and >90% of the amino acid is below the confidence score of 0.5 for disordered condition, suggested the lowest possibility of distortion and conferred the high stability to the predicted protein.

Binding sites were predicted using predict protein server, where 14 different protein binding sites were identified at positions, viz.: 30–31; 41; 43; 52–54; 81; 83; 92–94; 104; 110–114; 114; 117; 120–122; 124–124; and 133–136 (Fig. 2). Gene ontology predicted and categories the functional aspects as cellular, molecular, and biological, where this Helicase C-terminal domain protein found to be subcellular or the part of host cell or membrane; metabolic processes such as primary and cellular metabolic processes including cyclic, heterocyclic, and aromatic compound metabolism processes. Molecular function including binding (Score: 49) involves heterocyclic (Score: 49), organic (49), cyclic compound binding; small molecule (Score: 49), and nucleotide binding (Score: 49) whereas and nucleic acid binding (Score: 32) include nucleoside-triphosphatase activity (Score: 32) showed in Table 3.

Helicase C-terminal domain target sequence was inserted as input (FASTA format) in Swiss-Model workspace. The Swiss-MODEL template library (SMTL) was searched with HHBlits [17] resulted in total 32 templates. Among the seven most favorable template (2qeq.1.A; 2qeq.1.A; 2qeq.2.A; 2qeq.2.A 2z83.1.A; 2z83.1.A; 2v80.1.A) are those who have GMQE value greater than 0.90. Target sequence was selected based on the Qualitative Model Energy Analysis (QMEAN) score (−1.33), Global model quality estimate (GMQE) 0.96, percentage of sequence identity (96.39), similarity (0.61) and coverage (1.00). The model was based on target template alignment using ProMod3, where insertion, deletions remodeled, and side chains were then rebuilt.

Our model showed resemblance with 2qeq.1.A. So that ProMod3 Program selected 2qeq.1.A. as template structure which belongs to Flavivirin protease NS3 catalytic subunit. Template list for homology modeled structure is showed in Table 4. The model was generated and saved in PDB format (Fig. 3 left). PROCHECK, another

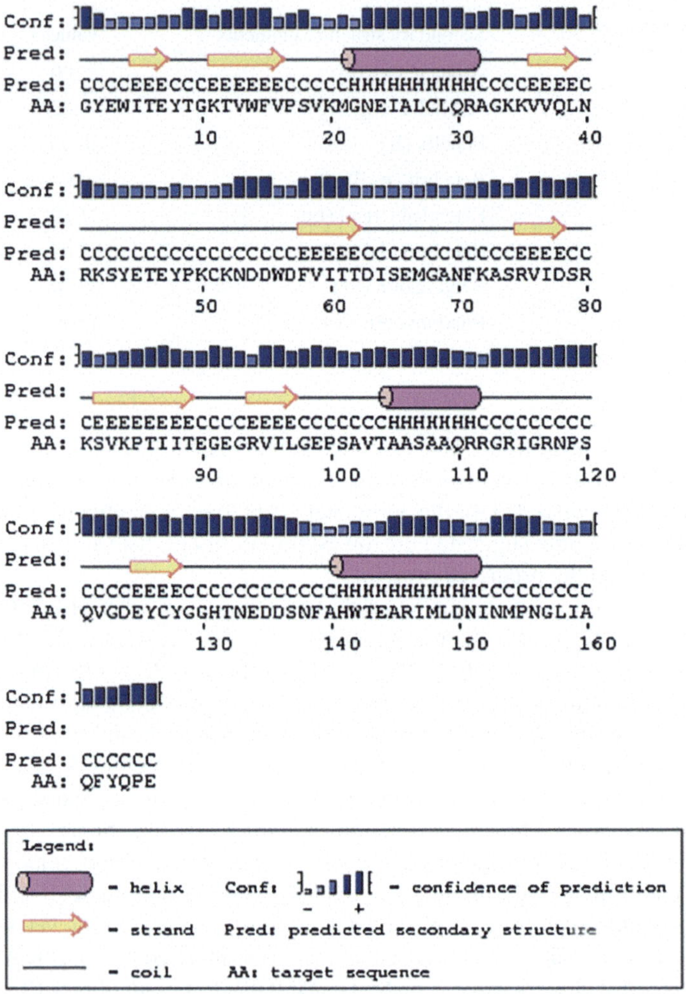

Fig. 1 Secondary structure prediction by PSIPRED

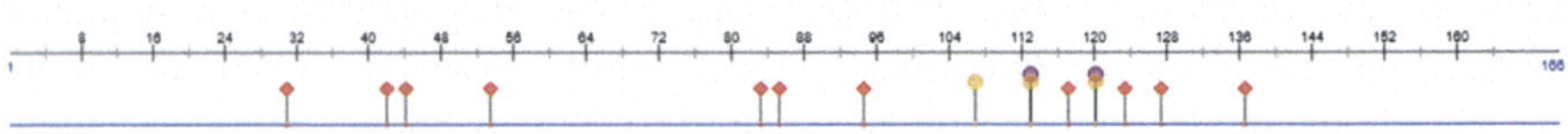

Fig. 2 Protein–protein and protein-polynucleotide binding sites

tool was used to measure the accuracy of protein models. Ramachandran plot statistics of Helicase C-Terminal revealed that 91.7% of amino acid residues were found in the most favored regions. Thus, the selected protein model was very good as all of the amino acid residues were within the limits of Ramachandran plot (Fig. 5). Verification was also done by ERRAT and QMEAN server. The respective values Z-scores of

Table 3 Molecular functional ontology

GO id	GO term	Reliability (%)
GO:0005488	Binding	49
GO:1901363	Heterocyclic compound binding	49
GO:0000166	Nucleotide binding	49
GO:1901265	Nucleoside phosphate binding	49
GO:0036094	Small molecule binding	49
GO:0097159	Organic cyclic compound binding	49
GO:0003676	Nucleic acid binding	36
GO:0003723	RNA binding	36
GO:0016787	Hydrolase activity	32
GO:0017111	Nucleoside-triphosphatase activity	32

Table 4 Alignment of selected template

Name	Title	GMQE	Identity (%)
2qeq.1.A	Flavivirin protease NS3 catalytic subunit	0.98	96.39
2qeq.2.A	Flavivirin protease NS3 catalytic subunit	0.77	96.39
2qeq.1.A	Flavivirin protease NS3 catalytic subunit	0.97	96.39
2qeq.2.A	Flavivirin protease NS3 catalytic subunit	0.63	96.39
2z83.1.A	Helicase/Nucleoside-triphosphatase	0.96	84.94
2z83.1.A	Helicase/Nucleoside-triphosphatase	0.94	84.85
2v8o.1.A	Flavivirin protease NS3	0.95	80.12

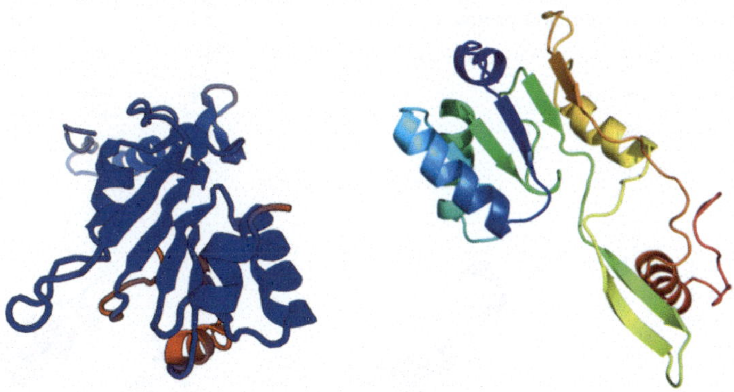

Fig. 3 Helicase C domain structure with helix, strands and coil predicted by Swiss Model [left], Helicase-C domain structure predicted by Phyre2 [right]

Cbeta interaction energy, torsion angle energy, solvation energy, secondary structure in case of Helicase C-Terminal Domain are −1.89, −0.79, −0.37, and −0.56. The overall QMEAN score for Helicase C-Terminal is −1.43. QMEAN generated results confers Helicase C-Terminal Domain as a qualified model for drug target scopes. Stereochemical quality of the Swiss model predicted Helicase C-terminal domain structure was evaluated by plotting Ramachandran map (PROCHECK) shown in Fig 5. 91.7% of the total residues (132) (Table 5) were found in the core (A; B; L; P) whereas 7.6% of residues were in the allowed (a; b; l; p) regions. Disallowed region constitutes 0.0% of the residues. PROCHECK analysis showed max deviation of 4.2 (residue properties), with bond length/angle of 5.1 and 93.9% planar groups within the limits (Fig. 4).

Similarly, the homology modeling of Helicase C-terminal domain was performed by Phyre2. Based on the six templates (d1yksa2, c2vbcA, c2wv9A, c2jlrA, c5aorA, and c5vheA), protein model was generated with 95% of the residues modeled at 100% confidence with the template structure d1yksa2 which belongs to Crystal structure of yellow fever virus NS3 helicase and the homology modeled protein is shown in Fig. 3 right with coordinates (A): X: 56.637, Y: 44.638, Z: 56.329 (based

Table 5 Ramachandran plot analysis

Ramachandran plot statistics	Residue	%
Residues in the most favored regions [A, B, L]	132	91.7
Residues in the additional allowed regions [a, b, l, p]	11	7.6
Residues in the generously allowed regions [a, b, l, p]	1	0.7
Residues in the disallowed regions [xx]	0	0.00
Number of non-glycine and non-proline residues	144	100.0
Number of end residues (excl. Gly and Pro)	1	–
Number of glycine residues	14	–
Number of proline residues	7	–
Total number of residues	166	–

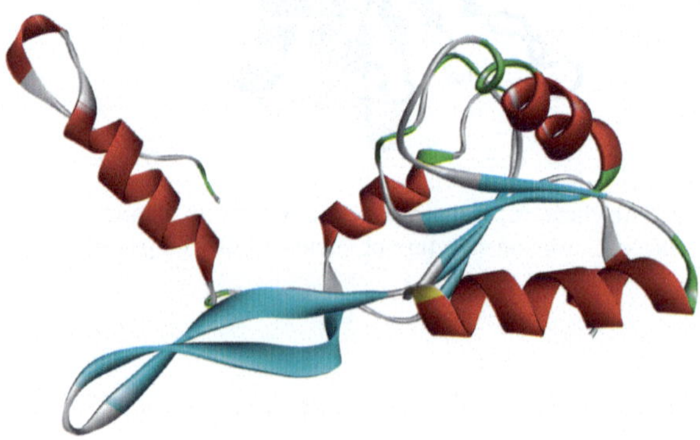

Fig. 4 Homology modeled hypothetical Helicase C-terminal domain visualized by discovery studio

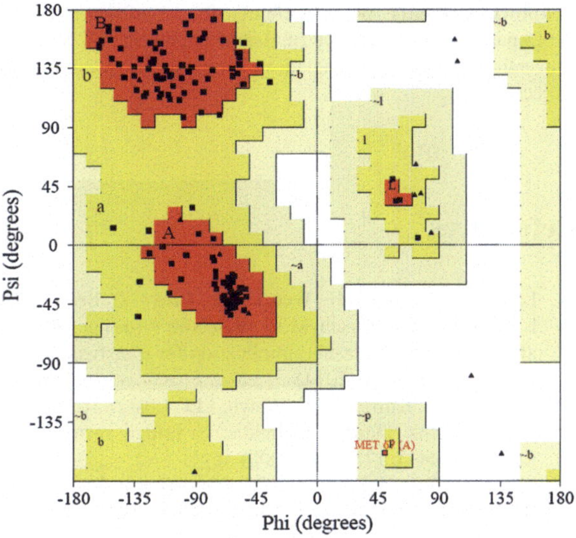

Fig. 5 Ramachandran plot analysis of Helicase C-terminal through Swiss Model workshop

on heuristics to maximize confidence, percentage identity, and alignment coverage). Secondary structure prediction by Phyre2 was described as Disordered (22%), Alpha helix (32%), and beta strand (25%) (data not shown). Phyre2 predicted structural model was evaluated for the stereochemical quality using Ramachandran map (PROCHECK). 88.8% of the residues were found in the core (A; B; L; P), whereas 8.4% of residues were in the allowed (a; b; l) regions. However, 2.8% residues were aligned in generously allowed region (~a, ~b, ~l ~p), whereas disallowed region constituted 1.5% of the residues. Among residual properties max deviation was 0.00%, bond length/angle 18.2 with 1 cis-peptides with 92.4% planar groups within limits (data not shown).

4 Conclusion

We have reported the structural model of Helicase C-terminal domain with predicted active site for ligand binding through in silico approach whereas we showed the position of amino acid in favored region by evaluating the structure. The physico-chemical parameters prediction and functional annotation are useful for understanding the action of this viral protein domain activity. Our homology-modeled protein domain provides insights into the functional role of Helicase C-terminal domain in viral pathogenesis which will help to design potential therapeutic drug against West Nile Infection.

Acknowledgements We are grateful to the department of Biochemistry and Molecular Biology and the Department of Pharmacy of Bangabandhu Sheikh Mujibur Rahman Science and Technology University, Gopalganj, Bangladesh for giving us a computational platform for completing this project.

References

1. Lanteri, M.C., Lee, T.H., Wen, L., Kaidarova, Z., Bravo, M.D., Kiely, N.E., Kamel, H.T., Tobler, L.H., Norris, P.J., Busch, M.P.: West Nile Virus nucleic acid persistence in whole blood months after clearance in plasma: implication for transfusion and transplantation safety. Transfusion **54**(12), 3232–3241 (2014)
2. Hamer, G.L., Kitron, U.D., Brawn, J.D., Loss, S.R., Ruiz, M.O., Goldberg, T.L., Walker, E.D.: Culex pipiens (Diptera: Culicidae): a bridge vector of West Nile virus to humans. J. Med. Entomol. **45**(1), 125–128 (2008)
3. Liu, Y., Wimmer, E., Paul, A.V.: Cis-acting RNA elements in human and animal plus-strand RNA viruses. Biochimica Et Biophysica Acta (BBA)-Gene Regul. Mech. **1789**(9–10), 495–517 (2009)
4. Yu, L., Nomaguchi, M., Padmanabhan, R., Markoff, L.: Specific requirements for elements of the 5′ and 3′ terminal regions in flavivirus RNA synthesis and viral replication. Virology **374**(1), 170–185 (2008)
5. Bollati, M., Alvarez, K., Assenberg, R., Baronti, C., Canard, B., Cook, S., Coutard, B., Decroly, E., de Lamballerie, X., Gould, E.A., Grard, G.: Structure and functionality in flavivirus NS-proteins: perspectives for drug design. Antivir. Res. **87**(2), 125–148 (2010)
6. Perera, R., Kuhn, R.J.: Structural proteomics of dengue virus. Curr. Opin. Microbiol. **11**(4), 369–377 (2008)
7. Brinton, M.A.: The molecular biology of West Nile Virus: a new invader of the western hemisphere. Ann. Rev. Microbiol. **56**(1), 371-402 (2002)
8. Jankowsky, E.: RNA helicases at work: binding and rearranging. Trends Biochem. Sci. **36**(1), 19–29 (2011)
9. Gasteiger, E., Hoogland, C., Gattiker, A., Wilkins, M.R., Appel, R.D., Bairoch, A.: Protein identification and analysis tools on the ExPASy server. In: The Proteomics Protocols Handbook, pp. 571–607. Humana press (2005)
10. Avn, S.: Insilico exploration of PYC gene from corynebacterium glutamicum atcc 13032
11. Graña, O., Baker, D., MacCallum, R. M., Meiler, J., Punta, M., Rost, B., Tress, M.L., Valencia, A.: CASP6 assessment of contact prediction. Proteins Struct. Funct. Bioinf. **61**(S7), 214–224 (2005)
12. Liu, X.: Preparation and functional investigation of protease inhibitors conjugated to soluble dextran (2018)
13. McKee, L.S., Brumer, H.: Growth of Chitinophaga pinensis on plant cell wall glycans and characterisation of a glycoside hydrolase family 27 β-L-arabinopyranosidase implicated in arabinogalactan utilisation. PLoS ONE **10**(10), e0139932 (2015)
14. Bennett, M.J., Huey-Tubman, K.E., Herr, A.B., West, A.P., Ross, S.A., Bjorkman, P.J.: A linear lattice model for polyglutamine in CAG-expansion diseases. Proc. Natl. Acad. Sci. **99**(18), 11634–11639 (2002)
15. Vikram, T., Kumar, P.: Analysis of Hepatitis E virus (HEV) X-domain structural model. Bioinformation **14**(7), 398 (2018)
16. Bellingham, J., Foster, R.G.: Opsins and mammalian photoentrainment. Cell Tissue Res. **309**(1), 57–71 (2002)
17. Sandal, M., Duy, T.P., Cona, M., Zung, H., Carloni, P., Musiani, F., Giorgetti, A.: GOMoDo: a GPCRs online modeling and docking webserver. PLoS ONE **8**(9), e74092 (2013)

Author Index

S. Kundu et al. (eds.), *Proceedings of the 2nd International Conference on Communication, Devices and Computing*, Lecture Notes in Electrical Engineering 602, https://doi.org/10.1007/978-981-15-0829-5

The manufacturer's authorised representative in the EU is Springer
Nature Customer Service Centre GmbH, Europaplatz 3, 69115 Heidelberg,
Germany. If you have any concerns regarding our products, please
contact ProductSafety@springernature.com

Printed and bound by CPI Group (UK) Ltd, Croydon, CR0 4YY
25/09/2026
02222886-0014